CURRENT AIR QUALITY ISSUES

Edited by **Farhad Nejadkoorki**

Current Air Quality Issues

http://dx.doi.org/10.5772/58743
Edited by Farhad Nejadkoorki

Contributors

Kazuo Shimizu, Farideh Golbabaei, Margherita Ferrante, Teresa Fortoul, Adriana Gonzalez-Villalva, Marcela Rojas-Lemus, Vianey Rodriguez-Lara, Patricia Bizarro-Nevares, Isabel García-Peláez, Martha Ustarroz-Cano, Sofía López-Zepeda, Silvana Cervantes-Yépez, Nelly López-Valdez, Nayeli Meléndez-García, Maribel Espinosa-Zurutuza, Gumaro Cano-Gutierrez, Laura Colín-Barenque, María Concepción Cano-Rodríguez, Giuseppe Petrone, Carla Balocco, Radim J. Sram, Danni Guo, Davidson Moreira, Chang Jin Ma, Blanka Mankovska, Alberto Mendoza, Edson Carrillo, Byeong-Kyu Lee, Thanh-Dong Pham, Chi Hyeon Lee, Minh Viet Nguyen, Zhenyi Chen, Elaine Araújo, Divânia Silva, Shirley Cavalcanti, Márbara Vilar Almeida, Edcleide Araujo, Marcus Vinícius Lia Fook, Anca Maria Moldoveanu, Jian Hai Sun, Ronny Brandenburg, Cristina Paradisi, Milko Schiorlin, Michael Schmidt, Ester Marotta, Agata Giardina, Ralf Basner, Florentina Villanueva, Sheikh Saeed Ahmad, Rosa Maria Ceron, Julia Griselda Ceron Breton, José Del Carmen Zavala, Manuel Muriel García, Carmen Guadalupe Carballo Pat, Roberto Carlos Escoffie Espinosa, Francisco Anguebes Franseschi, Miguel Angel Ramirez Elias, Eliane R. Rodrigues, Mario H. Tarumoto, Guadalupe Tzintzun, Claudia Cappello, Sabrina Maggio, Monica Palma, Daniela Pellegrino, Sandra De Iaco, Veronica Distefano, Donato Posa, Alessandra Spennato, Ivan Komar, Branko Lalić, Mark Wolters, Charmaine Dean, Ana Rowland Godson

CBS, Edition **2017**

Published by InTech

Janeza Trdine 9, 51000 Rijeka, Croatia

Publishing Process Manager Iva Simcic

Technical Editor InTech DTP Team

Cover Designer InTech Design Team

Additional hard copies can be obtained from orders@intechopen.com

Current Air Quality Issues, Edited by Farhad Nejadkoorki
p. cm.
ISBN 978-953-51-2180-0

INTECH

open science | open minds

Contents

Preface

While significant progress has been witnessed in air quality management, there are concerns that remain to be addressed in some key areas. This book is a collection of selected contributions in key fields of air quality. Each contribution comes as a distinct chapter complete in itself but directly related to the book's themes and purposes describing air pollution case studies. The book contains findings of air pollution scientists from around the globe on five areas of interest in the field: Air Pollution Modeling, Air Quality and Health, Air Pollution Risk Assessment, Air Pollution Monitoring, and Indoor Air Pollution. The book is recommended to professionals who wish to exercise applications of the concepts in air pollution discussed in the book.

Dr. Farhad Nejadkoorki
Associate Professor of Environmental Science
Department of Environmental Engineering
Yazd University,
Yazd, IRAN

Section 1

Air Pollution Monitoring

Chapter 1

Pollutant Degradation in Gas Streams by means of Non-Thermal Plasmas

Milko Schiorlin, Cristina Paradisi,
Ronny Brandenburg, Michael Schmidt,
Ester Marotta, Agata Giardina and Ralf Basner

Additional information is available at the end of the chapter

http://dx.doi.org/10.5772/60049

1. Introduction

In the industrialized society air pollution has become a major concern for the environment and public health. Pollutants such as particulate matter and harmful chemicals are causing disease or death to humans and other living organisms and strongly affect ecosystems (e.g. acidification, eutrophication). Although significant progress has been made in recent years in tackling the emissions of some air pollutants (e.g. sulphur dioxide), a large fraction of the population is still exposed to excessive concentrations of certain air pollutants, in particular volatile organic compounds, particulate matter, and ammonia. Therefore, environmental standards are continuously being raised and more effective technologies for depollution of gas streams are needed.

Air non-thermal plasmas (NTPs) are strongly oxidizing environments and thus useful means for the activation of advanced oxidation processes for air and water remediation. NTPs are used for the generation of ozone as an important oxidant for water or air cleaning and the removal of dust from flue gases in electrostatic precipitators [1, 2, 3]. Within the last years deodorization by means of NTP or NTP-supported methods has reached the level of commercialisation.

The term plasma denotes an ionised gas containing free electrons, ions and neutral species (atoms and molecules) characterized by collective behaviour. Often referred to as the "4th state of matter", plasmas have unique physical and chemical properties distinct from solids, liquids and gases. They are electrically conductive, respond to electromagnetic fields, contain chemically reactive species as well as excited species and emit electromagnetic radiation in various wavelength regions. Plasmas are generated artificially by supplying energy to gases, liquids or solids.

In principle, gaseous plasma depollution can be done by an increase of the gas enthalpy in so-called "translational" plasmas, which are non-equilibrium plasmas (electrons and heavy particles have different mean kinetic energies) but with gas temperatures reaching several thousand K. Plasma torches, arcs, arc jets or gliding arcs are examples for such plasma-based incineration sources [4]. However in the following we will focus on non-thermal plasmas which stays at moderate gas temperatures, also referred to as "cold" non-thermal plasmas. Another approach, which is also excluded in this chapter, is the electron beam injection, which is under development for very large combustion facilities (coal fired power plants) [5].

The most common method of producing "cold" non-thermal plasmas for technological applications is based on the application of an electric field to a gas. If the applied field exceeds a certain threshold value (breakdown field strength) a gas discharge and thus plasma are generated. The specific feature of so-called non-thermal plasmas is that most of the coupled energy is primarily released to the free electrons which exceed in temperature that of the heavy plasma components (ions, neutrals) by orders of magnitude. Thus, strong non-equilibrium conditions are achieved in which the gas temperature remains nearly at or slightly above room temperature ("cold") while "hot" electrons initiate chemical processes resulting eventually in the oxidation of pollutants.

This chapter is meant to provide some insight into the application of NTP for air remediation. After a brief introduction to summarize the principles of NTP generation, the main aspects of discharge physics and plasma chemistry involved in air treatment are described and discussed. Special focus is on the removal of volatile organic compounds (VOCs). The two major types of plasma sources for such applications, namely dielectric barrier discharges and corona discharges, are described in the next section. Various aspects of plasma generation and the ensuing chemical processes will be discussed in two separate sections based largely on work carried out by the Authors in Padova and in Greifswald. Finally, some conclusions and perspective outlook on the field are given.

2. Overview non-thermal plasma-reactors and processes for depollution of gases

The most common principles for the generation of "cold" non-thermal plasmas are dielectric barrier discharges (DBDs) and corona discharges. They are most suitable for treating exhaust gases from manufacturing processes and mobile emission sources, as they offer a compact design and good scalability.

DBD based devices consist of at least two electrodes enclosing a gas space which is filled with, or bound by, an insulating material [6]. Typically, dielectric materials such as glass, quartz, ceramics, enamel, plastics, silicon rubber or Teflon are used as barrier materials. There are many possible DBD arrangements. Traditionally, DBDs were generated in parallel plate reactor geometries or in coaxial cylindrical reactor geometries, as shown in Figure 1 (left). The dielectric barrier(s) can cover one or both electrodes entirely, but they can also be separated from both electrodes, forming two discharge gaps. The gap widths are typically in the range of 0.1 – 5 mm. When a sufficiently high voltage is applied between the electrodes, an electrical breakdown occurs and plasma is formed. Furthermore, both electrodes can be arranged in such a way that they are in direct contact with the barrier. In this case, the gas discharge is formed in the gas at the exposed electrode and propagates along the dielectric surface, and is therefore called 'surface discharge' or 'surface DBD'. In 'co-planar DBDs' both electrodes are embedded in the insulator. The so-called 'sliding DBD' is based on surface DBDs but with a third electrode which is placed opposite to the top electrode on the dielectric surface [7]. This arrangement allows the discharge 'to slide' over the dielectric. The so-called 'packed bed reactor' is also often classified as DBD-type plasma. Dielectric or ferroelectric pellets are packed in between the two electrodes. Due to polarization of the pellet material, regions with high electrical fields are generated, leading to gas discharges in the void spaces between the pellets and on their surfaces [8]. Porous ceramic foams can also be used instead of pellets beds [9]. Packed-bed reactors are relevant for depollution since the filling can feature catalytic properties.

The insulator suppresses large currents on the electrodes and thus keeps the plasma in the non-thermal regime [6]. Because of the capacitive coupling of the insulating material to the gas gap, DBD generation requires alternating or pulsed operating voltages. For the treatment of gas streams the gas is injected into the device flowing along the electrode arrangement. In industry DBDs are used for the generation of ozone, deodorization, surface treatment and many more.

Corona discharges are characterized by non-uniform electrical field geometries, e.g. needle-to-plate or wire-in-cylinder electrode arrangements, as shown in Figure 1 (right). DC and low frequency AC operated corona discharges expand from the needle or wire electrode in the outer regions towards the plate or cylinder electrode. The energy is mainly dissipated in the high-ohmic region of non-ionized gas in the outer drift region, where the electrical field is lower than in the plasma region around the wire/needle electrode. In this region the discharge is not supported anymore and is thus kept in the non-thermal regime [10]. DC-operated corona discharges are used in electrostatic precipitators. For environmental applications (e.g. VOC removal, water purification) corona discharges operated by pulsed high voltage are proposed since higher densities of reactive species can be achieved [11]. Pulsed corona discharges are characterized by plasma regions which fill a much larger fraction of the discharge gap than DC or low frequency corona discharges.

In molecular gases at atmospheric pressure, corona discharges and DBDs are typical examples of non-uniform, filamentary plasmas, consisting of many individual microdischarges or discharge channels. Each volume element of the flowing gas is repeatedly subjected to the action of these filaments as it passes through the reactor. Non-thermal plasma based remediation of air is due to chemical reactions with photons and active species created in the plasma, namely radicals or ions. The different physical and chemical processes associated with and induced by non-thermalizing discharges span a time range of about 12 orders of magnitude. The equilibrium of the electrons with the local electrical field is usually approached within

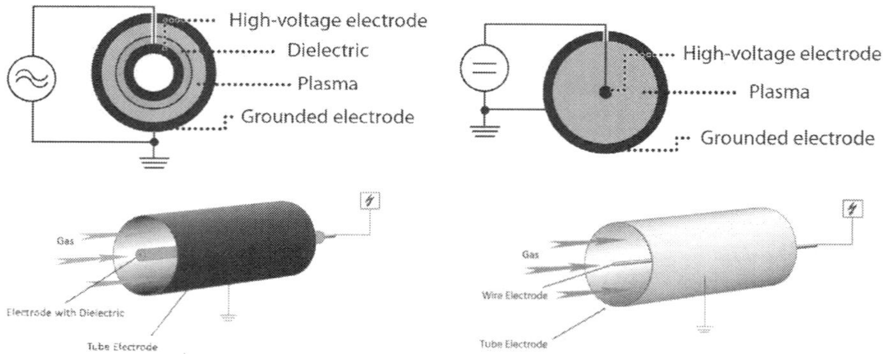

Figure 1. Top pictures: Cross sectional view of coaxial DBD (left) and DC corona discharge (right) arrangement; Bottom pictures: Coaxial DBD (left) and corona discharge (right) reactor for treatment of air streams.

picoseconds [6, 12, 13]. Ionization and electrical breakdown typically proceed at the nanosecond time scale via electron collisions. For example, DBD microdischarges in atmospheric air have duration of about 20 - 50 ns. The development of discharge channels and microdischarges is dominated by the build-up and spatio-temporal enhancement of volume space charges, resulting in propagating perturbation of the electric field, which has been investigated as a cathode directed streamer or ionization wave [12, 13]. In the filaments the highest electron density and electron temperature are achieved and electron-induced dissociation and thus formation of radicals occur. The pollutant degradation is initiated by secondary reactions with these free radicals and ions, mainly on a micro- to millisecond time scale, i.e. after the active discharge filament has faded [14]. Ion-molecule reactions occur on an intermediate time scale, typically in the range of 10 ns up to 1 μs. The pollutant molecules react with oxidizing atoms and radicals (e.g. O, OH) or with plasma-generated ozone (O_3). Water vapor may play an important role, as it acts as the precursor for hydroxyl radicals (OH) and hydroperoxyl radicals (HO_2). When hydrocarbons or other VOCs are present in the gas, other radicals are also produced and radical chain reactions occur. Beside electrons, photons or collisions with metastable or excited species can lead to ionization.

The chemical equilibrium including heat and mass transfer is commonly settled within milliseconds to seconds.

The application of NTP for pollutant degradation in gases in industry must comply with the demands on removal efficiency, energy efficiency and selectivity.

The removal efficiency is defined as the removed molar fraction of the pollutant related to the initial molar fraction C_{in}: $\eta=(C_{in}-C_{out})/C_{in}$ (C_{out} is the molar fraction of the pollutant after the plasma treatment) [2, 14]. Energy efficiency relates to the energy needed to achieve a given removal efficiency and can be expressed in several ways. For example the energy yield is the decomposed pollutant mass per dissipated energy. A high energy yield is not necessarily assuring sufficient removal efficiency as this is determined by the initial molar fractions of the pollutant [2]. The removal efficiency and the energy efficiency depend on the specific energy density of the plasma and are also determined by a number of conditions (gas composition, humidity and temperature; level of initial contamination) [2, 14].

Selectivity is defined as the fraction of the desired product of the plasma-chemical conversion to the total amount of products of the conversion process. The chemistry and fraction of desired products and undesired by-products can also be characterized by mass balances (e.g. carbon balance in the case of conversion of hydrocarbons). A high selectivity is required to achieve a reasonable performance in terms of energy efficiency and by-products [2]. High reactivity of radicals usually results in a poor selectivity, since competing reactions which result in the formation of undesired by-products happen simultaneously. One important reaction which consumes oxygen atoms alongside the reactions with pollutant molecules is the generation of O_3. For some pollutants (e.g. NO_x, alkenes and other unsaturated VOCs) O_3 is an efficient oxidizer, but in other cases it constitutes an additional pollutant by-product.

In the following paragraphs the main aspects of pollutant degradation by means of DBDs and corona discharges are discussed using selected examples. The discussion is focussed on VOCs as a class of contaminants present in many different industries (e.g. semiconductor manufacturing, chemical processing, painting and coating) as well as in indoor air (outgassing of paint, carpets etc.). VOCs contribute to the generation of photochemical smog and to certain health diseases like nausea and skin irritation. Some are associated with high cancer risk [15, 16]. Conventional methods for VOCs removal are thermal oxidation, condensation, absorption and biofiltration. The thermal oxidation and condensation are economic only for situations in which VOCs are present in moderate to high concentrations. The absorption process does not destroy VOCs but only transfers them to another medium. In addition, this technology suffers from problems arising by deposits of dirt or clog on filters. Biofilters are useful only for VOCs that have some solubility in water and they are cost-effective if the volume of air to be treated is in the range of $10^4 – 10^5$ m^3/h. In NTP energy of about 10-30 eV are needed to produce an O-atom or an OH-radical in (humid) air, which makes the decomposition also energy consuming [14]. However, the total energy consumption can be low in case of small concentrations of pollutants. Thus NTP-based processes are feasible for low contamination levels. For VOCs, this level is about 100 mg/m_N^3 (N refers to standard conditions for pressure and temperature) [2, 14].

Among the different VOCs which are being routinely monitored for air quality, toluene (methylbenzene, C_7H_3) is one of the most important ones. Toluene is widely used as feedstock in the chemical industry for the synthesis, among others, of drugs, dyes, explosives, and as a solvent (e.g. thinner, paints, adhesives). Exposure to toluene is known to affect the central nervous system and may cause tiredness, confusion, weakness, memory loss, and nausea. Toluene is water-insoluble and thus cannot be scrubbed. For its wide use, diffusion and well known properties and reactivity, toluene has become sort of a standard for testing and comparing non-thermal plasma based air treatment for VOCs removal. Thus, the discussion in the following sections will focus largely on experiments with toluene. The conversion of toluene in hybrid systems in which NTP is combined with a catalyst is also being extensively studied and has been reviewed [17]. Such hybrid processes will not be covered in this chapter.

3. VOC removal by means of corona discharges

Air plasma produced by corona discharges and its performance in the oxidation of VOCs are being investigated in Padova using a prototypal large corona reactor [18–20]. The reactor and the auxiliary apparatus were designed in order to achieve stable and reproducible plasma regimes and experimental conditions, which are necessary for quantitative kinetic and product studies. Reproducibility and stability of experimental conditions in our set-up allow to test and compare the performance of different corona regimes, notably dc+, dc– and pulsed+ within the same apparatus and under otherwise identical experimental conditions. The experimental set-up, comprising the corona reactor, the gas flow line and instrumentation for *in line* and *off line* analysis of the treated gas, is schematically reproduced in Figure 2.

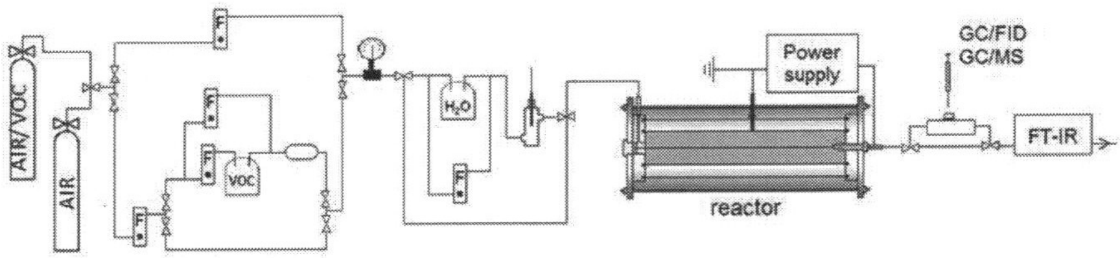

Figure 2. Schematics of corona reactor, gas flow line and instrumentation for *in line* and *off line* analysis of the treated gas.

The corona reactor has a wire/cylinder electrode configuration. The wire electrode (stainless steel, outer diameter 1 mm) is electrically connected to the high voltage supply and fixed along the axis of a stainless steel cylinder (38.5 mm i.d. x 600 mm) which is electrically grounded.

The reactor can be energized by dc or pulsed high-voltage power. The dc power supply has an output voltage of ± 25 kV and an output current of 0 – 5 mA. For generating pulsed corona, a pulsed high voltage with dc bias (PHVDC) was used, based on a spark gap switch with air blowing, with the following specifications: dc bias of 0 – 14 kV, peak voltage of 25 – 35 kV (with dc bias), peak current up to 100 A, maximum frequency 300 Hz, rise-time of the pulses less than 50 ns. To measure the power input two homemade current probes (*shunt*), one of 1.1 Ω for pulsed current, the other of 52 Ω for dc current, were used. The experimental apparatus was described in detail previously [18].

The reactor is connected to a gas flow line made of Teflon tubing (inner diameter 4 mm). The air/VOC mixture is prepared by bubbling synthetic air (80% nitrogen: 20% oxygen from AirLiquide) through a sample of liquid VOC and by diluting the outcoming flow with a second flow of synthetic air to achieve the desired gas composition (VOC mixing ratio in the 100 - 1000 ppm range) and flow rate (usually kept constant at 450 $mL_N \cdot min^{-1}$). The gas flow line is equipped with a loop for humidification and with a probe to measure the humidity. The treated gas exiting the reactor goes through a small glass reservoir equipped with a sampling port from which aliquots are withdrawn with a gastight syringe for off-line chemical analysis by

GC-MS (Agilent Technologies 5973) and GC-TCD/FID (Agilent Technologies 7890). *In line* IR analysis is performed with an FTIR Nicolet 5700 spectrophotometer using a 10 cm long gas cell with windows of NaCl (for experiments with dry air) or of CaF_2 (for experiments with humidified air). The determination of ozone, CO, and CO_2 were performed by integration of characteristic IR bands as described previously [18]. The determined conversion data, i.e. $[VOC]/[VOC]_0$ as a function of the specific input energy (SIE, also referred to as specific energy density SED) usually follow a first order exponential decay profile. The SIE was determined as described previously for dc [18, 21] and pulsed [20, 21] corona, respectively. The data are thus interpolated with the equation (1) to obtain the energy constant kE, which is a measure of the process energy efficiency.

$$[VOC] = [VOC]_0 \cdot e^{(-k_E \cdot SIE)} \tag{1}$$

Current/voltage characteristics of dc corona, both of positive and negative polarity, were monitored in synthetic air with and without VOC admixture (500 ppm VOC concentration). For each applied voltage, the mean current intensity was measured, after a stabilization time of 5 minutes, using a multimeter. The ions present in the air plasma produced by +dc and –dc corona were investigated using an APCI (*Atmospheric Pressure Chemical Ionization*) interfaced to a quadrupole mass analyzer (TRIO 1000 II, Fisons Instruments) [22, 23]. A schematic drawing of the arrangement and the gas inlet systems is shown in Figure 3.

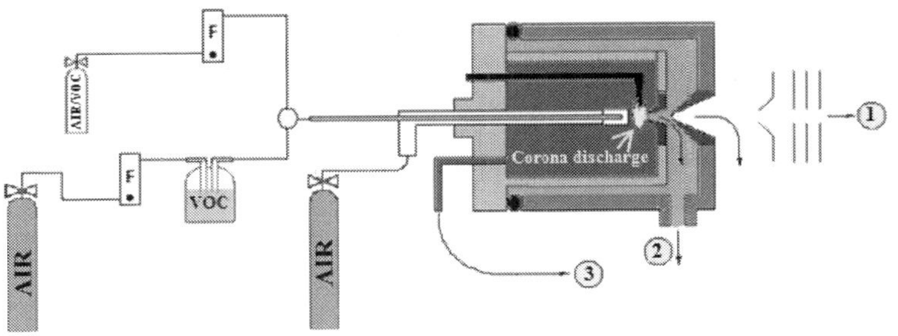

Figure 3. Schematics of APCI ion source and gas inlet system (1) quadrupole analyzer (2) rotary pump, (3) diaphragm pump.

The corona discharge is kept at atmospheric pressure by a flow of synthetic air (4–5 L·min^{-1}) introduced through the nebulizer line, a capillary of ca. 2 mm (inner diameter). Vapors of the desired VOC, stripped by an auxiliary flow of synthetic air (typically 5–50 mL·min^{-1}) from a liquid sample contained in a reservoir, enter the APCI source through another capillary (inner diameter 0.3 mm) placed coaxially inside the nebulizer line. A second line allows for the introduction of water vapors as desired. The needle electrode for corona discharge was kept at 3 kV. Ions leave the source through an orifice (50 μm in diameter) in the counter electrode, called the "sampling cone" and held at 0–150 V relative to ground. The ions then cross a low pressure region (down to ca. 10^{-2} Torr) and, through the orifice in a second conical electrode, called the "skimmer cone" and kept at ground potential, reach the low pressure region hosting the focusing lenses and the quadrupole analyser. Prior to running the experiments with the VOC, a preliminary analysis is routinely conducted to acquire the "background" spectra with only synthetic air and humidified synthetic air [a)].

The efficiency, products and mechanisms of VOC oxidation were studied systematically under variation of the corona type (dc or pulsed), the corona polarity (negative or positive), the VOC (a few hydrocarbons, halogenated and oxygenated organic compounds have been investigated), the VOC inlet concentration and the level of humidity. These studies have provided a large body of experimental results which give insights into corona induced chemical oxidation and useful hints for its application.

The type of corona has major impact on the process efficiency. In Figure 4 an example is shown, which is reporting a comparison of the decay profile of toluene concentration as a function of SIE under three different corona regimes: dc+, dc– and pulsed+ [21]. The much better efficiency of pulsed+ corona with respect to dc corona of either polarity is evident. Also evident is the better performance of dc– with respect to dc+ corona. Analogous results were obtained in similar experiments with other VOCs, including *n*-hexane [18, 20] and dibromomethane [24].

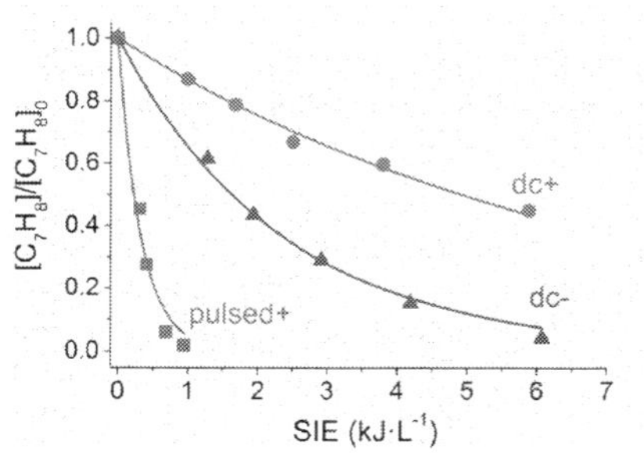

Figure 4. Decay profile of toluene (500 ppm in synthetic air) as a function of SIE in corona induced oxidation under the following regimes: pulsed+, dc– and dc+ [21].

The better efficiency of pulsed+ corona with respect to dc corona of either polarity is consistent with the results of an emission spectroscopy study which showed that at any specific input energy significantly greater average electron energy is obtained with pulsed corona than with dc coronas [25]. Correspondingly, a higher density of reactive O atoms is observed in pulsed + corona than with dc coronas [25]. Due to the filamentary nature of the plasma, not only the energy but also the spatial distribution of electrons and other short-lived reactive species is very different from that in glow dc coronas: the plasma is affecting a relatively larger volume thus accounting for a more efficient process.

A second important variable is the VOC initial concentration. Usually, the corona induced oxidation efficiency decreases as the VOC initial concentration $[X]_0$ is increased and often a linear correlation is observed between k_E and $1/[X]_0$ within a significant range of concentrations. An example is shown in Figure 5.

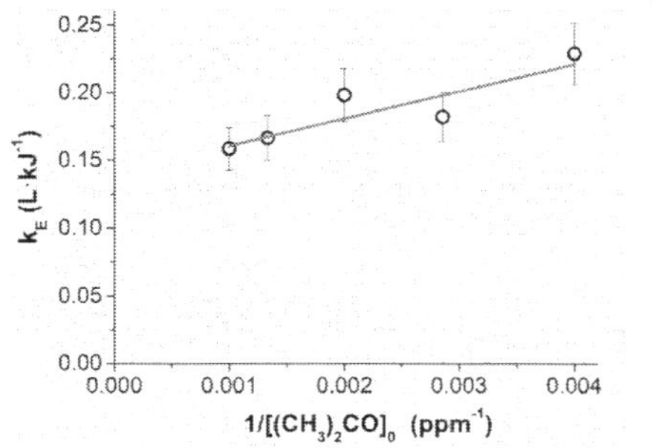

Figure 5. Dependence of process efficiency (k_E) on the reciprocal of VOC initial concentration for dc+ corona induced oxidation of acetone in dry synthetic air.

Other similar cases are reported in the literature [14, 24, 26–29] and have been interpreted based on a simple scheme of inhibition by the intermediates formed in the VOC reaction [30]. Finally, the VOC chemical composition and structure also matters and different VOCs are oxidized with different efficiencies under the same experimental conditions. A few representative data are reported in Table 1.

VOC[b)]	dc–		dc+	
	dry air	humid air	dry air	humid air
n-hexane [c)]	$7.7 \cdot 10^{-1}$	1.1	$2.0 \cdot 10^{-1}$	$1.8 \cdot 10^{-1}$
i-octane	$4.2 \cdot 10^{-1}$	$7.5 \cdot 10^{-1}$	$1.3 \cdot 10^{-1}$	$1.2 \cdot 10^{-1}$
toluene	$4.1 \cdot 10^{-1}$	$8.1 \cdot 10^{-1}$	$1.4 \cdot 10^{-1}$	$1.3 \cdot 10^{-1}$
CH_2Br_2	$2.1 \cdot 10^{-1}$	$2.6 \cdot 10^{-1}$	$6.4 \cdot 10^{-2}$	$5.1 \cdot 10^{-2}$
CF_2Br_2	$1.4 \cdot 10^{-1}$	$1.1 \cdot 10^{-1}$	$4.5 \cdot 10^{-2}$	$3.9 \cdot 10^{-2}$

[a)] Data are from ref. [27] unless otherwise specified. [b)] VOC initial concentration was 500 ppm. [c)] Data from ref. [25].

Table 1. Reaction efficiency data, expressed as k_E in L kJ^{-1} units, for corona processing of different VOCs in dry and in humid (40% RH) synthetic air.

This outcome might not have been anticipated *a priori* since the generally accepted notion is that plasma chemical processes proceed via radical reactions which are usually very fast and poorly selective. This is the case, for example, for the reaction of OH radicals with organic compounds which is viewed as a major contributor to VOC oxidation in humid air plasmas.

The data in Table 1 show instead that air plasmas are somewhat selective. This selectivity might originate from either of two circumstances (or possibly a combination of the two): within a given type of plasma, say that produced by dc–, different VOCs either react along different paths or react with the same species but at different rates. The vast available bibliography on rate constants for reactions of many VOCs with air plasma reactive species (atoms, radicals, ions) and on ionization energies and electron affinities provides tools to exclude some possibilities and sort out which reactions are most likely paths. Chemical knowledge and intuition help in providing model VOCs to be used as reactivity probes.

A most intriguing and informative response is found in studying the effect of humidity on the efficiency of VOCs oxidation. The data in Table 1 show that for all VOCs considered, except CF_2Br_2, the presence of humidity in the air produces an increase in efficiency with dc– corona and no effect or a slight decrease in efficiency with dc+ corona. The increase in efficiency observed with dc– is rather straightforwardly attributed to the OH radicals formed by corona discharges in humid air. OH radicals are among the strongest known oxidants of VOCs. Compare for example the rate constants for reaction of toluene with atomic oxygen (2) [31] and with OH radical (3) [32], a channel becoming more available in humid air plasma:

$$C_7H_8 + O \rightarrow Products \qquad k_{298} = 7.6 \cdot 10^{-14}\ cm^3 \cdot molecule^{-1} \cdot s^{-1} \tag{2}$$

$$C_7H_8 + OH \rightarrow Products \qquad k_{298} = 5.7 \cdot 10^{-12}\ cm^3 \cdot molecule^{-1} \cdot s^{-1} \tag{3}$$

Support for the conclusion that reaction with OH radicals is important in dc– corona induced oxidation of hydrocarbons and of CH_2Br_2 (Table 1) came from experiments with CF_2Br_2 (halon 1020) [27]. Like other perhalogenated saturated hydrocarbon, CF_2Br_2 is not attacked by OH and other atmospheric radicals: the reaction of CF_2Br_2 with OH radicals is more than 220 times slower [33] than that of CH_2Br_2. And indeed there was no increase in efficiency for dc– processing of CF_2Br_2 in humid air, but rather a slight decrease with respect to dry air. This slight decrease in efficiency was attributed to reaction (4) which contributes to reduce the average electron energy while producing OH radical, which is unable to attack this specific VOC.

$$H_2O + e^- \rightarrow OH + H + e^- \tag{4}$$

Less straightforward was to explain the decrease in efficiency observed with dc+. To make sure that OH radicals also form in dc+ corona regime and to compare their relative densities in dc + and dc– air plasmas the well known reaction of OH with CO to form CO_2 (eq. (5)) was used [34].

$$CO + OH \rightarrow CO_2 + H \qquad k_{298} = 2.41 \cdot 10^{-13}\ cm^3 \cdot molecule^{-1} \cdot s^{-1} \tag{5}$$

Indeed, in a control experiment CO did not react at all in dry air under the effect of either dc + or dc– corona. In contrast, in humid air (40% RH) reaction (5) occurs both with dc–, more efficiently, but also with dc+ corona, thus proving the presence of OH radicals in such plasmas. Since with dc+ oxidation of hydrocarbons is less efficient in the presence of OH radicals than it is in dry air (Table 1), it was concluded that reaction with OH radicals is not the dominant initiation channel for their oxidation in dc+ corona. Thus, it appears clearly that VOC oxidation induced by dc+ and dc– corona in air occurs by different mechanisms. For the investigated hydrocarbons (see Table 1) oxidation induced by dc+ corona is believed to be initiated by ion-molecule reactions. Support for this hypothesis comes from direct observation of the ions within the plasma achieved by APCI-mass spectrometry analysis and from comparison of current/voltage (I/V) profiles measured with only synthetic air and with VOC-containing synthetic air.

Figure 6 (a) reports *I/V* data monitored in experiments with toluene (500 ppm initial concentration). It is seen that for dc– the profiles determined with and without toluene are nearly superimposed, whereas in the presence of toluene for dc+ the current intensity measured is significantly lower, at any applied voltage, than found in pure synthetic air. Since corona current is due to ion transport across the drift region of the interelectrode gap, the results suggest that with dc+ corona different ions are present in pure air and in toluene containing air. Accordingly, different average ion mobilities are derived from the current/voltage characteristics of Figure 6 (a) [35–37]: 2.35 $cm^2 \cdot V^{-1} \cdot s^{-1}$ for pure air and 1.79 $cm^2 \cdot V^{-1} \cdot s^{-1}$ for toluene-containing air, respectively.

APCI-mass spectrometry is a powerful tool for monitoring and characterizing the ions formed by corona discharges and their reactions. The APCI-mass spectra reported in Figure 6 (b) show the ions present in the plasma produced in synthetic air by dc– and dc+ corona discharge, respectively. These ions are water clustered O_2^- and O_3^- ions ($O_2^-(H_2O)_n$ (n = 0-2: m/z 32, 50, 68), $O_3^-(H_2O)_n$ (n = 0-1: m/z 48, 66)) as well as $O_2^-(O_2)$ (m/z 64) for dc– corona and $H_3O^+(H_2O)_n$ (n = 2 - 3: m/z 55, 73) and $NO^+(H_2O)_n$ (n = 1 - 2; m/z 48, 66) for dc+ corona, respectively.

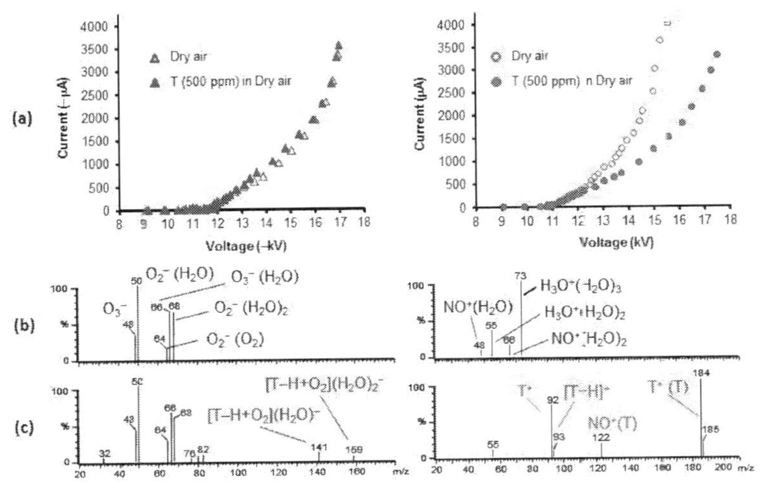

Figure 6. (a) Current/voltage profiles measured with dc– and with dc+ corona in pure air (open symbols) and in toluene (500 ppm) containing air (closed symbols). (b) APCI mass spectra recorded with dc– and dc+ corona in pure air. (c) APCI mass spectra recorded with dc– and dc+ corona in air containing toluene (500 ppm).

The APCI mass spectra recorded under the same experimental conditions except for the presence of a small amount of toluene (500 ppm) in the air are shown in panel (c) of Figure 6. The effects are significantly different for dc– and dc+ corona. Thus, with dc– corona the major ions observed in the plasma are the same regardless of whether toluene is present or not in the gas (compare the mass spectra on the left-hand side in panels (b) and (c) of Figure 6). These observations are fully consistent with and provide a rationale for the nearly identical *I/V* curves determined for dc– in pure air and in toluene containing air (panel (a) of Figure 6).

In contrast, in the case of dc+ the mass spectrum of toluene containing air is completely different from that of pure air (compare the mass spectra on the right-hand side in panels (b) and (c) of Figure 6). Thus, in air contaminated with toluene (500 ppm) the prevailing charged species are T^+ (m/z 92) and $[T+H]^+$ (m/z 93) (T stands for the toluene molecule), along with their ion-molecule complexes $T^+(T)$ (m/z 184) and $[T+H]^+(T)$ (m/z 185). These ions form [38-40] via exothermic charge- and proton transfer ion-molecule reactions (eq. 6 and 7), characterized by rate constants of $1.8x10^{-9}$ and $2.2x10^{-9}$ cm^3 $molecule^{-1}$ s^{-1}, respectively [38], followed by ion-molecule complex formation (eq. 6a and 7a).

$$\begin{aligned} T + O_2^+ &\rightarrow T^+ + O_2 \\ T^+ + T + M &\rightarrow T^+(T) + M \quad (a) \end{aligned} \tag{6}$$

$$\begin{aligned} T + H_3O^+ &\rightarrow [T+H]^+ + H_2O \\ [T+H]^+ + T + M &\rightarrow [T+H]^+(T) + M \quad (a) \end{aligned} \tag{7}$$

Finally, the $NO^+(T)$ (m/z 122) ion-molecule complex is also observed. Thus, mass spectroscopic ion analysis provides a rationale, at the molecular level, for interpreting *I/V* curves observed with dc+ (Figure 6 panel (a) right hand side). In addition, these results suggest that ionic reactions might be responsible for the initial stages of toluene decomposition induced by dc+ corona. This hypothesis is consistent with the observed insensitivity of the dc+ process efficiency to the presence of humidity, which rules out a significant role of the OH radical.

Analogous results were obtained with other hydrocarbons leading to the conclusion that the initial step of oxidation depends on the plasma regime applied: ion-molecule reactions are favored with dc+ whereas reactions with O atoms and OH radicals prevail in the case of dc– corona discharges.

The yield of the final oxidation product, CO_2, as a function of SIE has been determined and compared with the profile of VOC conversion (Figure 7). CO_2 production is clearly less energy efficient than VOC conversion as is reasonable to expect for a process which involves many steps and oxidation intermediates.

In comparing pulsed and dc coronas, at any given value of VOC conversion the yield of CO_2 increases in the order pulsed+ < dc– < dc+. This is evident from the data shown in Figure 8a concerning experiments with *n*-hexane. Comparing the results corresponding to a given

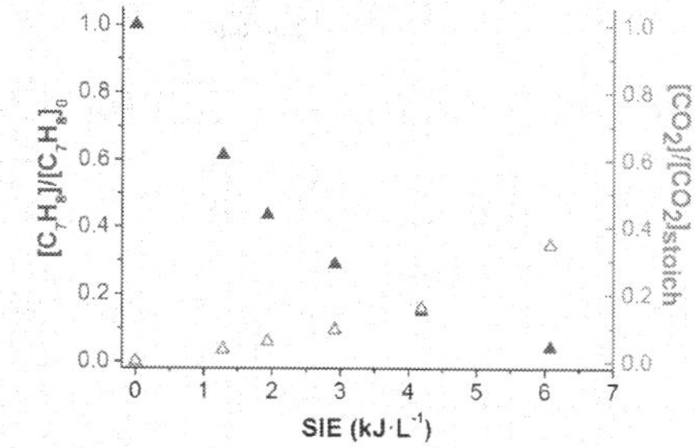

Figure 7. Profiles of VOC decay and CO_2 production as a function of SIE for treatment of toluene (500 ppm) with dc– corona in dry synthetic air.

decomposition fraction of the VOC, for example 0.7 (70% conversion), the corresponding amount of CO_2 released with dc+ is about 4 times larger than with pulsed+ and about 1.6 times larger than with dc–. Thus, among the different types of corona tested, dc+ has the poorest efficiency for VOC conversion but the best selectivity for CO_2 production. On the other hand, a consistently lower CO_2/CO ratio is found with dc+ than with dc– and pulsed+ (Figure 8b).

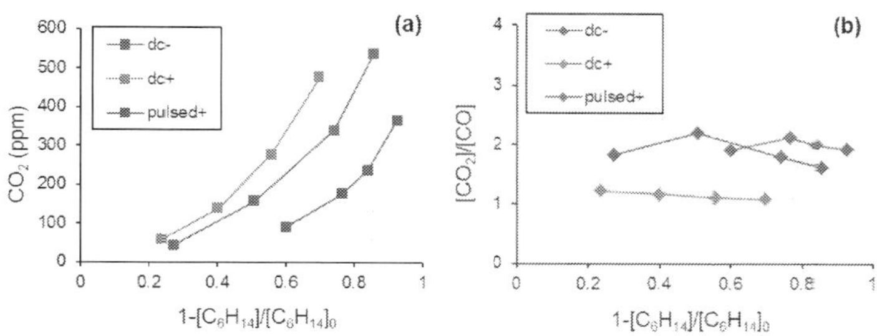

Figure 8. CO_2 production and CO_2/CO ratio for treatment of *n*-hexane (500 ppm) with dc+, dc– and pulsed+ corona in dry synthetic air. The data are displayed as a function of the fraction of decomposed *n*-hexane.

In search for the missing fraction of organic carbon several oxidation intermediates were identified and quantified by means of GC/MS and GC/FID analysis and proper standards. In the case of *n*-hexane the major detected intermediates were a few aldehydes and ketones, as shown in Figure 9. It is seen that the concentration of most of these intermediates reaches a maximum and then decays, showing that they are in turn oxidized in air non-thermal plasmas. It is also seen that, with the exception of acetaldehyde, the concentrations of these organic intermediates are very small under any of the applied conditions. The experimental data have been fitted according to a simple kinetic model for consecutive reactions [41] to obtain relative reactivity data of the intermediates with respect to that of the precursor, *n*-hexane [18].

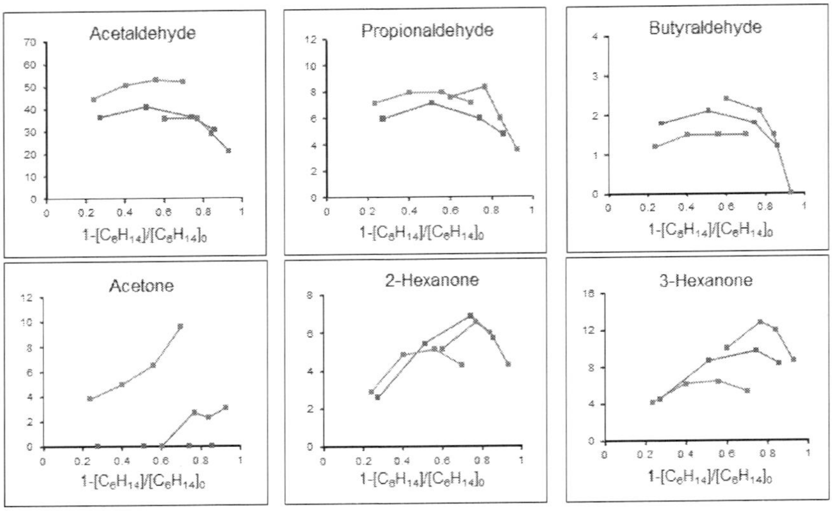

Figure 9. Aldehydes and ketones detected as intermediates in the oxidation of *n*-hexane (500 ppm in dry synthetic air) induced by pulsed+, dc– and dc+. The data are displayed as a function of the fraction of decomposed *n*-hexane [18].

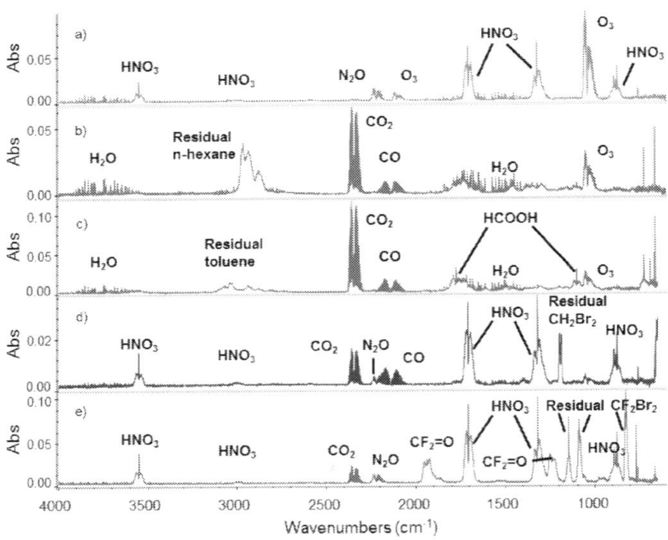

Figure 10. FT-IR spectra recorded in experiments with dc+ corona (+19 kV) in (a) pure synthetic air, used as reference spectrum, and in synthetic air containing 500 ppm of the following VOCs: (b) *n*-hexane; (c) toluene; (d) CH_2Br_2; (e) CF_2Br_2.

In line FT-IR spectroscopy gives a comprehensive overview of the composition of the treated gas analyzed at the outlet of the corona reactor. Besides CO_2 and CO, other species can be conveniently determined, including ozone, various nitrogen oxides and derivatives (N_2O, HNO_3, etc), and, depending on the specific VOC, also other volatile organic oxidation intermediates. A few examples, reported in Figure 10, show characteristic bands for VOC specific products such as formic acid in the case of toluene and CF_2O in the case of CF_2Br_2.

The production of HNO_3 is high and ozone is almost completely absent in corona discharge treatment of air containing bromo derivatives CH_2Br_2 and CF_2Br_2, (Figure 10 d and e). These observations have been explained considering reactions and catalytic cycles involving BrOx (x = 0, 1) and NOx (x = 1, 2) species [24], which have been extensively investigated as major contributors to the depletion of stratospheric ozone [42].

4. VOC removal by means of barrier discharges

The experimental results obtained with a surface DBD reactor developed at INP Greifswald are reported. The discharge arrangement consists of two metal woven meshes and a dielectric plate (mica) in between the two electrodes [43]. As shown in Figure 11 the surface DBD arrangement (110 x 80 x 3 mm) was installed in a gastight chamber made of poly(methyl methacrylate) (PMMA) where the gas mixture to be treated (mixed from gas cylinders by means of mass flow controllers) was conveyed with a total gas flow of 75 L_N/h. In order to vary the humidity of the gas mixture the partial gas flow from oxygen gas cylinder was directed through a water containing bubbler. Water content of around 0.5% was realized in this way. Most of the gas flows along the electrodes configuration instead of entering the active plasma region between the electrodes and the dielectric as depicted by the arrows in Figure 11 (right). An advantage of this configuration is the very small back pressure, which is desired for the treatment of large gas flows.

The plasma reactor was energized with a programmable high-voltage power source and a high-voltage transformer. The frequency of the applied voltage was ranged from 400 Hz to 1 kHz. The power dissipated into the plasma was analyzed by recording the high voltage operating the reactor via a high voltage probe. Additionally, the voltage drop over a capacitor (capacitance 100 nF), connected in series with the reactor between the grounded electrode and protected earth, was recorded. By multiplying the voltage drop over the capacitor by the capacitance the transferred charge was obtained.

Samples of the gas mixture were analyzed by Flame Ionization Detector (FID) (Testa FID-2010T), measuring the total amount of organic carbon present in the exhaust gas. Additionally, a Fourier Transform InfraRed spectrometer (FTIR) (Bruker Alpha, spectral resolution 1 cm^{-1}, optical path length 5 m) was used to monitor the processed gas mixtures. The gas cell of the spectrometer was heated up to 40°C in order to avoid water condensation. Higher temperatures would be desirable to avoid water condensation but cannot be used otherwise ozone will decompose and give rise to sort of a "post-plasma" contribution to toluene oxidation which should be avoided.

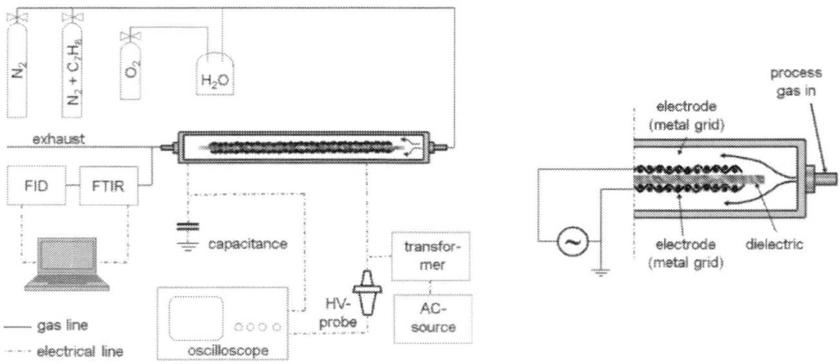

Figure 11. Experimental setup for toluene removal studies (left) and detailed horizontal cut view of the Surface DBD reactor (right).

In order to obtain comparable results of the measurements under the selected conditions the operating voltage was chosen as the electrical parameter to be set for every measurement. The electrical data recorded during the experiments were investigated during the analysis procedure. The power input into the plasma reactor was calculated by integrating the area of the charge-voltage plot (Q-V plot) and multiplying the resulting value with the frequency of the applied voltage [44]. By this, it was found that the Q-V plots for different frequencies at the same driving voltage were almost identical.

Examples are given in the left part of Figure 12. The black curves show the Q-V plots recorded at a frequency of 400 Hz, whereas the red curves display that one recorded at 1 kHz. The operating voltage was 8.3 kV and 6.9 kV, respectively. The equality of the Q-V plots implies that the energy transferred per cycle into the plasma is almost identical. Thus, the power input for a fixed operating voltage should depend only on the frequency in a linear relation [44]. Using the frequency ratio 2.5 (1000 Hz divided by 400 Hz) and multiplying it with the power input measured at 400 Hz, one gets the calculated power at 1 kHz. In the right part of Figure 12 the power measured at 400 Hz (black boxes), the power calculated for 1 kHz (blue triangles) and the power measured at 1 kHz (red circles) are shown. The values calculated and measured at 1 kHz are in good agreement and can be taken as another evidence of the proportionality of frequency and power, as already mentioned by Kogelschatz [44] and Manley [45].

Further investigations showed that the slope from the bottom right corner to the upper right one of the charge-voltage plot, which gives the capacitance of the plasma reactor during the discharge period, increases with increasing the operating voltage (Figure 13, left). The reason is suggested to be the increase in the active area of the electrode, which means the surface of the electrode covered with plasma. Photographs of the plasma were taken (Nikon D5100, aperture 5.6, exposure time 30 s) and reworked with an image manipulation program (Gimp 2.8, color correction) to make the plasma visible (Figure 13, right). An analysis of the extension of the visible plasma was performed. The obtained values were normalized as well as the values of the measured capacitance. The result is given in the left part of Figure 13. The normalized capacitance (red circles) increases linearly with the increasing driving voltage. The normalized active area of the plasma (black boxes) increases almost linearly, except the value at 7.3 kV. The linear relation between the dielectric capacitance and the amplitude of the applied voltage is further confirmed by the fact that the bottom right corner as well as the left top corner of the Q-V-plot (which both correspond to the inset of the discharge in every half period of the applied voltage) are not sharp. This would be the case for a uniform breakdown of the gas discharge.

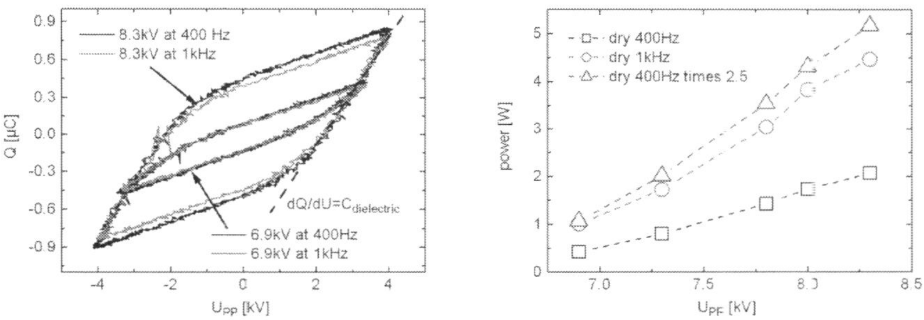

Figure 12. Left: Q-V plots recorded at 8.3 kV and 6.9 kV at 400 Hz and 1 kHz under dry conditions. Right: Power input under dry conditions at 400Hz and 1 kHz and at 1 kHz calculated based on the power of 400 Hz.

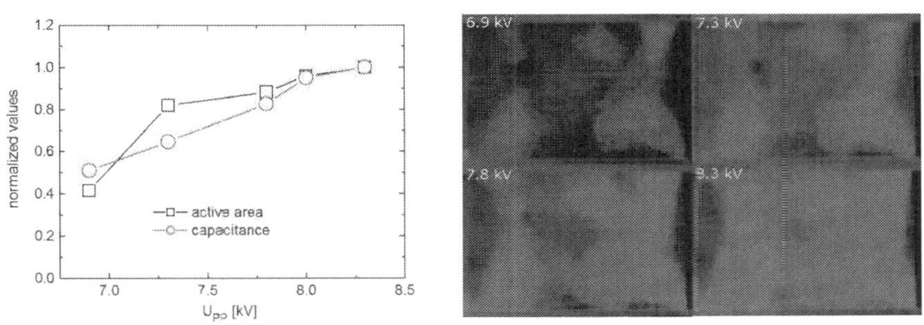

Figure 13. Left: Increase in active plasma area and increase in capacitance with respect to the operating voltage. Right: Photographs of the plasma operated at different voltages.

In a first step the overall plasma chemistry of toluene removal was investigated by FTIR. In Figure 14 samples of selected spectra are presented showing the so-called fingerprint region (i.e. wavenumber 700–1500 cm^{-1}). The left graph shows infrared-spectra taken at an applied voltage amplitude of 8.3 kV$_{PP}$ and a frequency of 400 Hz under dry conditions. In the untreated gas mixture (black curve) the absorption band related to toluene is the only detectable band.

With plasma (red curve) the strong absorption band of ozone appears at 1053 cm^{-1}, which is the main stable by-product of an NTP operated under ambient air conditions. Additionally, nitric acid HNO_3 is detected. It is assumed that HNO_3 is formed by the reaction of intermediate NO_x with hydroxyl radicals produced by the toluene decomposition. The toluene absorption band is replaced by a broad absorption band whose origin could not be identified. Because no infrared absorption spectrum of the known products or intermediates (formaldehyde, benzaldehyde, benzoic acid, benzene, nitrobenzene, phenol, formic acid, and acetic acid) of the toluene removal process fits to the measured spectra it is assumed that it is a compound emitted from the material of the reactor housing (PMMA). According to the results given by the FID this analyzed gas mixture does not contain any hydrocarbons at all.

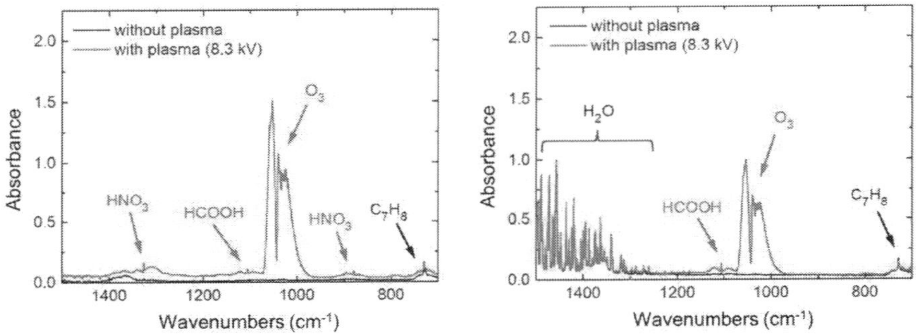

Figure 14. FTIR absorption spectra without plasma (black lines) and with plasma (frequency 400 Hz, operating voltage 8.3 kV, red lines) under dry (left) and wet conditions (right) of 50 ppm toluene in synthetic air.

Under wet conditions (right graph, same electrical parameters) there is no nitric acid detectable. Moreover, the absorption of ozone is much smaller than under dry conditions. Both phenomena are attributed to the consumption of energetic electrons, which under dry conditions are used to produce NOx as a necessary intermediate for the production of nitric acid. These changes result in a considerable production of formic acid. The lower energy efficiency in toluene removal under wet conditions is also assumed to be due to the consumption of high energy electrons for the vibrational excitation of water.

FID is used to study the toluene removal since it is not sensitive to the main by-products of toluene removal that were identified by FTIR. With the concentration of toluene at about 50 ppm in the untreated gas mixture the molar fraction was calculated and plotted against the SIE. The results are shown in Figure 15. The removal efficiency increases with the SIE up to total removal at around 55 $J \cdot L^{-1}$ under dry conditions (black boxes). The same efficiency is achieved for 400 Hz and 1 kHz. Under wet conditions (red circles) the removal efficiency is smaller and about twice as much energy is needed to achieve complete removal of toluene which is only obtained at 1 kHz for this conditions. This dependency on the frequency is only found under wet conditions.

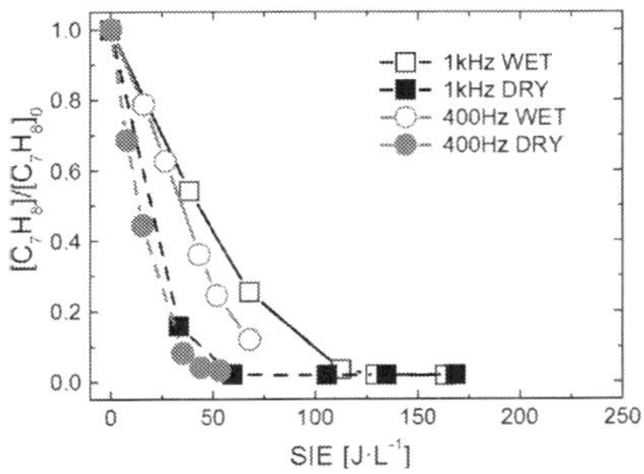

Figure 15. Molar fraction of toluene under dry (black boxes) and wet (red circles) conditions at 400 Hz (circle) and 1 kHz (square).

In order to discuss the energy efficiency the energy constant parameter k_E was evaluated according to the eq. (1). As reported in Table 2 the energy constants obtained under dry conditions are very similar. Those obtained under wet conditions are significantly smaller but also differ significantly.

Frequency	k_E L·J^{-1}	
	dry	wet
400 Hz	$5.85 \cdot 10^{-2}$	$3.03\ 10^{-2}$
1 kHz	$5.63 \cdot 10^{-2}$	$2.06\ 10^{-2}$

Table 2. Reaction energy efficiency data, expressed as k_E in L J^{-1} units, for DBD processing of toluene.

Quantitative analysis data for CO_2 and CO produced under different experimental conditions are shown in Figure 16 as a function of the SIE. At dry conditions the production of CO is favored compared to CO_2, while in humid conditions the amounts of both compounds are almost the same. The increased selectivity towards CO_2 in humid conditions could be explained as follows. In the presence of water vapor in the plasma area the production of OH radicals is higher. These radicals can react with the CO molecules to produce CO_2, according to the eq. (5).

Usually the formation of OH radicals in the plasma area is also accompanied by an enhancement of the energy efficiency in the VOC removal process (e.g. see **Section 3** of this Chapter). In the case of this setup, as discussed above, the presence of water vapor in the process gas is obviously responsible of a decrease in the energy efficiency of toluene removal.

In Figure 17 the selectivity to CO_2 of the plasma treatment for pollutant degradation is reported. The black vertical line is referred to a value of SIE of 55 J·L^{-1}, the energy value at which the toluene is completely decomposed, but as reported in Figure 17 there is selectivity of 40% in dry conditions and of 60% in humid conditions. The complete oxidation of the toluene is

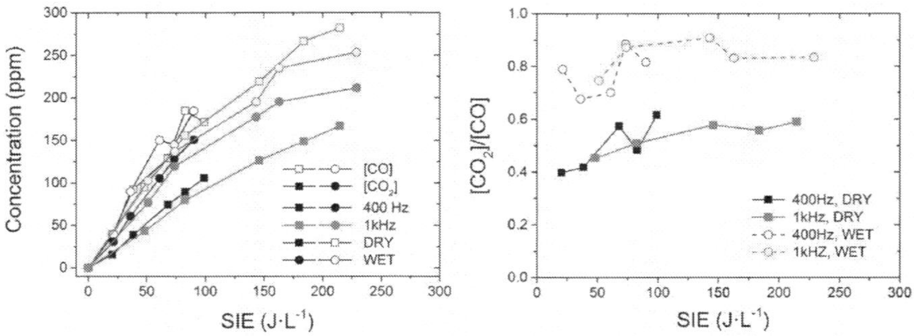

Figure 16. Left: CO (empty symbols) and CO_2 (full symbols) production during toluene decomposition experiments. Comparison at 400 Hz (black) and 1 kHz (red) and between dry conditions (full line) and wet conditions (dashed line). Right: CO_2/CO ratio as a function of SIE at 400 Hz (black) and 1 kHz (red). Comparison between dry condition (full symbols and straight lines) and humid conditions (empty circle symbols).

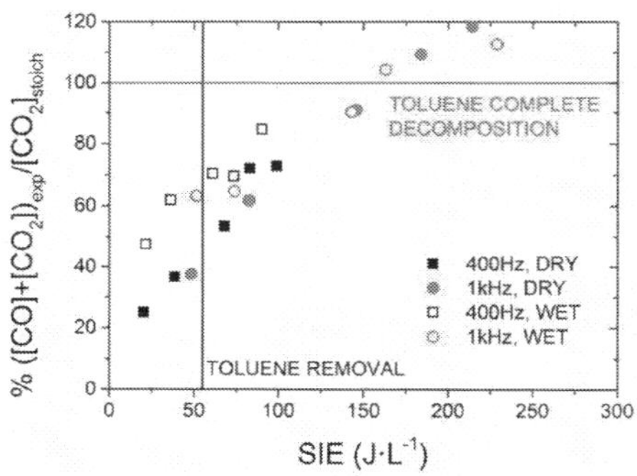

Figure 17. Carbon balance for the different experimental conditions being tested. 400 Hz (black symbols) and 1 kHz (red symbols) under dry (full square symbols) and wet (empty symbols and dashed lines) conditions.

achieved at a value of SIE of 150 J·L^{-1}, where all the VOC is decomposed to CO and CO_2. The carbon fraction which is missing is mainly formic acid that could be easily removed from the effluent gas by means of water scrubbing. Despite the reduction of the toluene removal efficiency made by the presence of water vapor, it is clear how the selectivity to CO_2 production is improved (Figure 16, right), but also the carbon balance is clearly improved at least until the energy value of 100 J·L^{-1}. Above the value of 150 J·L^{-1} the carbon balance is exceeding 100% (marked by the red horizontal line). This is due to some additional degradation of the acrylic housing. This effect was also noted in the IR spectra where an additional band around 700 cm^{-1} was recorded (see Figure 14).

Because of the construction of the NTP-reactor with the electrode configuration in the middle of the discharge reactor and, therefore, a huge gas volume not in direct contact with plasma, the toluene to be removed hardly comes in contact with plasma. Thus, electron dissociation cannot be the main process. As ozone is generated in the air plasma it is a possible oxidizer of toluene. The effect of ozone on the toluene removal has been studied in a separate experiment. Therefore the reactor was operated with a pure oxygen gas flow which was mixed with the toluene polluted air in a separate reaction chamber (volume about 250 mL; not shown in Figure 11). The results of this experiment are presented in Figure 18.

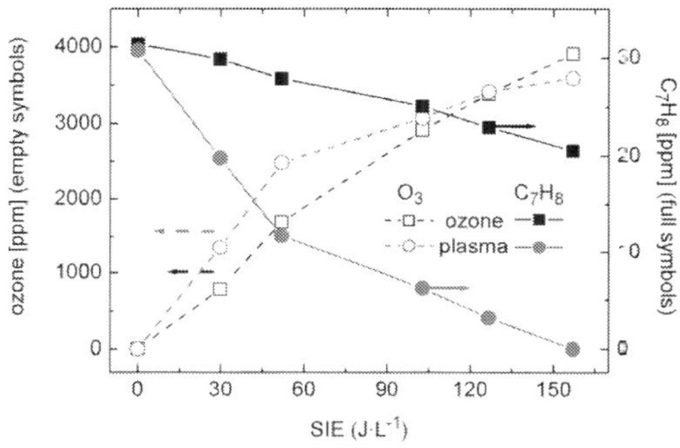

Figure 18. Ozone and toluene concentrations as a function of SIE (frequency 1 kHz) at the surface DBD arrangement. Comparison of the direct plasma treatment (red symbols) with the ozone injection experiment (black).

Ozone concentration increases and toluene concentration decreases with increasing SIE. The ozone production is similar in both direct discharge and ozonation treatments, but the toluene removal varies significantly. In the case of indirect treatment with ozone a small reduction of toluene is obtained. In additional experiments a similar reduction was obtained even without plasma operation. Thus it must be that toluene is adsorbed somewhere in the system (reaction chamber, pipes etc). In case of direct treatment at 157 JL^{-1} toluene is completely removed. These results show that reactions with ozone are not dominant, which is in agreement with the fact that the rate coefficient of the reaction of toluene with ozone is small (eq. (8) [46]) compared to the reaction with atomic oxygen (eq. (2)).

$$C_7H_8 + O_3 \rightarrow Products \qquad k_{298} = 1.5 \cdot 10^{-22}\ \mathrm{cm}^6 \cdot \mathrm{molecule}^{-1} \cdot \mathrm{s}^{-1} \qquad (8)$$

The reaction with atomic oxygen (eq. (2)) is considered to be the most important removal process.

Since for the production of these species energetic electrons are needed, the lower efficiency under wet conditions can be explained by their lower production efficiency due to the consumption of these electrons as well as a reduction of the electron temperature by water dissociation and vibrational excitation. More detailed investigation on the toluene by-products are needed to clarify the carbon balance and the selectivity of the process which remains to future investigations.

5. Conclusions and outlook

The feasibility of NTP for the removal of VOCs has been demonstrated by means of two different gas discharge concepts, namely DBD and corona discharge. There are evident peculiarities in the two different approaches and, notably, also among different types of coronas. Different discharge regimes create NTPs with distinct properties and compositions (type and density of reactive species), which reflect on the process chemical outcome. This knowledge is important for mastering the process product selectivity, i.e. the chemical composition of the treated gas. Likewise important is the process efficiency. Unfortunately, up to now it is possible to make a comparison between the two reactor arrangements only for the specific case of toluene. There are reviews in the literature in which the efficiency of different NTP processes is compared [17].

Regarding the main chemical reaction involved in the decomposition process of toluene a similar mechanism for DBD and dc– corona can be proposed. Here, the most important reactive species are radicals. The main difference in the overall efficiency of the decomposition process is when water vapour is present in the treated gas, but this opposite effect is totally in agreement with the physical properties of the DBD. In the case of the DBD configuration the volume of gas which is directly affected by plasma is small compared to dc– corona. Accordingly, in the case of the corona a much higher probability of collisions between reactive species and substrate molecules (toluene) exists, but also a high probability to generate OH radicals. In the case of DBD the presence of water vapour is quenching the high energetic electrons and reducing the chance to generate more reactive species with an overall result of a decrease in energy efficiency. Despite this, if we calculate the Energy Yield (EY) according to the equation reported in [47]:

$$EY = \frac{C_{in} \cdot \eta \cdot M \cdot 0.15}{SIE} \tag{9}$$

where C_{in} is the starting concentration of the pollutant to be treated (ppm), η is the removal efficiency and M is the molar mass of the pollutants. For the DBD configuration EY value is 13.6 g/kWh while for the corona reactor, in pulsed+ mode, the value is 6.9 g/kWh.

Plasma type	Concentration range (ppm)	Maximum removal efficiency (%)	Energy Yield (g/kWh)
DBD (this work, sec. 4)	**50**	**>99**	**13.6**
DBD packed with glass pellets [47]	1100	80	11.5
Pulsed Corona (this work, sec. 3)	**500**	**>99**	**6.9**
DBD packed with glass beads [47]	240	36	6.8
DBD [47]	400	23	5.2
DC corona [47]	5 – 200	93	0.4

Table 3. Selected results on toluene removal with NTP in dry air.

In Table 3 are summarized selected results on toluene removal with NTP taken from [47] to better evaluate the EY of the two setups reported above. In general, the DBD setups present higher values compared to the corona ones. What is really interesting to note from this table is that not only the values of EY obtained with the setups developed by the authors are among the highest ones, but also these are the only cases in which the complete removal of toluene is reached.

In contrast to many other established technologies of air cleaning, NTPs can be controlled more or less instantaneously by their electrical operation parameters, and they can thus be adjusted to fluctuating gas flow volumes and/or contamination levels. However, nearly all practical processes of pollutant degradation in gases by means of NTPs are hybrid processes or a combination of NTPs with other technologies. In such combinations the NTP acts as an oxidation stage. However, the combination is not only a processing by means of subsequent methods but also offer multiple process synergies. Therefore NTP can be coupled with catalysts, adsorbing agents or scrubbing. For example, the oxidation of non-soluble VOCs results in soluble by-products such as formaldehyde or formic acid. This can also be used for the removal of NO_x [48]. The so-called Plasma Enhanced Selective Catalytic Reduction (PE-SCR) of NOx offers many synergies between plasma and catalyst.

In this context, the combination of plasma treatment with adsorption methods has also been proposed for VOC abatement and deodorization [49]. Several manufacturers offer devices for deodorization which sometimes combine a NTP with an active carbon or molecular sieve stage. The odour reduction of so-called indirect plasma treatment was also demonstrated. Indirect treatment means that the plasma processed air is injected in the VOC containing off-gas. In such case short lived radicals and ions may be less involved in the decomposition processes but the operation lifetime of such system is much longer. During the direct plasma decomposition aerosols can be formed by nucleation of intermediate products and deposit as layers on the electrodes which interfere with the plasma generation. This is avoided by indirect treatment. Many of such DBD-based installations are worldwide used for deodorization in several factories for producing food for fattening, fish meal and flavouring substances. The installations are low-maintenance and need about one third to one fifth of the space as conventional technologies. In [50] the investment- and running cost of numerous waste air purification processes for a gas flow of 50,000 m_N^3/h and for <100 mg VOC/m^3 in the flavour processing industry were determined and compared. NTP installation had the lowest investment costs (about 400,000 € compared to at least 700,000 € for combustive methods, biological filter or molecular sieve filtration) and second lowest operating cost (about 8 €/h, compared to 70 €/h for combustion and biological filtration with 35 €/h). Although the applicability of NTPs is devoted to low-contaminated gas streams, these examples show the high economic relevance and potential of such technologies.

Furthermore, the combination of NTP with absorbers offers the possibility to establish cyclic processes for the removal of low-concentrated pollutants [49]. In such processes, the low-concentrated pollutants are adsorbed and thus concentrated on solid matter in a storage phase. In the subsequent plasma phase, the adsorbed molecules are desorbed and decomposed by plasma activity. Since the retention time of the pollutants in the plasma phase and their concentration are increased, less energy is consumed in such a plasma-enhanced adsorption process. Such processes have been established to decompose different VOCs and NO_x, as summarized in [51]. The decomposition of adsorbed ethanol on active carbon samples by means of ozone generated in the plasma has been investigated in [52]. The regeneration of clinoptilolite (a natural zeolite) loaded with NH_3 has recently been shown by means of a packed-bed DBD reactor. The adsorbed NH_3 is released at a relatively low temperature and low energy consumption [53]. A cycled adsorption and plasma process using mineral granulates consisting of 80 % halloysite in a packed-bed DBD reactor for the removal of formaldehyde CH_2O was investigated in [51]. Here, the adsorbed CH_2O molecules were decomposed into CO_X and hydrocarbons in N_2 plasma. The total amount of decomposed CH_2O and the selectivity towards CO_2 increased with N_2 gas space-times (i.e. the time required to process one packed bed volume of adsorbing material with gas) and with oxygen fraction in the carrier gas. The above examples demonstrate the high potential of plasma-enhanced techniques, which can increase efficiency and lower operational costs. However, more research and development are necessary in order to establish a wider industrial breakthrough.

Acknowledgements

The authors would like to thank Dr. Tomáš Hoder, Mr. Wolfgang Reich and Mr. Alexander Schwock for their support on this work.

The work is partly supported by the European Regional Development Fund, Baltic Sea Region programme 2007-2013 (project No 033, "Dissemination and Fostering of Plasma Based Technological Innovation for Environment Protection in The Baltic Sea Region", PlasTEP). The research leading to these results has received further funding from the European Union Seventh Framework Programme (FP7/2007-2013) under grant agreement n°316216.

Author details

Milko Schiorlin[1*], Cristina Paradisi[2*], Ronny Brandenburg[1], Michael Schmidt[1], Ester Marotta[2], Agata Giardina[2] and Ralf Basner[1]

*Address all correspondence to: milko.schiorlin@inp-greifswald; cristina.paradisi@unipd.it

1 Leibniz Institute for Plasma Science and Technology (INP Greifswald), Greifswald, Germany

2 Department of Chemical Sciences, University of Padova, Padova, Italy

References

[1] von Sonntag C, von Gunten U. Chemistry of ozone in water and wastewater treatment: from basic principles to applications, ed. Iwa Publ., London, 2012

[2] Kim HH. Nonthermal plasma processing for air-pollution control: A historical review, current issues, and future prospects. Plasma Process. Polym. 2004;1(2), 91–110. DOI: 10.1002/ppap.200400028

[3] Mizuno A. Industrial applications of atmospheric non-thermal plasma in environmental remediation. Plasma Phys. Contr. Fusion 2007;49, A1-A15. DOI: 10.1088/0741-3335/49/5A/S01

[4] Hammer T, Bröer S. Plasma Enhanced Selective Catalytic Reduction of NOx for Diesel Cars. SAE Technical Paper 982428, 1998, DOI:10.4271/982428.

[5] Chmielewski AG, Tymiński B, Licki J, Iller E, Zimek Z, Radzio B. Pilot plant for flue gas treatment-continuous operation tests. Rad. Phys. Chem. 1995;45(4-6) 1067–1070. DOI:10.1016/0969-806X(95)00322-O.

[6] Kogelschatz U, Eliasson B, Egli W. Dielectric-Barrier Discharges. Principle and Applications. J. Phys IV France 1997;7(C4), C4-47–C4-66. DOI: 10.1051/jp4:1997405

[7] Moreau E, Sosa R, Artana G. Electric wind produced by surface plasma actuators: a new dielectric barrier discharge based on a three-electrode geometry. J. Phys. D-Appl. Phys. 2008;41(11) 115204. DOI: 10.1088/0022-3727/41/11/115204.

[8] Holzer F, Roland U, Kopinke FD. Combination of non-thermal plasma and heterogeneous catalysis for oxidation of volatile organic compounds: Part 1. Accessibility of the intra-particle volume. Appl. Catal. B-Environ. 2002;38(3) 163–181. DOI:10.1016/S0926-3373(02)00040-1.

[9] Kraus M, Eliasson B, Kogelschatz U, Wokaun A. CO_2 reforming of methane by the combination of dielectric-barrier discharges and catalysis. Phys. Chem. Chem. Phys. 2001;(3) 294–300. DOI: 10.1039/B007015G.

[10] Nijdam S, van Veldhuizen E, Bruggeman P, Ebert U. An Introduction to Nonequilibrium Plasmas at Atmospheric Pressure. In: Parvulescu VI, Magureanu M, Lukes P. (ed.) Plasma Chemistry and Catalysis in Gases and Liquids. Weinheim: Wiley-VCH GmbH & Co. KGaG; 2012. p.1–44.

[11] Hoeben WFLM, Beckers FJCM, Pemen AJM, van Heesch EJM, Kling WL. Oxidative degradation of toluene and limonene in air by pulsed corona technology. J. Phys. D: Appl. Phys. 2012;45, 055202. DOI: 10.1088/0022-3727/45/5/055202.

[12] Kozlov KV, Wagner HE, Brandenburg R, Michel P. Spatio-temporally resolved spectroscopic diagnostics of the barrier discharge in air at atmospheric pressure. J. Phys. D: Appl. Phys. 2001; 34(21), 3164–3176. DOI:10.1088/0022-3727/34/21/309.

[13] Hoder T, Černák M, Pailliol J, Loffhagen D, Brandenburg R. High-resolution measurements of the electric field at the streamer arrival to the cathode: A unification of the streamer-initiated gas-breakdown mechanism. Phys. Rev. E. 2012;86, 055401. DOI: 10.1103/PhysRevE.86.055401.

[14] Fridman A. Plasma Chemistry. Cambridge: Cambridge University Press; 2008.

[15] Urashima K, Chang JS. Removal of volatile organic compounds from air streams and industrial flue gases by non-thermal plasma technology. IEEE Trans. Dielec. Electric. Insul. 2000;7(5), 602–614. DOI: 10.1109/94.879356.

[16] Isbell MA, Stolzberg RJ, Duffy LK. Indoor climate in interior Alaska: simultaneous measurement of ventilation, benzene and toluene in residential indoor air of two homes. Science Total Environ. 2005;345(1-3), 31–40. DOI: 10.1016/j.scitotenv.2004.11.016.

[17] Vandenbroucke AM, Morent R, De Geyter N, Leys C. Decomposition of Toluene with Plasma-catalysis: A Review. J. Adv. Oxid. Technol. 2012;15(2), 232–241.

[18] Marotta E, Callea A, Rea M, Paradisi C. DC corona electric discharges for air pollution control. Part 1. Efficiency and products of hydrocarbon processing. Environ. Sci. Technol. 2007;41(16), 5862–5868. DOI: 10.1021/es0707411.

[19] Marotta E, Callea A, Ren XW, Rea M, Paradisi C. DC corona electric discharges for air pollution control, 2-ionic intermediates and mechanisms of hydrocarbon processing. Plasma Process. Polym. 2008;5(2), 146–154. DOI: 10.1002/ppap.200700128.

[20] Marotta E, Callea A, Ren XW, Rea M, Paradisi C. A Mechanistic Study of Pulsed Corona Processing of Hydrocarbons in Air at Ambient Temperature and Pressure. Int. J. Plasma Environ. Sci Technol. 2007;1(1), 39–45. http://www.iesj.org/html/service/ijpest/vol1_no1_2007/7_Marotta.pdf.

[21] Schiorlin M, Marotta E, Rea M, Paradisi C. Comparison of Toluene Removal in Air at Atmospheric Conditions by Different Corona Discharges. Environ. Sci. Technol. 2009;43(24), 9386–9392. DOI: 10.1021/es9021816.

[22] Marotta E, Scorrano G, Paradisi C. Ionic reactions of chlorinated volatile organic compounds in air plasma at atmospheric pressure. Plasma Process. Polym. 2005;2(3), 209–217. DOI: 10.1002/ppap.200400047.

[23] Donó A, Scorrano G, Paradisi C. Abatement of volatile organic compounds by corona discharge, a study of the reactivity of trichloroethylene under atmospheric pressure ionization conditions. Rapid Commun. Mass Spectrom. 1997;11(15), 1687–1694. DOI: 10.1002/(SICI)1097-0231(19971015)11:15<1687::AID-RCM33>3.0.CO;2-Y.

[24] Schiorlin M, Marotta E, Dal Molin M, Paradisi C. Oxidation Mechanisms of CF_2Br_2 and CH_2Br_2 Induced by Air Nonthermal Plasma. Environ. Sci. Technol. 2013;47(1), 542–548. DOI: 10.1021/es303561n.

[25] Zaniol B, Schiorlin M, Gazza E, Marotta E, Ren XW, Puiatti ME, Rea M, Sonato P, Paradisi C. An emission spectroscopy study of atmospheric plasmas formed by DC and pulsed corona discharges in hydrocarbon contaminated air. Int. J. Plasma Environ. Sci. Technol. 2008;2(1), 65–71. http://www.iesj.org/html/service/ijpest/vol2_no1_2008/IJPEST_Vol2_No1_08_pp65-71.pdf.

[26] Rudolph R, Francke KP, Miessner H. Concentration dependence of VOC decomposition by dielectric barrier discharges. Plasma Chem. Plasma Process. 2002;22(3), 401–412. DOI: 10.1023/A:1015369100161.

[27] Marotta E, Schiorlin M, Rea M, Paradisi C. Products and mechanisms of the oxidation of organic compounds in atmospheric air plasmas. J. Phys. D: Appl. Phys. 2010;43(12), 124011. DOI: 10.1088/0022-3727/43/12/124011.

[28] van Veldhuizen EM., editor. Electrical Discharges for Environmental Purposes: Fundamentals and Applications. New York: Nova Science Publischers; 2000.

[29] Blin-Simiand N, Jorand F, Magne L, Pasquiers S, Postel C, Vacher JR. Plasma reactivity and plasma-surface interactions during treatment of toluene by dielectric barrier discharge. Plasma Chem. Plasma Process. 2008;28(4), 429–466 DOI: 10.1007/s11090-008-9135-1.

[30] Slater RC, Douglas-Hamilton DH. Electron-beam-initiated destruction of low concentrations of vinyl chloride in carrier gases. J. Appl. Phys. 1981;52(9), 5820–5828. DOI: 10.1063/1.329476.

[31] Baulch DL, Cobos CJ, Cox RA, Frank P, Hayman G, Just T, Kerr JA, Murrells T, Pilling MJ, Troe J, Walker RW, Warnatz J. Evaluated kinetic data for combusion modelling. Supplement I. J. Phys. Chem. Ref. Data. 1994;23, 847–1033. http://www.nist.gov/data/PDFfiles/jpcrd484a.pdf.

[32] Bohn B. Formation of Peroxy Radicals from OH–Toluene Adducts and O_2. J. Phys. Chem. A. 2001;105(25), 6092–6101. DOI: 10.1021/jp0033972.

[33] Lias SG. Ionization Energy Evaluation NIST ChemistryWebBook. NIST Standard Reference Database No. 69. Gaithersburg MD: National Institute of Standards and Technology; 2006. http://webbook.nist.gov/chemistry/

[34] Atkinson R, Baulch DL, Cox RA, Crowley JN, Hampson RF, Hynes RG, Jenkin ME, Rossi MJ, Troe J. Evaluated kinetic and photochemical data for atmospheric chemistry: Volume II – gas phase reactions of organic species. Atmos. Chem. Phys. 2006;6, 3625–4055. http://www.atmos-chem-phys.net/6/3625/2006/acp-6-3625-2006.html.

[35] Feng JQ. An analysis of corona currents between two concentric cylindrical electrodes. J. Electrostat. 1999;46(1), 37–48. DOI: 10.1016/S0304-3886(98)00057-6.

[36] Zebboudj Y, Hartmann G. Current and electric field measurements in coaxial system during the positive DC corona in humid air. Eur. Phys. J. Appl. Phys. 1999;7(2), 167–176. DOI: 10.1051/epjap:1999211.

[37] Skalny JD, Mikoviny T, Matejcik S, Mason NJ. An analysis of mass spectrometric study of negative ions extracted from negative corona discharge in air. Int. J. Mass Spectrom. 2004; 233(1-3), 317–324. DOI: 10.1016/j.ijms.2004.01.012.

[38] Sieck LW, Buckley TJ, Herron JT, Green DS. Pulsed Electron-Beam Ionization of Humid Air and Humid Air/Toluene Mixtures: Time-Resolved Cationic Kinetics and

Comparisons with Predictive Models. Plasma Chem. Plasma Process. 2001;21(3), 441–457. DOI: 10.1023/A:1011030627752.

[39] Španěl P, Smith D. Selected ion flow tube studies of the reactions of H_3O^+, NO^+, and O_2^+ with several aromatic and aliphatic hydrocarbons. Int. J. Mass Spectrom. 1998;181(1-3), 1–10. DOI: 10.1016/S1387-3806(98)14114-3.

[40] Arnold ST, Dotan I, Williams S, Viggiano AA, Morris RA. Selected Ion Flow Tube Studies of Air Plasma Cations Reacting with Alkylbenzenes. J. Phys. Chem. A. 2000;104(5), 928–934. DOI: 10.1021/jp9928199.

[41] Baker J, Arey J, Atkinson R. Rate Constants for the Gas-Phase Reactions of OH Radicals with a Series of Hydroxyaldehydes at 296 ± 2 K. J. Phys. Chem. A. 2004;108(34), 7032–7037. DOI: 10.1021/jp048979o.

[42] Seinfeld JH, Pandis SN. Atmospheric Chemistry and Physics: From Air Pollution to Climate Change. Hoboken NJ: John Wiley & Sons, Inc; 2006.

[43] Müller S, Zahn RJ. Air Pollution Control by Non-Thermal Plasma. Contrib. Plasma Phys. 2007;47(7), 520–529. DOI: 10.1002/ctpp.200710067.

[44] Kogelschatz U. Dielectric-Barrier Discharges: Their History, Discharge Physics, and Industrial Applications. Plasma Chem. Plasma Process. 2003;23(1), 1–46. DOI: 10.1023/A:1022470901385.

[45] Manley TC. The Electric Characteristics of the Ozonator Discharge. Trans. Electrochem. Soc. 1943;84(1), 83–96. DOI: 10.1149/1.3071556.

[46] Pate CT, Atkinson R, Pitts JN Jr. The Gas Phase Reaction of O_3 with a Series of Aromatic Hydrocarbons. J. Environ. Sci. Health A. 1976;11(1), 1–10. DOI: 10.1080/10934527609385750.

[47] Vandenbroucke AM, Morent R, De Geyter N, Leys C. Non-thermal plasma for non-catalytic and catalytic VOC abatement. J. Hazard. Mater. 2011;195, 30–54. DOI: 10.1016/j.jhazmat.2011.08.060.

[48] Stamate E, Chen W, Jørgensen L, Jensen TK, Fateev A, Michelsen PK. IR and UV gas absorption measurements during NO_x reduction on an industrial natural gas fired power plant. Fuel. 2010;89(5), 978–985. DOI: 10.1016/j.fuel.2009.11.044.

[49] Kim HH, Ogata A, Futamura S. Oxygen partial pressure-dependent behavior of various catalysts for the total oxidation of VOCs using cycled system of adsorption and oxygen plasma. Appl. Catal. B Environ. 2008;79(4), 356–367. DOI: 10.1016/j.apcatb.2007.10.038.

[50] Rafflenbeul R. Non Thermal Plasma Plants: Experiences from the Industrial Praxis of Air Purification. Presentation at PlasTEP workshop;2010. Available from http://www.plastep.eu/fileadmin/dateien/Downloads/Downloads__Minutes_etc_/100910_Workshop/Vortrag_PPT__INP_Greifswald.pdf

[51] Saulich K, Müller S. Removal of formaldehyde by adsorption and plasma treatment of mineral adsorbent. J. Appl. Phys. D: Appl. Phys. 2013;46(4). 045201. DOI: 10.1088/0022-3727/46/4/045201.

[52] Basner R, Akimalieva A, Brandenburg R. Effect of atmospheric surface plasma on the adsorption of ethanol at activated carbon filter element. Surf. Coat. Tech. 2013;234, 126–134. DOI: 10.1016/j.surfcoat.2012.11.028.

[53] Müller S, Saulich K, Krueger T. Regeneration of Clinoptilolite Loaded with Ammonia by a Non-Thermal Plasma Method. Chem. Ing. Tech. 2012,84(12), 2190–2197. DOI: 10.1002/cite.201200044.

Chapter 2

Air Pollution in Welding Processes — Assessment and Control Methods

Farideh Golbabaei and Monireh Khadem

Additional information is available at the end of the chapter

http://dx.doi.org/10.5772/59793

1. Introduction

Welding is a very common operation in many industries and workplaces [1, 2]. According to American Welding Society, it is defined as "a metal joining process wherein coalescence is produced by heating to suitable temperature with or without the use of filler metal" [3]. There is a variety of welding processes that are used in different working conditions. According to some reports, from 0.2 to 2.0% of the working population in industrialized countries are engaged in welding activities [4]. Worldwide, over five million workers perform welding as a full time or part time duty [5, 6]. These welders, depending on conditions, work in outdoor or indoor workplaces, in open or confined spaces, underwater, and above construction sites. Welding operators face various hazards resulting in different injuries, adverse health effects, discomfort and even death. Furthermore, air pollution due to welding leads to certain consequents on humans and environment. Therefore, there are strong reasons to deal with the welding processes and the working environment of the welder from different aspects. A large number of welders experience some type of adverse health effects. Other workers near the place where welding process is done may be affected by the risks generated by it [1, 7]. Totally, welding risks can be classified as risks deriving from physical agents and risks related to the chemical components. The main hazards related to welding include electricity, radiation, heat, flames, fire, explosion, noise, welding fumes, fuel gases, inert gases, gas mixtures and solvents. Welders may be exposed to other hazards not directly related to welding, such as manual handling, working at height, in confined spaces, or in wet, hot or humid situations, and working with moving equipment, machinery and vehicles. Welding in a static awkward or horizontal posture may result in musculoskeletal injuries, such as strains and sprains. Prolonged use of a hard hat and a helmet can cause strain on the neck. Furthermore, long-term exposure, repetitive motions with arms and hands, and tasks inducing high force may lead to cumulative effects, increasing risk of injury. The main components of welding emissions are oxides of metals due to contact between the oxygen in the air and the vaporized metals. Common chemical hazards include particulates (lead, nickel, zinc, iron oxide, copper, cadmium, fluorides, manganese, and chromium) and gases (carbon monoxide, oxides of nitrogen, and ozone). Recently, nanoparticles emitted by welding operations are considered as an important group of air pollutants and there is a need to assess particle sizes and size distributions when risk assessment is done. Each welding technique produces a distinctive range of particulate composition and morphology. Different and complex profiles of exposures may be related to various welding environments [8-10].

HAZARD	WELDING PROCESS			
	PAW/PAC Carbon Arc Processes	SMAW GTAW GMAW FCAW	SAW	Oxyfuel
Ergonomic	+	+	+	+
Electric Shock	+	+	+	x
Bright light	+	+	-	+
Ultraviolet radiation	+	+	-	x
Toxic fumes and gases	+	+	-	+
Heat, Fire, and Burns	+	+	+	+
Noise	+	x	x	x

x No hazards, + Hazard present, - Hazard present if SAW flux is absent [11]

Table 1. The hazards associated with welding Processes

2. Welding technology

2.1. Applications

Welding is used extensively in various manufacturing industries including shipyards, automobile factories, machines, home appliances, computer components, bridge building and other constructions. Welding is used for manufacturing pressure vessels, heat exchangers, tanks, sheet metal, prefabricated metal buildings and architectural work. Also, welding is an applicable technique in maintenance operations and repair shops. It is used in mining, oil and gas transmission companies, piping systems, heavy equipment manufacturing, aerospace, electronics, medical products, precision instruments, electric power, and petrochemical industries. Perhaps artists and sculptors are the smallest group who use welding techniques to create artworks. Therefore, many things that people use in daily lives are welded or made by welded parts [12].

2.2. Workplace conditions

Welders, depending on conditions, work in outdoor or indoor workplaces, in open or confined spaces, underwater, and above construction sites. In some conditions, welding processes are carried out in confined spaces where the welding work area is surrounded on most sides by walls and there is no sufficient space for the installation of a conventional exhaust hood [1, 7].

Working in indoor environments includes all works which are done in buildings like workshops, repairing shops, storages, office, and any closed area in industries, factories, and other places. Welders may work in indoor areas to do welding tasks full time or part time. An important benefit of indoor workplaces is the protection against environmental factors such as rain, wind and sunshine. Outdoor workers spend long periods of time working in open areas. They are exposed to different hazards depending on their type of work, as well as geographic region, season, and the period of time they are outside. Outdoor works include agriculture, construction, mining, oil and gas transmission through pipelines, transportation, warehousing, utilities, and service sectors. Sometimes welders should work in such workplaces to do their tasks. Some workplace hazards related to outdoor areas include unpredictable weather conditions, bugs and wild animals, extreme heat, extreme cold, and ultraviolet (UV) radiation.

Many workplaces contain spaces that are considered "confined" because their configurations hinder the activities of employees who must enter, work in, and exit them. A confined space has limited or restricted means for entry or exit. Confined spaces include underground vaults, tanks, storage bins, manholes, reactor vessels, silos, process vessels, and pipelines. Confined spaces have the following characteristics: limited space, entry, or exit; poor ventilation and lack of safe breathing air. Welders may experience various hazards when welding in confined spaces, such as fire, explosion, electric shock, asphyxiation, and exposure to hazardous air contaminants [13-16].

2.3. Types of welding processes

There are different welding processes (over 50 types) that differ greatly in some parameters such as heat, pressure, and the type of equipment used. Welding process can be classified into various types based on different literatures. Some common types of welding are listed in five categories each of which includes some subcategories (Figure 1). The most common and known types of welding include:

Shielded Metal Arc Welding: (SMAW) also is known as Manual Metal Arc welding (MMA) or stick electrode welding. It is one of the oldest, simplest, and most versatile arc welding processes used for carbon steel welding and low alloy welding. In SMAW, the electrode is held manually, and the electric arc flows between the electrode and the base metal. The electrode is covered with a flux material which provides a shielding gas for the weld to help minimize impurities. A wide range of metals, welding positions and electrodes are available based on intended requirements. This type of welding is especially suitable for jobs such as the erection of structures, construction, shipbuilding, and pipeline work. Contrary to the other methods requiring shielding gas which are unsuitable in wind, SMAW can be used outdoors in different weather conditions. However, owing to the time required for removing the slag after welding and changing the electrodes, its arc time factor is relatively low. As a disadvantage, forming fumes in SMAW makes the process control difficult.

Gas Metal Arc Welding: (GMAW) or metal inert gas (MIG) welding is used for most types of metal and is faster than SMAW. It may be applied to weld vehicles, pressure vessels, cranes, bridges and others. This process involves the flow of an electric arc between the base metal and a continuous and consumable wire electrode. Shielding gas (usually an argon and carbon dioxide mixture) is supplied externally; therefore, the electrode has no flux coating or core. MIG welding is used for mild steel, low alloyed and stainless steel, for aluminum, for copper, nickel, and their alloys. Some parameters can affect MIG welding process, such as:

- Electrode diameter
- Voltage
- Wire feed speed and current
- Welding speed
- Shielding gas and gas flow rate
- Torch and joint position

To perform an optimum welding, most of the mentioned parameters should be matched to each other. In addition to affecting the quality of welding, some of these parameters can influence the fumes and gases emitted from the process. However, the fume produced by MIG welding is less than that of SMAW. Unlike the SMAW that is discontinuous due to limited length of the electrodes, GMAW is a continuous welding process. There is no slag and no need for high level of operators' skill. Nevertheless, expensive and non-portable equipment is required, and also outdoor applications are limited because of the negative effects of weather conditions like wind on the shielding gas [17, 18].

Gas Tungsten Arc Welding: (GTAW) is also known as tungsten inert gas (TIG) welding. GTAW is used on metals such as aluminum, magnesium, carbon steel, stainless steel, brass, silver and copper-nickel alloys. This technique uses a permanent non-consumable tungsten electrode. The filler metal is fed manually, the weld pool and the electrode are protected by an inert gas (usually argon), and high electrical currents are used in this type. Welding of stainless steel, welding of light metals, such as aluminum and magnesium alloys, and the welding of copper are the main applications of TIG welding. GTAW welds are highly resistant to corrosion and cracking over long time periods. However, TIG welding is suitable to weld thin materials and produces a high quality weld of most of metals. There is no need for slag removal in GTAW process. The concentration of heat takes place in a small zone, resulting in the minimal thermal distortion of work piece. The TIG welding has some disadvantages including low welding rate, expensiveness, and need for high level of operators skill. Although during TIG welding operators are exposed to dangerous gases and fumes, the generation of these compounds is very little in comparison with other welding processes.

http://dx.doi.org/10.5772/59793

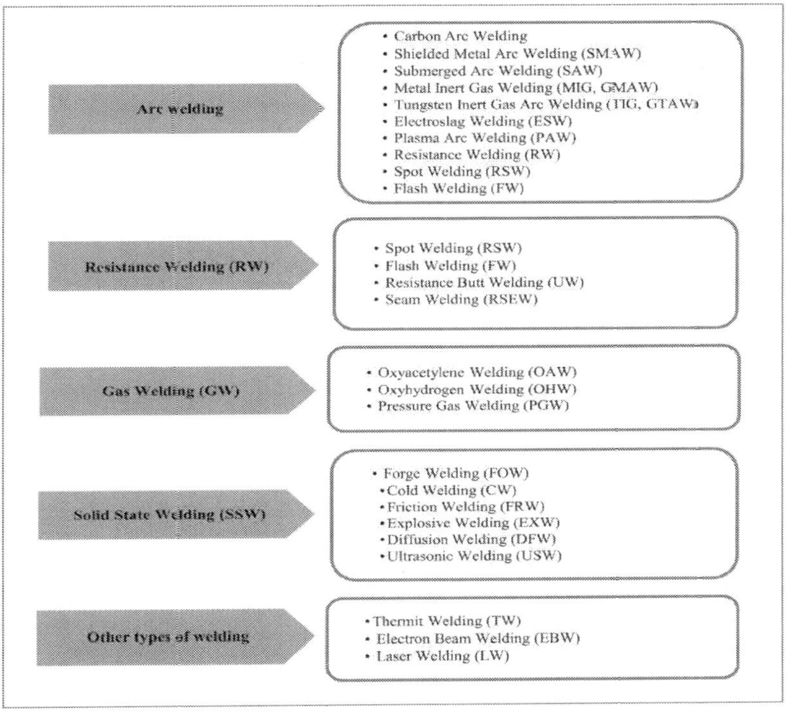

Figure 1. Classification of welding processes [18]

Submerged Arc Welding: (SAW) is a highly-productive welding method (4-10 times as much as the SMAW). SAW may be automatic or semi-automatic. It is used to weld thick plates of carbon steel and low alloy steels. In this welding process, the electric arc flows between the base metal and a consumable wire electrode; however, the arc is not visible since it is submerged under flux material. This welding process is usually used for large structures such as large tubes, cylindrical vessels, and plates in shipyards. Some parameters can affect SAW process such as welding arc voltage, arc current, the size and shape of the welding wire, and the number of welding wires. A low fume emission is produced during SAW process and there is a little ozone, nitric oxide and nitrogen dioxide generation because of the invisibility of the arc. Very high welding rate, suitability for automation, suitability for both indoor and outdoor works, and high weld quality are mentioned as advantages of SAW. Some limitations of this welding process include slag inclusion, limited applications often for welding in a horizontal position, and need for precise parameter setting and positioning of the wire electrode.

Plasma Arc Welding: (PAW) is an arc welding process in which arc is formed between an electrode and the workpiece. In PAW process, the plasma arc can be separated from the shielding gas cover by positioning the electrode within the body of the torch. It can be named as a key difference between GTAW and PAW. Two inert gases are used in the process, one forms the arc plasma and the second shields the arc plasma. Applying the plasma arc welding is being increased in industries, because it provides a high level of control and accuracy to produce high quality welds. Also, using the PAW leads to long electrode life for high production conditions. This welding process is suitable for both manual and automatic applications. It can be used for precise welding of surgical equipment, jet engine blades, and instruments required for food and dairy industry. There is a low level of fume generation during PAW, but welding gases especially ozone is often formed in this process. Need for less operator skill, high welding rate, high penetrating capacity, long electrode life, high accuracy and precision, and short weld time are considered as the advantages of PAW process. Its limitations include expensive process tools, needs for high power electrical equipment, more distortion and loss of mechanical properties due to the greater heat input.

Flux Core Arc Welding: (FCAW) is used for carbon steels, low alloy steels and stainless steels. This welding process has similarities to both SMAW and GMAW. This process is used in construction because of its high welding speed and portability. The consumable tubular electrode is continuously fed from a spool and an electric arc flows between the electrode and base metal. The electrode wire has a central core containing fluxing agents. There are a variety of cored wires; some of them require the use of shielding gas like carbon dioxide or the mixture of argon/carbon dioxide and the others (self-shielded flux cored wires) do not require additional shielding gas. The slag produced in FCAW process acts as an additional protection during cooling time but has to be chipped away after that. Like other welding process, FCAW has some advantages and limitations. No needs for skilled operators and pre-cleaning of metals, suitability for use in the outdoor or windy condition (it is true about self-shielded flux cored wires), suitability for use in all positions, and ease of varying the alloying constituents are mentioned as FCAW advantages. Its limitations include: emission of considerable amount of fumes in self-shielded wires, higher price of filler material and wire in comparison with GMAW, and needs for slag removal. Also, escaping of the shielding gas from the welded area leaves holes in welded metal, resulting in porosity in products [17, 18].

3. Air pollution out of welding

According to Flagan and Seinfeld definition, "the phenomenon of air pollution involves a sequence of events: the generation of pollutants at and their release from a source; their transport and transformation in and removal from the atmosphere; and their effects on human beings, materials, and ecosystems" [19]. Air pollution is indoor or outdoor contamination by particulates, biological molecules, or other harmful materials that changes the natural characteristics of the Earth's atmosphere. Household combustion devices, motor vehicles, forest fires, and industrial processes are common sources of air pollution. Major industrial sources of particulate matter include the metals, mineral products, petroleum, and chemical industries. Air pollution is considered as a threat to human health as well as to the Earth's ecosystems. Based on WHO report, around 7 million people worldwide died due to the air pollution in

2012 [20]. Welding, as an important operation in most industries, can considerably cause air pollution. In all types of welding processes, fume and gases are formed as air pollutants. Due to high temperature during the welding process, different substances in the arc are vaporized. Then, the vapor condenses and oxidizes in contact with the air, leading to the formation of fumes. The fume particles are so small and they can reach the narrowest airways of respiratory system (respiratory bronchioles). Some parameters like the welding type and consumables (filler metal and surface coatings) determine the kind and amount of generated particles and gases.

The composition of welding fumes and their generation rate is a function of different parameters. Welding fume particles are in the fine (<2.5 μm) to ultrafine (<100 nm) respirable size and can penetrate into the alveolar regions of the lungs. The generation of fumes depends on:

- -Amperage, voltage, gas and arc temperatures and heat input in the welding process
- -Consumables like electrodes
- -Materials
- -Welding duration [9, 21].

The most common gases emitted during welding are ozone, nitrous gases and carbon monoxide. Phosphine and phosgene are the other gases that may be produced during welding. Gases are generated due to the high temperature and ultraviolet (UV) radiation from the arc. Like fumes, some factors can affect the emission of gases during welding processes. For instance, ozone formation during welding depends on process type, used material, and shielding gases. Welding gases can also be generated when surface coatings or contaminants contact with hot surfaces or UV radiation.

Along with harming human health, air pollution may lead to various environmental impacts. Air pollution can adversely cause critical impacts on the atmosphere and natural environment in many ways. Welding, as an industrial process, causes serious impacts on the environment depending on its operation mode and the technological equipment. Environmental pollution in welding process is the result of some parameters, such as high percentage of heat that is released into the environment and materials including large amount of gases and fumes. Some factors needed to carry out the welding operation include: energy, mineral or organic substances (protective gases, cooling water, oils, grease and protective substances etc.). These consumables can be harmful for the environment. Furthermore, produced waste during the welding processes results in undesirable impact on the work or natural environment. To protect the welding region and prevent oxidation, inert gases like carbon dioxide and argon are used because of their availability and low cost. They are used as shielding gases and have undesirable impacts on the environment. To protect the environment and keep the resource for future, energy conservation and reducing greenhouse gas emissions should be considered. In this respect, the average consumption rate, usage rate and the purity of products and consumables are important factors [22, 23].

The generation of fumes and gases is directly related to the welding process. Fumes emitted during manual metal arc welding (MMA) and MIG welding is the same. In some conditions, the level of fume generated during MIG welding (with solid wire) may be much lower in comparison with the fumes produced by MMA. In TIG welding, a lower level of fumes is emitted compared to MMA and MIG welding. The composition of fumes is directly associated with the composition of used wire. MMA welding causes adverse health effects because of forming the hexavalent chromium (Cr (VI)) in the process. In addition, high rates of emission of toxic compounds generate in MMA-stainless steel (MMA-SS) welding [24]. During TIG welding, very little fume are generated. Welding fumes may be composed of oxides of chromium, nickel and copper, with very low specific limit values. The individual elements and also their synergetic effect must be considered when assessing fume toxicity. Lower ozone and nitrogen oxides are emitted during TIG welding than those in MIG/MAG welding. The amount of mentioned gases during TIG welding is dependent on current, arc length and the flow and type of shielding gas. High electrical currents cause the significant levels of ozone, nitric oxide and nitrogen dioxide. During MIG welding, significant levels of ozone and nitrogen oxides are produced because of intense current levels.

There is a little information concerning emissions during plasma arc welding (PAW). Due to the similarity of TIG and PAW welding techniques, they may probably emit air pollutants with the same magnitude. MIG welding of aluminum produces larger quantities of ozone than TIG welding of aluminum. Forming more nitrogen oxides in the latter process will keep the emitted ozone levels down [25, 26]. A study by Schoonover et al. showed that welders performing MIG and SMAW are exposed to higher fume concentrations than welders performing TIG. According to mentioned study, exposure to manganese during MIG was nearly two and ten times higher than in SMAW and TIG, respectively. In fact, not using a consumable electrode during TIG welding results in lower exposures. The highest average exposures occur in SMAW, followed by GMAW, and GTAW [21]. K. Fuglsang et al. investigated the Fume Generation Rates (FGR). This rate for MMA was 3-5 times higher than that found for MAG and MIG. The same FGR was found for TIG and MIG/MAG welding [27].

Various welding processes generate particles in different size distributions. Particles produced during MMAW, MAG, MIG, and laser welding are quite similar in size. Resistance Spot Welding (RSW) and TIG welding have a completely different structure for particle size distribution. These techniques produce particles smaller than 100 nm, in which, at least 90% are smaller than 50 nm. Particles generated during processes with high mass emission rates (MMAW, MAG, MIG, and Laser) have diameters about 100–200 nm and there are few nanoscaled particles between them. Processes with low mass emission rates (TIG and RSW) generate exclusively particles smaller than 50 nm; however, the number concentration of particles in these techniques is similar to the others. Although, welding types with low mass emission rates are called "clean techniques", their potential toxicological properties and health effects due to exposure to nanoscaled particles should be further studied [28].

A study by Keane M. introduced the pulsed axial spray method (from MIG process) as the best choice of the welding processes because of minimal fume generation (especially Cr (VI)) and cost per weld. The advantages of this method include usability in any position, high metal deposition rate, and simple learning and use. Totally, the highest amounts of fume are produced by the self-shielded cored wire electrodes. These electrodes are used without a shielding gas. Using solid wire electrodes results in emission of ozone and nitrogen oxides as in MAG welding [25, 29].

Airborne particles with diameter smaller than 100 nm are known as nanoparticles or ultrafine particles. According to researches, nanoparticles are more harmful to human health than larger particles. They can deeply penetrate inside the respiratory system and then enter the blood stream. The main character of nanoparticles is the high surface area, and their toxicity depends on the shape and penetration potential inside the respiratory system. In addition to the emission of fine particles with diameter less than 10 μm, nanoparticles may be emitted during welding operations. Some studies have indicated that the highest values of nanoparticles are related to MAG and TIG processes when applying the highest current intensities. Therefore, the higher amounts of nanoparticles are emitted by processes in which the higher energy intensities are used.

As it was stated, the emission of nanoparticles during welding operations increases with the increase of welding parameters like current intensity. Welding with short-circuit mode results in lower value of nanoparticles, because its low current intensity and tension causes an electric arc with lower temperature and thus emitting lower amounts of elements. Also, the high quantity of nanoparticles is generated by the stainless steel welding, which can be related to the presence of helium in the gas mixture of welding. Helium, due to high ionization energy, results in electric arc with high temperature that generates higher values of nanoparticles. Furthermore, the study of different base materials indicated that the higher quantity of nanosized particles is obtained for stainless steel compared to carbon steel. According to data from different investigations, the lowest level of ultrafine particles deposited in alveolar region of lungs was related to FSW, followed by TIG and MAG. Totally, all welding processes can result in deposition of a significant concentration of nanosized particles in lungs of exposed welders [30-32].

4. Welding health effects

Fume and gases emitted during welding pose a threat to human health while welding. The exposures may be varied depending on where the welding is done (on the ship, in confined space, workshop, or in the open air). The welding process and metal welded affect the contents of welding fumes. On the other hand, physical and chemical properties of the fumes and individual worker factors are effective on deposition of inhaled particles. In this respect, particle size and density, shape and penetrability, surface area, electrostatic charge, and hygroscopicity are the important physical properties. Also, the acidity or alkalinity of the inhaled particles are the chemical properties that may influence the response of respiratory tract. Welding gases can be classified into two groups; some gases are used as a shielding gas and the others are generated by the process. Shielding gases are usually inert, therefore, they are not defined as hazardous to health but they may be asphyxiants. Gases generated by welding processes are different based on welding type and may cause various health effects if over-exposure occurs. Welding emissions depending on some factors like their concentration, their properties, and exposure duration can lead to health effects on different parts of human body.

Hazards on Respiratory System - The inhalation exposures may lead to acute or chronic respiratory diseases in all welding processes. In the occupational lung diseases, the various reactions produced in respiratory tract depend on some parameters such as the nature of the inhaled matter, size, shape and concentration of particles, duration of exposure, and the individual workers susceptibility. Chronic bronchitis, interstitial lung disease, asthma, pneumoconiosis, lung cancer, and lung functions abnormalities are some hazardous effects on respiratory systems. The pulmonary disorders are various based on the differences in welding metals and their concentrations. Ozone, at low concentrations, irritates the pulmonary system and can cause shortness of breath, wheezing, and tightness in the chest. More severe exposures to ozone can lead to pulmonary edema. Exposure to nitrogen dioxide may cause lung function disorders like decrements in the peak expiratory [33, 34]. Kim JY in a study showed the PM2.5 concentration for welders (1.66 mg/m^3) was significantly greater than that for controls (0.04 mg/m^3), and the exposure of healthy working population to high levels of welding fumes resulted in the acute systemic inflammation [35].

Hazards on Kidney- Substantial exposure to metals and solvents may be nephrocarcinogenic. Chromium can deteriorate renal function because of accumulation in the epithelial cells of the proximal renal tubules and induce tubular necrosis and interstitial changes in animals and humans. Tubular dysfunctions have been identified in subjects occupationally exposed to Cr (VI) [33, 36]. Welders exposed to heavy metals like cadmium and nickel have also experienced kidney damage [7]. Pesch et al. indicated that there was an excess nephrocarcinogenic risk involved with soldering, welding, milling in females. So, it can be considered an evidence for a gender-specific susceptibility of the kidneys [37].

Hazards on Skin - Erythema, pterygium, non-melanocytic skin cancer, and malignant melanoma are the adverse health effects of welding on the skin among which erythema is a common one. The intense UV as well as visible and infrared radiations are produced by welding arc machines. Exposure to UV can lead to short- and long-term injuries to the skin [33, 38-40]. Some metals like beryllium, chromium and cobalt can cause direct effects (irritation and allergic impacts) on the skin. Also, they may be absorbed through the skin and cause other health effects such as lung damage. When the particles are small and there are cuts or other damages to the skin, the absorption through the skin is raised [7, 36]. Chromium (VI) may cause irritating and ulcerating effects when contacting with skin. An allergic response including eczema and dermatitis may be induced in sensitized individuals exposed to Cr (VI) [34].

Hazards on the visual systems - Most welding processes emit intense ultraviolet as well as visible and infrared radiations. Adverse effects on the eyes may be induced by these optical radiations. In addition, Tenkak reported that, welding may cause photokeratitis and some types of cataract. Erhabor et al. showed the most frequent symptoms among the welders were eye irritation (95.43%). Exposure to UV radiation can lead to short- and long-term injures to the eyes. Acute overexposure to UV radiation can result in the photokeratitis and photoconjunctivitis that are the inflammation of the cornea and the conjunctiva, respectively. These responses of the human eye to UV radiation are commonly known as snow blindness or welder's flash [33, 38, 41].

Hazards on Reproductive System - In the past, some studies have indicated the increased risk for infertility and reduced fertility rate in mild steel welders. There are some evidences that reduced fecundity can be related to exposure to hexavalent chromium and nickel. According to new investigations, damages to male reproduction system have been reported less than before, probably because of decreasing the exposure levels in the developed countries. However, some special tasks like stainless steel welding may impair welders' reproduction system [42-44]. A study by Bonde showed that mild steel welding, but not stainless one, resulted in significant effects on the fertility during years [45]. Mortensen [46] observed a greater risk for poor sperm quality among welders compared to controls, especially welders who worked with stainless steel. Therefore, welding in general, and specifically with stainless steel, may cause the reduced sperm quality. According to Sheiner, impaired semen parameters can be associated with the exposures to lead and mercury [47].

Hazards on the nervous system - Memory loss, jerking, ataxia and neurofibrillary degeneration have been attributed to exposure to aluminum. The accumulation of aluminum in the brain may develop some neuropathological conditions, including amyotrophic lateral sclerosis, Parkinsonian dementia, dialysis encephalopathy and senile plaques of Alzheimer's disease [36]. A review of literatures by Iregren suggests that occupational exposure to manganese results in the central nervous system damage that is generally irreversible [48]. Although there are multiple toxic agents in welding, more literatures have dealt with manganese as an important agent of toxicity. Welders are also exposed to high concentrations of carbon monoxide and nitrogen dioxide. Carbon monoxide can cause the neurological impairments of memory, attention, and visual evoked potentials. Both central and peripheral nervous system damages may be induced by exposure to welding fumes [49]. Some neurobehavioral impairments associated with exposure to lead and manganese have been indicated by Wang [50]. A study by Bowler (2003) showed there is a relation between welding and a decline in brain functions and motor abilities. In this survey, various questionnaire and tests like neuropsychological tests were used [49].

Carcinogenic effects - There are some concerns regarding the presence of carcinogens in the welding fumes and gases. Sufficient evidences for carcinogenicity of nickel, cadmium, and chromium (VI) have been reported through experimental and epidemiological studies. These three metals have been categorized as carcinogen "Class 1" by the International Agency for Research on Cancer [51-52]. Ozone has been introduced as a suspect lung carcinogen in experimental animals, but there are very few documents about its long term effects on welders. The ultraviolet emissions resulting from welding arc can potentially cause skin tumors in animals and in overexposed individuals, however, there is no definitive evidence for this effect in welders [53].

Other health problems - Welding on surfaces covered with asbestos insulation may lead to risk of asbestosis, lung cancer, mesothelioma, and other asbestos-related diseases in exposed welders. The intense heat and sparks of welding can cause burns. Eye injuries are possible because of contact with hot slag, metal chips, and hot electrodes. Lifting or moving heavy objects, awkward postures, and repetitive motions result in strains, sprains and musculoskeletal disorders. High prevalence of musculoskeletal complaints (back injuries, shoulder pain, tendonitis, carpal tunnel syndrome, and white finger) is seen in welders [54].

5. Exposure standards for welding emissions

Usually, exposure standards apply to long term exposure to a substance over an eight hour work per day for a normal working week, over an entire working life. Some organizations like American Conference of Governmental Industrial Hygienists (ACGIH), National Institute for Occupational Safety and Health (NIOSH), and Occupational Safety and Health Administration (OSHA) have published the exposure standards for various components in welding fumes and gases (table 2). According to Work Safe Australia exposure standards cannot be used as a fine dividing line between a healthy and unhealthy workplace. Adverse health effects below the exposure limits might be seen in some people because of individual susceptibilities and natural biological variation. ACGIH, however, recommends a TLV-TWA (Threshold Limit Value-Time Weighted Average) of 5 mg/m^3 for total welding fume, assuming that it contains no highly toxic components. Each metal or gas within the welding has its own exposure standard. As Table 2 indicates, biological media, Biological Exposure Index (BEI), and carcinogenicity class have been proposed for some welding emissions [55, 56].

Substance	OSHA PEL-TWA (mg/m3)	NIOSH REL-TWA (mg/m3)	ACGIH TLV-TWA (mg/m3)	ACGIH BEI	Carcinogenicity
Aluminum Fume	15 (Total) 5 (res)	5	5		
Arsenic	0.01	0.002 (Ceiling)	0.01	35 μg As/L	A1
Barium	0.5	0.5	0.5		
Beryllium	0.002	0.5 (Ceiling)	0.002		A1
Cadmium Fume	0.005	LFC (Ca)	0.01 (Total) 0.002 (Res)	5 μg Cd/g creatinine	A2
Cobalt	0.1	0.05	0.02	15 μg Co/L	A3
Chromium(VI)	--	0.001	0.05	25 μg Cr/L	A1
Chromium metal	1	0.5	0.5		A4
Copper Fume	0.1	0.1	0.2		
Iron Oxide	10 (as Fe)	5	5		A4
Lithium	--	--	--		
Manganese	5 (Ceiling)	1	0.2	range 0.5 to 9.8 mg/L; up to 50 mg/L for occupational exposure	
Molybdenum	5(Soluble) 15 (Insoluble)	--	5 (Soluble) 10 (Insoluble)		

Substance	OSHA PEL-TWA (mg/m3)	NIOSH REL-TWA (mg/m3)	ACGIH TLV-TWA (mg/m3)	ACGIH BEI	Carcinogenicity
Lead	0.05	0.1	0.05	30 µg /dL (whole blood)	A3
Nickel	1	0.015 (Ca)	1	10µmol/mol creatinine	Elemental (A5) Insoluble inorganic (A1)
Platinum	0.002 (Soluble)	1(Metal) 0.002 (Soluble)	1		
Selenium	0.2	0.2	0.2		
Silver	0.01	0.01	0.1		
Tellurium	0.1	0.1	0.1		
Thallium	0.1	0.1(Soluble)	0.1	50 µg Th/g creatinine	
Titanium Dioxide	15	LFC (Ca)	10		
Vanadium Pentoxide	0.1 (Ceiling)	0.05(Ceiling)	0.05	50 µg V/g creatinine	
Zinc Oxide	5	5	5		
Zirconium	5	5	5		
Total fumes	--	LFC (Ca)	5		
Carbon monoxide	50 ppm	35 ppm	25 ppm	3.5% of (Hemoglobin) 20 ppm (end-exhaled air)	
Nitrogen dioxide	5 ppm (ceiling)	5 ppm (ceiling) 1ppm (STEL)	3 ppm		
Ozone	0.1 ppm	0.1 ppm	0.08 ppm		

LFC=lowest feasible concentration; Res=Respirable; Ca=NIOSH potential occupational carcinogen [55, 57, 58]

Table 2. Exposure limit of each individual constituent of welding components

6. Welding monitoring and risk assessment

6.1. Monitoring of welding emissions

Managing the risks of pollutants generated by welding process is carried out in some steps inculing identifying hazards, assessing the risks arising from these hazards, eliminating or minimising the risks via proper control ways, and checking the effectiveness of controls. Monitoring the welder's exposure is a main component of risk management process. Welding process leads to chemical exposures to fumes and toxic gases in enormous quantity. The hazard identification and risk assessment are necessary to work safely in a welding environment. Enough information, education, training and experience are required in this respect. In addition to the full-time welders, a large number of part-time welders who work in small shops and workers in the vicinity of the welding process may also be exposed.

There is a greater potential for exposure due to welding in confined spaces with poor ventilation such as ship hulls, metal tanks and pipe, therefore, monitoring such welders should be seriously considered.

As it was stated previously, the level of welder's exposure to welding emission depends on some factors like the process type, process parameters, and consumables used. Materials and consumables used in welding determine the chemical composition of welding emissions. The specific toxicity of each element and the synergetic effect of generated constituents must be considered to evaluate the exposure status of welders. There are some other workplace specific factors, including the ventilation condition, welder position or posture, and the volume of welding room, that influence the exposure level. The emission rate and also its concentration in the breathing zone of the welder or in the work environment are directly related to the mentioned factors. When it is probable that the welders' exposure will be exceeded the prescribed limits, or when the workers' health and the environment are at risk, the monitoring of hazards and the risk assessment program are required. To evaluate the hazards caused by different welding emissions, collecting various information is recommended. Air monitoring and measuring related pollutants via personal and environmental sampling, biological monitoring, workplace assessment with regard to physical and chemical hazards, and occupational medical findings can be used to evalute the welder's exposure status compeletely [59-60].

Air Monitoring -Airborne pollutants generated by welding can threaten the worker's health and safety. Thus, during the health and safety program, air monitoring is used to identify and quantify welding emissions. To evaluate air contaminants, a sampling strategy is used for collection of exposure measurements. The choice of the best strategy is based on site-specific conditions. In a sampling strategy, some parameters like selection of workers for personal monitoring, sampling duration and required number of samples are important. The measurement of contaminants is carried out in the breathing zone of selected worker. The collected samples must be representative of the normal work activity and exposure of welder, because the sampling results are used to prevent overexposures. Air monitoring in welding processes includes the sampling and analysis of welding fumes and welding gases [61].

Within recent years, standard practices have been developed to monitor exposures considering the occupational exposure limits for elements. Most measurements are made using personal monitoring systems with a pump at a proper flow rate connected to a cassette containing a membrane filter for a suitable period of time. To obtain the accurate result, filter cassette must be placed inside the welding helmet. Time-weighted average concentrations of total fumes is obtained by weighing the filter before and after exposure; the concentrations of elements are determined by chemical analysis methods provided by related organizations like American Welding Society and British Standards Institution [51], NIOSH Manual of Analytical Methods (NMAM) for metals in air and urine and OSHA Sampling and Analytical Methods are used to monitor the welding workplaces. In these methods, analysis of metals is performed by Inductively Coupled Argon Plasma-Atomic Emission Spectroscopy (ICP-AES) after sample preparation by acidic ashing [61, 62]. It is worth mentioning that the microwave digestion can be used instead of acidic ashing to prepare samples, leading to reduction in ashing time up to 90 percent, as well as cost saving and providing a healthier work environment for laboratory operators. Golbabaei et al. used the microwave digestion to prepare urine samples before urinary metal analysis by graphite furnace atomic absorption spectrometry [52].

As it was stated previously, there are different workplace conditions for workers who are welding in confined spaces compared to other welders. Limited access and little airflow or ventilation are the characteristics of a confined space. Hazardous concentrations of welding emissions can accumulate very quickly in such small spaces. Hazardous concentrations of welding emissions can accumulate very quickly in such small space. Thus, confined spaces should be monitored for toxic, flammable, or explosive emissions to evaluate welders' exposure. In some situations, continuous air monitoring may be necessary when workers are welding in a confined space with special conditions. Golbabaei et al. conducted an investigation to assess the risk related to welding pollutants for welders who work in confined spaces. Almost for all analyzed metals, there were significant differences between back welders and controls. Back welding is a task that workers perform welding inside the pipe as a confined space. Based on risk assessment, back welding was a high risk task [16]. These authors in another study assessed the welder's exposure to carcinogen metals (Cr, Cd, and Ni). The NIOSH methods were used for sampling and measurement of metals. Back welders group had maximum exposure to total fume and mentioned elements [52].

Determination of occupational exposures to gases must be based on workplace measurements, because the local ventilation and workplace design can affect the actual concentrations of toxic gases (ozone, carbon monoxide, nitrogen oxides) in the welders' breathing zone. Hariri et al. surveyed the appropriate personal sampling methods to measure the welding emissions in small and medium enterprises. They proposed NIOSH methods to evaluate the fumes and direct reading instruments for measurement of gases. Also, they offered some guidelines for correct assessment of welding workplaces [60]. Choonover et al. showed welders were exposed to higher concentrations of NO_2 and O_3 than controls. These gases were collected on pre-treated filters with proper solutions. Then, NO_2 and O_3 were analyzed by spectrophotometry and ion chromatography (IC), respectively [21]. Azari et al. conducted a study to evaluate exposure of mild steel welders to ozone and nitrogen oxides during TIG and MIG welding. OSHA ID214 and NIOSH 6014 methods were used to evaluate ozone and nitrogen oxides, respectively. High exposure of welders to these gases was reported in the study [64]. Golbabaei et al. also used OSHA and NIOSH methods as well as direct reading instruments for sampling and measurement of various gases [65].

Although there are various techniques for monitoring of welding emissions (both fumes and gases) in air samples, selecting the proper ones depends on some parameters. Availability of sampling media, sample storage time, and the simplicity, cost, time and sensitivity of analytical technique are essential to planning proper sampling strategies. It is necessary to consider those workers who probably have the highest exposures due to used materials and processes, the characteristics of their tasks, their postures during welding, the conditions of work environment, and other pollutants from processes in the vicinity of welding environment. It is known that high concentrations of some welding fumes and gases can also be explosive; therefore, the workplace should be tested to ensure a safe working environment [61, 66].

Biological Monitoring - Biological monitoring means the measurement of the concentration of a contaminant, its metabolites or other indicators in the tissues or body fluids of the worker. In some cases, biological monitoring may be a supplementary monitoring for the personal assessment [53]. Another advantage of the biological monitoring is the detection of biological effects of the chemical by monitoring reversible and irreversible biochemical changes. It can be used in the medical treatment to identify the real exposures of chemicals absorbed into the body of employees suspected of over-exposing to a chemical [58]. Airborne contaminants measurement and biological monitoring are complementary procedures used to prevent occupational disease, assess the risk to workers' health, and evaluate the effectiveness of control ways. Biological monitoring must be conducted based on a proper strategy. Careful considerations are required to select the best biological matrix for each component. To obtain valid results, timing sample collection, sample preparation and analytical method used to determine the concentration of components are critical. There are different methods for biological monitoring of some welding emissions. As it is indicated in Table 2, biological media and biological exposure indices (BEIs) have been recommended for some metals and gases emitted by welding processes. Totally, complete information can be provided by biological monitoring and air monitoring to assess the worker exposure to welding emissions.

Ellingsen et al. studied the concentration of manganese in whole blood and urine in welders. Concentration of Mn in whole blood (B-Mn) was about 25% higher in the welders compared to the controls. The increase in B-Mn and the dose-response relation between air-Mn and B-Mn in the welders are strong indicators of Mn. Long-term high exposure to welding fumes may lead to alterations of the urinary excretion of certain cations that are transported through the DMT1 transport system (divalent metal transporter 1 that is found on the surface of the lung epithelial cells) [67]. Kiilunen study showed the metal concentration in post shift urine samples were correlated with the personal air monitoring results. There were statistical significant correlations between urinary concentrations of chromium and nickel and the related total metal concentration in air in wire welding processes. Also, in MIG/MAG welding, chromium is accumulated in the body with a long half life. There is an association between the airborne concentration of nickel and its post shift urinary concentration. In welding, the nickel concentration in post shift urine samples can indicate the body burden [68]. In a study conducted by Hassani et al. the correlation between airborne Mn and urinary Mn was significant for all exposed subjets. The obtained result can introduce the urinary Mn as a biomarker for exposure to this element [69]. Azari et al. measured the serum level of malondialdehyde in welders. Serum MDA of welders was significantly higher than that of the control group. A significant correlation was detected between ozone exposure and level of serum MDA, but the correlation was not observed for nitrogen dioxide exposure [64]. Rossbach recommended the determination of Al in urine for biological monitoring because of the higher sensitivity and robustness of this marker compared to Al in plasma [70]. Golbabaei et al. analyzed the urinary metals among the different groups of welders. According to the results, exposure of welders to fume components leads to more accumulation of them at welders' bodies [52]. Based on different studies, the soluble metal compounds are accumulated in the body, affecting the critical organs. Urinary concentration of metal is used as a biomarker of metal exposure. Therefore, biomonitoring serves as an appropriate tool to monitor both the recent and past exposure and it can be related to the total chemical uptake through all exposure routes [69].

Health monitoring - In addition to the assessment of the airborne concentration of a particular contaminant and its comparison with standard limit, health monitoring may also be done for some hazardous chemicals to assess risks to exposed workers. Health monitoring means monitoring workers exposed to hazardous pollutants to identify changes in their health status and evaluate the effects of exposure. Health monitoring can provide effevtive information to implement proper ways for eliminating or minimizing the risk of exposure and improving control measures. Health monitoring considers all routes of exposure to contaminants [9, 66, 71]. Some tests including spirometry (lung function), audiometry (hearing), biochemical tests (e.g. kidney or liver function), cardiac function tests (heart function), nerve conduction velocity and electromyography tests (nerve and muscle function), and neurobehavioural tests (nerve and brain function) may be used in health monitoring. The type of test used will depend on the occupational hazards that the employee are exposed to [58]. Donaldson [72] and Antonini [73] surveyed lung functions in exposed welders and showed that exposure to welding fumes is associated with both pulmonary and systemic health endpoints, including decrease in pulmonary function, increased airway responsiveness, bronchitis, fibrosis, lung cancer and increased incidence of respiratory infection. In addition to these pulmonary effects, metal fume fever is frequently observed in welders. Exposure to metal fumes and irritating gases cause chronic obstructive pulmonary disease (COPD). Health monitoring of welders can help detect breathing problems and reduced lung functions in early stages, resulting in prevention of further damages. Spirometric tests are used by an occupational phisycian to assess lung functions [74]. Totally, health monitoring may include simple observation of the worker's skin to complicated tests in special cases. Health monitoring must be done by the experienced medical practitioner. An occupational physician can provide specialist services and testing such as spirometric tests, respiratory screening and chest X-rays. It is necessary to do the health monitoring before beginning work with a hazardous chemical to provide enough information for following changes in the worker's health during periods of exposure.

6.2. Risk Assessment of welding emissions

Risk is defined as the possibility of occurance of an event leading to clear concequences. Evaluating risks to workers' safety and health is conducted in risk assessment process. It is performed in some steps including:

- Hazards identification and those at risk
- Evaluating the risks (qualitative or quantitative)
- Elimination or minimization of risks via implementing control measures and taking actions
- Monitoring and reviewing the effectiveness of adopted controls

The severity of hazard and the exopsure level determine the health risk and the type of chemical and nature of work are important factors in this regard. All workers in the vicinity of a special activity should be considered to assess the risk associated with chemical hazards, because they may potentially be at risk of chemicals emmitted by that activity.

In welding environments, employers are resposible to ensure the safety and health of welders and take proper measures for their protection. Although, preventing the occupational risks is the main purpose of risk assessment, it is not possible in all situations; therefore, risks should be reduced using control measures. There are different hazards related to welding process resulting in risks to welders. Chemical hazards, physical hazards, and those associated with ergonomics threaten the health of welders. Since this text deals with air pollution, the risk assessment of welding emissions i.e. fumes and gases is considered. Hazardous chemicals in the workplace result in different risks to workers.

There are different methods to do risk assessment of chemicals in which some principles should be considered. These principles include addressing all relevant hazards and risks and beginning the elimination of risks, if it is possible.

The ministry of manpower of Singapore has published a guideline intitled "semi-quantitave method to assess occupational exposure to harmful chemicals"[75]. This method may be useful to assess the risks resulting from welding emissions. Risk assessment is conducted for following purposes:

- Identifying the hazards related to each harmful chemical
- Evaluating the degree of exposure to chemical of interest
- Determining the likelihood of chemical adverse effects

A risk rating to different tasks can be designate using the mentioned method. After that, using risk rating matrix, hazards are ranked as negligible, low, medium, high and very high (legends 1 to 5) and required actions are prioritized to select appropriate controlling plans. This guideline deals with the health risk to workers exposed to chemicals via inhalation. There are eleven steps for hazard identification and rating, exposure evaluation, and assessing risk. The actual exposure level is required for determination of exposure rating and risk level. A step by step flow chart for assessing the risk, forms needed for completing some steps, and different tables and equations for evaluating the risk have been provided by guideline. All components to assess the risks are available in guideline and it can be used for risk assessment of welding emissions in a simple and fast way. Following, the process flow chart has been presented to understand the consept of risk assessment.

Golbabaie et al. used mentioned guideline to assess the health risks arising from metal fumes on back welders. Risk assessment was performed according to the steps previously explained. Cadmium concentration was ranked as "very high" group. Also, total fumes, total chromium, and nickel were ranked as "high" legend. Findings indicated back welding is a high risk task. High concentration of metals confirmed that working in confined spaces creates a great risk for welders. In some cases as in cadmium despite the rather low concentration of the pollutants, the risk is ranked as "very high" due to the carcinogenisity nature of this element. Therefore, it is not always possible to judge the health hazards of the pollutants based on their concentrations.

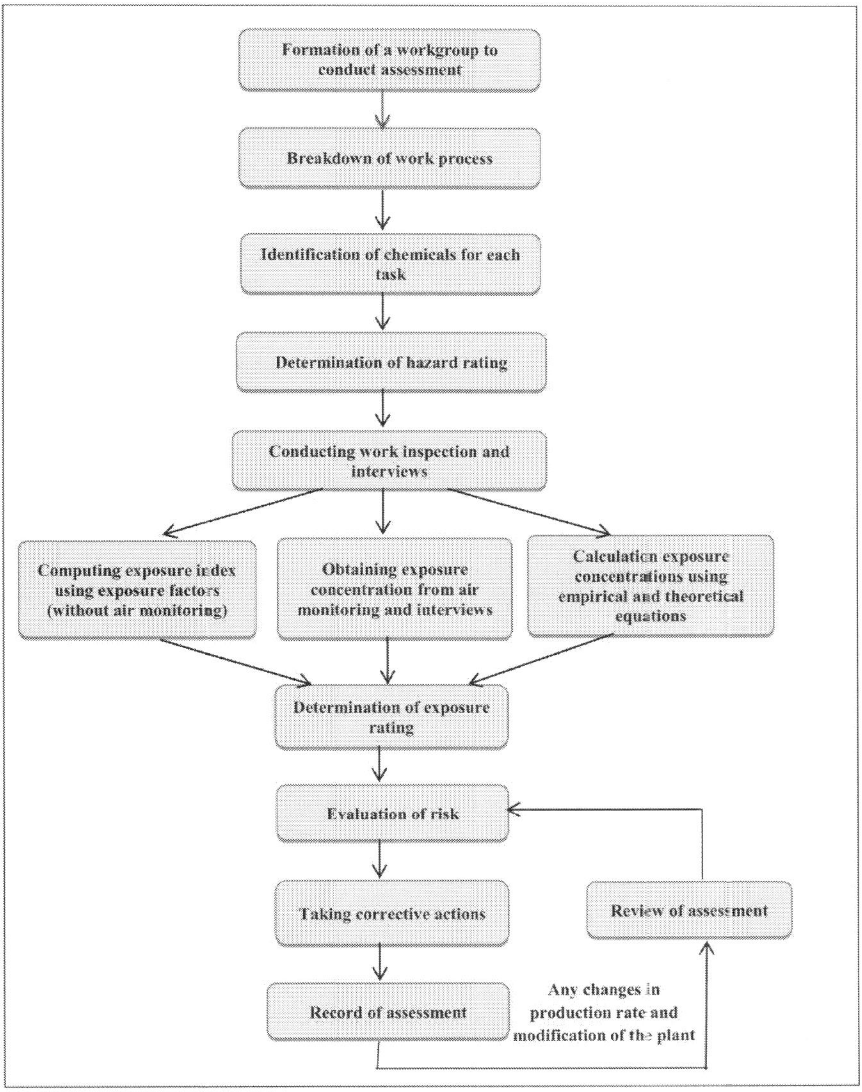

Figure 2. Process flow chart of semi quantitave method for chemicals risk assessment [75]

Following the risk assessment, employers can decide on required preventive measures, the working and production procedures, and also improving the level of welder protection. To complete risk assessment of welding chemicals, data related to air monitoring, biological monitoring, and health monitoring may be required for true judgement. Totally, risk assessment in workplace can result in some advantages. Workers do their tasks in a safe manner; employers provide appropriate programs to prevent high exposure and increase job satisfaction; regulators and related organizations can reliably present health and safety standards. The process of risk assessment is a basis for risk management to reduce welding hazards by choosing correct actions [76-77].

7. Occupational control

Air pollution control deals with the reduction of air pollutants emitted into the atmosphere using different technologies. Sometimes, managing the production process is used to control air pollutant emisstion, therefore, checking the production process can be useful for beginnig the air pollution control. Elimination of a hazard is the first aim to control related risk. In essence, keeping the pollutant emission at the minimum level during the process is the main purpose of controling the air pollution. Based on the risk assessment results, employers can decide for control of risk using proper ways. There are various ways to control the risk of chemicals like welding emissions. If the hazard elimination in not reasonably practicable, other approaches are used to minimize the risk. Substitution, isolation, engineering controls, work practices, and personal protective equipment (PPE) are used to reduce risks to the lowest practicable level in order of priority. Using personal protective equipment is the least recommended control way. To provide a layered safety net, a combination of several control ways may be adopted for preventing risks [66, 76, 78]. In the case of welding, if the elimination of fumes is not practicable, other controling measures should be applied. Modifying the welding process, improving working practices, ventilation, and using PPEs are considered in order to control of fumes.

7.1. Choosing or modifying the welding process

Employers can choose the welding type for production process based upon its efficiency, weld quality, available equipment, and economics. For instance, TIG welding generates less fume compared to MMA, MIG and FCAW processes, so, it can be a proper choice for welding operations. In order to modify the welding process, selecting consumables with minimum fume emissions and considering the welding parameters to minimize the emissions are recommended to employers. The generation of welding fumes is minimized using the lowest acceptable amperage. To optimize the process modification, paying attention to consumables, equipment, and control system is necessary. Selecting proper consumbles leads to minimizing the environmental impacts and controling risks to welders. Welding on non-painted or coated surfaces can also reduce the production of emissions. Process modification in welding results in decreasing needs for administrative controls and other expensive procedures, and also simplifying the process of risk assessment.

7.2. Improvement of working practices

Working practice, the way used to do work, can be improved for control of workers' exposure. Safe work practices are provided by company or organization to perform a task with minimum risk to workforce, environment, and process. Such practices control the manner of performing work and complete engineering measures. Placing the workpiece, as an improving measure, can keep the welders away from plume rising above the weld. Minimizing the welding in confined or enclosed spaces leads to reduction of exposure to pollutants. Proper training programs, housekeeping, maintenance, and doing task on time are the safe welding habits to reduce exposure. Consequently, welding based on safe practices and instructions results in healthier workplace and diminishing the risks of exposure to hazardous emmisions [79, 80].

7.3. Ventilation

Ventilation is the most effective way for removing welding emissions at source to reduce exposure to fumes and gases in welding operations. Designing the ventilation system in accordance with the types of hazardous emissions results in providing a safe atmosphere in the workplace. This control procedure is classified into dilution (general) ventilation and local exhaust ventilation (LEV). The most efficient method to control welding emissions is the combination of LEV and dilution ventilation.

General or Dilution Ventilation -This type of ventilation uses the flow of air into and out of a working environment to dilute contaminants by fresh air. The required fresh air can be supplied by natural or mechanical ways. Dilution ventilation may not be sufficient to control exposure to welding emissions, because it cannot provide enough air movement to prevent the entry of fumes and gases into the welder's breathing zone before removing them from welding environment. In fact, the general ventilation is not suitable for controlling the toxic substances, specially when the worker is downstream of contaminant. To ensure the efficiency of the system, measuring airflow regularly and sampling contaminants to assess exposure are required. A well designed dilution system can be approprite for situations in which welding is done on clean, uncoated, mild steels. In dilution ventilation, draft fans or air-movers, wall fans, roof vents, open doors and windows may be used to move air through the work environment. Totally, if the generated contaminant is in low concentration and can be controlled to the standard exposure level, dilution systems will be effective enough as a control measure [66, 80-82].

Local Exhaust Ventilation - Local exhaust ventilation (LEV), as a primary engineering control, is used to remove contaminants before entering the breathing zone of workers. LEV can be used to control welding emissions close to the generation source. To be effective, LEV system should be well designed and installed, used correctly and properly maintained. Type of generated contaminants and characteristics of the process and work environment are crucial to design LEV [81]. To design a suitable system in welding process, some parameters should be considered, such as fume generation rate, arc- to-breathing zone distance, work practices and worker's exposure. Various parameters related to type of welding have important roles in the fume generation rate and fume composition. Therefore, considering these parameters is necessary to design LEV system [83-85].

For welding processes like stainless steel or plasma arc welding in which fumes containing heavy metals are generated, the LEV system can effectively be used to control worker's exposure. A local exhaust ventilation consists of a hood, fan, duct, and air cleaner. All parts of LEV system must be designed according to correct rules and requirements to remove air pollutants with appropriate efficiency. For instance, the ducting material and structure, air velocity through ducts, the number of branches, and the probability of the leakage and corrosion are important factors related to duct that can affect the LEV system. There are some considereations to select a suitable fan for the system. Some variables such as pressure, flow rate, power, noise, and rotation speed are the main characteristics influencing on the fan performance. Air cleaner is a device to capture welding emissions before it can escape into the ambient air. To select an appropriate air cleaner, some design considereations need to be addressed. Size and shape of welding space, pollutants generation rate, pollutant composition, cost of devices, process type, and the availability of equipment may be effective factors in this respect. In welding processes, source capture systems can be the ideal choise to control fume contaminants using the least air flow rate. In some situations, a source capture system cannot be used. For example situations in which worker has to work on mobile positions; there are a large number of small welding points producing hazardous emissions; welding must be done in confined spaces; and there are some obstructions like overhead cranes leading to problems with ducting installation. Dust collectors (filtration units) and electrostatic precipitators (ESP) can also be used as air cleaners to capture welding emissions before escaping into the environment. ESPs are ideal to collect submicron particles, especially in carbon steel welding. Although the efficiency of ESP is lower than filtration system, it needs very little maintenance and also there is no cost for filter replacement. ESPs are not recommended for stainless steel welding.

Some general considereations should be addressed to design a LEV system. Ducting system should be resistant to the captured emissions; the risks of contaminants accumulation and fire propagation in ducting system should be taken into account; exhausted air containing welding emissions should not be discharged where other workers or people are present; any draught from open doors or windows should be considered because of interference with hood performance. In addition, a maintenance program is required to ensure that control measures remain effective. For instance, regular inspections of LEVsystems should be carried out to check their effectiveness. As an other maintening plan, periodic air monitoring is done to ensure the system has proper performance. Therefore, as well as correct and completed design of LEV system, other elements like employee training, proper use, cleaning, and maintenance are required to achieve the effective protection.

Portable Systems - In some situations, portable systems may be used. These systems are used where welding is infrequently performed and the existing sysrem can be shared between working stations. Also, small mobile units may be used in confined spaces where installing the usual systems is not practical. In these cases, installing the hood close to the emissions point of origin, the hood placement and its distance from the source of welding emissions should be considered. Adequate ventiltion is essential in confined spaces, because the accumulation of hazardous emissions may lead to oxygen deficiency and also adverse effects related to generated fumes and gases. Commercially, there are different portable ventilation systems to use in confined spaces. Flexible air ducts and different kinds of portable fans are available for a variety of ventilation applications. In general, approximately 10 air exchanges per hour should be provided by ventilation in confined spaces. The volume of space and the flow rate of fan determine the time of each exchange. Before entry into the confined space for welding, that space should be ventilated for a minimum of five minutes. It is important to select a proper fan with enough capacity and position it in correct place. Some related organizations have provided procedures and instructions related to working in confined spaces, including ventilation equipment, confined spaces entry, emergency action plan, permit forms, and other requirements for working in these spaces [66, 81, 84, 86].

7.4. Respiratory protection equipments

Personal protective equipment (PPE) should not be used instead of other control measures, but sometimes they may be required along with engineering controls and safe work practices. Respiratory Protection Equipments (RPEs) are used to protect the workers against inhalation of hazardous emissions in the workplace, where exposures cannot adequately be controlled by other ways.

Using a respirator not selected appropriately leads to a false sense of protection for wearer and exposure to hazardous substances. It must be specific to the pollutant and fitted, cleaned, stored and maintained based on provided standards and guidelines for respirators. Each RPE has a protection factor (PF) that is determined as the ratio of the concentration of the pollutant outside the respirator to that inside the respirator. There is a wide range, from low to high, for protection factors. Some organizations like NIOSH have provided required equations and tables to calculate protection factors for respirators. There are different types of respirators and it is possible to select the most appropriate type for existing circumstances. In welding processes, respirators should be selected in accordance with generated emissions, welding type, welding task, and working conditions. For example, NIOSH recommends a self-contained breathing apparatus for welding in confined spaces because the oxygen concentration in the space may be reduced due to welding. Also, a combination of particulate/vapour respirator may be used because of the generation of both of fumes and gases during welding. A standard program is needed for using raspiratory protection devices. Some requirements are followed in this program including hazard assessment, selecting the appropriate respirators in respect of pollutants, respirator fitting test, worker training on how to use respirator correctly, inspection and maintenance of respirator, and recordkeeping. There are two types of RPE. The first type is respirators that clean workplace air before being inhaled and the second type is air-supplied respirators in which air supply is separate from workplace atmosphere. Totally, the suitable RPE for welding processes should be selected by an expert and based on fume concentration, presence of toxic gases, and the probability of oxygen deficiency. Selecting air-purifying respirators with correct filtration cartridge results in protection of welders from low levels of metal fumes and welding gases [87, 88].

8. Conclusion

Air pollution is contamination of the indoor or outdoor environment, leading to changes in the natural characteristics of the atmosphere. In all welding processes, various types of air pollutants are generated. Air pollutants created by welding include fumes and gases whose composition and emission level depend on some factors such as the welding method, welding parameters (current, voltage, shielding gas and shielding gas flow), base metal and other consumables. Exposure to excessive levels of fume and gases can cause different adverse health effects on workers. Since a large number of workers are exposed to welding emissions and also the generated pollutants have negative impacts on environment, a risk assessment program is required to protect workers and environment by suitable procedures. In an effective program, worker's safety and health is considered by management as a fundamental value.Taking different precautions can improve the welder's work situation. There are various techniques for evaluating and monitoring welding pollutants in air samples and biological matrices and also different procedures for their control. Selecting the proper engineering controls can lead to protection of workers and environment. During the risk assessment program and selection of control measures, it is necessary to consider nanoparticles emitted by welding operations. Particle sizes and size distributions of welding emission are critical to determine the efficient control devices. In some cases, breathing zone protection can be used. Health hazards can be reduced by choosing a correct welding helmet and by using the proper shielding gas and welding parameters. It is worth mentioning that proper information should be provided for workers about hazards of their tasks. The welder should be informed of operating techniques and all procedures that reduce welding fumes. The training programs should be included proper ways to perform tasks and proper work practices to reduce fumes. This program includes safety training, monitoring the good safety practices and good environmental practices. Also, the respirator and cartridge selection, fit-testing and respirator maintenance and storage are considered in a suitable training program. Furthermore, employers must be informed about industrial hygiene programs at workplaces and quantitative risk assessment for workers exposed to hazardous compounds. In recent years, different organizations have focused on climate change and environmental impacts of all industrial activities including welding. Various laws, instructions, and guidelines have been provided for protecting the air, environment, and water. Employers are responsible for the purchase of proper welding equipment to meet environmental requirements and choose more environmentally friendly processes.

Author details

Farideh Golbabaei* and Monireh Khadem

*Address all correspondence to: fgolbabaei@sina.tums.ac.ir

Department of Occupational Health Engineering, School of Public Health, Tehran University of Medical Sciences, Tehran, Iran

References

[1] National Institute for Occupational Safety and Health (NIOSH). Nomination of Welding Fumes for Toxicity Studies. In: U.S. Department of Health and Human Services, Public Health Service: NIOSH; 2002.

[2] El-Batanouny MM, Amin Abdou NM, Salem EY, El-Nahas HE. Effect of Exercise on Ventilatory Function in Welders. Egyptian Journal of Bronchology. 2009; 3(1): 103-11.

[3] American Welding Society (AWS). Standard Welding Terms and Definitions. In: Committee TA, 12 ed. Miami: American Welding Society; 2009.

[4] Stern RM. Process-Dependent Risk of Delayed Health Effects for Welders. Environmental Health Perspectives. 1981; 41: 235-53.

[5] Husgafvel-Pursiainen I, Siemiatycki J. IARC Monograph on Welding Fumes. 1990; 49.

[6] Erdely A, Antonini JM, Salmen-Muniz R, Liston A, Hulderman T, Simeonova PP, et al. Type I interferon and pattern recognition receptor signaling following particulate matter inhalation. Particle and fibre toxicology. 2012; 9 (1).

[7] Gonser M, Hogan, T., editor. Arc Welding Health Effects, Fume Formation Mechanisms, and Characterization Methods: InTech; 2011.

[8] Carter GJ. Risk Assessment and Control of Exposure During Arc Welding of Steel. Intrernational Conference on Health and Safety in Welding and Allied Processes; 9-11 May; Kopenhagen, Denmark 2005.

[9] Safe Work Australia. Guidance On The Interpretation Of Workplace Exposure Standards For Airborne Contaminants. 2012. p. 1-63.

[10] Finneran A, O'Sullivan L. Force, posture and repetition induced discomfort as a mediator in self-paced cycle time. International Journal of Industrial Ergonomics. 2010; 40 (3): 257-66.

[11] Canadian Centre for Occupational Health and Safety (CCOHS). Welding-Overview of Types and Hazards [Internet]; 2010. Available from: http //www.ccohs.ca/oshanswers/safety haz/

[12] Turan E, Kocal T, unlugencoglu K. Welding technologies in shipbuilding industry. The Online Journal of Science and Technology. 2011; 1 (4).

[13] Occupational Safety and Health Administration. Safety and Health Topic Criteria for a Recommended Standard: Working in Confined Spaces. Washington, DC Occupational Safety & Health Administration; 2007. Available from: www.osha.gov/SLTC/confinedspaces.

[14] American Welding Society (AWS). Confined Spaces, Safety and health Fact Sheet No. 11. 1995.

[15] NIOSH. Hazards to Outdoor Workers [Internet]. Centers for Disease Control and Prevention, Workplace Safety & Health Topics; 2013. Available from: http://www.cdc.gov/niosh/topics/outdoor/.

[16] Golbabaei F, Khadem M, Hosseini M, et.al. Exposure to metal fumes among confined spaces welders. Ital J Occup Environ Hyg. 2012; 3 (4): 196-202.

[17] Weman K. Welding processes handbook. bington Hall, Abington Cambridge, England: Woodhead Publishing Ltd and CRC Press LLC; 2003.

[18] Kopeliovich D. Classification of welding processes: substech; 2012 [updated 2014/09/03]. Available from: http://www.substech.com/dokuwiki/doku.php

[19] Flagan RC, Seinfeld JH. Fundamentals of Air Pollution Engineering. Englewood Cliffs, New Jersey Prentice Hall; 1988.

[20] World Health Organization (WHO). Air Pollution; 2014. Available from: http://www.who.int/topics/air pollution/en/.

[21] Schoonover T, Conroy L, Lacey S, Plavka J. Personal Exposure to Metal Fume, NO_2, and O_3 among Production Welders and Non-welders. Industrial Health. 2011; 49: 63-72.

[22] Amza G, Dobrota D, Dragomir MG, Paise S, Apostolescu Z. Research on Environmental Impact Assessment of Flame Oxyacetylene Welding Processes. METALURGIJA 2013; 52 (4): 457-60.

[23] Nakhla H, Shen JY, Bethea M. Environmental Impacts of Using Welding Gas. The Journal of Technology, Management, and Applied Engineering 2012; 28 (3): 1-11.

[24] Kalliomaki PL, Hyvarinen HK, Aitio A, Lakoma EL, Kalliomaki K. Kinetics of the metal components of intratracheally instilled stainless steel welding fume suspensions in rats. British journal of industrial medicine. 1986; 43 (2): 112-9.

[25] AGA. FACTS ABOUT Fume and gases. AGA (a member of Linde Group), Report No. 110199 0912 – 1.3 HL.

[26] Spear JE. Welding Fume and Gas Exposure. Magnolia, Texas J.E. Spear Consulting, LLC, 2004.

[27] 27. Fuglsang K, Gram LK, Markussen JB, Kristensen JK. Measurement of ultrafine particles in emissions from welding processes. 16th International Conference on Joining of Materials; 10-13 May 2011; Elsinore, Denmark 2011.

[28] Brand P, Klaus L, Uwe R, Kraus T. Number Size Distribution of Fine and Ultrafine Fume Particles From Various Welding Processes. Ann Occup Hyg. 2013; 57 (3): 305-13.

[29] Keane M, Siert A, Stone S, Chen B, Slaven J, Cumpston A, et al. Selecting Processes to Minimize Hexavalent Chromium from Stainless Steel Welding. WELDING JOURNAL. 2012; 91: 241s-246s.

[30] Gomes J. Albuquerque P, Miranda R, Vieira T. On the toxicological effects of airborne nanoparticles from welding processes. IIW European-South American School of Welding and Correlated Processes 18 –20th May 2011, Brazil: 2011.

[31] Gomes J, Miranda R. Emission of airborne ultrafine particles during welding of steel plates. Ciencia & Tecnologia dos Materiais. 2014; 26 (1) :1-8.

[32] Gomes JFP, Albuquerque PCS, Miranda RMM, Vieira MTF. Determination of airborne nanoparticles from welding operations. Journal of Toxicology and Environmental Health, Part A. 2012; 75 (13-15): 747-55.

[33] A-Meo S, Al-Khlaiwi T. Health Hazards of Welding Fumes. Saudi medical journal. 2003; 24: 1-25.

[34] De Flora S. Threshold Mechanisms and Site Specificity in Chromium (VI) Carcinogenesis. Carcinogenesis. 2000; 21 (4): 533-41.

[35] Sowards JW, Ramirez AJ, Lippold JC, Dickinson DW. Characterization Procedure for the Analysis of Arc Welding Fume. Welding Research. 2008; 87: 76-83S.

[36] Keegan GM, Learmonth ID, Case CP. Orthopaedic metals and their potential toxicity in the arthroplasty patient. The Journal of Bone and Joint Surgery. 2007; 89-B (5): 567-73.

[37] pesch B, Haerting J, Ranft U, et.al. Occupational Risk Factor for renal Cell Carcinoma: Agent-Specific Results from a Case Control Study in Germany. International Journal of Epidemiology. 2000; 29: 1014-24.

[38] Erhabor GE, Fatusi S, Obembe OB. Pulmonary functions in ARC-welders in Ile-Ife, Nigeria. Ast Afr Med J. 2001; 78 (9): 461-4.

[39] Tenkate TD. Optical radiation hazards of welding arcs. Rev Environ Health. 1998; 13 (3): 131-46.

[40] Mariutti G, Matzeu M. Measurement of ultraviolet radiation emitted from welding arcs. Health Phys. 1988; 54 (5).

[41] Tenkate TD. Occupational exposure to ultraviolet radiation: a health risk assessment. Rev Environ Health. 1999; 14 (4): 187-209.

[42] Ernst E, Bonde JP. Sex hormones and epididymal sperm parameters in rats following sub-chronic treatment with hexavalent chromium. Hum Exp Toxicol. 1992; 11 (4): 255-8.

[43] Jensen TK, Bonde JP, Joffe M. The influence of occupational exposure on male reproductive function. Occupational Medicine. 2006; 56: 544-53.

[44] Bonde JP. Subfertility in relation to welding. A case referent study among male welders. Dan Med Bull. 1990; 37 (1): 105-8.

[45] Bonde JP, Hanse KS, Levine RJ. Fertility among Danish male welders. Scand J Work Environ Health. 1990; 16 (5): 315-22.

[46] Mortensen JT. Risk for reduced sperm quality among metal workers, withspecial reference to welders. Scand J Work Environ Health. 1988; 14 (1): 27-30.

[47] Sheiner EK, Sheiner E, Hammel RD, Potashnik G, Carel R. Effect of Occupational Exposures on Male Fertility: Literature Review. Industrial Health. 2003; 41: 55-62.

[48] Iregren A. Manganese neurotoxicity in industrial exposures: proof of effects, critical exposure level, and sensitive tests. Neurotoxicology. 1999; 20 (2-3): 315-23.

[49] Bowler RM, Gysens S, Diamond E, Booty A, Hartney C, Roels HA. Neuropsychological Sequelae of Exposre to Welding Fumes in a Group of Occupationally Exposed Men. Int J Hyg Environ Health. 2003; 206: 517- 29.

[50] [50]Wang X, Yang Y, Wang Xi, Xu S. The Effect of Occupational Exposure to Metals on the Nervious System Function in Welders. J Occup Health. 2006; 48: 100-6.

[51] International Agency for Research on Cancer (IARC). welding fumes and gases. lARC Monograhs 49; 1987.

[52] Golbabaei F, Seyedsomea M, Ghahri A, Shirkhanloo H, Khadem M, Hassani H, Sadeghi N, Dinari B. Assessment of Welders Exposure to Carcinogen Metals from Manual Metal Arc Welding in Gas Transmission Pipelines, Iran. Iranian J Publ Health. 2012; 41 (8): 61-70.

[53] National Occupational Health and Safety Commission. Welding: Fumes and Gases. In: Commonwealth of Australia, editor: Ambassador Press Pty Ltd; 1990.

[54] Division of Workers' Compensation. Welding Hazards Safety Program. In: Texas Department of Insurance, editor; 2012.

[55] American Conference of Governmental Industrial Hygienists (ACGIH). Threshold Limit Values for chemical substances and physical agents & biological exposure indices.: ACGIH, Cincinnati, OH; 2010.

[56] Safe Environment-Managing Property Risk. Welding Fume [Internet]. Safe Environment, Sydney; 2012. Available from: http://www.safeenvironments.com.au/welding-fume/.

[57] Ashby HS. Welding Fume in the Workplace, Preventing Potential Health Problems through Proactive Controls. Professional Safety. 2002; Apr: 55-60.

[58] Department of Consumer and Employment Protection. Risk-based health surveillance andbiological monitoring. Resources Safety, Department of Consumer and Employment Protection, Western Australia; 2008.

[59] Pires I, Quintino L, Miranda R, Gomes J. Fume emissions during gas metal arc welding. Toxicological and Environ Chemistry. 2006; 88 (3): 385-94

[60] Spiegel-Ciobanu V. Occupational health and safety regulations with regard to welding and assessment of the exposure to welding fumes and of their effect. Welding and Cutting. 2012; 11 (1): 61.

[61] Occupational Safety and Health Administration (OSHA). OSHA Technical Manual, Personal Sampling for Air Contaminants In: U.S. Department of Labor, Washington, DC OSHA; 1999.

[62] Education and Information Division. National Institute for Occupational Safety and Health. Occupational Exposure Sampling Strategy Manual. In: Department of Health Education and *Welfare*, Cincinnati: NIOSH; 1977.

[63] Hariri A, Yusof MZM, Leman AM. Sampling Method for Welding Fumes and Toxic Gases in Malaysian Small and Medium Enterprises (SMEs). Energy and Environment Research. 2012; 2 (2): 13-20.

[64] Azari MR, Esmaeilzadeh M, Mehrabi Y, Salehpour S. Monitoring of Occupational Exposure of Mild Steel Welders to Ozone and Nitrogen Oxides. Tanaffos. 2011; 10 (4): 54-9.

[65] Golbabaei F, Hassani H, Ghahri A, Arefian S, Khadem M, Hosseini M, Dinari B. Risk Assessment of Exposure to Gases Released by Welding Processes in Iranian Natural Gas Transmission Pipelines Industry. International Journal of Occupational Hygiene (IJOH). 2012; 4 (1): 6-9.

[66] Safe Work Australia. Managing Risks of Hazardous Chemicals in the Workplace. Australia; 2012.

[67] Ellingsen DG, Dubeikovskay L, Dahl K, et al. Air exposure assessment and biological monitoring of manganese and other major welding fume components in weldersw. J Environ Monit. 2006; 8: 1078-86.

[68] Kiilunen M. Use of biological monitoring for exposure assessment in welding Health and safety in welding and allied processes Brondby, Denmark; 2005.

[69] Hassani H, Golbabaei F, Ghahri A, et al. Occupational Exposure to Manganese-containing Welding Fumes and Pulmonary Function Indices among Natural Gas Transmission Pipeline Welders. J Occup Health. 2012; 54: 316-22.

[70] Rossbach B, Buchta M, Csanady GA, et al. Biological monitoring of welders exposed to aluminium. Toxicology Letters. 2006; 162: 239-45.

[71] Safe Work Australia. Health Monitoring for Exposure to Hazardous Chemicals, Guide for Persons Conducting -A Business or Undertaking. Australia; 2013.

[72] Donaldson K, Tran L, Jimenez LA, Duffin R, et al. Combustion-derived nanoparticles: A review of their toxicology following inhalation exposure. Particle and Fibre Toxicology. 2005; 2 (10): 1-14.

[73] Antonini JM, Lewis AB, Roberts JR, Whaley DA. Pulmonary Effects of Welding Fumes: Review of Worker And Experimental Animal Studies. American Journal of Industrial Medicine. 2003; 43: 350-60.

[74] Health and safety in welding and allied processes - assessing and controlling the exposure risk and complying with the COSHH regulations for welding fume [Internet]. Technical Knowledge 2014. Available from: http://www.twi-global.com/technical-knowledge/faqs/health-and-safety-faqs

[75] Ministry of manpower. A semi-quantitative method to assess occupational exposure to harmful chemical. In: Occupational Health and safety Division, Singapore; 2005.

[76] Europian Commision. Guidance on Risk assessment at Work. In: Directorate General Employment and Social Affairs, Luxembourg: Office for Official Publication; 1996.

[77] National Academy of Sciences. Occupational Health and Safety in the Care and Use of Nonhuman Primates. Washington DC: The National Academies Press; 2003.

[78] Overseas Environmental Cooperation Center. Air Pollution Control Technology Manual. Japan: Environmental Agency, Government of Japan; 1998.

[79] Hewitt PJ. Strategies for risk assessment and control in welding: challenges for developing countries. Annals of Occupational Hygiene. 2001; 45 (4): 295-8.

[80] The welding Institute (TWI). Control of welding fume, Health, safety and accident prevention. [Internet]. The Welding Institute: UK; 2014. Available from: http://www.twi-global.com/technical-knowledge/job-knowledge.

[81] Government of Alberta-Employment and Immigration. Welder's Guide to the Hazards of Welding Gases and Fumes Canada: Workplace Health and Safety Bulletin; 2009.

[82] Health and Safety Executive. General Ventilation in the Workplace, Guidance for Employers. Sheffield: Health and Safety Executive; 2000.

[83] American Welding Society (AWS). Fumes and gases in the welding environment. Miami, FL: AWS; 1979.

[84] Chevron MCBU. Confined Space Entry. USA: HSE, 2012 Contract No. CPL-HES 201

[85] Flynn MR, Susi P. Local exhaust ventilation for the control of welding fumes in the construction industry-a literature review. Annals of occupational hygiene. 2012: 1-13.

[86] OSHA. OSHA and Welding Exhaust In: General Industry Standards and Interpretations, Marine Chemist Service, Inc; 1979.

[87] Health and Safety Authority. A Guide to Respiratory Protective Equipment. Dublin; 2010.

[88] Employment and Social Development Canada. A Guide to Health Hazards and Hazard Control Measures with Respect to Welding and Allied Processes. Canada, Quebec; 2007.

Chapter 3

Aldehyde Measurements in Indoor and Outdoor Environments in Central-Southern Spain

Florentina Villanueva, Araceli Tapia,
Inmaculada Colmenar, José Albaladejo,
Beatriz Cabañas and Ernesto Martínez

Additional information is available at the end of the chapter

http://dx.doi.org/10.5772/60016

1. Introduction

Aldehydes are indoor and outdoor chemical pollutants of particular interest due to their potential adverse health effects on humans and to their important role in atmospheric chemistry. In fact, carbonyls are of critical importance since they are the most stable intermediate species in the oxidation of volatile organic compounds. Even in the indoor environments, aldehydes can be released from ozone reactions with unsaturated VOCs [1]. The most important global source of formaldehyde and acetaldehyde in the atmosphere is of secondary origin, and it is the oxidation of natural and anthropogenic hydrocarbons. However, motor vehicle exhausts represent the primary emission source of these pollutants in urban areas [2].

In occupational and residential indoor environments, predominant carbonyls are aldehydes, mainly formaldehyde and acetaldehyde. Formaldehyde, usually the most abundant aldehyde in air, is classified in Group 1 (human carcinogen) by the International Agency for Research on Cancer [3] and acetaldehyde is classified in Group 2B as a possible carcinogenic in humans [4, 5]. Also, benzaldehyde and acrolein are suspected carcinogens and mutagens, as well as other low-molecular-mass aldehydes, which reactivity and possible mutagenicity are similar to those of acetaldehyde [6-8].

Generally speaking, exposure to carbonyls is higher indoors than outdoors. Indoor aldehyde concentrations are usually 2-10 times higher than the outdoor ones [9, 10], indicating the presence of significant indoor sources such as direct emissions and indoor chemical formation although these higher values are also due to the low air exchange rates in the indoor environment [11]. Particularly, formaldehyde is widely used not only in construction (wood processing, furniture, textiles, and carpeting) but also in various industries [12, 13]. It is also a by-product of certain anthropogenic activities (e.g., smoking tobacco, burning automotive (and other) fuels, and residential wood burning) [12]. Formaldehyde is even a component of many consumable household products such as antiseptics, medicines, cosmetics, dish-washing liquids, fabric softeners, shoe-care agents, carpet cleaners, glues and adhesives, lacquers, etc. [14, 15]. Formaldehyde is also employed as food preservative [16, 17]. For these reasons, formaldehyde is generally found in higher concentrations indoors than outdoors.

Because aldehydes may contribute to different diseases and are mainly found in indoor environment, their measurements are of particular interest, especially when it is known that most people in developed countries spend up to 90% of their time indoors [18]. The objective of this study is to measure and analyze the levels of formaldehyde and acetaldehyde and even other carbonyls in indoor and outdoor air in this region from central-southern Spain. The measurements have been carried out in different sampling periods from 2010 to 2014. Very few studies have been carried out in the indoor air in Spain with the aim of determining different air pollutants [19-27] and most of which have been focused on particulate matter [20, 23-26]. In this paper, the levels of carbonyls in different indoor environments in our country are presented for first time. It consists of new field campaigns together with the results obtained from our previous studies [21, 29].

2. Experimental section

2.1. Sampling sites and their characteristics

Ciudad Real is an urban area with around 65,000 inhabitants and is located in the heart of La Mancha region in central-southern Spain (38.59°N, 3.55°W, at approximately 628 m above sea level) in a fairly flat area, 200 km south of Madrid. With a low presence of industry, traffic is the most important source of air pollution in this city [28]. Puertollano (38° 42"N 04° 07"W, at approximately 700 m above sea level) is a very important industrial area with almost 52,000 inhabitants located at 40 km to the southwest of Ciudad Real. The existent industries situated about 5-6 km southeast from the town center include an oil refinery, a petrochemical industry, a nitrogen fertilizer factory, two power plants, and a coal mine, which means an important source of air pollution in this city. Field measurements in the ecological area were conducted on the southwestern border of the park at about 6 km east of the village of Horcajo de los Montes (39.2°N, 04.4°W, 617 m above sea level) (see Fig. 1). Meteorologically, this zone is characterized by very hot and dry summer period with high insulation (Fig. 1).

The first series of field measurements was conducted in Ciudad Real from September 2008 to April 2009 and were focused on different places of the Faculty of Chemistry: two research laboratories, a laboratory used by practical classes, an office, and the bar of the campus. In addition, an outdoor sampling point was selected to measure levels of the aldehydes. Also, indoor and outdoor measurements were carried out in two private homes, one with smokers (home 1) [19]. Indoor measurements were carried out in two points of the homes, the kitchen and the living room.

Figure 1. Situation map of Ciudad Real, Puertollano, and the sampling point in Cabañeros National Park.

In a second series of measurements, aldehydes (formaldehyde, acetaldehyde, acrolein, acetone, propanal, crotonaldehyde, butanal, benzaldehyde, pentanal, tolualdehydes, and hexanal) were measured in six reading rooms of the campus library and also different common rooms of another building of the Faculty of Chemistry. The third series of field measurements was conducted in two classrooms of one school in Ciudad Real and Puertollano and in two homes (living rooms) in Puertollano. All these measurements were carried out in 2012 and 2013. Finally, during August-November 2010 and February-August 2011, carbonyl compounds were measured in the ecological area of Cabañeros National Park [29].

The first series of field measurements was carried out using Analyst® passive samplers, while Radiello® passive samplers were used in the other campaigns. The Analyst® passive sampler was used for assessment only for formaldehyde and acetaldehyde; for this reason and because both aldehydes produce adverse health effect and formaldehyde is the most abundant aldehyde, this study focuses especially on formaldehyde and acetaldehyde, although in some places other carbonyls have also been measured such as the campus library some places in the Faculty of Chemistry, or Cabañeros National Park.

2.2. Sampling and analytical methods

Radiello® (Fondazione Salvatore Maugeri, Padova, Italy) and Analyst® (Marbaglass, Palombara Sabina, Rome, Italy) passive samplers were used for monitoring carbonyl compounds. The Radiello® passive samplers for carbonyls consist of a stainless steel cartridge filled with

2,4-dinitrophenylhydrazine-coated Florisil® inside a diffusive body, while Analyst® sampler consists of three parts: a polyethylene cylinder, an antiturbulence net (made of silver for outdoor sampling or stainless steel for indoor sampling), and a 2,4-DNPH-coated adsorbent bed (Florisil or silica gel). Analyst® passive samplers were prepared in the laboratory. A detailed description of both passive samplers is given elsewhere [19].

Indoor devices were positioned at a height of 1.5-2.0 m above the floor, in the middle of the room when possible. Outdoor samples were taken simultaneously in the windows or balconies and protected from bad weather conditions by a mountable polypropylene shelter (for Radiello) or stainless steel shelter (for Analyst).

The sampling duration of the Analyst® samplers was 14-20 days, while Radiello® was 7 days. After exposure, the Radiello® cartridges were introduced in their sealed glass tubes, and the Analyst® was cap and stored in the dark and refrigerated until the analysis. Field blanks were transported together with samplers to the sampling point.

The extraction of the hydrazones and the analytical conditions have been described in previous works [19, 29]. Briefly, both Analyst® and Radiello® passive samplers were extracted with 2 ml of acetonitrile (HPLC grade), and the extract was filtered (PTFE 0.45 mm) and analyzed by HPLC (Varian prostar, CA, USA) coupled to a photodiode array detector. For this, 20 µl of the solution obtained after extraction was injected by a sampling loop into a reversed-phase column C-18 (a Varian Microsorb MV 100-5, 25 cm length × 4.6 mm i.d. and a Supelcosil™LC-18 25 cm × 4.6 mm × 5 µm for the Analyst® and for Radiello®, respectively) and detected at a wavelength of 365 nm, according to the literature. The program of the mobile phase was as follows: 0-7 min, 60% acetonitrile (HPLC grade), and 40% water (from a Milli-Q system); 7-20 min, a gradient up to 100% acetonitrile. The flow was 1 ml min^{-1}.

A series of standards (TO11/IP-6A Aldehyde/Ketone-DNPH Mix, Supelco, Bellefonte, USA) containing formaldehyde, acetaldehyde, acrolein, acetone, propanal, crotonaldehyde, butanal, benzaldehyde, isopentanal, pentanal, o-tolualdehyde, m-tolualdehyde, p-tolualdehyde, hexanal, and 2,5-dimethylbenzaldehyde in acetonitrile were used to obtain a five-point calibration curve for each compound in concentration ranges similar to the tested samples (0.2-4 µg ml^{-1}). There were very good linear relationships between concentration and instrumental response for all carbonyls measured ($R^2 > 0.99$).

2.3. Quality assurance

Blank samples, limits of detection (LOD), and reproducibility of the Analyst® and Radiello® passive samplers were assessed for quality assurance. Method detection limits (MDLs) were defined as three times the standard deviation of the blanks. The aldehyde amount in the Analyst® blank samples ranged from 0.06 to 0.11 µg for formaldehyde (estimated air concentration, 0.19-0.35 µg m^{-3}) and from 0.04 to 0.31 µg for acetaldehyde (estimated air concentration, 0.17-1.32 µg m^{-3}). LOD values for blank samples were 0.39 µg for formaldehyde and 0.09 µg for acetaldehyde, corresponding to a concentration of 1.2 and 0.4 µg m^{-3}, respectively, calculated for a period of 14 days. The coefficients of variation in the reproducibility test ($n = 5$) were 5.8% for formaldehyde and 4.5% for acetaldehyde, while for the Radiello® passive samplers ranged from 0.5% (acetaldehyde) to 4.5% (m/p-tolualdehyde) for carbonyl compounds except for acetone, which was 10%. Method detection limits calculated for a sampling period of 7 ranged from 0.01 μg m^{-3} for 2,5-dimethylbenzaldehyde and isopentanal to 0.26 μg m^{-3} for acetaldehyde.

3. Results and discussion

Table 1 shows a summary of the formaldehyde and acetaldehyde concentrations in all sampling sites investigated in Ciudad Real and Puertollano.

3.1. Levels of carbonyls in indoor environments

3.1.1. Faculty of chemistry

The Faculty of Chemistry consists of several buildings, and the laboratories sampled are in a different building to the rest of common areas listed in Table 1. These common areas in the case of Inorganic and Organic Chemistry Departments are halls where offices and laboratories are around or near them.

In the laboratories sampled, the formaldehyde levels varied between 4 and 23.5 μg m^{-3} depending on the laboratory, while those of acetaldehyde varied from 1.3 to 13.2 μg m^{-3}. The class practical laboratory showed the highest formaldehyde and acetaldehyde concentrations, while the lowest concentrations were found in Research Laboratory 1. The difference between Research Laboratory 1 and the other two laboratories (Research Laboratory 2 and class practical lab) was the ventilation, a fact that could explain the differences in the concentrations of the two aldehydes [19]. Although literature information is hardly available for carbonyl compounds measured in laboratories these measurements are in the range of those determined in research laboratories and class practical laboratories in an academic institute in Fortaleza, Brazil [30], and are lower than those determined in Rio Grande University in Brazil [31] for a research laboratory and class practical laboratories as shown in Table 2 for comparison.

Sampling site	*n*	Formaldehyde			Acetaldehyde			
		Average ± SD	Min	Max	Average ± SD	Min	Max	Reference
Ciudad Real (urban area) *Research Laboratory 1*								[19]
Indoor	11	7.7 ± 3.0	4.0	11.6	5.3 ± 2.1	1.3	7.6	
Outdoor	10	1.5 ± 0.4	0.9	2.3	1.7 ± 0.6	0.8	3.1	
Research Laboratory 2	2	18.8 ± 4.0	16.0	21.6	5.5 ± 0.8	4.9	6.0	[19]
Class practical lab	2	19.6 ± 5.6	15.6	23.5	1.2 ± 1.4	11.2	13.2	[19]

Sampling site	n	Formaldehyde			Acetaldehyde			
		Average ± SD	Min	Max	Average ± SD	Min	Max	Reference
Office	4	11.3 ± 2.0	8.5	13.3	5.7 ± 1.3	5.0	7.7	[19]
Common areas Faculty Chemistry								This study
Inorganic chemistry (2nd floor)	2	9.2 ± 0.6	8.8	9.7	2.0 ± 0.2	1.9	2.1	
Organic chemistry (2nd floor)	2	17.8 ± 0.6	17.4	18.2	2.9 ± 0.2	2.7	3.0	
Library (2nd floor)	2	12.7 ± 0.9	12.1	13.3	2.4 ± 0.07	2.3	2.4	
Reception (1st floor)	2	11.5 ± 0.3	11.2	11.7	2.3 ± 0.006	2.3	2.3	
Hall (1st floor)	2	6.0 ± 0.3	5.7	6.2	1.5 ± 0.03	1.5	1.5	
Coffee area (1st floor)	2	6.1 ± 0.3	5.9	6.2	1.6 ± 0.2	1.4	1.7	
Outdoor	2	1.7 ± 0.4	1.4	2.0	1.8 ± 1.7	0.6	3.0	
Bar	1	13.4	-	-	26.0	-	-	[19]
University library								This study
Ground floor	8	11 ± 1.9	8.4	14.1	5.2 ± 0.9	4.2	6.8	
First floor	8	10.6 ± 1.6	8.6	14.2	5.05 ± 0.8	4.0	5.9	
Second floor	8	11.1 ± 1.4	8.4	15.0	5.1 ± 0.7	4.2	6.1	
Outdoor	4	2.1 ± 0.4	1.6	2.6	2.2 ± 0.5	1.6	2.9	
Home 1								[19]
Living room	1	75	-	-	19.2	-	-	
Kitchen	2	43.7 ± 2.2	42.1	45.2	21.5 ± 4.4	18.4	24.6	
Outdoor	6	2.2 ± 0.5	1.3	2.7	2.4 ± 1.1	0.82	3.55	
Home 2								[19]
Living room	4	46.0 ± 5.6	41.6	53.3	57.4 ± 8.5	43.3	62.2	
Kitchen	2	38.2 ± 3.3	35.8	40.5	69.2 ± 8.2	63.4	75.0	
Outdoor	6	3.1 ± 0.8	1.9	4.2	3.8 ± 2.0	1.2	6.5	
School								This study
Indoor	4	18.9 ± 4.7	14.7	23.1	5.2 ± 1.8	3.6	6.7	
Outdoor	2	1.8 ± 0.8	1.2	2.3	2.1 ± 2.1	0.6	3.6	
Puertollano (industrial area)								
School								This study
Indoor	4	35.4 ± 14.6	21.5	49.2	8.5 ± 5.4	3.7	13	
Outdoor	2	2.0 ± 1.6	0.9	3.1	0.4 ± 0.1	0.3	0.5	

Sampling site	n	Formaldehyde			Acetaldehyde			
		Average ± SD	Min	Max	Average ± SD	Min	Max	Reference
Home 1								This study
Indoor	8	24.1 ± 7.5	17.4	37.1	15.2 ± 5.0	9.3	21.4	
Outdoor	8	3.1 ± 0.6	2.4	3.8	2.4 ± 0.9	1.5	3.9	
Home 2								This study
Indoor	8	74.8 ± 27.3	37.9	107.5	23.6 ± 10.7	9.6	36.3	
Outdoor	8	3.2 ± 0.7	2.2	4.3	2.4 ± 0.9	1.2	3.5	
Cabañeros (ecological area)								
Outdoor	46	0.96 ±0.54	b.d.l	2.56	0.79 ± 0.49	0.13	1.89	[29]

Table 1. Formaldehyde and acetaldehyde concentrations (μg m^{-3}) in indoor and outdoor environments

In the office, the levels of formaldehyde and acetaldehyde were 11.3 and 5.7 μg m^{-3}, respectively. Formaldehyde and acetaldehyde concentrations were quite similar to those obtained for the offices with no smokers of a public building in Rome [19] and were lower than the levels found in the offices of Rio Grande University [31]. The values for formaldehyde are also in the range of those measured in offices from different countries from Europe [32].

Regarding common areas, formaldehyde and acetaldehyde concentrations ranged from 5.7 to 18.2 μg m^{-3} and from 1.4 to 3.0 μg m^{-3}, respectively. The lowest levels for both aldehydes were found on the first floor (hall and coffee area). The coffee area is located on the right side of the hall, and it is not separated from this by any wall. The hall has an extension of approximately 350 m^2, and there is not any furniture; there are some tables and chairs and two vending machines in the coffee area only. Because of this, it seems reasonable that the low levels of both aldehydes are found here. The highest levels were registered on the second floor especially in the Organic Chemistry Department, where the levels were similar to those obtained in the research laboratory 2 and the class practical laboratory placed in another building. This fact could be due to the reactions between the variety of organic solvents employed and ozone, but ozone levels are low usually in indoor environments, and they have not been measured in this study. Therefore, care must be taken when assuring this. The library is located beside the Organic Chemistry Department, which could influence the levels of formaldehyde found here. The reception is located on the first floor and presents levels of formaldehyde and acetaldehyde higher than those found on the other two sampling points of the first floor, maybe due to the presence of a big copying machine in the room and the emission of ozone with the consequent reaction with the unsaturated VOCs in the air (such as cleaning products) [1].

Not only formaldehyde and acetaldehyde have been measured in these common areas of the Faculty of Chemistry but also other carbonyls have been identified and quantified and are shown in Table 3. These are acrolein-acetone, crotonaldehyde, benzaldehyde, pentanal, p-tolualdehyde, and hexanal. The most abundant carbonyl was acrolein-acetone (both

Site	Formaldehyde ($\mu g\ m^{-3}$)	Acetaldehyde ($\mu g\ m^{-3}$)	Activity	Reference
Rio Grande University, Brazil	32.3-41.0	18.3-26.2	Office	[31]
Rome	8.9-9.4	4.2-4.7	Office	[19]
Europe	3-33	-	Offices	[32]
Fortaleza, Brazil	0.32-81.6	1.2-4.43	Research laboratories	[30]
Rio Grande University, Brazil	96.5	79.4	Research laboratory	[31]
Fortaleza, Brazil	3.78-60.75	1.4-3.18	Class practical laboratories	[30]
Rio Grande University, Brazil	56.5-161.5	38.1-91.2	Class practical laboratories	[31]
Modena, Italy	1.7-67.8	-	Libraries	[36]
Strasbourg, France	8.6-94.5	3.7-25.9	Libraries	[37]
Helsinki, Finland	8.1-77.8	3.7-41.5	Living room	[45]
Strasbourg, France	6-93	0-66	Living room	[44]
10 European cities	14.4 ± 4.9-30.7 ± 17.8	9.6 ± 2.3-15.8 ± 5.4	Homes	[43]
Bari, Italy	3.2-49	1.3-23.5	Kitchen	[46]
Paris, France	21.7±1.9	10.1±1.8	Kitchen	[39]
Lisbon, Portugal	6.3-23.8	-	Schools	[49]
11 European cities	5.6-49.7	1.4-29.1	Schools and kindergarten	[32]

Table 2. Formaldehyde and acetaldehyde concentrations measured in different indoor environments in previous studies

appear together in the chromatogram) with a range of concentration between 14.2 and 58.3 $\mu g\ m^{-3}$. The rest of carbonyls present concentrations between 0.2 $\mu g\ m^{-3}$ for benzaldehyde and 4.1 $\mu g\ m^{-3}$ for hexanal (without taking into account formaldehyde and acetaldehyde explained above).

The indoor/outdoor ratio is generally used to infer penetration to indoor environments and indoor sources. An I/O ratio lower and close to one indicates more outdoor sources [33]. In all sites, air concentrations for formaldehyde inside the buildings were between 5 and 13 times higher than outside, which appears to indicate that strong indoor sources exit, clearly determining indoor air concentrations. The same conclusion can be obtained for the pair acrolein-acetone or p-tolualdehyde (I/O = 8-20) and hexanal (I/O = 13-34), but the indoor levels of these two last carbonyls are low in all sampling sites. For the rest of compounds, the main source could be from outdoor even acetaldehyde with a ratio of 2.4-4.4.

Carbonyl	Inorganic hall	Organic hall	Library	Reception	Hall first floor	Coffee area	Outdoor
Formaldehyde	10.1	19.5	13.9	12.6	6.5	6.6	1.5
Acetaldehyde	2.1	3.1	2.5	2.4	1.6	1.7	0.7
Acrolein-acetone	58.3	16.4	14.2	40.1	30.4	22.4	<LOD
Propanal	1.4	0.3	0.2	<LOD	0.9	0.2	<LOD
Butanal	<LOD	<LOD	<LOD	<LOD	<LOD	<LOD	<LOD
Croton-aldehyde	0.5	1.3	0.7	0.9	0.6	0.4	0.4
Benzaldehyde	0.3	0.3	0.2	-	0.2	0.2	<LOD
Pentanal	1.0	0.7	0.8	2.1	0.4	0.4	0.2
p-tolualdehyde	0.8	1.0	0.6	0.5	0.7	0.4	0.05
hexanal	1.6	3.4	2.8	4.1	1.8	1.6	<LOD

Table 3. Carbonyls measured and their concentrations (in µg m^{-3}) in the common areas of Faculty of Chemistry

3.1.2. Bar

In the bar of the campus (about 200 m^2), the concentration of acetaldehyde was much higher than that of formaldehyde. The higher acetaldehyde value could be due to the cigarette smoke and combustion processes such as those carried out in the kitchen of the bar, although acetaldehyde is also present in many foods and alcoholic drinks [3, 5]. These levels of carbonyls are lower than expected. Hodgson et al. [34] estimated that cigarette smoking contributed from 57% to 84% to the total formaldehyde levels in the smoking areas of cafeterias where the formaldehyde concentrations varied between 5 and 42 µg m^{-3} when 20-100 cigarettes were being smoked. The combustion of 5-10 cigarettes in a room of 30 m^2 increases the formaldehyde concentrations from 240 to 600 µg m^{-3} [35]. However, formaldehyde levels measured in the bar were low and similar to those found in the teacher's office and in Research Laboratory 1 probably due to mechanical ventilation and the great surface of more than 200 m^2 of the bar.

3.1.3. University campus library

Carbonyl concentrations were also measured in the University Campus Library since paper can emit large amounts of formaldehyde [36]. Six reading rooms of the library and a sampling point outdoor were monitored during four consecutive weeks from February 6 to March 5, 2011. Two passive samplers were placed in two reading rooms on the ground floor and the same on the first and the second floor. The total samples were 28 between indoor and outdoor sampling.

The carbonyl compounds identified in the library were formaldehyde, acetaldehyde, acetone-acrolein, propanal, crotonaldehyde, butanal, benzaldehyde, pentanal, p-tolualdehyde, and hexanal, and their average concentrations were as follows (in µg m^{-3}): 10.9 ± 1.3, 5.1 ± 0.2, 12.4 ± 2.4, 2.5 ± 0.1, 0.4 ± 0.1, 15.4 ± 1.3, 0.55 ± 0.05, 2.9 ± 0.4, 0.8 ± 0.2, and 6.6 ± 1.1, respectively (quoted errors correspond to the standard deviation). Butanal appeared overlapped with an interference (no carbonyl compound) in the samples. Their high levels obtained with respect to the rest of carbonyls together with the UV spectrum indicate that butanal appears overlapped with other carbonyl compound may be 2-butanone. Both compounds appear together in the chromatogram when a standard is introduced under the analytical conditions. Therefore, butanal has not been considered for the discussion. The most abundant carbonyls were acetone-acrolein and formaldehyde. Their concentrations in the individual samples varied from 8.6 to 17.3 μg m^{-3} and from 8.4 to 15 μg m^{-3}, respectively.

Carbonyl	Ground floor	First floor	Second floor	Outdoor
Formaldehyde	11.0 ± 1.9	10.6 ± 1.9	11.1 ± 2.2	2.1 ± 0.4
Acetaldehyde	5.2 ± 0.9	5.1 ± 0.8	5.1 ± 0.7	2.0 ± 0.5
Acrolein-acetone	15.0 ± 1.9	11.1 ± 1.7	11.2 ±1.8	3.8 ± 0.8
Propanal	2.4 ± 0.3	2.6 ± 0.5	2.6 ± 0.4	1.0 ± 0.1
Crotonaldehyde	0.4 ± 0.1	0.5 ± 0.3	0.4 ± 0.3	0.7 ± 0.1
Benzaldehyde	0.5 ± 0.1	0.6 ± 0.1	0.6 ± 0.1	0.3 ± 0.06
Pentanal	3.3 ± 1.0	2.6 ± 0.4	2.7 ± 0.6	1.0 ± 0.1
p-tolualdehyde	0.9 ± 0.5	0.7 ± 0.1	0.8 ± 0.3	0.7 ± 0.3
hexanal	7.7 ± 2.0	5.9 ± 0.9	6.1 ± 0.9	1.7 ± 0.07

Table 4. Levels of carbonyls (in μg m^{-3}) measured in the Campus University Library together with standard deviations

Generally, the concentrations of the carbonyl compounds found in the different reading rooms of the library were low. These values are in the range of those obtained in 20 university libraries in Strasbourg (France) [37], where the levels found ranged from 8.6 to 94.5 for formaldehyde, from 3.7 to 25.9 μg m^{-3} for acetaldehyde, from 2.1 to 58.8 μg m^{-3} for hexanal, from 0.2 to 5.3 μg m^{-3} for benzaldehyde, and from 0.7 to 16.3 μg m^{-3} for propanal. Our measurement values for formaldehyde are also in the range of those reported by Fantuzzi et al. [36] for 16 libraries of the university of Modena (Italy), which ranged from 1.7 to 67.8 μg m^{-3} with an average value of 32.7 ± 23.9 μg m^{-3} (see Table 2). However, the average concentrations for formaldehyde calculated in this study are lower than the studies mentioned above. Formaldehyde, acetaldehyde, propanal, benzaldehyde, and hexanal have usually been detected in libraries [37].

Table 4 shows the concentrations of all carbonyls measured on the different floors of the library. There are not differences between the concentrations measured of the carbonyl compounds in the different reading rooms of the different floors of the library. The indoor/outdoor ratios vary from 0.65 crotonaldehyde, which indicates mainly outdoor sources to 5.1 for formaldehyde indicating mainly indoor sources. Other I/O ratios that could suggest indoor sources but are not strong are 3.3 for acetone and 3.8 for hexanal.

3.1.4. Private houses

Formaldehyde and acetaldehyde measurements were performed simultaneously in the living room and outdoor in Ciudad Real and Puertollano; aldehydes in the kitchens were also measured only in Ciudad Real (see Table 1). Homes were ventilated as usual by opening the windows during the sampling. Two samples were collected at the same time in Ciudad Real, while the samples were collected during 8 consecutive weeks in Puertollano. All houses were located in an area of moderated traffic. Approximate home ages are between 1 and 8 years old in Ciudad Real and Puertollano; only home 2 in Ciudad Real is more than 10 years. This information is important since building materials can emit formaldehyde, but these emissions generally decrease with the age of the house. Smoking people are present in one house of each town.

Formaldehyde indoor air concentration was in average 24.5 and 15.3 times higher than the concentration outdoor in Ciudad Real and Puertollano, respectively, while acetaldehyde was 11.5 and 7.9 times higher in Ciudad Real and Puertollano, respectively. These ratios are similar to those found by Geiss et al. [32] for a new manufacturer house (20.2 for formaldehyde and 9.2 for acetaldehyde), however, the homes in Ciudad Real are more than five years old and between 1 and 8 in Puertollano. This fact indicates that strong indoor sources existed, clearly determining indoor air concentrations and denoted that indoor sources were dominant for these compounds. Indoor materials such as consumer products, furniture, and decorations are important indoor sources [38-41]. Carbonyls such as acetaldehyde are generated from smoking [42]. A smoker person lives in home 2 (Ciudad Real) and presented the highest acetaldehyde levels in both kitchen and living room. Home 1 (Puertollano) is also the home of one smoker, but the frequency of smoking is low; in this case, the levels of acetaldehyde are even lower than the levels obtained in the homes of nonsmokers. Except for home 2 from Ciudad Real, formaldehyde levels were higher than the acetaldehyde ones. Bruinen de Bruin et al. [43] measured formaldehyde and acetaldehyde concentrations in residential environments in ten European cities, demonstrating mean levels in the range of 14.4 ± 4.9 μg m^{-3} (Dublin) to 30.7 ± 17.8 μg m^{-3} (Arnhem) for formaldehyde and from 9.6 ± 2.3 to 15.8 ± 5.4 μg m^{-3} (Budapest) for acetaldehyde, thus confirming low concentrations of aldehydes in European homes. Another recent study measured formaldehyde and acetaldehyde inside the homes of Strasbourg (France) [44]. The indoor levels were in the range of 6 to 93 μg m^{-3} for formaldehyde and from 0 to 66 μg m^{-3} for acetaldehyde. Therefore, the mean concentration of formaldehyde in the indoor air in the present work is in the range of those found in other European cities such as Strasbourg or Helsinki (8.1-77.8 μg m^{-3}) [45] (see Table 2). Indoor acetaldehyde had also a mean concentration similar to that obtained in other European cities (see Table 2), except for home 2 in Ciudad Real wherein concentration is the highest with an average of 57.4 μg m^{-3}. This concentration is in agreement with some living rooms in Strasbourg [44], where the range of concentrations was from 0 to 66 μg m^{-3} for acetaldehyde.

In the case of concentrations of aldehydes measured in the kitchens of Ciudad Real, formaldehyde presents practically the same level for both homes. Our mean value is approximately 40% higher than that reported in previous studies [39] and is in the range of that reported in reference [46] for a study performed in Bari (Italy). For acetaldehyde, the mean concentration registered in the kitchen home 2 is higher than those reported in other European cities.

According to the World Health Organization, the lowest formaldehyde concentration that has been associated with nose and throat irritation after short-term exposure is 100 μg m^{-3} [47]. Therefore, it represents the recommended maximal value. In our study, there is no such place with formaldehyde values exceeding this level. Despite that, it is important to note that the sampling period was 14 days for the Analyst® passive sampler (homes in Ciudad Real) and 7 days for Radiello® (homes in Puertollano). Thus, our results express an average concentration over a long period and do not provide information about exposure peaks, which must be higher than the average concentration measured. However, other recommended values exist, which were exceeded in our study. For example, the indoor formaldehyde concentrations in the homes monitored exceeded the guideline value of 30 μg m^{-3} proposed for the prevention of irritant effects [48] referred to a period of 30 min.

Hence, it is necessary to carry out an exhaustive research with more homes monitored in order to assess the health risk. In addition, taking into account the low outdoor pollution levels observed, an adequate airing of the rooms could reduce the indoor pollution by formaldehyde and acetaldehyde. Nevertheless, this study only presents preliminary results of the private homes in Ciudad Real and Puertollano due to the low number of samples. Measurements in Ciudad Real were performed only to check the viability of the Analyst® passive sampler in the indoor environment [19]. Regarding homes in Puertollano, this study presents the preliminary results of a further research. There are not clear differences between the levels of formaldehyde and acetaldehyde registered in the urban and industrial area.

3.1.5. Schools

Two schools were sampling for the present study, one in Ciudad Real and one in Puertollano located in the same area as the homes sampled. Samples were simultaneously collected in two classrooms from each school and outside the building. Every classroom has an area of about 50 m^2, and there are between 20 and 27 children. The school in Puertollano is more than 20 years old, and the classrooms have not been renovated recently, while in Ciudad Real, the school is also old but the classrooms were renovated 7 years ago. The ventilation in both schools is natural by opening the windows during 15 min per day.

The concentration of formaldehyde and acetaldehyde was slightly higher for the industrial area 35.4 ± 14.6 and 8.5 ± 5.4 μg m^{-3}, respectively, versus 24.3 ± 4.7 and 6.2 ± 1.8 μg m^{-3}, respectively, for the urban area. The concentrations of formaldehyde were more abundant than the acetaldehyde ones. In both schools, the air concentrations for both aldehydes inside the buildings were higher than outside. The indoor/outdoor ratios for formaldehyde and acetaldehyde in Ciudad Real were 13.5 and 2.9, respectively, and were higher for both aldehyde in the school from the industrial area 17.7 and 21 for formaldehyde and acetaldehyde, respectively. Except acetaldehyde in the school from Ciudad Real, formaldehyde and acetaldehyde are mainly indoor pollutants derived from indoor sources.

http://dx.doi.org/10.5772/60016

Our data for formaldehyde are similar or a bit higher than the levels registered in fourteen schools in Lisbon (Portugal) [49]. The levels of both aldehydes are in agreement with the concentrations registered in eleven European cities [32] (see Table 2).

The levels of formaldehyde exceeded the guideline value of 30 μg m^{-3} proposed for the prevention of irritant effects [48] in the industrial area. High levels of formaldehyde are likely associated with the age of building and renovating activities of old buildings. However, both schools are more than 10 years old, and they did not have recent renovation therefore, these levels are due to other sources.

3.2. Levels of carbonyls in outdoor environments

In all sampling sites, formaldehyde and acetaldehyde were present in both indoor and outdoor air. Generally, in all sites, air concentrations for these aldehydes inside the buildings were higher than outside. Mean values for formaldehyde in Ciudad Real varied between 1.5 and 3.1 μg m^{-3}, while acetaldehyde varied between 1.7 and 3.8 μg m^{-3} for the sampling period of December-February. Outdoor acetaldehyde concentrations were similar to those of formaldehyde. In the case of the industrial area, formaldehyde varied in the range of 2 (in March) and 3.2 μg m^{-3} (May-June) and acetaldehyde varied between 0.4 (in March) and 2.4 μg m^{-3} (May-June), being formaldehyde levels slightly higher than the acetaldehyde ones. Similar values were found in samplings conducted in rural and semirural areas [50-52].

Formaldehyde/acetaldehyde ratio has been proposed as an indicator of the biogenic source of formaldehyde [53] and can vary between 1 (urban area) and 10 (deciduous forest). This ratio for Ciudad Real and Puertollano is between 0.9 and 1.3 reflecting typical values for urban air. The formaldehyde/acetaldehyde ratio for the sample taken in the school in Puertollano gave a value of 5, which could imply a biogenic contribution in the early spring.

A study about levels of carbonyls was carried out in the ecological area of Cabañeros National Park [29]. Twelve compounds were identified and quantified: formaldehyde, acetaldehyde, acetone-acrolein, propanal, crotonaldehyde, butanal, benzaldehyde, isopentanal, pentanal, o-tolualdehyde, m/p-tolualdehyde, and hexanal (the sum of m/p-tolualdehyde and acetone-acrolein was reported because they could not be well separated by the analytical method). The most abundant carbonyls were hexanal, acetone-acrolein, formaldehyde, and acetaldehyde.

Because of the study covered all seasons, August 2010 was the most carbonyl-polluted month followed by July 2011 and September 2010. October and November 2010 were the months with lower concentration of carbonyl compounds showing an apparent seasonal variation with maximum values observed in summer months. The concentrations of the carbonyls were as follows: acetone-acrolein mixing ratios ranged from 0.35 in February to 4.52 μg m^{-3} in June 2011, with an average of 1.78 μg m^{-3}, while ambient levels of hexanal varied from 0.67 μg m^{-3} in October to 1.72 μg m^{-3} in April; the average concentration during the sampling period was 1.06 μg m^{-3}. Formaldehyde was in the range of values below detection limit and 2.56 μg m^{-3} in October and June, respectively, with an average concentration of 0.96 μg m^{-3} and acetaldehyde varied from levels below detection limits in February, March, and April to 1.89 μg m^{-3} in June. The concentrations of other carbonyls ranged from non-detected to 1.18 μg m^{-3}.

The seasonal cycles obtained for formaldehyde, acetaldehyde, and acetone-acrolein with respect to the other carbonyls suggest different formation mechanism and sinks compared to the others. This could be due to photochemical processes (oxidation of biogenic and even anthropogenic hydrocarbons that arrive to the area under determined meteorological conditions) and also the direct emission from vegetation. For example, two Mediterranean tree species such as *Quercus ilex* (holm oak), one of the most abundant oak in the study area, and *Pinus pea* (Italian stone pine) emit formaldehyde and acetaldehyde [54, 55].

The formaldehyde and acetaldehyde concentrations measured at Cabañeros National Park are in the same range or lower than the levels reported in other forests or rural areas (Germany [56] and Brazil [57]). Our data are similar to those reported in the small village of Covelo in Portugal, considered as a rural/forest site [58].

On the other hand, the levels found in the ecological area during the months of February and March for formaldehyde (0.51 and 0.59 μg m^{-3}) and acetaldehyde (0.33 and 0.29 μg m^{-3}) are lower than those registered in the urban (outdoor samples collected at the same time as common areas of Faculty of Chemistry or University Library) and industrial area (outdoor samples collected in the school) during the same period. Also, the levels of both aldehydes during the months of May and June were lower (1.15-1.37 μg m^{-3} for formaldehyde and 0.70-1.09 μg m^{-3} for acetaldehyde) than the levels registered in the industrial area during the same months in the outdoor samples collected in the homes.

3.3. Health risk evaluation

Formaldehyde is classified in Group 1 by IARC. It was based on inhalation causing squamous cell carcinoma in rats and nasopharyngeal cancer in humans [3]. Recently, the classification has been expanded with formaldehyde causing leukemia and limited evidence of sinonasal cancer in humans [59]. The lowest concentration reported to cause sensory irritation of the eyes in humans is 0.38 mg m^{-3} for 4 h. Increases in eye blink frequency and conjunctival redness appear at 0.6 mg m^{-3}, which is considered equal to the no-observed-adverse-effect level (NOAEL). There is no indication of accumulation of effects over time with prolonged exposure, and there is no evidence indicating an increased sensitivity to sensory irritation to formaldehyde among people often regarded as susceptible (asthmatics, children, and older people) [60].

The formaldehyde exposure-response relationship is highly nonlinear and biphasic, supporting a NOAEL that allows setting a guideline value [60]. This means that it cannot be assumed a low-dose linear relationship for the carcinogenic effects, and therefore, the calculation of the incremental lifetime risk of cancer for formaldehyde, assuming a low-dose linear relationship, would give a value highly overestimated and it would be meaningless.

As commented above, WHO established a guideline value for formaldehyde of 100 μg m^{-3} that should not be exceeded for any 30-min period of the day [47]. This short-term guideline will also prevent effects on lung function as well as long-term health effects, including nasopharyngeal cancer and myeloid leukemia [60].

We can compare our data with the short-term guideline value of 100 μg m^{-3}. In our study, the maximum values registered in the different environments were below 100 μg m^{-3}, except for home 2 in the industrial area that registered a maximum value of 108 µg m^{-3} during one of the 8 weeks sampled. The average concentrations for formaldehyde found in this study ranged from 6 to 75 µg m^{-3} in indoor environments, while outdoor concentrations were considerably lower. These values are below the guideline value considered preventive of carcinogenic effects; however, our results represent an average value during 1 or 2 weeks of exposition, and this guideline value is referred to a period of 30 min. Nevertheless, as long as formaldehyde concentrations are below 100 µg m^{-3}, there should be no chance of developing cancer.

On the other hand, the most important way to control the indoor formaldehyde concentration is the air exchange rate and the use of low-emitting materials and products. Environmental tobacco smoke and ozone-initiated reactions of alkene compounds may contribute to temporary peak levels [60]

4. Conclusion

This paper presents an overview of the concentrations of priority aldehydes observed in indoor and outdoor air in an urban and an industrial area of central-southern Spain. Very few studies have been carried out in the indoor air of Spain. Therefore, we have tried to gather the data previously published about formaldehyde and acetaldehyde in this region and other new data to obtain a general view of the levels in different outdoor and indoor environments.

From the results, we can conclude that the highest concentrations of formaldehyde and acetaldehyde are found inside the homes and schools. The levels of formaldehyde in laboratories or in rooms near laboratories, where many kinds of solvents are used, are much lower. Overall indoor and outdoor concentrations of formaldehyde and acetaldehyde in all sites sampled were below the threshold limit of 100 µg m^{-3} proposed by WHO [47] associated with nose and throat irritation and also to prevent all types of cancer. Therefore, it represents the recommended maximal value. In our study, there is no such place with formaldehyde values exceeding this level. Despite that, it is important to note that the sampling period was 14 and 7 days for Analyst® and Radiello®, respectively. Thus our results express an average concentration over a long period and do not provide information about exposure peaks, which must be much higher than the average concentration measured. On the other hand, the indoor formaldehyde concentrations in the homes and one school monitored exceeded the recommended guideline value of 30 µg m^{-3} proposed by Kotzias et al. [48] for the prevention of irritant effects. Also this value is referred to a period of 30 min.

Therefore, although the indoor sources are dominant for all sites sampled, the most important to be studied are at homes and schools. This is only a preliminary study about the levels of formaldehyde and acetaldehyde in homes and schools due to the low number of samples, and an exhaustive research is necessary in order to better characterize the chemical composition of the air that people breathe daily and to assess the health risk.

Author details

Florentina Villanueva[1,2*], Araceli Tapia[3], Inmaculada Colmenar[3], José Albaladejo[3], Beatriz Cabañas[1] and Ernesto Martínez[1]

*Address all correspondence to: florentina.vgarcia@uclm.es

1 Atmospheric Pollution Laboratory, Research Institute for Combustion and Atmospheric Pollution, University of Castilla La Mancha, Ciudad Real, Spain

2 Castilla La Mancha Science and Technology Park, Spain

3 Physical Chemistry Department, Faculty of Chemical Science and Technologies, University of Castilla La Mancha, Ciudad Real, Spain

References

[1] Zhang J, Wilson WE, Lloy PJ. Indoor air chemistry: formation of organic acids and aldehydes. Environmental Science and Technology 1994;28 1975-1982.

[2] Granby K, Carsten SC, Lohse C. Urban and semi-rural observations of carboxylic acids and carbonyls. Atmospheric Environment 1997;31 1403-1415.

[3] IARC. IARC monographs on the evaluation of carcinogenic risks to human. Formaldehyde, 2-butoxyethanol and 1-tert-butoxypropan-2-ol. IARC, Lyon, France, 2006;88 39-325.

[4] IARC. IARC monographs on the evaluation of carcinogenic risks to human. Re-evaluation of some organic chemicals, hydrazine and hydrogen peroxide: acetaldehyde. IARC, Lyon, France, 1999;71.

[5] WHO, Acetaldehyde. Environmental Health Criteria, No. 167, World Health Organization, Geneva, 1995.

[6] Liu W, Zhang J, Korn LR, Zhang L, Weisel CP, Turpin B, Morandi M, Stock T, Colome S. Predicting personal exposure to airborne carbonyls using residential measurements and time/activity data. Atmospheric Environment 2007;41 5280-5288.

[7] Tsai SW, Chang TA. Time-weighted average sampling of airborne n-valeraldehyde by a solid-phase microextraction device. Journal of Chromatography A 2002;954 191-198.

[8] US EPA 2003. Integrated Risk Information System. Washington, DC. http://cfpub.epa.gov/ncea/iris/

[9] Zhang F, He Q, Lioy PJ. Characteristics of aldehydes: concentrations, sources, and exposures for indoor and outdoor residential microenvironments. Environmental Science and Technology 1994;28 146-152.

[10] Baez A, Padilla H, Garcia R, Torres MDC, Rosas I, Belmont R. Carbonyl levels in indoor and outdoor air in Mexico City and Xalapa, Mexico. Science of the Total Environment 2003;302 211-226.

[11] Salthammer T, Fuhrman F, Kaufhold S, Meyer B, Schwarz A. Effect of climate parameters on formaldehyde concentrations in indoor air. Indoor Air, 1995;5, 120.

[12] ATSDR. Agency for Toxic Substances and Disease Registry. Toxicological Profile for Formaldehyde. Agency for Toxic Substances and Disease Registry, US Department of Health and Human Services, 1999.

[13] Bolt HM, Degen GH, Hengstler JG. The carcinogenicity debate on formaldehyde: how to derive safe exposure limits? Archives of Toxicology 2010;84(6) 421-422.

[14] Bracamonte BG, Ortiz FJ, Diez LI. Occupational allergic contact dermatitis due to formaldehyde and textile finish resins. Contact Dermatitis 1995;33 139-140.

[15] CARB. California Air Resources Board. Determination of Formaldehyde and Toluene Diisocyanate Emissions from Indoor Residential Sources. Research Division, California Air Resources Board, Sacramento, CA, 1996.

[16] Cai Y, Lin S, Jin Y, Chen H, Zheng J, Chen B. Investigation of formaldehyde level in marketed marine products. Strait Journal of Preventive Medicine 2000;6(5) 32-39.

[17] Naya M, Nakanish J. Risk assessment of formaldehyde for the general population in Japan. Regulatory Toxicology and Pharmacology 2005;43(3) 232-248.

[18] WHO. Environment and Health Risks: A Review of the Influence and Effects of Social Inequalities. World Health Organization, Copenhagen, Denmark, 2010.

[19] Villanueva F, Colmenar I, Mabilia R, Scipioni C, Cabañas B. Field evaluation of the Analyst® passive sampler for the determination of formaldehyde and acetaldehyde in indoor and outdoor ambient air. Analytical Methods 2013;5 516-524.

[20] Rivas I, Viana M, Moreno T, Pandolfi M, Amato F, Reche C, Bouso L, Alvarez-Pedrerol M, Alastuey A, Sunyer J, Querol X. Child exposure to indoor and outdoor air pollutants in schools in Barcelona, Spain. Environment International 2014;69 200-212.

[21] Nadala M, Inzab I, Schuhmachera M, Figuerasb MJ, Domingo JL. Health risks of the occupational exposure to microbiological and chemical pollutants in a municipal waste organic fraction treatment plant. International Journal of Hygiene and Environmental Health 2009;212 661-669.

[22] de Blas M, Navazo M, Alonso L, Durana N, Gomez MC, Iza J. Simultaneous indoor and outdoor on-line hourly monitoring of atmospheric volatileorganic compounds in

an urban building. The role of inside and outside sources. Science of the Total Environment 2012;426 327-335.

[23] Moreno T, Rivas I, Bouso L, Viana M, Jones T, Alvarez-Pedrerol M, Alastuey A, Sunyer J, Querol X. Variations in school playground and classroom atmospheric particulate chemistry. Atmospheric Environment 2014;91 162-171.

[24] Viana M, Rivas I, Querol X, Alastuey A, Sunyer J, Álvarez-Pedrero M, Bouso L, Sioutas C. Indoor/outdoor relationships and mass closure of quasi-ultrafine, accumulation and coarse particles in Barcelona schools. Atmospheric Chemistry and Physics 2014;14 4459-4472.

[25] Minguillón MC, Schembari A, Triguero-Mas M, de Nazelle M, Dadvand P, Figueras F, Salvado JA, Grimalt JO, Nieuwenhuijsen M, Querol X. Source apportionment of indoor, outdoor and personal PM2.5 exposure of pregnant women in Barcelona, Spain. Atmospheric Environment 2012;58 426-436.

[26] Horemansa B, Cardell C, Bencs L, Kontozova-Deutsch V, De Wael K, Van Grieken R. Evaluation of airborne particles at the Alhambra monument in Granada, Spain. Microchemical Journal 2011;99 429-438.

[27] Parra MA, Elustondo D, Bermejo R, Santamaría JM. Exposure to volatile organic compounds (VOC) in public buses of Pamplona, Northern Spain. Science of the Total Environment 2008;404 18-25.

[28] Villanueva F, Notario A, Albaladejo J, Millán MC, Mabilia R. Ambient air quality in an urban area (Ciudad Real) in central-southern Spain. Fresenius Environmental Bulletin 2010;19b 2064-2070.

[29] Villanueva F, Tapia A, Notario A, Albaladejo J, Martínez E. Ambient levels and temporal trends of VOCs, including carbonyl compounds, and ozone at Cabañeros National Park border, Spain. Atmospheric Environment 2014;85 256-265.

[30] Cavalcante RM, Campelo CS, Barbosa MJ, Silveirab ER, Carvalho TV, Nascimento RF. Atmospheric Environment 2006;40 5701-5711.

[31] Cavalcante RM, Seyffert B, Montes D'Oca M, Nascimento RF, Campelo CS, Pinto IS, Anjos FB, Costa AHR. Exposure assessment for formaldehyde and acetaldehyde in the workplace. Indoor Built Environment 2005;14 165-172.

[32] Geiss O, Giannopoulos G, Tirendi S, Barrero-Moreno J, Larsen BR, Kotzias D. The AIRMEX study-VOC measurement in public buildings and schools/kindergartens in eleven European cities: statistical analysis of the data. Atmospheric Environment 2011;45 3676-3684.

[33] Edwards RD, Jurvelin, J, Koistinen K, Saarela K, Jantunen M. VOC source identification from personal and residential indoor, outdoor, and workplace microenvironment samples in EXPOLIS-Helsinki, Finland. Atmospheric Environment 2001;35 4829-4841.

[34] Hodgson, AT, Beal D, McIlvaine JER. Sources of formaldehyde, other aldehydes and terpenes in a new manufactured house. Indoor Air 2002;12 235-242.

[35] Grimaldi F, Botti P, Bouthiba M, Gouezo F, Viala A. Study of indoor air pollution by carbonyl compounds. Pollution Atmospherique 1996;38(149) 58-67.

[36] Fantuzzi G, Aggazzotti G, Righi E, Cavazzuti LGP, Franceschelli A. Indoor air quality in the university libraries of Modena (Italy). Science of the Total Environment 1996;19349-56.

[37] Allou L, Marchand C, Mirabel Ph, Le Calve S. Aldehydes and BTEX measurements and exposures in university libraries in Strasbourg (France) Indoor and Built Environment 2008;17(2) 138-145.

[38] Baumann MGD, Lorenz LF, Batterman SA, Zhang GZ. Aldehyde emissions for particleboard and medium density fiber board products. Forest Product Journal 2000;50 75-82.

[39] Clarisse B, Laurent AM, Seta N, Moullec LY, Hasnaoui EA, Momas I. Indoor aldehydes: measurement of contamination levels and identification of their determinants in Paris dwellings. Environmental Research 2003;92 245-253.

[40] Fjällström P, Andersson B, Nilsson C. Drying of linseed oil paints: the effects of substrate on the emission of aldehydes. Indoor Air 2003;13 277-282.

[41] Knudsen HN, Kjaer UD, Nielsen PA, Wolkoff P. Sensory and chemical characterization of VOC emissions from building products: impact of concentration and air velocity. Atmospheric Environment 1999;33 1217-1230.

[42] Baek SK, Jenkins RA. Characterization of trace organic compounds associated with aged and diluted sidestream tobacco smoke in a controlled atmosphere—volatile organic compounds and polycyclic aromatic hydrocarbons. Atmospheric Environment 2004;38 6583-6599.

[43] Bruinen de Bruin Y, Koistinen K, Kephalopoulos S, Geiss S, Tirendi S, Kotzias D. Characterisation of urban inhalation exposures to benzene, formaldehyde and acetaldehyde in the European Union. Environmental Science and Pollution Research 2008;15 417-430.

[44] Marchand C, Le Calvé S, Mirabel Ph, Glasser N, Casset A, Schneider N, de Blay F. Concentrations and determinants of graseous aldehydes in 162 homes in Strasbourg (France). Atmospheric Environment 2008;42 505-516.

[45] Juverlin J, Vartiainen M, Jantunen M. Personal exposure levels and microenvironmental concentrations of formaldehyde and acetaldehyde in the Helsinki Metropolitan Area, Finland. Journal of the Air and Waste Management Association 2001;51 17-24.

[46] Lovreglio P, Carrus A, Iavicoli S, Drago I, Persechino B, Soleo L. Indoor formaldehyde and acetaldehyde levels in the province of Bari, South Italy, and estimated health risk. Journal of Environmental Monitoring 2009;11 955-961.

[47] WHO. Air Quality Guidelines for Europe. World Health Organization, Regional Office for Europe, Copenhagen, Denmark, 2nd ed., 2001.

[48] Kotzias D, Koistinen K, Kephalopoulos S, Schlitt C, Carrer P, Maroni M, Jantunen M, Cochet C, Kirchner S, Lindvall T, McLaughlinc J, Molhavec L, Fernandes E, Seifert B. The INDEX Project. Critical Appraisal of the Setting and Implementation of Indoor Exposure Limits in the EU, European Commission, Directorate General, Joint Research Centre, EUR 21590 EN, 2005

[49] Pegas PN, Alves CA, Evtyugina MG, Nunes T, Cerqueira M, Franchi M, Pio CA, Almeida SM, Cabo Verde S, Freitas MC. Seasonal evaluation of outdoor/indoor air quality in primary schools in Lisbon. Journal of Environmental Monitoring 2011;13 657-667.

[50] Christensen CS, Skov H, Nielsen T, Lohse C. Temporal variation of carbonyl compound concentrations at a semi-rural site in Denmark. Atmospheric Environment 2000;34 287-296.

[51] Gaffney JS, Marley NA, Martin RS, Dixon RW, Reyes LG, Popp CJ. Environmental Science and Technology 1997;31 3053-3061.

[52] Possanzini M, Tagliacozzo G, Cecinato A. Ambient levels and sources of lower carbonyls at Montelibretti, Rome (Italy). Water, Air and Soil Pollution 2007;183 447-454.

[53] Shepson PB, Hastie DR, Schiff HI, Polizzi M, Bottenheim JW, AnlaufK, Mackay GI, Karecki DR. Atmospheric concentrations and temporal variations of C_1-C_3 carbonyl compounds at two rural sites in central Ontario. Atmospheric Environment 1991;25A 2011-2015.

[54] Schäfer L, Gabriel R, Müller H, Wolf A, Kesselmeier J. Emission of short chained organic acids and aldehydes in relation to physiological activities and apoplastic ion concentration in Mediterranean tree species during the B.E.M.A. field campaign in May 1994. In: Biogenic Emissions in the Mediterranean Area, BEMA-Project, Report on the 1st BEMA measuring campaign at Castelporziano, Rome (Italy), May 1994, EUR 16293 EN, Brussels, 1995;233-247.

[55] Kesselmeier J, Bode K, Hofmann U, Müller H, Schäfer L, Wolf A, Ciccioli P, Brancaleoni E, Cecinato A, Frattoni M, Foster P, Ferrari C, Jacob V, Fugit JL, Dutaur L, Simon V, Torres L. The BEMA-Project: emission of short chained organic acids, aldehydes and monoterpenes from *Quercus ilex* L. and *Pinuspinea* L. in relation to physiological activities, carbon budget and emission algorithms. Atmospheric Environment 1997;31 119-134.

[56] Müller K, Haferkorn S, Grabmer W, Wisthaler A, Hansel A, Kreuzwieser J. Biogenic carbonyl compounds within and above a coniferous forest in Germany. Atmospheric Environment 2006;40 S81-S91.

[57] Custódio D, Guimaraes CS, Varandas L, Arbilla G. Pattern of volatile aldehydes and aromatic hydrocarbons in the largest urban rainforest in the Americas. Chemosphere 2010;79 1064-1069.

[58] Evtyugina MG, Nunes T, Pio C, Costa CS. Photochemical pollution under sea breeze conditions, during summer at the Portuguese West Coast. Atmospheric Environment 2006;40 6277-6293.

[59] IARC, IARC monographs on the evaluation of carcinogenic risks to human. Formaldehyde. IARC, Lyon 100F:1-36. Monographs.iarc.fr/ENG/Monographs/vol100F/mono100F-29.pdf. Accessed June 17, 2015.

[60] WHO. Guidelines for Indoor Air Quality: Selected Pollutants. World Health Organization, Regional Office for Europe, Copenhagen, Denmark, 2010.

Chapter 4

Air Pollution Monitoring and Prediction

Sheikh Saeed Ahmad, Rabail Urooj and Muhammad Nawaz

Additional information is available at the end of the chapter

http://dx.doi.org/10.5772/59678

1. Introduction

One of the most important emerging environmental issues in Asian cities is air pollution. Air pollution is an atmospheric condition in which the concentration and duration of certain substances present in the air produce injurious and destructive effects on both man and the surrounding environment [1]. The most common pollutants in air are sulfur oxide, nitrogen dioxide, carbon monoxide and dioxide, and particulate matter.

Geographical Information Systems (GISs) are computer-based applications used for mapping and analyzing the earth and related spatially distributed phenomena. GIS applications integrate unique visualizations with common databases, which make it possible to capture, model, manipulate, retrieve, analyze, and present the geographically referenced data. Compared to other information systems, GIS systems have advantages, including the high power of analyzing spatial data and handling large spatial databases.

GIS applications can be used in air quality management and for controlling pollution, for handling and managing large amount of data. GIS systems manage spatial and statistical data, which facilitates depiction of the association between the frequency of human activities leading to bad environmental health and poor air quality. GIS modeling and statistical analysis also enables to examine and predict the impact of climatic variables on air pollution. In this way, GIS systems help in monitoring air pollution and emissions of pollutants from different sources.

Air pollution mapping is a helpful method for determining the concentration of pollutants. As the result of air pollution mapping, overviews of pollution in cities can be created and their sources of pollution emission can be identified, which help in controlling emissions. Different studies have been executed on air pollution in conjunction with GIS [2-11]. Consequently, GIS applications in air monitoring are necessary to determine air quality to reduce pollution to such a level at which harmful impacts on human health and the environment is reduced.

With the help of GIS applications, an output report of pollutants in Air Quality Management Systems (AQMSs) can be achieved in the form of three-dimensional (spatial) records. In AQMS emission time, concentration and place of air pollutants are regulated in order to achieve the predefined air quality standards of ambient air. It encompasses the estimation of the pollutants' emission schedule in a way to determine the consequences to air quality and the design of alternative programs for emission control in order to meet air quality standards, which are subject to some limitations, for example, technological viability and lowest charges. For environmental modeling with GIS applications, AQMSs are considered to locate monitoring stations, for development of geospatial model for air quality, and for spatial decision-support systems. However, the most significant step in an AQMS is data mining. The data mining method is a skill, which is used to analyze the data, uncover hidden patterns, and find interesting information from large amounts of data or huge databases. The most commonly used technique in data mining is artificial neural networks [12].

The human brain consists of a large number of neurons connected to each other by synapses to make networks, and these networks of neurons are called neural networks, or natural neural networks. Similarly, the artificial neural network (ANN) is basically a mathematical model of a natural neural network. The ANN uses a mathematical or computational model based on connectionist approach for solving the given problem. The concept of ANN is derived from biological neural network systems. The key applications of neural networks are control systems, classification systems, and prediction and vision systems.

Three basic components are important in order to make functional model, like: synapses of neuron; an added that sum all input in form of weights; and activation function. In Figure 1, synapses are shown by weights. Basically, a strong connection between input and neuron is noted by synapses or value of weight. Negative values reflect inhibitory connections, whereas excitatory connections are shown by positive values. Activation functions regulate the output of neurons within an acceptable range from -1 to 1.

1.1. Sources

Air pollution takes place due to natural and anthropogenic activities. But air pollution as the result of man-made activities like fossil fuel combustion, construction, mining, agriculture, and warfare are the most significant and cause problems in the atmosphere [13].

Basically, two types of pollution sources have been categorized, i.e., Stationary and Mobile. The stationary source is a type of source that is fixed or is a preset pollutant emitter, for example, fossil fuel burning power plants and refineries. The mobile source is a nonstationary type of pollutant emitter, for example, vehicles. The most emerging and leading cause of air pollution is the motor vehicle [14]. Pollutants that are emitted directly from the source into the air are known as primary pollutants, for example, carbon dioxide, carbon monoxide, sulfur dioxide, etc. When these primary pollutants react in atmosphere with each other to form another type of pollutants, they are called secondary pollutants, which are not directly emitted

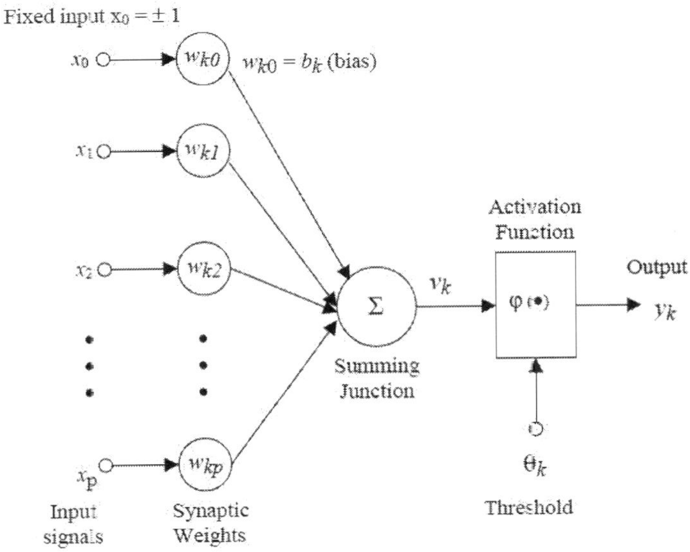

Figure 1. Model of a neuron

but formed as a result of primary pollutants' reaction in the atmosphere. For example, ozone forms when nitrogen oxides react with hydrocarbons in the presence of sunlight, and the resulting nitrogen dioxide reacts further with oxygen and forms ozone as pollutant.

1.2. Health effects

Air pollution and its effects in rural and urban areas are directly related to the ongoing activities. For example, in cities, pollution is related to the products of combustion in industries and vehicles. Many large cities all over the world exhibit excessive levels of air pollutants. Among all dangerous pollutants, nitrogen dioxide (NO_2) is important due to its capacity of causing dangerous effects on humans and the environment, which results in photochemical oxidation and acid rain.

The effects of air pollution cannot be ignored even within homes. Many air pollutants can cause cancer and other diseases among inhabitants. In 1985, it was reported that indoor toxic chemicals are three times more potent in causing cancer than outdoor air pollutants [15]. In America, health issues caused by buildings are called "sick building syndrome"[16].

1.3. Case study

In Pakistan, air pollution is emerging as a serious problem in its mega cities, which needs to be monitored and addressed at the root level in order to reduce the lethal impacts of pollutants on man and environmental health. The present study of Pakistan focuses on the most important twin cities of Pakistan, which are Rawalpindi and Islamabad. Both cities are commonly viewed as one unit and are 15 km apart. The study area with 135 sampling locations is shown in Figure 2. The climatic condition of Rawalpindi and Islamabad is sub-humid to tropical, with hot and long summers (May to August) accompanied by a monsoon season (July to August) followed by short and mild winters (October to March). The average low temperature is 12.05 °C in January and average high temperature is 31.13 °C in July.

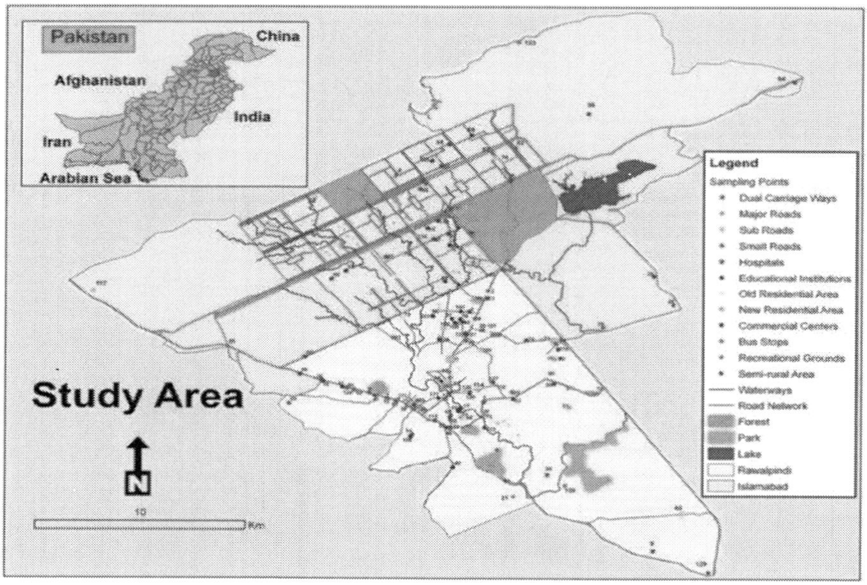

Figure 2. Base map

For the monitoring campaign, the maximum area (135 sampling sites) was covered in order to represent different traffic intensity and congestion levels in the urban area of Rawalpindi and Islamabad, for sampling. These sites included dual carriageways, major, linking, and small roads, healthcare centers, educational institutes, commercial areas, old residential areas, modern residential areas, recreational spots and semi-rural areas.

Research was carried out in order to monitor the NO_2 concentration in the ambient air of Rawalpindi city. Passive samplers were used within the city from January to December in 2008. The average concentration found was 27.46±0.32 ppb. The highest concentration was recorded near the main roads and in the vicinity of schools and colleges due to the large number of transport vehicles, which exceeded the set limit concentration value given by the World Health Organization.

2. Experimental design

2.1. Passive Sampling of NO_2

The most frequent method in monitoring studies for passive sampling of NO_2 is using diffusion tubes described by Atkins [17]. This method for NO_2 measurement is reliable, easy to handle, and it is an inexpensive method for screening air quality. Moreover, passive samplers are preferably appropriate for extensive spatial measurement of NO_2, and they have been reported in many studies of NO_2 monitoring of air in many countries like the United Kingdom, USA, France, Turkey, Argentina, and China [18].

Basically, passive samplers are designed on the principle of air diffusion having an efficient absorber at one end of the tube, and the flow rate (sampling rate) at constant temperature can be measured by using Flick's Law [19]. For that, the length and diameter of diffusion tubes are known, whereas sampling by using diffusion tubes is independent of air pressure.

2.2. Neural network design

From different sampling sites covering the whole study area, data was collected for neural network analysis. Collected data was fed to the neural network that has area_id, season_id, temperature, humidity, rainfall, and the respective concentrations as columns. For the neural network, the marked value was set to predict concentrations and rests were used as input to the neural network.

Neural network has two phases: training and testing. In the first phase (training), the network is trained by providing the complete information about the characteristics of data and observable outcomes to perform a particular task.

A neural network can develop a model that learns the relationship between input data and the desired outcome in the training phase. In the testing phase, testing data are provided as input. The performance of the testing phase depends upon the training phase (it depends on the number of samples that are provided during the training phase and also on the number of times that the network is accurately trained. However, it is impossible that the output is 100% precise for any network input. MS Access was used as the database engine because it is easy to use for all.

For testing the neural network, the cross validation method is used by using holdout method in which data was divided into testing and training data. The database consisted of two tables: training_ data and testing_data. The function of training_data is to train the ANN by adjusting weights in order to maximize the predictive ability of ANN and minimize error during forecasting. Testing data was used to test the prediction accuracy of ANN on new data. The structure of training data and testing data is given in Table 1.

In Table 1, the first key "id" is primary key, which contains the number that indicates row number and the second key "loc_id" contains the number that indicates location from where data is gathered, loc_name indicates the name of location and the next six fields indicate position of location with respect to north and east. The next two indicate temperature and humidity levels.

The 13th and 14th fields indicate concentration of NO_2 and level of concentration value. The last field of dataset contains week number, which indicates the number of weeks in which data is gathered from particular location. The attribute for testing data are the same in the testing data structure.

Field Name	Data type	Primary key	Field size
Id	Number	Yes	Long Integer
loc_id	Number		Long Integer
loc_name	Text		50
map_id	Number		Long Integer
north_d	Number		Long Integer
north_m	Number		Long Integer
north_s	Number		Long Integer
east_d	Number		Long Integer
east_m	Number		Long Integer
east-s	Number		Long Integer
Temp	Text		50
Humidity	Text		50
Concentration	Number		Long Integer
con_level	Number		Long Integer
Week	Number		Long Integer

Table 1. Structure of training data

For designing a network, we need to specify the architecture of a neural network by designing a number of hidden layers and units in each layer along properties of network that describe error function and network activation.

For optimal generalization of collected data, two types of architectures: the rtNEAT (real-time neuro evolution of augmented topologies) architecture with evolution algorithm and the feed forward architecture with back propagation algorithm of ANN are used in order to ensure high accuracy of ANN prediction about impacts of NO_2 concentration achieved in future. This rtNEAT architecture is used to train neural network with evolutionary algorithm, which has three steps, i.e., selection, mutation, and reinsertion. But before the training of neural network, the topology has to be created in the design of the neural network. A neural network is a connection of neurons, which contains three types of nodes: input, output, and hidden node. All nodes are randomly created during its execution.

Table 2 describes the properties of network, which contains an error function and network activation parameters. These properties are functional to all tested networks by the architecture search method and manually selected network.

Parameter	Value
Input activation FX	Logistic
Output name	Concentration
Output error FX	Sum-of-squares
Output activation FX	Logistic

Table 2. Network properties

The logistics function has a sigmoid curve and sum of squares. The sum of squares is the most frequent function error, which is used for the classification problem. The error is the sum of the square differences between the real input value and neural network target value.

2.3. Architecture search

A heuristic search is used to search the dataset for the best networks. Heuristic methods are used to speed up the process of finding a satisfactory solution. The architecture search for the designed neural network NO_2 is given in Table 3.

ID	Architecture	# of Weights	Fitness	Train Error	Validation ErrorError	Test ErrorError	AIC	Correlation	R-Squared
1	[8-1-1]	11	0.079965	11.371084	12.151164	12.505404	-4220.122343	0.652911	0.424949
2	[8-20-1]	201	0.080369	11.295746	12.093373	12.442678	-3846.484024	0.653839	0.427147
3	[8-12-1]	121	0.080841	11.017718	12.044772	12.37002	-4030.333807	0.668927	0.446734
4	[8-7-1]	71	0.080593	11.182193	12.108417	12.407984	-4116.153129	0.662147	0.438331
5	[8-16-1]	161	0.081507	10.941978	11.986108	12.26894	-3956.935279	0.670637	0.448777
6	[8-18-1]	181	0.080474	10.917611	12.026946	12.4264	-3919.068903	0.676823	0.457557
7	[8-14-1]	141	0.080839	11.105445	12.044827	12.370266	-3982.744071	0.666178	0.44366

Table 3. Heuristic architecture search for NO2

2.4. Training of neural network

The next step is to train the neural network for the NO_2 dataset by using the propagation algorithm. Weight change is calculated by the quick propagation algorithm by utilizing the quadratic function $f(x) = x^2$. In neural networks, several layers contain neurons in each layer that are connected with each other like neurons in the input layer connected to one or more neurons of the hidden layer, which are further connected to the output layer's neuron. With each presentation in neural network, error is computed as the difference between network output and observable output. The combination of randomly assigned weight (giving low error) replaces weights that are at the first location. This is called training to adjust the connection weights to enable the network to produce the expected output. Two different weights having two different error values are two points of a secant. Relating this secant to a quadratic function, it is possible to calculate its minimum $f(x) = 0$. The x-coordinate of the minimum point is the new weight value.

$$S(t) = \frac{\partial E}{\mathbb{R}w_i(t)}$$

$$\Delta w_i(t) = \alpha \cdot \frac{\partial E}{\partial w_i(t)} \text{(Normal back propagation)}$$

$$\frac{\Delta w_i(t)}{\alpha} = \frac{\partial E}{\partial w_i(t)}$$

$$S(t) = \frac{\partial E}{\partial w_i(t)} = \frac{\Delta w_i(t)}{\alpha}$$

$$\Delta w_i(t) = \frac{S(t)}{S(t-1) - S(t)} \cdot \Delta w_i(t-1) \text{(Quick propagation)}$$

Here w =weight, i =neuron, E =error function, t =time (training step), α= learning rate, and μ= maximal weight change factor

The quick propagation coefficient was set to 1.75, learning rate was 0.1, and iterations were 500. The training graph for dataset errors for NO_2 is shown in Figure 3.

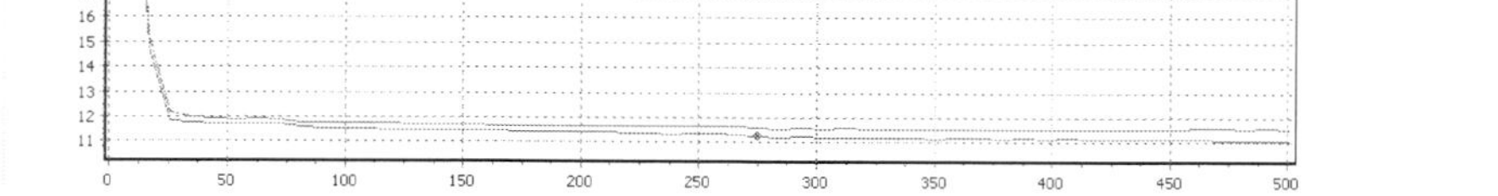

Figure 3. Dataset errors for the NO_2 dataset

The training graph of correlation for NO_2 is shown in Figure 4.

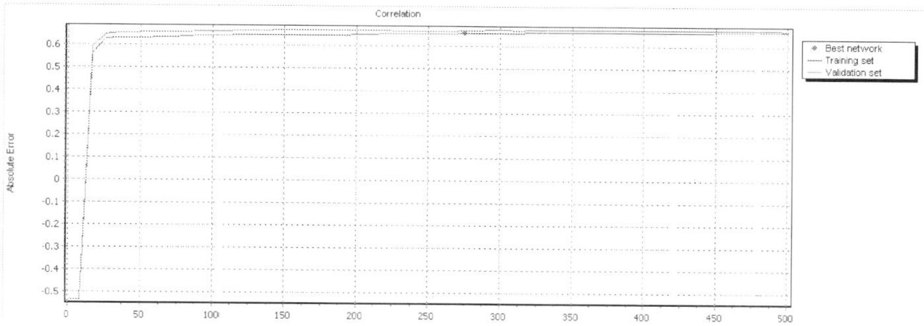

Figure 4. Graph of correlation for NO_2

The graph of error improvement – network errors for NO_2 is shown in Figure 5.

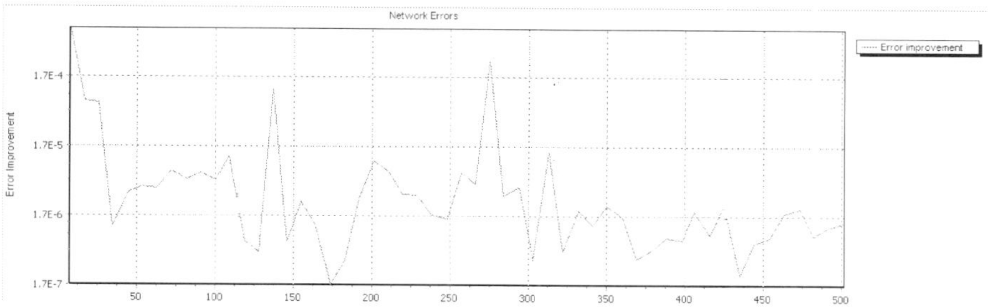

Figure 5. Network errors for NO_2

The error distribution of network statistics obtained after training of neural network is shown in (Figure 6).

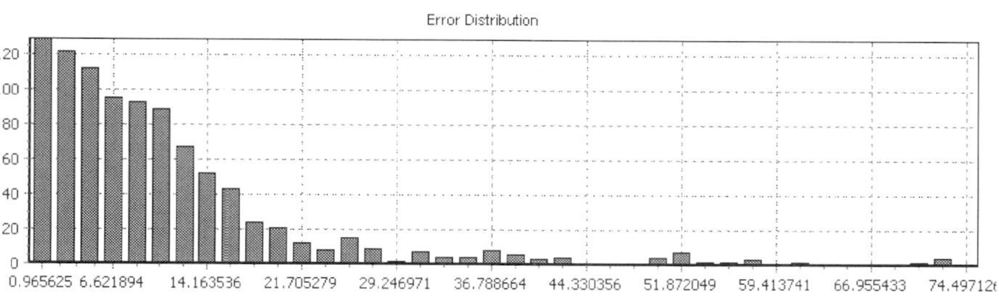

Figure 6. Error distribution for NO_2

3. Data analysis

In order to determine the seasonal variation and statistical significance, results are presented in tabular format. Tables 4 a and 4 b show the average concentration level of NO_2, season-wise, along standard deviation (SD) values measured at different sampling sites of study.

Table 4 a shows average values of NO_2 concentration in different seasons of 12 major sampling categories in urban Rawalpindi and Islamabad from November 2009 to July 2010.

Table 4 b shows the seasonal average concentration of NO_2 of 12 major sampling categories in urban Rawalpindi and Islamabad from September 2010 to March 2011.

Table 5 presents NO_2 concentration for each selected category, as described in study area profile, to understand the general trends of NO_2 concentration levels among different categories during the course of experimental period.

	Sampling Categories	Mild Winter (Nov)	Winter (Dec to Jan)	Early Spring (Feb)	Spring (Mar)	Mild Summer (April)	Summer (Pre-Monsoon) (May to June)	Monsoon (July to August)
	NO_2 Conc. (weekly basis)	(ppb)	(ppb)	(ppb)	(ppb)	(ppb)	(ppb)	(ppb)
Rawalpindi	Dual Carriage Ways (5)	87±19.78	98±26.87	63±12.29	53±6.49	44±10.64	22±4.22	18±1.91
	Major Roads (10)	60±12.19	68±9.56	52±13.52	45±10.23	36±8.97	26±5.88	19±4.74
	Sub-roads (6)	74±20.50	86±24.47	60±16.49	50±11.05	38±12.65	33±13.01	21±4.39
	Small Roads (3)	55±9.78	63±4.89	47±5.57	40±8.24	31±3.40	25±4.68	18±4.81
	Public Hospital (5)	48±18.71	63±18.40	37±0.74	29±2.29	22±2.24	18±0.79	14±0.96
	Private Hospitals (8)	61±14.47	75±14.19	38±1.16	32±2.0.3	25±2.29	20±5.57	14±3.98
	Public EI (11)	85±30.58	95±32.94	75±23.75	63±17.94	47±17.37	31±10.14	20±1.94
	Private EI (17)	55±9.71	66±9.54	45±4.56	43±9.65	38±10.89	26±4.54	18±3.18
	Old Residential Areas (5)	83±15.24	95±16.09	55±13.32	51±6.66	37±6.44	26±2.54	19±1.05
	Modern Residential Areas (5)	65±20.07	73±14.89	69±24.49	59±12.55	36±7.13	28±5.08	21±2.61
	Commercial Area (2)	75±0.83	82±17	61±6.69	51±7.11	36±4.29	21±6.20	18±4.78
	Bus Stops (9)	74±20.26	83±31.47	69±33.78	58±17	39±17.32	28±8.41	20±5.25
	Recreational Spots (9)	75±38.40	87±40.76	62±36.39	56±21.88	43±19.97	31±11.12	19±2.37

	Sampling Categories	Mild Winter (Nov)	Winter (Dec to Jan)	Early Spring (Feb)	Spring (Mar)	Mild Summer (April)	Summer (Pre-Monsoon) (May to June)	Monsoon (July to August)
	NO_2 Conc. (weekly basis)	(ppb)	(ppb)	(ppb)	(ppb)	(ppb)	(ppb)	(ppb)
Islamabad	Dual Carriage Ways (3)	84±28.73	95±33.64	66±23.78	57±12.31	45±16.69	24±5.98	19±4.16
	Major Roads (3)	50±3.72	60±2.04	40±0.81	32 ±2	26±4 42	21±2.97	15±2.16
	Sub-roads (4)	54±6.06	67±6.39	49±6.49	43±7.24	38±12.79	25±1.38	18±1.26
	Small Roads (3)	59±12.65	64±6.33	51±9.60	44±8.93	35±4 53	26±3.66	20±3.08
	Public Hospitals (3)	44±0.58	57±0.29	39±0.29	32±0.58	23±1.47	19±0.51	15±1.71
	Private Hospitals (1)	42	56	38	30	24	19	14
	Public EI (5)	53±13.34	64±9.32	46±9.30	39±10.76	34±14.19	25±6.01	18±1.28
	Private EI (6)	58±11.23	63±7.18	49±7.72	39±9.93	31±5.85	24±2.66	17±1.77
	Commercial Area (1)	61	68	57	50	35	25	16
	Bus Stops (12)	72±14.25	78±16.23	65±7.51	55±5.23	34±6.22	25±3.21	19±2.56
	Recreational Spots (2)	62±5.97	69±4.58	57±2.45	48±1.59	38±2.48	25±3.15	17±1.56
	Semi-Rural Areas (7)	46±8.98	59±5.64	42±6.41	33±7.87	31±5.29	24±3.22	18±3.19

(a)

Rawalpindi	Dual Carriage Ways (5)	30±4.00	51±7.52	88±22.28	100±26.42	63±18.52	50±20.32
	Major Roads (10)	27±6.57	49±9.72	61±10.33	68±9.34	48±3.76	37±3.79
	Sub-roads (6)	32±7.31	53±12.97	74±23.42	87±26.60	58±13.69	39±6.36
	Small Roads (3)	28±3.65	37±2.53	53±6.30	62±2.58	54±12.39	43±11.13
	Public Hospital (5)	20±0.98	32±5.46	48±18.01	64±18.03	40±6.13	29±3.94
	Private Hospitals (8)	23±3.90	38±7.19	60±15.37	73±14.03	40±4.05	31±1.95
	Public EI (11)	44±16.98	81±36.87	86±31.78	96±34.20	73±21.26	63±18.23
	Private EI (17)	31±4.85	42±6.10	55±8.91	66±9.82	45±7.08	35±5.90

	Sampling Categories	Mild Winter (Nov)	Winter (Dec to Jan)	Early Spring (Feb)	Spring (Mar)	Mild Summer (April)	Summer (Pre-Monsoon) (May to June)	Monsoon (July to August)
	NO_2 Conc. (weekly basis)	(ppb)	(ppb)	(ppb)	(ppb)	(ppb)	(ppb)	(ppb)
Islamabad	Dual Carriage Ways (3)	31±5.72	50±11.40	82±21.11	99±32.70	67±19.78	49±12.84	
	Major Roads (3)	22±0.80	37±1.93	53±4.33	65±0.30	44±2.30	33±2.00	
	Sub-Roads (4)	26±2.48	39±4.60	54±5.74	65±4.08	46±3.02	35±9.17	
	Small Roads (3)	30±5.94	41±4.12	54±4.24	63±6.67	47±2.60	38±2.79	
	Public Hospitals (3)	22±2.14	34±2.66	45±0.80	60±1.41	45±0.22	34±0.82	
	Private Hospitals (1)	22	31	40	55	38	30	
	Public EI (5)	31±7.41	40±3.60	52±7.94	64±8.39	46±9.34	37±9.62	
	Private EI (6)	29±11.58	41±8.65	54±10.14	63±7.03	47±7.93	36±9.17	
Twin Cities	Old Residential Areas (5)	27±2.97	61±14.74	84±14.18	95±16.51	58±12.41	48±10.06	
	Modern Residential Areas (5)	32±7.86	49±11.70	66±20.07	75±16.16	60±19.16	48±16.53	
	Commercial Area (3)	32±1.23	46±6.09	63±1.00	71±3.57	56±7.02	48±8.41	
	Bus Stops (11)	32±9.11	53±20.30	76±20.07	87±32.40	69±31.34	54±19.54	
	Recreational Spots (10)	37±18.55	52±25.23	71±37.63	84±39.83	57±29.71	46±24.78	
	Semi-Rural Areas (7)	31±9.47	41±7.44	53±6.51	62±6.21	44±7.50	36±6.99	

(b)

Table 4. (a): Seasonal mean values of NO_2 from November 2009 to July 2010 (b): Seasonal mean values of NO_2 from September 2010 to March 2011

In Table 5 most of the sampling sites of study area showed nearly similar average concentration from month of November 2009 to March 2011. Maximum concentration of NO_2 shown on dual carriage ways.

The possible cause of such elevated levels of NO_2 concentration is extensive increase in number of vehicles, increase in population, busy roads, fuel inefficient vehicles, driving ways, and traffic jams. Gilbert reported that NO_2 is considerably related to both the distance from the nearest highway and the traffic count on the nearest highway [20].

The rest of the categories showed nearly the same average concentration. Major roads and sub-roads showed average NO_2 concentration levels of 53.56 ppb and 51.78 ppb, respectively. Sub-roads, bus stops, recreational spots, and educational institutions showed similar concentration levels of approx. 51 ppb.

http://dx.doi.org/10.5772/59678

Sampling Categories	No. of Sites	Average NO_2 Conc. (ppb)
Dual Carriage Ways	8	55.23
Major Roads	13	53.56
Sub-roads	10	51.78
Bus Stops	11	51.62
Educational Institutions	39	51.26
Recreational Spots	10	51.18
Old Residential Area	5	48.97
Small Roads	6	48.23
Commercial Area	3	47.59
Hospitals	17	47.44
Modern Residential Area	5	46.25
Semi-Rural Area	7	37.65

Table 5. Average NO_2 concentration levels in twin cities from November 2009 to March 2011

Educational institutions and recreational spots, being present close to the dual carriage ways, also experience elevated concentration levels. Old residential areas (48.97 ppb) showed slightly higher NO_2 concentration levels as compared to modern residential areas (47.59 ppb).

Narrow road, enclosing architecture, and congestion among the old residential areas result in traffic emission being trapped and buildup leading to higher NO_2 concentration levels, whereas in modern residential areas increased vehicular number is the major cause of elevated NO_2 levels. The minimum NO_2 concentration levels were indicated in semi-rural areas, that is 37.65 ppb. A study in Vilnius commented the same phenomena; NO_2 average rates depend upon traffic and are highest in cross roads and lowest at the background suburban areas [21].

For annual average concentration level of nitrogen dioxide, a spatial interpolation map has been developed by using inverse distance weighted (IDW). IDW in Figure 7 is clearly depicted as the areas of higher and lower concentration level of NO_2 in Rawalpindi and Islamabad.

Higher concentration levels are represented by darker shades while the lower concentration levels are shown with lighter shades. The maximum NO_2 values were found at the center of the city, where they reached the concentration of 83–110 ppb. Values were low on the outskirts of the city, with the lowest concentration in north (31–44 ppb).

A study in Vilnius commented the same phenomena; NO_2 average rates depend upon traffic and are highest in cross roads and lowest at the background suburban areas. Dual carriage ways, sub roads, major roads, commercial areas, old residential areas, and areas where schools and colleges are existing have higher concentration levels of NO_2. Intense traffic flow and congestion were the major reasons for these elevated levels of nitrogen dioxide concentration in those areas as vehicular emission is the predominant source of NO_2.

Vehicle growth rate in twin cities is extensively high. Load of traffic is continuously increasing with growing population rate and demand of motor producing industry. Due to this, traffic congestion is also increasing day by day with growing vehicle population, resulting in highest emission rates per vehicle.

The higher emission rate of NO_2 can also be attributed to the type of fuel and quality of fuel [22]. In Figure 7 Rawalpindi showed more concentration levels than Islamabad due their building patterns.

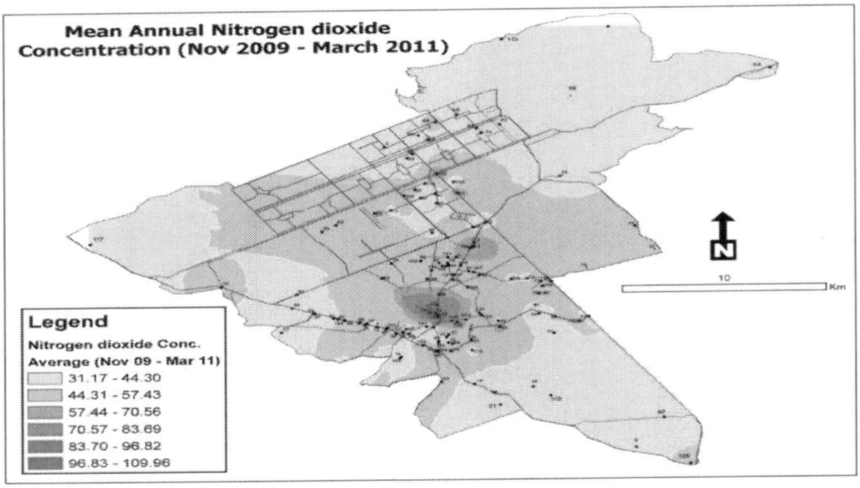

Figure 7. Spatial distribution of NO_2 concentration

3.1. Neural network data analysis

Based on the design of neural network, with the neural architecture and properties discussed, the data space is searched by using heuristic search method with 500 iterations and fitness criteria is set to Inverse Test error. The best top 5 networks explored from the space by the heuristic search are graphically shown (Figure 8).

Heuristic search is a problem-solving method that analytically searches a space of problem states. The best network is obtained when the absolute error gets minimum in the initial iterations so the best network out of the 5 best networks is shown (Figure 9).

Results for all data sets produced after training and testing data. Real vs. target graph represented a line graph of real- and network-predicted target values for record displayed in Table 6. X-axis shows the selected input column values and Y-axis represents network-predicted output values. Table 6 presents the summary of the real vs. output table after training.

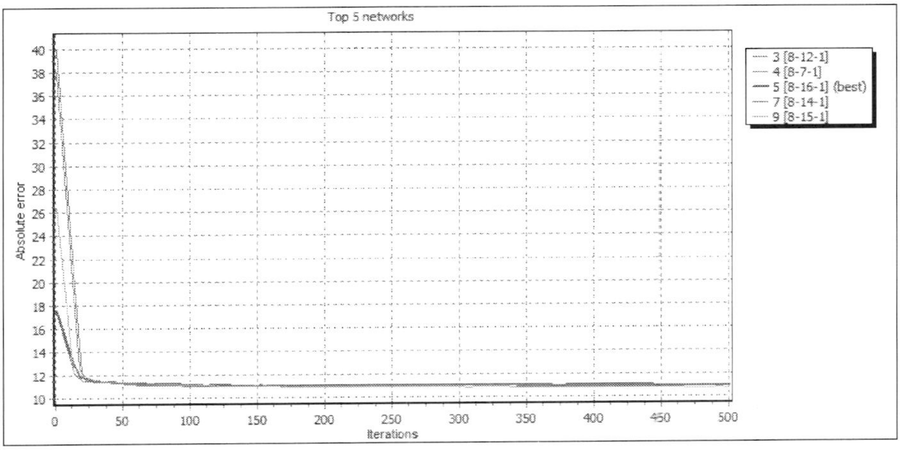

Figure 8. The top five networks explored by heuristic search approach

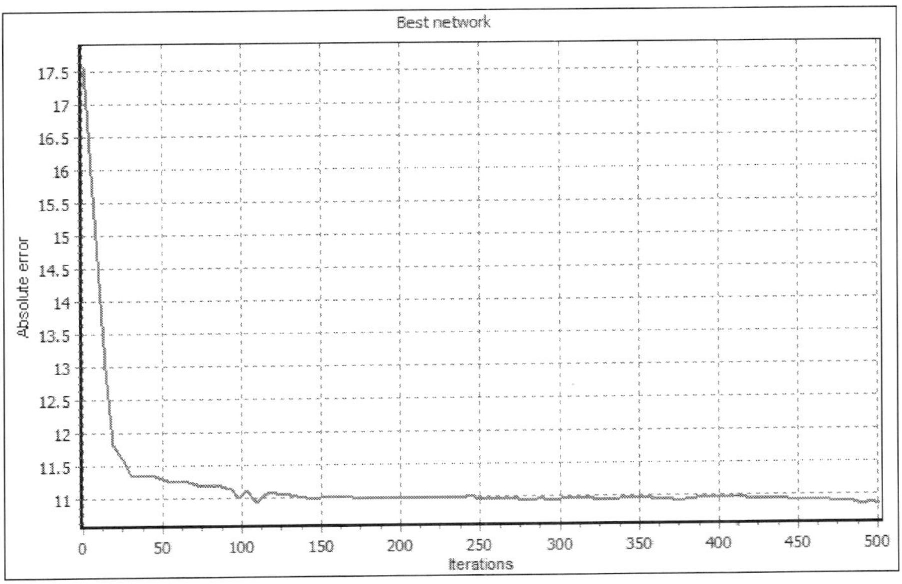

Figure 9. Network explored by heuristic search

	Target	Output	AE	ARE
Mean:	45.237265	45.09091	11.341221	0.250292
StdDev:	20.98552	13.97879	11.112997	0.180871
Min:	11.3	20.16986	0.004569	0.000171
Max:	132.72	63.353673	73.986765	1.096446
Correlation, 0.653989; R-squared, -0.290243				

Table 6. Summary of real vs. target

The visualization for real vs. output with row number on x-axis and target/output (area_id) on y-axis is shown (Figure 10).

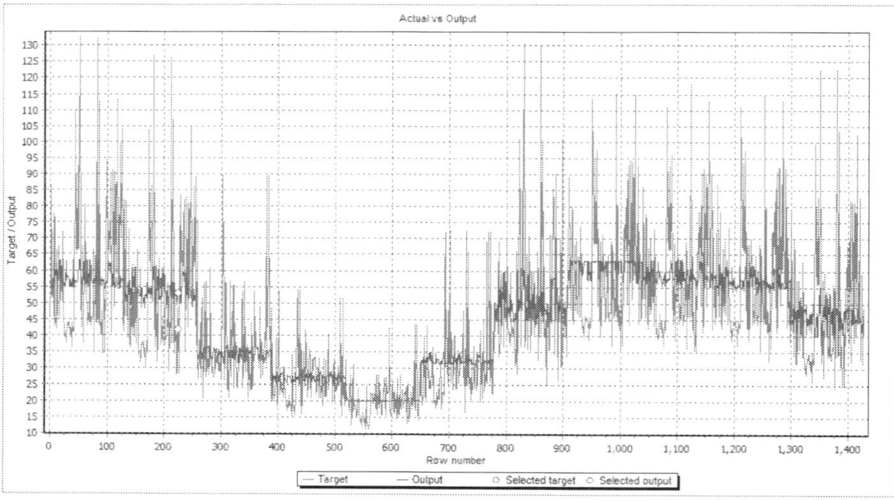

Figure 10. Real vs. network output

Figure 11 shows a scatter plot of the real and forecasted output values. X- axis presents the real values and Y-axis shows predicted network values.

Graph in Figure 12 shows the Network Error Dependence on values, which are numerically input in columns of data sheet. Through graph of Error Dependence, the ranges of the selected input column that can produce network error can be identified.

The last phase after the neural network is trained and tested is to query the network. The concentration is the output value for the neural network. So the input queries are subjected area_id, season_id, temperature, relative humidity, and rainfall (Figure 13).

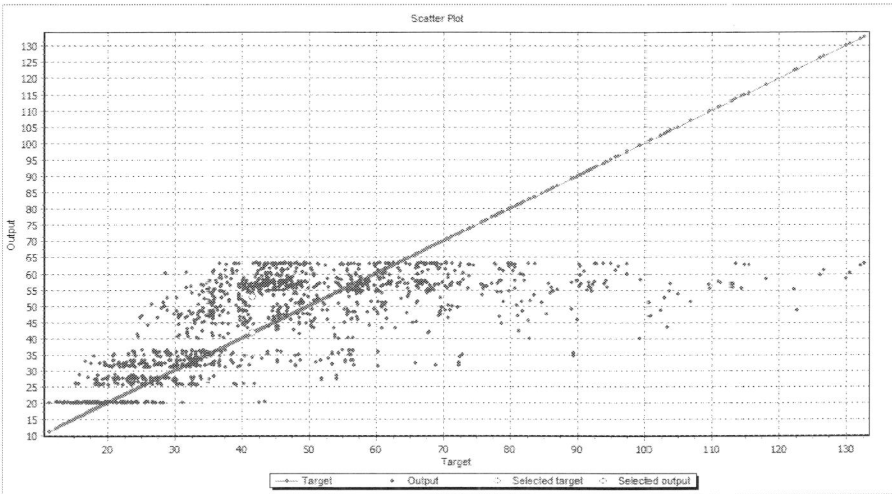

Figure 11. Scattered plot of real and output network values

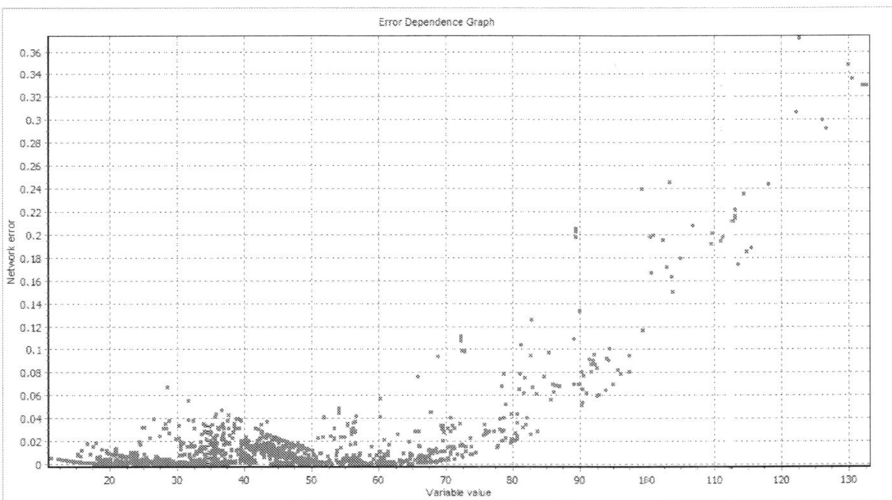

Figure 12. Graph of error dependence

The input Excel sheets are prepared for the GIS mapping. Sheets include area_id, their latitude, longitude, and their concentrations. With the help of interpolation, maps are created for the service.

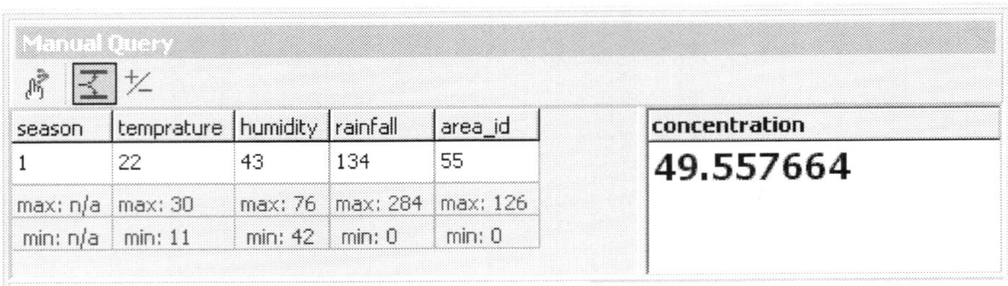

Manual Query

season	temprature	humidity	rainfall	area_id	concentration
1	22	43	134	55	49.557664
max: n/a	max: 30	max: 76	max: 284	max: 126	
min: n/a	min: 11	min: 42	min: 0	min: 0	

Figure 13. Excel sheet presenting manual query

Temporal variation can be explained through meteorological recorded conditions. However, most of the variations on a local scale are due to the impact of air pollutants.

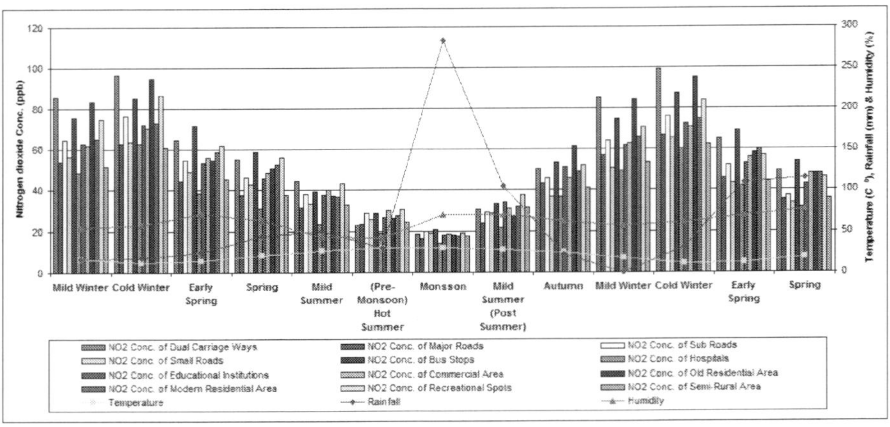

Figure 14. Relationship of rainfall, temperature, and humidity with NO_2 concentration (November 2009–March 2011)

Figure 14 indicates the positive association of NO_2 concentration level with humidity (RH in %) and negative association with the temperature. Figure 15 shows the concentration of NO_2 during summer when recorded temperature, rainfall, humidity are 31°C, 67, and 17mm, respectively.

Figure 16 shows the concentration of NO_2 during the winter season at 11 °C, 68% humidity, and 9mm rainfall.

http://dx.doi.org/10.5772/59678

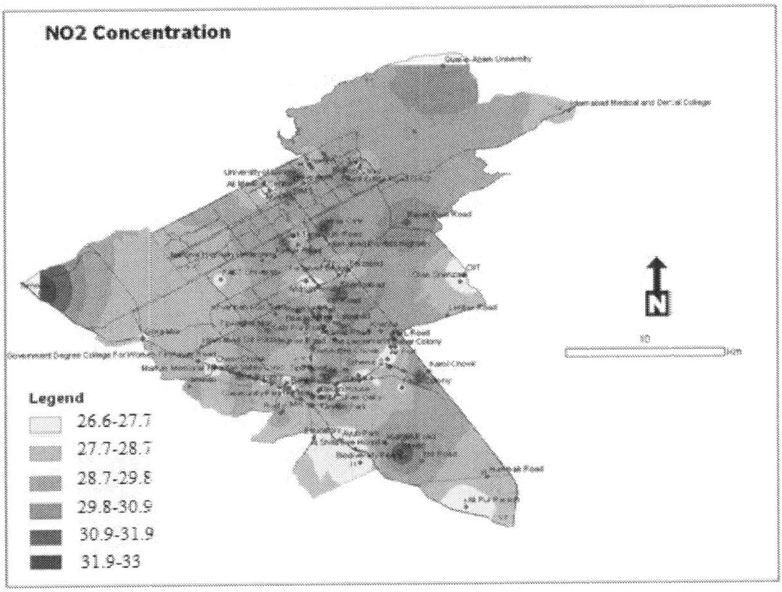

Figure 15. NO_2 concentration in summer

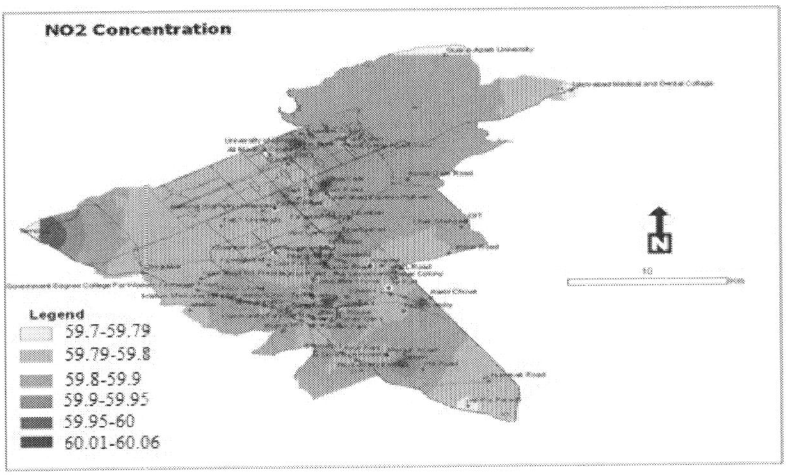

Figure 16. NO_2 concentration in winter

Concentration of NO_2 during the spring season, shown in Figure 17, when recorded temperature is 35°C, humidity is 58%, and rainfall is 60 mm.

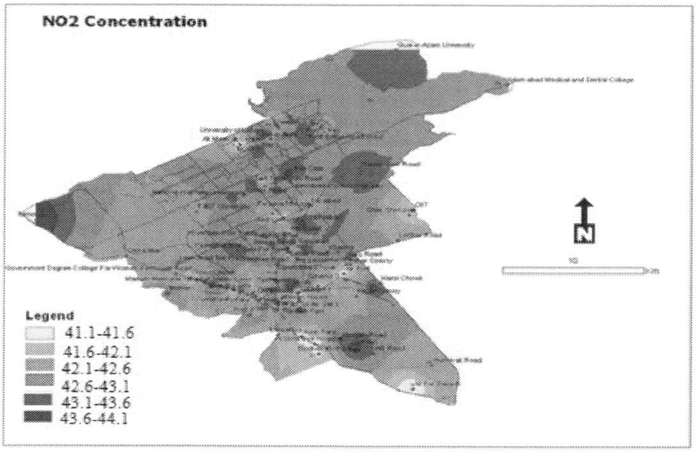

Figure 17. NO_2 concentration in spring

Figure 18 shows predicted concentration of NO_2in autumn season when recorded temperature, humidity, and rainfall are 29 °C, 69, and 22 mm, respectively.

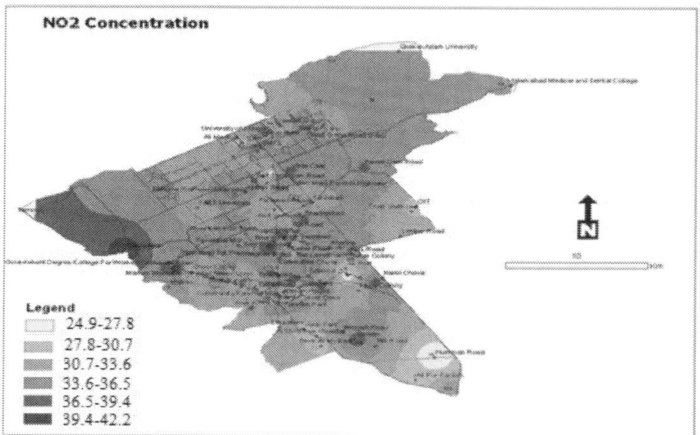

Figure 18. NO_2 concentration in autumn

http://dx.doi.org/10.5772/59678

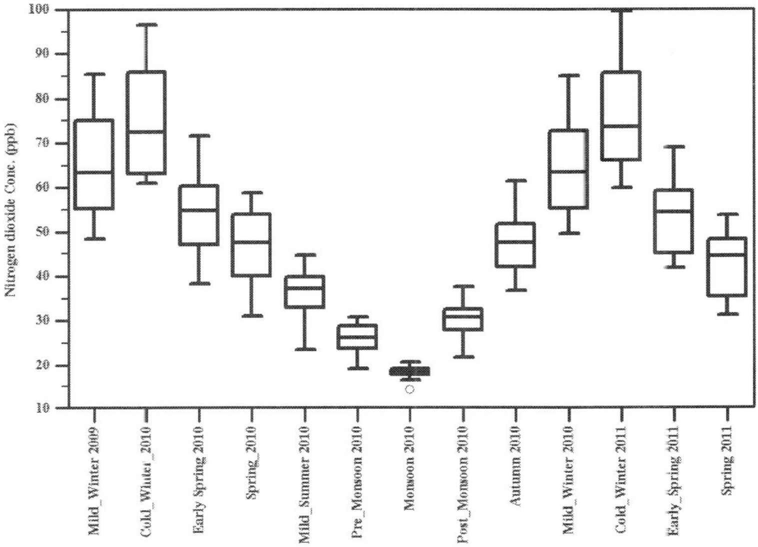

Figure 19. Seasonal variation in NO_2 concentration levels (November 2009 – March 2011)

Figure 19 shows that concentration of NO2 varies in different seasons. The months from May to August were months in which the minimum value of NO2 was recorded, and the maximum concentration was measured in the winter season from December to January.

4. Conclusion

NO_2 concentration levels were recorded on hourly and weekly basis in Rawalpindi and Islamabad city by using diffusion tubes. Artificial neural networks were trained to generalize the process of air pollutant spread over three dimensions. Prediction capabilities of ANN were analyzed through generalization by using hold-out evaluation method of classification. Results showed the advantage of using rtNEAT-like architecture of ANN where a neural network can modify its architecture to reduce the error up to the maximum possible limit. Results showed that annual average concentration of NO_2 concentration was 44 ± 6 ppb. However, the highest concentration was recorded in winter season near the dual carriage ways, schools, and colleges because of the higher number of transport vehicles on the road. This endorsed the fact that the reduced photolysis leads to the accumulation of NO_2 during winter due to less solar radiation. This is again attributed by the results of correlation, which reveal the negative correlation of nitrogen dioxide concentration levels with rainfall and temperature and the positive correlation with humidity. Moreover, the results of correlation reveal that the measured NO_2 concentration levels at different sampling areas exceeded the set limit of concentration value of the World Health Organization and Pak-EPA standard policy. This type of investigative study of artificial neural networks in the area of air pollution modeling shows promising applications for advanced machine learning algorithms in the emerging area of research called eco-informatics.

Author details

Sheikh Saeed Ahmad[1,2*], Rabail Urooj[1,2] and Muhammad Nawaz[1,2]

*Address all correspondence to: drsaeed@fjwu.edu.pk

1 Department of Environmental Sciences, Fatima Jinnah Women University, Mall Road, Rawalpindi, Pakistan

2 BZ University, Multan, Pakistan

References

[1] Mulaku C. Mapping and analysis of air pollution in Nairobi, Kenya. International conference on spatial information for sustainable development, Kenya, 2001.

[2] Gualtieri G and Tartaglia M. Predicting urban traffic air pollution: a GIS framework. Transportation Research – D; 1998; 3(5): 329–336.

[3] Pummakarnchana O, Tripathi N and Dutta J. Air pollution monitoring and GIS modeling: a new use of nanotechnology based solid state gas sensors. Science and Technology of Advanced Materials 2005; 6: 251–255.

[4] Afshar H and Delavar MR. GIS-based air pollution modeling in Tehran. Environmental Informatics 2007; 5: 557–566.

[5] Barnes J, Parsons B and Salter L. GIS Mapping of nitrogen dioxide diffusion tube monitoring in Cornwall, UK. Air Pollution 2005; 13: 157–166.

[6] Elbir T, Mangir N, Kara M, Simsir S, Eren T and Ozdemir S. Development of a GIS-based decision support system for urban air quality management in city of Istanbul. Atmospheric Environment 2010; 44: 441–454.

[7] Veen AVD, Briggs DJ, Collins S, Elliott S, Fischer P, Kingham S, Lebret E, Pryl K, Reeuwijk HV and Smallbone K. Mapping urban air pollution using GIS: a regression-based approach. International Journal of Geographical Information Science 2010; 11(7): 699–718.

[8] Vienneau D, de Hoogh K and Briggs D. A GIS-based method for modeling air pollution exposures across Europe. Science of the Total Environment 2009; 408: 255–266.

[9] Banja M, Como E, Murtaj B and Zotaj A. Mapping air pollution in urban Tirana area using GIS. International Conference SDI, Skopje, 15–17 September 2010, 105–114.

[10] Jensen SS. Mapping human exposure to traffic air pollution using GIS. Journal of Hazardous Material 1998; 61(3): 385–392.

[11] Kim JJ, Smorodinsky S, Lipsett M, Singer BC, Hodgson AT and Ostro B. Traffic-related air pollution near busy roads. American Journal of Respiratory and Critical Care Medicine 2004; 170(5): 520–526.

[12] Alexander SM. Data mining 2005. hhp//www.eco.utexas.edu/~norman/BUS, FOR/course.mat/Alex/ (Accessed 12th February 2008).

[13] United Nation. Conference on the Human Environment, Sewden, 1972.

[14] United Nation. Environmental Performance Annual Report, New York, 2001.

[15] US-EPA. Health affects of different air quality index (AQI) levels caused by nitrogen dioxide, 2008.

[16] Miller D. Potential hazards of future volcanic eruptions. California, 1989.

[17] Atkins DHF, Sandallas J, Law DV, Hough AM and Stevenson K. The measurement of nitrogen dioxide in the outdoor environment using passive diffusion tube samplers. AEA Technology, 1986.

[18] Varshney CK and Singh AP. Passive samplers for NO_x monitoring: a critical review. The Environmentalist 2003; 23: 127–136.

[19] Palmes ED, Gunnison AF, Dimattio J, and Tomczyk C. Personal sampler for nitrogen dioxide. American Industrial Hygiene Association Journal 1976; 37: 570–577.

[20] Gilbert NL, Goldberg MS, Beckerman B, Brook JR and Jerrett M. Assessing spatial variability of ambient nitrogen dioxide in Montreal, Canada, with a land-use regression model. Journal of the Air & Waste Management Association 2005; 65: 1059–1063.

[21] Lozano A, Usero J, Vanderlinden E, Raez J, Contreras J, Navarrete B and Bakouri HEI. Air quality monitoring network design to control nitrogen dioxide and ozone applied in Granada Spain. Ozone: Science & Engineering 2011; 33(1): 80–89.

[22] Heywood JB. Internal combustion engine fundamentals. McGraw-Hill, New York, 1998.

Chapter 5

PM_{10} Time Series Analysis Through Geostatistical Techniques

Claudia Cappello, Sabrina Maggio,

Daniela Pellegrino and Donato Posa

Additional information is available at the end of the chapter

http://dx.doi.org/10.5772/60115

1. Introduction

Particulate matter (PM) is an air pollutant comes from vehicular traffic, industrial activities and street dust, or from the atmosphere, by transformation of the gaseous emissions. In recent years the interest in the health effects of this pollutant have increased, since high concentration levels in urban area have been measured.

Several studies suggest an association between fine particulate air pollution and the increase of the mortality rate [1]. In particular, PM up to 10 micrometers in size (PM_{10}) could cause negative health effects such as respiratory illness or cardiovascular problems. Hence, the analysis of temporal evolution of this pollutant could be useful in decision-making process for environmental policy.

Typically, in time series analysis, the Box-Jenkins methodology is widely applied and the autocorrelation function (ACF) is used as a standard exploratory tool to identify the model structure [3, 4]. In this context, the use of geostatistical techniques could also be convenient, nevertheless these techniques are usually applied to analyze, through the variogram, spatial relationships among sample data measured at some locations in a domain and to predict the corresponding spatial phenomena [6, 18, 22, 29]. In particular, the variogram could represent a complementary exploratory tool for assessing stationarity in time series [2, 19] and it has the considerable advantage that it is defined in much wider circumstances than the autocovariance and the autocorrelation. Moreover, this analytical tool is appropriate to identify trends and periodicity exhibited by the data and to obtain kriging predictions of the variable under study, either for temporal intervals with missing values (interpolation mode) and in time points after the last available data (extrapolation mode).

Different studies have suggested the use of geostatistical methods in time domain [7, 19]. In particular, De Iaco et al. [12] illustrated the role of variogram in this context for different purposes.

The aim of this paper is to analyze PM_{10} air pollution in an area of South Italy characterized by high levels of industrial emissions and vehicular traffic, through geostatistical techniques.

Thus, after a brief review on stochastic processes and geostatistical methods in time series analysis, the temporal evolution of PM_{10} daily concentrations, for the period 2010-2013 has been assessed. After the identification of trend and periodicity, the reconstruction of the analyzed time series by estimation of missing values has been discussed, and predictions of PM_{10} daily concentrations at some unsampled points have been produced. Moreover, the probability distributions of the variable under study have been estimated for future time points.

For interpolation and prediction purposes, a modified version of *GSLib* kriging routine has been used.

2. Theoretical framework

In time series analysis the observed values of a variable for different time points or intervals can be reasonably considered as a finite realization of a real-valued random process, denoted with $\{X_t, t \in T \subseteq \mathbb{R}\}$.

Besides the common second-order moments used to describe the random process $\{X_t, t \in T\}$, such as the autocovariance function and the autocorrelation function, the variogram can also be considered and even preferred with respect to covariance function [10, 19].

Given a stochastic process $\{X_t, t \in T\}$ over a temporal domain $T \subseteq \mathbb{R}$, the corresponding variogram is defined as follows

$$\gamma(t, t+h_t) = 0.5Var\left[X_t - X_{t+h_t}\right], \qquad t, t+h_t \in T. \tag{1}$$

Note that a function $\gamma(\cdot)$ is a variogram if and only if it is conditionally strictly negative definite [23].

As known in the literature [3–5], time series analysis is based on the theory of stationary processes. It is worth highlighting that the second-order stationarity implies the intrinsic stationarity, but the converse is not true [18, 22].

In particular, the stochastic process $\{X_t, t \in T\}$ is intrinsically stationary if its variogram $\gamma(t, t+h_t)$ depends solely on the temporal lag h_t and the expected value of the difference $(X_t - X_{t+h_t})$ is constant.

The variogram, widely used in geostatistical context, could be applied efficiently in time series analysis [14, 15], since

- it can describe a wider class of stochastic processes, i.e. the class of intrinsic stochastic processes, which includes the class of second-order stationary stochastic processes,

- its estimation does not require the knowledge of the expected value of the associated stochastic process,
- it is appropriate to identify trend and periodicity exhibited by data,
- it can be used for prediction purposes.

Regarding this last aspect, geostatistical techniques provide different parametric and nonparametric prediction methods, among these the sample and ordinary kriging, the universal kriging and the indicator kriging. Further details can be found in the specialized literature [7, 12, 19]. Thus, the estimation of the unknown value x_t of the stochastic process $\{X_t, t \in T\}$, using the data observed in the past (extrapolation mode), or the data observed before and after the time point t (interpolation mode) can be easily supported by geostatistical tools.

In the following, the ordinary kriging method and the indicator kriging approach are briefly reviewed, since these geostatistical tools are used for analyzing the variable under study.

Let $\widehat{X}_t$ the linear predictor of the intrinsic stationary process $\{X_t, t \in T\}$:

$$\widehat{X}_t = \sum_{i=1}^{n} \lambda_i(t) X_{t_i}, \tag{2}$$

where $\lambda_i(t), i = 1, 2, \ldots, n$, are unknown real coefficients and X_{t_i} are random variables of the process X at the sampled time points t_i. The unknown weights $\lambda_i(t), i = 1, 2, \ldots, n$, of (2) are obtained by solving the following kriging system

$$\begin{bmatrix} \gamma_{11} & \cdots & \gamma_{1n} & -1 \\ \gamma_{21} & \cdots & \gamma_{2n} & -1 \\ \vdots & \ddots & \vdots & \vdots \\ \gamma_{n1} & \cdots & \gamma_{nn} & -1 \\ 1 & \ldots & 1 & 0 \end{bmatrix} \begin{bmatrix} \lambda_1 \\ \lambda_2 \\ \vdots \\ \lambda_n \\ \mu \end{bmatrix} = \begin{bmatrix} \gamma_{10} \\ \gamma_{20} \\ \vdots \\ \gamma_{n0} \\ 1 \end{bmatrix}, \tag{3}$$

where $\gamma_{ij} = 0.5\, Var(X_{t_i} - X_{t_j})$, $\gamma_{i0} = 0.5 Var(X_{t_i} - X_t)$, μ is the Lagrange multiplier. If γ is conditionally strictly negative definite, then the above system presents one and only one solution.

The ordinary kriging [22] requires only the knowledge of the variogram model and it is used when the expected value of the process is constant and unknown. Since the kriging system can be expressed in terms of the variogram, as in (3), the kriging predictor can be used even when the stochastic process under study satisfies the intrinsic hypothesis. Moreover, using a predictor based on a variogram, rather than on a covariance, avoids the estimation of the expected value, if this last is unknown.

The usefulness of geostatistical techniques in time series analysis can be appreciated through nonparametric estimation of the variable under study.

The kriging approach, based on the knowledge of variogram, leads naturally to nonparametric estimation [17]. Indicator kriging is a nonparametric approach to estimate the posterior cumulative distribution function (c.d.f.) of the variable under study at an unsampled point [16, 25, 26].

In this context, given the observed time series x_{t_i}, t_i, $i = 1, 2, ..., n$, the conditional probability $Prob\{X_t \leq x|\mathcal{H}_n\}$, with $\mathcal{H}_n = \{x_{t_i}, t_i, i = 1, 2, ..., n\}$, is interpreted as conditional expectation of an indicator random field $I(t; x)$ [27], that is

$$Prob\{X_t \leq x|\mathcal{H}_n\} = E[I(t; x|\mathcal{H}_n)]$$

where

$$I(t; x) = \begin{cases} 1, & if \ X_t \leq x \\ 0, & if \ X_t > x. \end{cases}$$

In the case study presented hereafter, ordinary kriging and indicator kriging are applied for interpolation and prediction purposes of an environmental variable. Note that a *GSLib* routine for kriging, named "KT3DP" [12], has been used in order to define appropriate temporal search neighborhoods in presence of periodicity, since environmental time series, such as the ones for air pollution data, usually are characterized by a periodic behavior. Hence, the use of periodic and nonperiodic variogram models have been proposed through two different approaches:

- the periodic component has been factored out using the moving average method [5] and nonperiodic variogram model has been fitted;
- the periodicity has been retained and described by a periodic variogram model.

3. PM_{10} time series

In the present case study, the analysis of daily concentrations of PM_{10} ($\mu g/m^3$), measured at one of the monitoring stations of Brindisi district during the period 2010-2013, has been conducted through geostatistical techniques.

These data have been collected by the Environmental Protection Agency of Apulian region (ARPA Puglia) which controls the air quality of urban, suburban, and industrialized areas of the region.

Note that PM_{10} monitoring stations are classified in the following three categories:

- traffic stations, located in areas with heavy traffic;
- industrial stations, located close to industrialized areas;
- background stations, located in peripheral areas.

The analyzed station, named "Torchiarolo" is located in the municipality of Torchiarolo (Brindisi district), as shown in Fig. 1. It is classified as industrial station, since it is strictly close to an industrial site, i.e. the thermoelectric power station "Enel-Federico II" in Cerano (Brindisi district).

http://dx.doi.org/10.5772/60115

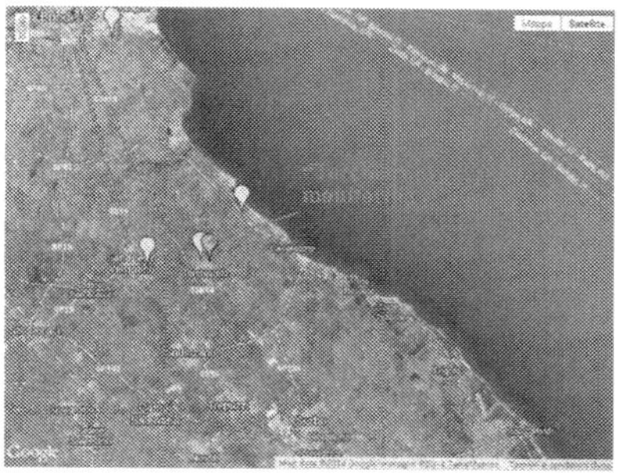

Figure 1. "Torchiarolo" monitoring station belonging to the Environmental Protection Agency of Apulian region (ARPA Puglia)

3.1. Exploratory Data Analysis

In order to assess the statistical properties of PM_{10} measured at the "Torchiarolo" station in the period 2010-2013, an exploratory data analysis has been performed. Some results are shown in Tab. 1.

Year	Min	Max	Mean	Standard Deviation	Number of exceedances
2010	8	114	35.10	20.09	67
2011	8	147	36.08	21.57	66
2012	10	108	32.83	17.01	49
2013	8	146	35.83	22.71	61
2010-2013	8	147	34.97	20.47	243

Table 1. Descriptive statistics of PM_{10} ($\mu g/m^3$), measured in the period 2010-2013 at the "Torchiarolo" monitoring station

According to the National Law concerning the human health protection, PM_{10} daily average concentrations cannot be greater than 50 $\mu g/m^3$ for more than 35 times per year. During the period under study, the PM_{10} daily values exceeded the threshold 243 times; in addition, the station has measured more than 35 exceedances per year.

The box plot in Fig. 2 shows that the observed time series is characterized by a seasonal component. During summertime, particle pollution shows lower levels compared to those recorded during wintertime; in particular, in summertime, PM_{10} doesn't exceed the limit value fixed by the National Law.
On the other hand, in wintertime changes in the lower layer of the troposphere determine PM_{10} stagnation. Hence, high levels of this pollutant are recorded.
In the following sections, the study of the temporal evolution of PM_{10} at the analyzed station has been conducted by performing

- structural analysis,
- estimation of some consecutive values assumed as missing,
- prediction of PM_{10} daily averages,
- estimation of the c.d.f. of PM_{10} daily concentrations at some unsampled time points.

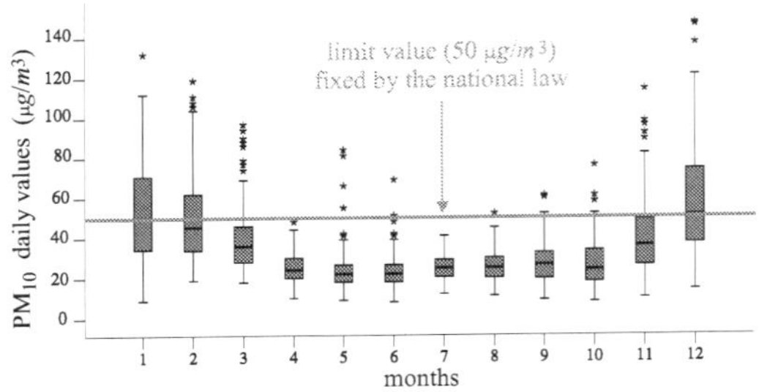

Figure 2. Box plot of PM_{10} daily concentrations, grouped by month, and limit value fixed by the National Decree (Decree-law 60/2002)

4. Structural Analysis

As previously pointed out, the variogram can describe a wider class of stochastic processes, that is the class of intrinsic stochastic processes and is usually preferred to the use of the covariance function.

In structural analysis, before modeling the temporal correlation described by the variogram, its estimation from data is required. The following classical estimator [9] is often used:

$$\hat{\gamma}(r_t) = \frac{1}{2\,|M(r_t)|} \sum_{M(r_t)} \left[X_{t+h_t} - X_t\right]^2 \tag{4}$$

where r_t is the temporal lag, $M(r_t) = \{t + h_t \in H \text{ and } t \in H, \text{ such that } \|r_t - h_t\| < \delta_t\}$, δ_t is the tolerance, H is the set of data at different time points (not necessarily equally-spaced) and $|M(r_t)|$ is the cardinality of this set.

In the present case study the variogram has been used as an exploratory tool to assess stationarity and periodicity. In particular, sample temporal variogram for PM_{10} daily observations, shown in Figure 3-a), reproduces the seasonal behavior of the variable under study, which presents an annual periodicity at 365 days. In equation (5) the analytic expression of the periodic variogram model, fitted to the sample variogram for the observed values, is proposed:

$$\gamma(h_t) = 265\,Exp(|h_t|; 10) + 130\,Cos(|h_t|; 365); \tag{5}$$

where $Exp(\cdot)$ and $Cos(\cdot)$ are the exponential and the cosine variogram models [29], respectively.

On the other hand, since the variable under study, is characterized by periodicity, this seasonal component could be factored out. Moving average and monthly averages techniques have been applied in order to obtain PM_{10} residuals. Note that the FORTRAN program "REMOVE" [11] has been used to apply moving average techniques.

http://dx.doi.org/10.5772/60115

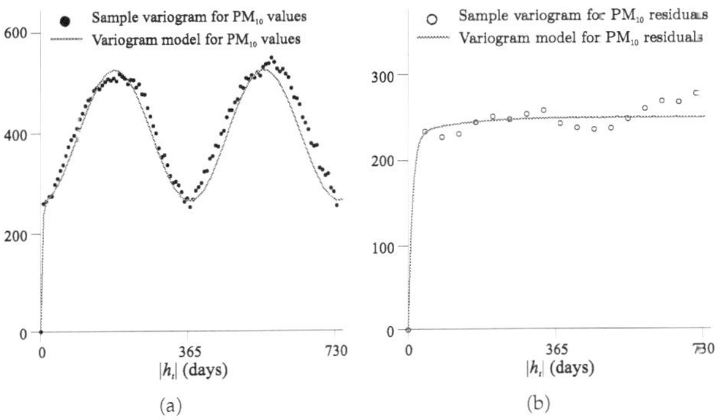

Figure 3. Sample temporal variograms and fitted models. (a) Variogram for PM_{10} daily concentrations (b) Variogram for PM_{10} residuals

The sample variogram of the residuals has been computed and modelled and the following nonperiodic variogram model has been chosen:

$$\tilde{\gamma}(h_t) = 348\,Exp(|h_t|;30) + 30\,Exp(|h_t|;365); \tag{6}$$

where $Exp(\cdot)$ is the exponential variogram model [8].

The sample temporal variogram for PM_{10} daily residuals and the corresponding nonperiodic fitted model (6) are illustrated in Fig. 3-b).

In both cases (original data and residuals), the behavior of the variogram functions near the origin is assumed to be linear with no nugget effect.

The goodness of variogram models (5) and (6) has been evaluated through cross-validation, which allows the estimation for PM_{10} daily concentrations and PM_{10} residuals, respectively, at all data points. Figure 4 shows the scatter plots of PM_{10} observed values (a) and PM_{10} residuals (b) towards the corresponding estimated values. The high values of the linear correlation coefficients (0.783 and 0.780, respectively) confirm the goodness of the above fitted models.

It is important to point out that the variogram model (5) has been validated using a modified version of the *GSLib* program "KT3D" [13], named "KT3DP". This program has been developed in order to properly define the neighborhood, i.e. the subset of available data used in the kriging system.

By taking into account the main features of the analyzed pollutant and its temporal behavior (periodicity at 365 days), the kriging routine has been modified and the value at time t is estimated by considering data observed

- at the two adjacent time points $(t-1)$ and $(t+1)$,
- at the same day of the year before and/or later, $(t-d)$ and $(t+d)$, with $d = 365$ and some days before and/or later, $(t-d\pm k)$ and $(t+d\pm k)$, with $k = 1,2,3$,

- at the same day of two years before and/or later, $(t - 2d)$ and $(t + 2d)$, with $d = 365$ and some days before and/or later, $(t - 2d \pm k)$ and $(t + 2d \pm k)$, with $k = 1, 2, 3$,

up to a maximum number of eight values.

The variogram model (6), which describes the temporal correlation for PM_{10} residuals, has been validated using the *GSLib* program "KT3D".

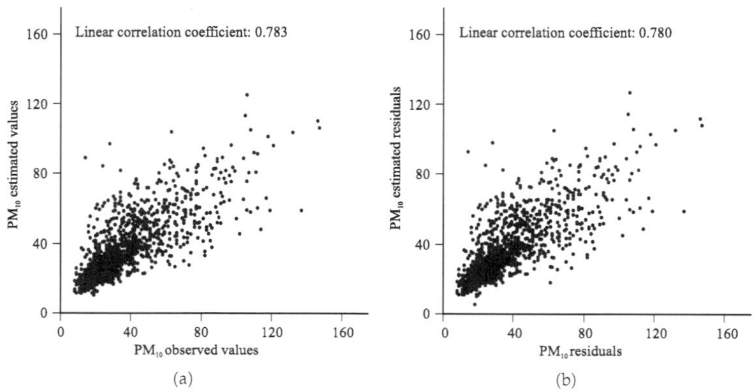

Figure 4. Scatter plots between observed and estimated values. (a) Diagram of PM_{10} daily concentrations towards the estimated ones (b) Diagram of PM_{10} residuals towards the estimated ones

5. Estimation of missing values

In this section the reconstruction of PM_{10}, by using the kriging technique, has been discussed [20, 21, 30, 31].

The reconstruction of temporal data is required if a time series is incomplete. This problem could be due to a malfunction of the monitoring station or the presence of invalid data.

With this aim, six consecutive PM_{10} values from the 12th to the 17th of June 2011, have been considered as missing, both for the observed time series with a 365-day periodic behavior and the deseasonalized values.

Kriging daily estimations for these missing values have been obtained using, alternatively

1. the periodic variogram model (5), which describes the temporal correlation for PM_{10} daily concentrations,
2. the nonperiodic variogram model (6), which describes the temporal correlation for PM_{10} daily residuals.

Since the time series of the observed values is characterized by a periodic behavior, *GSLib* routine "KT3DP", properly modified in order to define an appropriate neighborhood, has been used with the aim to estimate PM_{10} daily measurements.

On the other hand, for the deseasonalized time series, PM_{10} residuals have been estimated by the original version of "KT3D". Finally the periodic component, previously estimated by the moving average and monthly averages techniques, has been added to the estimated residuals, in order to obtain estimates of PM_{10} daily concentrations

Time series of estimated missing values, obtained with the periodic variogram model (5) and the nonperiodic variogram model (6), are shown in Fig. 5, together with the time series of PM_{10} values, observed on June 2011.

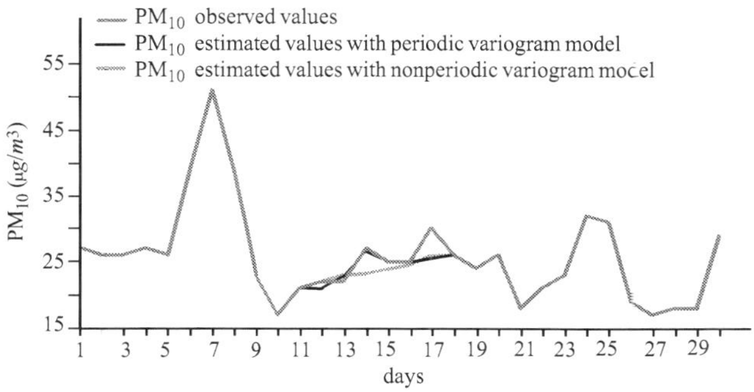

Figure 5. Time plot of PM_{10} estimated missing values and PM_{10} daily concentrations ($\mu g/m^3$), from the 12th to the 17th of June 2011

In order to test the validity of the estimation procedure, the linear correlation coefficients have been computed. In particular, the linear correlation coefficient between the PM_{10} observed values and the corresponding estimates, obtained with the periodic variogram model, is equal to 0.805. On the other hand, the linear correlation coefficient between the residuals and the corresponding estimates, obtained with the nonperiodic variogram model, is equal to 0.831. These results confirm the goodness of the kriging technique as estimator of missing values.

In Table 2 some results of estimation procedure are shown. Note that the mean value of the kriging standard error is lower if the periodic variogram model is used, compared with the kriging results based on the nonperiodic variogram model.

Therefore, the flexibility of kriging to reconstruct the time series has been demonstrated even when the periodic component is not factored out and the temporal correlation is described by a periodic variogram model.

6. Prediction of PM_{10} values

In this section, predictions for the variable under study in time points after the last available data are discussed [23, 24, 28].

The periodic variogram model (5) of PM_{10} concentrations and the nonperiodic variogram model (6) of PM_{10} residuals, have been used in order to predict six time points after the last available data, i.e. the 31st of December 2013. In particular, kriging predictions have

June 2011	PM_{10} value	PM_{10} Est. value[a]	Est. Error[a]	Est. value[b]	Est. Error[b]
12th	22	20.970	-1.030	22.029	0.029
13th	22	22.821	0.821	22.940	0.940
14th	27	26.437	-0.563	23.178	-3.822
15th	25	24.867	-0.133	23.878	-1.122
16th	25	24.926	-0.074	24.473	-0.527
17th	30	25.466	-4.534	25.825	-4.175
Mean values	21.167	24.248	-0.919	23.720	-1.446

[a] Results obtained by using the periodic variogram model (5) [b] Results obtained by using the nonperiodic variogram model (6)

Table 2. Kriging estimations of a sequence of 6 missing values, from the 12th to the 17th of June 2011 and corresponding errors for periodic and nonperiodic variogram models

been computed for the period ranging from the 1st to the 6th of January 2014, by using, alternatively

1. the available data, the variogram model (5) and the modified *GSLib* routine "KT3DP" which builds the searching neighborhood taking into account the periodicity exhibited by the data,
2. the deseasonalized PM_{10} observations, the variogram model (6) and the original *GSLib* routine "KT3D" which produces PM_{10} predicted residuals at which the diurnal component of the day before has been added to obtain predictions of PM_{10} daily concentrations.

In Fig. 6, the time series of PM_{10} daily concentrations measured from the 9th of December

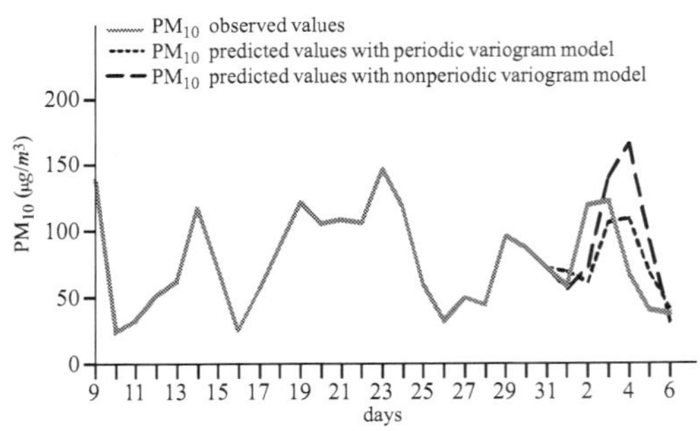

Figure 6. Time plot of PM_{10} predicted values and PM_{10} daily concentrations ($\mu g/m^3$), from the 1st to the 6th of January 2014

2013 to the 6th of January 2014 is shown together with the predicted PM_{10} values for the period ranging from the 1st to the 6th of January 2014. Note that the kriging procedure using the nonperiodic variogram model (6) related to PM_{10} residuals has produced overestimates of the pollution levels.

Moreover, in Table 3 some results of the performance of the prediction procedure are presented. The mean value of the kriging standard error is lower if the periodic variogram model is used, compared with the case of kriging based on the nonperiodic variogram model.

January 2014	PM_{10} Obs. value	PM_{10} Est. value[a]	Est. Error[a]	Est. value[b]	Est. Error[b]
1st	58	69.415	11.415	55.774	-2.226
2nd	119	61.031	-57.969	72.985	-46.015
3rd	122	106.051	-15.949	137.731	15.731
4th	67	108.397	41.397	164.909	97.909
5th	40	67.545	27.545	91.090	51.090
6th	37	41.498	4.498	30.738	-6.262
Mean values	73.833	75.656	1.823	92.205	18.371

[a] Results obtained by using the periodic variogram model (5) [b] Results obtained by using the nonperiodic variogram model (6)

Table 3. Kriging predictions of a sequence of six days, from the 1st to the 6th of January 2014 and corresponding errors for periodic and nonperiodic variogram models

It is important to highlight that in the period 1-5 January 2014 predicted values greater than 50 $\mu g/m^3$ (i.e. the limit value fixed by the National Law) have been obtained.

Note that in the period 1-4 January 2014, PM_{10} values greater than this threshold have been measured. On the other hand, at day 5th, the kriging procedure produces overestimate of the variable under study.

7. Estimation of the c.d.f.

For a given time series of PM_{10}, it might be useful to estimate the probability that the variable under study exceeds a fixed limit, so that appropriate and prompt solutions might be adopted if necessary.

In this section, estimation of c.d.f. of PM_{10} daily concentrations ($\mu g/m^3$) has been conducted.

In particular, the c.d.f. of PM_{10} at unsampled time points has been estimated by indicator kriging [17].

Six threshold values for PM_{10} (22, 35, 50, 78, 98, and 108 $\mu g/m^3$) have been properly chosen, and six indicator variables according to the fixed thresholds have been defined as follows

- $I_1(t;22) = \begin{cases} 1, & if\ PM_{10} \le 22 \\ 0, & if\ otherwise \end{cases}$
- $I_2(t;35) = \begin{cases} 1, & if\ PM_{10} \le 35 \\ 0, & if\ otherwise \end{cases}$
- $I_3(t;50) = \begin{cases} 1, & if\ PM_{10} \le 50 \\ 0, & if\ otherwise \end{cases}$
- $I_4(t;78) = \begin{cases} 1, & if\ PM_{10} \le 78 \\ 0, & if\ otherwise \end{cases}$
- $I_5(t;98) = \begin{cases} 1, & if\ PM_{10} \le 98 \\ 0, & if\ otherwise \end{cases}$
- $I_6(t;108) = \begin{cases} 1, & if\ PM_{10} \le 108 \\ 0, & if\ otherwise \end{cases}$

with $t \in T$. Note that indicator data are equal to 1 if the values of the variable under study are not greater than the considered threshold and they are equal to 0 otherwise. For each threshold, the temporal indicator variogram has been computed and modelled (Figs. 7, 8).

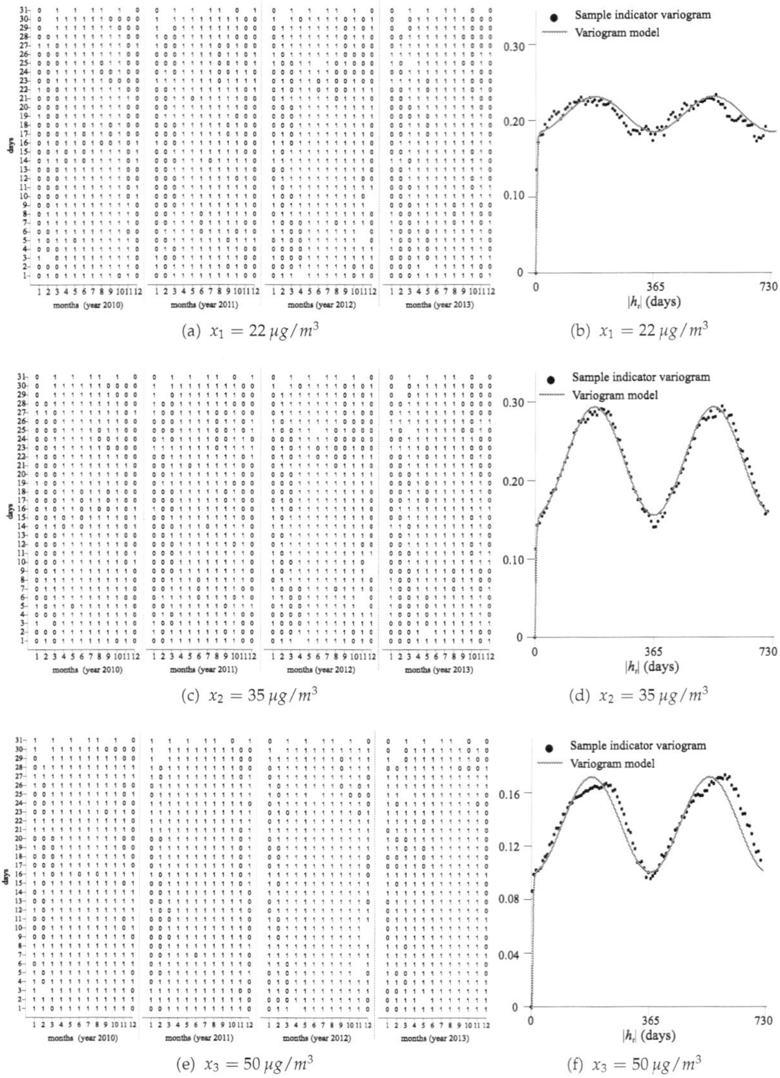

(a) $x_1 = 22\,\mu g/m^3$ (b) $x_1 = 22\,\mu g/m^3$

(c) $x_2 = 35\,\mu g/m^3$ (d) $x_2 = 35\,\mu g/m^3$

(e) $x_3 = 50\,\mu g/m^3$ (f) $x_3 = 50\,\mu g/m^3$

Figure 7. Indicator maps of PM_{10} daily concentrations and their sample indicator variograms with the fitted models, for three threshold values. (a) Indicator map and (b) sample variogram indicator for the threshold $x_1 = 22\,\mu g/m^3$. (c) Indicator map and (d) sample variogram indicator for the threshold $x_2 = 35\,\mu g/m^3$. (e) Indicator map and (f) sample variogram indicator for the threshold $x_3 = 50\,\mu g/m^3$

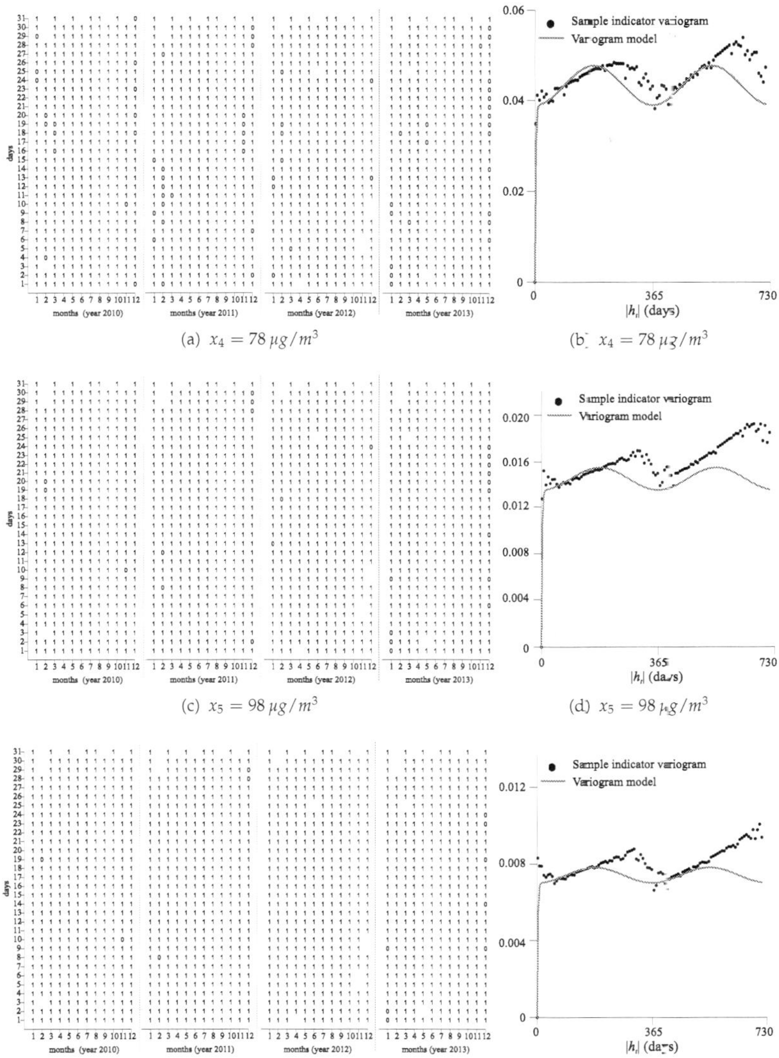

(a) $x_4 = 78\,\mu g/m^3$ (b) $x_4 = 78\,\mu g/m^3$

(c) $x_5 = 98\,\mu g/m^3$ (d) $x_5 = 98\,\mu g/m^3$

(e) $x_6 = 108\,\mu g/m^3$ (f) $x_6 = 108\,\mu g/m^3$

Figure 8. Indicator maps of PM_{10} daily concentrations and their sample indicator variograms with the fitted models, for three threshold values. (a) Indicator map and b) variogram for the threshold $x_4 = 78\,\mu g/m^3$. (c) Indicator map and (d) variogram for the threshold $x_5 = 98\,\mu g/m^3$. (e) Indicator map and (f) variogram for the threshold $x_6 = 108\,\mu g/m^3$

In particular the following models have been fitted

- $\gamma_{I_1}(h_t;22) = 0.185\, Exp(|h_t|;10) + 0.023\, Cos(|h_t|;365)$,
- $\gamma_{I_2}(h_t;35) = 0.15\, Exp(|h_t|;10) + 0.07\, Cos(|h_t|;365)$,
- $\gamma_{I_3}(h_t;50) = 0.102\, Exp(|h_t|;10) + 0.036\, Cos(|h_t|;365)$,
- $\gamma_{I_4}(h_t;78) = 0.039\, Exp(|h_t|;10) + 0.0004\, Cos(|h_t|;365)$,
- $\gamma_{I_5}(h_t;98) = 0.013\, Exp(|h_t|;10) + 0.001\, Cos(|h_t|;365)$,
- $\gamma_{I_6}(h_t;108) = 0.007\, Exp(|h_t|;10) + 0.0004\, Cos(|h_t|;365)$.

Thus, the c.d.f.s corresponding to six different unsampled time points, i.e. the days 1-6 of January 2014, have been estimated by using the "KT3DP" routine.

For each day of interest, the c.d.f. has been estimated by solving as many kriging systems as the number of threshold values considered. For each threshold, the corresponding indicator variogram model has been used for the kriging procedure.

Figure 9 shows the c.d.f.s estimated at days 1-6 of January 2014. It is clear that the probability of not exceeding a fixed threshold increases gradually from the 1st to the 6th of January 2014. For example, the estimated probability that PM_{10} concentrations, on the 1st of January 2014, do not exceed 22 $\mu g/m^3$ is lower than the estimated probability on the 3rd or the 6th of the considered month.

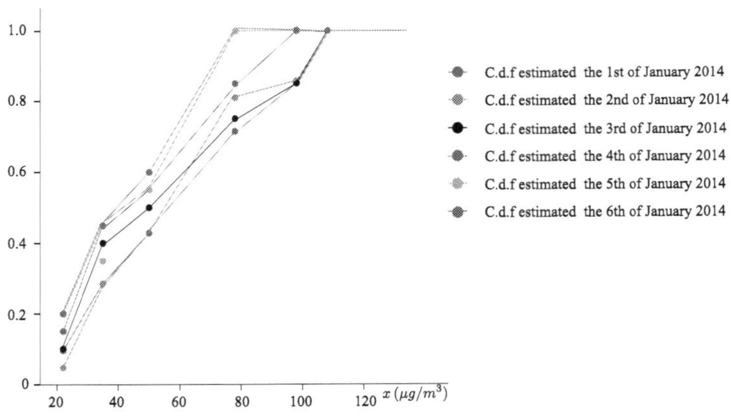

Figure 9. C.d.f.s estimated for PM_{10} daily concentrations ($\mu g/m^3$) at days 1-6 of January 2014

http://dx.doi.org/10.5772/60115

Thresholds	January 2014					
	1st	2nd	3rd	4th	5th	6th
22	0.095	0.047	0.100	0.150	0.200	0.200
35	0.285	0.286	0.400	0.450	0.350	0.450
50	0.429	0.429	0.500	0.550	0.550	0.600
78	0.714	0.810	0.750	0.850	0.900	0.900
98	0.857	0.857	0.850	0.900	0.950	1
108	1	1	1	1	1	1

Table 4. Estimated values for c.d.f. at days 1-6 of January 2014, for fixed thresholds

Moreover, note that it is almost sure that PM_{10} concentrations do not exceed the cutoff 78 $\mu g/m^3$ at days the 5th and 6th (Table 4).

The probability that the variable under study doesn't exceed the threshold fixed by the National Law (50 $\mu g/m^3$) is high (equal to 60%) for the last day of interest. In fact, the 6th of January is a non-working day (Epiphany) characterized by low traffic and low industrial emissions.

The local government could use these results in order to carry out environmental policies for the control of high levels of PM_{10}, since it is well known that high concentrations of this pollutant are dangerous for the human health.

Indeed, the estimation of the c.d.f. is a very powerful tool since any action of environmental protection might be adopted in advance by taking into account the actual likelihood of dangerous PM_{10} exceeding. For example, decisions about traffic limitation in high traffic urban area might be supported by the knowledge of the probability that a hazardous pollutant exceeds the level of attention.

8. Conclusions

In this paper, PM_{10} time series analysis, by using geostatistical techniques, has been discussed and the importance of appropriate tools of Geostatistics to study the temporal evolution of this environmental phenomena has been highlighted.

The seasonal behavior of PM_{10} levels has been evaluated through the variogram, that is the basic tool of Geostatistics. Moreover, estimation and prediction problems in the analysis of the time series of this pollutant, characterized by a periodic behavior, have been solved through kriging geostatistical techniques.

The computational aspects have been performed through the use of "KT3D" for the observed values and "KT3DP" for the residuals obtained after removing the periodic component.

Finally, the indicator approach and its capability for assessing the probability that PM_{10} exceeds the specific threshold values have been demonstrated.

The results obtained in this paper by applying geostatistical techniques to analyze PM_{10} time series could be useful to support national policies for environmental and health protection. Governments' activity must be oriented to control that the concentrations of the analyzed pollutant don't exceed specific thresholds according to national or international directives, since it has been demonstrated that particulate matter is dangerous for the human health.

Author details

Claudia Cappello, Sabrina Maggio*, Daniela Pellegrino and Donato Posa

*Address all correspondence to: sabrina.maggio@unisalento.it

University of Salento, Dept. of Management, Economics, Mathematics and Statistics, Italy

References

[1] Arden Pope III C., Dockery D. W. Health effects of fine particulate air pollution: lines that connect. Journal of Air and Waste Management Association 2006; 56 709–742.

[2] Bisgaard S., Khachatryan D. Asymptotic confidence intervals for variograms of stationary time series. Quality and Reliability Engineering International 2010; 26(No. 3) 259–265.

[3] Bloomfield P. Fourier analysis of time series: An Introduction. USA:Wiley & Sons Inc.; 2000.

[4] Box GEP., Jenkins GM. Time series analysis: Forecasting and Control. San Francisco:Holden Day; 1976.

[5] Brockwell P.J., Davis R.A. Time Series: Theory and Methods. New York: Springer; 1987.

[6] Chilés J.P., Delfiner P. Geostatistics: Modeling Spatial Uncertainty. New York: Wiley; 1999.

[7] Cressie, N. (1988). A graphical procedure for determining nonstationarity in time series, Journal of the American Statistical Association 1988; 83(No. 44) 1108–1116.

[8] Cressie, N. (1993). Statistics for Spatial Data, Wiley Series in Probability and Mathematical Statistics, Wiley, New York.

[9] Cressie N., Hawkins D.M. Robust estimation of the variogram. Mathematical Geology 1980;12(2) 115–125.

[10] Cressie N., Grondona M.O. A comparison of variogram estimation with covariogram estimation. In: Mardia K.V. (ed.) The Art of Statistical Science. Chichester: Wiley; 1992.

[11] De Cesare L., Myers D.E., Posa D. FORTRAN 77 programs for space-time modeling. Computers & Geosciences 2002;28(2) 205–212.

[12] De Iaco S., Palma M., Posa, D. Geostatistics and the role of variogram in time series analysis: a critical review. In: Montrone S. & Perchinunno P. (ed.) Statistical Methods for Spatial Planning and Monitoring. Italia: Springer-Verlag; 2013. p47–75.

[13] Deutsch C.V., Journel, A.G. GSLib: Geostatistical Software Library and User's Guide. New York: Oxford University Press; 1998.

[14] Gevers M. On the use of variograms for the prediction of time series. Systems & Control Letters 1985;6(1) 15–21.

[15] Haslett J. On the sample variogram and sample autocovariance for non-stationary time series. Statistician 1997;46(4) 475–485.

[16] Janis M.J., Robeson S.M. Determining the spatial representativeness of air-temperature records using variogram-nugget time series. Physical Geography 2004;25(6) 513–530.

[17] Journel A.G. Nonparametric estimation of spatial distributions. Mathematical Geology 1983;15(3) 445–468.

[18] Journel A.G., Huijbregts C.J. Mining Geostatistics. London: Academic; 1981.

[19] Khachatryan D., Bisgaard S. Some results on the variogram in time series analysis. Quality and Reliability Engineering International 2009;25 947–960.

[20] Little R.J.A., Rubin D.B. Statistical Analysis with Missing Data. New York: Wiley; 2002.

[21] Luceno A. Estimation of missing values in possibly partially nonstationary vector time series. Biometrika 1997;84(2) 495–499.

[22] Matheron G. Principles of Geostatistics. Economic Geology 1963;58(8) 1246–1266.

[23] Matheron G. Les variables régionalisées et leur estimation. Paris: Masson; 1965.

[24] Matheron G. The intrinsic random functions and their applications. Advances in Applied Probability 1973;5(3) 439–468.

[25] Myers D.E. Interpolation with positive definite functions. Science de la Terre 1988;28 251–265.

[26] Myers D.E. Kriging, cokriging, radial basis functions and the role of positive definiteness. Computers & Mathematics with Applications 1992;24(12) 139–148.

[27] Posa D. The indicator formalism in spatial data analysis. Journal of Applied Statistics 1992;19(1) 83–101.

[28] Posa D., Rossi M. Applying stationary and non-stationary kriging. Metron XLVII 1989;(1-4) 295–312.

[29] Posa D., De Iaco S. Geostatistica: teoria e applicazioni. Torino: Giappichelli editore; 2009.

[30] Uysal M. Reconstruction of time series data with missing values. Journal of Applied Sciences 2007;7(6) 922–925.

[31] Weerasinghe S. A missing values imputation method for time series data: an efficient method to investigate the health effects of sulphur dioxide levels. Environmetrics 2010;21(2) 162–172.

Chapter 6

The Integrated Mini GC-PID System for Monitoring Air Pollution

J.H. Sun, F.Y. Guan. X.F. Zhu, Z.W. Ning and T.J. Ma

Additional information is available at the end of the chapter

http://dx.doi.org/10.5772/59725

1. Introduction

Volatile organic compounds (VOCs) are the premise of pollutants for forming PM 2.5 or heavy haze and choking smog, so monitoring VOCs is a very important measure for preventing environmental pollution. In addition, rapid detection of VOCs is also very necessary in some on-site detection, including indoor air quality, industrial emissions control, air pollution emergency in urban area, rapid detection of explosives in public place, and breath analysis. Therefore, fast, on-site, high-precision, and portable systems are the pressing need for providing real-time monitoring for quantification and identification analysis of environmental samples.

Conventional gas chromatography (GC), which is a powerful and commonly used analytical technique, has become the premier technique for separation and analysis of harmful compounds due to its high resolving power. However, in some cases, gas samples need to be collected in the field and analyzed in the laboratory using conventional GC, which makes this process inefficient and inconvenient in field use.

As MEMS technology matures, micro gas chromatographic systems (μGC) [1-3] have a prospect future with some advantages of small volume, rapid analysis, and less power consumption compared with conventional GC systems, which are very urgent to be used for analyzing the complicated gas mixtures. Therefore, these attributes make these systems based on microelectromechanical systems (MEMS) technology attractive for a number of applications involving on-site monitoring of environmental samples, and this research direction is under intensive research.

In this chapter, we reported the design, fabrication, and characterization of micro preconcentrator, micro GC column, and mini photoionization detector (mini PID). Then a mini

GC system based on mini PID was integrated with the micro GC column and the micro pre-concentrator.

2. Schematic representation of the mini GC system

A mini (portable) GC system usually includes an injection unit, a micro pre-concentrator, a micro GC column, and a mini PID. Figure 1 shows schematic of the mini GC system for environmental monitoring and analysis.

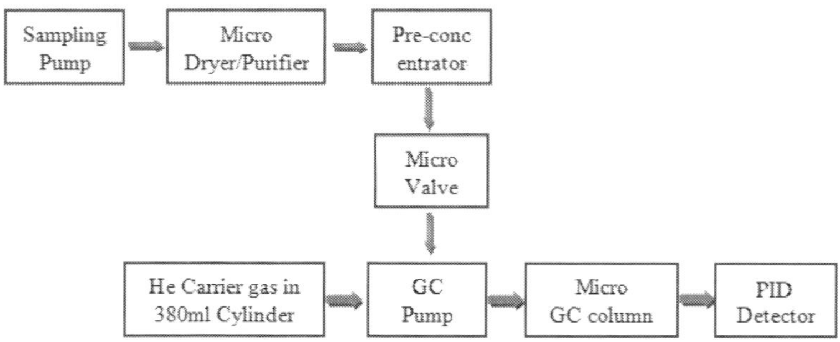

Figure 1. A schematic representation of the mini GC system architecture

The injection unit supplies sample to the pre-concentrator or the GC separation column. The pre-concentrator is a sampling and concentrating device that can significantly improve detection limit of mini GC system with 1~3 orders. The GC separation column separates the complicated mixtures on the basis of their affinity with the stationary phase distributed all along the column length. The detector acquires the signals provided by the separated components at the GC column outlet.

The mini GC system is characterized by a series of innovations. 1) The micro dryer and purifier can remove vapor and particulates in environmental sample. 2) The integrated pre-concentrator can make the system easy to deploy trace gas and suitable for in-field use. 3) A micro GC column instead of a conventional column was used to separate mixtures, which can greatly reduce volume of the system.

3. Micro dryer and purifier

Environmental samples do not only have a very low concentration but also contain vapor and various particles, which will make GC column and detector inactive or even fail. Therefore, the proposed mini GC system was integrated with a micro dryer and purifier which can clear up the vapor and micro pollutants in the samples.

In order to rapidly and effectively eliminate the vapor from the environmental samples, two kinds of micro-pillars were fabricated in micro flow channel of the dryer. The shape of the former pillar is shown in Figure 2(b) (refer to the I pillar), which can break and rapidly atomize the water droplets in the samples. The second pillar (refer to the II pillar in Figure 2(b)) has a relatively large surface area, which can completely come into contact with the micro droplets and then quickly vaporize them.

Fabrication of the chip used silicon wafer and Pyrex 7740 glass wafer as the substrates. The patterns of the dryer and purifier were formed by lithography and fabricated through a series of lithography, etching, and bonding process, where the channels and pillars were fabricated using a deep reactive-ion etching technology. For details on the fabrication process of the micro dryer and purifier, refer to the work in [4]. The length of the chip is 10 mm and the depth and width of the dryer channel were 400 μm and 500 μm, respectively. In order to clear up the vapor, molecular sieve 5A (60–80 mesh) was used as adsorption material and filled in the channel of the micro purifier. In order for the dryer to have a fast self-heating capacity, a micro heater was integrated on the backside of the dryer. The heater with resistance of 8 Ω was fabricated as a 20 nm/250 nm Cr/Pt stack deposited by a magnetron sputtering technology and patterned by a lift-off technology, which can increase the temperature at a speed of 10℃ per second and the highest temperature can be raised up to 200℃in less than 30 s Figure 2 shows the proposed micro dryer and purifier.

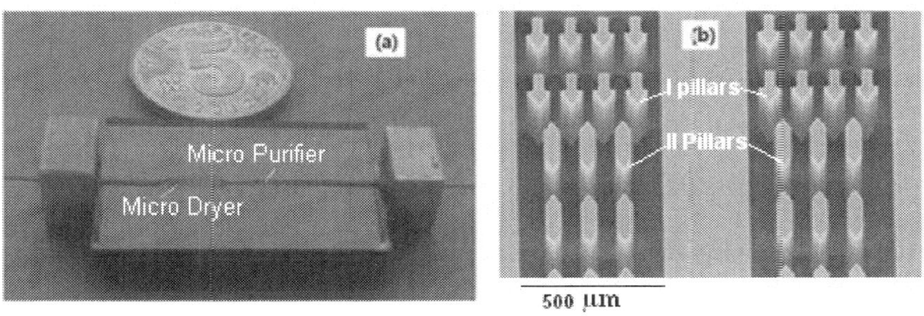

Figure 2. The fabricated micro dryer and purifier (a) and micro-pillars in channel (b)

In order to evaluate the performance, one end of the device is connected to a sampling pump, and the other end is connected to a transparent pipe filled with anhydrous $CuSO_4$. Firstly, an environmental sample containing a certain amount of water vapor was passed through the device which is in working state. After being processed by the device, the anhydrous $CuSO_4$ in the pipe did not change in color, which verifies that vapor was completely absorbed by the device.

4. Micro pre-concentrator

After removing vapor and particulates in environmental samples by the micro dryer and purifier, the sample can be introduced into a pre-concentrator for further processing.

The principle of the gas concentrator was detailed as follows: Analytes were transported into the pre-concentrator through a sampling pump, and the analytes were absorbed by the adsorption material. After the adsorption capacity was saturated, the pre-concentrator was heated to release the absorbed components. Finally, the released components with high concentration were transported into the detector.

The pre-concentrator can concentrate trace gas, that is, makes its concentration reach or exceed detection limit of the detector and makes the detector yield a good response for trace environmental sample. Therefore, pre-concentration of sample prior to analysis as it enables detection of trace gas and also improves detection sensitivity of mini-instruments, and sample pretreatments occupied in an important position for on-site rapid detection.

The conventional pre-concentrator configurations consist of a trap metal tube that is cooled by a flow of cold gas during sample collection and successively heated in order to release the sample through a rapid thermal desorption mechanism. However, the typical pre-concentrator needed a high power due to its larger thermal mass. Moreover, the conventional pre-concentrator is very difficult to be integrated into a portable GC system, resulting in an inconvenient in-field application.

Micro pre-concentrator with reduced thermal mass can raise temperature much faster at lower power compared with conventional desorption tubes, realizing a much higher concentration factor. Micro pre-concentrators with a small size and high concentration efficiency were easily integrated into micro/portable chromatographic systems, which make these systems very suitable for on-site or online analysis. Therefore, micro pre-concentrator received an unprecedented development with the development of MEMS technology.

Many efforts to improve concentration efficiency have been made in many works [5-13], and these efforts included the use of promising materials for high adsorption capacity, fabrication of high aspect ratio of the channels, improvement of sample capacity by fabricating multi-channels, and fabrication of micro channels with embedded micro-pillars which are able to fill more sorbent material and increase high aspect ratio of the channels. Although these efforts have yielded important progress, the released components during thermal desorption process should be controlled and the peak broadening also should be compressed.

4.1. Consideration of the micro pre-concentrator

As concentration of components in environmental gas is very low (ranged from several ppb to hundreds ppb) and the available analytical instruments are not sufficiently sensitive, the development of micro pre-concentrators for effectively concentrating components is very important and necessary.

According to theories of the pre-concentrator, there are several methods for improving concentration factor: (1) optimization of device structure for increasing the adsorption capacity; (2) the use of efficient adsorption material for improving specificity and adsorption capacity; (3) integration of cooler and heater on the pre-concentrator for improving adsorption capacity and desorption efficiency, respectively; and (4) integration of a micro valve behind the pre-concentrator for closing the released components during the desorption stage.

Therefore, in order to improve the concentration factor, in this work, we optimized the above considerations including measures as follows: the use of nanomaterials as adsorbent materials, integration of a heater and cooler on the pre-concentrator, and integration of a valve behind the pre-concentrator.

4.2. The adsorption material

The adsorption material is a key factor for affecting performance of a pre-concentrator device. Firstly, the material characteristics determine the types of concentrated gas. Secondly, material characteristics also determine the adsorption capacity of samples. Generally speaking, in order to improve the concentration factor, the first consideration is the selectivity and specificity of adsorption material, and then the adsorption capacity of the material is the next consideration.

With the development of technology, there are more adsorption materials that are available than before. In addition to the traditional adsorption material, such as Tenax-TA, Carbopack X, and activated carbon, recently some nanomaterials were developed and widely used as adsorption material, such as carbon nanotubes, molecularly imprinted polymers, nanoparticles, and so on.

Single-walled carbon nanotubes (SWCNTs) have a lot of advantages, such as large aspect ratio, high effectiveness of surface area, chemical and thermal stability, a large affinity to nonpolar compounds, and higher adsorption capacity than Carbopack X. Moreover, SWCNTs cannot be dissolved in water and organic solvents. Therefore, SWCNTs were the preferred material used as adsorbent material.

4.3. Fabrication of micro pre-concentrator

In this work, in order to improve the concentration factor of the pre-concentrator, a micro pre-concentrator with four parallel channels filled with SWCNTs as adsorbent materials was developed for concentrating VOCs in environmental samples. The details of the fabrication process of the pre-concentrator are the following (refer to Figure 3). Firstly, configuration of the micro pre-concentrator was drawn by AutoCAD software. Subsequently, rectangular micro channels were etched on silicon and glass wafer by laser dicing saw according to the configuration, and the length, depth, and width of the channels were 20 mm, 400 μm, and 1000 μm, respectively. Then these channels on silicon were aligned and bonded to these channels on glass wafer.

In order to increase adsorption capacity as much as possible in adsorption process, a cooler was integrated upside of the pre-concentrator, and the temperature of the pre-concentrator

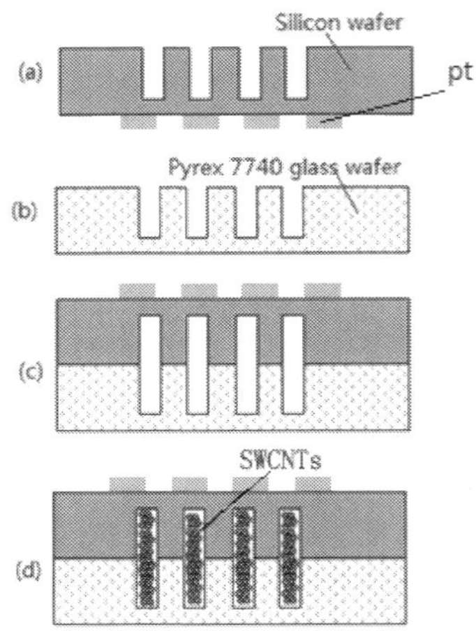

Figure 3. The details of the fabrication process

can be cooled down to 5 °C in 100 s. In the desorption stage, the faster the increase of the temperature, the higher concentration factor of the pre-concentrator. Therefore, a micro heater was integrated on the backside of the pre-concentrator, which can quickly release the analytes. The heater with resistance of 8 Ω was fabricated as a 20 nm/250 nm Cr/Pt stack deposited by a magnetron sputtering technology and patterned by a lift-off technology. The heater can rapidly heat the pre-concentrator at a speed of 10℃ per second, and the highest temperature can be raised up to 200℃ in less than 30 s, which the instantaneous concentration factor can be significantly improved. Furthermore, to compress peak broadening and further improve the concentration factor, a micro valve (purchased from Shenzhen Keyto Fluid Control Co., Ltd) was integrated behind the pre-concentrator (refer to Figure 4), which can close the released components during thermal desorption process, thus increasing the concentration factor as maximum as possible. In addition, four micro filters integrated in the end of the channels can prevent SWCNTs out of the pre-concentrator from the outlet. A photograph of the fabricated micro pre-concentrator is shown in Figure 5.

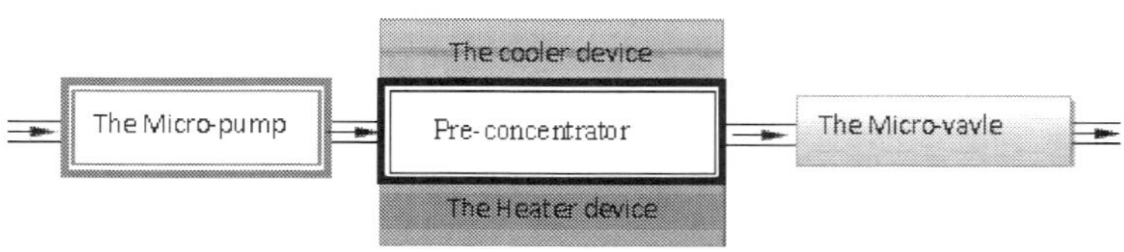

Figure 4. The diagram of the pre-concentrator

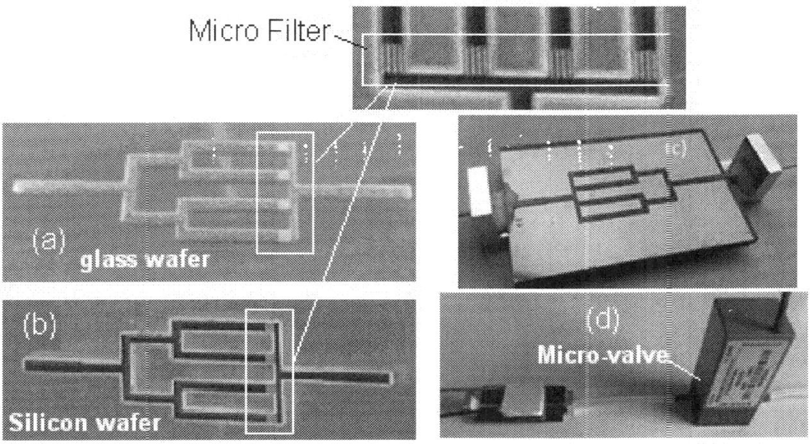

Figure 5. The channels on the glass wafer (a) and silicon wafer (b), the fabricated micro pre-concentrator (c)(d)

4.4. Filling of the adsorbent material

The details of filling process were reported as follows: (1) SWCNTs were added into TNWDIS solvent for forming suspending agent; (2) one end of the column was connected to a capillary which was submerged into a SWCNT solution, and the other end of the pre-concentrator was connected to a micro-pump. (3) The SWCNT solution was transported into the channels after the pump was turned on, and then nitrogen gas was continuously delivered through the channel to completely evaporate the TNWDIS solvent. (4) Then the pre-concentrator was put into a temperature-programmed oven under a nitrogen flow inside, and the temperature was increased gradually by 5 °C/min until 200 °C and kept for at least 4 h.

4.5. Characterization of the micro pre-concentrator

To improve the concentration factor, a cooler and a heater were, respectively, integrated on the upside and backside of the micro pre-concentrator. The cooler and heater cover the whole pre-concentrator chip region (i.e., the channel region) which has a temperature distribution over the chip surface as uniform as possible. The low temperature can increase adsorption capacity as much as possible in adsorption state, and the high temperature can rapidly and completely release the analytes from the adsorbent materials.

To evaluate the temperature characteristics of the micro pre-concentrator, a measurement cycle of 200 s was considered, and the result is reported in Figure 6. As can be disclosed, temperature of the pre-concentration chip can be dropped down to 5 °C in 100 s in the state of cooling and then can be raised up to 180 °C after 150 s and successively settles to a steady value of 200 °C in the heating state.

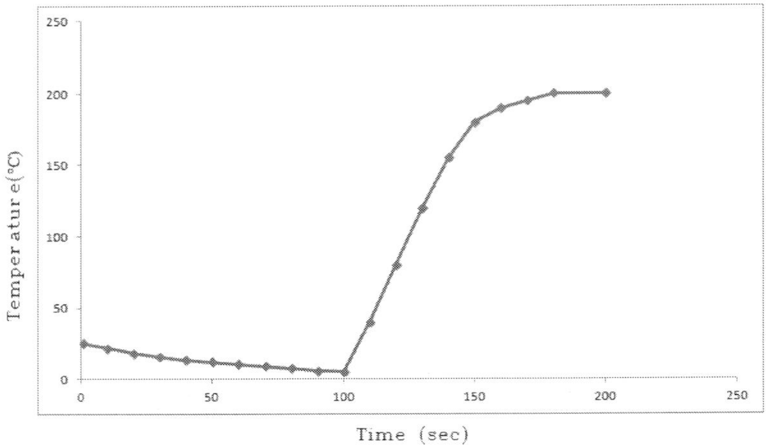

Figure 6. Temperature characteristics of the micro pre-concentrator

In order to evaluate the concentration factor of the fabricated micro pre-concentrator, tests of adsorption and desorption of analytes were carried out (refer to Figure 7). Firstly, the analytes were introduced into the micro pre-concentrator using a sampling pump, and the analytes were absorbed by the SWCNTs at the temperature of 5 ºC (the cooler is at cooling state) until the adsorption saturation. Then, the micro pre-concentrator was heated, but the desorbed analytes from SWCNTs were closed in the channels by the control of the valve. When the analytes were completely desorbed, the analytes were consequently carried to a laboratory-made portable photoionization detector (PID) [14] by carrier gas after the valve was opened.

In order to evaluate capabilities of the micro pre-concentrator, the sample composed of benzene with concentration of 500 ppb was used to be concentrated. The setup of the experiment is shown in Figure 7, where the outlet of the pre-concentrator was connected with a micro valve. Firstly, an amount of sample was transported into the pre-concentrator by opening the sampling pump. In this adsorption state, the cooler was working and the pre-concentrator was in cooling state; the sample was adsorbed in the SWCNTs (acted as adsorbent material). After the sample was completely adsorbed, the micro valve was closed and the micro heater was in working state; the analytes were quickly released from the SWCNTs by rapidly increasing the temperature of the pre-concentrator. After the analytes were completely released, the micro valve was opened, and the released analytes were consequently transported into the PID by carrier gas. Compared with the response without the proposed pre-concentrator (refer to Figure 8), the response with the pre-concentrator was significantly improved over 20 times more. Moreover, the chromatography peak broadening was greatly compressed, and peak tailing was satisfactorily solved. Therefore, the micro-fabricated pre-concentrator can be easy to deploy and suitable for a number of applications involving on-site monitoring of environmental samples.

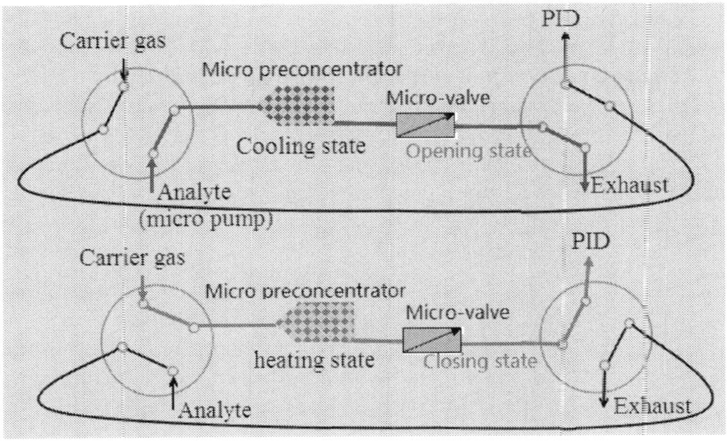

Figure 7. The setup for evaluating performance of the micro pre-concentrator

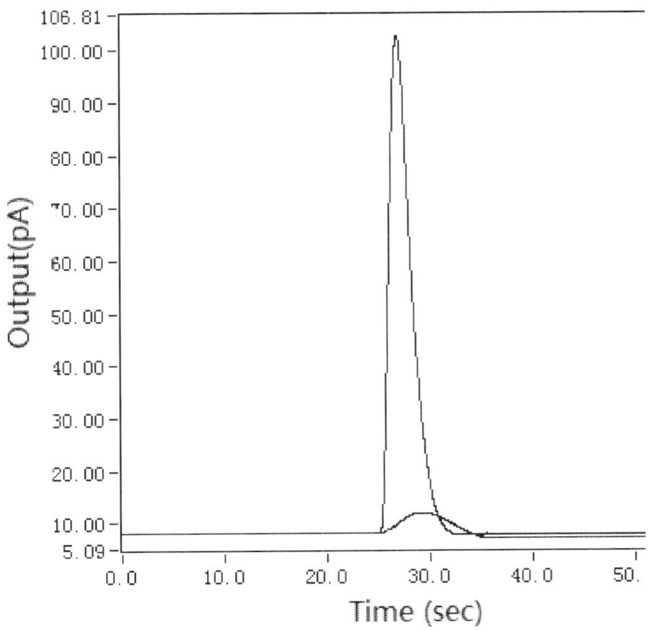

Figure 8. The performance of the pre-concentrator

5. Micro GC column

With more and more demands for online monitoring of environmental sample, high sensitivity of portable GC instruments were very urgent to be used. However, the development of the portable GC systems was limited due to the large volume, high power consumption of traditional GC column. As MEMS technology matures, the micro GC columns [15-19] have a prospect future with some advantages of small size, rapid analysis, batch production, and less power consumption. The portable system integrated with micro GC columns can provide real-time monitoring for quantification and identification analysis of environmental sample. Therefore, these attributes make micro GC columns attractive for a number of applications.

In this work, a high-separation-efficiency micro-fabricated GC column embedded with micro-pillars was developed. These embedded pillars can significantly improve the overall surface area of the columns and reduce the effective width of the column, which can enhance the sample capacity and obtain an excellent separation performance.

5.1. Column fabrication

In this work, an (100) n-type silicon wafer and Pyrex 7740 glass wafer were used as the substrates. The proposed column was fabricated through a series of lithography, etching, and bonding process, where the channels and pillars were fabricated using a deep reactive-ion etching technology. For details on the fabrication process of the micro column, refer to this work [20]. Figure 9 shows a photograph of the fabricated GC column and SEM of the channel and pillars. The diameter of pillars is 50 μm; the depth and the width of the micro channels were 350 μm and 300 μm, respectively; and the length of the micro column is 2 m. Therefore, the overall surface area of the columns, which was able to support the stationary phase, is over three times larger than that of open rectangular columns with the same dimensions, leading to higher column efficiency.

In order to separate the sample, OV-101 was acted as the stationary phase, and the stationary phase was coated via a static coating procedure. The coating process was shown as follows:

1. Preparation of the coating solution. Stationary phase OV-101 with a mass of 10 mg was dissolved in n-pentane and dichloromethane solvent (the volume ratio of n-pentane and dichloromethane solvent is 1:1), and the coating solution concentration is 15 mg/ml.
2. Coating of the column. The coating solution was transported into the column by a micro-pump, and the column was kept in a water bath at 40 °C. After the column was filled with coating solution, one end of the column was sealed by wax, and the excess solvent is slowly evaporated by the pump from the other end of the column.
3. Aging of the column. The fabricated column was put into a temperature-programmed oven which was protected by a nitrogen flow, and the temperature of the column was successively increased to 80 °C and 120 °C for 2 h to harden the stationary phase film; then the temperature of column was heated up to 220 °C for 4 h to cure the stationary phase film, leaving behind a layer of thin and uniform stationary phase film with a thickness of about 0.20 μm on the column walls.

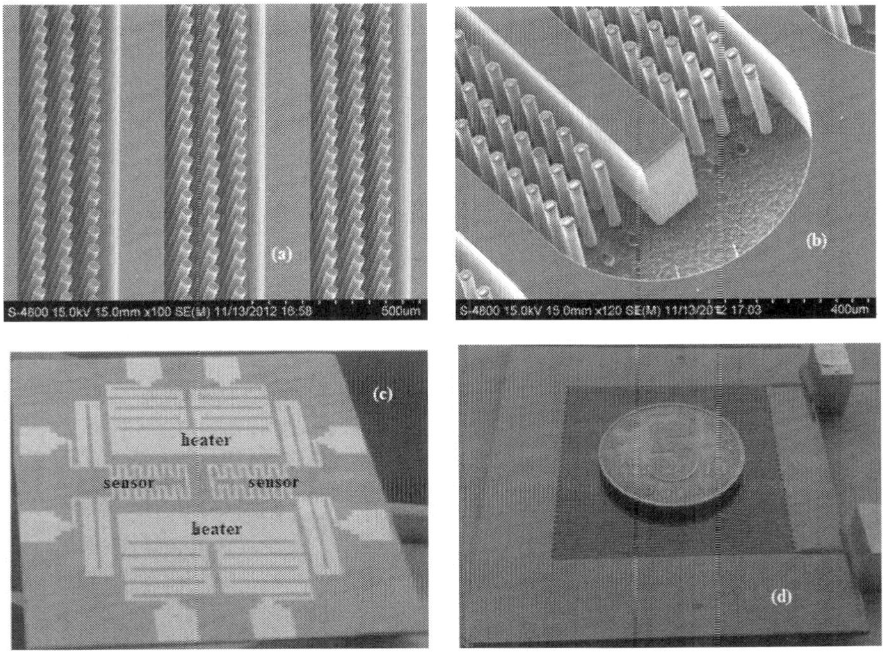

Figure 9. (a) (b) The channels embedded with micro-pillars, (c) the heaters integrated on the GC column, and (d) photograph of the fabricated GC column

5.2. Column efficiency

There are many factors to evaluate the performance of GC column, such as the separation resolution, theoretical plate number, and separation speed, but the theoretical plate number can basically determine the separation performance of GC column. According to theory of chromatography, theoretical plate number and height equivalent to a theoretical plate (*HETP*) have a reciprocal relationship. Therefore, in order to evaluate separation efficiency of the proposed GC columns, we carried out a detailed theoretical analysis on *HETP*. According to previous reports, the *HETP* can be expressed in Equation 1 [21].

$$\mathrm{HETP}=2\frac{D_g}{u}f_1f_2+\left(\frac{1+9k+25.5k^2}{105(1+\mathrm{k})^2}\frac{\mathrm{w}^2}{D_g}\frac{f_1}{f_2}+\frac{2\mathrm{kd}_f(\mathrm{w}+\mathrm{h})^2}{3D_s(1+\mathrm{k})^2h^2}\right)u \tag{1}$$

where D_g and D_s are the binary diffusion coefficients in the mobile and stationary phases, respectively; d_f is the stationary phase thickness; w and h are the channel width and height, respectively; and f_1 (varies between 1 and 1.125) and f_2 (varies between 0 and 1) are the Gidding-Golay and Martin-James gas compression coefficients, respectively.

The theoretical analysis of curve in Figure 10 shows height equivalent to a theoretical plate versus average carrier gas velocity from equation 5.1. The D_g and D_s were considered as 0.093 cm2/s and 6.4 x 10-6 cm2/s, respectively. k, h, and w were 3, 350 μm and 150 μm in the calculations, respectively. The minimum HETP value, H_{min}, found at the optimal average carrier gas velocity, u_{opt}, gives the maximum number of theoretical plates N. The column yielded a minimum HETP of 0.011 cm (9,100 plates/m) at a linear gas velocity of 18 cm/s.

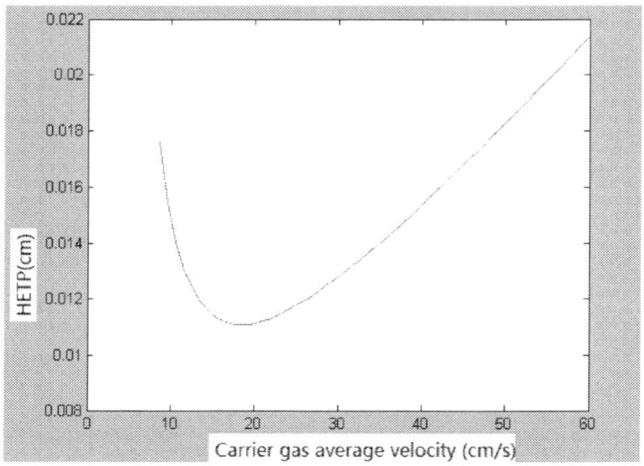

Figure 10. Height equivalent to a theoretical plate versus average carrier gas velocity

6. High-sensitivity PID

In the developed portable GC system, the PID is the core elements, which can provide qualitative and quantitative analysis for air pollution. Therefore, it is very important for developing a high-sensitivity mini PID.

6.1. Performance optimization of PID

In general, there are several factors affecting its sensitivity, such as background noise, volume of the ionization chamber, and electron collection efficiency.

Firstly, it is common knowledge that background noise will affect the detection limit and detection sensitivity of the sensor; even large noise will cause failure of the detector. Therefore, reduction of background noises is the primary consideration in PID configuration.

In terms of ionization chamber, the large ionization chamber cell will increase the dead volume of the detector, which will greatly reduce the response sensitivity of the detector. In addition, larger ionization chamber will consume more carrier gas, which will greatly reduce the working time of portable GC system in field use.

Electron collection efficiency directly influences the response sensitivity of the detector, and the low collection rate will greatly reduce the sensitivity and detection limit of detector. Therefore, maximum collection rate is an important consideration for designing PID.

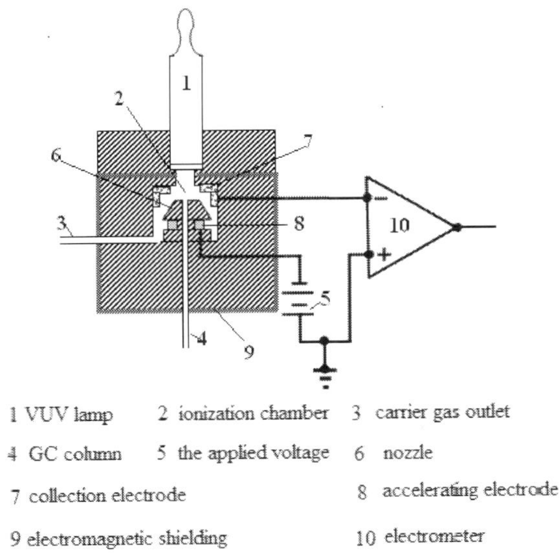

Figure 11. Schematic of the proposed PID

In order to develop a high-sensitivity PID, in this work, a mini PID with a lower background noise and a shorter response time was proposed. These important factors mentioned above were optimized for fabricating PID sensors. Figure 11 shows a side view of the PID proposed in this work. In this setup, a nozzle with a volume slightly less than the ionization chamber is parallel to the VUV lamp and positioned in the center of the ionization chamber, and thus the volume of the ionization chamber is substantially reduced to 10 μl, decreasing the response time and improving the response sensitivity.

In order to shield from the photoelectric effects produced by the VUV light radiation on the electrodes, an annular ion collection electrode that is perpendicular to the direction of energy radiated by the VUV lamp was positioned close to the inner wall of the ionization chamber, and an annular accelerating electrode was embedded in the groove of the nozzle. In addition, the ionization chamber was surrounded by an electromagnetic shield, which could prevent the external electromagnetic noises including the electromagnetic radiation of the VUV lamp from penetrating the ionization chamber. Therefore, these designs enable the PID to demonstrate a very low background noise and a small baseline drift.

In order to reduce the recombination of ions and electrons before collection, the collection electrode was closely positioned in the lamp window, in which the ions and electrons can be immediately captured by the collection electrode after gas molecule ionization. Moreover, a high voltage (which ranged from 100 V to 300 V) was applied between the collection and accelerating electrodes to capture all the ionized species resulting from exposure to VUV radiation. In the end, the output signal of the PID was amplified by an electrometer and monitored by a personal computer.

6.2. Background noise level

In this paper, a few key steps (e.g., shielding the electrodes from the VUV light radiation, surrounding the ionization chamber with an electromagnetic shield) have been used to reduce the background noise level, producing a one- to two-order decrease in the noise magnitude.

After a series of experiments, the results recorded a background current of 2×10-14 A; the output of signal acquisition system was 7.98 pA when the VUV lamp was off (see Figure 12 (a)). However, the output was increased to an upper limit of 8.00 pA when the VUV lamp was turned on (see Figure 12 (b)), which was still the lowest value reported in publications [22]. Moreover, the background noise of the PID was less than 1×10-14 A, lower than its commercial PID counterparts [23]. In addition, the baseline drift was extremely small, which was negligible in the test period.

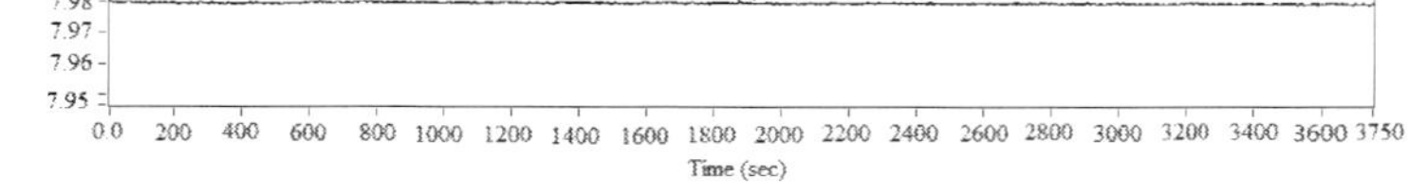

Figure 12. The background noise and current of the PID when the VUV lamp was off (a) and on (b)

6.3. The response time of the PID

In this setup, the ionization chamber of the proposed mini PID with a cell volume of 10 μl was much smaller than that of its commercial PID counterparts (from 40 μl to 200 μl [22,23]). The relatively large volumes of commercial PIDs lead to a large dead volume which can distort GC elution profiles and compromise the device performance. With reduction of the ionization chamber, the response time can be dramatically reduced.

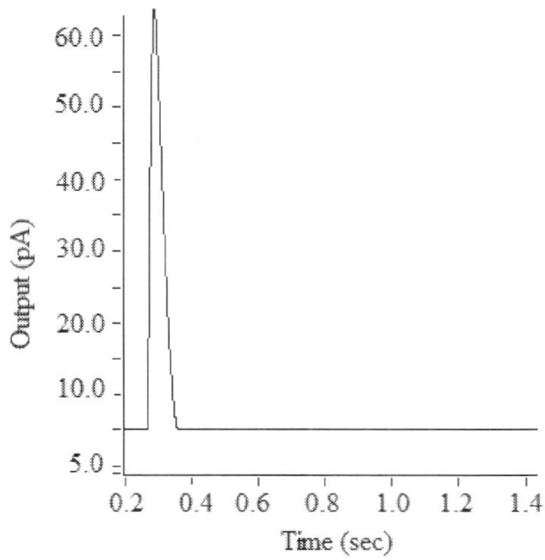

Figure 13. The response time of the proposed PID

In order to evaluate the response time of the mini PID, a capillary (20 cm×0.32 mm) instead of the GC column was used to directly connect the sample injector to the ionization chamber of the mini PID. The flow rate was set to 30 sccm so that the elapsed time in the capillary can be ignored. Figure 13 indicates the response time of the proposed mini PID. The response time was less than 30 ms from the sample entering into the capillary to 90 % response of the sample. To the best of the authors' knowledge, the response time reported in this study was the shortest among all the reported commercial counterparts.

7. Detection of harmful gases

Because of a large number of emissions of industrial waste gas and a substantial increase in automobile, air pollution is more and more serious. At present, PM 2.5, heavy haze, and choking smog have seriously affected people's health. In order to solve the problem of air pollution, real-time monitoring of air quality is very urgent and important. Volatile organic compounds (VOCs) are the premise of pollutants for forming the PM2.5 or heavy haze and choking smog, so monitoring VOCs is a very important measure. Therefore, we use the developed portable PID system for detecting these target pollutants released into the atmosphere, and these analytes mainly include benzene, toluene, styrene, phenol, etc.

Due to the lack of judgment on the qualitative components, the standard chromatogram of these main pollutants in the environmental air must be formulated. In this work, the standard sample with concentrations close to these components in environment was used to develop the standard curve, which can provide a scientific basis for environmental analysis (such as the city mobile monitoring air quality).

The experiments were performed under isothermal conditions at 40 °C with a carrier gas linear velocity of 18 cm/s. The pure He was used as carrier gas and the standard sample was composed of four components, namely, benzene, toluene, styrene, and phenol, with the concentrations of 5 ppm, 10 ppm, 8 ppm, and 10 ppm, respectively. The sample was diluted ten times and injected by a micro-pump and a six-port external sample injector, and carrier gas velocity was controlled by a gas flow controller. Figure 14 shows the chromatogram of the standard sample.

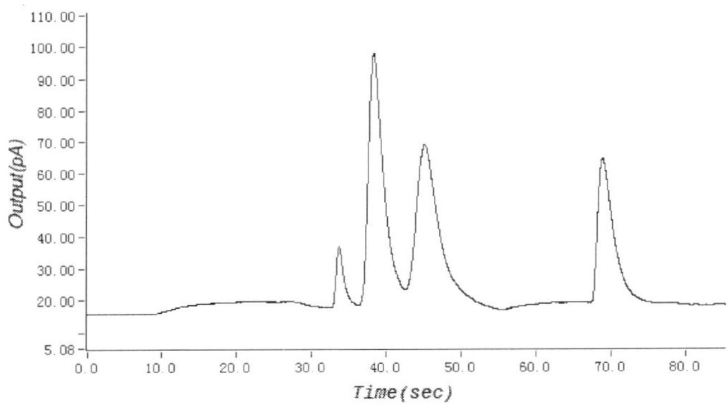

Figure 14. The chromatogram of standard sample

The result shows that the proposed system demonstrated a good separation and detection of these volatile organic compounds. The smallest response amplitude of the chromatographic peak (benzene) is also over 10 pA. Moreover, the minimum resolution of the closest two components is over 1.4. These experimental data are sufficient to show that the developed portable PID system can be widely applied to the trace detection of environment analysis.

8. Conclusion

The work here demonstrates that it is possible to fabricate a mini GC system integrated with a micro dryer and purifier, a micro pre-concentrator, a micro GC column, and a mini PID. The micro dryer and purifier can remove vapor and particulates from environmental samples, which makes the mini GC system suitable for field use. The micro pre-concentrator can concentrate the trace gas, which enables the mini GC system to detect environmental samples and also improves detection sensitivity of mini-instruments. In additional, the micro GC column can reduce the volume of the system and overcome low-resolution and poor anti-interference ability of other instruments. Based on the above experimental results, the mini GC system can effectively separate and detect the air pollutants. Therefore, the developed portable PID system can be widely applied to the trace detection of environment analysis. However, the standard sample instead of the actual sample was used in performing the experiments. In the following work, we will carry out the city mobile monitoring air quality based on this experimental basis, and the results will be reported in the next works.

Acknowledgements

The authors greatly acknowledge the financial support from the National Science Foundation of China under grant numbers 61176112 and 60976088. The authors greatly acknowledge the financial support from the Beijing Science and Technology Plan project under grant number Z141100003414003.

Author details

J.H. Sun[1*], F.Y. Guan[2], X.F. Zhu[3], Z.W. Ning[3] and T.J. Ma[1]

*Address all correspondence to: jhsun@mail.ie.ac.cn

1 State key laboratory of transducer technology, Institute of Electronics, Chinese Academy of Sciences, Beijing, China

2 International Centre for Bamboo and Rattan, Key Laboratory of Bamboo and Rattan, Beijing, China

3 Beijing Municipal Institute of Labour Protection, Beijing, China

References

[1] H. L. Wang, X. J. Cao, C. Bao, Z. B. Meng, G. X. Zhao, S. K. Sun, Chinese Journal of Analytical Chemistry. 39(10) (2011) 1513-1516.

[2] D. P. Neil, D. F. William, N. F. Robert, Open Access Emergency Medicine. (2009) 15-19.

[3] J. H. Sun, D. F. Cui, X. Chen, L. L. Zhang, H. Y. Cai, H. Li, Sens. Actuators B: Chem. 160(1) (2011) 936-941.

[4] J. H. Sun, D. F. Cui, F. Y. Guan, et al., Nanoscale Research Letters. 9 (2014) 576.

[5] S. M. Gordon, J. P. Szidon, B. K. Krotoszynski, et al., Clin Chem. 31(8) (1985) 1278-1282.

[6] P. Gang, T. Ulrike, A. Orna, H. Meggie, et al., Nature Nanotechnology. 4 (2009) 669-673.

[7] K. Schwarz, W. Filipiak, A. Amann, J Breath Res. 3 (2009) 027002.

[8] M. Ligor, T. Ligor, A. Bajtarevic, C. Ager, et al., Clin Chem Lab Med. 475(5) (2009) 50-560.

[9] V. Olavi, M. S. Florian, M. Markus, H. Lauri, Current Analytical Chemistry. 9 (2013) 463-475.

[10] D. N. Corrado, M. Antonella, M. Eugenio M, et al., Biosensors and Bioelectronics. 18(10) (2003) 1209-1218.

[11] R. F. Machado, D. Laskowski, O. Deffenderfer, et al., Am J Respir Crit Care Med. 171(11) (2005) 1286-1291.

[12] H. P. Chan, C. Lewis, P. S. Thomas, Lung Cancer. 63(2) (2009) 164-168.

[13] B. Buszewski, M. Kesy, T. Ligor, A. Amann, Biomed Chromatogr. 21(6) (2007) 553-566.

[14] J. H. Sun, F. Y. Guan, D. F. Cui, X. Chen, L. L. Zhang, J. Chen, Sensors & Actuators: B. Chemical. 188 (2013) 513-518.

[15] A. Bhushan, D. Yemane, E. B. Overton, J. Goettert, M. C. Murphy, J. Microelectromech. Syst. 16 (2007) 383.

[16] P. Zellner, L. Renaghan, Z. Hasnain, M. Agah, J. Micromechanics Microengineering. 20(4) (2010) 1.

[17] G. E. Spangler, Journal of Microcolumn Separations. 13 (2001) 285.

[18] H. S. Noh, P. J. Hesketh, G. Frye-Mason, Journal of Microelectromechanical Systems. 11(6) (2002) 718.

[19] J. H. Sun, D. F. Cui, X. Chen, L. L. Zhang, H. Li, J. Chromatogr. A. 1291(24) (2013) 122-128.

[20] J. H. Sun, D. F. Cui, F. Y. Guan, et al., Sensors & Actuators: B. Chemical. 201 (2014) 19-24.

[21] M. J. E. Golay, J Chromatogr. 216 (1981) 1.

[22] N. Ostojić, Z. Šternberg, Chromatographia. 7(1) (1974) 3-5.

[23] J. N. Driscoll, M. Duffy, Chromatographia. 2(4) (1987) 20-27.

Chapter 7

Spatial and Temporal Distribution of Throughfall Deposition of Nitrogen and Sulfur in the Mangrove Forests Associated to Terminos Lagoon

R.M. Cerón, J.G. Cerón, M. Muriel, F. Anguebes,
M. Ramirez, J. Zavala, C. Carballo and R.C. Escoffie

Additional information is available at the end of the chapter

http://dx.doi.org/10.5772/59726

1. Introduction

Air pollution is a major environmental concern in most urban areas. Atmospheric emissions of gaseous and particulate pollutants have caused profound environmental and health implications in last years. Sulphur dioxide (SO_2), nitrogen oxides (NO_x) and ammonia (NH_3) are gaseous pollutants commonly present in both, man- made and natural emissions in the lower atmosphere. Industrial processes, vehicular traffic, open biomass burning (including forest fires), dairy farming, intensive animal husbandry, and other anthropogenic activities are the main sources for these gaseous pollutants in the troposphere.

These pollutants may be removed from the atmosphere by two ways: wet deposition, which includes all pollutant material reaching the earth's surface by precipitation; and dry deposition, comprising the processes of adsorption of particulate and gaseous material by land or water surfaces [1]. In a general way, both processes govern the transfer of beneficial and toxic chemicals from the atmosphere on to surfaces.

Sulfur and nitrogen deposition (N and S) occurs as a result of these removal processes and is associated with acidification of soils and surface waters. Some environmental indicators have demonstrated that acidification related to air pollution may have already occurred in developing countries [2- 3]. The deposition of sulphur compounds cause changes in the chemistry and biology of the soils. These changes include decreasing pH and alkalinity, elevated concentrations of soluble aluminium and an imbalance in nutrient cycling leading to a change of ecosystem diversity [4]. There are evidences that nitrogen compounds deposition causes changes in ecosystems through eutrophication and acidification of soils and waters, decline in trees growth and losses of vitality in forest [5- 6]. In addition, N and S deposition can cause deterioration to historical monuments and diverse materials and can cause damages to human health.

In Mexico, N and S deposition monitoring, its spatial and temporal distribution, and its effects on historical and cultural heritage and sensitive vegetation species have not been sufficiently studied. Comparing current conditions with ten years ago, S emissions' scenarios have decreased as a result of reformulation of fuels and the application of rigorous regulations focused to reduce them; however, it has been reported that the relative importance of N deposition has increased [7].

The coast of the Gulf of Mexico is characterized to be a complex region where the co- existence of oil refineries, offshore platforms and other facilities for oil and gas exploration and production, historical monuments, archaeological zones, and many valuable aquatic and terrestrial ecosystems (this region has the widest mangrove forests cover in the country) occurs. Campeche State is located at the southeast of the Gulf of Mexico and it constitutes the most important oil and gas producer in the country. A total annual emission of 205.64 Gg of NO_x and 336.79 Gg of SO_2 has been reported for this state, mainly produced by urban and industrial sources. Consequently, it is expected that N and S deposition, in ecosystems located downwind from these sources, would be high. This region has important natural reserves, national parks, protected natural areas, historical monuments and archaeological sites that constitute the main support of the eco- tourism industry which generates significant economical resources for Campeche State. Only in Campeche State, mangrove cover accounts for 29.98% of the total country cover, which is, approximately 196,552 ha. Therefore, the potential ecological effects derived from atmospheric deposition on these sites constitute a key concern for this region.

Critical loads estimation method allows to quantify the grade of damage derived from atmospheric pollutants deposition on ecosystems. A critical load can be defined as the input of one pollutant at levels below of which harmful ecological effects do not occur in the long term. To establish these critical load values and estimate their exceedances in a given site, atmospheric deposition measurements in field are required. N and S levels, their spatial and temporal distribution and their deposition fluxes are used in order to determine the actual inputs of these pollutants to ecosystems. This information is commonly represented in deposition maps, most of them are based on three- dimensional chemical transport model results which need to be validated by comparison with field measurements [8- 10]. Such studies represent an opportunity for policy- makers to identify the potential impacts associated with different emission sources and their spatial and geographical distribution. Sensitivity mapping and the critical load approach are methods that may allow analysis of these risks.

Particularly, in Mexico, measurement- based maps are required to assess the current deposition fluxes and the vulnerability of the ecosystems. Considering that studies about critical loads and their exceedances are scarce in tropical humid forests, this research work had the following aims: 1) To establish a solid base line about throughfall deposition of N and S in mangrove ecosystems, and 2) To assess the temporal and geographical distribution of N and S deposition along Carmen Island in order to identify critical zones and seasons in which N and S deposition can be high as a result of the prevailing meteorological conditions.

2. Study area

This study was carried out in Carmen Island, Campeche, Mexico. This site constitutes a complex area due to its closeness to the most important offshore zone for gas and oil exploration and production. In addition, the urban area is located within the buffer zone and in the surroundings of the Natural Protected Area of "Terminos Lagoon". Climate in this site is sub-humid warm with rains occurring along the summer. Annual mean precipitation is 1300 mm and annual average temperature is 27 °C. Prevailing winds blew from NE from March to October (when the island is under the influence of cold fronts called "Norths"); and from SE during the rest of the year (from April to September) when this site is influenced by tropical maritime air as a result of the trade winds. In Figure 1. is presented a typical wind rose and backward trajectory for Carmen Island during the study period. Additionally, Carmen Island is all time under the influence of sea and land breezes as a consequence of differential heating. In this research, a multiple transect sampling schema was designed (13 sampling points were considered along the island) in order to assess N and S deposition temporal and spatial distribution along Carmen Island. Specific sampling points are presented in Figure 2.

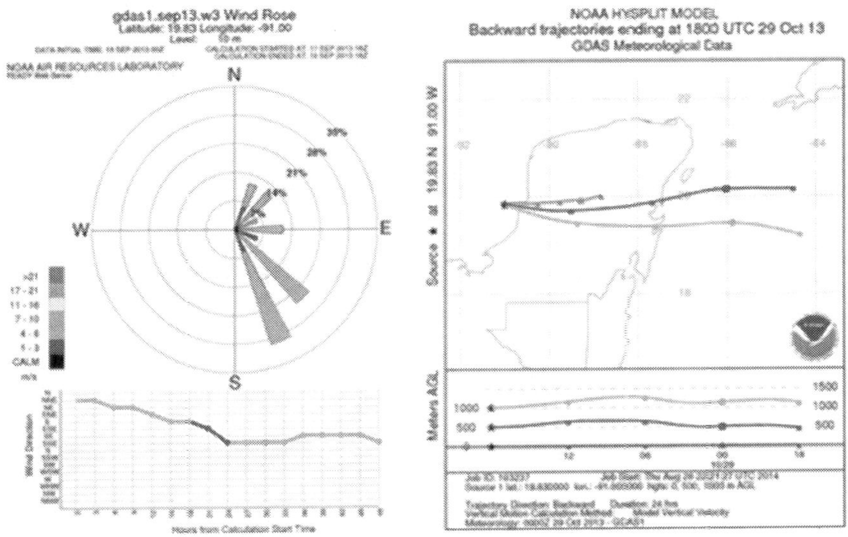

(a) (b)

Figure 1. Typical wind rose and backward trajectory for Carmen Island during the study period.

Figure 2. Sampling sites location along Carmen Island, Campeche, Mexico.

3. Methodology

3.1. Sampling

Throughfall deposition consists of collected solutes in wet deposition under the forest canopies. This method is widely used to estimate atmospheric deposition inputs to the forest ecosystems, since it includes both, wet and dry deposition. Because of the high cost and difficulty of the measurements of dry deposition fluxes in forest stands, throughfall collectors constitute a good choice to obtain a reliable estimation of N and S atmospheric inputs in a given ecosystem [11]. However, in regions where N deposition is chronic, throughfall collectors underestimate the total N fluxes (wet plus dry) [12]. This underestimation is attributed to the uptake and retention of atmospheric N compounds in tree canopies, especially during dry deposition [13]. In spite of this restriction, throughfall estimations of N and S deposition are useful to establish a base line in sites where atmospheric deposition data are not available as is the case of the tropical forests in Mexico. Since, a worldwide database of atmospheric N and S inputs is available; inputs quantified in a given region can be compared with these reference values to obtain a diagnosis about the severity of N and S deposition and the possible vulnerability of the ecosystems.

Automatic wet/dry collectors are expensive and need to satisfy certain requirements for their installation; therefore, passive sampling devices as throughfall collectors constitute a good sampling alternative in a given region; in addition, it is possible to increase the density of the sampling grid at a low cost. Throughfall deposition can be defined as the hydrologic flux of ions to floor contained within a solution. This work used passive throughfall collectors developed and tested by [14], constituted by an ionic exchange resin mixed bed within a column.

Throughfall deposition in Carmen Island was collected on one- year basis, from July 2013 to June 2014; passive sampling devices were exposed during two months for six periods throughout the year in each of the thirteen sampling points. Samples were collected at the end of each period.

Samples were collected with a funnel; the solution was channeled to the mixed resin bed through the column (particularly, in this work, a mixed Amberlite™ IRN150 ion exchange resin bed was used).. where ions were retained (Figure 3). At the end of the sampling period (two months), retained ions within the same column were extracted by using extraction solutions to recover the sampled elements. Sulfate and nitrate retained within the resin column were selectively extracted with a 2 N KCl solution and analyzed by turbidimetry and colorimetry, respectively.

Figure 3. Throughfall sampling devices used in this study.

Extraction efficiency was calculated as the percentage of the loaded ions on the columns to the recovered in sequential extractions. It was found that recovered N and S percentage from the third extraction was insignificant, therefore; only two extractions were considered for the recovery process, reaching an efficiency of approximately 96.8%. Similar extraction efficiencies have been reported by other authors [15].

3.2. Chemical analysis

NH_4^+ was analyzed by molecular absorption spectrometry using the blue indophenol method [16]. Sulfate was determined by turbidimetric method [17], whereas NO_3^- was analyzed colorimetrically by using the brucine method [18]. The weight of the extractant was converted to volume by using the specific density of the extractant solution (2N KCl = 1.05 g ml^{-1}). The amount of each collected ion in a given column was determined as the factor of extractant volume multiplied by ionic concentration in the extractant. The surface area of the funnel opening and the sampling period were used to estimate the deposition fluxes of N and S per land area per year (Kg ha^{-1} yr^{-1}).

3.3. Data interpolation and mapping

All methods based on point measurements (e.g., wet deposition, micrometeorological and dry deposition measurements, throughfall deposition), cannot be directly connected to emission inventories. Maps can only be produced directly from these measurements if the network is enough dense to account for spatial and temporal variations. This may be the case for networks measuring compounds with little spatial variation or for measurements of deposition in areas of simple terrain as the case of Carmen Island. A point measurements network should be interpolated using statistical methods as the kriging technique, which allows including monitoring data from neighboring stations for interpolation in a national scale [19]. Considering the characteristics of measured data and the expected interpolation results, diversified methods have been employed with atmospheric deposition network data [10, 20- 22].

To obtain N and S deposition maps in Carmen Island, a geostatistical procedure was used to interpolate field measurements into a continuous spatially pattern, where data were interpolated using the kriging interpolation technique. Kriging is a statistical method that provides unbiased estimates of variables in regions where the available data exhibit spatial autocorrelation and these estimates are obtained in such a way that they have minimum variance. The first step was to obtain the statistics summary of primary data. From this, a descriptive report of data was carried out, including mean, maximum, minimum values and their frequency distribution. This analysis was carried out in order to determine if it was necessary to make a data transformation since the results of the kriging interpolation are more reliable when data are normally distributed [20].Data interpolation was used in this work to obtain a continuous variable by isolines to increase the number of points in the sampling grid. Threshold values definition helps to obtain isolines or imaginary lines in which studied variables become continuous, this method is useful since it takes into account the spatial behavior of the variable from distance and the variability between points.

Values were obtained from a linear combination of the original points with known data. As a result, greater compact areas around the variable maximum values were estimated. Once additional points and isolines were obtained, deposition data were mapped to assess their spatial and temporal distribution along the Island.

4. Results and discussion

Critical loads have been estimated in several regions around the world. A critical load value of 5 Kg N ha^{-1} yr^{-1} has been reported for alpine ecosystems [23], whereas for New Mexico and California, values of 3- 8 Kg N ha^{-1} yr^{-1} and 4- 7 Kg N ha^{-1} yr^{-1}, respectively, have been found [24]. On the other hand, a critical load value of 3 Kg S ha^{-1} yr^{-1} for very sensitive areas and a range of 2 - 5 Kg S ha^{-1} yr^{-1} for natural forests have been proposed [25].

In Mexico, critical loads data are not available and only few studies have been carried out in Mexico Valley, mainly in pine forests. It has been reported an input of 15 Kg N ha^{-1} yr^{-1} for pine stands in Desierto de Los Leones in the surrounding of Mexico City [26], whereas in Zoquiapan (a site located at the east and upwind of Mexico City), the reported inputs were 5.5 and 8.8 Kg ha^{-1} yr^{-1} for N and S, respectively [27]. A research work carried out in Central Veracruz for several land- cover types reported inputs of 8- 17 Kg ha^{-1} yr^{-1} and 2- 4 Kg ha^{-1} yr^{-1} for S and N, respectively [28]. Additionally, some authors have studied the acidification in developing countries and have proposed a critical load approach on a global scale [3]. They assigned a relative sensitivity class of 3 to acidic deposition for terrestrial ecosystems in Carmen island region, and the preliminary critical load assigned to this site ranges from 50 to 100 meq m^{-2} yr^{-1}.

In this study, mean fluxes of throughfall deposition for N (as N- NH_4^+ + N- NO_3^-) and S (as SO_4^{2-}) at Carmen Island were 2.15 and 4.7 Kg ha^{-1} yr^{-1}, respectively. N mean deposition flux did not exceed critical loads proposed for very sensitive ecosystems, however, S mean deposition flux is already in the threshold limit value proposed for natural forests and it is greater than those reported for very sensitive areas.

Sulfur dioxide is oxidized to sulfate, and the oxidation rate determines its lifetime in the atmosphere. Sulfuric acid is produced from the oxidation of sulfur oxides, which in turn form sulfate particles. However, even in the atmosphere of rural or non- industrialized sites, significant levels of sulfate particles have been found, concluding that sulfate in these sites is related to atmospheric reactions from anthropogenic SO_2 [29]. SO_2 has an atmospheric residence time of 13 days, and may be transported great distances from anthropogenic sources [30]. On the other hand, the oxidation of NO_2 at atmospheric conditions is almost 10 times faster than the oxidation of SO_2 to SO_4, resulting in a residence time of approximately 1 day [31]. These chemical properties make nitrate or its parental gaseous precursors NO_x, be commonly known as local pollutants. Since the dry oxidation of SO_2 to SO_4, or wet oxidation via the bisulfate (HSO_3) intermediate at the ambient atmospheric conditions is much slower than that NO_x, SO_2 and its intermediate oxidative products have much longer residence time in the atmosphere. This makes them more susceptible to be transported by the movement of the air masses in comparison with NO_x. For this reason SO_4 or its parental gaseous precursors are known as regional pollutants.

Additionally, during the rainy season, when the mixing layer is very high, most of pollutants in precipitation derive from the rain- out process of condensation nuclei that have been transported long distances into the region. In contrast, during the dry season, the reduced mixing layer only concentrates ionic species of local origin. Dry deposition is, in general, greater than wet deposition near emission source.

To infer this local or regional influence, the sulfate: nitrate ratio in throughfall deposition was estimated. A ratio of 4.7 was obtained, suggesting that this site was under the influence of long-range transport. NH_4^+ and NO_3^- had a similar pattern in their deposition fluxes, with the highest fluxes occurring during the dry season (Figures 4a and 4b). It is agree with the local character of the emissions of NO_x.

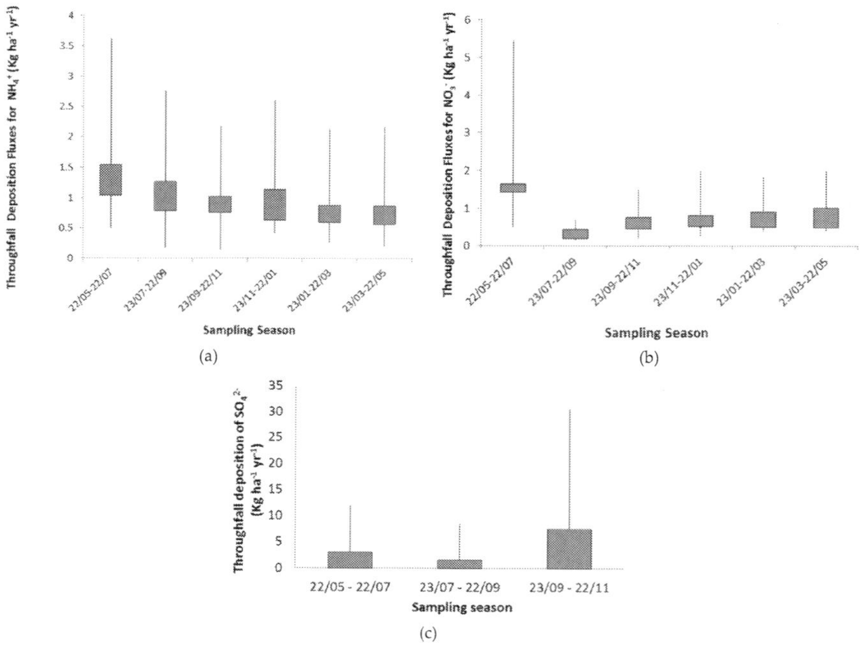

Figure 4. Atmospheric deposition fluxes for NH_4^+, NO_3^- and SO_4^{2-} for Carmen Island for sampling season.

On the other hand, SO_4^{2-} had a completely different behavior, with the highest levels occurring during the plenitude of the rainy season and at the beginning of the cold fronts season (Figure 4c). This is in agreement with the regional character of SO_2 emissions which are more connected to wet deposition.

Before applying Krigging interpolation, the thirteen sampling points in which throughfall deposition was collected, were grouped in three zones, considering the land- use along the Island. Identified zones were the following: *Industrial zone* (sampling points 1,2,3 and 4) located at the east edge of the island, *Zone with the greatest mangrove cover* (sampling points 5, 6, 7 and 8) located at the middle part of Carmen Island, and *Urban zone* (sampling points 9, 10, 11, 12 and 13) located at the west side of the island.

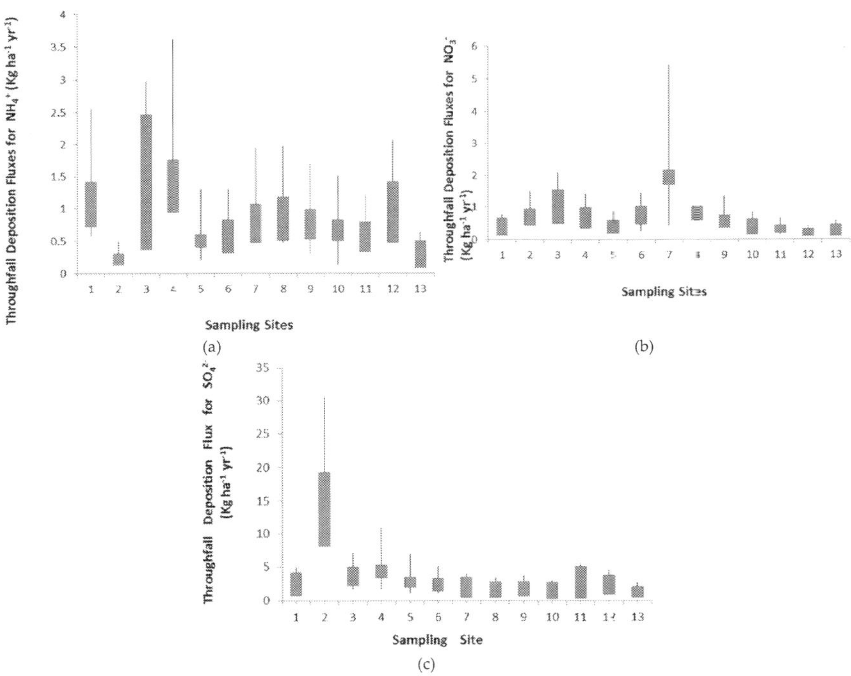

Figure 5. Atmospheric deposition fluxes for NH_4^+, NO_3^- and SO_4^{2-} for Carmen Island for sampling site.

Mean throughfall deposition fluxes for NH_4^+, NO_3^- and SO_4^{2-} were the highest in sampling points labeled as 3, 7 and 2, respectively (Figures 5a, 5b and 5c). Sampling points 7 and 8 are located at the limit of the urban zone; both points are within a complex area at the transition zone between urban area and mangrove forest. Sampling point labeled as "7" is located within an area characterized by a high vehicular density, small geographical extent and few circulation ways, so in peak hours, traffic vehicular is intense, resulting in high NO_x emissions of that are deposited as NO_3^- in the surroundings of the emission points.

Figure 6b illustrates that NO_3^- deposition was higher in the area adjacent to mangrove ecosystem, whereas the highest deposition of NH_4^+ and SO_4^{2-} occurred in the island industrial

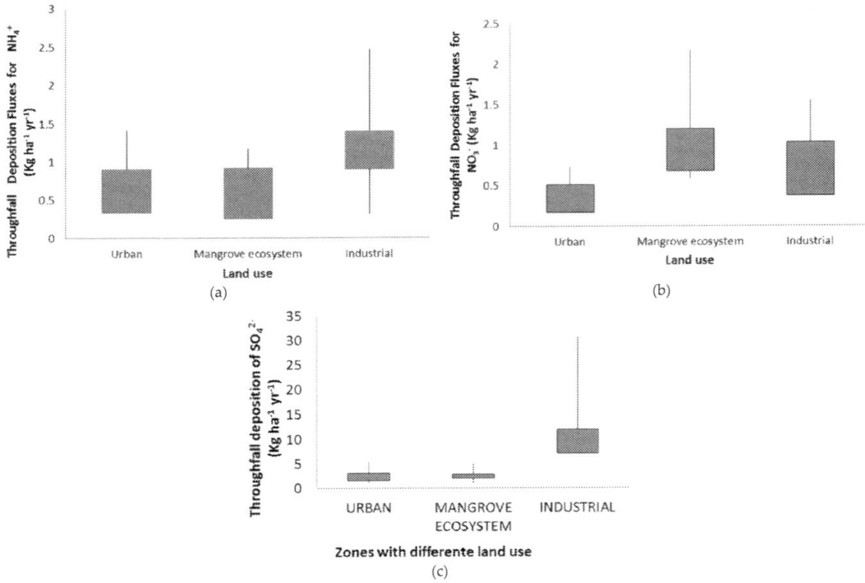

Figure 6. Atmospheric deposition fluxes for NH_4^+ , NO_3^- and SO_4^{2-} for Carmen Island for different land use: Urban, Mangrove ecosystem and industrial.

zone (Figures 6a and 6c). On the other hand, mean throughfall deposition fluxes for NH_4^+ and SO_4^{2-} were higher in sampling points 3 and 2, respectively. It is important to mention that the points 1, 2 and 3 are located in the mangrove forests boundaries. These points are placed at the east edge of the island, where some industrial facilities could contribute to deposition of local NH_4^+ and SO_4^{2-} . Since sampling points 2 and 3 were located along the Federal Highway 180, so that NH_3 could be also emitted from light- duty vehicles. Many authors have reported on NH_3 and amine emissions from gasoline- powered automobiles or engines with and without exhaust catalysts in dynamometer experiments [32, 33]. The production of NH_3 emissions depends on the vehicle's ability to produce NO in the presence of a catalytic convertor that has enough stored hydrogen to reduce the NO to NH_3. However, considering prevailing winds, a great proportion of NH_4^+ could also came from rural areas at the east of Carmen Island, specifically located crossing the bridge "La Unidad" in Isla Aguada and Sabancuy municipalities, where agriculture activities are developed. In addition, sulfate levels in throughfall deposition collected in Carmen Island could be enhanced by the long- range transport of SO_2 emissions from offshore platforms in the Gulf of Campeche where sour gas is burned in elevated flares. These SO_2 emissions could be washed- out during the rainy and cold fronts seasons since the wind roses and backward air mass trajectories pointed out that air masses followed this direction during this climatic period.

4.1. Mapping deposition fluxes of acidic compounds over the study region

Since successive monitoring of precipitation chemistry at the same station is scarce in Carmen Island and the data collected from various sources are highly discrete at the temporal scale, then all the concentration data obtained in this work were employed to produce continuous contours for spatial analysis. One- year mean results for all sites were interpolated to produce N and S deposition loads isopleths (Figures 7, 8 and 9).

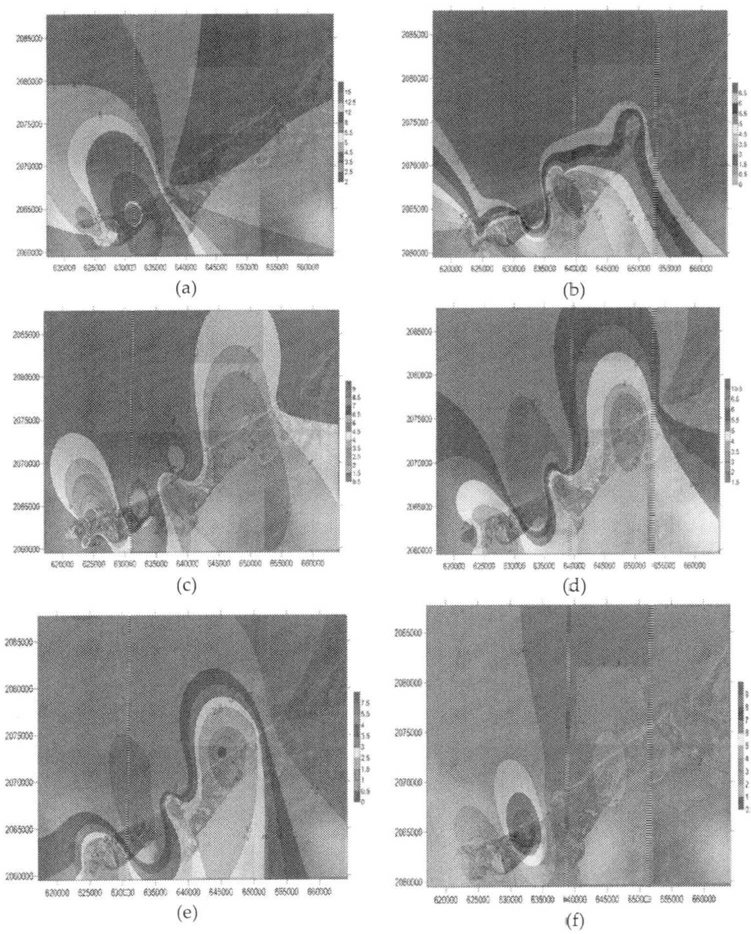

Figure 7. Spatial and temporal patterns of NH_4^+ throughfall deposition fluxes (Kg ha^{-1} yr^{-1}) in Carmen Island: (a) July, (b) September, (c) November, (d) January, (e) March, (f) May.

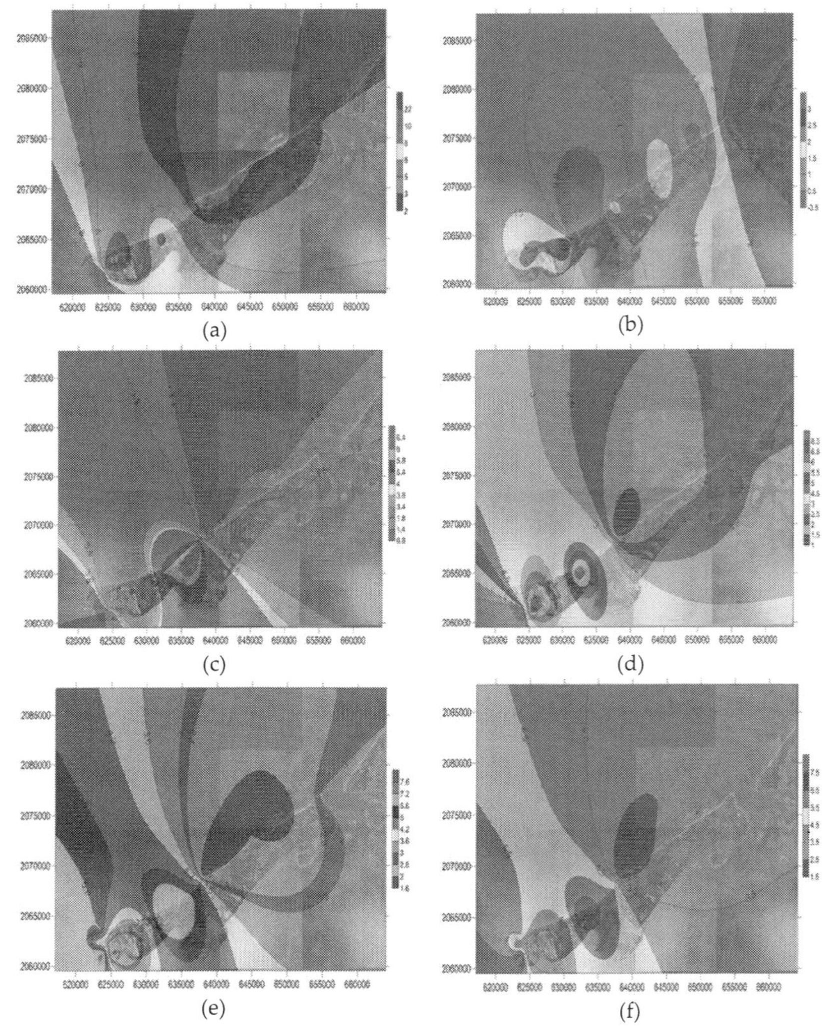

Figure 8. Spatial and temporal patterns of NO_3^- throughfall deposition fluxes (Kg ha^{-1} yr^{-1}) in Carmen Island: (a) July, (b) September, (c) November, (d) January, (e) March, (f) May.

NH_4^+ deposition maps indicate a clear seasonal pattern, positioning the highest values during the period September- January, just when the Island is under the influence of cold fronts (Norths). Highest fluxes associated to the spatial distribution, were in the surroundings of point 7 and in the east edge of the Island. This fact demonstrates that NH_4^+ probably could

http://dx.doi.org/10.5772/59726

have two main sources: light- duty vehicles circulating on the road with the highest traffic (point 7) and along the Federal highway 180 (points 1, 2 and 3) at the east edge of the Island. However, NH_4^+ deposition fluxes could be also enhanced by transport of emissions related to agricultural activities in Isla Aguada and Sabancuy municipalities.

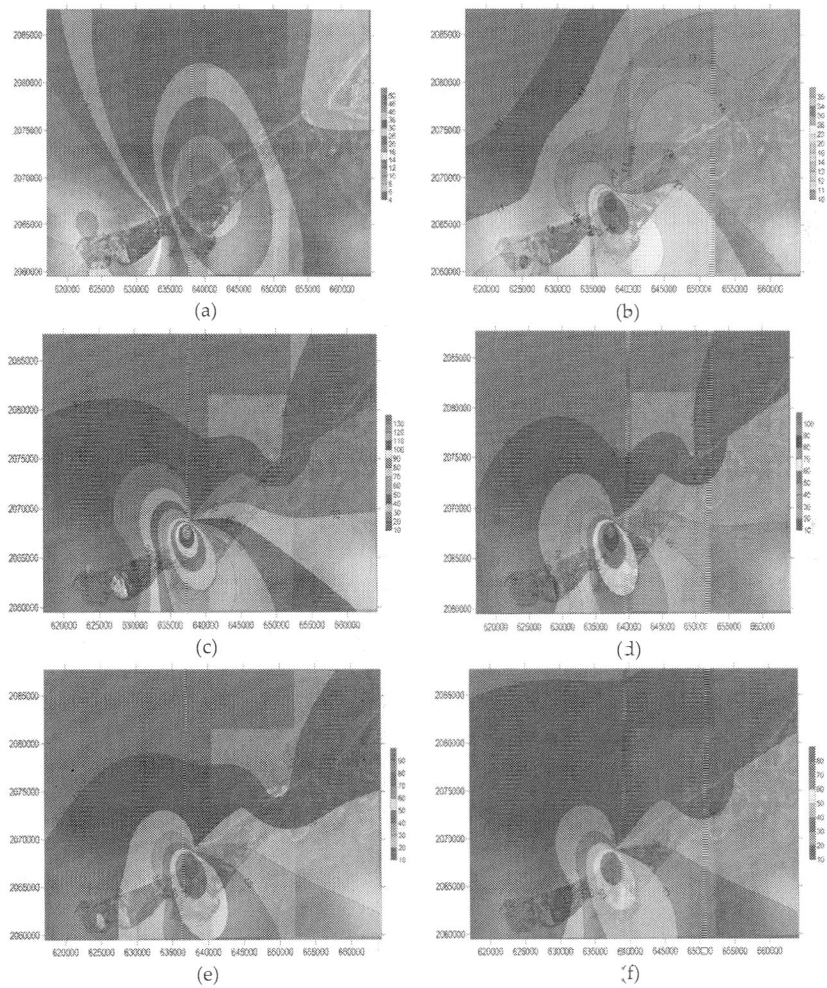

Figure 9. Spatial and temporal patterns of SO_4^{2-} throughfall deposition fluxes(Kg ha^{-1} yr^{-1}) in Carmen Island: (a) July, (b) September, (c) November, (d) January, (e) March, (f) May.

Nitrate did not show a clear seasonal trend and presented a similar pattern along the year representing its local character. However, it was observed a clear spatial pattern with peak values in the surroundings of points 6, 7 and 8. This zone is located at the mangrove forest boundaries and it is characterized by high vehicular traffic since there are not enough roads. Moreover, NO_3^- deposition fluxes showed a slight dilution effect, with relatively higher values during dry months, decreasing as rainy season progressed.

In the case of SO_4^{2-} , deposition fluxes were higher in the east edge of the island with a clear spatial trend which decreased progressively westward. Peak values were centered on the points 1, 2, 3 and 4. An evident seasonal pattern was identified since deposition fluxes were higher during Norths season; it suggests that sulfate levels could be enhanced by large- scale transported emissions from offshore platforms in the Gulf of Campeche demonstrating its regional character.

5. Conclusions

Results found in this research work suggest that NO_3^- atmospheric deposition in the study site has a local origin, and is mainly distributed along the urban zone, being its deposition higher during the dry season. On the other hand, NH_4^+ highest deposition fluxes were distributed at the east edge of the island just during the dry season, suggesting a local origin, probably from light- duty vehicles emissions and agricultural activities developed in nearby rural areas. Regarding SO_4^{2-} , its deposition has its origin in both, local and regional sources, being this contribution higher when the Island was subjected to the influence of "Norths". It was quite evident that in addition to local industrial sources in Carmen Island, background levels were enhanced by the long- range transport of SO_2 emissions from offshore platforms in the Gulf of Campeche during the "Norths" season when prevailing winds came from NE.

Exceedance of critical loads represents a potential risk indicator of acidification as well as its impacts on ecosystems; therefore, in order to assess the vulnerability of the mangrove ecosystems in Carmen Island, it is necessary to establish the critical loads and to estimate their exceedances. Only then, it will be possible to develop emissions control programs and propose prevention policies and local regulations focused to protect these ecosystems.

Interpolation of field measurements is one of the most convincing methods to explore N and S budget, even though the accuracy is limited by the characteristics of measured data and uncertainties associated with the interpolation technique.

This research work constitutes the first step to quantify the current status of N and S deposition in the study site. A long- range monitoring network over Mexico will be established in a short-time for estimating nitrogen and sulfur deposition fluxes and their distribution along the country. A series of monitoring stations should be set down in order to keep the uniformity in the measured data, allowing comparability and considering not only urban and industrialized sites but also rural and remote sites. Maps based on measurements are also required in Mexico to validate and calibrate three- dimensional transport models.

Acknowledgements

The authors acknowledge the financial support given to this research work by Autonomous University of Carmen through its Postgraduate Department. Thanks to the people, organizations, institutions and companies that provided a secure location for the sampling devices used in this study.

Author details

R.M. Cerón[1*], J.G. Cerón[1], M. Muriel[2], F. Anguebes[1], M. Ramirez[1], J. Zavala[1], C. Carballo[1] and R.C. Escoffie[3]

*Address all correspondence to: rceron@pampano.unacar.mx

1 Chemistry Faculty. Autonomous University of Carmen. Ciudad del Carmen, Campeche, Mexico

2 Marine Zone. Mexican Petroleum Institute. Ciudad del Carmen, Campeche, Mexico

3 Research Center of Environmental Sciences. Autonomous University of Carmen. Ciudad del Carmen, Campeche, Mexico

References

[1] Fowler, D. 1980. Wet and dry deposition of sulphur and nitrogen compounds from the atmosphere. In: "Effects of acid precipitation on terrestrial ecosystems", NATO Conference Series, T.C. Hutchinson and M. Haras Eds., 4: 9- 27.

[2] Zhao D, Zhang X, Yang J, Mao J and Xiong J. 1995. Critical load of sulphur deposition for ecosystem and its application in China. J. Environ. Sci, 7: 325–337.

[3] Kuylenstierna, J.C.I; Rodhe, H; Cinderby, S; Hicks, K. 2001. Acidification in Developing Countries: Ecosystem Sensitivity and the Critical Load Approach on a Global Scale. Ambio, 30: 20- 28. doi: 10.1579/0044- 7447- 30.1.20

[4] Nilsson, J. 1988. Critical loads for Sulphur and Nitrogen. In: "Air Pollution and Ecosystems", P. Mathy Ed., Proceedings of an International Symposium held in Grenoble, France, 18- 22 May, 1988: 85- 91.

[5] Goulding, K.W.T; Balley, N.J; Bradbury, N.J; Hargreaves, P; Howe, M; Murphy, D.V; Poulton, P.R; and Willison, T.W. 1998. Nitrogen deposition and its contribution to nitrogen cycling and associated soil processes. New Phytologist 139: 49- 58.

[6] Bishop, K.A. and Hultberg, H. 1995. Reversing acidification in a forest ecosystem: the Gårdsjön covered catchment. Ambio, 24: 85–91. doi:10.1021/es903722p.

[7] Fenn, M.E; de Bauer, L.I; Zeller, K; Quevedo, A; Rodríguez, C; Hernández- Tejeda, T. (2001). Nitrogen and sulfur deposition in the Mexico City Air basin: Impacts on forest nutrient status and nitrate levels in drainage waters. In: M.E. Fenn et al (eds.Urban Air Pollution and Forests. Springer- Verlag, New York. Pp. 319.

[8] Li,C.S;, Mosier,A; Wassmann, R;, Cai, Z; Zheng,X; Huang,Y; Tsuruta,H; Boonjawat, J; Lantin, R. 2004, Modeling greenhouse gas emissions from rice- based production systems: sensitivity and upscaling, Global Biogeochem. Cycles, 18. GB1043, doi: 10.1029/2003GB002045.

[9] Luo, C., John,J.C; Zhou, X; Lam, K.S; Wang,T;Chameides, W.L. 2000. A nonurban ozone air pollution episode over eastern China: Observations and model simulations, J. Geophys. Res, 105:1889–1908.

[10] Levy, H., Moxim, W. J. 1989. Simulated global distribution and deposition of reactive nitrogen emitted by fossil fuel combustion. Tellus, 41: 256–271. doi: 10.1111/j.1600-0889.1989.tb00305.x

[11] Butler, T.J; and Likens, G.E. 1995. A direct comparison of throughfall plus stemflow to estimates of dry and total deposition for sulfur and nitrogen. Atmos. Environ, 29: 1253- 1265. doi: 10.1016/1352- 2310(94)00339- M

[12] Fenn, M.E; and Bytnerowicz, A. 1997. Summer throughfall and winter deposition in the San Bernardino Mountains in Southern Californa. Atmos Environ, 31: 673- 683. doi:10.1016/S1352- 2310(96)00238- 5

[13] Garten, C.T; Schwab, A.B; and Shirshac, T.L. 1998. Foliar retention of ^{15}N tracers: Implications for net canopy exchange in low- and high- elevation forest ecosystems. For. Ecol. Manage, 103: 211- 216.

[14] Fenn, M.E; and Poth, M.A. 2004. Monitoring nitrogen deposition in throughfall using ion exchange resin columns: A field test in the San Bernardino Mountians. J. Environ. Qual, 30: 2007- 2014.

[15] Fenn, M.E; and Poth, M.A. 1999. Temporal and spatial trends in streamwater nitrate concentrations in the San Bernardino Mountains in Southern California. J. Environ. Qual, 28: 822- 836. doi: 10.2134/jeq1999.00472425002800030013x

[16] Fresenius, W., K.E. Quentin, & W. Schneider, (Eds.), 1988. Water Analysis. Springer Verlag, Berlín: 804 p.

[17] NMX- AA- 074- SCFI- 1981, 1981. Análisis de aguas. Determinación del ión sulfato en aguas naturals, residuales y residuales tratadas.

[18] NMX- AA- 079- SCFI- 2001, 2001. Análisis de aguas. Determinación de nitratos en aguas naturales, potables, residuales y residuales tratadas.

[19] Fowler, D; Smith, R.I; Weston, K.J. 1995. Quantifying the spatial distribution of surface ozone exposure at the 1kmx1km scale. In: Fuhrer, Achermann (eds.) (1995): 196-205.

[20] Van Leeuwen, E. P., Draaijers, G. P. J; Erisman, J. W. 1996. Mapping wet deposition of acidifying components and base cations over Europe using measurements. Atmospheric Environment, 30:2495–2511.

[21] Park, S. U., Lee, Y.H. 2002. Spatial distribution of wet deposition of nitrogen in South Korea, Atmos. Environ, 36: 619–628. doi:10.1016/S1352- 2310(01)00489- 7

[22] Holland, E.A; Braswell, B.H; Sulzman, J; Lamarque, J. F. 2005. Nitrogen Deposition onto the United States and Western Europe: Synthesis of Observations and Models. Ecological Applications, 15: 38- 57. doi: 10.1890/03- 5162

[23] Hiltbrunner, E; Schwikowski, M; Korner, C. 2005. Inorganic nitrogen storage in alpine snow pack in the Central Alps (Switzerland). Atmos. Env, 39: 2249- 2259. doi: 10.1016/j.atmosenv.2004.12.037

[24] Fenn ME, Geiser LH (2011) Temperate sierras. Chapter 15: 175- 180. In: Pardo LH, Robin- Abbott MJ, Driscoll CT, eds. Assessment of nitrogen deposition effects and empirical critical loads of nitrogen for ecoregions of the United States. Gen. Tech. Rep. NRS- 80. Newton Square, PA: U.S. Department of Agriculture, Forest Service, Northern Research Station.

[25] Grennfelt P, Nilsson J, Critical loads for sulphur and nitrogen. Report from a workshop hel at Skokloster, Sweden. March 19- 24, 1988. The Nordic Council of Ministers Report 1988: 15. Copenhagen, Denmark. ISBN: 87- 7303- 248- 4.

[26] Fenn ME, De Bauer LI, Hernandez- Tejeda T (2002) Summary of air pollution impacts on forest in the Mexico City Air Basin. In: Fenn ME, Bauer LI, Hernandez- Tejeda T eds., Urban Air Pollution and Forests: Resources at Risk in the Mexico City Air Basin. Ecological Studies Series, Volume 156. Springer- Verlag, New York, NY.

[27] Pérez- Suárez M, Fenn ME, Cetina- Alcalá VM, Alderete A (2008) The effects of canopy cover on throughfall and soil chemistry in two forest sites in the Mexico City Air Basin. Atmósfera 21: 83- 100.

[28] Ponette- Gonzalez AG, Weathers KC, Curran LM (2010) Tropical land- cover change alters biogeochemical inputs to ecosystems in a Mexican montane landscape. Ecol. Applic. 20: 1820- 1837. doi:10.1890/09- 1125.1. 96

[29] Khoder, M.I. 2002. Atmopsheric conversion of sulfur dioxide to particulate sulfate and nitrogendioxide to particulate nitrate and gaseous nitric acid in an urban area. Chemosphere, 49: 675- 684. doi: S0045- 6535(02)00391- 0

[30] Hitchcock, D.R; Spiller, L.L; Wilson, W.E. 1980. Sulfuric acid aerosols and HCl release in coastal atmospheres: Evidence of rapid formation of sulfuric acid particulates. Atmospheric Environment, 14: 165- 182. doi: 0004- 6981/80/0201- 0165

[31] Howells, G. 1995. Acid rain and Acid water. 2nd Ed. Ellis Horwood Ltd.

[32] Pierson W.R, Brachaczek, W.W (1983) Emissions of ammonia and amines from vehicles on the road. Environ. Sci. Technol. 17: 757- 760. doi: 0013- 936X/83/0917- 0757

[33] Bishop, G.A; Peddle, A.M; Stedman, D.H. 2010. On- road emission measurements of reactive nitrogen compounds from three California cities. Environ, Sci. Technol, 44: 3616- 3620. doi: 10.1021/es903722p

Chapter 8

Sea Transport Air Pollution

Ivan Komar and Branko Lalić

Additional information is available at the end of the chapter

http://dx.doi.org/10.5772/59720

1. Introduction

The aim of this chapter is to provide an overview of the air pollution generated by diesel engines of the ocean-going ships and the technologies as well as methodologies available to reduce these emissions. This chapter begins with general significant information of the air pollutant emission from ships followed by a summary of the International Maritime Organization (IMO) regulatory MARPOL Annex VI being developed to control marine shipping emissions as well as information on the various types of the ocean-going ships and their prime movers with particular emphasis on marine diesel engines as sources of air pollution from ships. For better understanding of the formation of air pollutants from marine diesel engines, authors gave a brief overview of the working principles of marine diesel engines as well as their combustion process and chemistry of the pollutant formation during that process. Finally, the chapter concludes with an analysis of several control methods that can effectively reduce harmful pollutant emissions from marine diesel engines.

Climate change on Earth is one of the largest civilised problems at the beginning of the twenty-first century. Anthropogenic impact on the Earth's climate became one of the crucial environmental issues of modern civilisation in the late twentieth century. Therefore, nowadays the ecology and preservation of human environment have become two of the very important human activities all over the world. Besides primary pollution from the land, nowadays attention is being paid to the pollution from the ships. In recent decades, shipping industry and maritime traffic have rapidly developed. From the economic point of view, this trend, which continues today, has a very positive impact on economic development but on the other hand, a very negative impact on the environment in terms of air pollution. Exhaust gases from marine diesel engines are the primary source of emissions from ships and contribute significantly to environmental pollution. Ocean-going ships are the major contributors to global emissions of several hazardous air pollutants such as nitrogen oxides (NO_x), sulphur oxide

(SOx), fine particulate matter (PM), hydrocarbons (HCs), carbon monoxide (CO) and greenhouse gas carbon dioxide (CO_2).

The presence of these pollutants has local and global impact. Impacts on local (or regional) air quality are mainly linked to pollutants such as PM, NOx and sulphur, while CO_2 has a global impact on climate [1]. The amount of gases emitted from marine engines into the atmosphere is directly related to the total fuel oil consumption. While pollutant emissions from land-based sources are gradually decreasing, those from shipping show a continuous increase. It is estimated that by 2020, the emissions NOx and SOx from international shipping around Europe are expected to equal or even surpass the total emissions from all land-based mobile, stationary and other sources in the 25 EU member states combined (see Figure 1). It should be noted that these figures refer only to ships in the international trade and do not include emissions generated from shipping in countries' internal waterways or from ships plying harbours in the same country, which are given in the domestic statistics of each country [2].

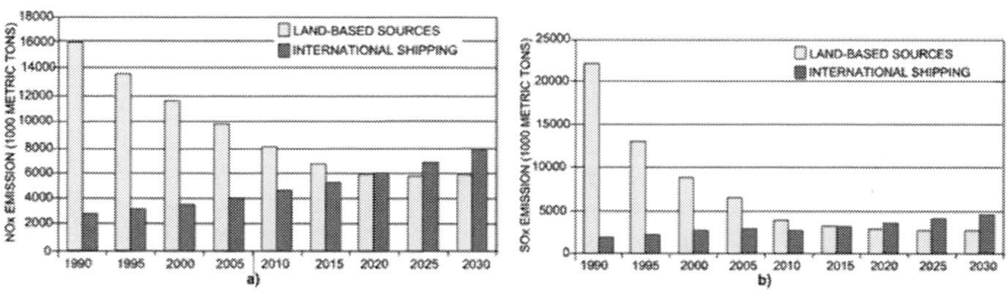

Figure 1. Emissions NOx (a) and SOx (b) 1990–2030 from land-based sources and international shipping

It has been estimated that about 90 % of the total sulphur dioxide (SO_2) and NOx emissions from ships in the North Sea, including the English Channel, originate from a zone of approximately 50 nautical miles from the coast line. International shipping was estimated to be a source of 97 % of the total SO_2 and NOx emissions in the North Sea within a distance of 100 nautical miles from the seaboard, as in [3]. The emissions of sulphur dioxide (SO_2), particulate matter (PM) and greenhouse gases (GHGs) from global shipping were increased from 585 to 1096 million tons between 1990 and 2007. The CO_2 emissions from international shipping are estimated at 943.5 million tons for the year 2007 and CO_2 emissions from global shipping are about 1 billion tons for the year 2006. International shipping is responsible for 3 % of global CO_2 emissions as in [4].

2. International regulation concerning air pollution from merchant shipping

Regulations concerning air pollution from merchant shipping are developed at the global level. Because shipping is inherently international, it is vital that shipping is subjected to uniform regulations on issues such as air emissions from ships. The shipping industry is principally regulated by the International Maritime Organization (IMO), which is a UN agency based in London and responsible for the safety of life at sea and the protection of the marine environment. IMO ship pollution rules are contained in the 'International Convention on the Prevention of Pollution from Ships' known as MARPOL 73/78, which represents the first set of regulations on marine exhaust emissions. The original MARPOL Convention was signed on 17 February 1973, but did not come into force. The current Convention is a combination of the 1973 Convention and the 1978 Protocol. MARPOL73/78 contains 6 annexes concerned with preventing different forms of marine pollution from ships:

- Annex I deals with regulations for the prevention of pollution by oil.
- Annex II details the discharge criteria and measures for the control of pollution by noxious liquid substances carried in bulk.
- Annex III contains general requirements for issuing standards on packing, marking and labelling.
- Annex IV contains requirements to control pollution of the sea by sewage.
- Annex V deals with different types of garbage, including plastics, and specifies the distances.
- Annex VI deals with gaseous emissions of ship engines and installations: the Convention regulates sulphur oxide, nitrogen oxide and particulate matter emissions from ship exhausts and prohibits deliberate emissions of ozone-depleting substances. It also contains provisions allowing for the creation of special Emission Control Areas (ECA) with even more stringent controls on air pollutant emissions. Annex VI also forbids any (deliberate) emission of an ozone-depleting substance, such as halons and chlorofluorocarbons (CFCs) as well as any new installation of equipment using these gases. Annex VI entered into force on 19 May 2005 and sets limits on nitrogen oxides, sulphur oxides and volatile organic compound (VOC) emissions from ship exhausts and prohibits deliberate emissions of ozone-depleting substances [5]. Annex VI regulation 13, nitrogen oxides (NOx), applies to diesel engines over 130 kW installed on ships built on or after 1 January 2000, excluding engines for emergency purposes such as emergency generator engine, lifeboat engine, etc.

Three different levels (tiers) of NOx control apply based on the ship construction date as follows:

- *Tier 1* entered into force in 2005 and applies to marine diesel engines installed in ships constructed on or after 1 January 2000 and prior to 1 January 2011.
- *Tier 2* entered into force in 1 January 2011 and replaced the Tier 1 NOx emission standard globally. It applies globally for new marine diesel engines installed in ships constructed on 1 January 2011 or later. Tier 2 NOx emission levels correspond to about 20 % reduction from the Tier 1 NOx emission standard. Tier 2 is applicable outside the Tier 3 designated Emission Control Areas (ECA).[1]
- *Tier 3* will enter into force in the year 2016 and it will by then apply for new marine diesel engines > 130 kW installed in ships constructed on 1 January 2016 or later when operating

inside the ECA. The Tier 3 NOx emission level corresponds to an 80 % reduction from the Tier 1 standard. The NOx emission limits are expressed as dependent on engine speed (n) in revolution per minute (RPM). These are shown in Table 1 and Figure 2 [6,7].

Tier	Effective date	NOx limit (g/kWh)		
		n < 130	130 < n < 2000	n > 2000
Tier I	2000	17	$45\ n^{-0,2}$	9,8
Tier II	2011	14,4	$44\ n^{-0,2}$	7.7
Tier III	2016	3,4	$9\ n^{-0,2}$	1,96

Table 1. NO_X limits according to MARPOL Annex VI

For engines with an engine speed lower than 130 RPM, the Tier III level is 3.4 g/kWh. When operating outside an ECA, the engine must meet the Tier II limit of 14.4 g/kWh. Engines with an engine speed higher than 130 RPM must meet even lower limits (see Table 1 and Figure 2). Any abatement technology reducing the NOx emission to the required level can be accepted.

Furthermore, MARPOL Annex VI has set a maximum global fuel sulphur limit of currently 3,5 % in weight (from 1 January 2012) for any fuel used on board a ship. Annex VI also contains provisions allowing for special SOx Emission Control Areas (ECA) to be established with more stringent controls on sulphur emissions. In an ECA, the sulphur content of fuel oil used on board a ship must currently not exceed 1 % in weight. The MARPOL Annex VI has undertaken a review with the intention to further reduce emissions from ships. The current and upcoming limits for future fuel oil sulphur contents are presented in Table 2 and Figure 2 [6,7].

Fuel sulphur cap	Area	Date of implementation
Max. 1 % S in fuel	ECA	1 July 2010
Max. 3,5 % S in fuel	Globally outside ECA	1 January 2012
Max. 0,1 % S in fuel	ECA	1 January 2015
Max. 0,5 % S in fuel	Globally outside ECA	1 January 2020

Table 2. Sulphur limits in fuel according to MARPOL Annex VI

The rules of SOx apply to all ships, no matter the date of ship construction. Although the SOx requirements can be met by using a low-sulphur fuel, the regulation allows alternative methods to reduce the emissions of SOx to an equivalent level.

1 Designated Emission Control Areas (ECA) that are defined by the IMO currently comprise the Baltic Sea, the North Sea, the English Channel, the US Caribbean Sea and the area outside North America (200 nautical miles – see Figure 3) The first sulphur Emission Control Area (ECA) was established in the Baltic and came into force internationally on 19 May 2005, and all ships were required to either use the 1.5 % low-sulphur fuel or fit an exhaust gas cleaning system as required by regulation 14 of Annex VI 12 months from this date which was 19 May 2006.

http://dx.doi.org/10.5772/59720

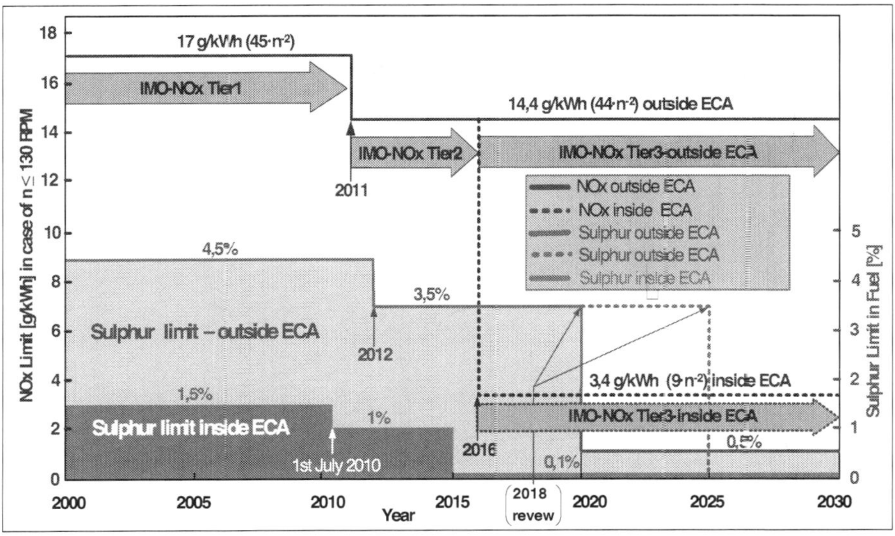

Figure 2. Reduction of NOx and sulphur in fuel on the global and ECA limit

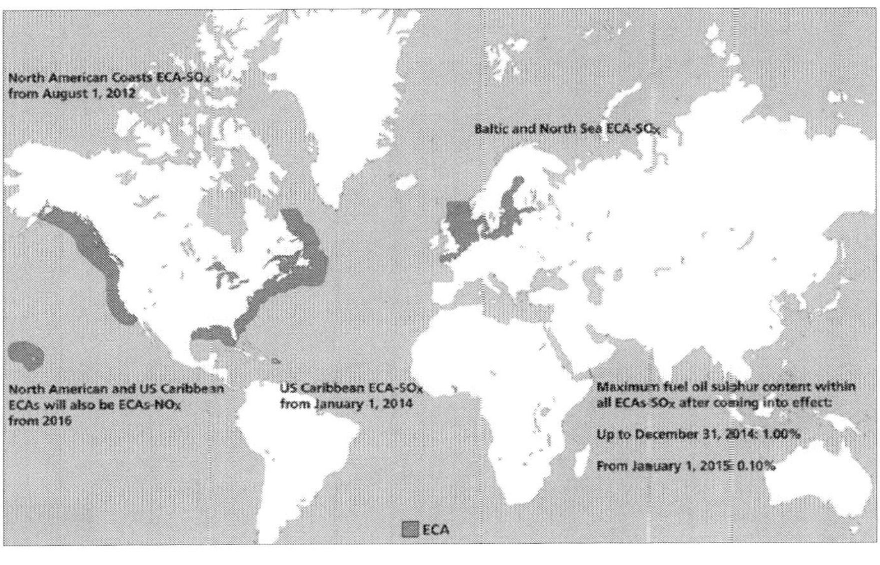

Figure 3. Existing Emission Control Areas as per MARPOL Annex VI (source: www.marlink.com)

The sulphur ECA limit which entered into force in 1 January 2015 can be met using a low-sulphur fuel with sulphur below 0.1 %. The global limit, outside sulphur ECA, can be met using fuel with sulphur content below 0.5 %, which will be required from 2020. The date of the global limit reduction may be changed to 2025 as a result of a feasibility review to be conducted no later than 2018 (see Figure 2).

In order to comply with the requirements of Regulation 14 of MARPOL, the burning of low-sulphur fuel oils (LSFOs) was introduced. There is also an EU regulative about the sulphur content in marine gas oil. Namely, in accordance with EU's marine fuel sulphur directive (1999/32/EG, Article 4 with amendment as per directive 2005/33/EC), the sulphur content in marine gas oil within the territorial waters of an EU member state may not exceed 0.1 % by weight. This applies to all ships regardless of flag. As of 1 January 2010, the sulphur content of any marine fuels may never exceed 0.1 % by weight for ships in port with the exception of short stays in port (up to 2 h).

For ships continuously operating on low-sulphur fuel oil and for those that may be fitted with an exhaust cleaning system, there is no issue. However, for ships which burn heavy fuel oil with 3.5 % sulphur content and are not equipped with scrubbing equipment, the problem of compliance is much greater as large quantities of fuel are being mixed during the oil change-over to low-sulphur fuel oil. MARPOL Annex VI Regulation 14 requires those ships using separate low-sulphur fuel to comply with this regulation and in entering or leaving an Emission Control Area, shall carry a written procedure showing how the fuel oil changeover is to be done, allowing sufficient time for the fuel oil service system to fully flushed out all fuel oils exceeding the applicable sulphur content prior to entry into a sulphur Emission Control Area. The volume of low-sulphur fuel oils in each tank as well as the date, time and position of the ship when any fuel oil changeover operation is completed prior to the entry into an ECA or commenced after exit from such an area shall be recorded in such logbook as prescribed by the Annex VI Regulation.

Furthermore, ships are also an important source of greenhouse gas (GHG) pollutants. According to the Green House Gas study by the IMO consensus, international shipping emitted 843 million metric tonnes of carbon dioxide, 2.7 % of the global CO_2 emissions in 2007. Including domestic shipping and fishing ships larger than 100 gross tonnes (GT),[2] the amount would increase to 1.019 billion tonnes, i.e. 3.3 % of the global CO_2 emissions. At the present trend, this percentage could go two or three times higher from the present by 2050 emissions [8, 9].

In order to control this CO_2 emission from shipping, the first formal CO_2 control regulations were adopted by the IMO at the 62nd session of the Marine Environment Protection Committee (MEPC) in July 2011. The amendments to MARPOL Annex VI included the addition of Chapter 4 on regulations on energy efficiency for ships to make mandatory the Energy Efficiency Design Index (EEDI) for new ships and the Ship Energy Efficiency Management Plan (SEEMP)

---

2 Gross tonnage (GT) determined under the IMO 1969 Tonnage Convention represent the total volume of of ship's enclosed spaces measured in register tonne (RT) where 1RT=100 f3 or 2,83 m3, conversely. Net tonnage (NT) determined under the IMO 1969 Tonnage Convention represents the volume of cargo spaces (cargo holds and cabins) in register tonne (RT).

for all ships. The regulations apply to all ships of 400 GT and above and are entered into force in 1 January 2013.

The basic formulation of EEDI is based on the ratio of total CO_2 emission per tonne mile. As CO_2 depends upon fuel consumption and fuel consumption depends upon the total power requirements, eventually this EEDI formulation has certain impact on ship design parameters and hydrodynamics. The SEEMP establishes a mechanism for a shipping company and a ship to improve the energy efficiency of ship operations. The SEEMP provides an approach for monitoring ship and fleet efficiency performance over time using. The results from the study that IMO ordered from Lloyd's Register and Det Norske Veritas to estimate the impact of the new requirements show that the EEDI will, as new ships are built, gradually reduce the emissions from the world fleet with 3 % in 2020, 13 % in 2030 and 30 % in 2050. The SEEMP will not directly mandate an emission reduction, but by increased awareness of costs and reduction potentials, the study estimated the reduction to be between 5 and 10 % from 2015 onwards [9,10].

3. Sea ship classification and quantification

As per rules of the Shipping Classification Societies, ship is defined as 'a floating unit intended for sea-going service with length greater than 12 meters and with GT greater than 15, or which carries more than 12 passengers. The present definition does not apply to ships of war and troopships'. Marine ocean-going ships are generally very large ships designed for deepwater navigation. Depending on the nature of their cargo, ships can be divided into different categories, classes and types. A majority of these ships can be classified as one of the following: tanker, bulk carrier, container ship, ro-ro ship, general cargo ship, reefer ship and passenger ship. There are also smaller ship types, which are not included in the largest categories of ship, as fishing ships intended and equipped for fishing or exploiting other living resources of the sea; tugs, a ship specially constructed and equipped for towing and/or rescuing and salvage of ships or other floating units; ships used by authorities which include the following types: pilot boats, rescue ships, police boats, custom boats, etc.; training ships provided for training of marine personnel gaining training and practical marine experience to develop seafaring skills suitable for a professional career at sea and provided with special equipment and arrangements suitable for that purpose (teaching rooms, accommodation spaces for teachers and trainees, etc.); research ship, a ship without cargo spaces, engaged in scientific research, noncommercial expeditions and surveys, carrying scientists, technicians and members of expeditions, and provided with special equipment and arrangements suitable for that purpose (i.e. laboratories, accommodation for research personnel, etc.); supply ship, a ship mainly intended and equipped for the carriage of special personnel, special materials and equipment which are used to provide facilities to offshore units and other marine installations, as well as to provide assistance in performing special activities; and icebreakers and recreational ships such as yachts classified as recreational craft for personal or commercial use, having hull length greater than 12 m, having facilities and accommodation for extended navigation, authorised to carry not more than 12 passengers, excluding crew. The following is a brief description of the characteristics of the main types of ocean-going ships:

Tanker is a merchant ship designed to transport liquids or gases in bulk. The major types of tanker ship include the oil tanker, the chemical tanker and gas carrier.

Oil tanker is a ship which is constructed primarily to carry oil in bulk and comes in two basic types: the crude carrier, which carries crude oil, and the clean product tanker, which carries the refined products, such as petrol, gasolene, aviation fuel, kerosene and paraffin. Tankers also include ship types such as combination carriers. Combination carrier is a general term applied to ships intended primarily to carry oil or dry cargoes, including ore, in bulk (ore/oil ships, oil/bulk/ore – OBO). These cargoes are not carried simultaneously. Generally they are constructed with a single deck, two longitudinal bulkheads and a double bottom throughout the cargo length area and intended primarily to carry ore cargoes in the centre holds or of oil cargoes in centre holds and wing tanks.

Chemical tankers are ships which are constructed generally with integral tanks and intended primarily to carry chemicals in bulk.

Gas carrier can be divided into two types: the *LNG tanker* carries liquified natural gas and the *LPG tanker* carries liquified petroleum gas. Tankers can range in size of capacity from several hundred deadweight tons (DWT),[3] which include ships for servicing small harbours and coastal settlements, to the real giants of several hundred thousand DWT: the VLCC (very large crude carrier) of between 200,000 and 300,000 DWT and the ULCC (ultra large crude carrier) of over 300,000 DWT.

Bulk carriers are sea-going self-propelled ships which are constructed and intended primarily to carry dry cargoes in bulk such as ore, coal, pulp, rock, cement, scrap metal, grain, flour, rice, fertilisers, sugar or any cargo that travels in bulk. Bulk carriers range from about 25,000 DWT ('handysize') through the medium-size ('Panamax') ships of about 75,000 DWT, to the giant ('capesize') ships of over 200,000 DWT.

Container ships are a type of dry cargo ships specially designed and equipped with the appropriate facilities for carriage of containers. They carry standardising container at 20-feet long (TEU – twenty-foot equivalent unit) or 40-feet long (FEU – forty-foot equivalent unit). Today's container ships are being built to take up to18,000 TEU.

Ro-ro (roll on/roll off) is a cargo ship (ferry) specifically designed for the carriage of vehicles, which embark and disembark on their own wheels, and/or goods on pallets or in containers which can be loaded or unloaded by means of wheeled vehicles. Another type of ro-ro is a passenger ship (ROPAX). The acronym ROPAX (roll-on/roll-off passenger) describes a ro-ro ship built for freight vehicle transport along with passenger accommodation. Ro-ro ships have built-in ramps that allow the cargo to be efficiently rolled on and off the ship when in port. The ramps and doors may be stern-only, bow and stern or side for quick loading.

3 Deadweight tonnage (also known as deadweight abbreviated to DWT) is a measure of how much weight a ship is carrying or can safely carry. It is the sum of the weights in metric tonnes of cargo, fuel, freshwater, ballast water, provisions, passengers and crew.

General cargo ship is a ship intended for the carriage of general cargo which will not be carried in containers.

Refrigerated cargo ship or reefer ship is a ship (excluding liquefied gas carriers and fishing ships) specially intended to carry permanently refrigerated cargoes such as fruits, vegetables, dairy products, fish and meat and has fixed refrigeration installations and insulated holds. Excluding the temperature control, the reefers are similar to other dry cargo ships or containers.

Passenger ship as per rules of the ship's classification society is a self-propelled ship with a permission to carry more than 12 passengers, specially designed and equipped for that purpose, with a single or multi-deck hull and superstructure and with cabin accommodation for passengers.

As of 1 January 2013, there were 48,732 merchant ships (597,709,000 GT) involved in international trade, registered under the flags of over 150 nations (see Figure 4).

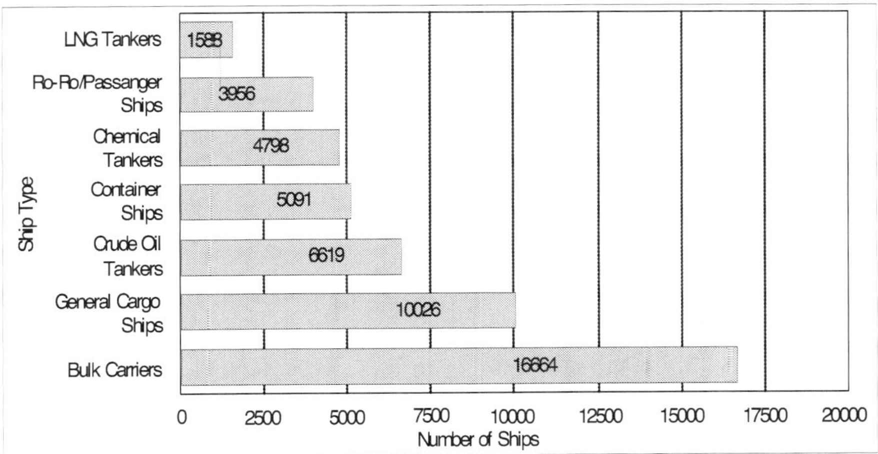

Figure 4. Number of the main types of ocean-going ships in the world merchant fleet as of 1 January 2013 (source: http://www.statista.com)

3.1. Ship engines

The power needed on ships is generated through main and auxiliary engines often called prime mover in the literature and can be sorted as diesel engine, gas turbine, steam turbine and electric motor. The diesel engine is the most common prime mover in the merchant marine, mainly due to its low fuel consumption in comparison with other prime movers. Power ranges between 0.25 MW for the smallest high-speed engines and 100 MW for the for the biggest low-speed marine diesel engines. The main advantages of diesel engines are the following: it is relatively insensitive to fuel quality; it can be operated by light fuel as well as the heaviest residual fuels; and it has high reliability and high efficiency. In the other hand, the main disadvantages of diesel engines are pollutant emissions, low power-to-weight ratio if compared with gas turbine and vibration and noise.

From the application viewpoint, three main types of diesel engines are available: low-speed diesel engines (rpm<200), medium-speed diesel engines (200<rpm<1000) and high-speed diesel engines rpm>1000).

From the construction viewpoint, two types can be distinguished: two-stroke low-speed engines and four-stroke engines (medium or high speed). Low-speed diesel engines are dominant in the deep sea tanker, bulk carrier and containership sectors. Such types of engines are used as ship's propulsion engine without gearbox, i.e. directly connected to the propeller shaft system. These engines are currently the most efficient in terms of the specific fuel consumption, but NOx emission level from these engines is very high in comparison with medium- or high-speed marine diesel engines (see Figure 5b). The diesel engines have specific fuel consumptions around 160–185 g/kWh, against the 220–240 g/kWh of gas turbines and 300 g/kWh of steam turbines (see Figure 5a).

Medium-speed engines are used for auxiliaries such as alternators and for the main propulsion engine, with a gearbox to provide a propeller speed reduction, for smaller cargo ships, ferries, passenger cruise liners, ro-ro carriers, supply ship, icebreakers, etc., while high-speed engines are used as propulsion and for auxiliaries in smaller ships as fishing ships, tugs, pilot ships, recreational ships, etc.

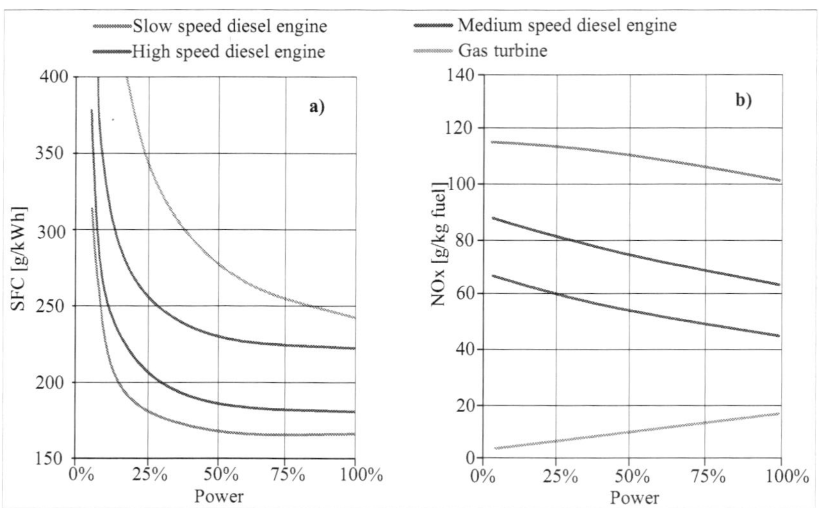

Figure 5. a) Specific fuel consumption of prime movers b) NOx emission ratio for prime movers

http://dx.doi.org/10.5772/59720

3.2. Marine diesel engine: Type and working principles

The diesel engine is reciprocating internal combustion engine where chemical energy of the fuel is converted into thermal energy by means of combustion reactions of the fuel, and then the thermal energy is converted into mechanical work. The actual cycle inside the engine can be done either in four strokes (two crank revolutions) or in two strokes (one crank revolution) [11]. A stroke is defined as the distance travelled by the piston between the top dead centre (TDC) and the bottom dead centre (BDC). Cycles in the four-stroke engine are compression, power, exhaust and intake strokes. During compression stroke, the piston moves upwards from BDC to TDC. Inlet and exhaust valves are closed, and the combustion air is compressed and thus increases air pressure and temperature (110 to 220 bar and 600 °C to 800 °C depending on the type of engine). Fuel is injected several crank degrees before TDC and ignited by the high temperature of the compressed air. Combustion starts at the end of the compression stroke. The combustion is continued over a considerable crank angle after TDC, while the combustion gases expand and perform work on the piston, forcing it down. This is power stroke. Towards the end of the stroke, the exhaust valve opens which releases the combustion gases into the exhaust manifold or exhaust gas receiver. Thereafter during the exhaust stroke, the piston moves from BDC to TDC. The exhaust valve is open and the rest of the combustion gases are forced out of the cylinder by the upward stroke of the piston. The gases that remain in the cylinder are dispelled by a scavenging process; the inlet valve is opened early, whereas the exhaust valve is closed late, so that both are open at the same time (overlap period). Thereafter during the intake stroke, the piston moves downwards from TDC to BDC. The inlet valve is open and the exhaust valve closed, while the cylinder is filled with a charge of fresh air and thereafter is ready for the compression stroke. A complete cycle takes four strokes. However, the power stroke is the only useful one, which suggests that there is a useful stroke every two crank revolutions.

Slow-speed marine diesel engine (65–200 rpm) operates on the two-stroke cycle. This means that this engine has one working or power stroke per every crank revolution. The main difference comparing two-stroke with the four-stroke cycle is that charging and exhaust take place without the piston enforcing the process. During the compression stroke, the inlet ports and exhaust valve are closed and a volume of air is trapped into the cylinder. The piston moves upwards to TDC, thus compressing this combustion air and causing a temperature and pressure rise (110 to 160 bar and 600 °C to 800 °C depending on the type of engine) that is sufficient to ignite the fuel that has been injected few degrees before TDC. At the end of the compression stroke, combustion starts and continues several degrees during power stroke. The combustion gases expand and perform work on the piston, forcing it down from TDC to BDC. At the end of expansion, exhaust valve opens and combustion gases blow down to exhaust receiver pressure. By the time the inlet ports are open, the cylinder pressure will have reached a pressure lower than that of the scavenging air, so scavenging starts. These processes take place in one stroke. Scavenging, which starts while the piston moves downwards, is completed while the piston moves upwards. Both the inlet ports and exhaust valve are opened, and fresh air from scavenging air receiver enters the cylinder, forcing the exhaust gases out. The scavenging and compression processes take place in another stroke. The process described here is for uniflow scavenging. It is necessary to compress the scavenging air by the turbocharger in order to scavenge the cylinder. The principal scheme of a turbocharged marine diesel engine is shown in Figure 6 as in [12]. Turbocharger (TC) is composed of the compressor (C) driven by the gas turbine (T) that receives its power from the heat energy of the exhaust gases flowing through. The compressor and the turbine are directly coupled and they are built together in a common housing. The gas turbine, usually one-stage axial type, is located after an exhaust gas receiver (EGR) which collects the exhaust gases from all the engine cylinders. The one-stage centrifugal-type compressor feeds compressed air through scavenging air cooler into scavenging air receiver (SAR) that supplies all the engine cylinders.

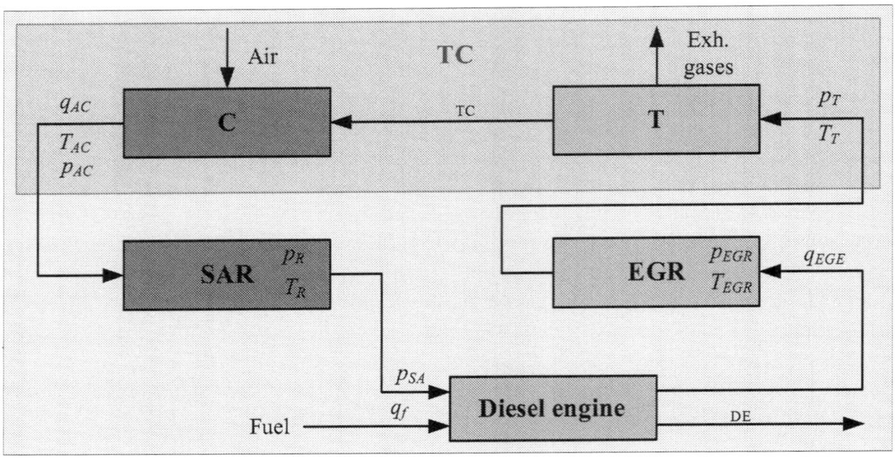

Figure 6. Principal scheme of turbocharging marine diesel engine

3.3. Types of fuel oils for marine diesel engine

Marine diesel engines use three types of liquid fuels standardised by ISO 8217 fuel standard for marine distillate fuels as gas oil (GO) with max. density of 890 kg/m^3 at 15 °C, marine diesel oil (MDO) with max. density from 890 to 900 kg/m^3 at 15 °C and heavy fuel oil (HFO) with max. density from 920 to 1010 kg/m^3 at 15 °C as in [13]. ISO 8217 fuel standard for marine distillate fuels also defines the values of the essential properties of each type of the fuel. The most important properties of marine fuels are as follows:

- Kinematic viscosity, expressed as mm^{2}s^{-1}, is a measure for the fluidity of the fuel at 50 °C. The viscosity of a fuel decreases with increasing temperature. The moment the fuel leaves the injectors, the viscosity must be within the limits prescribed by the engine manufacturer in order to obtain an optimal spray pattern; otherwise, it will lead to poor combustion, deposit formation and energy loss.

http://dx.doi.org/10.5772/59720

- Density expressed as kg/m^3 at 15 °C is an indicator of the ignition quality particularly for the low-viscosity HFOs.
- The sulphur content which depends on the crude oil origin and the refining process. When a fuel burns, sulphur is converted into sulphur oxides which has environmental implications. Maximum permitted sulphur content in the marine fuels is 3,5 %.
- Flash point is the temperature at which the vapours of a fuel ignite. The minimum flash point for marine fuels is 60 °C as per the IMO SOLAS Convention.
- Carbon residue is determined by a laboratory test performed under specified reduced air supply. It gives an indication of the amount of hydrocarbons in the fuel which have difficult combustion characteristics.
- Water in fuel is a contaminant and does not yield any energy. The percentage of water in the fuel can be translated into a corresponding energy loss for the customer. Water is removed on board the ship by centrifugal purification.
- Ash content is a measure of the metals present in the fuel (for GO and MDO max. 0,01 mass % and for HFO from 0,04 to 0,15 mass %), either as inherent to the fuel or as contamination.
- Vanadium and nickel are elements found in some heavy fuel oil molecules. Upon combustion, vanadium oxides are formed, and some have critical melting temperatures. The most critical are the double oxides/sulphates with sodium. The maximum permitted quantity is 50 to 450 mg/kg.
- Cetane index is a measure for the ignition quality of the fuel in a diesel engine. The higher the speed of the engine required, the higher the cetane index.
- Heating value (calorific value) is defined as the amount of heat that is released during combustion of 1 kg of fuel. Assumed that after combustion the water content in the fuel is present as vapour, the condensation heat is not included in the heating value and it is referred to as the net calorific value or lower heating value. Standard net calorific value for GO is 44,000 kJ/kg, for MDO 42,000 kJ/kg and for HFO 40,500 kJ/kg.

4. Combustion process in the marine diesel engines and formation of the air pollutants

4.1. Combustion stoichiometry

The aim of the combustion stoichiometry is to determine the required amount of air and fuel in order to achieve complete combustion. A stoichiometric mixture contains the exact amount of fuel and oxidiser, so that after combustion is completed, all the fuel and oxidiser are consumed to form combustion products. This ideal stoichiometric mixture approximately yields the maximum flame temperature, as all the energy released from combustion is used to heat the products. As in references [14], combustion stoichiometry for a general hydrocarbon fuel (CαHβOy) can be expressed by equation (1) and it can be applied only for single-component hydrocarbons (HC):

$$C_\alpha H_\beta O_y + \underbrace{\left(\alpha + \frac{\beta}{4} - \frac{y}{2}\right) \cdot \left(O_2 + 3{,}76N_2\right)}_{\text{stoichiometric air amount}} \rightarrow \alpha CO_2 + \frac{\beta}{2} H_2O + 3{,}76 \cdot \left(\alpha + \frac{\beta}{4} - \frac{y}{2}\right) \tag{1}$$

Typical approaches for multiple-component hydrocarbon fuels develop the stoichiometric combustion using the general principle of atomic balance, making sure that the total number of C, H, N and O atoms is the same in the products and the reactants (e.g. multiple-component mixture of a 95 % methane (CH_4) and 5 % hydrogen (H_2)):

$$0{,}95CH_4 + 0{,}05H_2 + 1{,}925\left(O_2 + 3{,}76N_2\right) \rightarrow 0{,}95CO_2 + 1{,}95H_2O + 7{,}238N_2 \tag{2}$$

If less air than the stoichiometric amount is used, the mixture is described as rich fuel or rich mixture, and if excess air is used, the mixture is described as lean fuel or lean mixture. For this reason, it is appropriate to determine the amount of the combustible mixture using one of the following methods: a)Fuel-air ratio (FAR), b) equivalence ratio (Φ) and c) percent excess air (% AE).

a. *Fuel-air ratio (FAR)* or f is the actual ratio of fuel mass m_f and air mass m_a and it is expressed as

$$f = \frac{m_\mathrm{f}}{m_\mathrm{a}} \tag{3}$$

and it is usually bounded by 0 and ∞. For a stoichiometric mixture, equation (3) becomes

$$f_s = \left.\frac{m_f}{m_a}\right|_{ST} = \frac{M_f}{\left(\alpha + \frac{\beta}{4} - \frac{y}{2}\right) \cdot 4{,}76 \cdot M_a} \tag{4}$$

where M_f is the molar mass of fuel and M_a molar mass of air which is approximately 28.96 kg/kmol. The stoichiometric mixture fuel-air ratio of the most hydrocarbon fuels is bounded by 0.05 and 0.07. Air-fuel ratio (AFR) is reciprocal of FAR and it is expressed as AFR=f^{-1}.

b. *Equivalence ratio (Φ)* is the actual ratio of fuel-air ratio f to the stoichiometric fuel-air ratio f_s:

$$\Phi = \frac{f}{f_s} = \frac{m_{as}}{m_a}, \tag{5}$$

and its value is bounded by 0 and ∞. $\Phi < 1$ is a lean mixture; $\Phi = 1$ is a stoichiometric mixture; and $\Phi > 1$ is a rich mixture. The fuel in the combustion process must be mixed with a greater amount of air than in stoichiometric mixture because it is not possible to bring the ideal amount of air to each fuel molecule in order to mix them perfectly so that complete combustion is achieved. In the combustion analysis, an alternative variable lambda (λ) is often used by engineers. Lambda is the ratio of the actual air-fuel ratio to the stoichiometric air-fuel ratio defined as

$$\lambda = \frac{AFR}{AFR_s} = \frac{1}{f / f_s} = \frac{m_a}{m_{as}} = \frac{1}{\Phi} \tag{6}$$

c. *Percent excess air* (% AE) is the amount of air in excess of the stoichiometric amount and it is defined as

$$\%EA = 100 \cdot \frac{m_a - m_{as}}{m_{as}} = 100 \cdot \left(\frac{m_a}{m_{as}} - 1 \right) \% \tag{7}$$

4.2. Combustion process in marine diesel engine

Process of fuel combustion is comprised of the following steps: entry of fuel jet into the combustion chamber, disintegration of the jet into droplets, decomposition of larger droplets into smaller, droplet heating, droplet evaporation, mixing of fuel vapour with the surrounding air, simultaneous auto-ignition of fuel mixture in several places, continued evaporation of the droplets and burning around (diffusion combustion), formation of soot during combustion in an area near droplets, temperature drop and slowing reaction due to expansion in the cylinder.

While the combustion temperature is still high, it is necessary that the soot particulate finds their reactants (oxygen) to complete combustion reaction. The phases until the simultaneous ignition of fuel mixture represent a delayed auto-ignition in several places which can be defined as the time or engine crank angle that elapses from the beginning of fuel ignition to the auto-ignition of the mixture.

A good spatial distribution of fuel affects the proper and economical operation of the engine. To achieve a good spatial distribution of fuel, it must be injected at a rate of about 150–400 ms^{-1}, which requires a pressure of over 80 MPa.

Dispersion quality is determined by the injection speed, fuel surface tension, fuel viscosity, density of air in the cylinder, turbulence and cavitation in the nozzle. Better turbulence, mixing with air and combustion can be achieved by better penetration and propagation of jet fuel.

In marine diesel engine, injection is performed by injectors with the nozzles that direct the fuel into the cylinder space. Under the influence of aerodynamic forces of compressed air, fuel jet expands and breaks down into small droplets. The quality of fuel atomisation is defined by a mean diameter of the droplets and their uniformity. Better fuel dispersion is achieved with the smaller diameters of the nozzle holes, greater injection pressure and higher compression pressure inside the cylinder. The combustion process in a diesel engine can be divided into four phases (see Figure7).

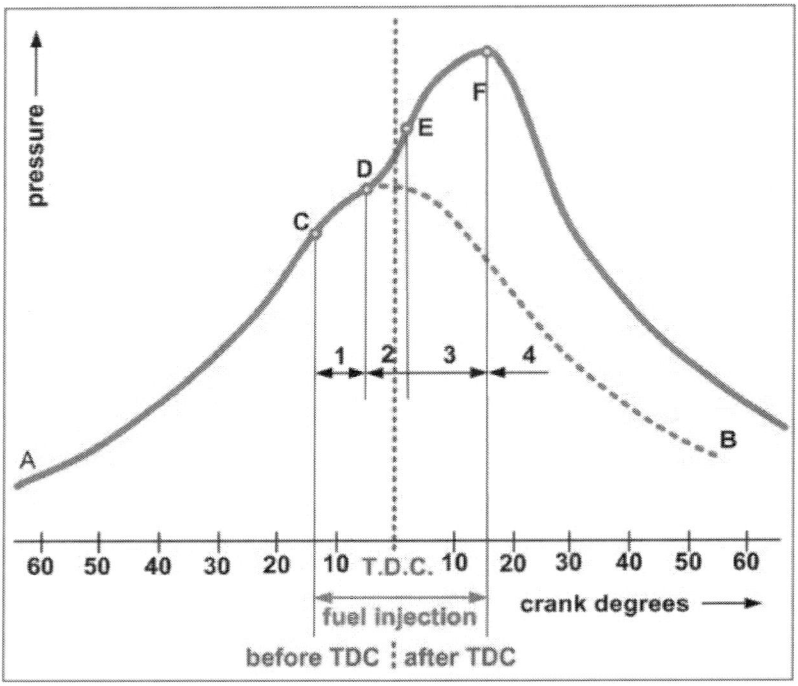

Figure 7. Phases of the combustion process

The first phase, 'ignition delay, curve C-D', defines the period from the beginning of injection until the ignition starts and has an impact on the pollutant formation. This period defines fuel atomisation, evaporation, mixing and the reaction beginning. At sufficiently low turbulence, local flame fronts are created and produce high temperature without soot.

The second phase, 'uncontrolled combustion, curve D-E', is a homogeneous phase of combustion. At this stage, there are sudden ignition and combustion of the already prepared fuel mixture during the delayed ignition phase. Combustion begins simultaneously in several places and conducts intensively, and there is a sudden increase in the pressure and the temperature.

The third phase, 'partially controlled combustion, curve E-F', is diffusion combustion when the fuel droplets vaporise from the surface. Evaporated fuel is mixed with air, and combustion speed is limited by the rate of fuel evaporation and the speed of creating fuel mixture.

The fourth phase, 'after burning, curve F till the end', is the final part of the combustion and it takes about half of the total combustion time duration. During that phase, reaction slows due to expansion and decrease amounts of reactants, and a part of soot that is created during combustion leaves the cylinder as portion of emissions.

4.3. Adiabatic flame temperature

One of the most important features of a combustion process is the highest temperature of the combustion products that can be achieved. The temperature of the combustion products will be the highest when there are no heat losses to the surrounding environment and when all energy released from combustion is used to heat the products. Constant pressure adiabatic temperature calculation, using a mean specific heat capacity method, can be performed for the lean and the rich combustion mixture as in [14]:

a. *For a lean mixture ($\Phi < 1$):*

$$T_{\mathrm{AFT}} = T_{\mathrm{R}} + \frac{\Phi \cdot f_s \cdot LHV}{\left(1+\Phi \cdot f_s\right) \cdot \bar{c}_{p,\mathrm{P}}} \tag{8}$$

where T_R represents the temperature of the reactants, i.e. fuel which has the compression temperature (T_2) after injection and ignition delay, and $\bar{c}_p$ is an average specific heat capacity of the mixture.

b. *For a rich mixture ($\Phi > 1$):*

$$T_{ATP} = T_R + \frac{f_s \cdot LHV}{\left(1+\Phi \cdot f_s\right) \cdot \bar{c}_{p,\mathrm{P}}} \tag{9}$$

5. Formation of the air pollutants during combustion process in marine diesel engine

The major pollutants in diesel exhaust emissions are a direct result of the diesel combustion process itself. Typical concentrations of exhaust gas emissions from marine diesel engine largely comprise nitrogen approximately 76 %, oxygen abt. 13 %, carbon dioxide (CO_2) abt. 5 % and water vapour abt. 5 %, with smaller quantities of pollutants: nitrogen oxide (NO_X) abt. 1200 ppm, sulphur oxide (SO_X) abt. 640 ppm, carbon monoxide (CO) abt. 60 ppm, partially reacted and non-combusted hydrocarbons (HC) abt. 180 ppm and particulate matter (PM) abt. 120 mg/Nm^3 [15]. The composition of this gas mixture, liquids and solids that are actually emitted into the air will vary depending on engine type, engine power, operating conditions as well as fuel and lubricating oil type and also depends on whether the emission control system is present. Pollutant formation in marine diesel engine is discussed below.

5.1. Nitrogen oxides (NOx)

Nitrogen oxides (NOx) generate thermally from atmospheric nitrogen oxygen in the intake or scavenging air. The oxidation of atmospheric nitrogen is influenced by local conditions in the combustion chamber, such as the maximum cylinder pressure, local peak temperatures and local air- fuel ratio. The primary reaction product is nitric oxide (NO) by approximately 90 % of the volume, but about 5 % of it is converted into nitrogen dioxide (NO_2) later in the combustion cycle, during expansion and during the flow through the exhaust system. At the same time, a very limited proportion of nitrous oxide N_2O is also formed. Further oxidation of NO to NO_2 subsequently continues at ambient temperatures after the exhaust gases have passed out to the atmosphere. Nitrogen oxide is of particular concern because of its detrimental effects on respiration and plant life, as well as its significant contribution to acid rain. In addition, NOx, together with volatile organic compounds (VOC), is also involved in a series of photochemical reactions that lead to an increase in troposphere ozone which, in turn, adversely affects human health and natural vegetation. These problems are only pronounced on land and especially in urban areas.

Analysis of the combustion process in the cylinder and the reactions which are involved in formation of NO has identified three main sources of NO formation of which, as mentioned above, some is converted to NO_2 to give the NOx mixture. These sources are thermal NO formation, prompt NO formation and fuel source. A majority of the NO emission is generated by internal combustion engines through the thermal process.

a. *Thermal nitric oxide (NO) formation*

During the combustion process in diesel engine, high temperatures are reached. Around 1700 K, and above up to 2500 K, sufficient thermal energy is available to dissociate oxygen, nitrogen and also other molecules formed during the combustion process itself. The recombination of the elements leads to the formation of NO. The reaction processes are quite slow so that most nitrogen oxides are formed during the mixing of the stoichiometric combustion gases with excess air in the cylinder. In low- and medium-speed diesel engines, by far the most important part of NOx is generated in the thermal NO process.

Formation of nitric oxide can be represented with three chemical reactions based on Zeldovich mechanism as in [16]:

$$N_2 + O \rightarrow NO + N \tag{10}$$

$$N + O_2 \rightarrow NO + O \tag{11}$$

$$N + OH \rightarrow NO + H \tag{12}$$

The first two reactions show the formation of nitric oxide for the lean mixture and the third for the rich mixture. The first reaction is the rate-limiting step due to its very high temperature activation. The high activation energy is required to break the triple bond in the nitrogen molecule (:N≡N:), which occurs at high combustion temperature; this is named thermal nitrogen monoxide (NO). The formation rate of thermal NO is practically insignificant if the temperature is below 1700 K. On the other hand, if the temperature rises, especially over 2000 K, the formation of thermal NO is strongly accelerated. The formation of thermal NO may be reduced by lowering and controlling the temperature peaks and minimising flue gas residence at high temperatures. As in [17], the equation for the total formation rate of thermal nitrogen oxides (NOx) is

$$\frac{d\left[NO_x\right]}{dt}=\frac{6\cdot 10^{16}}{T^{0,5}}\cdot e^{\left(\frac{-69090}{T}\right)}\cdot\left[N_2\right]\cdot\left[O_2\right]^{0,5}, \tag{13}$$

where T is absolute flame temperature (K), N_2 nitrogen molecule concentration (molcm^{-3}), O_2 oxygen molecule concentration (molcm^{-}) and dNOx/dt nitrogen oxide speed formation (molcm^{-3}).

b. *Prompt nitric oxide (NO) formation*

Prompt nitric oxide can be formed promptly at the flame front by the presence of hydrocarbon radicals produced only at the flame front at relatively low temperature. Nitric oxide generated via this route is named 'prompt nitric oxide (NO)'. Hydrocarbon (HC) radicals react with nitrogen molecules with the following sequence of reaction steps:

$$CH+N_2\rightarrow HCN+N \tag{14}$$

$$HCN+N\rightarrow\cdots\rightarrow NO \tag{15}$$

Nitrogen reacts with an HC radical to produce hydrogen cyanide HCN, and further, HCN reacts with nitrogen to produce nitric oxide via a series of intermediate steps. In contrast to thermal NO mechanisms that have activation temperature above 1700 K from (160), prompt NO can be formed starting at low temperature, around 1000 K as in [17].

c. *Fuel sources of NO formation*

NO formation from fuel becomes important when using heavy fuel oil because such fuels contain more organic nitrogen than marine diesel oil and other distillate fuels. Heavy fuel oil can contain up to 0.5 % nitrogen which increases the total NOx emission by as much as 10 %.

5.2. Sulphur oxides (SOx)

Formation of sulphur oxides (SOx) in the exhaust gases is caused by the oxidation of the elemental sulphur in the fuel into sulphur monoxide (SO), sulphur dioxide (SO_2) and sulphur trioxide (SO_3) during the combustion process. SOx emissions in diesel engine exhaust gas mostly comprise of sulphur dioxide and a small amount of sulphur trioxide. The stable products such as sulphur dioxide (SO_2), hydrogen sulphide (H_2S), carbon disulfide (CS_2) and disulfide (S_2) are created during the combustion of the rich mixtures. The radical sulphur monoxide (SO) reacts with oxygen (O_2) to produce sulphur dioxide (SO_2) at high temperatures. The amount of sulphur dioxide emissions depends on the sulphur content of the fuel used and cannot be controlled by the combustion process. Furthermore, sulphur trioxide (SO_3) cannot be created in the combustion under fuel-rich conditions, even when the combustion is near the stoichiometric point. However, if there is even a 1 % air excess, sulphur trioxide rapidly increases in its quantity. Typically, the amount of SO_3 is 5 % of the amount of sulphur oxides (SO_2 and SO_3). For example, if the fuel contains 3 % sulphur, the volume of SOx generated is around 64 kg per tonne of fuel burned; if fuel with 1 % sulphur content is used, SOx emission amount is about 21 kg per tonne of fuel burned as in [18]. SOx formed from diesel exhaust is corrosive and is partly neutralised by an engine's lubricating oil which is used as a typical base. Moreover, sulphur oxides (SOx) combine with moisture to form sulphuric acid (H_2SO_4), which is then excreted in the form of acid rain. It has a harmful effect on plants and human health and can damage many objects including buildings. Sulphur dioxide emissions also negatively impact human health; sulphate particles particularly can induce asthma, bronchitis and heart disease.

5.3. Carbon monoxide (CO)

The formation of carbon monoxide (CO) is a result of incomplete combustion of organic material, which is due to a lack of oxygen or low temperature at some points in the combustion chamber. Also the same reasons lead to the formation of hydrocarbons (HC). Hydrocarbons can also be formed from evaporation of the lubrication oil towards the end of the firing period. In diesel engines, the formation of carbon monoxide is determined by the air/fuel mixture in the combustion chamber, and since diesel fuel has a consistently high fuel-air ratio and the efficient combustion process, formation of this toxic gas is minimal. Nevertheless, insufficient combustion can occur if the fuel droplets in a diesel engine are too large or the level of turbulence is insufficient or swirl is created in the combustion chamber. When burning heavy fuel oil, the hydrocarbon emissions are lower than from the light fuel oil combustion due to lower evaporating level.

5.4. Hydrocarbon (HC)

Hydrocarbon (HC) emissions as fraction of the exhaust gases from diesel engines predominantly consist of unburned or partially burned fuel and lubricating oils as a result of insufficient temperature. This often occurs near the cylinder wall where the temperature of the air/fuel mixture is significantly less than in the centre of the cylinder. In the atmosphere, the hydrocarbons are subjected to photochemical reactions with nitrogen oxides forming the ground-level ozone and smog. Hydrocarbon (HC) emissions are represented as total hydrocarbons (THC) or as non-methane hydrocarbons (NMHC), as in [19].

5.5. Particulate matter (PM)

Particulate matter is a mixture of organic and inorganic substances largely comprising elemental carbon, ash minerals, heavy metals, condensed sulphur oxides, water and a variety of unburned or partially burned hydrocarbon components of the fuel and lubricating oils. More than half of the total particulate mass is soot (inorganic carbonaceous particles), of which the visible evidence is smoke. Some of the fuel particles do not burn completely, and they are emitted as droplets of heavy liquid or carbonaceous material. The incomplete burning is a result of locally low quantities of excess air. A mistimed or otherwise poorly operating fuel injection and poor mixing of fuel within the cylinder also result in incomplete combustion and increased the particulate matter emissions. Soot particles (unburned – elemental carbon) are not themselves toxic, but they can cause the build-up of aqueous hydrocarbons (HC), and some of them are believed to be carcinogens. Particulates constitute no more than around 0.003 % of the engine exhaust gases. Almost the entire diesel particle mass is in the fine particle range of 10 microns or less in diameter (PM_{10}). Approximately 94 % of the mass of these particles are less than 2.5 microns ($PM_{2.5}$) in diameter. Diesel PM is of specific concern because it poses as a lung cancer hazard for humans as well as a hazard from noncancer respiratory effects such as pulmonary inflammation. Because of their small size, the particles are readily respirable and can effectively reach the lowest airways of the lung along with the adsorbed compounds, many of which are known or suspected mutagens and carcinogens. Secondary reactions of NOx and SOx can also produce PM.

The most effective method of reducing particulate emissions is to use lighter distillate fuels; however, this leads to added expense. Additional reductions in particulate emissions can be achieved by increasing the fuel injection pressure to ensure that optimum air/fuel mixing is achieved. The third method of reducing particulate emissions is to use cyclone separators, which are effective for particle sizes greater than 0.5 μm.

5.6. Carbon dioxide (CO_2)

Carbon dioxide is one of the basic products of combustion and is not toxic; however, it has been linked to the 'greenhouse effect' and global warming. Diesel engine exhaust gases containing CO_2 as a result of carbon and oxygen O_2 combustion. The maximum concentration of carbon dioxide will be generated during stoichiometric combustion, i.e when complete amount of fuel reacts with oxygen from the air during combustion. The actual concentration of CO_2 depends on the relative contents of carbon (C), hydrogen (H) and other combustible elements in the fuel. The maximum values of carbon dioxide for common types of marine fuel are shown in Table 3 [20], assuming that the exhaust is dry.

The maximum value of carbon dioxide (CO_2 max) can be calculated according to the following expressions:

$$CO_2\,\text{max} = \frac{\text{No. of } CO_2 \text{ molecules produced by complete combustion of fuel}}{\text{Total no. of molecules of combustion products}} \tag{16}$$

Fuel	CO_2 max (%)
Natural gas	11.9
Light fuel oil	15.5
Heavy fuel oil	15.8

Table 3. CO_2 max values for marine fuel, assuming the gases are dry

For 'wet' exhaust gases

$$CO_2\,\mathrm{max} = \frac{c}{c + \frac{h}{2} + \frac{79,1}{20,9}\cdot\left(c + \frac{h}{4}\right)}\ \%, \tag{17}$$

For 'dry' exhaust gases

$$CO_2\,\mathrm{max} = \frac{c}{c + \frac{79,1}{20,9}\cdot\left(c + \frac{h}{4}\right)}\ \% \tag{18}$$

Carbon dioxide (CO2) concentration can be calculated in the exhaust gas emissions according to equation (19), provided that oxygen concentration (O_2), maximum concentration of carbon dioxide (CO_2) max and fuel type are known, as in [20]:

$$\left[CO_2\right] = \frac{CO_2\,\mathrm{max}\cdot\left(20,9 - \left[O_2\right]\right)}{20,9}. \tag{19}$$

Emission	Source
NOx	Function of peak combustion temperature, oxygen content and residence time (function of engine speed in rpm)
SOx	Function of fuel oil sulphur content
CO	Function of the air excess ratio, combustion temperature and air/fuel mixture
HC	Function of the amount of fuel and lubricating oil left unburned during combustion process
PM	Originates from unburned fuel as well as ash content in fuel and lubricating oil
CO_2	Function of combustion

Table 4. Summary of pollutants

Reduction of carbon dioxide emissions can be achieved by reducing specific fuel oil consumption (SFOC) since the amount of CO_2 produced is directly proportional to the volume of fuel used and therefore to the engine efficiency. An alternative is to use fuel with a low carbon ratio relative to hydrogen, which greatly increases the price of marine fuel oils. Table 4 provides a summary for the pollutants discussed above.

6. Methods for reducing harmful pollutant emissions from marine diesel engines

Different methods of reducing the pollutant emission from ship's diesel engines are briefly described in the following text. The focus is more on NOx and SOx than on other emissions, but some attention will be given to them as well.

6.1. NOx emission reduction technologies

Nitrogen oxides (NOx) are an important air pollutant created as a by-product of combustion. Air contains primarily nitrogen (N_2) and oxygen (O_2). The heat generated during combustion causes nitrogen (N_2) and oxygen (O_2) to react to form NO_X which is in direct proportion to peak combustion temperature and pressure. Therefore, NO_X emissions can be mitigated by engine controls that decrease combustion temperature and/or aftertreatment of the exhaust gas. NOx reduction methods are generally categorised as primary methods or internal measure and secondary methods or aftertreatment. The primary methods include changes to the combustion process within the engine and can be divided into three main categories: combustion optimisation, water-based controls and exhaust gas recirculation. Many of the mentioned methods aim to reduce NOx emissions by reducing peak temperatures and pressures of the combustion process in the engine cylinder. Secondary methods or aftertreatment implies post-combustion abatement in which the exhaust gas is treated in order to remove NO_X, either passing it through a catalyst or plasma system. Each of these methods is discussed below.

6.1.1. Primary methods or internal measure

The primary methods include changes to the combustion process within the engine and can be divided into three main categories. The first category is combustion optimisation; the second one is water-based control which consists of water injection, water/fuel emulsion and humidification; and the third category is exhaust gas recirculation. These methods have generally low impact on fuel consumption. Another drawback is to retrofit the system of existing engines. Tier II limits under MARPOL Annex VI, Regulation 13, can be achieved using primary controls.

a. *Combustion optimisation*

There are a number of ways to modify the combustion process, each aimed to reduce NOx emissions. Optimisation of the engine combustion process includes modifying the spray pattern by modification of the fuel valve design, injection timing, intensity of injection and injection rate profile (injection rate shaping), compression ratio, scavenge air pressure and scavenge air cooling. Delayed injection timing is very effective in reducing NOx but increases fuel consumption and smoke. It is usually combined with increased compression pressure and decreased injection duration to minimise or avoid increase in fuel consumption. Other operational modifications that could be made to reduce emissions are combustion chamber optimisation, variable valve timing, increasing the turbo efficiency, the use of a fuel injection system that can be easily adjusted (e.g. electronically controlled injection system) and decrease in the engine air intake temperature using Miller supercharging.[4] Modification of the fuel valve design means replacing the conventional injectors with fuel efficient valves (e.g. slide valves) that optimise the fuel injected into the cylinder. These valves differ from conventional valves in their spray patterns, and they are designed to reduce the dripping of fuel from the injector into the postinjection combustion zone. This fuel entering late the combustion zone is subjected to lower temperatures and therefore results in the emission of unburnt fuel (PM) and VOCs. Changing the conventional fuel valves with slide valves has a significant impact on NOx reduction and PM emissions since PM is a product of incomplete combustion and unburnt fuel. Currently, the slide fuel valves are only applicable for slow-speed two-stroke diesel engines. However, all new engines of this type are supposed to have these valves fitted as the standard. The fuel nozzle was optimised for NOx simultaneously with the development of the slide valve. Tests on a 12K90MC engine (55MW at 94 rpm) at 90 % load showed a 23 % reduction in NOx emissions for a slide-type valve compared with a standard valve and nozzle and with a 1 % fuel consumption increase. Furthermore, increasing the number of injectors per cylinder enables the combustion process to be better controlled and therefore more efficient combustion. However, additional injectors, piping and associated equipment are associated with a cost penalty. Nowadays, modern slow-speed engines use three fuel injectors located near the outer edge of the combustion chamber. With sequential injection, each of the three nozzles in a cylinder is actuated with different timings. Pulsed injection gives about 20 % NOx reduction with about 7 % increase in fuel consumption. Sequential and pre-injection gave less NOx reduction and less fuel consumption increase. The effects are the result of changes in the overall pressure development and interaction between fuel sprays. Pre-injection can be used to shorten the delay period in medium-speed engines and thus decreases temperature and pressure during the early stages of combustion, resulting in reduced NOx. Pre-injection can reduce particulates which are increased by other NOx control measures, thus allowing greater flexibility in NOx control. Delayed injection combined with increased compression ratio have effect on reducing the maximum combustion pressure and hence temperature. By using this simple technique, a reduction of up to 30 % can be achieved. However, delayed injection increases fuel consumption up to 5 % in specific fuel oil consumption due to later burning, as less of the combustion energy release is subjected to the full expansion process and gas temperatures remain high later into the expansion stroke, resulting in more heat losses on the

4 Reduced scavenge air temperature reduces combustion temperatures and thus NOx. For every 3 OC reduction, NOx may decrease by about 1 %. On four-stroke engines, the Miller concept can be applied to achieve low scavenge air temperature. Using a higher-than-normal pressure turbocharger, the inlet valve is closed before the piston reaches bottom dead centre on the intake stroke. The charge air then expands inside the engine cylinder as the piston moves towards bottom dead centre, resulting in a reduced temperature. Miller supercharging can reduce NOx by 20 % without increasing fuel consumption.

walls. In some engines, the timing adjustment can be made while in service. Smoke and emission of PM also increase due to reduced combustion temperatures and thus less oxidation of the soot produced earlier in the combustion, as in [21]. Furthermore, increasing injection pressure leads to better atomisation of the fuel and therefore to reduction in particulates and CO. Since combustion is cleaner, this technique tends to increase NOx reduction. There is also the new generation of the electronically controlled camshaftless engines that allow great flexibility for optimisation of the combustion process over the full range of operating conditions. These computer-controlled engines have allowed greater operational flexibility. As far as NOx is concerned, the main features are computer control of variable injection timing (VIT), injection rate shaping, variable injection pressure and variable exhaust valve closing (VEC). Variable exhaust closing gives the ability to change the effective compression ratio. With variable injection timing and variable exhaust valve closing, it is possible to optimise the injection timing delay and increased compression ratio over the whole load range to maintain peak pressures at low load while avoiding excessive peak pressures at high load. Computer-controlled camshaftless engines are equipped with common rail injection techniques which give high injection pressures and thus good spray characteristics even at low loads granting of NOx reduction emissions.

b. Water-based control

The second category is water-based controls consisting of water injection, water/fuel emulsion and humidification. Water-based controls reduce emissions of the NOx from diesel engines by introducing freshwater at different stages of the combustion process. Introduction of water into the combustion chamber reduces maximum combustion temperature due to the absorption of energy for evaporation and the increase in the specific heat capacity of the cylinder gases and thus reduce emissions of the NOx [23]. The in-cylinder evaporation of the water also improves the atomisation of the fuel and causes it to burn more completely. Freshwater can be introduced in the charge air (humidification), through direct injection into the cylinder or through water/fuel emulsion. Water/fuel emulsion is the process of introducing water into the fuel prior to injection into the combustion cylinder and can reduce smoke, while humidification can increase smoke. Direct water injection is the process of introducing water directly into the combustion cylinder at pressures of 200–400 bar. The water is injected into the cylinder by a combined injection valve and nozzle that allow injection of water and fuel oil. The process is electronically controlled. Direct injection of water and water/fuel emulsions place the water more directly in the combustion region, where it has maximum effect on NOx production. Generally, direct water injection or water/fuel emulsions will yield about 1 % reduction in NOx for every 1 % of water-to-fuel ratio. This one-to-one ratio is consistent up to about 30 % water content, at which point the combustion temperature decreases too much, resulting in an increase in PM emissions. Alternative to water injection and water/fuel emulsion is the scavenge air humidification as a second category of the primary methods for NOx emission reduction that implies injection of very fine water mist in scavenge airstream after the turbocharger using special nozzles (Scavenge Air Saturation System). The fine water droplets evaporate fast, and further heat is introduced in the scavenge air. Humidification requires about twice as much water for the same NOx reduction compared with direct injection of water and water/fuel emulsions. Humidification can reduce NOx levels down to 2 to 3 g/kWh without fuel consumption penalty. Similar technique that is used for humidification of the engine scavenge air is the so-called humid air motor (HAM). In this system, hot compressed air from the turbocharger is led to a humidification tower and exposed to a large surface area and flushed with hot water. The water can be heated by a heat exchanger connected to the jacket cooling system or using an exhaust gas boiler. One manufacturer claims considerable success in service in reducing NOx emissions with the added claim of increasing the indicated power of the engine at certain loads, therefore reducing fuel consumption hence proportionally reducing CO_2 emissions. The actual degree of NOx reduction varies from 10 % to over 60 %, depending on the engine type and which of the above reduction methods are adopted. For example, the experiment carried out on the Viking Line's MS Mariella has shown a NOx emission reduction from 17 to between 2.2 and 2.6 g/kWh and a decrease in fuel consumption of 2–3% using the HAM system [24].

c. *Exhaust gas recirculation (EGR)*

The exhaust gas recirculation (EGR) system is based on lowering of the combustion temperature and oxygen concentration thus lowering NOx. EGR reduces combustion temperatures by increasing the specific heat capacity of the cylinder gases and by reducing the overall oxygen concentration, taking away a part of the exhaust gases and mixing it into the engine intake air. Some of the exhaust gas is cooled and cleaned before recirculation to the scavenge air side. The usage of the exhaust gas as intake air reduces the oxygen content in intake air from 21 % to 13 % which limits the NOx that can be formed and reduces the amount of combustion products that can take place. In engines operating on poor-quality fuel, external EGR can lead to fouling and corrosion problems. The residue from cooling and cleaning the exhaust gas on ships using heavy fuel oil contains sulphur in a form which is difficult to dispose of. The relative changes in measured emission parameters as a function of the recirculation amount at 75 % engine load show that at increased recirculation amounts, the HC and PM emissions are reduced corresponding to the reduction of the exhaust gas flow from the engine. Increased recirculation amount leads to increase in CO emissions due to lower cylinder excess air ratios and thus lack of oxygen in the combustion chamber. Furthermore, EGR tends to increase smoke by reducing the O_2 concentration, increasing the combustion duration and decreasing the combustion temperature. All of that may be controlled using additional techniques such as water in fuel to achieve an optimum balance between NO_X, CO and PM. Test engine work by MAN Diesel & Turbo has shown that, with 40 % recirculation, EGR has the potential to reduce NO_X down to Tier III levels on a two-stroke low-speed marine engine and that increased fuel consumption, carbon monoxide emissions and PM emissions resulting from reduced combustion efficiency are manageable with engine adjustments. It is also reported that specific fuel consumption is greatly improved when using EGR to reduce NO_X down to Tier II limits, when compared with using engine adjustments to achieve the same level of emissions, particularly at part load as in [25]. There are many different components to an EGR system such as high pressure exhaust gas scrubber fitted before the engine turbocharger, cooler to further reduce the temperature of the recirculated gas, water mist catcher to remove entrained water droplets, high-pressure blower to increase recirculated gas pressure before reintroduction to the engine scavenge air and automated valves for isolation of the system. The scrubber in the EGR system is used to remove sulphur oxides and particulate matter from the recirculated exhaust, to prevent corrosion and reduce fouling of the EGR system and engine components. The system requires the use of an electrostatic precipitator and catalysts to remove the particulates from the exhaust gas before injecting it as intake air. There is also a need for wet-scrubbing technology to remove the sulphur components of the exhaust stream prior to reintroduction into the engine. A cooling unit is also needed to reduce the temperature of the exhaust gas before it returns to the engine.

6.1.2. Secondary methods or aftertreatment

Secondary methods, or aftertreatment, are based on treating the engine exhaust gas itself by passing it through a catalyst or plasma system. There has been much development in selective catalytic reduction (SCR) and nonthermal plasma (NTP) reduction systems over the last few years. Using these methods, NOx emission reductions of over 95 % can be achieved [26].

Selective catalytic reduction (SCR) is an exhaust gas treatment method by which the NOx generated in a marine diesel engine exhaust gas can be reduced to a level in compliance with the NOx Tier III requirements. The method involves mixing of ammonia as a reducing agent with the exhaust gas which is passed over a catalyst where more than 90 % of the NOx can be removed to below 2 g/kWh. The SCR system converts nitrogen oxides into harmless nitrogen and water, by means of a reducing agent injected into the engine exhaust stream before a catalyst. Hydrocarbons are also reduced. Exhaust emission abatement systems using SCR technologies usually use an ammonia (NH_3) reductant introduced as a urea/water solution ($(NH_2)_2CO$) into the exhaust stream, prior to the catalyst blocks. For marine systems, a 40 % solution of urea in de-ionised water is typically used for safe handling and toxic risk reasons [27, 28]. The use of urea in the system breaks down the NOx emissions to N_2 and H_2O. The degree of NOx removal depends on the amount of ammonia added. A NOx reduction efficiency of 90 % can be achieved using a urea injection rate of 15 g/ kWh NOx is reduced according to the following overall reaction scheme [26,27,29]:

Urea decomposition before entering the reactor:

$$(NH_2)_2CO\ (\text{urea}) \rightarrow NH_3\ (\text{ammonia}) + HNCO\ (\text{isocyanic acid}) \tag{20}$$

$$HNCO + H_2O \rightarrow NH_3 + CO_2 \tag{21}$$

The resulting quantity of CO_2 is minor when compared with that resulting from fuel combustion.

NO_X reduction at the catalyst:

$$4NO + 4NH_3 + O_2 \rightarrow 4N_2 + 6H_2O \quad (22)$$

$$2NO + 2NO_2 + 4\ NH_3 \rightarrow 4N_2 + 6H_2O \quad (23)$$

$$6NO_2 + 8NH_3 + O_2 \rightarrow 3N_2 + 6H_2O \quad (24)$$

Equation 22 shows the main SCR reaction as nitric oxide dominates in the exhaust. The reaction shown at equation 23 occurs at the fastest rate up to an NO_2:NO ratio of 1:1. However, at higher ratios, the excess NO_2 reacts slowly as per equation 24. The rate of urea injection must be sufficient to reduce NO_X emissions to the required level but not so great to avoid ammonia slip. Control is based on the load and speed of the engine with active feedback provided on some systems by NO_X and ammonia emission monitoring. At engine start-up, urea injection is initiated once the catalyst reaches operating temperature, which is key for effective NO_X reduction performance, deposit prevention and avoidance of ammonia slip. Catalysts have considerable heat capacity so the time taken to reach the injection trigger temperature is dependent on a number of factors including the minimum catalyst operating temperature recommended for the fuel type, the period of cooldown since the engine was last operated, the size of the catalyst and the engine load pattern at start-up. Injection can begin up to 30 min after a fully cold start, whereas it may begin within 10–15 min if the catalyst is still warm from running in the previous 6–10 h. SCR units are typically installed in the exhaust system of a diesel engine, if applicable, before the exhaust gas economiser and as close as possible to the engine because of the relatively high exhaust gas temperatures required by the catalysts for effective NOx reduction reactions. The SCR catalysts may also be integrated with the engine by close coupling to the engine, typically applicable to small high-speed diesel engines. For slow-speed diesel engines with inherently low relative exhaust gas temperatures, this may necessitate the integration of the SCR reaction chamber and catalysts before the turbocharger exhaust turbine. Depending on the engine load, the exhaust gas temperature on this side is 50–175 °C higher than on the low pressure side. Even though the reactor is placed before the turbine, the exhaust gas temperature will normally still be too low at low loads. To increase the temperature, a cylinder bypass from the scavenge air receiver to the turbine inlet is installed. The bypass is controlled by the cylinder bypass valve.

When opening the bypass, the mass of air through the cylinders will be reduced without losing the scavenge air pressure, and, accordingly, the exhaust gas temperature will increase. This system makes it possible to keep the temperatures above the required level. However, the cylinder bypass will increase the SFOC depending on the required temperature increase. Selective catalytic reduction is the only technology currently available to achieve compliance with the Tier III NOx standards for all applicable engines. Another option is selective non-catalytic reduction (SNCR), which works in a similar way with selective catalytic reduction but without the use of a catalyst. A reducing agent (ammonia or urea) injected during the combustion process converts the nitrogen oxides into nitrogen and water, reducing NOx emissions by 50 % [30]. The drawback of this system is that it is less efficient than the SCR method, because only 10–12 % of ammonia reacts with NOx. Since the cost of ammonia is relatively high and since the system requires extensive modification to the engine, the SNCR option does not appear to be competitive. Plasma reduction systems are based on the use of plasma. This is a partially ionised gas composed of a charge of a neutral mixture of atoms, molecules, free radicals, ions and electrons. Electrical power is converted into electron energy, and the electrons create free radicals, which destroy pollutants in exhaust emissions. Experiments have shown that plasma reduction systems can reduce NOx by up to 97 % [30]. It seems to be flexible in terms of size and shape and should be at relatively low cost. However, for marine use, it is still in the development phase.

6.2. SOx reduction technologies

The emission of sulphur dioxide is directly proportional to the content of sulphur in fuel. To meet the restrictions on emissions of sulphur oxides (SOx) and PM (particulate matter) that are determined by the MARPOL convention Annex VI which specifies a global and a local (ECA) limit on the sulphur content in marine fuel, there are only two possibilities for reducing SOx emissions: either use fuels with low sulphur content (there will be no restrictions in the use of heavy fuel oil) or apply an exhaust gas cleaning system to reduce the total emission of SOx. Although the SOx requirements can be met by using a low-sulphur fuel, the regulation allows alternative methods to reduce the emissions of SOx to an equivalent level. The process of exhaust gas cleaning is performed in a scrubber unit. There are two main types of SO_X scrubber: wet scrubbers that use water (seawater or fresh) as the scrubbing medium and dry scrubbers that use a dry chemical. The wet scrubbing technology is based on the fact that sulphur oxides dissolve in water. This means that when the exhaust gas is sprayed with the alkaline water in the scrubber, the SOx will dissolve in the scrubbing water and be cleaned from the exhaust gas. The water is injected into the exhaust gas stream and is discharged from the bottom of the scrubber. The alkalinity in the scrubbing water will neutralise the SOx emissions. The scrubbing water must be cleaned of particulate matter and other contaminants before being discharged out into sea. Wet scrubbing systems which are normally fitted on marine engines may be categorised as open-loop system, closed-loop system or hybrid system.

Open-loop systems – In an open-loop technology, the water comes from the sea and goes directly to the scrubbers. After the scrubbing process, the water goes through water treatment and to the sea again. This system takes advantage of the natural alkalinity of seawater to buffer the acidity of SO_X gases. The seawater flow rate in open-loop systems is approximately 45 m^3/MWh. Sulphur oxide removal rate is close to 98 % with full alkalinity of the seawater, meaning emissions from a 3.5 % sulphur fuel will be the equivalent of those from a 0.10 % sulphur fuel after scrubbing [29]. The sulphur oxides generated in the combustion process are dissolved and removed by the scrubber water. Sulphur dioxide (SO_2) is dissolved and ionised to bisulphite and sulphite, which is then readily oxidised to sulphate in seawater containing oxygen. As in reference [29], similarly sulphuric acid, formed from SO_3, and hydrogen sulphate dissociate completely to sulphate according to chemical reactions:

For SO_2:

$$SO_2 + H_2O \rightarrow H_2SO_3\left(sulphurous\,acid\right) \rightarrow H^+ + HSO_3\left(bisulphite\right) \tag{25}$$

$$HSO_3^-\left(bisulphite\right) \rightarrow H^+ + SO_3^{2-}\left(sulphite\right) \tag{26}$$

$$SO_3^{2-}\left(sulphite\right) + 1/2\,O_2 \rightarrow SO_4^{2-}\left(sulphate\right) \tag{27}$$

For SO_3:

$$SO_3 + H_2O \rightarrow H_2SO_4\left(sulphuric\,acid\right) \tag{28}$$

$$H_2SO_4 + H_2O \rightarrow HSO_4^-\left(hidrogen\,sulphate\right) + H_3O^+ \tag{29}$$

$$HSO_4^-\left(hidrogen\,sulphate\right) + H_2O \rightarrow SO_4^{2-}\left(sulphate\right) + H_3O^+ \tag{30}$$

Closed-loop systems use freshwater treated with sodium hydroxide (NaOH) as the scrubbing media for the neutralisation of SO_X. This results in the removal of SO_X from the exhaust gas stream as sodium sulphate according to the following chemical reactions as in [29]:

For SO_2:

$$NA^+ + OH^- + SO_2 \rightarrow NaHSO_3\left(aq\,sodium\,bisulphite\right) \tag{31}$$

$$2NA^+ + 2OH^- + SO_2 \rightarrow Na_2SO_3\left(aq\,sodium\,sulphite\right) + H_2O \tag{32}$$

$$2NA^+ + 2OH^- + SO_2 + 1/2\,O_2 \rightarrow Na_2SO_4\left(aq\,sodium\,sulphate\right) + H_2O \tag{33}$$

For SO_3:

$$SO_3 + H_2O \rightarrow H_2SO_4\left(sulphuric\,acid\right) \tag{34}$$

$$2NaOH + H_2SO_4 \rightarrow Na_2SO_4\left(aq\,sodium\,sulphate\right) + 2H_2O \tag{35}$$

http://dx.doi.org/10.5772/59720

In a closed-loop technology, absolutely no water comes from the sea. The freshwater comes from a buffer tank and is cooled by the seawater. The freshwater is composed of NaOH and leaves the buffer tank to go to the scrubber. After the scrubbing process, the water comes back to the buffer tank, cleaned by a filter. The black water goes to a sludge tank and the clean water goes back to the scrubbing cycle. A big storage tank fills up the buffer tank. Closed-loop systems can also be operated when the ship is operating in enclosed waters where the alkalinity would be too low for open-loop operation. Closed-loop systems typically consume sodium hydroxide in a 50 % aqueous solution. The dosage rate is approximately 15 l/MWh of scrubbed engine power of a 2.70 % sulphur fuel is scrubbed to equivalent to 0.10 %. Using a closed-loop technology can have some advantages. First, there is a possibility to increase the pH level in order to reduce more SOx. Also there is no corrosion of the parts and less discharge water to clean. The running costs of the closed-loop technology are relatively high because it uses NaOH which is 0.2 €/kg and its required monitoring units. Also the sludge tanks have to be discharged at the harbour which costs a lot of money. The closed-loop reduces about 98 % SOx in the exhaust gas.

Hybrid system – A hybrid system is a mixture of both open loop and closed loop. In harbours and ECA, the system can operate with freshwater without generating any significant amount of sludge to be handed at port calls. At open sea, the system switches to the seawater open loop. Using a hybrid technology can have some advantages. First, if the ship is running at open sea, after switching to open loop, the accumulated water of the buffer tank can slowly be removed back to the sea. Also, the tank is slowly filled up again to prepare for the arrival at sensitive areas.

Dry SO_Xscrubber known as an 'absorber' brings the exhaust gas from diesel engine in the multistage absorber where contact with calcium hydroxide ($Ca(OH)_2$) granules reacts with sulphur dioxide (SO_2), forming calcium sulphite as in reference [29]:

$$SO_2 + Ca(OH)_2 \rightarrow CaSO_3\left(calcium\,sulphite\right) + H_2O \tag{36}$$

The sulphite is then oxidised and hydrated in the exhaust stream to form calcium sulphate dihydrate, or gypsum:

$$2CaSO_3 + O_2 \rightarrow 2CaSO_4\left(calcium\,sulphate\right) \tag{37}$$

$$CaSO_4 + 2H_2O \rightarrow CaSO_4 \cdot 2H_2O\left(calcium\,sulphate\,dihydrate - gypsum\right) \tag{38}$$

Similarly, chemical reactions take place for SO_3:

$$SO_3 + Ca(OH)_2 + H_2O \rightarrow CaSO_4 \cdot 2H_2O\left(calcium\,sulphate\,dihydrate - gypsum\right) \tag{39}$$

Using a dry scrubber can have some advantages. First, the good point of this technology is that the desulphurisation unit requires, aside from electrical energy, only $Ca(OH)_2$ in the shape of spherical granulates. Also the dry scrubber further operates as a silencer. Dry scrubbers typically operate at exhaust temperatures between 240 °C and 450 °C. Calcium hydroxide granules are between 2 and 8 mm in diameter with a very high surface area to maximise contact with the exhaust gas. Within the absorber, the calcium hydroxide granules ($Ca(OH)_2$) react with sulphur oxides to form gypsum ($CaSO_4$ $2H_2O$). To reduce SO_X emissions to those equivalent to fuel with a 0.1 % sulphur content, a typical marine engine using residual fuel with a 2.70 % sulphur content would consume calcium hydroxide granules at a rate of 40 kg/MWh (i.e. a 20 MW engine would require approximately 19 tonnes of granulate per day) [29, 30]. The dry scrubber reduces up to 99 % SOx in the exhaust gas. It will be absolutely no problem to fulfil all the IMO 3 requirements for 2016. The dry scrubber compared to the wet scrubber has lower investment costs and higher running costs and requires a lot of space which reduces the benefits. The efficiency of the SOx scrubber systems depends on the sulphur content in the fuel and generally ranges up to 97 %. Anyhow, the efficiency system must be sufficient to achieve a SOx emission level that is equal to or lower than the required limit.

On the other hand, it should be noted that the reduction of both NOx and SOx emissions from marine diesel engines be achieved by replacing conventional fuels with alternative fuels, e.g. liquefied natural gas (LNG). In case of using LNG, NOx emission can be reduced by 60 % and SOx emission by 90–100 %.

There is also another option to use onshore power supply at ports which is especially beneficial for local air quality. In this case, NOx and SOx emissions can be reduced by 90 %, while CO_2 reduction depends on the source of electricity. The total CO_2 emission reduction depends on how the used electricity is produced. In the European Union, the use of shore-side electricity rather than electricity generated by a ship using low-sulphur fuel will cut CO_2 emissions by an average of 50 %.

One of the main benefits of shore connection systems stems from the fact that electricity generated on land by power plants has a smaller adverse impact on the ecosystem than that produced by ship engines. Namely, the main cause of air pollution from ships in ports is the use of auxiliary diesel engines to generate electricity on ships. Furthermore, experiments with wind and solar power, biofuels and fuel cells are ongoing and could be useful in the future to reduce air pollution from ships.

In Table 5, overview of different technologies and their potential for reduction of emissions from marine diesel engines are summarised as in [31].

Category	Technology aimed to reduce	NOx	SOx	PM	CO_2
Engine modification	NOx				
	Optimise combustion	20–40%	0%	25–50%	0%
Water-based control	Direct water injection	50%	0%	0%	0%

Category	Technology aimed to reduce	NOx	SOx	PM	CO_2
	Water in fuel (30 % emulsion)	30%	0%	0%	0%
	Scavenge air humidification	30–60%	0%	0%	0%
	Humid air motor	10–70%	0%	40–60%	0%
Exhaust gas recirculation	ERG	20–85%	0%	0%	0%
Aftertreatment	Selective catalytic reduction	90–95%	0%	0%	0%
	SOx	0%	0%	0%	0%
	Scrubber	0%	90–97%	80–85%	0%
(Alternative) fuels	Low-sulphur fuel 2,7 %S to 0,1 %S	0%	97%	20%	0%
	Both NOx and SOx	0%	0%	0%	0%
	LNG	60%	90–100%	72%	0–25%
	Onshore power supply (in harbour only)	97%	96%	89%	Depending on energy source, average 50%

Table 5. Different technologies and their reduction potential

7. Conclusion

Owing to rapidly developed shipping industry and maritime traffic in recent decades, air pollution emissions from ocean-going ships are continuously growing. Exhaust gases from marine diesel engines are the primary source of emission harmful pollutants such as nitrogen oxides (NOx), sulphur dioxide (SO_2), carbon monoxide (CO) and particulate matter (PM) which contribute significantly to environmental pollution, especially in port areas that are often located in or near urban areas, and a significant number of people are exposed to these emissions. The increased air pollutant concentrations and deposition have several negative effects. Nitrogen oxide and particulate matter can contribute to many serious health problems and increased morbidity and mortality (especially from cardiovascular and cardiopulmonary diseases). Nitrogen oxides also contribute to the formation of ground-level ozone, which has a harmful effect on plants and human health. Furthermore, sulphur dioxide and nitrogen oxide emissions increase acidification of sensitive forest ecosystems along the coastal areas and have a harmful effect on plants, aquatic animals and infrastructure by accelerating the deterioration process of various materials. Finally, ships are also a source of greenhouse gas, a pollutant which contributes to global warming. Recent studies indicate that the emission of CO_2 by ship corresponds to about 3 % of the global anthropogenic emissions. If things remain the same, by 2020, shipping will have been the biggest single emitter of air pollution especially in areas of the dense maritime traffic such as Europe, North America and East Asia which surprisingly surpasses the emissions from all land-based sources together. Since harmful pollutant emissions from ships have great impact on the human health and the environment, it is required to tighten uniform regulations at the global level, bearing in mind that shipping is inherently international. The International Maritime Organization (IMO) responsible for the safety of life at sea and the protection of the marine environment reacts on NOx, SOx, PM and CO_2 emissions from a ship by adoption of Annex VI of the MARPOL 73/78 Convention, titled 'Regulations for the Prevention of Air Pollution from Ships'. MARPOL Annex VI sets limits on NOx and SOx emissions from ship exhausts and prohibits deliberate emissions of ozone-depleting substances. Furthermore, the IMO marks out Emission Control Areas (ECAs) in cooperation with national governments with more stringent controls on sulphur emissions. These areas currently comprise the Baltic Sea, the North Sea, the English Channel, the US Caribbean Sea and the area outside North America (200 nautical miles). Ships are currently being permitted to burn fuel oils with sulphur content of less than 3.5 % while operating outside an ECA but must ensure that they burn fuel with a sulphur content of less than 1 % while within the sulphur Emission Control Areas. In accordance with EU's marine fuel sulphur directive, the sulphur content in marine gas oil within the territorial waters of an EU member state may not exceed 0.1 % by weight. This applies to all ships regardless of flag. Regarding reductions in nitrogen oxide emissions from marine engines, Annex VI introduced Tier I, II and III NOx emission standards for new engines. NOx emission limits are set for diesel engines depending on the engine's maximum operating speed. Tier I and II limits are global, while the Tier III standards apply only in NOx Emission Control Areas. Tier II NOx standards are currently being in force. In order to control CO_2 emission from shipping, the first formal CO_2 control regulations were adopted by the IMO, introducing a new chapter, Chapter 4, to Annex VI. Chapter 4 introduces two mandatory mechanisms intended to ensure an energy efficiency standard for ships: the first is the Energy Efficiency Design Index (EEDI), for new ships, and the second the Ship Energy Efficiency Management Plan (SEEMP) for all ships. The regulations apply to all ships of 400 gross tonnage and above and are entered into force on 1 January 2013. Energy Efficiency Design Index (EEDI) is the first globally binding climate change standard. It is anticipated that global CO_2 reductions of 10 to 20 % could be obtained by implementation of EEDI and SEEMP. Detailed descriptions of the emission restrictions prescribed by Annex VI of the MARPOL 73/78 Convention are listed in section 2 entitled 'International regulation concerning air pollution from merchant shipping'. To meet these restrictions on emissions of harmful pollutants from marine diesel engines, different methods and technical solutions can be implemented. Nitrogen oxide reduction methods are generally categorised as primary methods or internal measure and secondary methods or aftertreatment. Primary methods include changes to the combustion process within the engine and can be divided into three main categories: combustion optimisation, water-based control and exhaust gas recirculation, while secondary methods, or aftertreatment, is based on treating the engine exhaust gas itself by passing it through a catalyst system. MARPOL Annex VI will reduce global ship sourced

NOx emissions at a small rate because it only applies to new installations or major conversions. For NOx levels below MARPOL Annex VI, or for retrofitting, the main measures available now are water/fuel emulsions, direct water injection, inlet air humidification and catalyst system. Technical measures to reduce the sulphur oxide emission from ship's diesel engines include the adoption of low-sulphur fuels, the easiest way of reducing sulphur oxides emission. Usage of an exhaust gas cleaning system, i.e. scrubbers, is a possible alternative to low-sulphur fuels to reduce the total emission of SOx and considerably reduces emissions of other polluting particles. A detailed description of these methods is given in section 5 entitled 'Methods of reducing harmful pollutant emissions from marine diesel engines' As the air pollution emissions from ships are continuously growing, it is necessary to constantly improve and implement the efficient technologies and methods in order to reduce pollutant emissions from marine diesel engines and maintain them within the limits prescribed by MARPOL Annex VI as well as by other national and regional regulations. Some of these technologies and methods include the use of shore connection systems of which the main benefits stem from the fact that electricity generated on land by power plants has a smaller adverse impact on the ecosystem than that produced by ship engines, owing to that of particular concern is the pollution generated by ships at berth. Ships equipped with a green technology receive a higher grade. Furthermore, other technical measures for reducing air pollution from ships include the adoption of liquefied natural gas (LNG) as alternative fuel for marine engines. Wind, solar power, biofuels and fuel cells, the world of alternative energy is ongoing and could be useful for reducing air pollution from ships in the future. Finally, harmful pollutant emissions from ships require a stringent international standard due to their impact on the human health and the environment. It includes extending the SOx Emission Control Areas in the EU (e.g. in the Mediterranean, in the Black Sea, in the Irish Sea and in the North East Atlantic) and designating NOx Emission Control Areas as soon as possible.

Author details

Ivan Komar* and Branko Lalić

*Address all correspondence to: ivan.komar@pfst.hr

University of Split - Faculty of Maritime Studies - Marine Engineering Department, Split, Croatia

References

[1] Wahlström J, Karvosenoja N, Porvari P. Ship emissions and technical emission reduction potential in the Northern Baltic Sea, Finnish Environment Institute, Helsinki 2006 http://cleantech.cnss.no/wp-content/uploads/2011/06/2006-Wahlstrom-ships-

emissions-technical-emission-reduction-potential-in-northern-Baltic-Sea.pdf (accessed 17 July 2014)

[2] Friedrich A et al. Air pollution and greenhouse gas emissions from ocean-going ships, The International Council on Clean Transportation (ICCT), 2007 https://www.georgiastrait.org/files/share/PDF/MarineReport_Final_Web.pdf (accessed 17 July 2014)

[3] Air pollution from ships, The European Environmental Bureau, The European Federation for Transport and Environment, Seas At Risk, The Swedish NGO Secretariat on Acid Rain, Updated November 2004 http://www.airclim.org/sites/default/files/documents/shipbriefing_nov04.pdf (accessed 22 July 2014)

[4] Saracoglu H, Deniz C, Kılıç A. An Investigation on the Effects of Ship Sourced Emissions in Izmir Port, Turkey, The Scientific World Journal, Vol. 2013 (2013), Article ID 218324, http://dx.doi.org/10.1155/2013/218324 (accessed 5 August 2014)

[5] Komar I, Lalić B, Dobrota Đ. Air Pollution Prevention from Croude Oil Tankers with Volatile Organic Compound Emission, International Journal of Maritime Science & Technology "Our Sea", Vol. 57, No. 3-4, 138–145, Dubrovnik, 2010. (ISSN: 0469-6255)

[6] Lalić B, Komar I, Nikolić D. Optimization of Ship Propulsion Diesel Engine to Fulfill the New Requirements for Exhaust Emissions, Transactions on Maritime Science (ToMS), April 2014, Vol. 3, No. 1.3 (2014), 1 Split, 2014, 20–31, DOI:10.7225/toms.v03.n01.003

[7] Herdzik J. Emissions from Marine Engines Versus IMO Certification and Requirements of Tier 3, Journal of KONES Powertrain and Transport, Vol. 18, No. 2 (2011), 161–167

[8] Rashidul SMH. Impact of EEDI on ship design and hydrodynamics master of science thesis, Department of Shipping and Marine Technology Division of Sustainable Ship Propulsion, Chalmers University of Technology, Gothenburg, Sweden, 2011, http://publications.lib.chalmers.se/records/fulltext/151284.pdf (accessed 5 August 2014)

[9] Bazari Z, Longva T. Assessment of IMO mandated energy efficiency measures for international shipping, project final report, Lloyd's Register, London, UK, DNV, Oslo, Norway, IMO, 2011, http://www.imo.org/MediaCentre/HotTopics/GHG/Documents/REPORT%20ASSESSMENT%20OF%20IMO%20MANDATED%20ENERGY%20EFFICIENCY%20MEASURES%20FOR%20INTERNATIONAL%20SHIPPING.pdf (accessed 6 August 2014)

[10] Marty P. Ship energy efficiency study: development and application of an analysis method, PhD Thesis, Universités, École Navale, Nantes, 2014.

[11] Woodyard D. Pounder's Marine Diesel Engines and Gas Turbines, Elsevier, 2004

[12] Komar I, Antonić R, Kulenović Z. Experimental Tuning of Marine Diesel Engine Speed Controller Parameters on Engine Test Bed, Transactions of FAMENA, 33 (2009) 2, ISSN: 1333–1124, Zagreb, 2009, 51–70

[13] ISO 8217:2010/2012 Fuel Standard, Fourth Edition, https://www.bimco.org/News/2010/07/19_ISO_8217.aspx (accessed 20 August 2014)

[14] McAllister S, Chen JY, Fernandez-Pello AC. Fundamentals of Combustion Processes, Springer Science+Business Media, 2011

[15] Emission Control Two Stroke Low Speed Diesel Engines, MAN B&W Diesel A/S, 1996 http://www.google.hr/url?sa=t&rct=j&q=&esrc=s&source=web&cd=1&ved=0CBoQFjAA&url=http%3A%2F%2Fwww.flamemarine.com%2Ffiles%2FMANBW.pdf&ei=7nQZVLyUDsXaPPmw-geAI&usg=AFQjCNFBv8IBVUn1vTNfggYmTaJpBLbbjw&bvm=bv.75558745,d.ZWU (accessed 20 July 2014)

[16] Lavoie JB, Heywood JB, Keck JC. Experimental and Theoretical Investigation of Nitric Oxide Formation in Internal Combustion Engines, Combustion Science Technology Vol. 1 (1975), 313–326

[17] Bowman CT. Kinetics of Pollutant Formation and Destruction on Combustion, Progress in Energy and Combustion Science, Vol. 1, No. 1 (1975), 33–45

[18] Kuiken K. Diesel Engines, Part I, Target Global Energy Training; Onnen, 2012

[19] Rakopoulos CD, Giakoumis EG. Diesel Engine Transient Operation - Principles of Operation and Simulation Analysis, Springer-Verlag Ltd., London, UK, 2009.

[20] Technical Bulletin - Combustion Calculations, Normalisations and Conversions, Q Instrument Services Limited, Bishopstown, Cork, Ireland. http://www.qlimited.com/pdf/Land-Combustion-Calculations-Normalisations-&-Conversions-Q.pdf (accessed 5 June 2014)

[21] Fournier A. Controlling Air Emissions from Marine Vessels: Problems and Opportunities, University of California Santa Barbara, February 2006, http://fiesta.bren.ucsb.edu/~kolstad/temporary/Marine_Emissions__2-11-06_.pdf (accessed 5 June 2014)

[22] Lalić B, Komar I, Antonić R. Water-in fuel emulsion as primary method to reduce NOx from diesel engine exhaust gas, Trends in the Development of Machinery and Associated Technology TMT 2009 Conference Proceedings, 2009, Zenica, Barcelona, Catalunya, Istanbul, 457–460.

[23] Miola A et al. Regulating Air Emissions from Ships, European Commission Joint Research Centre Institute for Environment and Sustainability, Luxembourg, 2010, http://ec.europa.eu/dgs/jrc/downloads/jrc_reference_report_2010_11_ships_emissions.pdf (accessed 5 July 2014)

[24] MAN Diesel and Turbo: Tier III Two-Stroke Technology, https://www.google.hr/?gws_rd=ssl#q=MAN±Diesel±and±Turbo%3A±Tier±III±TwoStroke±Technology (accessed 5 August 2014)

[25] Komar I, Antonić R, Matić P. Selective catalytic reduction as a secondary method to remove NOx from exhaust gas of the ship's propulsion diesel engine, Proceedings of the IFAC Conference on Control Applications in Marine Systems Vukić, Zoran; Sauro, Longhi (ed.), Zagreb: CEPOST, 2007.

[26] Emission Project Guide MAN B&W Two-Stroke Marine Engines, MAN Diesel & Turbo, December 2013 http://www.mandieselturbo.com/download/project_guides_tier2/printed/7020-0145-001web.pdf (accessed 5 August 2014)

[27] Ship power four - stroke, Wärtsilä Environmental Product Guide, Vaasa, June 2014, http://www.wartsila.com/file/Wartsila/en/1278528485383a1267106724867-wartsila-o-env-product-guide.pdf (accessed 5 August 2014)

[28]]Lloyd's Register: Understanding Exhaust Gas Treatment Systems - Guidance for Shipowners and Operators, London, June 2012, http://www.google.hr/url?sa=t&rct=j&q=&esrc=s&source=web&cd=2&ved=0CCQQFjAB&url=http%3A%2F%2Fwww.lr.org%2Fen%2F_images%2F213-35826_ECGS-guide1212_web_tcm155-240772.pdf&ei=tlMVVNj9F8PXaubw-gYAF&usg=AFQjCNGz4a8vCcjsE4e1rsDHx1Rj8Rk0rQ&bvm=bv.75097201,d.bGQ (accessed 30 August 2014)

[29] Rr32 Guide to Exhaust Emission Control Options, MS3026, Land & Sea Systems, Bristol September 99, http://wenku.baidu.com/view/f634f9b81a37f111f1855b5f. (accessed 10 August 2014)

[30] CNSS: A Review of Present Technological Solution for Clean Shipping, http://www.google.hr/url?sa=t&rct=j&q=&esrc=s&source=web&cd=2&ved=0CCIQFjAB&url=http%3A%2F%2Fcnss.no%2Fwp-content%2Fuploads%2F2011%2F10%2FSummary-brochure10.pdf&ei=YVkaVPj2KMTfOajQgYAP&usg=AFQjCNGjfvIe8i4zF_nosPD4Ra1XnSyqqg&bvm=bv.75097201,d.ZWU (accessed 5 August 2014)

Chapter 9

Artificial and Biological Particles in Urban Atmosphere

Chang-Jin Ma and Mariko Yamamoto

Additional information is available at the end of the chapter

http://dx.doi.org/10.5772/59631

1. Introduction

In urban areas including smaller towns, the presence of ambient artificial particulate matters (PM) (e.g., aged fine PM and asbestos) and biological particles has been one of the main causes of adverse effects on human health and air quality. Bioaerosols are airborne particles of biological origin (e.g., bacteria, fungi, pollen, viruses). Some biological contaminants trigger allergic reactions, including hypersensitivity pneumonitis, allergic rhinitis, and some types of asthma.

In Japan, there is peculiar annual variation in these ambient PM due to the sources as well as the changing weather. Most synergistic effects take place during spring because of windborne pollen from trees, especially Japanese cedar trees.

From 1949 to 1970, a large number of Japanese cedar trees were planted in many parts of Japanese Island because of a great demand for new housing after World War II. Planted cedar forests now cover 12% of Japan's total land area, which is more than 45,000 km^2. Vast amounts of pollen, especially from cedar trees, drifts over wide areas every spring [1] As a result, more and more Japanese over the last 30 years have been affected by pollen allergies.

The air quality in urban area, especially its $PM_{2.5}$ (less than 2.5 μm in aerodynamic diameter) level, has become of increasing public concern because of its importance and sensitivity related to health risks [2]. This inhalable $PM_{2.5}$ is generally emitted from both natural (e.g., volcanic eruptions, forest fires, etc.) and anthropogenic sources (combustion processes in industrial sector and automobiles), and it can also be formed when gases react in the air. Inhalable particles, particularly $PM_{2.5}$, have been demonstrated to have the greatest impact on human health, visibility reduction, and solar radiation change, especially in densely populated urban areas.

Asbestos, a naturally occurring fibrous mineral, can be found naturally in the outdoor atmosphere and in some drinkable water, including water from natural sources around the world [3]. As asbestos had been a popular building material since the 1950s, it is still found in many buildings, including hospitals, schools and homes. Inhaling asbestos fibers is known to cause several serious and even fatal lung diseases. Studies have shown that the non-occupationally exposed population have 10,000 - 999,999 asbestos fibers in each gram of dry lung tissue, which translates into millions of fibers and tens of thousands of asbestos bodies in every person's lungs [4]. However, most building material products manufactured today do not contain asbestos. In the industrialized countries, asbestos was phased out of building products mostly in the 1970s, with most of the remainder phased out by the 1980s [5]. In 2006, the Japanese Ministry of Health, Labour and Welfare issued a final ruling banning most asbestos-containing products with the exception of 5 kinds of materials (e.g., some sealing materials). Those 5 unbanned materials were also banned eventually in 2011. Ever since its initial phase out in 2006 and permanent ban in 2011, it can still be found today in some older buildings and consumer goods. Ambient asbestos fibers will finally be lost in the air and eventually precipitate on the ground. These pieces of asbestos are likely to settle on the soil but can be re-released again into the atmosphere [6].

To assess the impact of both artificial and biological PM on the environment, including air quality, ecosystems, and human health, it is necessary to know its concentration, chemical composition, and the interplay among their components. The ambient outdoor PM in urban areas has seldom been evaluated with respect to both artificial and biological components. In light of this situation, we undertook a field campaign to evaluate the artificial and biological PM in an urban environment in Japan during springtime.

2. Material and methods

2.1. Description of Fukuoka prefecture

Fukuoka Prefecture is located on Kyushu Island, Japan's third largest island, located southwest of the main island, Honshu. According to the latest estimates (June 1, 2013), its population and total area are 5,088,480 and 4,971 km^2, respectively [7].

Fukuoka Prefecture faces the sea on three sides, bordering on Saga, Oita, and Kumamoto prefectures and facing Yamaguchi Prefecture across the Kanmon Straits. Fukuoka Prefecture includes the two largest cities on Kyushu, Fukuoka and Kitakyushu, and much of Kyushu's industry. Fukuoka prefecture's main cities form one of Japan's main industrial centers, accounting for nearly 40% of the economy of Kyushu. Major industries include automobiles, transport equipment, electronic parts and machine, general machinery, iron and steel, semiconductors, steel, and food [8].

2.2. Sampling and monitoring of ambient PM

An intensive measurement of PM was conducted at four selected sites (A: 33.65°N; 130.45°E, B: 33.63°N; 130.42°E, C: 33.39°N; 130.26°E, D: 33.66°N; 130.40°E,) in the Fukuoka Prefecture. The location of each sampling site in Figure 1 is indicated by filled circles.

http://dx.doi.org/10.5772/59631

Site A has six-lane roads with heavy traffic conditions. Site B is a residential area and surrounded by a residential area without any known point sources. Site C is an industrial area with roughly 60 manufacturing companies including small-to-medium sized enterprises. Most of them are distributed within a radius of one kilometer from site C. Site D is a desolate area with resort beach without the influx of people, as our study was conducted during an off-season period.

An intensive measurement of PM was simultaneously conducted at four-sites for two days beginning on April 18. 2007 and ending in April, 2007. For the sampling of size-classified PM, four impactor samplers (Tokyo Dylec Co.) were simultaneously operated. Particles were collected directly on the filter arranged behind the jet-nozzles of the 1st stage of sampler. For the collection of particle samples, airflow was maintained at 20 l min^{-1}. The size fraction of PM and filter kind at each stage are coarse fraction (> 2.5 μm) at the first stage (a 47 mm diameter, non hole Nuclepore® polycarbonate filter) and fine fraction ($PM_{2.5}$) at a back-up stage (a 47 mm diameter, 0.01 μm hole Nuclepore® polycarbonate filter), respectively.

Although the influx value (i.e., deposition rate, grains cm^{-2}) is a widely used method for studies of airborne pollen, in this study, for the purpose of measuring the ambient concentration (grains m^{-3}), airborne pollen grains were also collected on the 1st stage of impactor samplers. The directly collected pollen grains on the natural filter surface are more easily observed by a SEM without filter pretreatment compared to those collected on a traditional plastic tape coated by adhesive.

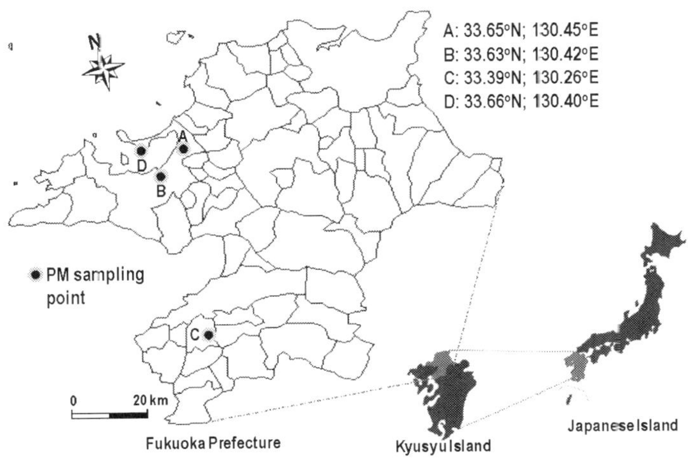

Figure 1. Locations of sampling stations for PM in Fukuoka Prefecture

In order to assess $PM_{2.5}$ mass concentration, 4 Dust scan Scouts (Rupprecht & Patashnick Co. Model 3020) were simultaneously operated at each site. This $PM_{2.5}$ monitoring system makes use of near-forward light scattering to measure the concentration of particulate matter in ambient air. The light source is a safety-interlocked laser that operates at a wave length of 670 nm. The scattered light caused by the presence of particles is received by a sensor, forming the basis of the monitor's computations. During sampling period, there was no Asian dust event. The wind speed was measured at 2.4 - 4.7 m s^{-1}, with the temperature at around 11.2 - 19.4 °C and the average relative humidity at 56%.

2.3. Sample pretreatment and analysis

2.3.1. Sample pretreatment

Figure 2 schematically illustrates the procedures of sample pretreatment and analysis. After sampling, filters were placed in sterilized airtight petridishes and stored in a refrigerator until analysis.

For the laboratory analysis, the filters capturing $PM_{2.5}$ were extracted with deionized water by ultrasonic treatment. And then the extracted water was filtrated through a 25 mm diameter Nuclepore® filter with 0.08 μm pore size to separate into the soluble and insoluble fractions. After filtration, the filtrate was considered to be soluble fractions. Blank filters were handled in the same manner as the samples.

Meanwhile, the coarse particles (>2.5 μm) deposited on the first stage were progressed to single particle analysis for identification of asbestos and pollen. Samples for observation of pollen were coated with a very thin layer of platinum by a machine called a sputter coater.

2.3.2. Analysis of major ionic species in $PM_{2.5}$

In general, sulfate, nitrate, and ammonia had the greatest ambient concentrations in particles [9,10]. The concentrations of major ions in $PM_{2.5}$ (ammonium, nitrate, and sulfate) were determined by Ion Chromatography (IC) (Dionex DX-100). To attain the reliability of the analyzed data, the quality assurance and quality control (QA/QC) was conducted by analyzing a set of known standard species. The data obtained by 5-time repeated IC analyses were tested for precision by checking the relative standard deviation (% RSD, (SD / mean) x 100) of each concentration in standard solution (0.5, 2, and 5 mg/L). As a result, all three ionic species maintained low % RSD levels (i.e., ammonium: 4.63 - 7.59%, nitrate: 0.04 - 9.03%, and sulfate: 1.96 - 9.72%). This high reproducibility is a clear indication of a methodological soundness.

2.3.3. Identification of asbestos and pollen

The most common method of identifying asbestos fibers in ambient PM is by polarized light microscopy. However, a Scanning Electron Microscopy (SEM) can be also usefully applied to the observation of asbestos. The advantage of using a SEM for asbestos identification is that it has better resolution through higher magnification and a greater depth of focus than polarized light microscopy. Most of all, a SEM equipped with an energy dispersive X-ray spectrometer (EDX) (i.e., SEM-EDX) can quantify the elemental components in asbestos. In Figure 3, the diagram of a SEM-EDX is schematically illustrated.

http://dx.doi.org/10.5772/59631

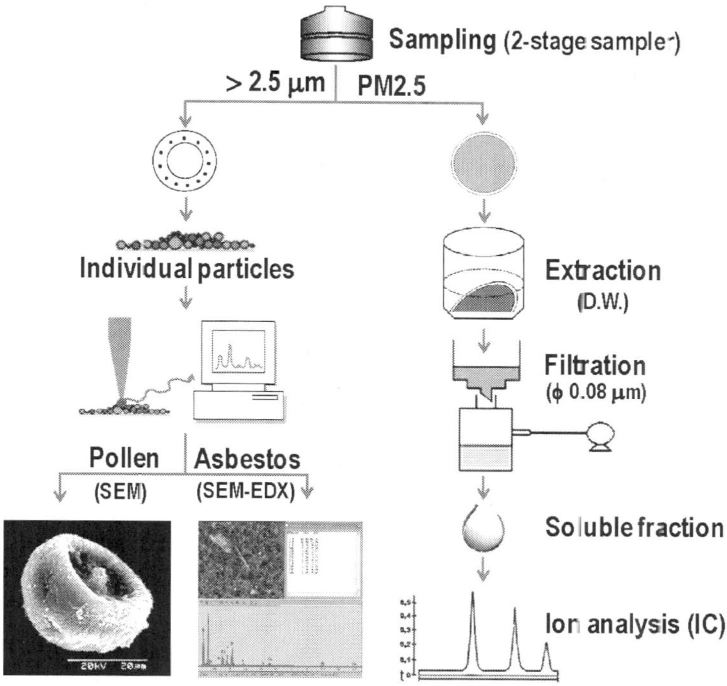

Figure 2. Procedures of sample pretreatment and analysis

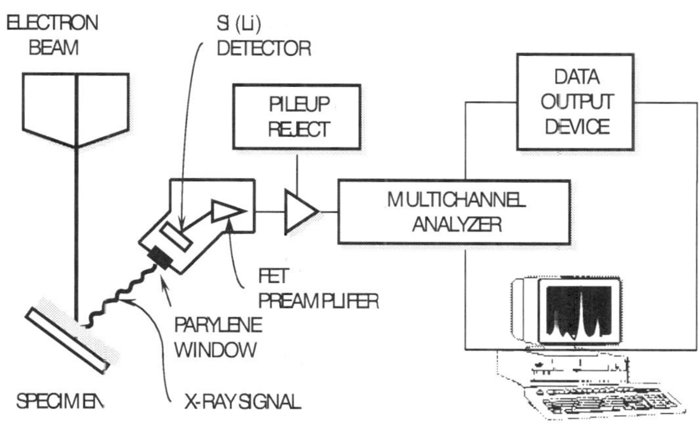

Figure 3. Schematic diagram of a SEM-EDX

Using an EDX and a computer system, information about the elemental properties of asbestos fibers can be gathered and graphed in their appropriate relative ratios. Computation of the exact ratios of the elemental compositions in asbestos fiber allows the researcher to distinguish not only one type of asbestos fiber from another but also asbestos fibers from non-asbestos fibers.

In this study, for the purpose of observing and analyzing the morphological and chemical properties of airborne asbestos and pollen grains, an SEM (JEOL JSM-5400) equipped with an EDX (Philips, EDAX DX-4) was employed. The samples were placed inside the SEM's vacuum column (10^{-6} Torr) through an air-tight door. Pollen species were also distinguished and countered under 3000 x magnification and 15 - 20 kV working conditions.

Calculation of the airborne asbestos fiber concentration on the filter sample was carried out using the following formula:

$$F_{con} = \frac{A_{eff} \times N_{total}}{a_{field} \times n_{field} \times V_{air}}$$

where F_{con} is airborne fiber concentration (fiber/L, f L^{-1}), A_{eff} is effective collecting area of filter (cm^2), N_{total} is the total number of fibers in an SEM field area, a_{field} is an SEM field area (cm^2), n_{field} is total number of fields counted on the filter, and V_{air} is total air volume (L) calculated by sample collection time (min) and pump flow rate (L/min).

3. Results and discussion

3.1. Spatial distribution of PM $_{2.5}$ and major water-soluble ions

Air pollution, especially $PM_{2.5}$, has become a serious issue in East Asia, and there is rising public criticism regarding its effects. Concern in Japan is also increasing as winds transfer the pollution into domestic areas. Figure 4 shows the distribution of $PM_{2.5}$ monitoring data at four-sampling sites (i.e., a heavy traffic area (A), a residential area (B), an industrial area (C), and a desolate area (D)).

Site C, an industrial area, showed the heighest $PM_{2.5}$ (65.3 μg m^{-3}) followed by site A (35.3 μg m^{-3}), site B (22.0 μg m^{-3}), and site C (11.3 μg m^{-3}). As might be expected, the highest $PM_{2.5}$ level was monitored at site C where there is a compact mass of manufacturing companies. Among the four monitoring sites, two sites (i.e., site A and C) exceeded the Japan's $PM_{2.5}$ criteria (a daily average of 35 μg m^{-3}). The reason for the relatively high $PM_{2.5}$ in both heavy traffic and industrial areas might be that much of fine particles and their precursors came from the vehicle emissions and fuel combustion and manufacturing processes in factories. The $PM_{2.5}$ data monitored in this study are compared with those measured in other urban and rural areas in Asia during the springtime. Zhang *et al.* [12] reported that $PM_{2.5}$ was 145.3 μg m^{-3} in Beijing during a non-Asian dust period in the springtime. Meanwhile, in Gosan, a typical rural area in Korea, $PM_{2.5}$ during non-Asian dust period was measured at 26.1 μg m^{-3} [13].

http://dx.doi.org/10.5772/59631

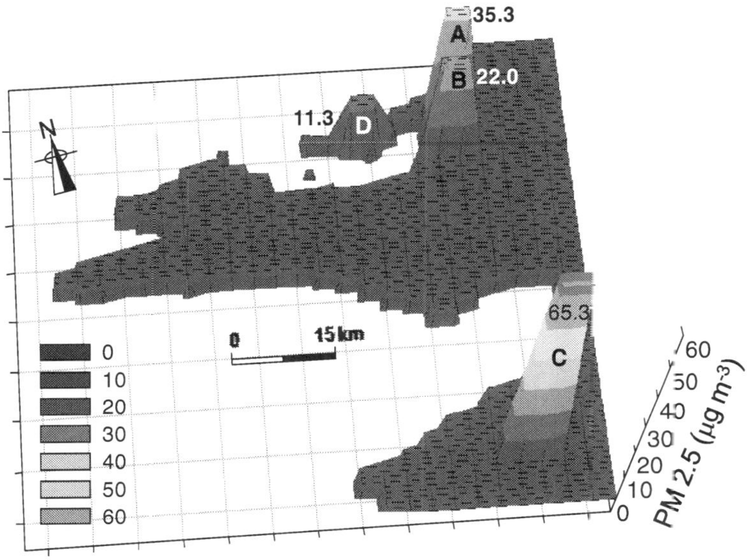

Figure 4. distribution of PM $_{2.5}$ monitoring data at 4-sampling site

It is a matter of course that a large number of factories contributed to the high $PM_{2.5}$ level and caused the regional worsening air pollution in a residential area (B). However, although the "yellow dust" warning did not issue during our field measurement, there is a possibility of the long-range transport of $PM_{2.5}$ and its precursors from the Asian continent to the local study site in Japan. Thus, it is required to clear the uncertainties regarding the linkage between locally high $PM_{2.5}$ at site C and its long-range transport under the springtime meteological conditions. In this study, in order to determine the source region of aerosols at the site C, the atmospheric backward dispersion model was applied.

Figure 5 displays the backward aerosol dispersion simulated by the NOAA Air Resources Laboratory (ARL) HYSPLIT dispersion-trajectory model started from site C. The area scales mean the integrated mass concentration [mass m^{-3}] at 100 - 1000 m height of site C (33.39°N; 130.26°E).

A detailed model description of HYSPLIT was given in reference [14]. According to the result of the HYSPLIT model, a high value of aerosol concentration at the present receptor (measurement location) was not driven from the Chinese continent but was generated from a local area.

Ambient concentrations of major ionic species (i.e., nitrate, sulfate, and ammonium) associated with $PM_{2.5}$ collected at four-sampling sites were overlapped with a map of Fukuoka Prefecture (Figure 6). As shown in Figure 6, the concentrations of major ionic species turned out to be of

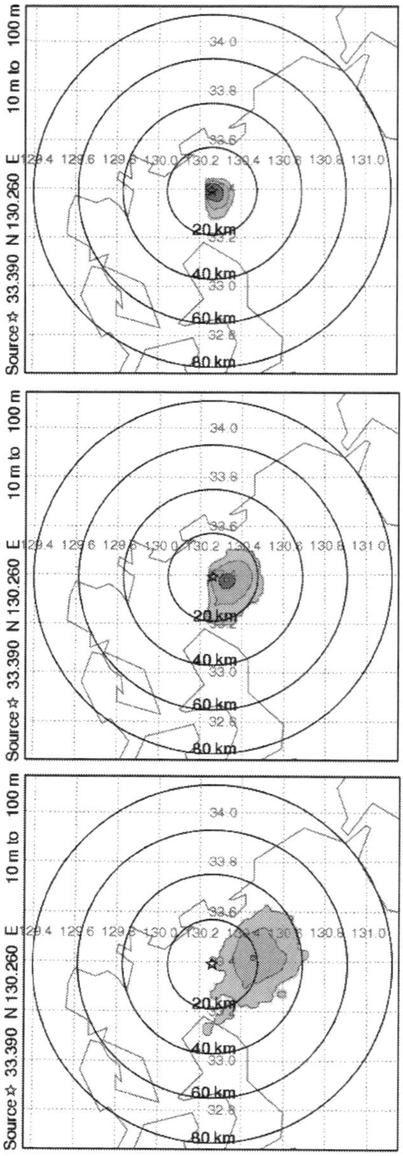

Figure 5. Backward aerosol concentrations simulated by the NOAA ARL HYSPLIT model. The area scales mean the integrated mass concentration [mass m-3] at 10-100 m height of site C (33.39oN; 130.26oE)

considerable variation among four sites. Sulfate was the most abundant species to record the highest concentrations in all urban areas of Fukuoka Prefecture (a heavy traffic area (14.1 μg m^{-3}), an industrial area (30.5 μg m^{-3}), and a residential area (2.8 μg m^{-3}). The sum concentrations of NH_4^+, NO_3^-, and SO_4^{2-} varied in a similar way, as PM $_{2.5}$ (i.e., site C (38.4 μg m^{-3}) > site A (18.3 μg m^{-3}) > site D (5.4 μg m^{-3}) > site B (4.9 μg m^{-3})).

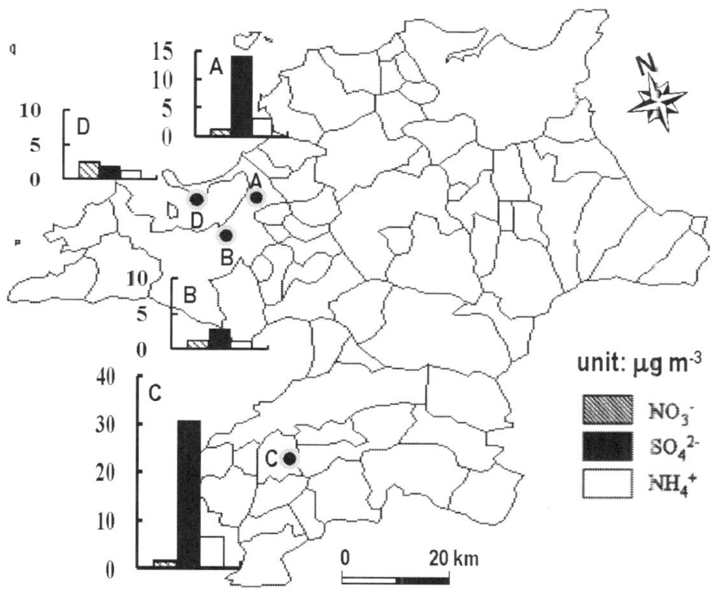

Figure 6. Ambient concentrations of major water-soluble ions associated with $PM_{2.5}$ collected at 4-sampling site in Fukuoka Prefecture

This suggests that the overwhelmingly high level of site C should be associated with the local pollution emissions. In the case of site C and A, the sum of three ionic components correspond to 58.8% and 51.8% of $PM_{2.5}$, respectively. Therefore, it seems reasonable to say that $PM_{2.5}$ in sites C and A was mainly composed by the secondary inorganic aerosol, which was formed by a gas-to-gas reaction in the atmosphere.

The concentrations of three kinds of ionic species in $PM_{2.5}$ in this study are comparable to those of urban areas in Beijing, China, and Durg, India. Gao *et al.* [10] reported that the concentrations of NH_4^+, NO_3^-, and SO_4^{2-} in $PM_{2.5}$ collected in Beijing were 20.5, 15.2, and 42.3 μg m^{-3}, respectively. On the other hand, those concentrations in Durg, a heavy traffic and industrial area, were marked as 2.1, 3.16, and 6.75 μg m^{-3}, respectively [9]. The occupation ratio of the sum concentration of three ionic components in this area (0.89%) is greatly dissimilar to those of this study.

3.2. Assessment of airborne asbestos fiber

Even though asbestos is closely regulated in the present, the deposited and accumulated asbestos fibers for the past several tens of years can be resuspended from the ground near the distributing and manufacturing shops of asbestos products.

Asbestos from natural geologic deposits is known as "naturally occurring asbestos" (NOA) [14]. Health risks associated with exposure to NOA are not yet fully understood. As air quality associated with asbestos in ambient outdoor air has seldom been evaluated in Fukuoka Prefecture, we evaluated airborne asbestos in the urban environment there. Six mineral types are defined by the United States Environmental Protection Agency as asbestos [15]. Among them, crocidolite and amosite are commonly known to cause negative health effects such as lung cancer and mesothelioma [4].

Figure 7 shows an example of asbestos image of SEM, the elemental wt% list, and elemental spectrum obtained from EDX analysis.

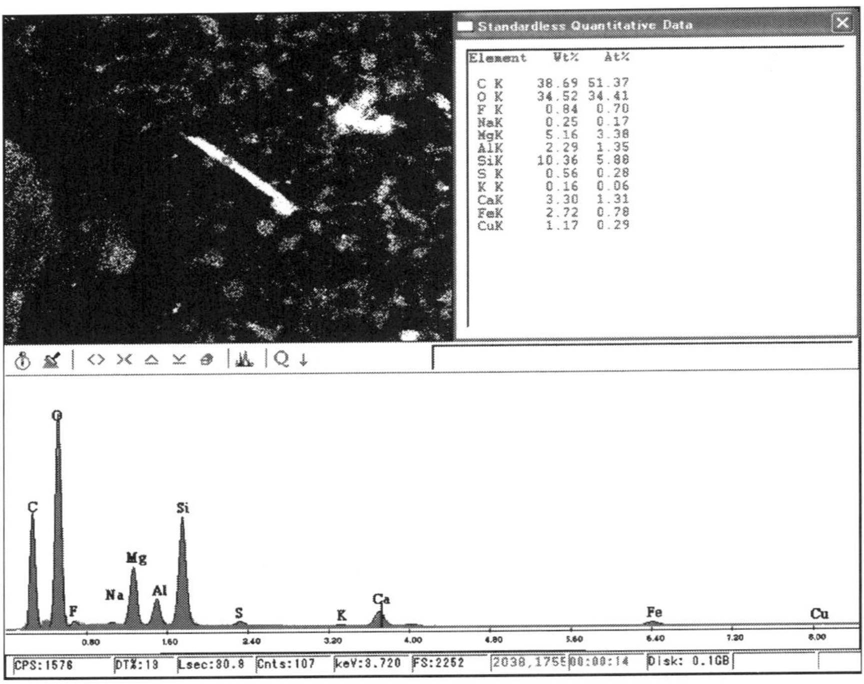

Figure 7. An example of asbestos image of SEM (left upper) and the elemental wt% list (right upper) and spectrum (bottom) obtained from EDX analysis

A part of results of SEM-EDX analysis for asbestos fibers and the processes of classifying asbestos types based on the SEM-EDX elemental wt% data are illustrated in Figure 8.

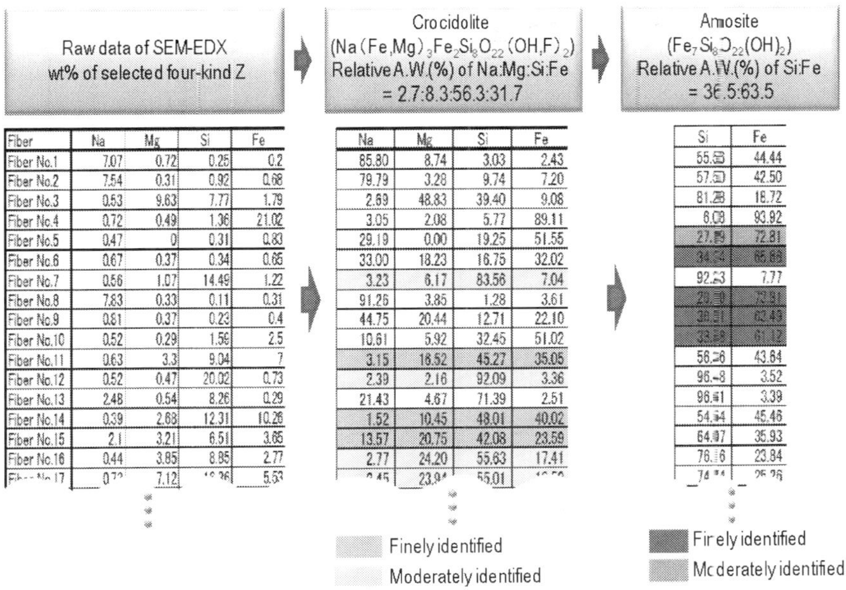

Figure 8. The raw data of SEM-EDX analysis for several asbestos fibers and the processes of classifying asbestos types based on the SEM-EDX elemental wt% data are illustrated in

Figure 9 illustrates the number concentration of asbestos fiber (a sum of crocidolite ($Na(Fe,Mg)_3Fe_2Si_8O_{22}(OH,F)_2$) and amosite ($Fe_7Si_8O_{22}(OH)_2$) at four-sampling sites and distribution of point sources of asbestos fiber in Fukuoka Prefecture. The highest airborne asbestos fiber concentration was recorded at site C (14.4 f L^{-1}), followed by site A (5.9 f L^{-1}), site D (3.4 f L^{-1}), and site B (2.5 f L^{-1}). The average concentration of the airborne asbestos fiber at all sites was 6.14 f L^{-1}. This average concentration level is similar to those (7 f L^{-1}) measured at asbestos abatement sites in Korea [16].

In 1989, the Japanese Air Pollution Control Law and related orders were revised to classify asbestos as a "specified dust" and to set up 10 fibers/liter (for including all type of asbestos) as the regulation guideline Concentration at the Boundary of the Asbestos Dusts Generation Facilities (i.e. asbestos products manufacturing facilities) [6]. Asbestos fiber concentration of site C (14.4 f L^{-1}) is considerably higher than the regulated levels of asbestos of the Concentration Standard at the Boundary of the Asbestos Dusts Generation Facilities.

To explore the possible reasons for the high asbestos fiber concentration at site C, the forecast atmospheric dispersion was simulated by the NOAA ARL Gaussian model. Figure 10 shows

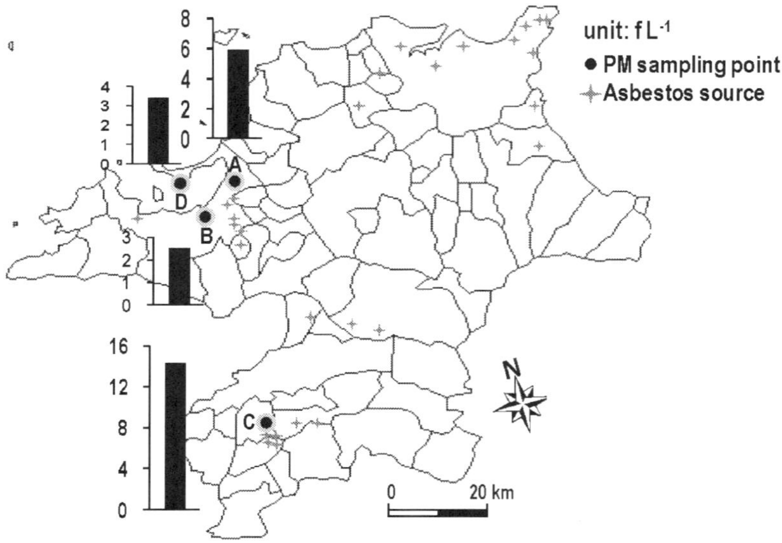

Figure 9. Number concentration of asbestos fiber (crocidolite and amosite) at 4-sampling site and distribution of point sources of asbestos fiber in Fukuoka prefecture.

sampling site C in Figure 9 having a dense asbestos point source and the area distribution of the air flume started from intensively distributed companies producing asbestos related products.

According to the result of NOAA's ARL Gaussian model, the C sampling site was in an downwind position. It is therefore suggested that the relative high asbestos concentration at site C was strongly influenced by the clumped distribution of point sources of asbestos fiber. Meanwhile, although the concentration of asbestos measured at site A (heavy traffic area) was lower than the national Concentration Standard at the Boundary of the Asbestos Dusts Generation Facilities, asbestos was detected at levels around 2 times higher than those of sites B and D.

NOAA's ARL Gaussian model shown in Figure 10 indicates that the point sources of asbestos fiber situated densely near site A did not exert a direct influence on the asbestos concentration at site A. For several years, automobile parts that needed insulation from heat and friction were manufactured from dangerous asbestos, due to its excellent heat-resistant qualities. Such parts included brake linings, clutch facings, transmission components, disc brake pads, drum brake linings, and brake blocks [5].

According to both the result of NOAA's ARL Gaussian model (Figure 10) and the information on extensive use of asbestos in automobiles, the asbestos concentration at site A was probably affected by automobiles.

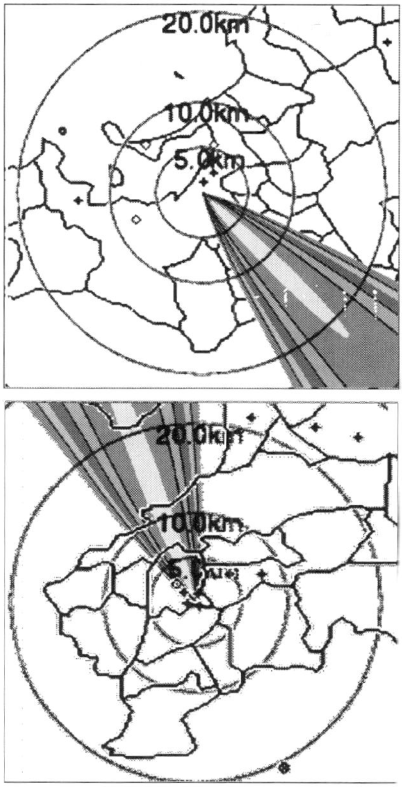

Figure 10. Map showing sampling site C in Figure 9 having dense asbestos point source and the area distribution of air flume started from the intensively distributed companies producing asbestos

3.3. Spatial distribution of ambient pollen

It can be considered that pollen contributes to the organic carbon fraction in totoal suspended particles (TSP). In addition, some plant types can produce pollen in huge quantities. For example, a single ragweed plant can generate a million grains of pollen a day. Therefore, though pollen granules are small and light, they can also attribute to the mass concentration of ambient aerosol particles during the main pollen season [17]. From this point of view, to study the species of pollens, their distribution and concentration can be helpful to understand their ambient behavior and public health, etc.

SEM images of several types of pollens are shown in Figure 11. Classification of pollen grains using SEM was based on morphological characteristics such as shape, size, apertures and ornamentation. The combination of these characteristics makes some pollen grains easily identifiable. However there were also pollen grains sharing several common characteristics that makes identification difficult.

Figure 12 shows the spatial variation of the number concentration of three-kind of airborne pollen at the four sites in Fukuoka Prefecture. Three-type mainly identified pollen grains are reported here. There was a noticeable spatial difference in the concentrations among three pollen types. Cedar (Cryptomeria, also called the Sugi tree in Japanese) pollen was distributed in advance of other pollens at sites A, B, and C, with the maximum concentration at site A. This cedar pollen is the most common allergen for seasonal allergic rhinitis in Japan. Cedar showed a higher concentration than other types of pollen and it was probably generated from the Sugi tree that is the most important timber tree in Japan.

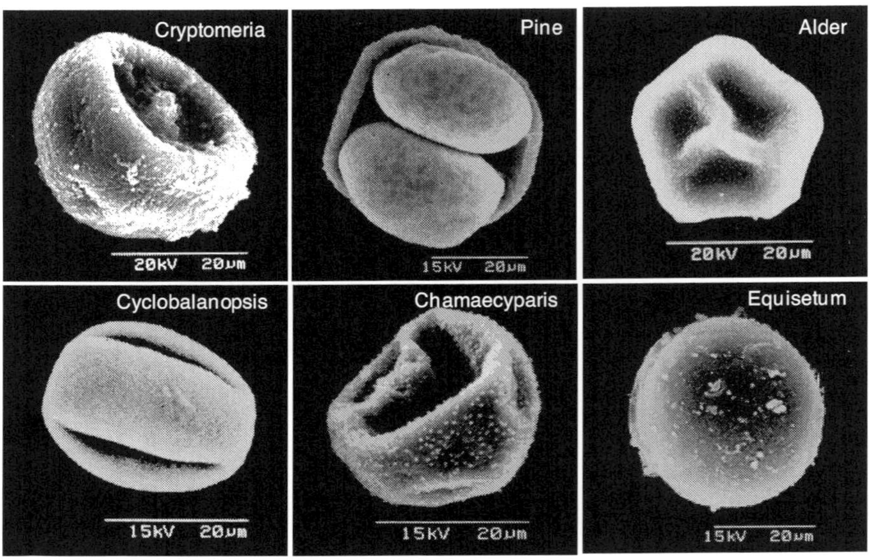

Figure 11. Morphologies of several types of pollen grains (Cryptomeria, Pine, Alder, Cyclobalanopsis, Chamaecyparis, and Equisetum) identified by SEM observation

The sum of the number concentration of three pollen types is still high at site A. Although the atmospheric presence of pollen grains is likely to vary depending on the local kind of plant, actual weather situations, and pollinating period, the vegetation of the surface is also important [18]. Cedar pollen easily absorbs water in the atmosphere and then settles down on surface. However, they can be easily resuspended into the atmosphere by the urban surface covered with asphalt and cement. Both hot-island and building wind can also promote the resuspension of pollen [19]. Therefore, site A, with its relatively high pollen concentrations, was probably affected by the typical urban surface and local plant.

http://dx.doi.org/10.5772/59631

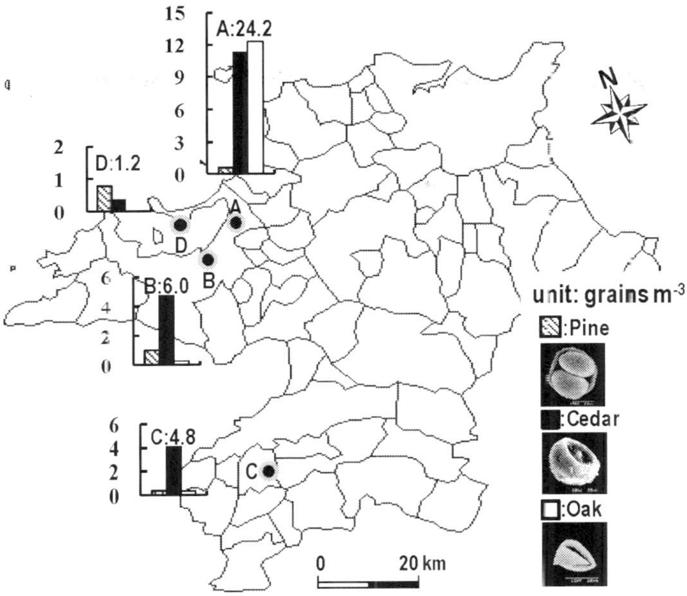

Figure 12. Pollens number concentration at each sampling site

4. Conclusion

A better knowledge of the impact of both artificial and biological PM on urban atmosphere can help establish improved the management strategies of urban air quality. This study focused on a comprehensive and detailed interpretation for the springtime air quality influenced by both artificial (particulate matter (PM) and asbestos) and biological (pollen) sources in Fukuoka Prefecture, Japan. An intensive measurement of PM was conducted at four characteristic sites (i.e., a heavy traffic area, a residential area, an industrial area, and a desolate area) in the Fukuoka Prefecture during spring of 2007. Analysis of major ionic species in $PM_{2.5}$ was performed by an Ion Chromatography, and asbestos and pollen were identified by Scanning Electron Microscopy with an energy dispersive X-ray spectrometer (EDX). $PM_{2.5}$ concentration (65.3 μg m^{-3}) measured in an industrial area (site C) was extraordinarily high compared to those monitored in other areas; it greatly exceeded the Japan s $PM_{2.5}$ criteria (a daily average of 35 μg m^{-3}). NOAA's HYSPLIT dispersion model suggests that this high level of $PM_{2.5}$ monitored at site C is unlikely to affect the Asian continent. The ambient concentrations of $PM_{2.5}$-related anions (NH_4^+, NO_3^-, and SO_4^{2-}) and their relative contributions to $PM_{2.5}$ were also investigated in four study areas. The concentrations of these major water-soluble ions exhibit not only strong spatial dependence but also different ratios to each other. Asbestos fiber (crocidolite and amosite) concentration values changed in the range of 2.5 to 14.4 f per liter of air. The number of pollen grains showed that Cedar ranked higher in concentration than other types of pollen, with the maximum concentration at site A.

The results of our intensive field measurement suggested that synergic biological effects induced by ambient allergenic pollen and urban fine PM in atmosphere are associated with a peculiar springtime air quality in the Fukuoka Prefecture. Our study also indicates that the specific artificial and natural sources are regionally distributed to influence local air quality and public health. We should be mindful of the fact that in order to improve the understanding of urban air quality regarding the environmental load and repercussions on human health, it will be necessary to continue the monitoring of not only particulate matters but also gaseous materials.

Acknowledgements

The HYSPLIT aerosol dispersion model (http://www.arl.noaa.gov/ ready.html) developed by the National Oceanic Atmospheric Administration (NOAA) for backward trajectory was very helpful to data interpretation. The authors also express sincere thanks to Prof. G.-U. Kang, Department of Medical Administration, Wonkwang Health Science University for his ion analysis.

Author details

Chang-Jin Ma[1*] and Mariko Yamamoto[1,2]

*Address all correspondence to: ma@fwu.ac.jp

1 Department of Environmental Science, Fukuoka Women's University, Fukuoka, Japan

2 Fukuoka Prefectural Munakata High school, Munakata, Japan

References

[1] Takahashi Y., Tokumam K., Kawashima S. Distribution Chart of Cryptomeria Japonica Forest Through Data Analysis of Landsat-TM. Japanese Journal of Palynology 1992; 38 140-147.

[2] Schwartz J., Neas L.M. Fine Particles Are More Strongly Associated than Coarse Particles with Acute Respiratory Health Effects in Schoolchildren. Epidemiology 2000(11) 6-10.

[3] Cook P.M., Glass G.E., Tucker J.H. Asbestiform Amphibole Minerals: Detection and Measurement of High Concentrations in Municipal Water Supplies. Science 1974; 185 853-855.

[4] Camus M., Siemiatycki J., Meek M.Sc.B. Nonoccupational Exposure to Chrysotile Asbestos and the Risk of Lung Cancer. New England Journal of Medicine 1998(338) 1565-1571.

[5] Lemen R.A. Asbestos in Brakes Exposure and Risk of Disease. American Journal of Industrial Medicine 2004(45) 229-237.

[6] Monthly Archives. Fukuoka-ken in Japan Encyclopedia. http://www.asbestosremovalsaustralia.com.au/blog/2013/04/Nussbaum (accessed 2 May 2014)

[7] Fukuoka Prefecture. About Fukuoka http. http://www.pref.fukuoka.lg.jp/somu/multilingual/english/about.html (accessed 2 May 2014).

[8] Nussbaum L.F. and Käthe R. Japan encyclopedia. Cambridge: Harvard University; 2005.

[9] Deshmukh D.K., Deb M.K., Tsai Y.I., Mkoma S.L. Water Soluble Ions in $PM_{2.5}$ and PM_1 Aerosols in Durg City, Chhattisgarh, India Aerosol and Air Quality Research 2011(11) 696-708.

[10] Gao X., Nie W., Xue L., Wang T., Wang X., Gao R., Wang W., Yuan C., Gao J., Ravi K.P., Wang J., Zhang Q. Highly time-resolved Measurements of Secondary Ions in $PM_{2.5}$ During the 2008 Beijing Olympics: The Impacts of Control Measures and Regional Transport. Aerosol and Air Quality Research 2013(13) 367-376.

[11] Vezey E.L., Skvarla J.J., Vanerpool S.S. Characterizing Pollen Sculpture of Three Elosely-related Capparaceae Species Using Quantitative Image Analysis of Scanning Electron Micrographs, Clarendon: Oxford; 19991.

[12] Zhang W.J., Sun Y.L., Zhuang G.S., Xu D.Q. Characteristics and Seasonal Variations of $PM_{2.5}$, PM_{10}, and TSP Aerosol in Beijing. Biomedical and Environmental Sciences 2006(19) 461-468.

[13] Stone E.A., Yoon S.C., Schauer J.J. Chemical Characterization of Fine and Coarse Particles in Gosan, Korea During Springtime Dust events. Aerosol and Air Quality Research 2011(11) 31-43.

[14] Interim Guidance. Naturally occurring asbestos (NOA) at school sites http://www.dtsc. ca.gov/Schools/upload/SMBRP_POL_Guidance_Schools_NOA.pdf. (accessed 3 July 2014).

[15] El Dorado County. Naturally occurring asbestos and dust protection. http://www.co.el-dorado.ca.us/emd/apcd/PDF/Naturally_Occuring_Asbestos_June_12.pdf (accessed 3 July 2014).

[16] Kim J.Y., Lee S.K., Lee J.H., Lim M.H., Kang S.W., Phee Y.G. A Study on the Factors Affecting Asbestos Exposure Level from Asbestos Abatement in Building Demolition Sites. Korean Industrial Hygiene Journal 2009 (19) 8-15.

[17] Farn P.P., Clarence T.N., Patrick J.S. Aerosol characteristics of Arctic haze sampler during AGASP 2. Atmospheric Environment 1990 (24A) 937-939.

[18] Mandrioli P., Negrini M.G., Scarani C., Tampieri F., Trombetti F. Mesoscale Transport of Corylus Pollen Grains in Winter Atmosphere. Grana 1980 (19) 227-233.

[19] Saito Y., Ide T., Murayama K. Science of Pollenosis. Kagakudouzinn co; 2006. p.14-42.

## Chapter 10

Emission Control Technology

Thanh-Dong Pham, Byeong-Kyu Lee,
Chi-Hyeon Lee and Minh-Viet Nguyen

Additional information is available at the end of the chapter

http://dx.doi.org/10.5772/59722

1. Introduction

The chapter presents various promising methods to control air pollution emissions. Although several previously published books have examined this field [1-6], most of them presented only the methods to control air pollutants generated from stationary sources. These days, however, mobile sources are also important sources contributing to air pollution. Among the six air criteria pollutants listed by the United States Environmental Protection Agency (EPA), including CO, particulate matter (PM), SO_2, NO_2, O_3, and Pb, mobile sources can contribute to 81% of all CO emissions, 37% of PM, 4% of SO_2, 45% of NO_2, 47% of volatile organic compounds (VOCs are precursors of O_3, their estimate is a surrogate of the O_3 concentration), and 72% of Pb [6]. Therefore, promising methods to control the emission of air pollutants generated from mobile sources should be included in the chapter to provide the readers with innovative ideas about the emission control of air pollution. Because of the variety of mobile sources and their mobility from one location to another, the methods applied to control the emission of air pollutants generated from stationary sources may not be useful for controlling air pollutants generated from mobile sources. Therefore, in this chapter, upstream control strategies, which try to control air pollutants from upstream emission processes, will be presented as promising methods to control the emission of air pollutants generated from mobile sources. Policies and regulations applying to control air pollutants emitted from transportation activities, agricultural activities, and construction fields will be presented as the main strategies to air pollution from upstream processes.

On the other hand, the chapter also presents downstream technologies to control air pollutants emitted from stationary sources. Based on the characteristics of target pollutants, the downstream control technologies will be classified into two categories: particulate matter and gaseous pollutant control technologies. The technologies of cyclone, wet scrubber, electrostatic precipitators, and fabric filter will be introduced in the chapter as methods to control particulate matter pollutants. To control gaseous pollutants, the chapter will present methods such as adsorption, absorption, condensation, incineration, applications of biological system, and photocatalyst. Each technology will be presented in detail from definition to principle and applications. The advantages and disadvantages of each technology will be described and compared in the chapter.

2. Upstream control strategies

2.1. Control transportation activities

Strategies to control air pollutants emitted from transportation activities include regulations to control precursor pollutants in raw materials; the application of catalytic converters to reduce NO_x, CO, and hydrocarbon emissions; the control of lubricant consumption; the reduction of motorized transportation demand; and the improvements to road quality and traffic flow.

2.1.1. Raw materials

Air pollutants can be controlled based on regulations that control precursor pollutants in raw materials such as sulfur and lead. Sulfur oxides (SO_x) in aerosol are formed during the combustion process of sulfur in fossil fuels such as gasoline and diesel. Therefore, the use of low-sulfur-content fuels can be considered as a control strategy to reduce SO_2 emissions. Many regulations have been issued to reduce the sulfur content in all transportation fuels. For example, Figure 1 presents a selection of a few of the gasoline and diesel sulfur specifications in major countries and the regulatory timetable associated with the introduction of these specifications [7].

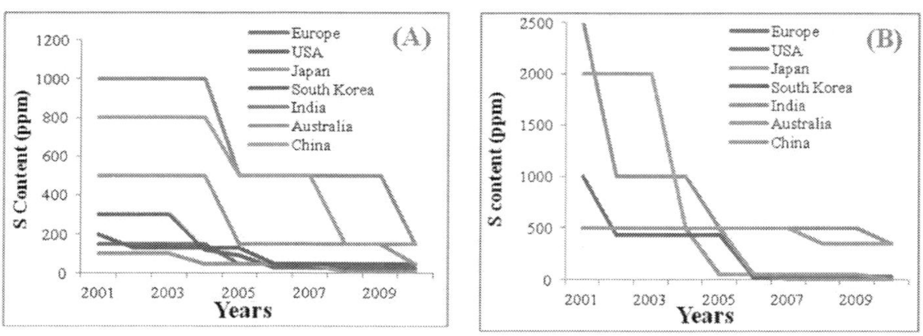

Figure 1. Regulations and regulatory timetable for sulfur content in gasoline (A) and diesel (B) in several countries.

In the 1920s, refiners started adding lead compounds to gasoline in order to increase octane levels and improve engine performance by reducing engine "knock" and allowing higher engine compression. However, the burning of leaded fuel introduced massive quantities of atmospheric lead leading to many adverse effects on human health. Therefore, regulations to eliminate lead from leaded fuel began to be issued worldwide from the 1970s. For example, the US EPA scheduled performance standards requiring refineries to decrease the average lead content of all fuels beginning in 1975, but these were postponed until 1979 through a series of regulatory adjustments. By the early 1980s, the lead content in fuels had declined by about 80% due to both the regulations and the fleet turnover [8]. In August 1984, the US EPA proposed a further reduction of lead to 0.1 grams per liter gallon (gplg) by January 1, 1986. However, several refineries were not able to achieve this time scale, so the US EPA postponed the deadline until January 1, 1988. Lead was banned as a fuel additive in the United States beginning in 1996. Figure 2 shows the decline over time in the lead content of leaded fuel in the United States [9].

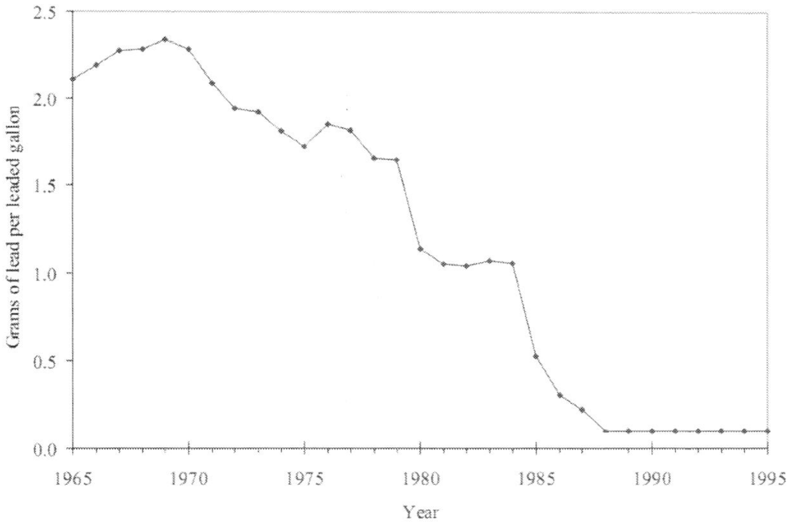

Figure 2. Temporal lead content in leaded fuel in the United States.

Regulations concerning oxygenated gasoline and reformulated gasoline were also issued in the United States in 1990 to reduce the vehicular emissions of CO and VOCs. The oxygenated gasoline regulation required a higher oxygen content in gasoline in order to ensure more complete burning of the gasoline and thereby reduce harmful tailpipe emissions from motor vehicles. In this respect, oxygen dilutes or displaces the precursor pollutant components in gasoline such as aromatics (e.g., benzene) and sulfur and thus decreases their content in gasoline. The reformulated gasoline regulation is specially blended gasoline. Therefore, the gasoline can be burned more cleanly and can be prevented from evaporating as quickly as conventional gasoline, thereby reducing the emission of smog-forming and toxic pollutants. To reduce the emission of benzene (VOCs) into an aerosol, the US EPA required all refiners to meet an annual average gasoline benzene content standard of 0.62% by volume (vol %) in all their gasoline, both reformulated and conventional, nationwide, from 2011.

In addition, the use of new-generation fuels such as biofuels and natural gas can be a valuable strategy to reduce air pollutant emission. Feedstocks for biofuels, mostly plants, are much more environmental-friendly and evenly distributed around the world than the feedstocks for traditional fuels such as oil and gas. Two of the most commonly used biofuels are ethanol and biodiesel. Biodiesel fuels are oxygenated organic compounds of methyl or ethyl esters derived from a variety of renewable sources such as vegetable oil, animal fat, and cooking oil. Therefore, the use of biofuels for transportation activities can significantly reduce the atmospheric emissions of CO, hydrocarbon, and lead.

2.1.2. Catalytic converters

A catalytic converter is a vehicle emission control device that converts toxic pollutants in exhaust gas to less-toxic pollutants by catalyzing a redox reaction (oxidation or reduction). A catalytic converter comprised usually of the following three main parts is used in internal combustion engines: substrate, washcoat, and catalytic materials. The substrate material is usually a ceramic monolith with a honeycomb structure. A washcoat, usually aluminum oxide, titanium dioxide, silicon dioxide, or a mixture of silica and alumina, is used as a carrier for the catalytic materials. The catalytic material is often a mix of precious metals such as platinum, palladium, and rhodium. The catalytic converter can reduce oxides of nitrogen (NO_x) into nitrogen gas (N_2), combine or oxidize carbon monoxide (CO) with/or unburned hydrocarbons (HC) to produce carbon dioxide (CO_2) and water (H_2O). A schematic diagram illustrating the role of the catalytic converter in reducing air pollutants is shown in Figure 3.

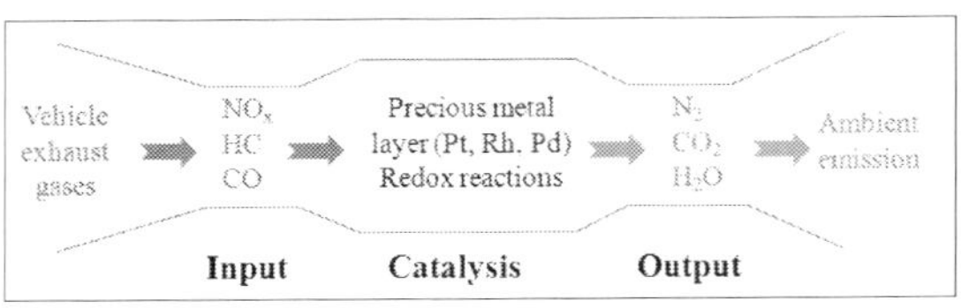

Figure 3. A schematic diagram of a catalytic converter.

The relevant catalysts applied to reduce the pollutants emitted from specific engines are listed in Table 1 [10].

To reduce NO_x, a controlled amount of the reactive chemical reductant such as anhydrous ammonia, aqueous ammonia, and urea is added to a stream of fuel or exhaust gas and adsorbed onto the catalyst. Due to the role of the catalysts, NO_x can react with the reductant to produce nitrogen gas and water according to the following reactions:

http://dx.doi.org/10.5772/59722

Engines	Pollutants	Catalysts
Gasoline engines	Nitrogen oxides (NO_x), Hydrocarbons	Pt/Pd/Rh/$Ce_xZr_{1-x}O_2$/(La, Ba)-Al_2O_3 on ceramic
Diesel engines (light vehicles)	(HC_s), and Carbon monoxide (CO)	and metallic monoliths
Diesel engines (heavy vehicles,	NO_x, HC_s, CO	Pt/Pd/Rh/BaO/Al_2O_3 on ceramic and metallic
truck, and bus)	NO_x, HC_s, CO	monoliths
Diesel engines	Particulate matter	V_2O_x/TiO_2 on ceramic monolith
		Cerium and iron oxides
		Pt/Al_2O_3
		Cu, V- and K-based catalysts

Table 1. The relevant catalysts for specific engines

$$4NO + 4NH_3 + O_2 \rightarrow 4N_2 + 6H_2O$$
$$2NO_2 + 4NH_3 + O_2 \rightarrow 3N_2 + 6H_2O$$
$$NO + NO_2 + 2NH_3 \rightarrow 2N_2 + 3H_2O$$

The reaction typically takes places at an optimal temperature range between 630 and 720 K, but can operate from 500 to 720 K with longer residence times. To operate an effective process, the engine requires an external urea tank and dosing system. The specific NO_x/ammonia and ammonia/catalyst ratios can be designed to optimize a specific application. At optimal conditions, the application of a catalyst in the downstream of engine can reduce NO_x emissions from vehicles by 70–90%.

The reactions to oxidize carbon monoxide and unburned hydrocarbon are described by the following reactions:

$$2CO + O_2 \rightarrow 2CO_2$$
$$C_xH_{2x+2} + [(3x+1)/2]O_2 \rightarrow xCO_2 + (x+1)H_2O$$

2.1.3. Lubricant consumption

Lubricants are composed of a base fluid and additives. The base fluid is the major part of the lubricant formulation and is mainly made from petroleum-based oils. The additives are used to obtain desirable properties. Lubricants are very important substances for reducing the wear and tear of machine parts. Lubricants also reduce friction, which in turn reduces heat loss. The worldwide consumption of lubricants is more than 41 million tones [11]. They have a soluble organic fraction of 60%, which contributed between 20% and 90% of the total particulates in air that were generated from engine lubricant consumption. Therefore, the particulate emission rate can be significantly reduced by controlling engine lubricant consumption. The strategies to control engine lubricant consumption include changing the piston-ring design and manipulating the operation conditions of the engine such as intake air pressure. The use of biolubricants is also a valuable strategy to reduce the adverse effects of traditional lubricants to the atmospheric emissions of particulate matter. Biolubricants are described in many ways such as eco-friendly lubricants, green lubricants, biodegradable lubricants, recyclable, nontoxic, and reusable.

2.1.4. Reduce motorized transportation demand

Strategies applied to encourage the use of nonmotorized transport, discourage nonessential trips, shorten trip lengths, and restrain the use of private cars can reduce the overall demand for motorized transport and thus minimize the emission of air pollutants from transportation activities. The strategies could be instigated based on the following regulations:

- Provide safe and comfortable conditions for walking and other forms of nonmotorized transport
- Improve public transportation quality and efficiency
- Increase fuel taxes
- Increase parking charges
- Increase road pricing
- Compact design of retail and entertainment centers with workers and public transport
- Limit use of private vehicles both by pricing and by administrative regulation

2.1.5. Road quality and traffic flow

Road quality also directly affects the air pollutants emitted from transportation activities. For example, the operation of vehicles on unpaved roads can introduce a significant amount of atmospheric particulate matter. Therefore, road qualities need to be improved to reduce air pollutant emission. The strategies to improve road quality include:

- Try to pave unpaved roads
- Sweep roads frequently (can reduce concentrations of PM up to 20%)
- Flush roads with water in the dry season
- Investigate new types of asphalt and concrete, which are cheap and environmental-friendly
- Cover operating trucks

Strategies to improve traffic flow and thereby minimize unnecessary braking and reduce congestion can result in high efficiency of vehicle operation and reduce undesired pollutant emissions. These strategies can be obtained based on the following methods:

- Control traffic signals
- Design road systems by use of ring roads and bypasses
- Increase infrastructure capacity
- Reduce congestion by congestion charging
- Reduce vehicle speeds because fast moving vehicles stir up dust (a reduction in speed from 40 miles per hour (m/h) to 20 (m/h) reduces dust emissions by 65%)

http://dx.doi.org/10.5772/59722

2.2. Control agricultural activities

Air pollutants emitted from the agricultural sector are mainly methane (CH_4), nitrous oxide (N_2O), and ammonia (NH_3). Agriculture is also a major source of PM, both primary and secondary in origin [12]. Agricultural pollutants are mainly generated from livestock production and the application of fertilizers and pesticides. Therefore, strategies to control air pollutants emitted from agriculture activities are strongly linked to the activities including strategies to control livestock feeding, animal housing systems, manure storage systems, application of manure for crops, and application of fertilizers and pesticides.

2.2.1. Livestock feeding

Because the quantity of nitro compounds, such as ammonia and nitrous oxide excreted from animal feces and urine, is linearly dependent on the intake of nitrogen in food (protein), the strategies to reduce the oversupply of protein in animal feedstock can reduce nitrogen excretions and thus decrease the emissions of nitrogen-containing compounds [13]. Such strategies involve adapting the amount of proteins in the food to the needs of the animals. For instance, young animals and high-productive animals require more protein than older and less-productive animals. On average, this measure leads to a NH_3 emission reduction of 10% for a 1% reduction in the mean protein content in the diet, but efficiencies depend strongly on the animal categories. It has no implications on animal health as long as the requirements for all amino acids are ensured. It is most applicable to housed animals while the practical applicability of feeding strategies to grazing animals is limited.

2.2.2. Animal housing systems

The available strategies to reduce NH_3 emissions from animal housing systems have been well known for decades and apply one or more of the following principles [13]:

Principles	NH_3 emission reduction
· Decrease the surface area fouled by manure	15–25% in pig housing
· Rapid removal of urine and rapid separation of feces and urine	25–46%
· Reduce pH of the manure	up to 20%
· Reduce temperature of the manure	up to 60% in pig and cattle housing systems
· Dry the manure (e.g., poultry litter)	45–75%
· Scrubbing ammonia from exhaust air	up to 70%
· Decrease housing time by increasing grazing time	70–95%
	10–50%, but some emission swapping

2.2.3. Manure storage

The strategies to eliminate air pollutants emitted from manure storage can be based on the following principles:

- Decreasing the surface area where emissions can take place, i.e., by covering the storage, encouraging crusting, and increasing the storage depth
- Reducing the pH and temperature of the manure
- Minimizing disturbances such as aeration

2.2.4. Manure used for crops

The application of manure for crops can emit a significant amount of atmospheric pollutants. The strategies or application techniques to control the emissions can be based on the following principles:

- Decrease the exposed surface area of slurries applied to surface soil through band application, injection, and incorporation
- Decrease the time that emissions can take place, i.e., bury the slurry or solid manures through injection or incorporation into the soil
- Decrease the source strength of the emitting surface, i.e., through lowering the pH and NH_4 concentration of the manure (through dilution)

2.2.5. Fertilizer application

The strategies to reduce emissions of pollutants from the application of fertilizers are based on one or more principles including:

- Decrease emission sites by decreasing the surface area via band application, injection, and incorporation
- Decrease the emission periods of pollutant via rapid incorporation of fertilizers into the soil or via irrigation
- Decrease the emitting source surface strength via urea inhibitors and blending
- Ban use of pollutant precursors such as ammonium (bi) carbonate

For example, the techniques for the application of urea and ammonium-based fertilizers can reduce levels of ammonia emission as follows [13]:

Fertilizer type	Application techniques	Emission reduction %
Urea	Injection	>80
	Urea inhibitors	>30
	Incorporation following surface application	>50
	Surface spreading with irrigation	>40
	Ban	~100
	Injection	>80
Ammonium carbonate	Incorporation following surface application	>50
Ammonium-based fertilizers	Surface spreading with irrigation	>40

http://dx.doi.org/10.5772/59722

2.3. Control construction fields

The strategies to control air pollutants emitted from construction fields can be classified into four categories: control of site planning, construction traffic, demolition works, and site activities [14].

2.3.1. Site planning

Regulations applied to control air pollutants emitted from site planning include:

- Erect effective barriers around dusty activities or entire site boundary
- Do not burn any material in entire site planning
- All site personnel to be fully trained
- Trained and responsible manager on site during working times to maintain logbook and carry out site inspections
- Hard surface all major haul routes through the site (e.g., use recycled rubber blocks, concrete blocks or tarmac)
- Use nearby rail or waterways for transportation to/from site

2.3.2. Construction traffic

Regulations applied to control air pollutants emitted from construction traffic activities include:

- No vehicles or plant will be left idling unnecessarily
- Wash or clean all vehicles effectively before leaving the site
- All loads entering and leaving site should to be covered
- On-road vehicles comply to emission standards
- Use ultra-low sulfur tax-exempt diesel for all nonroad mobile machineries
- Should minimize construction traffic activity around site
- Cover the haul routes with hard surface combining with frequently cleaning the surface and give an appropriate control of speed limit around site

2.3.3. Demolition works

Regulations applied to control air pollutants emitted from demolition works include:

- Use water as dust suppressant during demolition works
- Use enclosed chutes and covered skips
- Wrap building(s) before demolition works
- Bag and remove any biological debris or damp down before demolition
- Avoid explosive blasting where possible and consider using appropriate manual or mechanical alternatives

2.3.4. Site activities

Regulations applied to control air pollutants emitted from site activities include:

- Minimize dust-generating activities. For example, when a worker cuts concrete slabs or bricks with a power tool without extraction or suppression, a second worker can pour water from a plastic bottle over the material leading to reduce the great amount of generated dust.
- Use water as a dust suppressant where applicable
- Cover, seed, or fence stockpiles to prevent wind whipping
- Re-vegetate earthworks and exposed areas
- If applicable, ensure the concrete crusher or concrete batcher has a permit to operate
- Minimize drop heights to control the fall of materials

2.4. Miscellaneous

Coal, the most abundant solid fuel and widely used for power plant and other industrial activities, is the largest source of air pollutant emissions. Coal combustion produces a significant amount of air pollutants such as SO_x, heavy metals, and PM. For example, sulfur in coal occurs both as inorganic minerals (mainly pyrite and marcassite) and organic compounds incorporated in the combustible part of coal. The sulfur content can be converted into SOx during the coal combustion. Therefore, reducing the sulfur content in coal before the combustion processes is a great strategy to reduce SO_x emissions from the upstream coal combustion process. Inorganic sulfur in coal can be removed by coal washing and the organic sulfur by using chemical hydrogenation and gasification processes.

Mining activities can produce significant air pollutants such as heavy metals (in PM form), SO_x and NO_x. Strategies to reduce air pollutants emitted from mining activities from the upstream process include enclosure or cover mine, mining area, and transfer areas; water spraying mining area; and stabilizing unpaved traffic areas.

Indoor activities can also be a significant source of air pollution. Strategies to reduce air pollutants emitted from indoor activities include improvement of cooking devices, use of alternative fuels for cooking and reducing the need for fire. Strategies to improve cooking devices include stabilization of stove materials and improvement of stove chimneys, in particular, biomass stoves. Uses of alternative fuels for cooking including charcoal, biogas, liquid petroleum gas, and electricity can significantly reduce air pollutant emissions. For example, the transition from wood to charcoal for cooking can reduce PM10 emissions by more than 80% (although the wider environmental impacts of charcoal production must be considered). The need for fire can be reduced based on the use of solar heating or electric devices.

The change of building materials from high-polluting materials such as paint, linoleum, and gypsum to low-polluting materials such as PVC and polyolefin can also control air pollutants from upstream emissions.

http://dx.doi.org/10.5772/59722

3. Downstream control

3.1. Particulate matter control

3.1.1. Cyclone

The cyclone is a well-known device used primarily for the collection of medium-sized and coarse particles. The cyclone works by forcing a gaseous suspension downward. The particles move outward by centrifugal force and collide with the outer wall and then slide downward to the bottom of the cyclone. At the bottom of the cyclone, the gas reverses its downward spiral and moves upward in a smaller, inner spiral. The cleaned air exits from the top of the cyclone and the particles are expelled from the bottom of the cyclone through a pipe sealed by a spring-loaded flapper valve or a rotary valve. The cyclone collector is shown schematically in Figure 4.

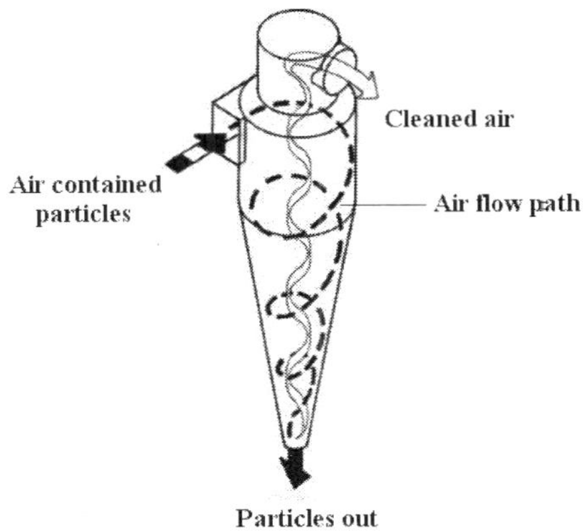

Figure 4. Schematic diagram of a cyclone.

Cyclones have a wide range of industrial applications in gaseous cleaning and product recovery. They are relatively inexpensive, easy to set up and maintain, and can work at high temperature and pressure. They can be used as a precollector for removing larger particles before next treatment. When well designed, the cyclone can collect particles larger than 10 μm with an efficiency of more than 90%. For smaller particles, however, the well-designed cyclone would have a considerable pressure drop with relatively lower collection efficiency [15]. In addition, the cyclone method cannot be used for removing sticky particles with high moisture content.

3.1.2. Wet scrubber

A wet scrubber system can be used to control fumes, mists, acid gasses, heavy metals, trace organics, and suspended dusts. An individual wet scrubber can usually be used to control a targeted pollutant. Therefore, a well-designed wet scrubber system often contains two or more single scrubbers leading to a multistage wet scrubber, which affords higher total removal efficiencies than that of a single-stage scrubber [1]. A schematic diagram of a wet scrubber is shown in Figure 5.

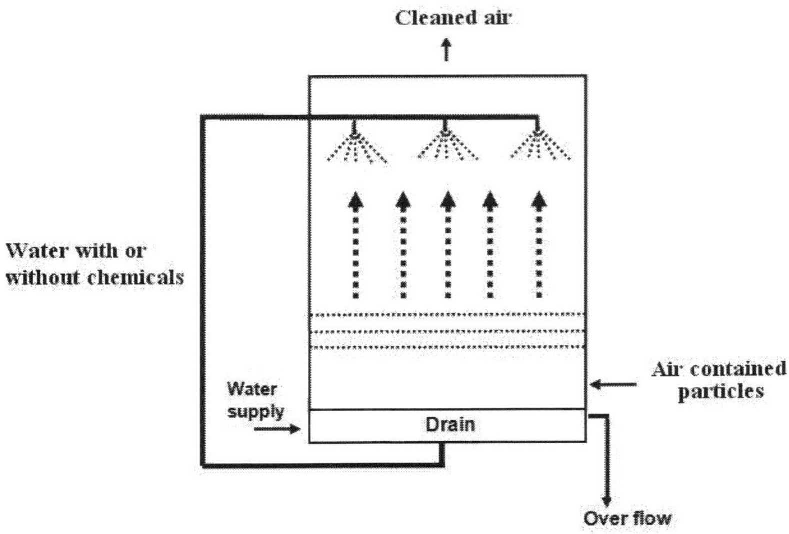

Figure 5. Schematic diagram of a wet scrubber chamber.

The wet scrubber system works based on direct interaction between the adsorbent liquid and the particles. The adsorbent liquid is usually water; however, several chemicals are also added to water to increase the adsorption ability of the liquid phase with the particles. Based on the interaction between the particles and the liquid phase, the particles can diffuse out of the gas phase and be absorbed in the liquid phase and the particle-loaded air can be cleaned. The absorption of particles into the liquid phase can be both physical and chemical absorption, depending on the particle and liquid phase and gaseous properties. The particle removal efficiency depends on:

- The solubility of the pollutant in the chosen scrubbing liquor
- Pollutant concentration in the gas phase being treated
- Flow rate of the gas and liquid phases
- Gas–liquid phase contact surfaces

http://dx.doi.org/10.5772/59722

- Stripping efficiency of the liquor and recycling of the solvent
- The ability to increase gas–liquid phase contact surfaces will result in higher absorption efficiency in a wet scrubber system. The reduced temperature and increased liquid-to-air ratio improve the absorption efficiency of a wet scrubber system.
- The pH of the scrubbing liquid is an important factor directly affecting the purification efficiency. The pH of the liquid may need to be low for ammonia scrubbing, while a neutral or high pH liquid is required for acid gas scrubbing. When scrubbing trace organics, a wet scrubber with alkaline pH is often used.

3.1.3. Electrostatic precipitators

Electrostatic precipitation (ESP), which is one of the most popular and efficient particle control systems in the United States, is defined as a particle control method that uses electrical forces to move the particles out of the flowing gas stream and onto collector plates [16]. The ESP processes include:

1. The ionization of particles, which can be dry dusts or liquid droplets, in contaminated air (particle charging)
2. The charged particles are deposited on an oppositely charged plate
3. The removal of the deposited particles from the plates

The particles are charged when the particles in the air stream pass through a corona, a region of gaseous ions flow. In the corona, the ions bombard the surface of the particles leading to charging particles. When these charged particles pass through the surface of the collecting electrodes, oppositely charged plates, they are trapped on the collected electrodes by the electrostatic field. The charged particles are accelerated toward the collecting electrodes by Coulomb forces, but inertial and viscous forces can resist the motion. When the plates (electrodes) collected a certain particle amount, the collected particles must be removed from the plates to prevent their re-entrainment into the gas stream. The plates could be knocked to let the collected layer of particles to slide down into a hopper from which they are evacuated [16]. The plates could also be continuously washed with water to remove the collected particles. A schematic diagram of an electrostatic precipitator is shown in Figure 6.

The principal difference between the ESP and other scrubbing methods are that in the ESP, the separation forces are electrical and are applied directly to the particles or droplets themselves while in others the separation forces are usually applied indirectly through the contaminated air system [17]. Therefore, the ESP could remove small particles or liquid droplets at a high efficiency with low energy consumption or low cost and small pressure drop through the gas cleaning system [17]. ESPs are built in either single-stage or two-stage versions. In the single-stage precipitator, the ionization and collection of particles or liquid droplets are achieved in a single stage and the corona discharge and precipitating field extend over the full length of the device. In the two-stage precipitator, the ionization of particles or liquid droplets is carried out in the first stage confined to the region around the corona discharge wires, followed by the particle collection in the second stage which provides an electrostatic field whereby the previously charged particles migrate onto the surface of the collecting electrodes [2].

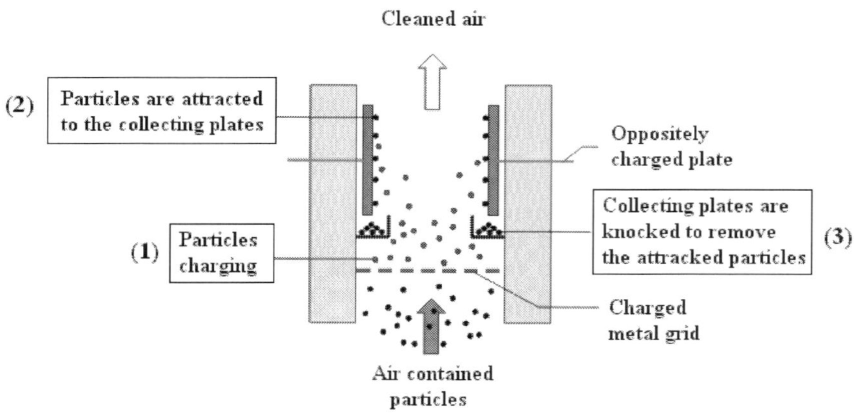

Figure 6. Schematic diagram of an electrostatic precipitator.

3.1.4. Fabric filtration

Fabric filtration is a well known and accepted physical technique in which a gas stream containing mainly solids passes through a porous fabric medium which retains the solids. This process may operate in a batch or semicontinuous mode for removing the retained solid particles from the filter medium. Filtration systems may also be designed to operate in a continuous manner.

In air fabric filtration, the contaminated gas flows into and passes through a number of filter bags placed in parallel, leaving the solid particles retained by the fabric filter. The fabric filter can be classified into two basic groups, depending on the fabric properties: felt and woven. Felt media are normally used in high-energy cleaning systems, while woven media are used in low-energy systems. Felt fabrics are tighter in construction and they can be considered to be more of a true filter medium and should be kept as clean as possible to perform satisfactorily as a filter. The woven fabric is merely a site upon which the true filtering occurs as the dust layer builds up, through which the actual filtering take place.

Particles are collected on the fabric surface through four mechanisms including:

- Inertial collection – the fibers, which are placed perpendicular to the gas flow direction, could collect the particles in the stream without changing gas flow direction.

- Interception – particles are trapped in the filter matrix.

http://dx.doi.org/10.5772/59722

- Brownian movement – submicron particles are collected on the surface of the filter.
- Electrostatic forces – the particles on the gas stream were captured because of electrostatic interaction between the particles and the filter.

The particles were captured on the filter leading to formation of a dust cake on the filter. The formation of the dust cake could increase the resistance to gas flow. Therefore, the filter containing the dust cake must be frequently cleaned.

Fabric filters are extremely efficient solid removal devices and operate at nearly 100% efficiency. The efficiency depends on several factors including:

- Particle properties
  - Size: Particles between 0.1 and 1.0 μm in diameter may be more difficult to capture.
  - Seepage characteristics: Small, spherical solid particles tend to escape.
  - Inlet dust concentration: The deposit is likely to seal over sooner at high concentrations.
- Fabric properties
  - Surface depth: Shallow surfaces form a sealant dust cake sooner than napped surfaces do.
  - Weave thickness: Fabrics with high permeability, when clean, show lower efficiencies.
  - Electrostatics: Particles, fabrics, and gas can all be influenced electrostatically and proper combination.
- Dust cake properties
  - Residual weight: The heavier the residual loading, the sooner the filter is apt to seal over.
  - Residual particle size: The smaller the base particles, the smaller (and fewer) are the particles likely to escape.
- Air properties
  - Humidity: With some dusts and fabrics, 60% relative humidity is much more effective than 20% relative humidity. Increased humidity or moisture level can be a frequent cause of clogging pores of the filter medium and increasing filter pressure drop.
- Operational variables
  - Velocity: Increased velocity usually lowers the efficiency, but this can be reversed depending on the collection mechanisms, for example, impaction and infusion.
  - Pressure: Probably not a factor, except that an increase in pressure after the dust cake has been formed can fracture the filter medium and greatly reduce efficiency until the cake reseals.
  - Cleaning: Without frequent or periodical cleaning, the air filtration system cannot be operated.

The advantages and disadvantages of methods to control particulate matter including cyclone, wet scrubber, ESP, and fabric filtration are summarized in Table 2.

Advantages	Disadvantages
Cyclone	
· Low capital cost · Simple and insignificant maintenance problems · Ability to operate at high temperature · Require small spaces	· Relatively low efficiencies for collection particles which are smaller than 10 mm · Could not use for sticky materials
Wet scrubber	
· No secondary production · Require small space · Operation to collect both gases and sticky particles · Operation at high-temperature as well as high-humidity gas streams · Low capital cost · Operation with flammable and explosive dust with little risk · High effective to collect fine particles	· Treatment issue concerning with water disposal/effluent · Corrosion problems · High pressure drop problems · Solid buildup problems at the wet–dry interface · Relatively high maintenance costs
Electrostatic precipitation (ESP)	
· Very high collection efficiencies of coarse as well as fine particulates with low energy consumption · Collection dry dust · Low pressure drop · Relatively low operation and maintenance costs · Operation capability at high temperatures as well as high pressure or under vacuum · High collect capacity	· High capital cost · High sensitivity to fluctuations in gas stream · Problems with particles, with extremely high or low resistivity · Require large space for installation · Produce ozone as by-product · Require highly trained maintenance personnel
Fabric filtration	
· Very high collection efficiencies of coarse as well as fine particulates · Relative insensitivity to gas stream fluctuations and large changes in inlet dust loadings · Recirculation of filter outlet air · No corrosion issues · Simple maintenance, flammable dust collection · High collection efficiency of submicron smoke and gaseous contaminants · Many application types · Simple operation	· Difficult to operate at high temperature · Need for fabric treatment after removal process · Require high maintenance costs · Explosion problems · Shortened fabric life at elevated temperatures and in the presence of acid or alkaline particulate or gas · Respiratory protection requirement for fabric replacement · Medium pressure-drop requirements
Source: Bounicore and Davis 1992 [18]	

Table 2. Advantages and disadvantages of particulate control methods

3.2. Gaseous pollutants control

3.2.1. Adsorption

Adsorption is the phenomenon via which molecules of a fluid adhere to the surface of a solid material (adsorbent). Gas adsorption is used for industrial applications such as odor control, recovery of volatile solvents such as benzene, toluene, and chloroflurocarbon, and drying of process gas streams. During this process, the molecules or particles (adsorbate) in airstream gases and liquids can be selectively removed or captured despite being at low concentrations. There are two distinct adsorption mechanisms: physisorption and chemisorption. Physisorption or physical adsorption, also called van der Waals adsorption, involves a weak bonding of gas molecules with the adsorbent. The bond energy is similar to the attraction forces between molecules in the stream. The adsorption process is exothermic and the heat of adsorption is slightly higher than the heat of the vaporization of the adsorbate. The forces holding the adsorbate to the adsorbent are easily overcome by either the application of heat or the reduction of pressure, which are methods that can be used to regenerate the adsorbent. Chemisorption or chemical adsorption involves an actual chemical bonding by reaction of the adsorbate with the adsorbent, leading to new chemical bonds such as covalent bonding generated at the adsorbent surface.

When a stream comes into contact with an adsorbent, one or several components of the stream are adsorbed by the adsorbent. At all adsorbent interfaces, adsorption can occur, but often at a low level unless the adsorbent is highly porous and possesses fine capillaries. For an effective solid adsorbent, it should have a large surface-to-volume ratio, and a preferential affinity for the individual component of concern. The adsorption occurs by a series of steps. In the first step, the adsorbate diffuses from the stream to the external surface of the adsorbent. In the second step, the adsorbate molecule migrates from the relatively small area of the external surface to the pores within each adsorbent. The bulk of the adsorption occurs in these pores because of the majority of available surface area. In the final step, the adsorbate adheres to the surface in the pores of the adsorbent [19].

Most industrial adsorbents could be divided into three classes including:

- Oxygen-containing compounds such as silica gel and zeolite
- Carbon-based compounds such as graphite or activated carbon materials
- Polymer-based compounds, which include functional groups in a porous polymer matrix

Silica gel, which is usually prepared by the reaction between sodium silicate and acetic acid, is a chemically inert, nontoxic, polar, and dimensionally stable amorphous form of SiO_2. It is used for the drying of processed air and the adsorption of polar hydrocarbons from natural gas.

Zeolites are natural or synthetic crystalline aluminosilicates, which have a repeating pore network and release water at high temperature. Zeolites are applied in the drying of processed air, CO_2 removal from natural gas, CO removal from reforming gas, air separation, catalytic cracking, and catalytic synthesis and reforming.

Activated carbon is a highly porous, amorphous solid consisting of microcrystallines with a graphite lattice, usually prepared in small pellets or a powder. It is nonpolar and cheap. Activated carbon is used for the adsorption of organic substances and nonpolar adsorbate. Activated carbon is also usually used for waste gas (and waste water) treatment. It is the most widely used adsorbent because its chemical and physical properties such as surface groups, pore size distribution, and surface area can be tuned as required. Its usefulness also derives from its large microspore (and sometimes mesoporous) volume and the resulting high surface area.

3.2.2. Absorption

Absorption is a physical or chemical process in which atoms, molecules, or ions enter some bulk phase – gas, liquid, or solid material. As compared to the adsorption process, in which the molecules are adhered on the surface of the adsorbent, the absorption process takes place when the volume takes up molecules.

Gas absorption is the removal of one or more pollutants from a contaminated gas stream when the gas stream passes through a gas–liquid interface and ultimate dispersion in the liquid. Absorption is a process that may be chemical (reactive) or physical (nonreactive). Physical absorption is formed based on the interaction of two phases of matter including a liquid absorbs a gas or a solid absorbs a liquid. When a liquid solvent absorbs a part or all of a gas mixture, the gas mass could move into the liquid volume. The mass transfer could take place at the interface between the gas and the liquid. The mass transfer rate depends on both the liquid and the gas properties. The solubility of gases, the pressure and the temperature are the main factors affecting to this type of absorption. In addition, the absorption rate also depends on the surface area of the interface and its duration in time. When a solid absorbs a part or all of a liquid mixture, the liquid mass could move into the solid volume. The mass transfer could take place at the interface between the liquid and the solid. The mass transfer rate depends on both the solid and the liquid properties. Chemical absorption or reactive absorption is a chemical reaction between the absorbed and the absorbing substances. Sometimes, it is combined with physical absorption. This type of absorption depends upon the stoichiometry of the reaction and the concentration of its reactants.

Gas absorption is usually carried out in packed towers. The contaminated gas stream enters the bottom of the column and passes upward through a wetted packed bed. The absorbing liquid enters from the top of the column and is distributed over the column packing. The column packing may have one or more commercially available geometric shapes designed to maximize the gas–liquid contact and minimize the gas–phase pressure drop [20]. The requirements for the packing column include high wetted area per unit volume, minimal weight, sufficient chemical resistance, low liquid holdup, and low pressure drop.

3.2.3. Condensation

Condensation is a separation process to convert one or more volatile components of a vapor mixture to a liquid through saturation process. Any volatile components can be converted to liquids by sufficiently lowering their temperature and increasing their pressure. The most common process is reducing the temperature of the vapor because increasing the vapor pressure is expensive. The condensation process is primarily used to remove VOCs from gas streams prior to other control methods, but sometimes it can be used alone to reduce emissions from high-VOCs concentration gas streams [2, 21].

The simple and relatively inexpensive condenser uses water or air to cool and condense the vapor stream to the liquid. Since these devices are not required to reach or capable of reaching low temperature, high removal efficiencies of most vapor pollutants cannot be obtained unless the vapor will condense at high temperature. That is why condensers are typically used as a pretreatment device. They can be used together with adsorption, absorption, and incinerators to reduce the gas volume to be treated by other expensive methods.

A typical condenser device includes condenser, refrigeration system, storage tanks, and pumps. The condensation process includes:

- The contaminated gas stream is compressed as it passes through the blower.
- The existing hot gas stream flows to an after-cooler, which is constructed of copper tubes with external aluminum fins. Air is passed over the fins to maximize the cooling effect. Some condensation occurs at this step.
- The gas stream continues to cool in an air-to-air heat exchanger.
- The condenser cools the gas to below the condensing temperature in an air-to-refrigerant heat exchanger.
- Finally, the cold gas passes to a centrifugal separator where the liquid is removed to the collecting vessel. The gas stream typically requires further treatment before being emitted to the atmosphere.

A condensing system usually contains either a contact condenser or a surface condenser. Contact condensing systems cool the contaminated gas stream by spraying ambient or chilled liquid directly into the gas stream. A packed column is usually used to maximize the surface area and contact time. The direct mixing of the coolant and contaminant necessitates separation or extraction before coolant reuse. This separation process may lead to a disposal problem or secondary emissions. Contact condensers usually remove more contaminated air as a result of greater condensate dilution. In the surface condensing systems, the coolant does not mix with the gas stream, but flows on one side of a tube or plate in the surface condensing systems. The condensing vapor contacts the other side, forms a film on the cooled surface, and drains into a collection vessel for storage, reuse, or disposal. Surface condensers require less water and generate 10–20 times less condensation than contact condensers do.

The advantages of the condensation method include lower installation cost, little required auxiliary equipment, and less maintenance requirement. However, the remaining disadvantages of the method include problems of water disposal, low efficiencies, and the need for further treatment.

3.2.4. Incineration

Incineration or thermal oxidation is a broadly used method to control air pollutants such as VOCs, using oxidation at high temperature. Incineration is considered as an ultimate disposal technique in which VOCs are converted to carbon dioxide, water, and other inorganic gases. The two popular incineration methods are thermal incineration and catalytic incineration.

In thermal incineration, organic compounds in the contaminated gas are burned or oxidized at a high temperature with air in the presence of oxygen [22]. The thermal oxidizer involves specifying a temperature of operation along with a desired residence time and then optimum sizing the device to achieve the desired residence time and temperature with proper flow velocity. Selection of the proper piece of equipment depends on the mode of operation, oxygen content, and concentration of the organic gases. They are very important when trying to minimize the overall cost of the incineration and reduce the volume of the gas stream to be treated as much as possible. Depending on the types of heat recovery, incinerators can be classified into two categories: recuperative and regenerative. The recuperative incinerator uses a shell and tube heat exchanger to transfer the heat generated by the incinerator to the preheat of the feed stream. The recuperative incinerator can recover about 70% of the waste heat from the exhaust gases [21]. The regenerative incinerator includes a flame-based combustion chamber that connects two or three fixed beds containing ceramic or other inert packing. The input gas enters over these beds where it is preheated before passing into the combustion chamber and being burned. Then, the hot flue gases pass through the packed beds where the heat generated during incineration is recovered and stored. The packed beds keep the heat during one cycle and release it as the beds preheat the input organic gases in the second cycle. This regenerative incinerator method can recover up to 95% of the energy from the flue gas [23].

In catalytic incineration, the organic compounds in the contaminated gas are converted into carbon dioxide and water by using a catalyst that facilitates incineration at low temperature. Thus, the requirement incineration temperature can be decreased by hundreds of degrees. Therefore, the application of catalyst incineration can save a large amount of energy to heat up the gas stream containing pollutants for combustion. The contaminated gases are heated by a small auxiliary burner, and then the gases passed through the catalyst bed. The space requirement for operation of catalytic incineration is much smaller than that of thermal incineration. Thanks to the catalytic activity, the degree of oxidation of the pollutants is greatly increased compared with that in the incineration system without any catalyst. The catalyst activity refers to the degree of the chemical reaction rate. The catalyst can also be selective with higher activity for some compounds. Such activity and selectivity enable a lower operating temperature while still achieving the desired destruction efficiencies. In air pollution control, the catalyst is usually a noble metal (Pd, Pt, Cr, Mn, Cu, Co, and Ni) deposited on an alumina support in a configuration to minimize the pressure drop, which is often critical for incinerator designs [23, 24].

3.2.5. Biological system

The biological system for controlling air pollutants such as VOCs and odor uses microbes or microorganisms, immobilized on a biologically active solid support, to treat the gas pollutants.

The principle of the method is that the gaseous pollutants are used by the microbes as a food or energy source and thus destroyed and converted into innocuous metabolic end products such as carbon dioxide and water. The processes via which microbes destroy or convert pollutants contain:

- First, the pollutant gas must be absorbed into the liquid film in which the microbes are growing.
- The pollutant is absorbed into the cells of the microbes and metabolized.
- Finally, the end products (mainly CO_2) must be expelled from the cell and diffuse outward through the liquid film.

The process requires careful attention to design and operation in order to ensure firstly good contact between the contaminated gases and the microbes contained on the solid support; and secondly, that the microbe population is sustained and maintained in a healthy state. The key concerns in the design and operation of a biological system to control air pollutants contain [25]:

- Identify the type of the contaminant and its concentration in the air stream
- Find the suitable microbial population and combining with a compatible medium
- Maintain sufficient moisture
- Design the bed to ensure suitable residence time for a given airflow rate
- Control the bed conditions including pH, nutrient levels, and temperature

The biological technology is most suitable for high volumetric flow rate air streams containing low pollutant concentrations. The two most common biological systems are biofilter and bioscrubber. Biofilter is a biological system which uses an organic or synthetic media to host and nourish the microorganism without the requirement for an aqueous flush system. Bioscrubber is a biological system which uses an inorganic or synthetic media to provide a structural base for physically hosting the microorganisms requiring a continuous water flush or an intermittent containing carbon nutrient that supports the microorganism. The use of a biological system to control air pollutant offers several advantages including effective removal of compounds, little or no by-product pollutants, uncomplicated installations, and low costs. However, the method retains several disadvantages including a reduced suitability to high concentration streams, large area requirement for installation, need for careful attention to moisture control, and the possibility of becoming clogged by particulate matter or biomass growth.

3.2.6. *Application of photocatalyst*

An alternative technology, which offers a number of advantages over the above-mentioned technologies, for controlling organic air pollutants is to use photocatalysts. The use of a photocatalyst supports the operation of a low- or room-temperature photocatalytic oxidation process that can degrade a broad range of organic contaminants into innocuous final products such as CO_2 and H_2O without significant energy input. When the photocatalyst, for example, TiO_2, is irradiated with ultraviolet (UV) radiation that has energy higher than the band gap energy of the TiO_2, an electron could be excited from the valence band to the conduction band of TiO_2, leaving holes behind on the TiO_2 surface. The leaved holes could react with surrounding H_2O to produce hydroxyl radicals ($^*OH$), while the excited electrons could react with O_2 to produce superoxide radical anions $\left(o_2^-\right)$. These oxy radical species can participate in the oxidation reaction to destroy many organic contaminants ($C_xH_yO_z$) completely [26]. The photocatalytic process mainly follows the following reactions:

$$
\begin{aligned}
& TiO_2 + h\nu \rightarrow e^- + h^+ \\
& h^+ + H_2O \rightarrow H^+ + {}^*OH \\
& e^- + O_2 \rightarrow O_2^- \\
& O_2^- + e^- + 2H^+ \rightarrow H_2O_2 \\
& H_2O_2 \rightarrow 2{}^*OH \\
& {}^*OH + C_xH_yO_z \rightarrow xCO_2 + yH_2O + \ldots
\end{aligned}
$$

The three common reactor types designed to use a photocatalyst for air purification purposes are the honeycomb monolith, fluidized-bed, and annular reactors [27]. A honeycomb monolith reactor contains a certain number of channels, each of which typically has an internal dimension of the order of 1 mm. The cross-sectional shapes of the channels are square or circular. The photocatalyst is coated onto the walls of channels in a very thin wash coat. Fluidized-bed reactors are designed to treat a high gas feed rates directly passing through the catalyst bed. Based on reactor design, the solid photocatalyst could directly contact with the UV irradiation as well as gaseous reactants. The fluidized-bed reactors generally consisted of two concentric cylinders, which form an annular region with a certain gap. The photocatalyst is deposited onto the interior wall of the outer cylinder. The light source is usually located at the center. The thickness of the deposited photocatalyst film is sufficiently thin ensuring that all of the photocatalyst could be illuminated by UV irradiation [28].

The applications of the photocatalyst for photocatalytic oxidation processes to reduce air pollutants have been considered as alternatives to conventional air pollution control technologies. However, they have yet to overcome the problems of low energy efficiency and poor cost competitiveness. Therefore, numerous methods for modifying photocatalysts have been developed and investigated to accelerate the photo-conversion, enable the absorption of visible light, or alter the reaction mechanism to control the products and intermediates [29]. In this regard, metals or nonmetals were used as doping agents to implant or coprecipitate on the surface or in the lattice of TiO_2. Electron donors or hole scavengers have been added to such photocatalytic systems. In addition, another semiconductor was integrated with TiO_2 to establish a suitable two-semiconductor system [29]. The modifications not only change the mechanism and kinetics of the photocatalytic processes under UV irradiation but also enhance the photocatalytic activities of the photocatalyst, thereby enabling the photocatalytic oxidation processes to proceed even under visible light [30].

http://dx.doi.org/10.5772/59722

4. Strategies to control climate change

Global climate change, also called global warming or the greenhouse effect, may be the most significant problem ever faced by humankind. Global climate change is caused by adding certain gases including carbon dioxide (CO_2), methane (CH_4), nitrous oxide (N_2O), and many chlorofluorocarbons (CFCs). The added gases absorbed infrared radiation leading to excess thermal energy within the earth's biosphere. The largest driver of global warming is carbon dioxide (CO_2) emissions from fossil fuel combustion, cement production, and land use changes such as deforestation. Therefore, the strategies to control CO_2 emission are also the main strategies to control global climate change.

4.1. Upstream control

Upstream strategies to control CO_2 emission include energy conservation, alternative and renewable fuels, and oxy-fuel combustion.

4.1.1. Energy conservation and efficiency use

Energy conservation refers to reducing energy consumption through using less of an energy service. The strategies concerned to energy conservation include energy taxes, building design, transportation, and consumer products.

- Energy taxes

Some countries employ energy or carbon taxes to motivate energy users to reduce their consumption.

- Building design

Energy conservation in building could be improved by using of an energy audit, which is an inspection and analysis of energy use and flows in the building, process, or system to reduce the amount of energy input into the system without negatively affecting the output(s). In addition, a passive solar building design in which windows, walls, and floors are made to collect, store, and distribute solar energy in the form of light or heat in the winter and reject solar heat in the summer. The design, leading to decreased use of mechanical and electrical devices, is also a solution for energy conservation.

- Transportation

The zoning reform and designs for walking and bicycling could allow greater urban density leading to reduce energy consumption concerning to transportation. The application of telecommuting is also a sufficient opportunity to conserve energy. For example, with people who work in service jobs, they could work at home instead of commuting to work each day.

- Consumer products

Because the consumers usually lack the information concerning to saving by energy-efficient products, we must inform the consumer in understanding the problems. For example, many consumers choose cheap incandescent bulbs, failing to take into account their higher energy costs and lower lifespan. However, as compared to modern compact fluorescent and LED bulbs, these products have a higher upfront cost, with their long lifespan and low energy.

4.1.2. *Alternative and renewable fuels*

Use of alternative energy sources could prevent CO_2 emission from fossil fuel. The alternative energy sources include wind, solar, hydropower, biomass, and geothermal energy.

- Wind energy

Airflows can be used to run wind turbines to produce electric energy. Globally, the long-term technical potential of wind energy is believed to be five times total current global energy production, or 40 times current electricity demand.

- Solar energy

Solar energy, radiant light and heat from the sun, is harnessed using a range of ever-evolving technologies such as solar heating, photo-voltaic, concentrated solar power, solar architecture, and artificial photosynthesis.

- Hydropower

Hydropower is the power derived from the energy of falling water and running water, which may be harnessed for useful purposes. Hydropower could be also captured from ocean surface waves and tidal power.

- Biomass

Biomass is the biological material derived from living or recently living organisms. Biomass could be used as energy source by either used directly via combustion to produce heat or indirectly after converting it to various forms of biofuels.

- Geothermal energy

Geothermal energy could be generated from thermal energy, which is stored in the earth. Because of the difference between temperature of the core of the planet and its surface, it drives a continuous conduction of thermal energy in the form of heat from the core to the surface.

4.1.3. *Oxy-fuel combustion*

When burning coal or other fossil fuel using ambient air, the air contains a huge amount of nitrogen as well as the oxygen needed for combustion (4:1). Raising the temperature of the nitrogen to the combustion temperature requires a great deal of heat. Therefore, reducing nitrogen content in the air input could be a good strategy to reduce fuel consumption, leading to reduced CO_2 emission. In the strategy, oxy-fuel combustion, a process of burning a fuel using pure oxygen instead of air as the primary oxidant, is applied. There are several researches being done in firing fossil-fueled power plants with an oxygen-enriched gas mixture instead of air. Almost all of the nitrogen is removed from input air, yielding a stream that is approximately 95% oxygen.

http://dx.doi.org/10.5772/59722

4.2. Downstream control

Technologies to downstream control CO_2 or remove CO_2 from atmospheric include biological capture, wet scrubbing, CO_2 absorption, and mineral carbonation.

4.2.1. Biological capture

Biological capture of CO_2 is a process in which photosynthetic organisms are used to absorb the CO_2 gas from air and convert it into a solid carbonaceous compound. The strategies to conduct for biological capture of CO_2 include:

- Reforestation

Reforestation means that tree could be replanted on marginal crop and pasturelands leading to incorporate atmospheric carbon (CO_2) into biomass. For a successful process, the incorporated carbon could not return to the atmosphere from burning or rotting when the trees die. Finally, the trees grow in perpetuity or the wood from them must itself be sequestered in the forms of biochar or bio-energy with carbon storage or landfill.

- Agriculture

Agricultural activity to capture CO_2 is also called as "capture" energy of the sun. Under solar light, the artificial plants could capture and convert the CO_2 in atmosphere into biomass, which can be storage or used as food or also used as raw material to make biofuels to replace the use of fossil fuel. Land-based plants such as corn and soybeans can be grown as energy crops, in particular to make biodiesel. Because of the limitations of land-based plants, there has been much interest over the years in developing systems that utilize microalgae for engineered biological CO_2 capture systems. Microalgae can fix CO_2 up to ten times faster than trees, and utilize sunlight much more efficiently than the land-based energy crops.

4.2.2. Wet scrubbing of CO_2

A carbon dioxide scrubber, which uses various amines such as monoethanolamine as absorbents, could absorb CO_2 to capture them from the atmosphere. The design and principle of wet scrubber have been presented in Section 3.2.2. Amines could be used as absorbents to absorb CO_2 based on following reactions:

$$2RNH_2 + CO_2 + H_2O \rightarrow (RNH_3)_2CO_3$$
$$(RNH_3)_2CO_3 + CO_2 + H_2O \rightarrow 2RNH_3HCO_3$$
$$2RNH_2 + O_2 \leftrightarrow RNHCOONH_3R$$

4.2.3. Mineral carbonation

Many chemical processes, known as carbon sequestration by mineral carbonation or mineral sequestration, could capture and store CO_2 from the atmosphere in stable carbonate mineral forms. In the process, CO_2 could react with abundantly available metal oxides such as MgO or CaO to form stable carbonates. These reactions are mostly exothermic and occur naturally.

$$CaO + CO_2 \rightarrow CaCO_3$$
$$MgO + CO_2 \rightarrow MgCO_3$$

CO_2 could also react with calcium and magnesium silicates including forsterite and serpentinite by the following the reactions:

$$Mg_2SiO_4 + 2CO_2 \rightarrow 2MgCO_2 + SiO_2$$
$$Mg_3Si_2O_5(OH)_4 + 3CO_2 \rightarrow 3MgCO_3 + 2SiO_2 + 2H_2O$$

These reactions are slightly more favorable at low temperatures. This process occurs naturally over geologic time frames and is responsible for much of the earth's surface limestone.

5. Conclusion

In the chapter, we described the current and emerging technologies and strategies that are being used or proposed to control air pollutants. Control technologies and strategies could be classified into upstream and downstream controls. Upstream technologies and strategies are usually being used to control air pollution emitted from mobile sources such as transportation, agriculture, and construction activities. Downstream technologies are usually applied to control particulate matter and gaseous pollutants emitted from stationary sources. The upstream control exhibited more advantages concerning to cost and efficiency than the downstream control.

The air pollutants including SO_x and Pb emitted from transportation activities could be significantly reduced by controlling the precursor pollutants in raw materials. The NO_x, CO, and hydrocarbon emissions from transportation activity could be reduced by the application of catalytic converters and the control of lubricant consumption. The strategies to reduce motorized transportation demand and improve the road quality and traffic flow decrease the energy consumption demand, leading to decrease in air pollution emission. The strategies to control livestock feeding, animal housing systems, manure storage systems, application of manure for crops and application of fertilizers and pesticides significantly reduced the emission of methane (CH_4), nitrous oxide (N_2O), and ammonia (NH_3). The strategies to control construction activities including control of site planning, construction traffic, demolition works, and site activities significantly reduced the emission of particulate matter.

To control particulate matter pollutants, the technologies including cyclone, wet scrubber, electrostatic precipitators, and fabric filter have been efficiency applied. Among them, electrostatic precipitator is the most advantageous method and widely applied in the United

http://dx.doi.org/10.5772/59722

States. The traditional methods such as adsorption, absorption, condensation, incineration to control air pollutants emitted from stationary sources exhibited many certain disadvantages including high cost and efficiency. Therefore, they should to be modified, enhanced, or replaced by modern technologies such as biological and photocatalytic technologies.

Acknowledgements

This work was supported by a grant from the National Research Foundation of Korea (NRF), funded by the Ministry of Science, ICT, and Future Planning (2013R1A2A2A03013138).

Author details

Thanh-Dong Pham, Byeong-Kyu Lee*, Chi-Hyeon Lee and Minh-Viet Nguyen

*Address all correspondence to: bklee@ulsan.ac.kr

Department of Civil and Environmental Engineering, University of Ulsan, Daehakro , Namgu, Ulsan, Republic of Korea

References

[1] Cooper CD, Alley FC. Airpollution Control: A Design Approach. Waveland Press Inc, USA, 2011.

[2] Davis WT. Air Pollution Engineering Manual. John Wiley & Son, Inc, USA, 2000.

[3] Heinsohn RJ, Kabel RL. Sources and Control of Air Pollution. Prentice-Hall Inc, UK, 2000.

[4] Wang LK, Pereira NC, Hung Y-T. Air Pollution Control Engineering. Humana Press Inc, USA, 2010.

[5] Vallero D. Fundamentals of Air Pollution. 4 ed., Elsevier Inc, USA, 2008.

[6] Arcadio PS, Gregoria AS. Environmental Engineering: A Design Approach. Prentice-Hall, Inc, USA, 1996.

[7] Eric Y. Alternative sulfur management solutions to help refiners meet clean fuel and environmental challenges. In: 10th Biennial Sulfur Market Symposium. Bejjing, China, 2006.

[8] Nichols AL. Lead in gasoline. In: Economic Analyses at EPA: Assessing Regulatory Impact, Resources for the Future. 1997;1:49–86.

[9] Newell RG, Rogers K. The U.S. Experience with the Phasedown of Lead in Gasoline, in Washington, USA, 2003.

[10] Bradley MJ. Best available technology for air pollution control: Analysis guidance and case studies for North America in USA, 2005.

[11] Amit KJ, Amit S. Biotechnology: a way to control environmental pollution by alternative lubricants. Res Biotechnol 2013;4:38–42.

[12] Erisman JW, Bleeker A, Hensen A, Vermeulen A. Agricultural air quality in Europe and the future perspectives. Atmospheric Environment 2008;42:3209–17.

[13] Susana LA, Guerreiro C, Viana M, Reche C, Querol X. Contribution of agriculture to air quality problems in cities and in rural areas in Europe. In: The European Topic Centre on Air Pollution and Climate Change Mitigatio, 2013;10:1–26.

[14] Livingstone K. The Control of Dust and Emissions from Construction and Demolition. In: London, 2006.

[15] Leith D, Jones DL. Handbook of Power Science and Technology. 2 ed., Chapman & Hall, New York, USA, 1997.

[16] Treybal RE. Mass Transfer Operations, McGraw-Hill, New York, 1980.

[17] White HJ. Role of electrostatic precipitators in particulate control: a retrospective and prospective view. J Air Pollut Control Assoc 1975;25:102–7.

[18] Bounicore AJ, Davis WT. Air Pollution Engineering Manual. New York: Van Nostrand Reinhold, 1992.

[19] Glasstone S. Physical Chemistry. Princeton, USA, 1946.

[20] Calvert S, Englund HM. Handbook of Air Pollution Technology. Wiley-Interscience, New York, USA, 1984.

[21] Wang LK, Pereira NC, Hung YT. Air Pollution Control Engineering. Humna Press Inc, USA, 2010.

[22] Clausen CA, Copper CD, Hewett M, Martinez A. Enhancement of organic vapor incineraion using ozone. J Hazard Mater 1992;31: 75–87.

[23] Taylor PH, Delligner B, Lee CC. Development of a thermal stability based ranking of hazardous organic compound incinerability. Environ Sci Technol 1990;24:316–28.

[24] Buonicore WTDAJ. Air Pollution Engineering Manual, Air and Waste Management Association, Van Nostrand Reinhold, New York, 1992.

[25] Crocker BB. Air pollution control methods. In: Kirk-Othmer Encyclopedia of Chemical Technology. John Wiley & Sons, Inc, 2000.

[26] Pham TD, Lee BK. Feasibility of silver doped TiO2/Glass fiber photocatalyst under visible irradiation as an indoor air germicide. Int J Environ Res Public Health 2014;11:3271–88.

[27] Zhao J, Yang X. Photocatalytic oxidation for indoor air purification: a literature review. Building Environ 2003;38:645–54.

[28] Dezhi S, Sheng C, Chung JS, Xiaodong D, Zhibin Z. Photocatalytic degradation of toluene using a novel flow reactor with Fe-doped TiO2 catalyst on porous nickel sheets. Photochem Photobiol 2005;81:352–7.

[29] Park HW, Park Y, Kim W, Choi W. Surface modification of TiO2 photocatalyst for environmental applications. J Photochem Photobiol C: Photochem Rev 2013;15:1–20.

[30] Pham TD, Lee BK. Cu doped TiO2/GF for photocatalytic disinfection of Escherichia coli in bioaerosols under visible light irradiation: application and mechanism. Appl Surface Sci 2014;296:15–23.

## Chapter 11

A Comprehensive Research of Atmospheric Haze by Optical Remote Sensing in Central China Region (CCR)

Wenqing Liu, Zhenyi Chen, Pinhua Xie,
Jianguo Liu and Tianshu Zhang

Additional information is available at the end of the chapter

http://dx.doi.org/10.5772/59753

1. Introduction

With the changing pollution scenario, the task of improving air quality is now faced with increasingly serious challenges. Aerosols and gaseous pollutants have great impact on the radiative balance of the Earth, and thus on climatic changes. Furthermore, the radiative forcing of natural and anthropogenic aerosols exhibits a strong seasonal and geographical variability, which can affect the diffusion of atmospheric pollution locally [1]. The increase of aerosols or gaseous pollutants results in impairment of visibility. Besides, high aerosol concentration is deleterious to human health. The Chinese Academy on Environmental Planning considers that air pollution is responsible for 411,000 premature deaths in China in 2003, primarily from respiratory problems and heart diseases [2]. Therefore, obtaining information on the vertical distribution of aerosols and gaseous pollutants in the pollution haze is crucial to understanding their optical characteristics as well as their impact on air quality and on human health.

During the past 10 years, with its dense population, rapid economic growth, and dramatic rate of urbanization, China has been experiencing elevated levels of air pollution [3-5]. This is particularly so in the case of the Pearl River Delta (PRD), Yangtze River Delta (YRD), and Central China Region (CCR). Among these, the CCR region is where the frequency of occurrence of haze has been increasing rapidly from 2012. In CCR, NOx, SO_2 and VOCs discharged by pollution sources are the significant primary pollutants, which greatly influence the formation of particulate matter and ozone because of photochemical reaction. Besides, the prevalence of mild and humid climate throughout the year in CCR leads to transformation and accumulation of particulate matter and gaseous pollutants. In addition, regional transportation contributes to combined pollution problems of atmosphere, manifested by high-concentration fine particulate matter, low visibility, high-concentration NO_2 column, etc.

Thus far, extensive studies using different methods have been conducted in China on atmospheric pollution [6-9]. For a better understanding of aerosol properties and their effect on environment, two major aerosol experiments were conducted in mainland China since 2004: (i) East Asian Study of Tropospheric Aerosols -An International Regional Experiment (EAST-AIRE) [10] and (ii) East Asian Study of Tropospheric Aerosols and Impact on Regional Climate (EAST-AIRC) [11]. Information on the properties of aerosols and their temporal and spatial distribution was obtained from studies using both short-term intensive field experiments and long-term operations, as well as modeling studies. Besides these filed experiments, lidar (LIght Ranging and Detection) and DOAS (Differential Absorption Spectroscopy) studies were conducted at several sites in China[12-18]. However, knowledge on the properties of aerosols and trace gases, particularly in the vertical dimension, in CCR China is rather scanty. Also, because of lack of real-time data on the quantum of pollutants discharged by major industrial parks, it is hard to resolve the task of pollution prevention. The present research was designed to take advantage of advanced environmental monitoring technical equipment (including Lidar[19,20], multi-axis DOAS[21,22], vehicle-mounted DOAS /infrared SOF[23-25], and meteorological parameter analyzer, etc.) to establish a stereoscopic observing system for key sources of regional pollution in CCR, China. The campaign was carried out from Dec 2013 to April 2014 in three key cities, namely Anyang (AY), Jiaozuo (JZ) and Sanmenxia (SMX). Continuous and automatic stereoscopic monitoring of primary pollution sources, transportation of pollutants, emission characteristics of pollution from major industrial parks, as well as the transport conditions in the surrounding regions of pollution under specific meteorological conditions was carried out to obtain data that can help in air quality improvement and in development of forecast warning model, besides making up for the deficiency in current ground monitoring system.

The scope of research resolves mainly into two parts:

1. Real-time monitoring of spatio-temporal distribution of regional haze pollution:

Representative monitoring points were identified in selected cities and surrounding areas along the direction of pollution transportation to measure spatial and temporal distribution characteristics of SO_2, NO_2 and aerosol particulate matter with lidar, MAX-DOAS, etc.

2. Real-time monitoring of disorganized pollutant discharge in key industrial park sources:

VOC is an important precursor to the formation of photochemical smog. The type and volume of VOC discharge, as also of SO_2 and NO_2, play a significant role in pollution. Aiming at the key source areas of elevated discharge, vehicle-mounted DOAS and vehicle-mounted SOF-FTIR were utilized for mobile scanning measurement of discharged smoke plumes. Using the absorption characteristics of pollution gas molecules in infrared and ultraviolet spectrum, the concentration of exhaust gas was qualitatively and quantitatively determined through SOF-FTIR and DOAS spectrum analytical methods.

2. Instrument and system

2.1. Lidar

The Lidar system, used for the study of spatial and temporal distribution of atmospheric particles, is shown in Fig 1; it adopts 532m Mie scattering and 532 polarization channels. Laser transmitter launches linear polarization lasers at 532nm by shooting vertically into atmosphere through beam expander, inversed mirror and quartz skylight. The laser pulse is scattered through air molecules, atmospheric aerosols or cloud, on its transmission route. The backward-scattered light firstly passes through skylight glass and then as parallel light through the diaphragm hole and convergent lens of the telescope. The converged parallel light is split into two beams by dichroic mirror: one is at 532nm, received by photomultiplier C after passing through optical filter; the other is at 355nm, whose vertical and parallel components are received respectively by A and B photomultipliers, after decomposition by an analyzer prism. After the conversion of optical signal into electrical signal by photomultiplier, the signal will be delivered to data acquisition system, where its ASCII code data file will be recorded and saved by main control computer.

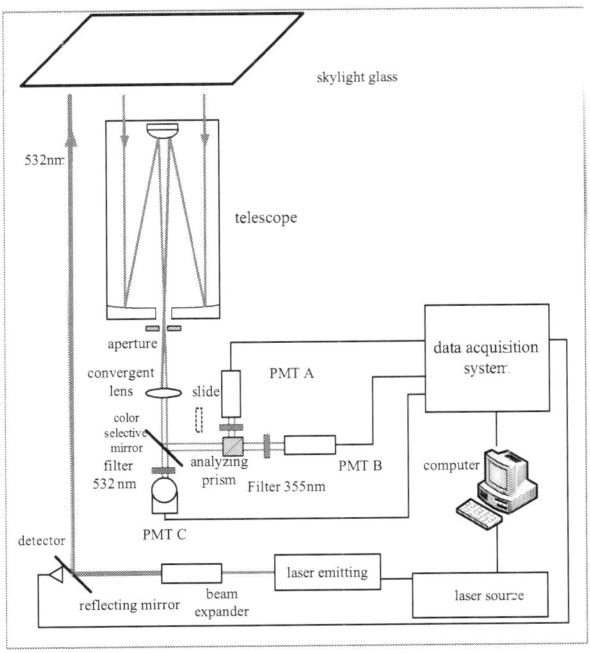

Figure 1. System structure of lidar

2.2. MAX-DOAS

MAX-DOAS is a special type of DOAS technology, which uses characteristic absorption structure of trace gases (such as NO_2, SO_2, O_3) in visible ultraviolet band for qualitative and quantitative analysis of gas. As a real-time and online measuring method, it is eminently suitable for remote sensing measurement of air pollution. Besides, the data on vertical column density of pollution gases acquired from this method can help in verifying satellite results [26]. MAX-DOAS receives scattered sunlight from telescopes at different elevations, as shown in Fig 2. Based on the characteristic absorption spectrums of pollution gas, it inverses the vertical column density of pollution gas and distribution profile of density. Besides, it acquires data on spatial distribution of pollution gas in atmosphere by measuring atmospheric absorption spectrums in zenith direction and several off-axis directions. During the monitoring process, the reflector orients itself to the north and is then driven by stepper motor to turn to 1°, 2°, 3°, 4°... 90° to finish one measurement cycle, thus realizing measurement of different angles. The schematic diagram of MAX-DOAS is presented below.

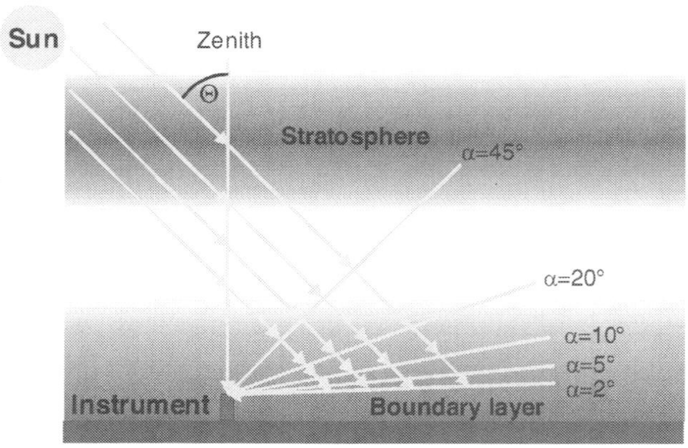

Figure 2. Schematic diagram of MAX-DOAS

2.3. Vehicle-mounted DOAS

With zenith-sky scattered light as the light source, vehicle-mounted DOAS technology is a kind of spectrum detection technology, which uses "fingerprint" absorption characteristics of optical radiation by gas molecule for qualitative and quantitative measurements of gas. The system includes mainly a spectrograph, a GPS system and a telescope [27]. As shown in Fig 3, the whole system is put on mobile platform and the telescope gathers solar spectrum along the measurement route. During operation, the band of spectrograph is fixed at 290nm~420nm (spectral region 130nm) as this band has strong SO_2 and NO_2 absorption structures. After the spectrograph delivers the gathered spectrum to computer for data processing, DOAS method is used to calculate the column density of pollution gas. Finally, density distribution of pollution gas on the measuring route can be acquired by combing the information from GPS on longitude and latitude.

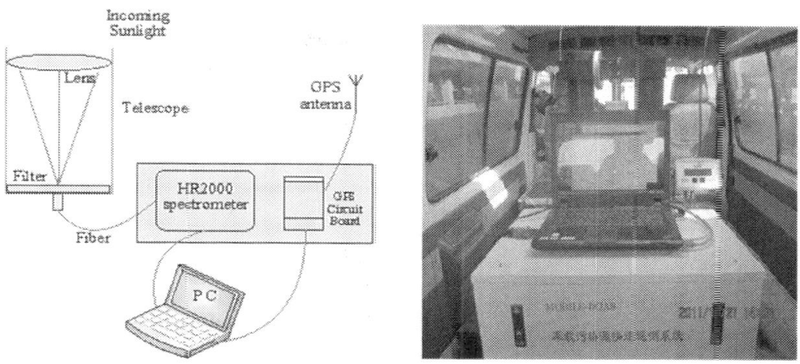

Figure 3. Structure and system of vehicle-mounted DOAS

2.4. Satellite

The spatial resolution of the OMI detector mounted on AURA satellite is 13 km×24 km [28]. NO_2 VCD of troposphere, below the satellite orbit, can be obtained through DOAS inversion of measured spectrum. After differential and meshing treatment, NO_2 VCD of each acquired orbit is integrated and then NO_2 density distribution over the region is obtained. Acquisition of regional density distribution data is significant for investigation of the processes of generation and transportation of pollution. For the present study, OMI Level 2 NO_2 VCD data was adopted. From the weekly average between Nov and Dec, it was found that CCR pollution reached its peak in the third week.

3. Results and discussion

3.1. Stereoscopic monitoring of regular pollution gas

1. Comparison of areal distribution at the three sites

The average of the values obtained during measurement period at the three observation sites (AY, JZ and SMX) were taken as the areal distribution level of NO_2 and SO_2 in CCR (Fig 4). It can be observed that the average level of NO_2 density at AY is $63.4*10^{15}$molec./cm^2, which is higher than the values at JZ and SMX. Density of NO_2 in SMX is the lowest at its average level of $35.3*10^{15}$molec./cm^2. SO_2, density at JZ is the highest at $149*10^{15}$molec./cm^2, which is higher than that at AY and SMX.

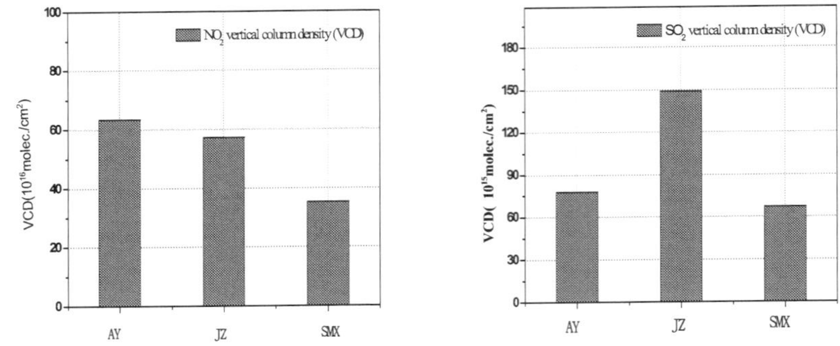

Figure 4. Characteristic contrasts about VCD (vertical column density) of SO_2 and NO_2 at the three sites

2. Results of monthly average variation

The results obtained during the measurement period were used to calculate the averages of different months and to analyze variation trends of pollutant concentration. It can be observed from the calculation results (see Fig 5) that NO_2 VCD (vertical column density) reached its peak in Dec and Jan, decreased from Feb onward, and then remained low between Mar and June. Monthly average of SO_2 VCD at each station is shown in the figure below. It is obvious from VCD of SO_2 is high in winter just as NO_2; it reached its peak in Dec at AY, in Jan at SMX, and in Feb at JZ. It obviously decreased from Mar and thereafter remained low up to June.

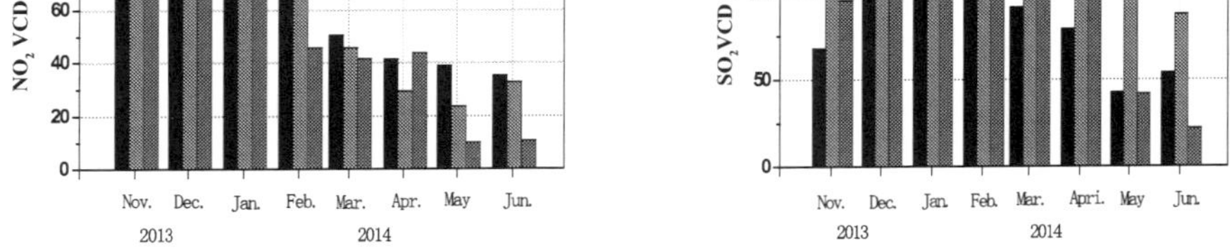

Figure 5. Monthly average variations about VCD (vertical column density) of SO_2 and NO_2 in observing sites

From an overview of the monthly averages discussed above, it can be concluded that the accumulation of pollutants had been accumulated rapidly and heavily because of high discharge and unfavorable weather conditions in winter. With the end of hot season after

February, change of wind direction and increase in rainfall would start favoring diffusion and sedimentation of pollutants, and consequently reduction of pollutant concentration.

3.2. Stereoscopic monitoring of aerosols

Monitoring of aerosols through lidar, from Nov 2013 to April 2014, (Fig 6) revealed that haze pollution was severe mainly from Nov. 2013 to Jan. 2014. Particularly from Dec 12 to Dec 20, and Jan 4 to Jan 17, the pollution lasted for almost half a month at AY and JZ. Overall air quality improved continuously, from Feb to June 2014, at AY, JZ and SMX. However, during wheat harvesting in June, because of straw burning, pollution levels increased at AY, JZ and SMX sites. The highest pollution levels were confined to the middle of June (14th-15th); the peak of aerosol extinction coefficients at AY, JZ and SMX sites are respectively 0.83km^{-1}, 0.91km^{-1} and 0.66km^{-1}.

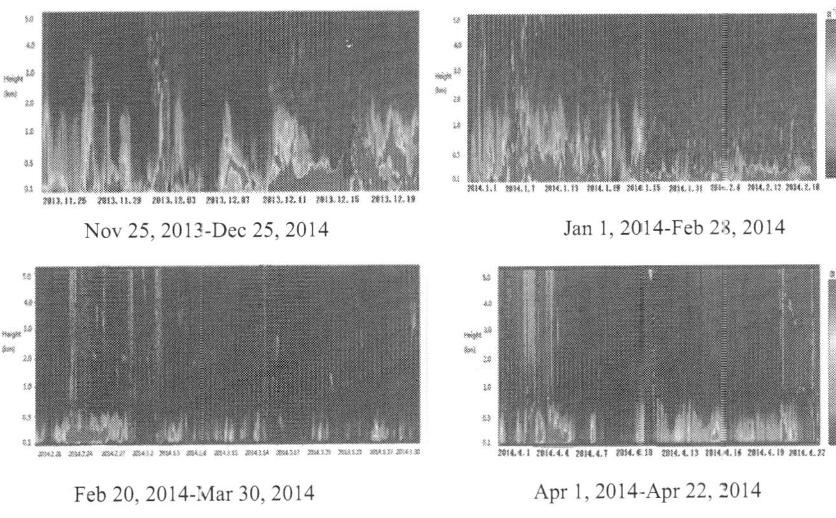

Figure 6. Monthly aerosol distributions from lidar observation

Pollution transportation (Nov 25, 2013-Nov 27, 2013) also occurred at AY, JZ and SMX, but the pollution characteristics and sources were different. Among these three sites, sand dust transportation in northwest direction affected SMX the most. Because of sand dust transportation, large particle concentration in air increased and RH decreased rapidly (at SMX, RH=78% on Dec 25 and RH=32% on Dec 27). Repeated monitoring of local pollutions (Dec 12, 2013-Dec 17, 2013; Jan 8, 2014-Jan 11, 2014; Jan 27, 2014-Jan 31, 2014; Feb 16, 2014-Feb 18, 2014; Apr 10, 2014-Apr 13, 2014) reveal that near-ground pollution at these three sites came mainly from local pollution (industrial emission, motor vehicle exhaust and coal combustion) and fire coal transportation from southern cities in the southeast direction. By the end of Jan 2014, local pollution aggravated because of discharge from firecrackers during spring festival. The pollution at higher levels in the air above 1km came mainly from the northwest.

In conclusion, it emerges that particulate pollution occurred almost simultaneously at the three sites AY, JZ and SMX. Local pollution (local source) is typical in AY, and near-ground pollution in JZ is higher due to the local industrial emission. Local pollution was relatively slighter in SMX, but it can be easily aggravated by northwest sand dust transportation. Sand dust pollution at AY manifests as superposed locally raised dust over northwest sand dust. In CCR, the ground weather conditions of uniform-pressure were conducive to pollutant diffusion. This weather had lower ground pressure field, lower ground and low-altitude wind speeds, and even calm wind sometimes. These were often accompanied by stronger radiation inversion or low-altitude temperature inversion that is difficult to eliminate. The structure of lower atmosphere layer was stable and might contribute to higher pollution.

3.3. Vehicle-mounted monitoring of pollutant discharge in some representative cities and key industrial parks

Quantitative modeling was done in some cities to evaluate the influence of key industries and industrial parks on ambient air quality, especially PM10. Vehicle-mounted DOAS system was used for monitoring industrial parks in both winter (Dec 2013) and summer (May-June, 2014) by way of acquiring data on spatial distribution of urban VOD of SO_2 and NO_2, emission characteristics of SO_2 and NO_2 in key industrial sources, as well as pollution transportation parameters under specific meteorological conditions.

At AY, the results obtained in winter and summer is similar. The concentration of SO_2 and NO_2 was caused mainly by the discharge from steelworks factories. However, observations during southeast/south winds in summer show that the concentration of $SO_{2,}$ in the southwest-northwest section of the measurement route, was high. This indicates that sporadic small-scale industrial discharge occurred beyond southwest-northwest section (south and southwest of AY), which was not found during observations in winter. Judging from the relative geographic positions of steelworks and urban area at AY, as well as forward track model of wind field (Fig 7), it can be concluded that during northwest winds, discharge of AY steelworks had great influence on the air quality of urban area in AY.

Mean values per hour (MV/H), obtained from ground monitoring of SO_2 at monitoring point of environmental protection agency in AY from May 18 to May 20 are shown in Fig 8. From these results, it can be observed that ground concentration of SO_2 in urban area increased during northwest winds. At the same time, it is also found that during southeast winds, ground concentration of SO_2 in urban area was high at about 11a.m. on May 18, when the observation time was just the same as that of third auxiliary chart in Figure 9. It is concluded that, under southeast wind field, sporadic pollution in the southwest and south, beyond measurement area, had great influence on the air quality of main urban area in AY.

Figure 7. Spatial distribution of SO_2 and wind field track superposition at AY under northwest wind field

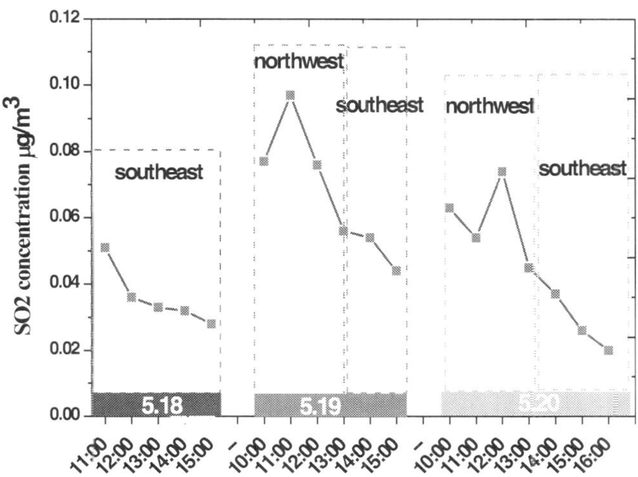

Figure 8. MV/H about ground concentration of SO_2 in monitoring point at AY

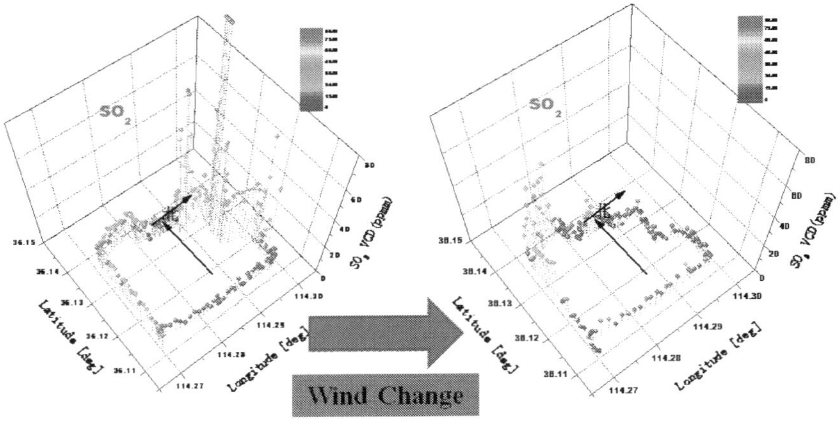

Figure 9. Spatial distribution changes of SO_2 during the wind filed transformation process at AY steelworks

3.4. Analysis of the typical haze process

Haze episode (Dec 12, 2013-Dec, 16, 2013)

From Dec 12, 2013 to Dec 16, 2013, a typical haze phenomenon was observed at AY, JZ and SMX, when the pollution was mainly local. Under the influence of regional weak wind field, the distribution range of local pollution was large and the duration long, but intermittent. Besides, the extinction coefficient of aerosol and concentration of NO_2 reflect a completely accumulative, rising, sustaining and stable process. Based on the distribution range and strength, two main phases were identified: generation phase of local pollution and pollution stabilization phase.

1. **Aerosols**

AY city: Since the afternoon of Dec 12, 2013, the pollutants gradually rose to 1.9 km and then began descending on Dec 13 (Fig 10). In the afternoon of Dec 14, it again started rising, but this time the height was lower than that on Dec 13. During Dec 15-16, near-ground pollution intensified and lasted for a long time, with the maximum value of extinction coefficient reaching 0.87km-1 (see Fig 11).

a. Formative period of haze (Dec 12-14)

b. Stable formation of pollution (Dec 15-16)

JZ city: The height of pollutants started rising from the afternoon of Dec 12 and the average of extinction coefficient was higher than that at AY (Fig 12). The height of daytime pollutants dropped on Dec 13, but it began rising again in the morning of Dec 14; near-ground pollution continued on Dec 15-16 (Fig 13).

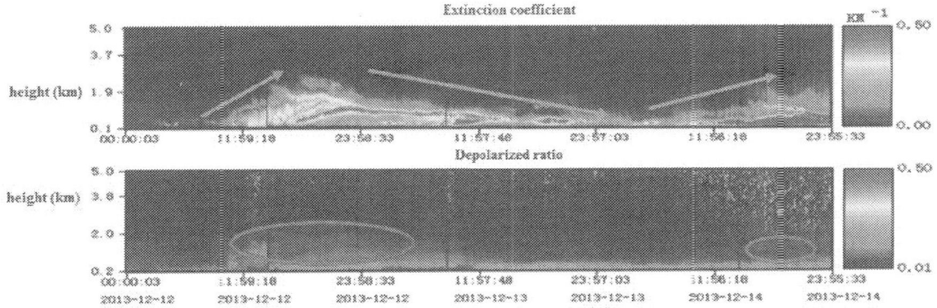

Figure 10. Spatial distributions of aerosols from Dec 12 to Dec 14 at AY, CCR

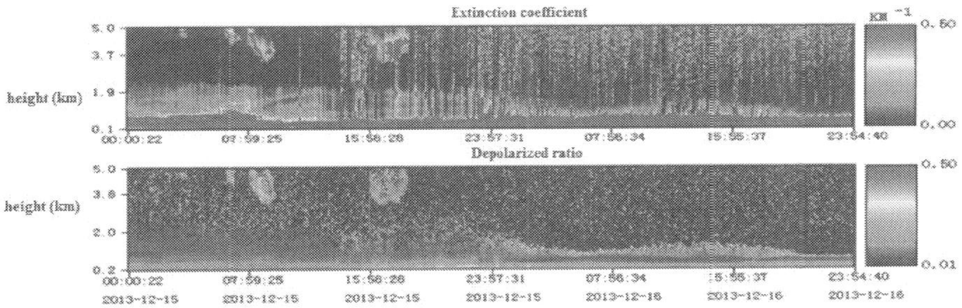

Figure 11. Spatial distributions of aerosols from Dec 15 to Dec 16 at AY, CCR

c. Formative period of haze (Dec 12-14)

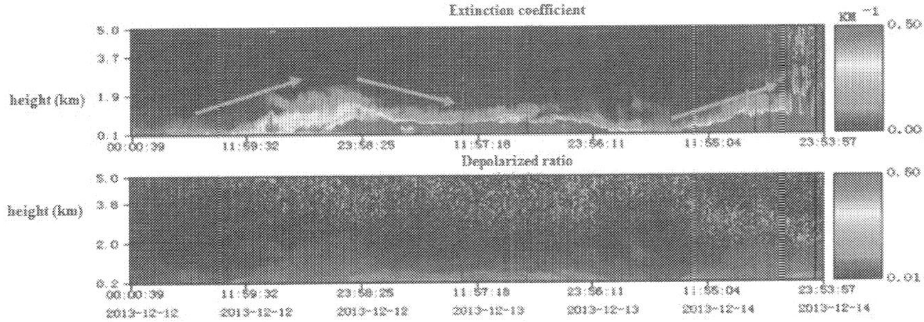

Figure 12. Spatial distributions of aerosols from Dec 12 to Dec14 at JZ, CCR

d. Stable formation of pollution (Dec 15-16)

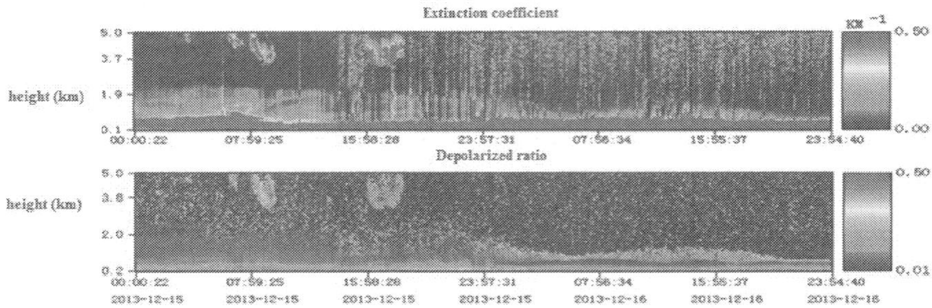

Figure 13. Spatial distributions of aerosols from Dec 15 to Dec16 at JZ, CCR

SMX city: Local pollution was noticed from the afternoon of Dec 12; the height of the pollutants rose gradually from the evening of Dec 13 and the average of extinction coefficient was 0.39km^{-1} (Fig 14). High concentration pollution was confined to the period from the evening of Dec 13 to evening of Dec 14. No obvious change of extinction coefficient profile is observed on Dec 15-16; the height of the pollutants was below 1.2km (Fig 15).

e. Formative period of local pollution (Dec 12-14)

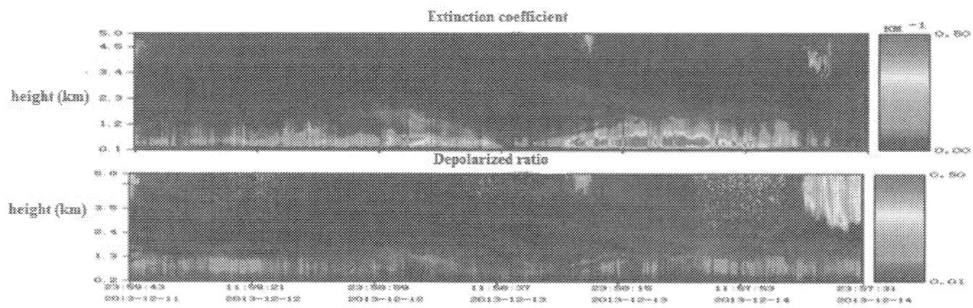

Figure 14. Spatial distributions of aerosols from Dec 12 to Dec 14 at SMX, CCR

f. Stable formation of pollution (Dec 15-16)

2. Pollution gas

During the haze period from Dec 14 to Dec 20, the whole process of NO_2 accumulation and dispersion, as observed by MAX-DOAS, is depicted in Figs 16 & 17. The changes in pollution concentration are most evident at AY, where the concentrate started rising since Dec 12 and

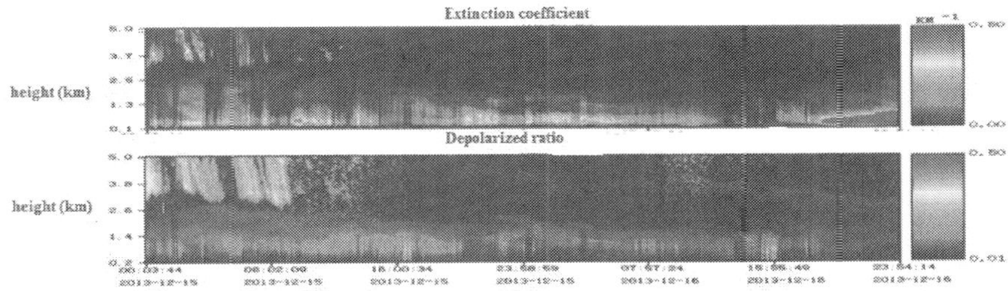

Figure 15. Spatial distributions of aerosols from Dec 15 to Dec16 at SMX, CCR

continued until Dec 14; thereafter, it remained at a high level. However, the increase at JZ was relatively slow. The concentration of both NO_2 and SMX NO_2 increased obviously on Dec 14 and again, and this time is more higher on December 18. On the contrary, the concentrations of NO_2 at AY and JZ began decreasing. The haze formation was caused mainly by internal and northern atmosphere transmission; besides, the obstruction to the airflow in the north militated against pollutant diffusion and instead favored accumulation. Air humidity, coupled with the convergence of cold and hot airflows, contributed to the process of haze formation.

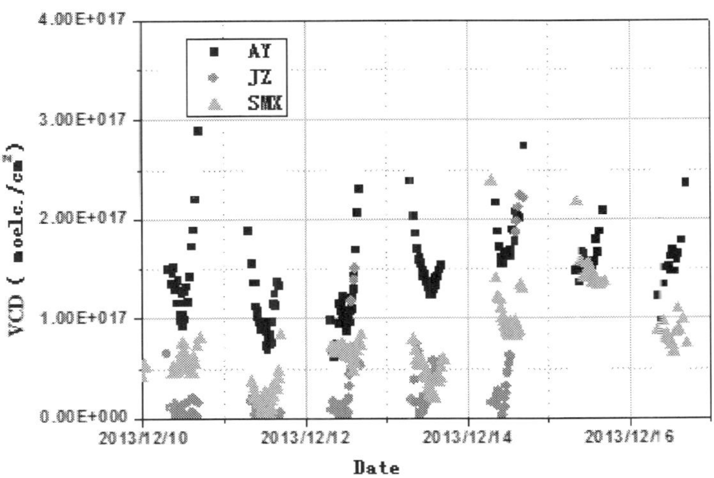

Figure 16. NO_2 VCD from Dec 10 to 16 in the observing sites

3. Meteorological conditions

On Dec 15, the average wind speed was 4.1 m/s and the temperature was1.8°C at the height of 1000m. The prevailing wind was northwest wind (Fig 18). The high-altitude west wind was

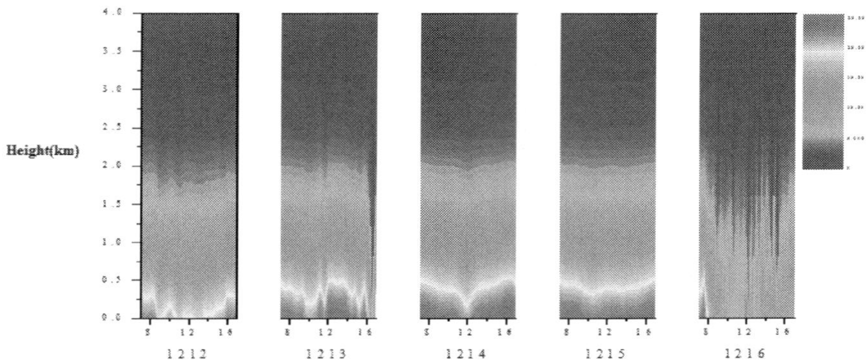

Figure 17. NO_2 profiles from Dec 12 to 16 at AY, CCR

affected by the downdraft in leeward slope terrain, leading to the convergence of the regional weak wind field pollutants, and finally to heavy pollution.

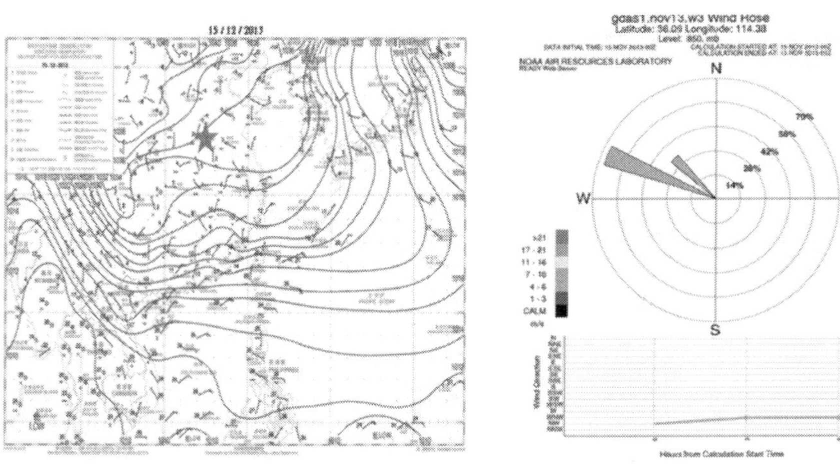

Figure 18. Meteorological conditions and wind rose diagram

From the HYSPLIT backward trajectory model in Fig 19 [29], the pollution at the three observation points was found to be mainly from the northwest. The main pollution source could be from burning of fossil fuels or bio-fuels, aided by the continental airstream. By noon on Nov 25, there was evident increase in pollutants at both AY and JZ. Subsequently, until the evening on Dec 26, at both the observation points, the pollutants decreased and then again

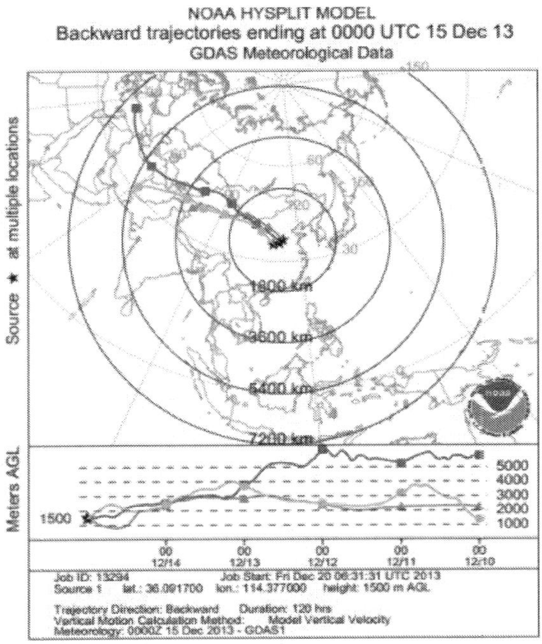

Figure 19. Backward trajectories from Hysplit

increased and stabilized. The overall aerosol extinction coefficient at AY was higher than that at JZ, and the peal values are 0.87km^{-1} and 0.51km^{-1}, respectively. The pollution layer was below 1.9 km, corresponding to the pollutant transmission layer in a low level jet (LLJ) [30] of stable or neutral boundary layer. For two days, it was observed that the height of the mixing layer grew linearly over time. When the height of the mixing layer grows to the height of the pollutant transmission belt in the nocturnal boundary layer, the vertical mixing effect would result in rapid concentration of pollutants on the ground. Therefore, there was increase in aerosol values, as well as in local pollutants discharge during the days; the downward transmission from the higher layers was also an important factor. Compared with AY and JZ, the air at SMX was cleaner and the height of pollutants was lower at 1.2 km below, but by the evening of June13 the pollution became serious.

4. Satellite

From the satellite results in Fig 20, it can be observed that high NO_2 VCD values existed during the haze period. From the daily variations in NO_2 VCD values between Nov and Dec at AY, JZ and SMX (seen in Fig 21), it can be seen that the results at AY clearly indicate a relation between NO_2 VCD values and haze. For example, at AY, the daily average value of NO_2 VCD

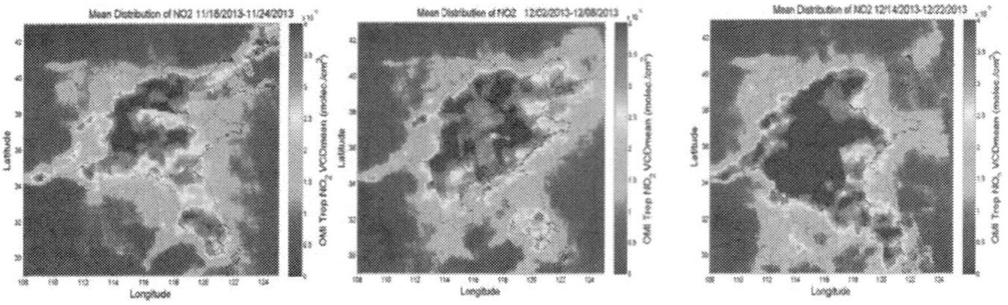

Figure 20. NO_2 VCD distribution result chart from Nov 11 to Dec 20 in CCR during haze period

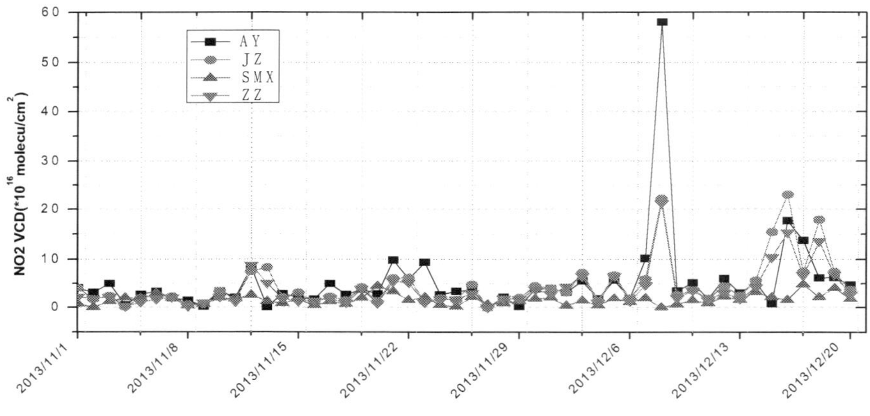

Figure 21. NO_2 VCD change chart detected by OMI satellite from Nov 1 to Dec 20

during Nov 1-20 was 3.32×10^{16}molec./cm^2. For the first haze period during Nov 21-23, the daily average value was 8.37×10^{16} molecule/cm^2, which was 2.5 times more than the normal average value. For the second haze period on Dec 8, it was 5.8×10^{17} molecule/cm^2, which was the highest and 18 times more than the normal average value. For the third haze period during Dec 16-17, it was 1.57×10^{17} molecule/cm^2, which was 4.7 times more than the average value. From the values of NO_2 during these three haze periods, which were multiple times more than the average value, it clearly follows that there exists certain relation between NO_2 and haze, and that high NO_2 VCD values contribute to hazy weather.

3.5. Typical straw burning process analysis (June 12-16, 2014)

1. **Aerosols**

Wheat harvesting from June 12 to June 16 in early summer of 2014, straw burning contributed to local pollution at all the three observation points in AY, JZ and SMX (Figs 22 & 23). At all

http://dx.doi.org/10.5772/59753

these points, pollution was at its peak during June 14-15. And the biomass burning, such as straw burning, which releases particulate matter of small grain size, is confirmed by the depolarization ratio values from lidar observations.

June 13, 2014-June 14, 2014 (initial phase)

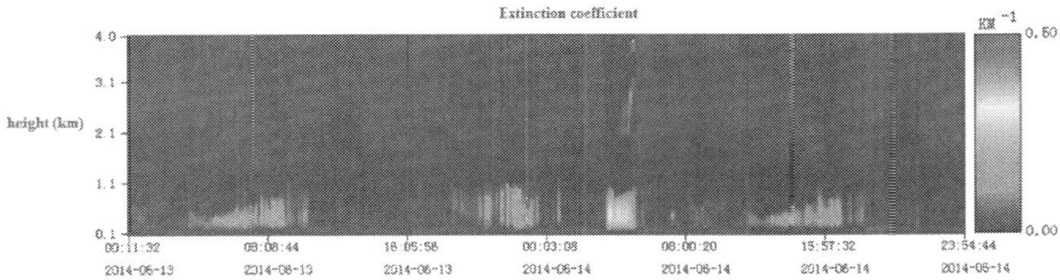

Figure 22. Spatial distribution of aerosol from June 13 to June 14, 2014 at AY, CCR

June 15, 2014-June 16, 2014 (serious pollution phase)

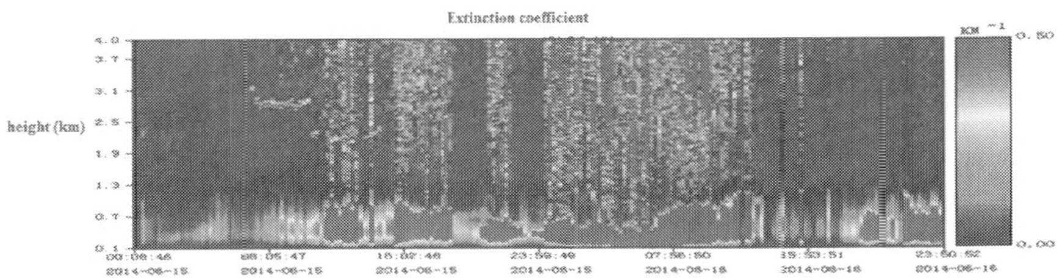

Figure 23. Spatial distribution of aerosol from June 15 to June16, 2014 at AY, CCR

During the straw burning period in CCR, the aerosol extinction coefficient fluctuated from 0.69 to 1.23 km^{-1}. The low depolarization peak value of 0.2 confirms the presence of fine particulates. Besides particulates, usually straw burning also releases various gaseous components such as CO_2, CO, NOx, VOC, sulfuric acid gas, hydrochloric acid gas, and alkaline gas ammonia. The retention time of ammonia in the air is short, because it quickly reacts with acidic gases or acidic particulate matter, and the resulting components can be precipitated on the particle surfaces of the particulate matter leading to the generation of new condensation nucleus. The aging process will increase the amount of particulate matter, which can lead to the increase in particulate emission factors. The burning of straw near CCR, together with surrounding pollution transmission, aging of particulate matter during long distance transmission, and the emission factors such as diameter aggravated local pollution.

2. Pollution gas

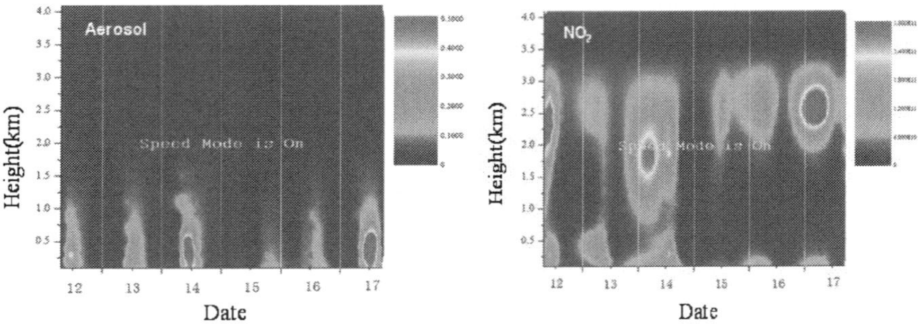

Figure 24. Aerosol and NO_2 profiles from June 15 to June16, 2014 at AY, CCR

NO_2 and SO_2 values were at low level for most of the time, but at some stages during the measurement period from the end of May to the middle of June, they were at high level. According to the results of aerosol and NO_2 for the period June 12 to 17 (see Fig 24), high aerosol extinction appeared at 300 m on June 12; it reached the peak on June 14 and diffused on June 15. It appeared on June 16 again and reached the highest level on June 17. NO_2 transmission process appeared at the height of 1.8 to 2.7 km on June 12; it weakened on June 13 although the transmission close to the ground was stronger. The transmission process appeared again at the height of 1.5 to 2.2 km on June 14, which reappeared, though as a weaker one, on June 15 and 16 at the height of 2.2 to 3 km. But the transmission remaining at the same height became stronger on June 17.

3. Fire spots and backward trajectory

To identify the sources of pollution transmission, local information was taken as the reference. But during the wheat harvesting season, the burning of straw on the surrounding farmlands was believed to be the main cause for high values of pollution. Therefore, the analysis was made by referring to the satellite fire spots diagram of Ministry of Environmental Protection, which is shown in Fig 25.

From these fire spots [31], it was inferred that multiple fire points existed around AY and JZ since June. However, by the middle of June they increased and many more fresh points appeared in the agricultural area of south CCR in the southeast. The backward trajectories, shown in Fig 26, indicate that concentrated burning of wheat straw brought about peripheral transmission.

From the foregoing results of aerosols and pollution gas, it emerges that the transportation brought by wheat straw burning manifested mainly as rapid increase of PM10, which is related to moisture absorption and expansion of particulate matter during transportation process. NO_2 and SO_2 were at their peaks on June 13 and 14 respectively; so, the amplification was not

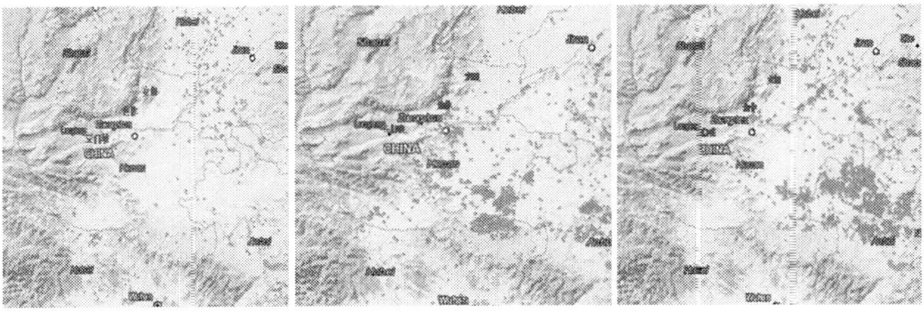

Figure 25. Fire spots from May 26 to June 1, June 2 to June 8, and June 9 to June 15

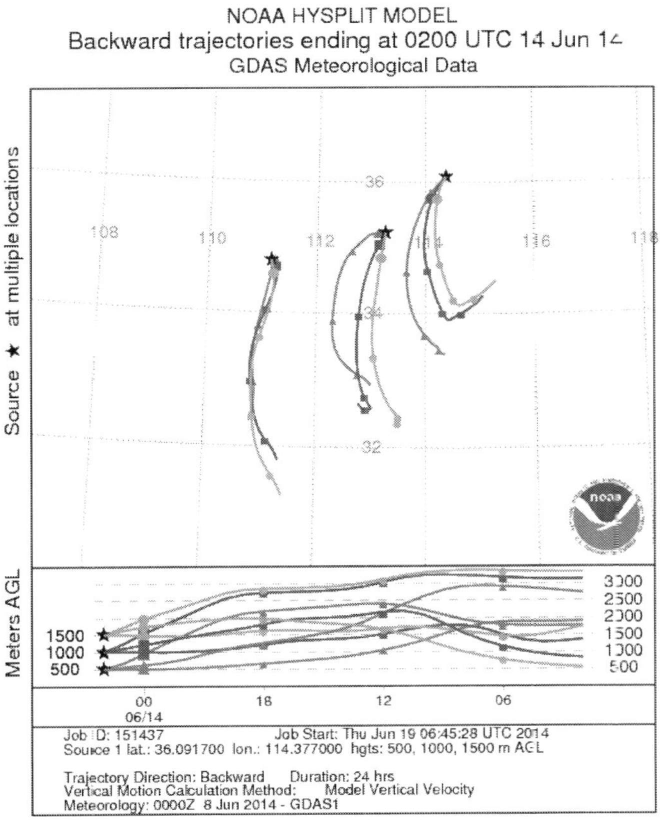

Figure 26. Backward trajectories from Hysplit

as obvious as that of PM10. For evaluating the influence of wheat straw burning on local air pollution, the values of NO_2 and SO_2 on June 11, before wheat straw burning, were taken as the background values. The NO_2 values obtained after wheat straw burning on June 13, and again the SO_2values obtained on June 14, were recorded and evaluated. From these evaluation results, it's concluded that from the end of May to the middle of June, concentration of gaseous pollutants (NO_2 and SO_2) and particulate matter increased obviously due to straw burning in CCR and surrounding cities.

4. Conclusions

CCR is an important industrial site in the mid- west region of China. Historical review of satellite data shows that the cities with relatively serious air pollution are confined mainly to the north because of their geographic location, weather conditions and industrial activity. Besides, the "accumulation zone" of pollutants exists in the region east of Tai-hang Mountains. Analysis of the data obtained from stereoscopic monitoring of the selected sites shows that the prevalence of pollution at the three sites, AY, JZ and SMX, was almost simultaneous. However, AY was characterized by local pollution (local source) and JZ by near-ground pollution, where the pollution concentration was also higher due to higher emission of industrial gases. Though local pollution was milder at SMX, it still can be easily aggravated by northwest sand dust transportation. Through statistical analysis of satellite and ground monitoring data, pollution was more serious during the 1st and 4th quarters (winter and spring), followed by the 2nd quarter and the 3rd quarter. During the heating period, the air pollution intensified obviously, because the urban energy consumption structure was still coal-oriented. During late May to mid-June, straw burning in the south CCR and surrounding provinces, such as Anhui and Jiangsu, contributed to the increase of aerosol, NO_2 and SO_2, in some cities of CCR.

As for the reasons for the complex process of haze formation, it was found that local pollution (local sources) was the main cause, besides the effect of superposition of pollution transportation in some time frames. Near-ground pollution at the three sites was caused mainly by local pollution (industrial emission, motor vehicle exhaust, coal combustion, etc.) and pollution transportation of some northern cities in the southeast direction. AY was more affected by local transportation and emissions from some industrial parks, and JZ by near-ground local pollution. Moreover, pollution concentration was higher at JZ than that at AY because of more industrial emissions. Local pollution at SMX was comparatively less obvious, but was lasting longer due to the unfavorable diffusion condition. In CCR, formation of ground weather situations of uniform-pressure field and North China low pressure, which are bad for pollutant diffusion, is easy. Such weather has lower ground pressure field, and smaller speeds of ground and low-altitude winds, which are sometimes calm. It is often accompanied with stronger radiation inversion or low-altitude temperature inversion that is difficult to eliminate. The structure of lower atmosphere layer is stable and might contribute to higher pollution.

So far many efforts were made from ground based and space borne observations to study aerosol distribution and properties, along with model simulations, but such integrated studies are still limited over the East Asian, especially in CCR region in China. Satellite observations provide time constrained observations of aerosol vertical distributions but ground based lidar can give information on the evolution of the distribution over a location. Thus the optical properties of aerosols or gaseous pollutants obtained from ground-based remote sensing technology, especially lidar and DOAS is being studied to determine the aerosol optical characteristics and source origin. In the future, the more integrated combination of lidar, satellite data, in-situ instruments, and sunphotometer data will help us to improve the characterization of a variety of aerosol events and get deeper insight into the mechanism of long-range transportation of pollution in China. Furthermore, the information of vertical distribution of haze layer during long-range transportation could permit us to make better assessment of the effect of aerosols on the radiative forcing and climate change.

Acknowledgements

This work was supported by the Funding from Henan Program (Researches on Atmospheric Haze pollution in Henan Province), Major State Basic Research Development Program of China (2014CB447900) and NSFC (41305126). The authors are also thankful to the NOAA Air Resources Laboratory (ARL) for providing the HYSPLIT transport and dispersion model used in this research.

Author details

Wenqing Liu, Zhenyi Chen*, Pinhua Xie, Jianguo Liu and Tianshu Zhang

*Address all correspondence to: zychen@aiofm.ac.cn

State Environmental Protection Key Laboratory of Optical Monitoring Technology, Ministry of Environmental Protection of the People's Republic of China, Anhui Institute of Optics & Fine Mechanics, Chinese Academy of Sciences, Hefei, Anhui, P.R. China

References

[1] Ramanathan, V., P. J. Crutzen,, J. T. Kiehl, and D. Rosenfeld, Atmosphere: aerosols, climate, and the hydrological cycle. *Science.*,2001, 294, 2119–2124.

[2] Ian G. C., Guard C. L. and Jami H., China's Growing Urban Health Inequalities: The Challenges Ahead, J. Management and Sustainability, 2013, 3(2):10-18

[3] Sun Y. L., Zhuang G. S., Wang Y., Han L. H., Huo G. J. and Dan M., The air-borne particulate pollution in Beijingconcentration, composition, distribution and sources. *Atmos. Environ.*, 2004,38(35): 5991–6004

[4] Tie, X., Brasseur G., Zhao C. S., Granier C., Massie S., Qin, Y., Wang P. C., Wang G. L., Yang P. C. and Richterg A., Chemical characterization of air pollution in Eastern China and the Eastern United States. *Atmos. Environ.*, 2006,40, 2607–2625

[5] Wu, D., Deng X. J., Bi X. Y., Li F., Tan H. B., Liao G. L., Study on the visibility reduction caused by atmospheric haze in Guangzhou area. *J. Tropical Meteorology*, 2006,13, 77–80

[6] Zhang, Y. H., Hu M., Zhong L. J., Wiedensoher A., Liu S. C., Andreae M. O. and Fan F. J., Regional integrated experiments on air quality over Pearl River Delta 2004 (PRIDE-PRD2004), *Atmos. Environ.*,2004,42,6157-6173

[7] Wang Y.S., Xin J. Y., Li Z. Q., Wang P. C., Wang S. G., Wen T. X., Sun Y., AOD and Ångström Parameters of Aerosols Observed by the Chinese Sun Hazemeter Network from August to December 2004, *Environ. Sci.*, 2006,27(9)

[8] Tesche, M., Ansmann A., Müller D., Althausen D. and Engelman R., Particle backscatter, extinction, and lidar ratio profiling with Raman lidar in south and north China, *Appl. Opt.*, 2007,46(25):6302-8

[9] Xiao, R., Takegawa N., Zheng M., Kondo Y., Miyazaki Y., Miyakawa T., et al.,Characterization and source apportionment of submicron aerosol with aerosol mass spectrometer during the PRIDE-PRD 2006 campaign, *Atmos. Chem. Phys.*, 2001,11, 6911-6929

[10] Li, Z., et al., Preface to special section on East Asian Studies of Tropospheric Aerosols: An International Regional Experiment (EASTAIRE), *J. Geophys. Res.*, 2007,112, D22S00, doi:10.1029/2007JD008853.

[11] Li, Z., et al., East Asian Studies of Tropospheric Aerosols and their Impact on Regional Climate (EAST-AIRC): An overview, *J. Geophys. Res.*, 2011,116, D00K34, doi: 10.1029/2010JD015257

[12] Atsushi, S., Nobuo S., Ichiro K. A., Itsushi U., Toshiyuki M., Naoki K.,et al., Continuous observations of Asian dust and other aerosols by polarization lidars in China and Japan during ACE-Asia, *J. Geophys. Res.*,2004 109, D19S17

[13] He, Q. S., Li C. C., Mao J. T. and Li P. R., A study on the aerosol extinction-to-backscatter ratio with combination of micro-pulse LIDAR and MODIS over Hong Kong, *Atmos. Chem. Phys.*, 2006, 3243–3256

[14] Müller, D., Tesche M., Eichler H., Engelmann R., Althausen D., Ansmann A., et al,Strong particle light absorption over the Pearl River Delta (south China) and Beijing (north China) determined from combined Raman lidar and Sunphotometer observations, *Geophys. Res. Lett.*, 2006, 29 (17)

[15] Mattias J., Bo G., Tong Y., Lin T., Chen D.L., Li H.J., et al., Quantification of total emission of air pollutants from Beijing using mobile mini-DOAS, 2008,*Atmos. Environ.*,42(29), 6926–6933

[16] Meng X.L., Fan X.Y., Meng X., Zheng J.P., Mason R.,Submicron-sized fluid inclusions and distribution of hydrous components in jadeite, quartz and symplectite-forming minerals from UHP jadeite–quartzite in the Dabie Mountains, China: TEM and FTIR investigation, 2009,*Applied Geochemistry*,24(4), 517–526

[17] Cheng, Y., Lee S. C., Ho K. F., Wang Y. Q., Cao J. J., Chow J. C. and Watson J. G., Black carbon measurement in a coastal area of south China, *J. Geophys. Res. Atmos.*, 2006, 111, D12310, doi: 10.1029/2005JD006663

[18] Franke, K., A. Ansmann, D. Muller, D. Althausen, C. Venkataraman, M. S. Reddy, F. Wagner and R. Scheele (2003),Optical properties of the Indo-Asian haze layer over the tropical Indian Ocean, J. Geophys. Res., 108, D2, 4059, doi: 10.109/2002JD002473

[19] Klett, J. D., Stable analytical solution for processing lidar returns, *Appl. Opt.*, 1981, 20, 211–220,

[20] Frederick G. Fernald. Analysis of atmospheric lidar observations: some comments, *Appl. Opt.*, 1984, 23(5), 211–220

[21] Xu J., Xie P.H., Si F.Q.,Li A.,et al., Comparison of NO2 slant columns between two ground-based Max-DOAS, *Spectroscopy and Spectral Analysis*, 2012,32(2),558-564

[22] Wang Y., Li A., Xie P.H., Chen H., Xu J.,Wu F. C., Liu J.G., Liu W.Q., Retrieving vertical profile of aerosol extinction by multi-axis differential optical absorption spectroscopy, *Acta Phys. Sin.*,2013, 62(18), 180705

[23] Wu J.G. Modern Fourier Transform Infrared Spectroscopy and Its Application, Beijing: Scientifical and Technical Documents Press,1994,1, 492-505

[24] Manne K, Ph. D. Dissertation, Sweden: Chalmers University of Technology.2005

[25] Fransson, K and Mellqvist, J., Measurements of VOCs at Refineries Using the Solar Occultation Flux Technique, Report to Preem Environmental foundation, Göteborg, 2002

[26] Wang T., Wang P.C., Yu H., Zhang X.Y., Zhou B.,Si F.Q., et al., Intercomparison of slant column measurements of NO2 by ground-based MAX-DOAS, *Acta Phys. Sin.* 2013, 62(5), 054206

[27] Wu F.C., Xie P.H., Li A.,Xu J., Shi P., Qin M.,Wang M.H., et al., Studies on the Remote Measurement of the Distribution ion of City Gaseous Pollutant by Mobile Passive Differential Optical Absorption Spectroscopy, *Spectroscopy and Spectral Analysis*, 2011,31(3),583-588

[28] Witte J.C., Duncan B.N., Douglassb A.R., Kurosuc T.P., Chancec K., Retscherd C.,The unique OMI HCHO/NO2 feature during the 2008 Beijing Olympics: Implications for ozone production sensitivity, *Atmos. Environ.*,2011, 45(18), 3103–3111

[29] Draxler, R. R., and Rolph G. D., HYSPLIT (Hybrid Single-Particle Lagrangian Integrated Trajectory) Model access via NOAA ARL READY, (2003) web site (http://www.arl.noaa.gov/ready/hysplit4.html), NOAA Air Resources Laboratory, Silver Spring, MD.

[30] Hu X.M., Klein P.M., Xue M., Zhang F.Q., Doughty D.C., Forkel R., Joseph E, Fuentes J.D., Impact of the vertical mixing induced by low-level jets on boundary layer ozone concentration, *Atmos. Environ.*, 2013,70,

[31] http://hjj.mep.gov.cn/stjc/

Chapter 12

Temporal and Spatial Trends (1990–2010) of Heavy Metal Accumulation in Mosses in Slovakia

B. Maňkovská, M. V. Frontasyeva and T. T. Ostrovnaya

Additional information is available at the end of the chapter

http://dx.doi.org/10.5772/59634

1. Introduction

The multielement biomonitoring surveys, using suitable plant biomonitors [1] can provide information about long-term and large-scale atmospheric deposition rates of elements. The large-scale biomonitoring programs using selected bioindicators were introduced in Slovakia in the end of 1980s. The bioindicators are commonly available elsewhere in the landscape, and the bioindicated air quality parameters can be related to the particular sampling sites within the ecosystems. mosses and foliage of forest tree species [2,3] as biomonitors of atmospheric deposition of heavy metals began in Slovakia more than 30 years ago, in connection with the problems of dying forests.

Moss species such as *Pleurozium schreberi, Hylocomium splendens*, and *Dicranum* sp. can effectively adsorb deposited air pollutants on pectine and cell structures. Bryomonitoring method was validated and tested to large-scale estimate current atmospheric deposition rates of elements between 1970 and 2000 [4-6]. Since 2000, in the frame of UN ECE ICP-Vegetation program, more than 30 European countries have monitored the current element content in mosses at about 7000 sampling sites in 5-year intervals. The Slovak national moss surveys since 1990–2010 have mapped elemental content distribution within the whole country (16x16 km net).

2. Material

Two complementary analytical techniques, instrumental neutron activation analysis (INAA) and atomic absorption spectrometry (AAS) were used for determination of the elemental concentrations in the samples of moss for year 2000. For INAA, moss samples of about 0.3 g were packed in aluminum cups for long-term irradiation or heat-sealed in polyethylene foil bags for short-term irradiation in the IBR-2 reactor, Dubna, described elsewhere [7]. The samples of mosses were not washed before analysis. Sulfur and nitrogen concentrations were determined using LECO corporation equipment (S: LECO SC 132 and N: LECO SC 228). Atomic absorption spectrometry (VARIAN SPECTRA A-300 and mercury analyzer AMA-254) was carried out in Forest Research Institute Zvolen (1990, 1995, 2000, 2005, 2010). The accuracy of data published in paper was verified by 109 individual laboratories and tested by the IUFRO program [8].

The monitoring studies have been undertaken in the framework of the international project Atmospheric Deposition of Heavy Metals in Slovakia Studied by the Moss Biomonitoring Technique Employing Nuclear and Related Analytical Techniques and GIS Technology. *Project REGATA (2003-2015).*

3. Results and discussion

The principal investigator of the project, Dr. Maňkovská (at that time working in the Forest Research Institute in Zvolen, Slovakia) was invited by Scandinavian specialists (Finland, UNIDO, 1986) to join the existing European biomonitoring program focused on monitoring of actual deposition of selected set of elements using analyses of mosses in 1990. The first collection of moss samples of *Pleurozium schreberi* and *Hylocomium splendens* at 58 permanent monitoring sites in Slovakia was made in the same year in accordance with the European network. The following elements were analyzed by atomic absorption spectroscopy (AAS): Cd, Cr, Cu, Fe, Mn, Ni, Pb, S, and Zn.

In the second European moss survey conducted in 1995, moss samples were collected at 78 permanent monitoring sites. In 1996, moss samples were collected at 69 and in 1997 at 74 permanent monitoring sites. The contents of As, Cd, Cr, Cu, Fe, Hg, Ni, Pb, V, and Zn were determined by AAS and Hg was determined by AMA-254.

The third moss survey at the European scale on actual levels of atmospheric deposition of elements was conducted within the ICP Vegetation in 2000. Collection of moss samples (*Pleurozium schreberi, Hylocomium splendens,* and *Dicranum* sp.) in Slovakia was performed at 86 permanent monitoring sites. NAA was carried out in the Frank Laboratory of Neutron Physics of the Joint Institute for Nuclear Research in Dubna, Russia. A total of 39 elements (Ag, Al, As, Au, Ba, Br, Ca, Ce, Cl, Co, Cr, Cs, Fe, Hf, I, In, K, La, Mg, Mn, Mo, Na, Ni, Rb, Sb, Sc, Se, Sm, Sr, Ta, Tb, Th, Ti, U, V, W, Yb, Zn, Zr) were determined. Varian Techtron atomic absorption spectrometer was used for determination of Cd, Cr, Cu, Hg, Ni, Pb, and Zn. Sulfur and nitrogen determination was performed by LECO corporation equipment as earlier (see above).

In the fourth European moss survey in 2005, moss samples (*Pleurozium schreberi, H. splendens, Dicranum* sp.) were collected at 77 permanent monitoring sites in Slovakia. They were analyzed for contents of Cd, Cu, Fe, Hg, N, Ni, Pb, S, V, and Zn by AAS Varian Techtron, AMA-2454, LECO SC 132, and LECO SP 228. Results from required monitoring elements were published in the European reports [9, 10]. Results from required and optionally monitored elements from Slovakia were evaluated in the context of neighboring countries of Visegrad Four [11].

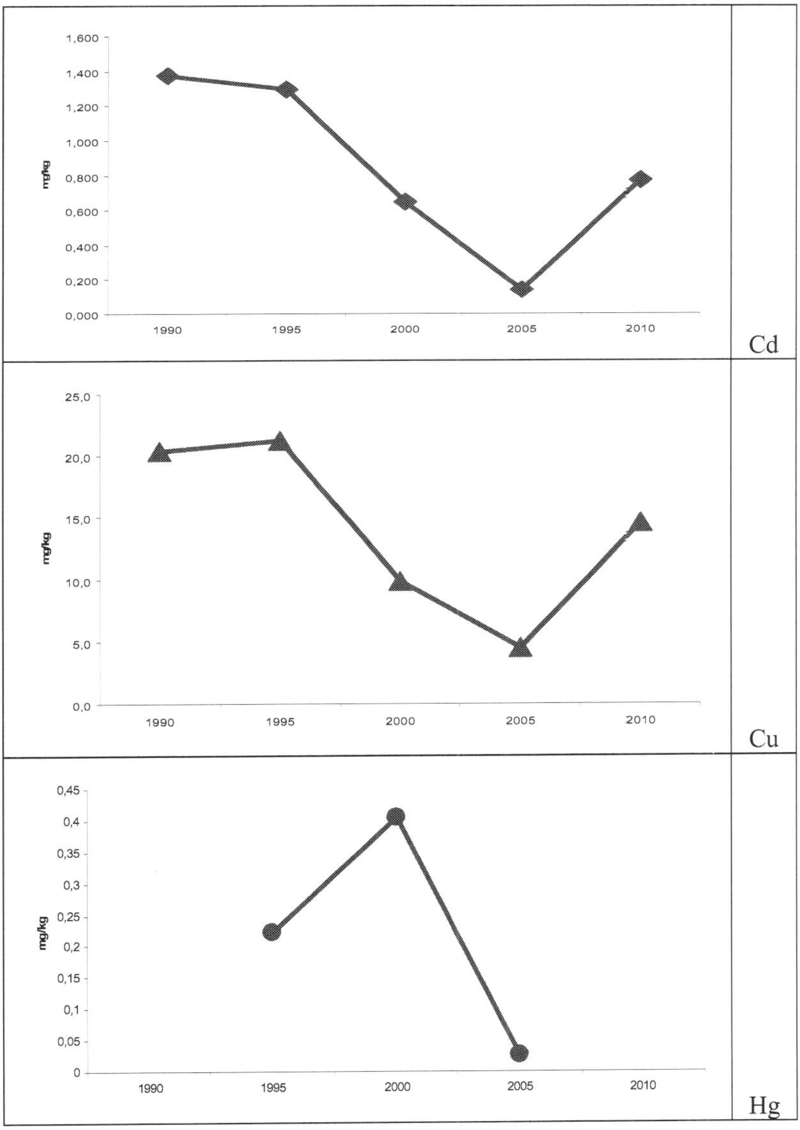

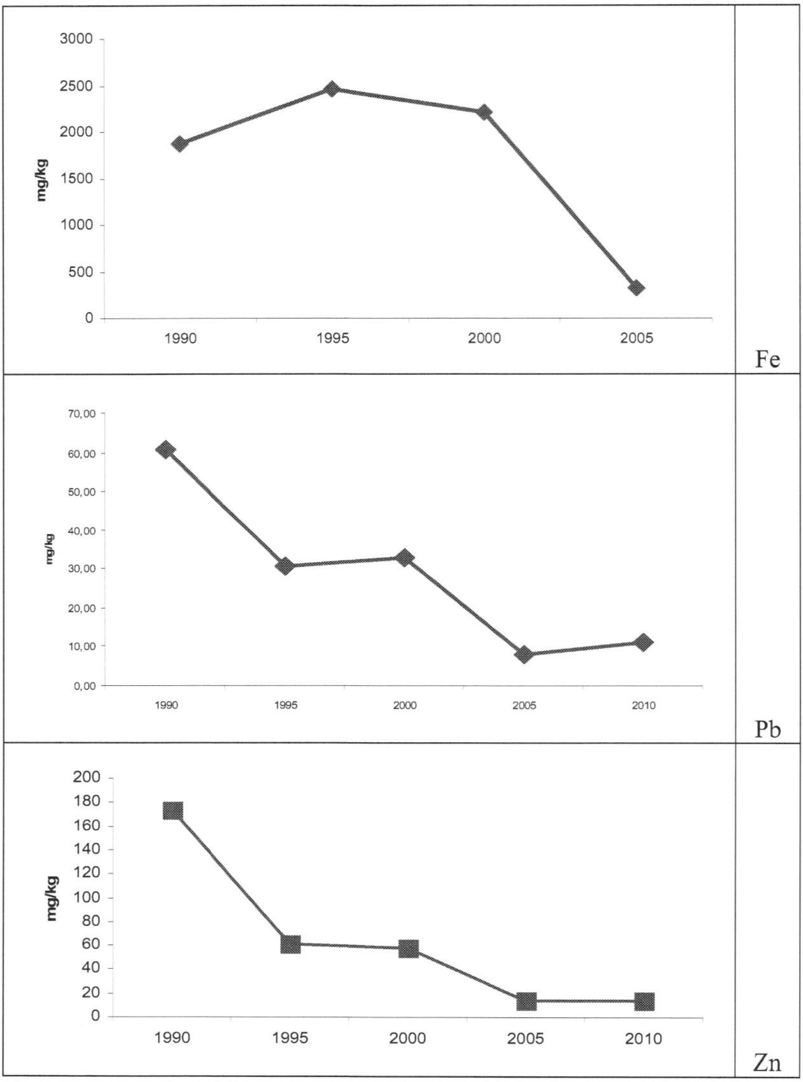

Note: Year (number of PMP): 1990(58);1995(79); 1996(69); 1997 (74); 2000 (86); 2005(82), 2010(67). PMP-permanent monitoring plots

Figure 1. Concentration of Cd, Cu, Hg, Fe, Pb, and Zn (average in mg/kg) in mosses for Slovakia in 1990, 1995, 2000, 2005, 2010.

http://dx.doi.org/10.5772/59634

So far, in the last, that is, the fifth European moss survey in 2010 in Slovakia, collection of moss samples was made at 68 permanent monitoring sites (*P. schreberi, H. splendens, Dicranum* sp.). They were analyzed for contents of Al, Ca, Cd, Cl, Cu, Dy, I, K, Mn, Pb, S, Ti, V by use of AAS Varian Techtron, LECO SC 132, and LECO SP 228, and by use of NAA in the Frank Laboratory of Neutron Physics in Dubna, Russia. Results from required monitoring elements were published in European reports [12].

The concentration of Cd, Cr, Cu, Fe, Hg, Ni, Pb, V, and Zn in mosses between 1990 and 2010 are shown in Fig. 1.

The moss biomonitoring technique is based on the fact that the concentration of heavy metals in mosses correlates with the atmospheric concentration. It was proven that it is possible between the concentration of the given element in mosses and the concentration of the same element in the atmosphere. The concentration of individual elements in precipitation was calculated to the time of exposure of mosses (3 years). In case of each element, there was a good linear relationship between the concentrations of a given element in mosses and in precipitation. There is a valid equation [concentration in moss] mg.kg^{-1} = [4x atmospheric deposition] mg.m^{-2}.year^{-1} [13]. The concentration of elements in mosses in comparison with Norway (Table 1 and Table 2) is expressed by means of the coefficient of loading by elements K_F and classified into 4 classes; class < 1 – elements are within norm and do not exceed the value 1; class 2 – slight loading (elements range from 1 to 10); class 3 – moderate loading (elements range from 10 to 50); class 4 – heavy loading (elements are higher than 50 times higher value).

Contamination factor K_F				
<1	1-2	2-5	5-10	>10
Br, I	Cl, Mn, Na, Ni, Se, Rb, U, Zn	Ba, Ca, Co, Cr, Cu, Fe, Hg, K, Sm, Tb, Th, Ti, V	Al, Au, Ce, La, Sb, Se, Sr, Yb, Pb	Ag, Cd, Mo, Ta, W

Note: K_F = contamination factor as the rate of median values of element in Slovak mosses vs. Norvay mosses (Steinnes et al., 2007). K_F Slovakia= 9.5: K_F Norway=1.

Table 1. The rate of median values of element in Slovak vs. Norway mosses in year 2000

The marginal 2 hot spots were shown in Central Spiš (metallurgical plants), Žiar basin (nonferrous ores processing and aluminium plant). The protected area of Morské oko (chemical industry) is also of great interest. In comparison with the mean Austrian and Czech values of heavy metal contents in moss, the Slovak atmospheric deposition loads of these elements were found to be 2–3 times higher on average. The transboundary contamination by Hg through dry and wet deposition from Czech Republic and Poland is evident in the bordering territory in the north-western part of Slovakia (Black Triangle II), known for metallurgical works, coal processing, and chemical industries. Spatial trends of heavy metal concentrations in mosses were metal-specific. Since 1990, the metal concentration in mosses has declined for cadmium, chromium, cooper, iron, lead, mercury, nickel, and zinc.

Sites	Coefficient of loading by elements K_F				K_F
	< 1	1 -10	10-50	>50	
Hot Spots					
Žiar basin	Au, Br,Cl, I, In, Mn,	Ag, Al, As, Ba, Ca, Cd, Ce, Co, Cr, Cs, Cu, Fe, Hg, K, La, Mg, Mo, Na, Ni, Rb, Sc, Se, Sm, Sr, Tb, Th, Ti, U, V, W, Zn	Hf, Pb, Sb, Ta, Yb	F	**6.2**
Central Spiš	Au	Br, Ca, Cl, In, K, Mg, Mn, Rb, Se,	Al, As, Ba, Cd, Co, Cr, Cs, Cu, Fe, Hg, I, La, Mo, Na, Ni Sc, Sr, Th, U, V, W, Zn	Ag, Hf, Pb, Sb Ta Tb, Yb	**45**
National Parks					
Nízke Tatry	Au, Br, I, Mg, S, Se, Sm,Ti	Ag, Al, As, Ba, Ca, Cd, Ce, Cl, Co, Cr, Cs, Cu, Fe, Hg, In, K, La, Mn, Mo, N, Na, Ni, Pb Rb, Sb, Sc, Sr, Ta, Tb, Th, U, V, W, Yb, Zn, Zr	Hf		**4.2**
Vysoké Tatry	Au, Br, Ca, I, Se	Ag, As, Ba, Cd, Ce, Cl, Co, Cs, Cu, Fe, Hg, In, K, La, Mg, Mn, Mo, N, Na, Ni, Pb, Rb, S, Sc, Se, Sm, Sr, Tb, Th, Ti, U, V, W, Zn	Al, Cr, Sb, Ta, Yb, Zr	Hf	**6.7**
Protected Area					
Veľká Fatra	Au, Br,In Sm	Ag, Al, As, Au, Ba, Ca, Cd,Ce, Cl, Co, Cs, Cu, Fe, Hg, I, K, La, Mg, Mn, Mo, N,Na, Ni, Pb Rb, S, Sc, Se, Sr, Ti, U, V, W, Zn	Cr, Sb, Ta, Tb, Th, Yb, Zr	Hf	**7.6**
Báb	Au, Br, In, Mg, N, S, Se	Ag, As, Ba,Ca, Cl, Co, Cr, Cs,Cu, Fe,Hg, I, K, Mn, Na, Ni, Rb, Sm, Sr, Ti, U, V, W, Zn	Al, Cd, Ce, La, Mo, Pb, Sb, Sc, Ta, Tb, Th,Yb, Zr	Hf	**8.8**
Slovenský raj	Au, Br, In Sm, Se	Al, As, Ba, Ca, Cd, Ce, Cl, Co, Cr, Cs, Cu, Fe, I, K, La, Mg, Mn, N, Na, Ni, Rb, S, Sc, Sr, Th, Ti, U, V, W, Zn	Ag, Hg, Mo, Pb, Ta, Tb, Yb, Zr	Hf, Sb	**11.8**
Poľana	Au	Br, Ca, Cl, Cu, In, K, Mg, Mn, Na, Rb, Se, Zn	Ag, Al, As, Ba, Cd, Co, Cr, Cs, Fe, Hg, I, La, Mo, Ni, Pb, Rb, Sc, Sr, Ta, Tb, Th, U, V, W, Yb	Sb, Hf	**19**
Morské oko	Au	Br,Ca, Cl, In, K, Mg, Mn, Rb, Se, Zn	Ag, As, Ba, Cd, Co, Cr, Cs, Cu, Fe, Hg, I, La, Mo, Na, Ni, Pb, Sr, U, V, W	Al, Hf, Sb, Sc, Ta, Tb, Th, Yb	**44**

Table 2. Coefficient of loading by elements K_F in the year 2000

The temporal trends in the concentration of Cd, Cr, Cu, Fe, Hg, Ni, Pb, V, and Zn between 1990 and 2010 were observed. In general, the concentration of Cd, Cr, Cu, Fe, Hg, Ni, Pb, V, and Zn in mosses decreased between 1990 and 2010; the decline was higher for Pb than for Cd. The observed temporal trends for the concentrations in mosses were similar to the trends reported for the modeled total deposition of cadmium, lead, and mercury in Europe. The level of elements determined in bryophytes reflects the relative atmospheric deposition loads of the elements at the investigated sites. Factor analysis was applied to determine possible sources of trace element deposition in the Slovakian moss. In the industrial area of Central Spiš, in comparison with the Norwegian limit values (Central Norway is considered a relatively pristine region), exceeded levels for Al, As, Ca, Cd, Cl, Co, Fe, K, Mn, Sb, Sm, Sr, W, and Zn were found.

4. Conclusion

Moss surveys can provide quick and cheap information about spatiotemporal changes of the current deposition rates of about 40 chemical elements across the country. Figures from the moss surveys may be the only data about elemental deposition rates that have not been determined at measurement stations of air quality (e.g., Be, Li, Se, Tl, Th, and REEs).

Moss biomonitoring is an effective tool for detecting effects of new technologies on deposition zones in the vicinity of emission sources. All results of the Slovak moss surveys were accepted and stored in the UN ECE ICP-Vegetation database for checking of deposition loads in Europe and their environmental effects.

On the basis of biomonitoring using 3-year-old segments of *Pleurozium schreberi*, *Hylocomium splendens*, and *Dicranum* sp. at 10 sites in Slovakia, it was determined that:

a. The concentration of elements (in parentheses) is more than 50 times higher at sites Báb (Hf), Poľana (Hf, Sb); Vysoké Tatry (Hf); Slovenský raj (Hf, Sb); Veľká Fatra (Hf); Central Spiš (Ag, Hf, Pb, Sb Ta Tb, Yb); Žiar basin (F), and site Morské oko (Al, Hf, Sc, Sb, Ta, Tb, Th, Yb) compared to the Norwegian values.

b. Air pollutants K_F varies in the range of 4–45 (4.2 – Nízke Tatry; 6.2 – Žiar basin; 6.7 – Vysoké Tatry; 7.6 – Veľká Fatra; Báb – 8.8; 11.8 – Slovenský raj; 19 – Poľana; 44 – Morské oko; and 45 – Central Spiš). Results of biomonitoring campaigns serve as a reliable basis for planning and long-term exploitation of the landscape of the country and for further environmental investigations.

Acknowledgements

This article was made possible with the financial support of grant APVV-0663-10, VEGA and by the grant of the Plenipotentiary of the Slovak Republic at the Joint Institute for Nuclear Research, Dubna, Russian Federation.

Author details

B. Maňkovská[1*], M. V. Frontasyeva[2] and T. T. Ostrovnaya[2]

*Address all correspondence to: bmankov@stonline.sk

1 Institute of Landscape Ecology, Slovak Academy of Sciences, Bratislava, Slovakia

2 Frank Laboratory of Neutron Physics, JINR, Dubna, Russian Federation

References

[1] Markert, B.A., Breure, A.M., Zechmeister, H.G., 2004: *Bioindicators and Biomonitors. Principles, Concepts and Applications*. 2nd Edition, 997 pp. Amsterdam – Tokyo: Elsevier.

[2] Maňkovská, B., 1995: Mapping of forest environment load by selected elements through the leaf analyses. *Ecology* (Bratislava) 14, 2, 205-213.

[3] Maňkovská, B., 1996: Geochemický atlas Slovenska - Lesná biomasa. Geologická služba Bratislava, ISBN 80-85314-51-7, 87 pp.

[4] Rühling, A., Tyler, G., 1970: Sorption and retention of heavy metals in the woodland moss *Hylocomium splendens* (Hedw.) *Br et Sch Oikos*, 21(1), 92–97.

[5] Ross, H.B., 1990: On the use of mosses (*Hylocomium splendens* and *Pleurozium schreberi*) for estimating atmospheric trace metal deposition. *Water Air Soil Poll*, 50(1-2), 63–76.

[6] Berg, T., Steinnes, E., 1997: The use of mosses (*Hylocomium splendens* and *Pleurozium schreberi*) as biomonitors of heavy metal deposition from: From relative to absolute deposition values. *Environ Poll*, 98(1), 61–71.

[7] Frontasyeva, M.V., 2011: Neutron activation analysis for the Life Sciences. A review: "Physics of Particles and Nuclei", 42, 2, p. 332-378 (in English). http://www.springerlink.com/content/f836723234434m27

[8] Maňkovská, B., Oszlányi, J., 2010: Concentration of 45 elements in moss and their temporal and spatial trends in Slovakia (1990-2005). In *Landscape Ecology - Methods, Applications and Interdisciplinary Approach*. Bratislava. Institute of Landscape Ecology Slovak Academy of Sciences, p. 341-351. ISBN 978-80-89325-16-0.

[9] Schröder, W., Pesch, R., Englert, C., Harmens, H., Suchara, I., Zechmeister, H.G., Thöni, L., Maňkovská, B., Jeran, Z., Grodzinska, K., Alber, R., 2008: Metal accumulation in mosses across national boundaries: Uncovering and ranking causes of spatial variation. *Environ Pollut*, 151: 377–388.

[10] Harmens, H., Frontasyeva, M., Maňkovská, B., et al., 2010: Mosses as biomonitors of atmospheric heavy metal deposition: spatial and temporal trends in Europe. *Environ Poll*, 158: 3144-3156.

[11] Suchara I., Florek M., Godzik B., Maňkovská B., Rabnecz G., Sucharova J., Tuba Z., Kapusta P., 2007: Mapping of Main Sources of Pollutants and their Transport in Visegrad Space. Silvia Taroucy Institute for Landscape and Ornamental Gardening Průhonice, CZ, ISBN 978-80-85116-55-7.

[12] Harmens, H., Ilyin, I., Mills, G., Aboal, J.R., Alber, R., Blum, O., Coskun, M., De Temmerman, L., Fernandés, J.Á., Figueira, R., Frontasyeva, M.V., Godzik, B., Goltsova, N., Jeran, Z., Korzekwa, S., Kubin, E., Kvietkus, K., Leblond, S., Liiv, S., Magnússon, S.H., Maňkovská, B., Nikodemus, S., Pesch, R., Pikolainen, J., Radnovič, D., Rühling, A., Santamaria, J.M., Schröder, W., Spiric, Z., Stafilov, T., Steinnes, E., Suchara, I., Tabors, G., Thöni. L., Turcsányi, G., Yurukova, L., Zechmeister, H.G., 2012: Country-specific correlations across Europe between modelled atmospheric cadmium and lead deposition and concentrations in mosses. *Environ Poll*, 166, 1–9.

[13] Steinnes, E., Berg, T., Uggerud, H., Vadset, M., 2007: Atmospheric deposition of heavy metals in Norway (in Norwegian). Nation-wide survey in 2005. State Program for Pollution Monitoring, Report 980/2007. Norwegian State Pollution Control Authority, Oslo 2007, 36 pp.

Chapter 13

Ambient Level of NO_x and NO_y as Indicators of Photochemical Activity in an Urban Center

Alberto Mendoza and Edson R. Carrillo

Additional information is available at the end of the chapter

http://dx.doi.org/10.5772/59752

1. Introduction

Atmospheric pollution is considered a severe problem, especially in large urban areas where anthropogenic emissions (e.g., emissions from domestic, industrial, and transportation activities, as well as from other productive sectors) mix with biogenic emissions (i.e., emissions with natural origins). Anthropogenic emissions include gas-phase primary air pollutants such as nitric oxide (NO), nitrogen dioxide (NO_2), carbon monoxide (CO), volatile organic compounds (VOCs), and sulfur dioxide (SO_2). Although these pollutants can produce harmful health effects, their ability to react as precursors of secondary air pollutants is one of their most relevant characteristics.

Ozone (O_3) is produced as a secondary pollutant by photochemical reactions occurring between nitrogen oxides (NO_x, where $NO_x = NO + NO_2$) and VOCs in the presence of sunlight [1, 2]. For a long time, interest in O_3 focused on its direct health effects as a major constituent of photochemical air pollution [3, 4] and its impacts on vegetation [5, 6]. However, O_3 also affects the energy budget of the atmosphere; thus, it has become part of a family of species referred to as short-lived climate pollutants (SLCP) [7].

Photochemical activity occurs in both natural and human-altered environments. Theoretically, in the presence of only NO_x in the troposphere, O_3 generation could be described by a simple mechanism known as the NO_x-O_3 photostationary state, summarized by the following reactions:

$$NO_2 + hv \rightarrow NO + O_3 \tag{1}$$

$$O^{\cdot} + O_2 + M \rightarrow O_3 + M \tag{2}$$

$$O_3 + NO \rightarrow NO_2 + O_2, \tag{3}$$

where hv represents a photon (sunlight energy), $O^{\bullet}$ denotes an oxygen free radical (an oxygen atom with an unpaired electron), and M is known as a third body (molecule) that acts as an energy sink. In the atmosphere, this third body is typically N_2 or O_2. Under this chemical reaction scheme, the net O_3 production is zero, and according to the photostationary state equation, the concentration of O_3 can be determined based on the concentration of NO_x and the amount of solar radiation:

$$\left[O_3\right]_{ps} = \left(\frac{j_{NO_2}}{k_{R3}}\right)\frac{\left[NO_2\right]}{\left[NO\right]}, \tag{4}$$

where k_{R3} is the kinetic reaction rate constant for reaction (3) and j_{NO2} is the photolysis rate of NO_2. As a demonstration, if a 10 parts per billion (ppb) value is considered for the term j_{NO2}/k_{R3}, the O_3 concentration would be 27 ppb, with an initial value of 100 ppb of NO [8]. An O_3 concentration ranging between 10–40 ppb is typical of rural areas around the globe [9].

When VOCs are added to the mixture of species present in the troposphere, the observed O_3 levels are higher than those predicted by the photostationary state formulation. In this case, interactions in the O_3-NO_x-VOC system are initialized by the hydroxyl ($HO^{\bullet}$) radical, through the photolysis of O_3:

$$O_3 + hv \rightarrow O_2 + O\left(^1D\right) \tag{5}$$

$$O\left(^1D\right) + H_2O \rightarrow 2HO^{\cdot}, \tag{6}$$

where $O(^1D)$ represents an excited singlet oxygen atom. The oxidation of the VOCs (represented here as RH, where R is an organic functional group, e.g., an alkyl group) continues in the presence of $HO^{\bullet}$ and NO_x to produce more O_3 and a variety of nitrogen-containing species. The process can be represented by the following generalized reactions [8]:

$$RH + HO^{\cdot} + O_2 \rightarrow RO_2^{\cdot} + H_2O. \tag{7}$$

The alkyl peroxide radical ($RO_2^{\bullet}$) reacts with NO to form aldehydes (R'CHO) and peroxide radicals ($HO_2^{\bullet}$):

http://dx.doi.org/10.5772/59752

$$RO_2^{\cdot} + NO + O_2 \rightarrow NO_2 + R'CHO + HO_2^{\cdot} \tag{8}$$

Reaction 8 is a relevant process because it represents an alternate route for the production of NO_2 from NO, but without destroying O_3 (in contrast to reaction [3]). A similar process occurs when the aldehydes continue reacting to eventually form acyl peroxy radicals ($R'C(O)O_2^{\bullet}$) that undergo a similar fate as the $RO_2^{\bullet}$ radicals in reaction (8). Thus, O_3 starts to accumulate in the system, as it is no longer destroyed by reaction (3). It continues to be produced by reactions (1) and (2) as NO_2 and NO cycle through this set of reactions. In addition, the $RO_2^{\bullet}$ radicals can produce nitrates when they react with NO_2:

$$RO_2^{\cdot} + NO + M \rightarrow RONO_2 + M \tag{9}$$

$$HO^{\cdot} + NO_2 + M \rightarrow HNO_3 + M. \tag{10}$$

This description is not exhaustive, as we have focused our attention on the main reactions of O_3 production by NO_x and VOCs. As indicated, NO_x acts as catalyst in these reactions, while the VOCs continue to undergo oxidation until they are converted to CO_2. In parallel, a variety of different inorganic (e.g., HNO_3) and organic (e.g., $RONO_2$) nitrogen-containing species are also produced. Some of these substances act as reservoirs of NO_x which are released to the reacting mixture upon decomposition (e.g., peroxy acetyl nitrate or PAN, an organic nitrate), and others act as sinks (most notably, HNO_3). The sum of NO_x and these additional inorganic and organic nitrates is referred to as total reactive nitrogen or total odd nitrogen oxides (NO_y).

In general, the O_3-NO_x-VOC system increases the NO_x available to react through reactions (1)-(2) (increasing O_3), thereby producing a complex mixture of partially oxidized VOCs that mixes with freshly emitted VOCs and different oxidized nitrogen species (HNO_3, HNO_2, NO_3, PAN, etc.). The study of the dynamics of O_3 production can be quite challenging, given the complexity of the chemical mixtures and their nonlinear response to emission changes and meteorological conditions. For this reason, comprehensive air quality models have been devised to study the complex physical and chemical processes that participate in gas-phase production of O_3 and other air pollutants [10].

Observational-based approaches have proven to be valuable for describing the conditions and regimes that foster air pollution. Some observational approaches use considerable amounts of data that can be obtained through networks of routine air quality and meteorological monitoring stations, and conclusions are then inferred based on the statistical analysis of these data [11-17]. The analyses of these databases can be enhanced if information is also available for additional indicator species that are typically not routinely monitored, such as NO_y.

In this study, we have analyzed data gathered by the routine air quality monitoring stations of the Monterrey Metropolitan Area (MMA; 25° 40′ N, 100° 18′ W). Monterrey is the third most populated urban center in Mexico (4.1 million inhabitants), second in size in the country in terms of industrial infrastructure, and one of the cities with the worst air quality problems in Mexico [18]. The MMA (see Figure 1) has been in violation of the 1-hr Mexican Air Quality Standard for O_3 (0.11 ppm) since the establishment of its routine air quality monitoring system in 1993. For example, the 1-hr O_3 standard was exceeded on 48 different days in 2011 [19]. Peak O_3 concentrations can reach 170 ppb and typically occur at the downtown or western air quality stations [20]. Some studies report that the MMA is the fifth most polluted Latin American city in terms of O_3 [21]. In addition to the large number of industrial facilities located inside or nearby the metropolitan area that contribute to the poor air quality, emissions from several gas-fired electric utilities and one of the six refineries that operate in the country (located less than 40 km to the east of downtown Monterrey) add to the anthropogenic burden imposed on the airshed. Despite the importance of the contribution of area and point sources to the total emissions inventory, the mobile sources represent approximately 75% of the total anthropogenic emissions released in the MMA [20]. This is a result of a relatively high proportion of vehicles per inhabitant registered in this urban center (approximately one vehicle for every two inhabitants).

2. Methods: Database and analysis tools

The study used data collected by the *Sistema Integral de Monitoreo Ambiental* (SIMA; Integrated Environmental Monitoring System) of the MMA. At the time of the study, valid data from six operational routine air quality stations were available: Downtown, Southeast, Southwest, Northeast, Northwest, and North (see Figure 1). Data archived for the period of August 2012 to August 2013 were retrieved. The database contained hourly-average concentrations of CO, NO, NO_2, and O_3, as well as meteorological parameters (relative humidity, atmospheric pressure, dry bulb temperature, solar radiation [SR], wind speed [WS], and wind direction [WD]). These data go through a quality assurance/quality control process defined by SIMA in compliance with international standards. Additional air quality parameters monitored by the stations (mainly, SO_2 and particulate matter with aerodynamic diameter less than or equal to 10 microns [PM_{10}] and less than or equal to 2.5 microns [$PM_{2.5}$]) were not used in this study, as the focus was on the relationship between O_3 and NO_x.

In addition to the above chemical parameters, NO_y was monitored in the Downtown station from August 2012 to August 2013, using a NO-NO_y (Thermo Scientific, Model 42i-Y) chemiluminescence continuous sampling device. NO_y is a chemical parameter that is not routinely measured by Mexican air quality stations, as it is not considered a criteria pollutant. The decision to deploy the NO-NO_y instrument at the location of the Downtown station was based on spatial homogeneity studies performed in the past for the MMA [22], which indicate that pollutant levels observed in the Downtown station are representative of the MMA, with the exception of the Southeast region. The data collected by the NO-NO_y device went through a validation process similar to that routinely conducted by SIMA on its own collected data. The main conditions to reject data included the following red flags: obstructed capillary tubing, low flow in the inlet, and concentrations outside the measurement range (i.e., above 200 ppbv). The data were then consolidated and analyzed on a daily basis, a monthly basis, and by seasons (Summer and Fall 2012, and Winter, Spring, and Summer 2013). Summer 2012 included only data from August and September of 2012; Summer 2013 did not include data for September 2013.

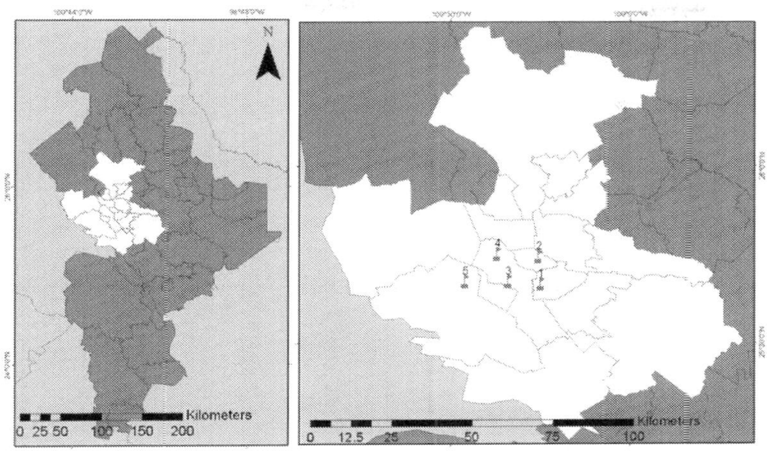

Figure 1. Municipalities that comprise the Monterrey Metropolitan Area (left panel) and location of the air quality stations monitoring stations used in this study (right panel): 1 Southeast, 2 Northeast, 3 Downtown, 4 Northwest, 5 Southwest; data from the North station was not used

Once the data were consolidated, an exploratory analysis of the data set was conducted to derive descriptive statistics. In addition, correlation analysis was conducted to explore relations between the chemical and meteorological parameters, with emphasis on the relationships among O_3, NO_x, and NO_y. This analysis was complemented by the use of polar plots and wind roses to determine the relationship between high pollutant levels and transport conditions (wind speed and direction).

Grouping techniques were then applied for further exploration of the data. Two methods were used: Principal Components Analysis (PCA) and Analysis of Variance (ANOVA). PCA is helpful in reducing the dimensionality of the dataset, and ANOVA can identify differences among data groups. Thus, PCA was used to identify the most important variables in the dataset which in turn merit further exploration. With the ANOVA, we attempted to determine differences in the weekday and weekend conditions that resulted in high O_3 levels. All statistical analyses were conducted using Minitab® 16, and plots were constructed using the R programming language.

3. Results

3.1. Descriptive statistics

Table 1 summarizes the descriptive statistics of the chemical and meteorological parameters retrieved from the Downtown air quality monitoring station. The values presented are based on the daily means for the August 2012-August 2013 period. The climatology of the MMA is portrayed in these results: a hot semi-arid region characterized by extreme weather conditions [23]. The long-term average temperature of Monterrey is around 23°C, with temperatures that can reach 40°C in summer and below 0°C in winter. Average humidity is 62%; the rainy season is between August and October, directly related to the occurrences of hurricanes in the Gulf of Mexico (the average mean annual precipitation is around 600 mm). Due to its geographic location, the MMA is influenced by anti-cyclonic systems from the Gulf of Mexico [24], which in some instances result in high atmospheric stability in the region, thereby inhibiting the vertical mixing of pollutants.

Species/Parameters	Mean ± Std. Dev.	Median	Maximum	Minimum
O_3 (ppb)	25.7 ± 9.0	24.4	50.7	2.6
NO_2 (ppb)	16.3 ± 5.9	15.3	39.2	3.3
NO (ppb)	6.5 ± 4.7	5.3	34.4	0.6
NO_y (ppb)	40.0 ± 20.3	35.2	188.2	12.1
CO (ppm)	0.78 ± 0.26	0.75	1.78	0.21
Solar radiation (kW/m^2)	0.24 ± 0.12	0.24	0.70	0.00
Temperature (°C)	24.4 ± 5.2	25.8	32.5	9.5
Wind speed (km/h)	1.84 ± 0.47	1.84	3.21	0.71
Relative Humidity (%)	62.9 ± 16.8	64.17	98.9	13.5

Table 1. Descriptive statistics for parameters monitored at the Downtown air quality station of the MMA (annual means, medians, maximum values, and minimum values of the daily averages)

Details of the monthly variation of the daily averages of the chemical and meteorological parameters are depicted in Figure 2. O_3 levels tend to be the highest during springtime, as well as during late summer and early fall. These periods correspond to the times when temperature and SR are high and wet precipitation is low, i.e., conditions that foster photochemical reactions. Another relevant feature presented in Figure 2 involves the inverse relationship observed between O_3 and NO_x. That is, during the cold months, when SR is low, NO_x levels tend to be high due to inhibition of the reactions between VOCs and NO_x that promote the conversion of NO to NO_2, as well as suppression of the photo-dissociation of NO_2 to produce O_3. We need to bear in mind that NO_x emissions are mainly in the form of NO. Thus, an accumulation of NO and NO_2 is observed in winter. At other times, temperatures and SR increase, promoting photochemical reactions. A meteorological component influences the observed NO_x concentrations: shallow mixing heights and low wind speeds during winter. The O_3 seasonal boxplots indicate that even though fall is the second most important season with respect to the frequency of peak O_3 levels, it is also the time when the average O_3 is the lowest (Figure 3). This can be explained by the incidence of rain events associated with this season.

http://dx.doi.org/10.5772/59752

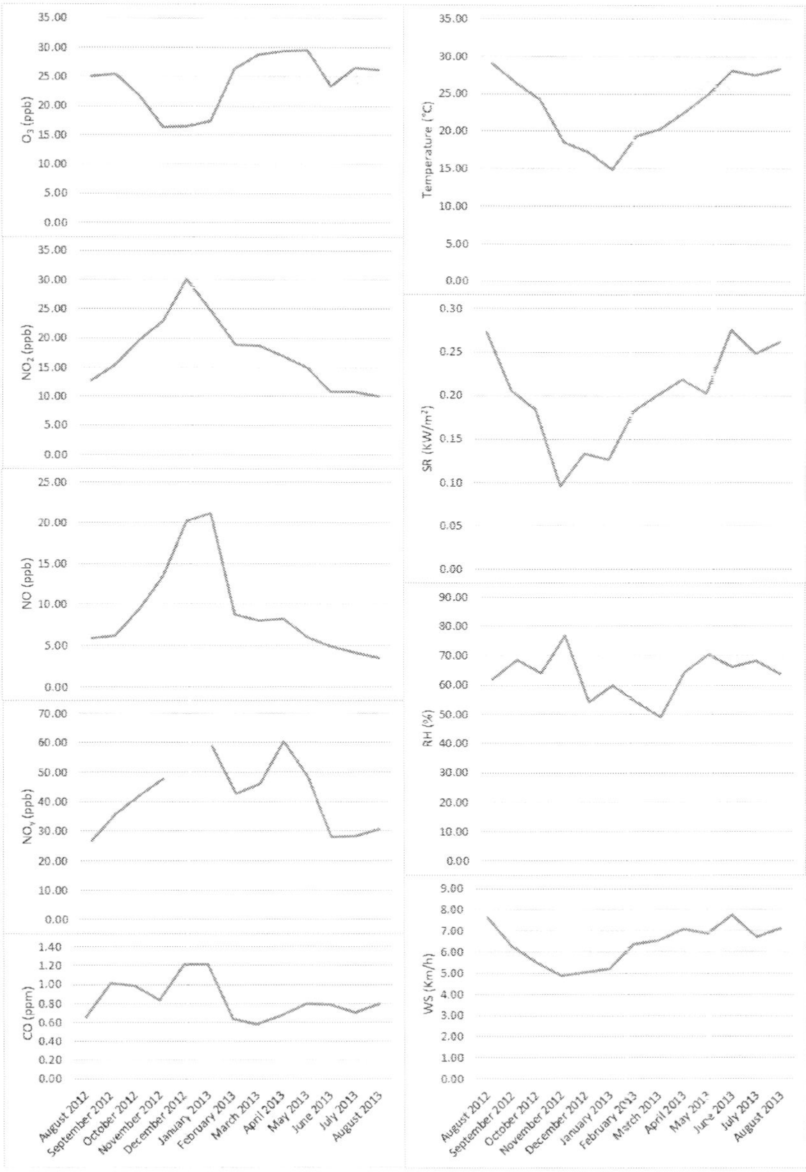

Figure 2. Time series plots of monthly averages of the mean daily values of chemical and meteorological parameters observed at the Downtown station

3.2. Spatial analysis

The exploratory analysis was complemented by the construction of bivariate polar plots of NO_x (see Figure 4) and wind roses (see Figure 5) for five of the six air quality monitoring stations, covering the main sub-regions of the MMA. The five selected stations correspond to typical upwind locations (Northeast and Southeast), Downtown Monterrey, and typical downwind locations (Northwest and Southwest), as illustrated in Figure 5. Seasonal variation is also presented in these plots.

The polar plots represent NO_x concentration as a function of wind speed and direction, and they help identify the possible occurrence/prevalence of horizontal transport conditions that lead to high NO_x events. Further away from the center of the plot, the wind speed is higher. The form of the polar plot can provide an indication of the wind direction and speed frequency; however, this information is better observed through the corresponding wind roses.

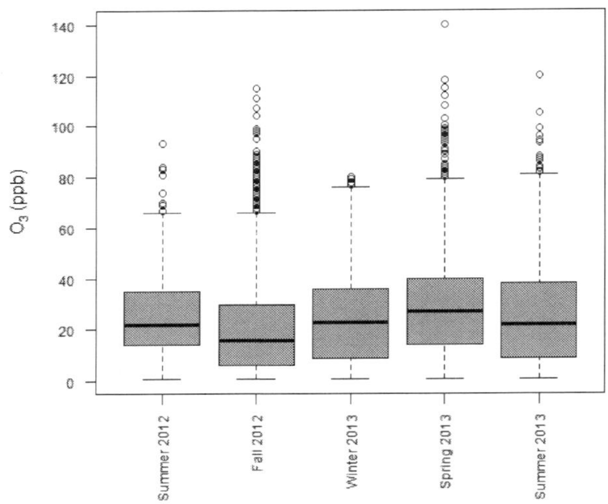

Figure 3. O_3 boxplots by seasonal periods for 1-hr average values reported at the Downtown station

In general, Figure 4 indicates higher transport of NO_x from the northern stations (this can be catalogued as "fresh" NO_x), which then is consumed as the air masses traverse the urban core. The transport patterns appear to be similar in all seasons. Thus, the change in NO_x by season appears to be more strongly linked to photochemical conditions (temperature and SR) than to wind patterns.

Using the same analysis as was used for NO_x, Figure 6 depicts the behavior of O_3 as a function of wind speed and direction exclusively at the Downtown station. In all cases, the higher concentrations begin to occur when the wind velocity was above 2 m/s, indicating the influence of transport and aging of air masses on the observed O_3 concentrations and weak O_3 production under stagnant conditions. Transport influences the levels of O_3; consequently, the highest O_3 levels correlate with the prevailing wind directions (wind blowing from the Northeast-East-Southeast and from the North-Northwest-West). The East-to-West air flow channels air masses through the long axis of the MMA, adding more precursors to the chemically aged mixtures as they traverse the urban core, thereby promoting O_3 production.

The frequency of high O_3 concentrations related to a specific wind direction cannot be inferred from the polar plots, however. That analysis was conducted by constructing boxplots for 1-hr average O_3 concentrations for the three main wind categories (classified based on direction from which the wind was blowing) and segregated by season (see Figure 7). The year around the peak O_3 concentrations and the maximum 1-hr average concentrations occur when the wind blows from the Northeast-East-Southeast. O_3 levels, on average, are lower when the wind blows from the North-Northwest-West, and they are the lowest when winds blow from the South-Southeast (with the exception of the summer, when this direction provides higher average concentrations than the North-Northwest-West direction).

Comparison of the O_3 levels presented by the polar plots (Figure 6) for spring and fall indicates that the influence of transport is less important during fall than in spring, as wind speeds tend to be lower; high O_3 events are mainly related to events when the speed is in the range of 2.0-3.5 m/s. In spring, O_3 levels can remain high at wind speeds above 4 m/s.

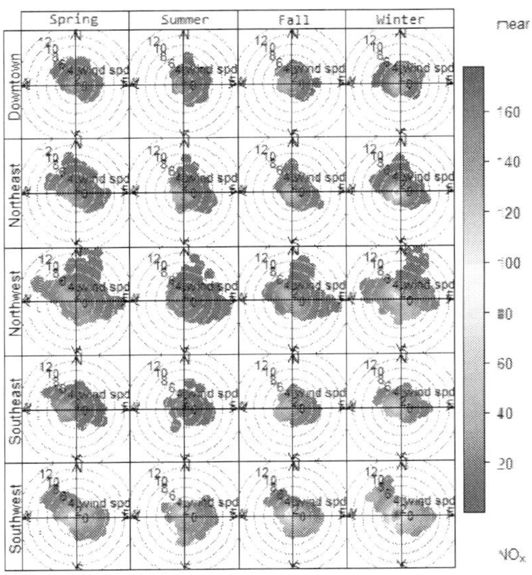

Figure 4. Polar plot for 1-h average NO_x concentrations at five different monitoring stations within the MMA (the radial dimension is an indication of increasing wind speed [m/s])

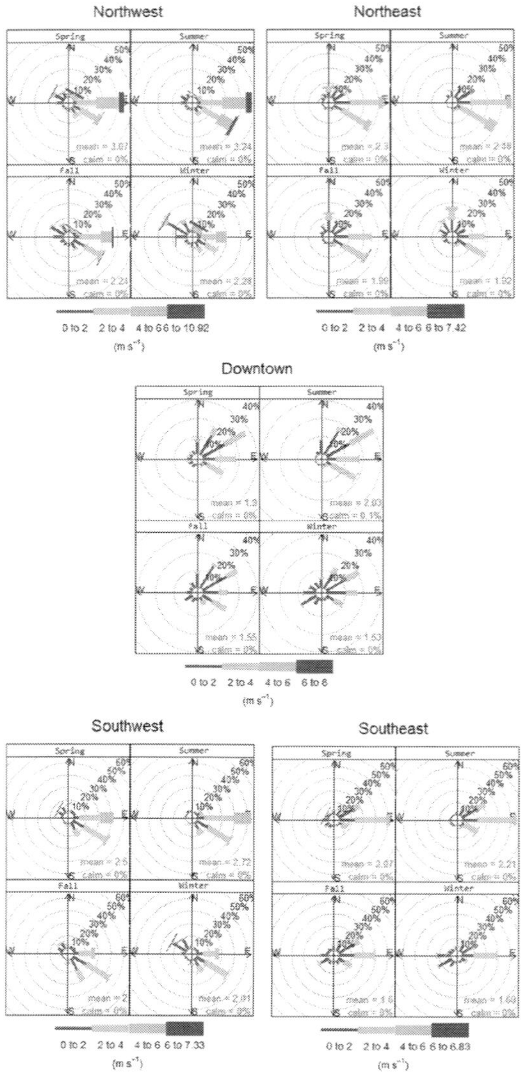

Figure 5. Wind roses at five different monitoring stations within the MMA

http://dx.doi.org/10.5772/59752

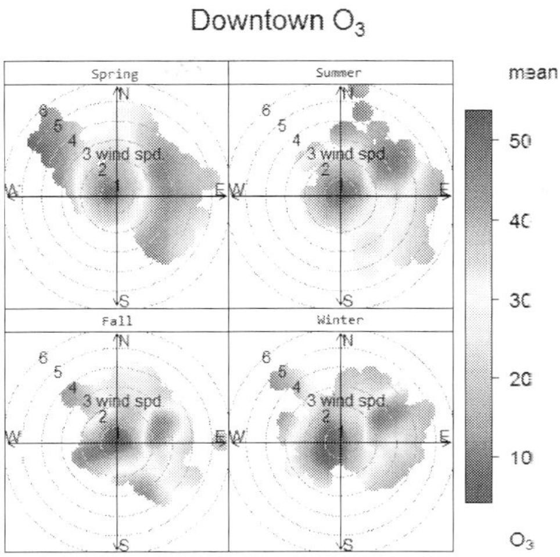

Figure 6. Polar plot for 1-h average O_3 concentrations at five different monitoring stations within the MMA (the radial dimension is an indication of increasing wind speed [m/s])

3.3. Regression analysis

Figure 2 gives an initial indication of a direct relationship between NO and NO_2 levels and an inverse relationship between NO_x and O_3 levels. This was explored further by means of scatter plots (not presented for brevity), which were used to derive linear regression expressions among NO_x constituents and between NO_x and O_3:

$$[NO_x] = 1.5\,[NO] + 13.0 \tag{11}$$

$$[O_3] = -0.37\,[NO_x] + 34 \tag{12}$$

where $[i]$ represents the molar concentration of species i. Here, the 1-hr average concentrations were used to derive equations (11) and (12). These expressions corroborate prior findings. The relevance of equation (12) is that it provides a measure of the expected, first order change in O_3 provided by a change in NO_x concentration (which can occur from a change in NO emissions). The results indicate that O_3 would, in fact, increase as NO_x tends to be reduced. This counterintuitive finding is a well-known result seen in urban areas dominated by a VOC-sensitive regime [25]. Other studies that have used complex photochemical air quality models have reached similar conclusions. Downtown Monterrey appears to be dominated by a VOC-sensitive regime in which VOC controls would work better for reducing O_3 levels [20].

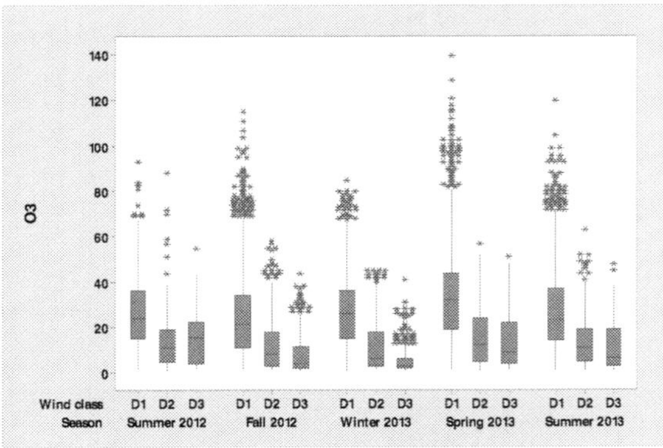

Figure 7. Boxplots of 1-hr averaged O_3 concentrations observed at the Downtown station, classified by prevailing wind direction: D1 (wind blowing from 20 to 120°), D2 (wind blowing from 270 to 360°), and D3 (wind blowing from 180 to 150°)

3.4. Analysis of NO_y levels

Up to this point, our analysis has focused on describing the relationship between NO_x and O_3 production based on the meteorological conditions that foster photochemical reactions. The generalized chemical mechanism that describes this relationship indicates that other reaction products can be of interest in describing the dynamics of O_3 production. In particular, the levels of NO_y provide additional information and serve as an indicator of the degree to which the air masses have been subjected to photochemical processing (aging), thereby providing information on O_3 production.

Table 2 presents the results of the correlation analysis of different chemical and meteorological parameters with NO_y. A correlation coefficient (R^2) of NO_2 with NO_y above 0.78 is noted for months in which O_3 production is not the highest, and between 0.40 and 0.68 for the months with the highest O_3 production. In general, a high percentage of NO_y corresponds to NO_x (Figure 2); thus, the response of one clearly depends on the other.

The difference between NO_y and NO_x (known as NO_z) includes the inorganic and organic nitrates derived from photochemical processing of the NO_x-VOCs mixtures. The typical composition of NO_z is approximately 50%-55% HNO_3, 35%-40% PAN, 1%-5% HNO_2, and a marginal contribution of other nitrates [26, 27]. During the cold months, the production of NO_z can be inhibited by meteorological conditions that do not foster photochemical reactions. Thus, the relative contribution of NO_x to NO_y is the highest, and, therefore, the responses of both species are expected to show high correlation. When conditions favor O_3 production, more NO_z is produced, thereby limiting the contribution of NO_x to the total NO_y. Consequently, the level of correlation drops as the nonlinear photochemistry and sinks of NO_y increase in importance. Of interest is the fact that during spring, the season with the highest peak O_3 1-hr concentrations, the correlation of NO_x species with NO_y is the lowest. A similar response is obtained between CO and NO_y, which can be explained by the typical strong correlation between CO and NO_x emissions, particularly in urban areas that are heavily influenced by mobile-source emissions. With respect to meteorological variables, the highest correlation was with wind speed, and, as in the case of NO_x and CO, the correlation was the lowest during spring.

	Summer 2012	Fall 2012	Winter 2013	Spring 2013	Summer 2013
NO_2 (ppbv)	0.811 (< 0.001)	0.676 (< 0.001)	0.819 (< 0.001)	0.411 (< 0.001)	0.783 (< 0.001)
NO (ppbv)	0.335 (< 0.001)	0.610 (< 0.001)	0.661 (< 0.001)	0.275 (< 0.001)	0.615 (< 0.001)
O_3 (ppbv)	0.031 (< 0.001)	0.106 (< 0.001)	0.151 (< 0.001)	0.065 (< 0.001)	0.070 (< 0.001)
CO (ppmv)	0.764 (< 0.001)	0.420 (< 0.001)	0.664 (< 0.001)	0.227 (< 0.001)	0.712 (< 0.001)
Solar radiation (kW/m^2)	0.011 (< 0.001)	0.001 (0.639)	0.004 (0.052)	0.001 (0.165)	0.009 (< 0.001)
Temperature (°C)	0.009 (0.048)	0.024 (< 0.001)	0.022 (< 0.001)	0.005 (0.002)	0.063 (< 0.001)
Wind speed (km/h)	0.252 (< 0.001)	0.247 (< 0.001)	0.203 (< 0.001)	0.112 (< 0.001)	0.225 (< 0.001)

Table 2. Correlation coefficients (R^2) of parameters monitored by the SIMA downtown station with NO_y (in parenthesis p-values at an α of 0.05)

The results of the correlation analysis suggest a relationship between NO_x and NO_y, but a correlation between NO_y and O_3 is less evident. Figure 2 shows two peaks in the annual cycle of the NO_y monthly average concentration (constructed from daily NO_y average concentrations). One peak occurs during the cold months and is associated with the high levels of NO_x (low NO_z component), while the second peak appears during spring, which is the period with highest photochemical activity (highest O_3 levels); thus, as NO_x is consumed, and its levels decrease and NO_y levels are high, NO_z must be contributing heavily to NO_y. The peaks of NO_y are associated with different conditions, so we looked in greater detail into the seasonal levels of NO_y. Figure 8 depicts the 1-hr NO_y concentrations by season. The highest levels (both of peak and season-average concentrations) occur in winter, followed by fall and spring (when O_3 is the highest), with summer registering the lowest levels. The O_3 levels presented in Figure 3 do not appear to track the NO_y levels.

We then attempted to determine a clearer relationship between O_3 and NO_y by conducting a Principal Components Analysis (PCA) (see Table 3). The first principal component (PC1) has to do with O_3 precursors in the form of primary emissions or intermediate products (NO_x, NO_y, and CO), in line with the results from the descriptive statistics. The second principal component (PC2) covers photochemical O_3 production (O_3 and solar radiation). The third principal component (PC3) refers to a combined meteorological effect that includes horizontal transport (wind direction) and temperature— an effect that was also observed in the descriptive statistics results. The first two principal components represent almost 60% of the variance.

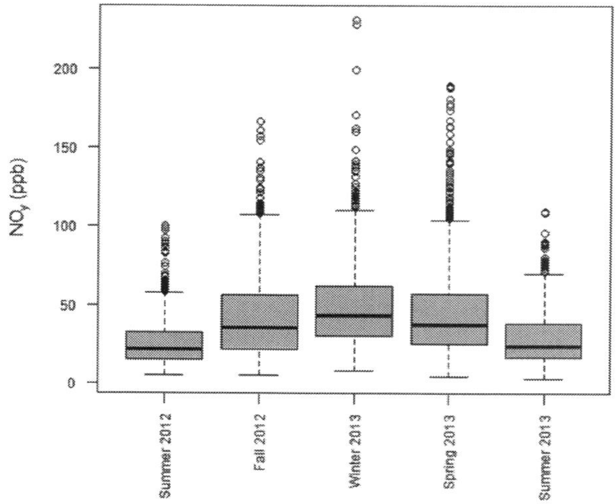

Figure 8. Annual average distribution of NO_y concentrations

Component	PC1	PC2	PC3
NO_y	**0.481**	0.055	–0.009
NO_2	**0.387**	0.087	–0.123
NO	**0.375**	0.029	0.151
O_3	–0.226	**0.447**	0.075
CO	**0.400**	0.110	0.081
Solar radiation	–0.090	**0.497**	0.031
Temperature	–0.197	0.370	**0.411**
Wind velocity	–0.288	0.276	0.084
Wind speed	0.175	0.339	**0.794**
Accumulated variance (%)	0.395	0.595	0.672

Table 3. Principal components analysis results

Given these results, we conducted an additional assessment on the effects of meteorological parameters on O_3 and NO_y levels, which showed the effects of horizontal transport. O_3, temperature, wind speed, and wind direction data were particularly correlated (see Table 4). In addition, complementary polar plots were constructed for NO_y (Figure 9). The highest correlation between O_3 and wind speed occurred during the winter of 2013. In this case, the slope of the linear regression was positive (i.e., as the wind speed increases, O_3 also increases). In a similar way, the highest correlation between O_3 and wind direction occurred during the winter of 2013. This indicates that transport is a major factor responsible for the average O_3 levels observed in the winter in the MMA. The polar plot for NO_y confirms that high levels of NO_y during winter (which are, as discussed, mostly NO_x) are associated with low wind speeds. This relationship between NOy and wind speed would indicate that NO_x is not being processed and translated to high local production of O_3 in the winter.

	Wind speed				
	Summer 2012	Fall 2012	Winter 2013	Spring 2013	Summer 2013
R^2	0.212	0.271	0.350	0.025	0.231
Slope	2.348	3.625	3.342	1.164	2.986
Intercept	6.836	0.937	3.319	35.560	4.850
p-value	< 0.001	< 0.001	< 0.001	< 0.001	< 0.001
	Winddirection				
	Summer 2012	Fall 2012	Winter 2013	Spring 2013	Summer 2013
R^2	0.001	0.022	0.216	0.002	0.010
Slope	–0.005	–0.031	–0.088	0.014	–0.027
Intercept	25.432	24.893	35.313	44.662	28.428
p-value	< 0.001	< 0.001	< 0.001	< 0.001	< 0.001
	Temperature				
	Summer 2012	Fall 2012	Winter 2013	Spring 2013	Summer 2013
R^2	0.454	0.327	0.341	0.344	0.536
Slope	2.834	1.877	2.017	2.021	3.792
Intercept	–57.407	–21.939	–14.324	–5.969	–81.202
p-value	< 0.001	< 0.001	< 0.001	< 0.001	< 0.001

Table 4. O_3 correlations with meteorological parameters (p-values at an α of 0.05)

An additional way to establish the influence of transport on the levels of air pollutants is to analyze the CO-to-NO_y ratio. Values close to 10 are indicative of an influence of local sources, and ratios larger than 100 indicate the influence of regional (remote) sources [11]. Figure 10 depicts the CO-to-NO_y ratios observed at the Downtown station as a function of the season. The average ratio tends to be the inverse with respect to the behavior of the average levels of NO_y, although one main difference exists: the lowest ratios are observed during winter but also during spring, intermediate values are observed during the fall, and the highest average values are registered during the summer. In a few cases, the CO-to-NO_y ratio reached values of 100 or more, and the 95th percentile, in general, was below 56. Thus, air pollution in the MMA during winter appears to be heavily influenced by low wind conditions (Figure 2 indicates the lowest average wind speeds during this season), and poor photochemical activity limits the levels of NO_z and O_3. The O_3 levels that are registered are characteristic of background amounts that are transported to the urban center when wind speeds are sufficiently strong. The spring level is also heavily influenced by local emissions, but meteorological conditions do favor photochemical activity, resulting in production of the highest levels of O_3. Fall would seem to be influenced partly by local emissions and partly by transport. Finally, summer is characterized by the influence of more regional (and aged) emissions. During the first two months of the summer season, the ambient conditions tend to be characterized by high temperatures and deep planetary boundary layers that promote vertical mixing. In addition, conditions favor the mixing of these vertical columns with pollutants that undergo long-range transport.

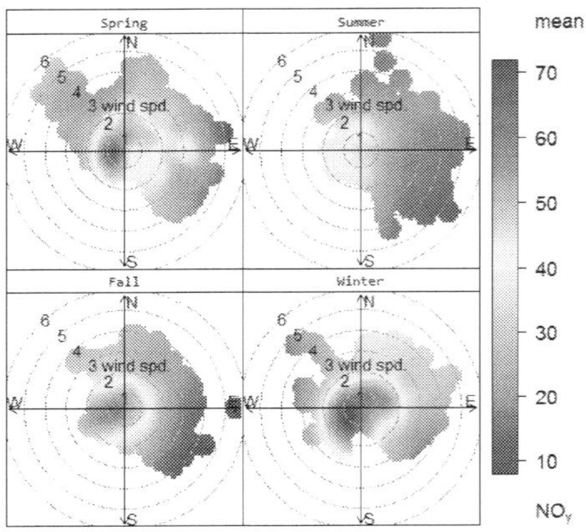

Figure 9. Polar plot for NO_y at the Downtown monitoring station

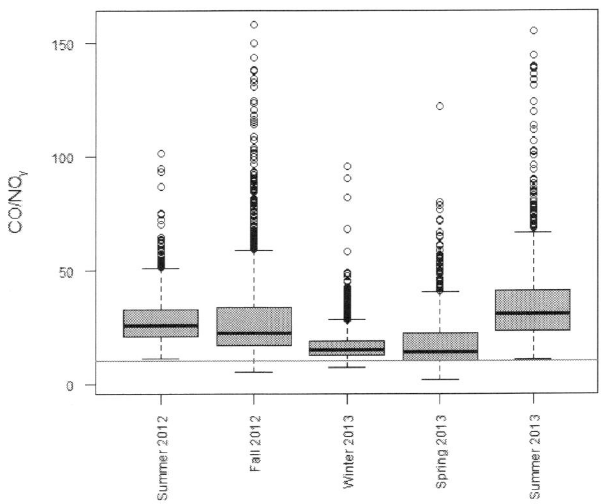

Figure 10. Seasonal variation of CO-to-NO_y ratios (the red line represents a value of 10 for the ratio)

3.5. Weekend-weekday effect

A final aspect of interest in this study was to define if a weekend-weekday effect could be established based on NO_x and NO_y readings. Previous studies have demonstrated that O_3 levels in urban centers can be higher during weekends [28, 29], a condition that might be thought of as counterintuitive, as emissions from anthropogenic sources tend to be lower during weekends. An explanation for the occurrence of this weekend-weekday effect can be obtained by analyzing the photochemical mechanism for the production of O_3. Urban centers tend to be VOC-sensitive; that is, NO_x is in excess and, thus, processes that remove NO_2 from the reactive mixtures are promoted. For example, NO_2 efficiently scavenges $HO^•$ radicals through reaction 9, producing HNO_3, which is a direct sink for NO_x. This limits the reaction of $HO^•$ radicals with VOCs, which, in turn, controls the production pathways for O_3. During weekends, the ratio at which VOCs and NO_x are emitted can change, preferentially reducing the emissions of NO_x. If this happens, less NO_x is available in the atmosphere, reaction 9 becomes less important, more $HO^•$ radicals become available to react with the VOCs, and, thus, O_3 tends to increase. Figure 11 illustrates the distribution of NO_y concentrations by day of the week. At a first glance, NO_y values tend to be quite constant throughout the day. However, Tuesdays and Wednesdays present the lowest peak values, while average levels on Sunday tend to be lower than during the rest of the week. If the average NO_x-to-NO_y ratio is examined, values tend to be lower during the weekends (Figure 12). The NO_x-to-NO_y ratio provides information on the level of photochemical processing of the observed air parcels. Ratios closer to 1 would indicate that most of the NO_y is in the form of NO_x, and thus, it could imply the presence of fresh emissions or the transport of unprocessed emissions. As the ratio drops, the relative contribution of NO_z increases, indicating that an aged (photochemically processed) air parcel is being observed.

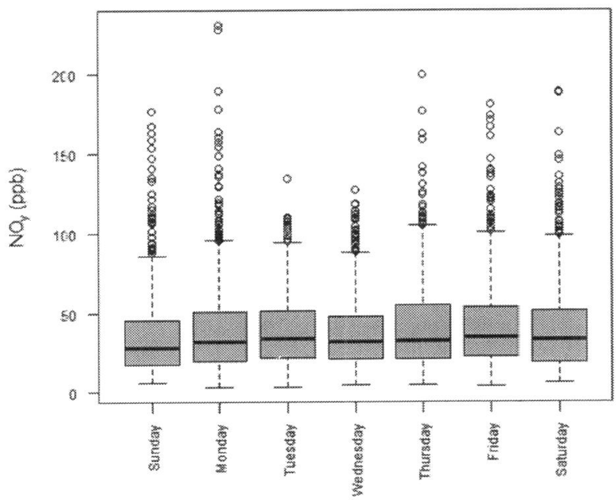

Figure 11. Boxplot for NO_y levels by day of the week

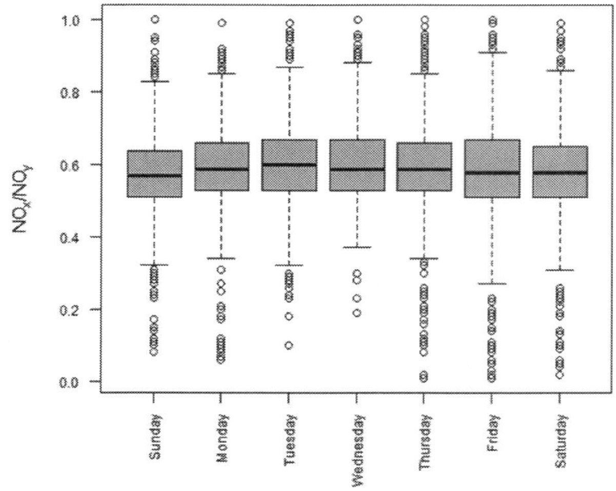

Figure 12. Boxplot for the NO$_x$-to-NO$_y$ ratio by day of the week

Finally, we conducted an Analysis of Variance (ANOVA) test for the NO$_x$-to-NO$_y$ ratio by season (see Table 5). For the hot months (spring and summer), lower ratios were observed during weekends compared to weekdays. In fact, the lowest average ratio was obtained during spring, the season with the highest average and peak O$_3$ concentrations. In contrast, the winter of 2013 showed a marginally higher ratio during the weekends than on the weekdays, but the difference was not statistically significant. This could be the result of an intensive use of fuels during the whole week in response to the low temperatures. For the fall of 2012, this statistical test also indicates no difference between weekdays and weekends. Fall, as the results from previous sections suggest, is influenced by days with high O$_3$ production and days with very low O$_3$ levels as a result of the transition from a period with high SR and temperature to the regional rainy season.

Season	Average ± Standard Dev. (Weekdays)	Average ± Standard Dev. (Weekends)	*p-value*	Difference between Weekdays and Weekends
Summer 2012	0.606 ± 0.139	0.582 ± 0.124	0.033	Yes
Fall 2012	0.671 ± 0.141	0.668 ± 0.107	0.657	No
Winter 2013	0.545 ± 0.086	0.549 ± 0.077	0.504	No
Spring 2013	0.553 ± 0.112	0.496 ± 0.142	< 0.001	Yes
Summer 2013	0.586 ± 0.110	0.561 ± 0.112	< 0.001	Yes

Table 5. Results of the ANOVA test for the NO$_x$-to-NO$_y$ ratio (p-values are at an α of 0.05)

http://dx.doi.org/10.5772/59752

4. Conclusions

Observational-based methods have proven valuable for analysis of the chemical interactions that give rise to high air pollution events in urban areas. In particular, the vast amount of information gathered by networks of air quality stations can provide insight into the production of secondary air pollutants such as O_3, as atmospheric conditions change throughout the year. This analysis can be enhanced if complementary observations are also made of species that are not routinely registered, such as NO_y.

In this study, we presented a statistical analysis performed on the air quality and meteorological data registered by the routine air quality stations of the MMA. The analysis included descriptive statistics, regression analysis, correlation analysis, PCA, and ANOVA, along with the interpretation of bivariate polar plots, wind roses, and boxplots. In addition, ratios of NO_x with NO_y and CO with NO_y provided additional information on the level of chemical processing of the air masses traversing the MMA. When used together, these techniques prove to be complementary, thus providing more robust results.

In the MMA, O_3 registers two distinct annual peaks: one in spring and the other in late summer-early fall. The analysis of meteorological conditions and air pollutant levels indicate that the O_3 concentrations during winter would be characteristic of background conditions that get transported to the urban center when wind speeds are sufficiently strong. O_3 production in winter is small because typical conditions that foster photochemical activity are not present, such as relatively high temperatures and strong solar radiation. During this season, NO_y is composed mainly of locally-emitted NO_x, which corroborates the low photochemical activity. That is, the ambient conditions do not favor the catalytic effect of NO_x to produce O_3. Spring is also heavily influenced by local emissions, but meteorological conditions favor photochemical activity, leading to production of the highest levels of O_3. In this season winds bring background O_3 to the MMA, and precursors traverse the long-axis of the urban core, allowing for chemical processing, injection of fresh emissions, and high photochemical production. Spring registers the highest amount of NO_z, indicating that NO_x is reacting efficiently with VOCs to produce photochemical oxidants, including O_3. Fall would seem to be influenced partly by local emissions and partly by transport. Even though fall has the second highest peak O_3 levels, the average O_3 in this season is relatively low. This can be explained by the fact that the rainy season occurs during this period. Finally, summer would be characterized by the influence of more regional (and aged) emissions. Deep planetary boundary layers are characteristic of this season, allowing mixing and dispersion of air pollutants that result in the lowest NO_y levels of the year. Thus, even though temperature and solar radiation levels could suggest high O_3 production, photochemical activity is limited by transport and mixing effects. Finally, for the hot months (spring and summer), a distinct weekday-weekend effect can be identified, as the NO_x-to-NO_y ratio tends to be higher during weekdays than during weekends. This would indicate changes in NO_x emission rates during the week, which could lead to higher O_3 events during the weekend, in line with the VOC-sensitive condition of the MMA atmosphere that has been suggested by others.

Overall, the analysis of NO_x and NO_y levels with other chemical and meteorological variables as well as the correlation and ratios between NO_x and NO_y provide indicators of the level of photochemical activity that fosters O_3 production.

Acknowledgements

This study was supported by the Mexican Council for Science and Technology (CONACYT) through grant No. CB-2010-1-154122. Additional support was received from Tecnológico de Monterrey through grant no. CAT-186. E. Carrillo appreciates the scholarship received from CONACYT during his research stay (MSc) at Tecnológico de Monterrey.

Author details

Alberto Mendoza* and Edson R. Carrillo

*Address all correspondence to: mendoza.alberto@itesm.mx

School of Engineering and Sciences, Tecnológico de Monterrey, Monterrey, Mexico

References

[1] Seinfeld JH. Urban Air Pollution: State of the Science. Science 1989; 243(4892) 745-752.

[2] Atkinson R. Atmospheric chemistry of VOCs and NOx. Atmospheric Environment 2000; 34(12–14) 2063-2101.

[3] Yang C, Yang H, Guo S, Wang Z, Xu X, Duan X, Kan H. Alternative ozone metrics and daily mortality in Suzhou: The China Air Pollution and Health Effects Study (CAPES). Science of The Total Environment 2012; 426: 83-89.

[4] Ha S, Hu H, Roussos-Ross D, Haidong K, Roth J, Xu X. The effects of air pollution on adverse birth outcomes. Environmental Research 2014; 134: 198-204.

[5] Vlachokostas C, Nastis SA, Achillas C, Kalogeropoulos K, Karmiris I, Moussiopoulos N, Chourdakis E, Banias G, Limperi N. Economic damages of ozone air pollution to crops using combined air quality and GIS modelling. Atmospheric Environment 2010; 44(28) 3352-3361.

[6] Feng Z, Sun J, Wan W, Hu E, Calatayud V. Evidence of widespread ozone-induced visible injury on plants in Beijing, China. Environmental Pollution 2014; 193: 296-301.

[7] Fang Y, Naik V, Horowitz LW, Mauzerall DL. Air pollution and associated human mortality: the role of air pollutant emissions, climate change and methane concentration increases from the preindustrial period to present. Atmospheric Chemistry and Physics 2013; 13(3) 1377-1394.

[8] Seinfeld JH, Pandis SN. Atmospheric Chemistry and Physics: From Air Pollution to Climate Change. 2nd ed. Hoboken, New Jersey: John Wiley & Sons.; 2006.

[9] Warnek P. Chemistry of the Natural Atmosphere. International Geophysics Series. New York: Academic Press. 757; 1988.

[10] Russell A, Dennis R. NARSTO critical review of photochemical models and modeling. Atmospheric Environment 2000; 34(12–14) 2283-2324.

[11] Kleanthous S, Vrekoussis M, Mihalopoulos N, Kalabokas P, Lelieveld J. On the temporal and spatial variation of ozone in Cyprus. Science of The Total Environment 2014; 476–477: 677-687.

[12] Nishanth T, Praseed KM, Kumar MKS, Valsaraj KT. Influence of ozone precursors and PM_{10} on the variation of surface O_3 over Kannur, India. Atmospheric Research 2014; 138: 112-124.

[13] Iqbal MA, Kim K-H, Shon Z-H, Sohn J-R, Jeon E-C, Kim Y-S, Oh J-M. Comparison of ozone pollution levels at various sites in Seoul, a megacity in Northeast Asia. Atmospheric Research 2014; 138: 330-345.

[14] Zhang Y, Mao H, Ding A, Zhou D, Fu C. Impact of synoptic weather patterns on spatio-temporal variation in surface O_3 levels in Hong Kong during 1999–2011. Atmospheric Environment 2013; 73: 41-50.

[15] Alghamdi MA, Khoder M, Harrison RM, Hyvärinen AP, Hussein T, Al-Jeelani H, Abdelmaksoud AS, Goknil MH, Shabbaj II, Almehmadi FM, Lihavainen H, Kulmala M, Hämeri K. Temporal variations of O_3 and NO_x in the urban background atmosphere of the coastal city Jeddah, Saudi Arabia. Atmospheric Environment 2014; 94: 205-214.

[16] Jenkin ME. Investigation of the impact of short-timescale NO_x variability on annual mean oxidant partitioning at UK sites. Atmospheric Environment 2014; 90: 43-50.

[17] Hassan IA, Basahi JM, Ismail IM, Habeebullah TM. Spatial Distribution and Temporal Variation in Ambient Ozone and Its Associated NO_x in the Atmosphere of Jeddah City, Saudi Arabia. Aerosol and Air Quality Research 2013; 13 1712-1722.

[18] González-Santiago O, Badillo-Castañeda CT, Kahl JDW, Ramírez-Lara E, Balderas-Renteria I. Temporal Analysis of PM_{10} in Metropolitan Monterrey, México. Journal of the Air & Waste Management Association 2011; 61(5) 573-579.

[19] Sistema Integral de Monitoreo Ambiental. Estadística SIMA: Gobierno del Estado de Nuevo León. http://www.nl.gob.mx/?P=med_amb_mej_amb_sima_estadisti (accessed 1 Oct 2013).

[20] Sierra A, Vanoye AY, Mendoza A. Ozone sensitivity to its precursor emissions in northeastern Mexico for a summer air pollution episode. Journal of the Air & Waste Management Association 2013; 63(10) 1221-1233.

[21] Green J, Sánchez S. Air Quality In Latin America: An Overview. Washington, DC: Clean Air Institute; 2013. http://www.cleanairinstitute.org/calidaddelaireamericalatina/ cai-report-english.pdf (accessed 22 September 2014).

[22] Mancilla Y. $PM_{2.5}$ Source Apportionment using Partial Least Squares Based on Organic Molecular Markers: Monterrey Metropolitan Area, Mexico. PhD thesis. Tecnológico de Monterrey; 2013.

[23] SEMARNAT, Gobierno del Estado de Nuevo León. Programa de Gestión para Mejorar la Calidad del Aire del Área Metropolitana de Monterrey 2008-2012. 2008. http://www.semarnat.gob.mx/archivosanteriores/temas/gestionambiental/calidaddelaire/Documents/Calidad%20del%20aire/Proaires/ProAires_Vigentes/6_ProAire%20AMM%202008-2012.pdf (accessed 23 September 2014).

[24] Rodwell MJ, Hoskins BJ. Subtropical Anticyclones and Summer Monsoons. Journal of Climate 2001; 14(15) 3192-3211.

[25] Sillman S. The relation between ozone, NO_x and hydrocarbons in urban and polluted rural environments. Atmospheric Environment 1999; 33(12) 1821-1845.

[26] Luke WT, Kelley P, Lefer BL, Flynn J, Rappenglück B, Leuchner M, Dibb JE, Ziemba LD, Anderson CH, Buhr M. Measurements of primary trace gases and NO_y composition in Houston, Texas. Atmospheric Environment 2010; 44(33) 4068-4080.

[27] Luecken DJ, Tonnesen GS, Sickles IJE. Differences in NO_y speciation predicted by three photochemical mechanisms. Atmospheric Environment 1999; 33(7) 1073-1084.

[28] Wang YH, Hu B, Ji DS, Liu ZR, Tang GQ, Xin JY, Zhang HX, Song T, Wang LL, Gao WK, Wang XK, Wang YS. Ozone weekend effects in the Beijing–Tianjin–Hebei metropolitan area, China. Atmospheric Chemistry and Physics 2014; 14 2419-2429.

[29] Warneke C, de Gouw JA, Edwards PM, Holloway JS, Gilman JB, Kuster WC, Graus M, Atlas E, Blake D, Gentner DR, Goldstein AH, Harley RA, Alvarez S, Rappenglueck B, Trainer M, Parrish DD. Photochemical aging of volatile organic compounds in the Los Angeles basin: Weekday-weekend effect. Journal of Geophysical Research: Atmospheres 2013; 118(10) 5018-5028.

Section 2

Air Pollution Modelling

Chapter 14

Modelling PM2.5 with Fuzzy Exponential Membership

Danni Guo

Additional information is available at the end of the chapter

http://dx.doi.org/10.5772/59617

1. Introduction

Air quality is a public and environmental issue that concerns people, whether in terms of global climate change or for health and the quality of life. The Environmental Protection Agency (EPA) regulates six primary air pollutants: Ozone, Particulate matter, Carbon Monoxide, Nitrogen Oxides, Sulfur Dioxide, and Lead [3]. Particulate matter (PM) refers to solid particles and liquid droplets found in air. Many manmade and natural sources produce PM directly, or produce pollutants that react in the atmosphere to form PM [9]. $PM_{2.5}$ are small particles or particulate matter, that are less than 2.5 micrometers in diameter. $PM_{2.5}$ can be produced by combustion from motor vehicles (esp. diesel powered buses and trucks), power plants, residential wood burning, forest fires, agricultural burning, and industrial processes [15]. They can also be formed in the air through when gases (air poullutants) and orgnaic compounds are transformed through chemical reactions.

These tiny particles $PM_{2.5}$ can cause health hazards for people and also for the environment. People with heart diesease and lung problems including asthma, and also the elderly and children, are particularily vulnerable and at high risk when exposed to high levels of $PM_{2.5}$. Due to the tiny size of these $PM_{2.5}$ particles, they can penetrate to the deepest parts of the lungs, which is very dangerous to the human health. The California Air Resources Board (CARB) scentifically conducts studies and reports on the impacts of air pollutant expcsures on public health, and the studies shows the negative impacts of $PM_{2.5}$ which is known to cause premature death [2]. Scientific studies has repeatedly found links between particulate matter and many health problems of people who has been exposed to high levels of $PM_{2.5}$, including asthma, bronchitis, respiratory problems, including shortness of breath and painful breathing, and premature deaths [3].

One must also be aware of the effects of high levels of $PM_{2.5}$ on the environment as well. Since $PM_{2.5}$ are tiny, they can be carried by wind and travel great distances, so that it can cause problems for areas downwind far from the actual source of air pollution. They have adverse effect on urban areas, agriculture, and the natural envrinonment. High levels of $PM_{2.5}$ can results in visibility problems, urban haze, and acid rain [3].

The U.S. Environmental Protection Agency has established standards requiring the annual average of the $PM_{2.5}$ to be not more than 15 micrograms per cubic meter [3]. The State of California monitors and reports on their air pollutants carefully, setting very high standards for their air quality (μg/m^3). From 1999-2011, there are 113 station locations monitoring $PM_{2.5}$. The site design originally planned was well spread statistically. See Figure 1. However, in reality, it is too costly in terms of time, finance, and manpower to keep all the 113 sites to be monitoring and recording every single year. Each year, only a part of the 113 sites were actually sampled, and each year at different locations.

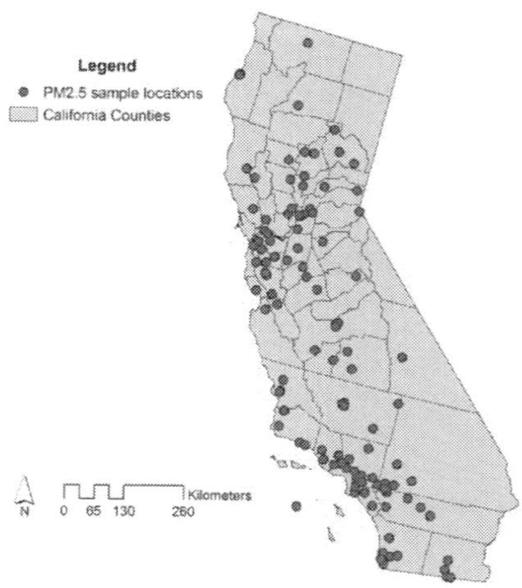

Figure 1. Complete 113 $PM_{2.5}$ Observational Sites in the California State

Comparisons of $PM_{2.5}$ between the years are difficult, due to "missing data" at sample sites [6, 9]. A site that does not have a recorded $PM_{2.5}$ value is referred to as "missing value", and since there are no patterns so that serious problems would twist the kriging map constructions.

Observing the dataset in Figure 2, the worst (in 1999) only 11 sites (9.73% of 113 sites) were used and at the best (in 2009) 65 sites (57.52% of 113 sites) were used. Over 13 years, 1469 annual arithmatic means (μg/m^3) should be recorded, but actually, 556 annual arithmatic means (μg/m^3) were reported, which occupied 37.85%. Sitewise looking, only one site, Site 2596 (Placer County APCD), was collected data annually and had 13 recored annual arithmatic means (μg/m^3), while 16 sites had one annual arithmatic mean (μg/m^3) only. The comparisons of $PM_{2.5}$ annual arithmatic means (μg/m^3) between years for a given site or between sites for a given year, i.e., the investigations of $PM_{2.5}$ annual arithmatic means (μg/m^3) patterns will be an

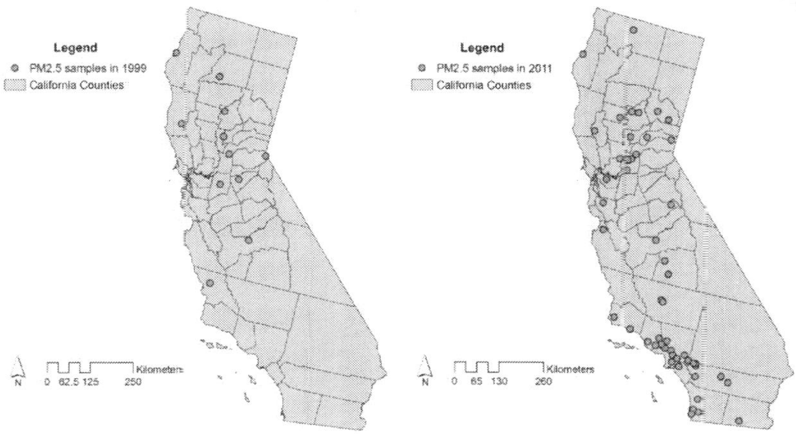

Figure 2. $PM_{2.5}$ Samples Collected in California in 1999 (11 sites) and 2011 (52 sites)

extremely difficult task due to data incompleteness. Therefore it is logical to engage fuzzy theory for treating the "missing" or scarce data.

2. Literature review and methodology

2.1. Fuzzy theory and membership kriging approach

Zadeh's fuzzy theory [22, 23] poineered a new mathematical branch. His membership approaches were quickly spreading and merging into many other mathematical branches, for example, engineering, business, economics, etc. and generating huge impacts in mathematical theories and applications. But it is aware that associated with Zadeh's fuzzy mathematical achivements, researchers gradually discovered three fundamental issues: self-duality dilemma, variable dilemma, and membership dilemma. Guo et al. [8] discussed those dilemmas in detail and pointed out Liu's credibility measure theory [11] is a solid mathematical treatment to address fuzzy phenomenon modelling. The credibility measure, similar to probability measure, assumes self-duality. Consequently, parallel to probability theory, a fuzzy variable and its (credibility) distribution can be defined. Furthermore, the membership function of a fuzzy variable can also be specified by its credibility distribution. Without any doubts, credibility measure theory is applying to practical situations sucessfully, say, Peng and Liu [14] considered parallel machine scheduling problems with processing times Zheng and Liu [24] studied a fuzzy vehicle routing optimization problem, Guo et al. [2008] proposed credibility distribution grade kriging for investigation California State PM_{10} spatial patterns, Wang et al. [20, 21] investigated a fuzzy inventory model without backordering, Sampath and Deepa [16] developed sampling plans containing fuzziness and randomness, and others. Nevertheless, the investigations in statistical estimation and hypothesis testing problems with fuzzy credibility distribution are very slowly progressed, for example, Li et al. [10], Sampath and Ramya [17, 18] in worked with fuzzy normal distributions, and studied the exponential credibility distribution function [1].

With statistically well-designed scheme, the collected spatial data can be analyzed and presented by kriging maps. If we are facing spatial data with "missing" or scarce fuzziness, it is impossible to construct kriging maps. It is noticed that the air pollutants were measured in different sites each year, even the site design originally planned was well spread statistically. We call a site that does not have a recorded $PM_{2.5}$ concentration as "missing value" site, as continued from Guo's research [4, 5]. To address the basic requirements in constructing kriging maps, Guo first proposed membership kriging in Zadeh's sense, see the MSc thesis [4] in which the linear, quadratic and hyperbolic tangent membership functions were used. Later Guo [5, 7] had developed the membership kriging under credibility theory. Following the membership kriging route of treating uncertainty, Shada et al. [19] and Zoraghein et al. [25] made considerable contributions in their papers. In this chapter, we will integrate exponential membership and kriging, to fill in the "missing data" based on existing sample data, and making a comparison of $PM_{2.5}$ concentrations in California from 1999-2011.

2.2. Fuzzy exponential distribution

It is necessary to introduce the basics of Liu's fuzzy credibility theory [11]. Let Θ be a nonempty set and M a σ- algebra over Θ. Elements of M are called events. $\mathrm{Cr}\{A\}$ denotes a number or grade associated with event A, called credibility (measure or grade). Credibility measure $\mathrm{Cr}\{\cdot\}$ satisfies the axioms normality, monotonicity, self-duality and maximality:

$$\begin{aligned} &\mathrm{Cr}\{\Theta\} = 1 \\ &\mathrm{Cr}\{A\} \le \mathrm{Cr}\{B\} \text{ for } A \subset B \\ &\mathrm{Cr}\{A\} + \mathrm{Cr}\{A^c\} = 1 \text{ for any event } A \\ &\mathrm{Cr}\left\{\bigcup_i A_i\right\} = \sup_i \left(\mathrm{Cr}\{A_i\}\right) \text{ for any events } A_i \text{ with } \sup_i\left(\mathrm{Cr}\{A_i\}\right) < 0.5 \end{aligned} \tag{1}$$

Definition 1: The set function Cr{} on M is called a credibility measure if it follows Axiom Normality, Axiom Monotonicity, Axiom Self-Duality and Axiom Maximality shown in Equation 1. The triplet (Θ, M ,Cr) is called a credibility space.

Definition 2: A measurable function,η, mapping from a credibility space (Θ, M,Cr) to a set of real numbers ф B? $\subset \mathbb{R}$.

Definition 3: The credibility distribution Ψ of a fuzzy variable η on the credibility measure space (Θ, M,Cr) is Ψ: $\mathbb{R} \rightarrow$ [0,1], where $\Psi(x) = \mathrm{Cr}\{\theta \in \Theta \mid \eta(\theta)\ x\}$, $x \in \mathbb{R}$.

Definition 4: The function μ of a fuzzy variable η on the credibility measure space (Θ, M,Cr) is called as a membership function: $\mu(x) = (2\mathrm{Cr}\{\eta = x\}) \wedge 1, \quad x \in \mathbb{R}$.

Theorem 5: (Credibility Inversion Theorem) Let η be a fuzzy variable on the credibility measure space (Θ, M,Cr) with membership function μ. Then for any set B of real numbers,

$$\operatorname{Cr}\{\eta \in B\} = \frac{1}{2}\left(\sup_{x \in B} \mu(x) + 1 - \sup_{x \notin B} \mu(x) \right). \tag{2}$$

Corollary 6: Let η be a fuzzy variable on a credibility measure space (Θ, M,Cr) with membership function μ. Then the credibility distribution Ψ is

$$\Psi(x) = \frac{1}{2}\left(\sup_{y \le x} \mu(y) + 1 - \sup_{y > x} \mu(y) \right), \text{ for } \forall x \in ¡ . \tag{3}$$

It is obvious the concept of credibility measure is very abstract although the credibility measure has normality, monotonicity, self-duality and maximality mathematical properties. The credibility measure loses the intuitive feature as Zadeh's membership orientation. Credibility Inversion Theorem and its corollary have just revealed the deep link between an abstract measure and intuitive membership. Such a link definitely paves the way for real-life applications. For example, the trapezoidal fuzzy variable has membership function μ:

$$\mu(x) = \begin{cases} 0 & x \le c - c_2 \\ \dfrac{c_2 + x - c}{c_2 - c_1} & c - c_2 < x \le c - c_1 \\ 1 & c - c_1 < x \le c + c_1 \\ \dfrac{c_2 - x + c}{c_2 - c_1} & c - c_1 < x \le c + c_1 \\ 0 & c + c_2 < x \end{cases} \tag{4}$$

where c, c_1, c_2 are all positive200, $c > c_1, c > c_2, c_2 > c_1$.

With the help of Equation 3, the fuzzy credibility distribution is thus,

$$\Psi(x) = \begin{cases} 0 & x \le c - c_2 \\ \dfrac{c_2 + x - c}{2(c_2 - c_1)} & c - c_2 < x \le c - c_1 \\ \dfrac{1}{2} & c - c_1 < x \le c + c_1 \\ \dfrac{2c_2 - c_1 + x - c}{2(c_2 - c_1)} & c - c_1 < x \le c + c_1 \\ 1 & c + c_2 < x \end{cases} . \tag{5}$$

Liu [12, 13], Wang and Tian [21] and [1] studied the exponential fuzzy distribution with a membership function, denoted as Exp(m):

$$\mu(x) = \frac{2}{1+e^{\frac{\pi x}{\sqrt{6}m}}}, x \geq 0, m > 0 \tag{6}$$

The support of an exponential membership function is set $[0,+\infty)$, the nonnegative part of the real line, $\mathbb{R}$. The expected value and second moment of exponential fuzzy variable are

$$\mathrm{E}[\eta] = \frac{\sqrt{6}\ln 2}{\pi} m \tag{7}$$

and

$$\mathrm{E}\left[\eta^2\right] = m^2 \tag{8}$$

where the parameter $m > 0$ determining the mean and variance of the exponential fuzzy variable. Thus it is intuitive to reveal how the shape of membership curve affected by the possible values of the parameter $m > 0$.

x	m=12	m=15	m=22	m=50
0	1.0000	1.0000	1.0000	1.0000
1	0.9466	0.9573	0.9709	0.9872
2	0.8935	0.9147	0.9418	0.9744
3	0.8410	0.8724	0.9128	0.9615
4	0.7894	0.8306	0.8839	0.9487
5	0.7390	0.7894	0.8553	0.9360
6	0.6899	0.7490	0.8269	0.9232
7	0.6424	0.7094	0.7987	0.9105
8	0.5968	0.6707	0.7709	0.8978
9	0.5530	0.6332	0.7435	0.8851
10	0.5113	0.5968	0.7165	0.8724
11	0.4717	0.5616	0.6899	0.8598
12	0.4342	0.5277	0.6638	0.8473
13	0.3990	0.4952	0.6382	0.8348
14	0.3660	0.4640	0.6132	0.8223
15	0.3351	0.4342	0.5887	0.8100
16	0.3063	0.4059	0.5647	0.7976
17	0.2796	0.3789	0.5414	0.7854

x	m=12	m=15	m=22	m=50
18	0.2549	0.3533	0.5187	0.7732
19	0.2320	0.3292	0.4966	0.7610
20	0.2110	0.3063	0.4752	0.7490
21	0.1917	0.2848	0.4544	0.7370
22	0.1739	0.2645	0.4342	0.7251
23	0.1577	0.2455	0.4147	0.7133
24	0.1428	0.2277	0.3959	0.7016
25	0.1293	0.2110	0.3777	0.6899
26	0.1170	0.1954	0.3602	0.6784
27	0.1057	0.1808	0.3433	0.6669
28	0.0955	0.1672	0.3270	0.6556
29	0.0863	0.1546	0.3114	0.6443
30	0.0779	0.1428	0.2964	0.6332

Table 1. Impacts of m in the Shape of Exponential Membership Curve

The value choice of parameter m is not aimless. m = 22 corresponds to the $PM_{2.5}$ annual arithematic mean 11.85 (μg/m^3), while m = 50 corresponds to the $PM_{2.5}$ annual arithematic mean 25.89 (μg/m^3). Therefore, the one-parameter exponential fuzzy variable may well cope to the modelling requirements of the California $PM_{2.5}$ annual arithematic mean dataset. Furthermore, using the one-parameter exponential fuzzy variable it may develop a delicate scheme of testing credibility hypothesis.

2.3. Credibility hypothesis testing with exponential membership

Similiar to the popular Neyman-Pearson Lemma in probability theory, likelihood ratio L_0/L_1, constant k, and critical region C of size α, are involved in the testing hypothesis H_0: $\theta = \theta_0$ against alternative hypothesis H_1: $\theta = \theta_1$. The likelihood is defined as the product of the densities for given sampled population. Hence we can call Neyman-Pearson testing criterion as likelihood ratio criterion. Inevitably, Type I error and Type II error concepts are also engaged in describing testing procedure. For hypothesis testing under credibility theory, Sampath and Ramya [17] proposed a membership ratio criterion. The membership criterion applies to any forms of credibility distributions, but exponential credibility distribution has its unique advantage. Therefore, the remaining descriptions will be focused on credibility hypothesis testing under exponential membership function [1].

Definition 7: Credibility hypothesis is a statement describing the possible rejection of a null hypothesis, H_0: $\mu = \mu_0$ with the credibility distribution of a fuzzy variable against an alternative hypothesis H_1: $\mu = \mu_1$ with another credibility distribution of a fuzzy variable

Definition 8: Credibility hypothesis testing is the rule describing reject or not reject a null hypothesis if the calculated values sampled from the fuzzy distribution defined by null hypothesis.

Definition 9: Credibility rejection region is the subset of the support under a fuzzy distribution, denoted as C, on which the null credibility hypothesis is rejected H_0: $\mu = \mu_0$, i.e., $C = \{\eta \in \Theta \mid H_0 \text{ is rejected}\}$

Definition 10: Type I error is the mistake by rejecting the null credibility hypothesis H_0: $\mu = \mu_0$ when it is true and Type II error the mistake by not rejecting the null credibility hypothesis H_0: $\mu = \mu_0$ when it is false.

Definition 11: Level of credibility significance is the maximal credibility to make a Type I error in testing a credibility hypothesis H_0: $\mu = \mu_0$, denoted as α.

Definition 12: The best credibility rejection region of credibility significance level α, C^*, if this region possesses the maximal power (measured by credibility) under alternative hypothesis K with all possible credibility rejection regions of level of credibility significance α, i.e.,

$$\mathrm{Cr}\{\eta \in C^* \mid K\} \geq \mathrm{Cr}\{\eta \in C \mid K\}, \tag{9}$$

where C is any region satisfying the condition $\mathrm{Cr}\{\eta \in C \mid H\} \leq \alpha$. The power of the credibility hypothesis testing is $\mathrm{Cr}\{\eta \in C^*\}$.

With the exponential membership function having single parameter $m > 0$, the best credibility rejection region of credibility significance level α, C^* should be an interval so that we name it as best credibility rejection interval of credibility significance level α.

Theorem 13: For testing the null credibility hypothesis H_0: $m = m_1$ against the alternative credibility hypothesis K: $m = m_2$, $(m_1 < m_2)$, under exponential fuzzy distributions $\mu(x) = \mathrm{Exp}(m)$ as Equation 6 specified, the membership ratio criterion is engaged. The criterion states that given credibility significance level $\alpha < 0.5$, the best credibility rejection interval C^*:

$$C^* = \left[\frac{\sqrt{6}m_1}{\pi}\ln\left(\frac{1-\alpha}{\alpha}\right), +\infty\right), \quad \alpha \in (0, 0.5). \tag{10}$$

The credibility of credibility rejection interval C^* under alternative hypothesis is greater than α.

α	0.10	0.15	0.20	0.25	0.30
x_0	37.56361	29.65463	18.78181	14.48535	10.58305

Table 2. x_0 and α under H_0: m_1=21.926

Table 2 illustrates relationship between the best credibility rejection interval boundary x_0 for selected credibility significance level $\alpha < 0.5$. For example, let $\alpha = 0.20$, $m_1 = 21.93$, then the best credibility rejection interval C^*= [18.78, +∞). We have to point out that the choice of credibility significance level α in credibility hypothesis testing should not follow the "thumb rule" the significance level α = 0.05 in probability hypothesis testing. Although the two significance levels have the same role in hypothesis testing, nevertheless, the practical meanings of credibility significance level α and the significance level α are quite different. From Table 2 and California $PM_{2.5}$ distribution patterns, it is is logical and practical selecting the credibility significance level α = 0.25, which gives x_0 =14.485.

3. Analysis and results

3.1. Exponential fuzzy membership kriging

Now having examined the methodology, we can now calculate the membership grades. But first let us have a quick overlook on California $PM_{2.5}$ 1999-2011 records.

Year	Number of Sites	Annual Maximum	Annual Minimum	Annual Average
1999	11	23.4190	8.3274	13.6146
2000	22	23.9229	7.7971	14.3465
2001	25	25.0446	4.1492	12.5643
2002	41	25.8332	6.3356	15.9160
2003	44	24.8348	4.3802	12.6993
2004	43	18.1962	4.4146	11.3249
2005	40	22.4178	4.8117	12.0750
2006	47	21.5991	5.1022	11.4211
2007	55	25.2073	3.2883	11.6240
2008	59	21.2357	6.7716	12.3861
2009	65	21.2423	3.3373	10.7292
2010	52	17.1908	2.9665	9.2255
2011	52	18.0950	3.3682	10.7934
Total	556			

Table 3. California $PM_{2.5}$ 1999-2011 Annual $PM_{2.5}$ Concentrations

From Table 3, we can see that the annual average $PM_{2.5}$ range between 9.2255 and 15.9160. Most of the averages are wondering about 11.0 and 12.7. Therefore, it is very reasonable to estimate $\overline{X}$ =11.85.

Exponential membership grade kriging scheme:

1. Calculating overall exponential membership parameter m. Based on Equation 7, we can estimate the exponential membership parameter m. Then in this paper, m = 21.93 is used for membership grade smoothing, kriging, and hypothesis testing.

2. Calculating every exponential membership grade $e_{ij} = 2.0/(1+\exp(\pi x_{ij}/m\sqrt{6}))$ for site j given year i. For any site, if "missing" value, x_{ij} = 1, otherwise, $x_{ij} = e_{ij}$ (> 0).

3. Performing membership grade interpolation. (i) Equal weights, for any given site j, if x_{ij} = 1, which is a "missing" value, let $x_{ij} \triangleq \hat{e}_{ij}$ where $\hat{e}_{ij}=(e_{i-1,j}+e_{i+1,j})/2$ (conditioning on the availability of the nearest neighbours $e_{i-1,j}$ and $e_{i+1,j} < 1$); (ii) Unequal weights, for a given site j, if the neighbour years are quite far, say, N between the gap, then, we may take linear interpolation for filling those "missing" value. that is given e_{ij} and $e_{i+N,j} < 1$, let $x_{i+l,j} \triangleq e_{ij} + l * (e_{i+N,j}-e_{ij})/N$, $l = 1,2,\ldots, N\text{-}1$.

4. Performing membership grade extrapolation, including forward and backward extrapolation. Equal weights (1/3) are used mostly. Unequal weights are also used.

5. Carrying the filling "missing" cells task until thirteen years 1999 to 2011, each year 113 membership grades are all calculated, $\{\hat{e}_{ij}, i=1, 2, \cdots, 13, j=1, 2, \cdots, 113\}$.

Now, it is ready to construct exponential membership grades kriging maps and use these 13 maps for comparisons.

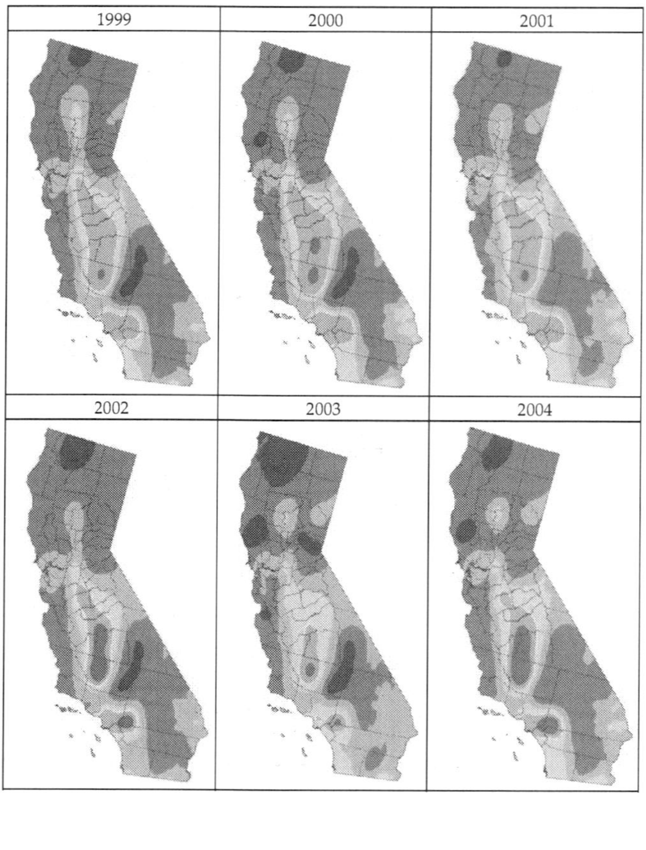

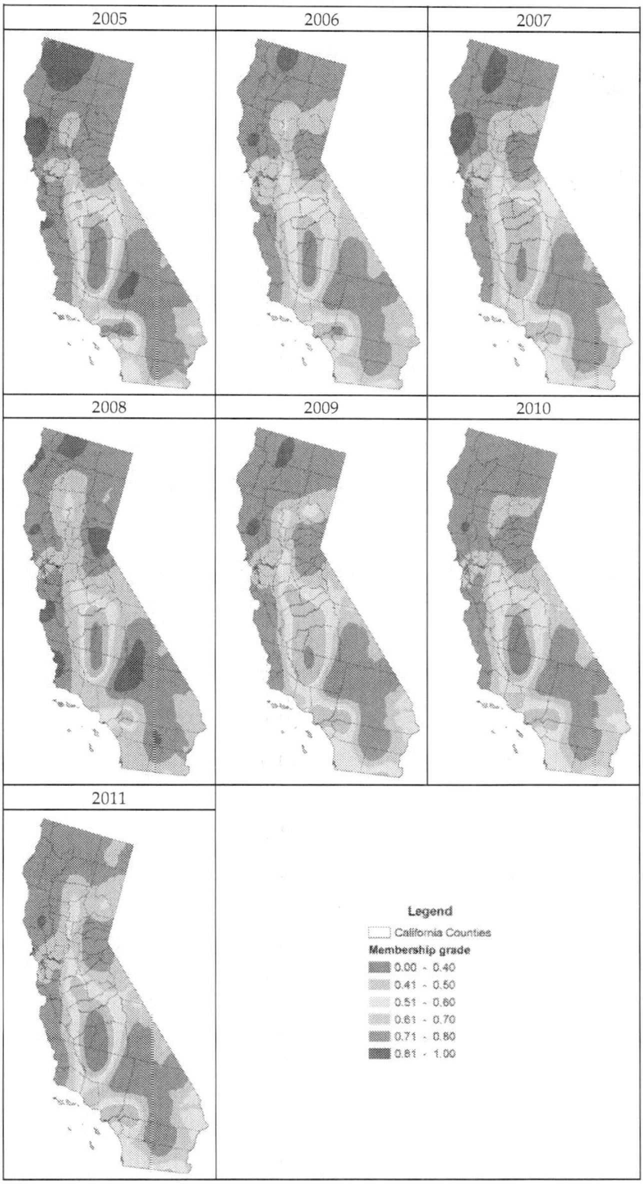

Figure 3. Kriging maps of Exponential membership grades for California 1999-2011

As one can observe from Figure 3, central California regions have very low membership grades, and the rest of Calfornia have higher membership grades.

3.2. Calculated $PM_{2.5}$ concentrations

It is impossible for the public and governmental officers to accept the membership grade kriging maps. Therefore, kriging maps of every year $PM_{2.5}$ concentrations (collected and estimated together) must be constructed. The conversion formula is

$$\tilde{x}_{ij} = \frac{\sqrt{6}m}{\pi}\ln\left(\frac{2-\hat{e}_{ij}}{\hat{e}_{ij}}\right). \tag{11}$$

Kriging maps with "completed" of every year $PM_{2.5}$ data $\{x_{ij}, i=1,2,...,13, j=1,2,...,113\}$. Comparison of kriging maps with the "completed" data can now be performed. See Figure 4. A complete dataset is now available, the 113 observation sites now have the full 13 year $PM_{2.5}$ concentration, containing 1469 data records. The 13 ordinary kriging prediction maps are generated, and one can clearly observe the changes in $PM_{2.5}$ year by year.

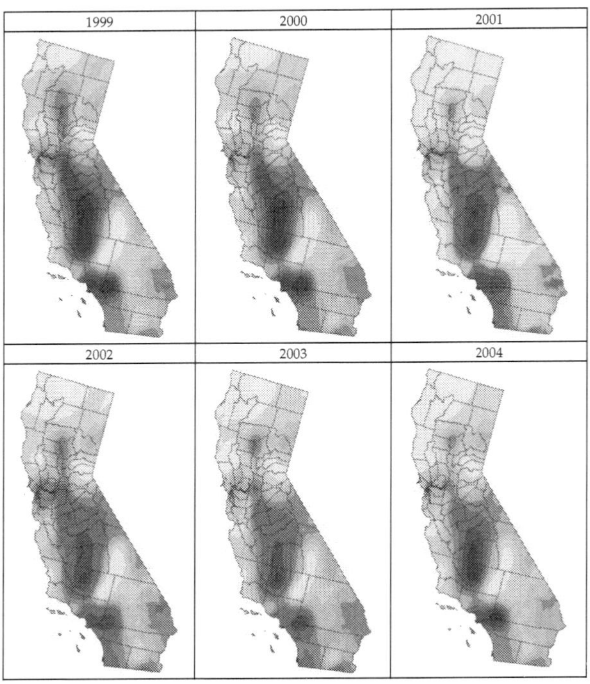

http://dx.doi.org/10.5772/59617

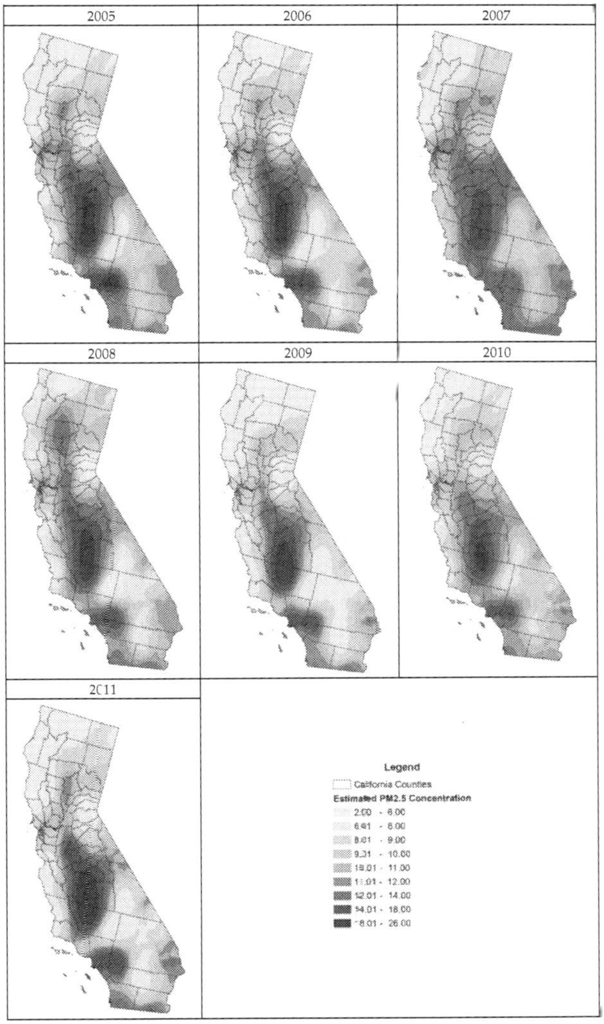

Figure 4. Kriging maps of completed estimated $PM_{2.5}$ concentrations for California 1999-2011

One can now observe and compare the year by year changes of $PM_{2.5}$ concentration, and note the regions of high and low $PM_{2.5}$ concentration. The dark brown colours represent high $PM_{2.5}$ concentrations, and light yellow colours represent areas with low $PM_{2.5}$ concentrations. Note that central California shows to have continual high levels of $PM_{2.5}$ concentration, year after year.

4. Interpretation and conclusion

In the results section, the dataset is now calculated and completed. However, it is now up to us to interpret the maps, and decide how to best make use of the calculated dataset, so that it provides us with easy to read information. And we can do this through a change map and 13 health safety maps.

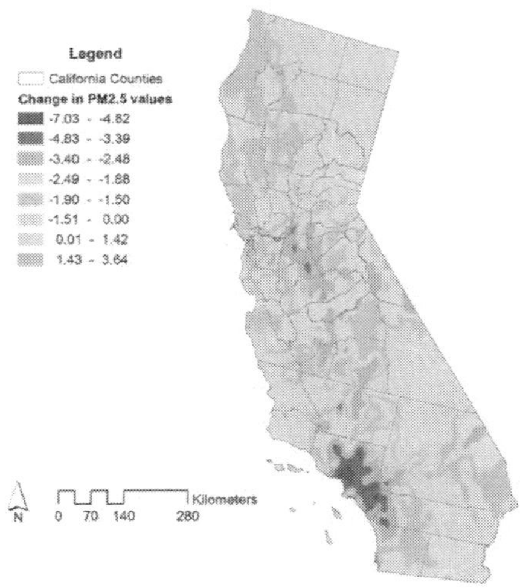

Figure 5. Changes in $PM_{2.5}$ Concentrations in California between 1999 and 2011

As one can clearly see from Figure 5, that $PM_{2.5}$ concentration has clearly decreased and air quality has improved remarkably over the years. The blue and green colours show negative changes, and orange shows positive changes. Counties such as Los Angeles and Orange show the highest decrease, and other counties such as Lassen, Plumas, Sierra, Inyo, and Imperial show some increase in $PM_{2.5}$ concentration. However, *a decrease in* $PM_{2.5}$ *concentration does not indicate safety in air quality.*

In terms of credibility hypothesis testing, say, with credibility significance level $\alpha = 0.25$, critical point for the best credibility rejection interval is $x_0 = 14.485$. The indictor λ_{ij} is defined as

$$\lambda_{ij} = \begin{cases} 1 & \tilde{x}_{ij} < 14.485 \\ 0 & \text{othewise} \end{cases}, \tag{12}$$

For comparisons of air quality safety, we generate $PM_{2.5}$ safety maps with two colours: blue colour if $\lambda_{ij} = 1$, orange colour otherwise, in total 13 safety maps. See Figure 6.

http://dx.doi.org/10.5772/59617

Figure 6. Air Quality $PM_{2.5}$ Safety areas in California 1999-2011

One can now observe that over the 13 years period 1999-2011, Stanislaus, Merced, Madera, Fresno, Kings, Tulare, Kern, Los Angeles, San Bernardino, Orange, and Riverside counties are the counties with the highest $PM_{2.5}$ safety problems. These areas have shown to be unsafe for public health safety, and especially for those with lung and heart problems, and for children and the elderly. These places are also sources of environmental and ecological concerns.

In conclusion, facing the difficult problem of "incomplete" $PM_{2.5}$ data in California from 1999-2011, we used the interpolation and extrapolation smoothing approaches for "filling" those "missing value" sites. For easy computation, the fuzzy exponential membership function is assumed. The treatment is based on an assumption that the smoothing is performed for a given site rather than over different sites for a given year. Such an assumption is emphasizing the fact: the data recorded are $PM_{2.5}$ concentration annual arithmetic means and they shouldn't change too dramatically over neighbour years. As to neighbour sites impacting, the membership grade kriging approach is adequate enough for generating smoothed maps. Furthermore, for utilizing credibility hypothesis testing theory, we perform parameter estimation of the fuzzy exponential membership function and in terms of membership ratio criterion for deriving the safety maps under 0.25 credibility significance level. Membership ratio criterion is very similar to likelihood ratio criterion in theoretical development. By comparing those 13 $PM_{2.5}$ concentration safety maps to 1999-2011 change map, it is quite justifiable to say the safety maps under the credibility hypothesis testing procedure are very intuitive and convenient to the public. Finally, interpreting the 13 safety maps will provide the public with knowledge of air quality in California.

Acknowledgements

I would like to thank the California Air Resources Board for providing the air quality data used in this paper. This study is supported financially by the National Research Foundation of South Africa (Ref. No. IFR2009090800013).

Author details

Danni Guo*

Address all correspondence to: d.guo@sanbi.org.za

Statistical Ecology Division, Biodiversity Research, Assessment and Monitoring, South African National Biodiversity Institute, Cape Town, South Africa

References

[1] Authors unknown. Credibility Hypothesis Testing of Fuzzy Exponential Distributions. International Journal of Fuzzy Computation and Modelling (IJFCM). (*accepted*)

[2] California Environmental Protection Agency Air Resources Board (CARB). Ambient Air Quality Standards (AAQS) for Particulate Matter. www.arb.ca.gov (accessed 3 July 2014)

[3] Environmental Protection Agency (EPA). National Ambient Air Quality Standards (NAAQS). U.S. Environmental Protection Agency. www.epa.gov (accessed 3 July 2014)

[4] Guo D. Integrating GIS with Fuzzy Logic and Geostatistics: Predicting Air Pollutant PM10 for California, Using Fuzzy Kriging. MSc thesis. University of Durham; 2003.

[5] Guo D. Contributions to Spatial Uncertainty Modelling in GIS: Small Sample Data. PhD Thesis. University of Cape Town; 2007.

[6] Guo D., Guo R., Cui YH., Midgley GF., Altwegg, R., Thiart, C. Climate Change Impact on Quiver Trees in Arid Namibia and South Africa. Blanco, Kheradmand (ed.)

Climate Change - Geophysical Foundations and Ecological Effects. InTech; 2011. p323-342.

[7] Guo D, Guo R, Thiart C. Predicting Air Pollution Using Fuzzy Membership Grade Kriging. Computers, Environment and Urban Systems. 2007 31(1) 33-51.

[8] Guo D, Guo R, Thiart C. Credibility Measure Based Membership Grade Kriging. International Journal of Uncertainty, Fuzziness and Knowledge-Based Systems. 2007 15(2) 53-66.

[9] Guo D., Guo R., Thiart C., Cui YH. Imprecise Uncertainty Modelling of Air Pollutant PM10. In: Nejadkoorki (ed.) Advanced Air Pollution. InTech; 2011. p193-212.

[10] Li X., Qin Z, Ralescu D. Credibilistic parameter estimation and its application in fuzzy portfolio selection. Iranian Journal of fuzzy systems. 2011 15(2) 57-65.

[11] Liu B. Uncertainty Theory: An Introduction to Its Axiomatic Foundations. Springer-Verlag Heidelberg, Berlin. 2004.

[12] Liu B. Uncertainty Theory. 3rd ed. Springer-Verlag Berlin. UTLAB; 2010.http://orsc.edu.cn/liu/ut.pdf

[13] Liu B. Uncertainty Theory. 4th ed. Springer-Verlag Berlin. UTLAB; 2012. http://orsc.edu.cn/liu/ut.pdf

[14] Peng J, Liu B. Parallel machine scheduling models with fuzzy processing times. Information Sciences. 2004 166(1-4) 49-66.

[15] Pittsburghtoday. Environment / PM2.5. Pittsburghtoday.org. www.pittsburghtoday.org/CurrentPM25US (accessed 3 July 2014)

[16] Sampath S, Deepa SP. Determination of Optimal Chance Double Sampling Plan using Genetic Algorithm. Model Assisted Statistics and Applications. IOS Press. 2013 8(4) 265-273.

[17] Sampath S, Ramya B. Credibility hypothesis testing of expectation of fuzzy normal distribution. Journal of Intelligent & Fuzzy Systems. 2014.

[18] Sampath S, Ramya B. Credibility hypothesis testing of fuzzy triangular distributions. Journal of Uncertain Systems. (*accepted*)

[19] Shada R, Mesgarib MS, Abkarc A, Shad A. Predicting air pollution using fuzzy genetic linear membership kriging in GIS. Computers, Environment and Urban Systems. 2009 33(6) 472-481.

[20] Wang X, Tang W, Zhao R. Fuzzy Economic Order Quantity Inventory Models without Backordering. Tsinghua Science and Technology. 2007 12(1) 91-96.

[21] Wang Z, Tian F. A Note of the Expected Value and Variance of Fuzzy Variables. International Journal of Nonlinear Science. 2010 9(4) 486-492.

[22] Zadeh LA. Fuzzy sets. Information and Control. 1965 8 338-353.

[23] Zadeh LA. Fuzzy sets as a basis for a theory of possibility. Fuzzy Sets and Systems. 1978 1 3-28.

[24] Zheng Y. Liu B. Fuzzy vehicle routing model with credibility measure and its hybrid intelligent algorithm. Applied Mathematics and Computation. 2006 176(2) 673-683.

[25] Zoraghein H, Alesheikh A A, Alimohammadi A, Vahidnia, M H. The utilization of soft transformation and genetic algorithm to model two sources of uncertainty of Indicator Kriging. Computers, Environment and Urban Systems 2012 36(6) 592-598.

Chapter 15

On the Analytical Formulations for Pollutant Dispersion Simulation in the Atmospheric Boundary Layer

Davidson Martins Moreira, Tiziano Tirabassi and Taciana Albuquerque

Additional information is available at the end of the chapter

http://dx.doi.org/10.5772/60023

1. Introduction

The management and preservation of air quality require information of the environmental status. Such information involves both cognitive and interpretative features Networks and measurements for monitoring, together with emission inventory of sources, are of great significance for the building of the cognitive representation, but not the interpretative one. Air quality control demands interpretative tools that are able to extrapolate in space and time the values measured by analytical instrumentation at field sites, while environmental improvement can only be obtained by means of emissions reduction from a systematic planning by using tools (mathematical models of dispersion) adequate to link the causes (sources) with the effects (pollutants).

The pollutants' transport and diffusion processes are complex and without mathematical models it is impossible to account for them. Such models therefore are an indispensable technical instrument of air quality management.

The theoretical approach to the atmospheric dispersion assumes different points of view. The diffusion in the K approach is proportional to the local gradient concentration from diffused material (at the fixed point in space). As a result, it is fundamentally Eulerian, once it assumes the motion of fluid within a fixed system of reference.

Lagrangian models differ from Eulerian ones in adopting a system of reference that follows atmospheric motions. This category contains all models that consider the pollutant cloud as discrete "elements" (puffs or computer particles). In Lagrangian approach, the dispersion is simulated through the motion of particles whose path allows the concentration field computation from the liberated material.

In this chapter, we consider the Eulerian approach.

2. K models

The Eulerian models takes into account the resolution of the mass conservation equation of pollutant chemical species [1]:

$$\frac{\partial c}{\partial t} + u\nabla c = D\nabla^2 c + S. \tag{1}$$

In (1), $\boldsymbol{u}$ is the wind speed vector, $D\nabla^2$c is the molecular diffusion, D is the molecular diffusion coefficient, S is the term source, is an operator, and ∇^2 is the Laplacian.

To obtain the solution of equation (1), it is inevitable to know the wind field, something that is not possible because it is changing in space and time. As a consequence, wind is divided into two parts:

$\bar{u}$, ensemble average

u', turbulent fluctuation

Therefore, the wind speed can be expressed as the sum of the two contributions:

$$u = \overline{u} + u' \tag{2}$$

with $\bar{u}'=0$.

Similar hypothesis can be assumed for the concentration:

$$c = \overline{c} + c' \tag{3}$$

with $\overline{c'} = 0$.

The new u and c are introduced into (1). Then, the following is obtained:

$$\frac{\partial \overline{c}}{\partial t} = -\overline{u} \cdot \nabla \overline{c} - \nabla \cdot \overline{c'u'} + D\nabla^2 \overline{c} + \overline{S}. \tag{4}$$

This equation possesses new unknown variables. This feature leads to a number of unknowns greater than the number of equations. By solving this problem it is possible to parameterize some terms with known quantities.

The widely used approach is the parameterization of second order moments. Such an approach is called as K-theory or flux-gradient theory:

$$\overline{c'u'} = -K\nabla\overline{c}, \tag{5}$$

where K is the well-known eddy diffusivity.

A great variety of K-formulations exists [2]. The major part them take into account the similarity theory, and give different results for the same atmospheric stability.

Applying the following approximations:

a. Tensor is diagonal

b. c represents the concentration of a non-reactive pollutant (thusS = S)

c. The molecular diffusion is insignificant

Then, equation (4) can be written as:

$$\frac{\partial\overline{c}}{\partial t} = -\overline{u}\cdot\nabla\overline{c} + \nabla\cdot K\nabla\overline{c} + S. \tag{6}$$

Equation (6) can be solved analytically or numerically, with data for u. K, and S and the initial and boundary conditions for$c$. This equation can be resolved in two ways:

i. With analytic methods, obtaining exact solutions

ii. With numerical methods, obtaining approximate solutions

In this chapter we take into consideration analytical solutions.

3. Analytical solutions

Analytical solutions of equations are very important to understand and describe physical phenomena. Moreover, they are fast, simple, and, generally, do not require complex meteorological inputs.

There are analytical solutions of two-dimensional advection-diffusion equation [3]:

$$u\frac{\partial\overline{c}(x,z)}{\partial x} = \frac{\partial}{\partial z}\left(K_z\frac{\partial\overline{c}(x,z)}{\partial z}\right) + S, \tag{7}$$

where u is longitudinal mean speed, $\bar{c}$is the mean concentration, S is the source term, and K_z is the vertical eddy diffusivity. Furthermore, the longitudinal diffusion can be neglected because it is considered less in respect to the advection. Although, very recently a steady-state analytical solution for dispersion of contaminants in low winds by taking into account the longitudinal diffusion in the advection-diffusion equation was formulated [4].

The best-known solution is the so-called Gaussian solution, where both wind and turbulent diffusion coefficients are constant with height. So it is not a realistic solution of the transport and diffusion equation in the atmosphere. In the so-called Gaussian models, the solution is forced to represent real situations by means of empirical parameters, referred to as "sigmas." The different versions of Gaussian models substantially differ in the techniques utilized to calculate the "sigmas" as a function of atmospheric stability and the downwind distance from the emission source. Gaussian models are fast, simple, and do not require complex meteorological inputs. For these reasons they are still widely used by all the environmental agencies over the world for regulatory applications. At this point, it is important to mention that there are models based on non-Gaussian analytical solutions.

[5] report a bi-dimensional solution considering cases where the wind speed and vertical eddy diffusivities follow power laws as a function of height:

$$u=u_1\left(z/z_1\right)^{\alpha} \tag{8}$$

$$K_z=K_1\left(z/z_1\right)^{\beta}, \tag{9}$$

where z_1 is the height when u_1 and K_1 are computed.

[6] report a bi-dimensional solution for linear profiles of the eddy diffusivity. [7] showed the bi-dimensional equation with u and K power functions of height with the exponents following the law of Schmidt.

[8] shows a solution considering u constant, but K as follows:

$$K_z = K_0 z^a \left(h-z\right)^b, \tag{10}$$

where $K_z$is a constant and a and b can be expressed as:

a³ 0 ; b = 0
a = 0 ; b > 0 for 0£ ≤ z ≤ h
a = 1 ; b > 0 for 0 ≤ z ≤ h
a = 1 ; b = 0 for 0 ≤ z ≤ h/2 ; a = 0 and b = 1 for h/2 ≤ z£ ≤ h,

where h is the ABL height.

[9] proposed a solution with constant u and K_z as:

$$K_z \equiv z \quad \text{for } 0 \le z \le z_s \tag{11}$$

$$K_z = K_z(z_1) \ \text{ for } z_s < z \le h, \tag{12}$$

where z_s is a predetermined height (generally, the height of the surface layer). This solution allows (as boundary conditions) a net flow of material toward the ground:

$$K_z \frac{\partial \bar{c}}{\partial z} = V_g \bar{c}, \tag{13}$$

where V_g is the deposition velocity.

[10, 11] report bi-dimensional solutions considering elevated sources with wind speed and vertical eddy diffusivity power profiles, but for an unbound atmosphere. That is:

$$K_z \frac{\partial \bar{c}}{\partial z} = 0 \text{ at } z = \infty. \tag{14}$$

[12] shows a solution considering identical conditions, but for a vertically limited boundary layer. That is:

$$K_z \frac{\partial \bar{c}}{\partial z} = 0 \text{ at } z = h\,. \tag{15}$$

The solutions [10, 11, 12] are used in the KAPPAG model [13, 14, 15].

[16] using the Monin-Obukhov similarity theory to diffusion, has derived a solution from continuous sources near the ground with u and K follow power profiles. SPM [17] is a model that uses this solution.

[18] report a solution, which was a particular case of [7, 8]. In the sequence, [19] extended the solution to the situation of a growing ABL. [20] extended the solution to the case of nonzero mean vertical wind profiles.

[21] presented a three-dimensional atmospheric dispersion-deposition model for a ground-level area source.

[22] extended the solution obtained in [12] with boundary conditions considering dry deposition to the ground.

In the work [23] were reported equations for point sources releases considering the first four moments of the vertical concentration distribution and the magnitude and downwind location of the maximum ground concentration.

[24] obtained a three-dimensional solution (with u and K_z constant) for low wind condition where the diffusion coefficients are function of down-wind distance from the source. [25] presented a two-dimensional solution with K_z constant and function of down-wind distance from the source, with a power low wind profile and for a finite boundary layer.

[26] obtained an analytical solution with dry deposition to the ground with any vertical function of wind and eddy coefficients but for a fixed vertical shape of contaminant concentrations.

$$\bar{c}(x,z)=F(x)\left(1-\frac{z}{h}\right)^2, \tag{16}$$

where $F(x)$ is any function of x.

Using ADMM (Advection-Diffusion Multilayer Method) approach, [27] proposed a general solution for any wind and eddy coefficient profiles, but represented by a stepwise function in z. Recently, using the ADMM approach, [28] found a solution for two-dimensional steady-state solution considering eddy coefficient profile depending of x and z variables.

Finally, [29, 30] applying GILTT technique (Generalized Integral Laplace Transform Technique), found a general two-dimensional steady-state solution for any profiles of wind and eddy coefficient diffusions and a limited ABL.

3.1. GILTT approach solution

The General Integral Transform Technique (GITT) is a well-known hybrid method that solved a wide class of direct and inverse problems mainly in the area of Heat Transfer and Fluid Mechanics. There is a vast literature about this subject, including papers and books that become impossible to mention all of them. But, for instance, we mention the works of [31, 32, 33, 34], etc. In this chapter, we restrict our attention to the linear problem, because for nonlinear problem the linear result is iteratively used after the linearization of nonlinear transformed equation. Indeed, for the linear problems, the GITT transformed problem is analytically solved by the Laplace Transform technique. Consequently, the GITT solution becomes an analytical solution in the sense that no approximations are made along its derivation. It is important to notice that the GITT solution for nonlinear problems is semi-analytical because, in each iteration, the solution is analytical. This methodology is known as GILTT (Generalized Integral Laplace Transform Technique).

Bearing in mind that the novelty of the GILTT approach relies on the analytical solution of GITT linear transformed problem, in what follows, we restrict our analysis to the ensuing standard GITT time-dependent transformed problem:

$$E\frac{dY(t)}{dt}+FY(t)=0 \qquad t>0, \tag{17}$$

subject to the initial condition,

$$Y(0) = Y_0. \tag{18}$$

Here, Y(t) is the unknown vector with N components, E and F are constant matrices of order (NxN), and Y_0 is a known vector with N components. We begin our analysis recasting equation (17) like,

$$\frac{dY(t)}{dt} + GY(t) = 0 \qquad t > 0, \tag{19}$$

where $G = E^{-1}F$ and E^{-1} denotes the inverse of the matrix E. Now, applying the Laplace transform technique to equation (19), we obtain,

$$s\overline{Y(s)} + G\overline{Y(s)} = Y_0, \tag{20}$$

where $\overline{Y(s)}$ denotes the Laplace transform of the vector $Y(t)$. Assuming that the matrix G is nondegenerate, we may write,

$$G = XDX^{-1}. \tag{21}$$

Here, D is the eigenvalue of the diagonal matrix, X is the matrix of the respective eigenfunctions, and X^{-1} its inverse.

Indeed, replacing equation (21) in equation (20) we get,

$$s\overline{Y(s)} + XDX^{-1}\overline{Y(s)} = Y_0, \tag{22}$$

or

$$\left(sI + XDX^{-1}\right)\overline{Y(s)} = Y_0, \tag{23}$$

where the matrix I is the (NxN) identity matrix. Recalling that $X X^{-1} = X^{-1}X = I$, equation (23) reads like,

$$\left(sXX^{-1} + XDX^{-1}\right)\overline{Y(s)} = Y_0, \tag{24}$$

or

$$X(sI+D)X^{-1}\overline{Y(s)}=Y_0, \tag{25}$$

which has the well-known solution,

$$\overline{Y(s)}=X(sI+D)^{-1}X^{-1}Y_0. \tag{26}$$

Performing the Laplace transform inversion of equation (26), we have,

$$Y(t)=X\,L^{-1}\left\{(sI+D)^{-1}\right\}X^{-1}Y_0. \tag{27}$$

Here, L^{-1} denotes the inverse Laplace transform operator. We must notice that the matrix $(sI + D)$ has the form,

$$(sI+D)=\begin{pmatrix} s+d_1 & 0 & \cdots & 0 \\ 0 & s+d_2 & \cdots & 0 \\ \vdots & \vdots & \ddots & \ddots \\ 0 & 0 & \cdots & s+d_N \end{pmatrix}, \tag{28}$$

where d_i are the eigenvalues of the matrix G. From the diagonal structure of the matrix G, we can straightly write down its inverse like:

$$(sI+D)^{-1}=\begin{pmatrix} \frac{1}{s+d_1} & 0 & \cdots & 0 \\ 0 & \frac{1}{s+d_2} & \cdots & 0 \\ \vdots & \vdots & \ddots & \ddots \\ 0 & 0 & \cdots & \frac{1}{s+d_N} \end{pmatrix}. \tag{29}$$

Performing the Laplace transform inversion of equation (29) by using the standard results of the Laplace transform theory we obtain:

$$L^{-1}\left\{(sI+D)^{-1}\right\}=E(t)=\begin{pmatrix} e^{-d_1 t} & 0 & \cdots & 0 \\ 0 & e^{-d_2 t} & \cdots & 0 \\ \vdots & \vdots & \ddots & \ddots \\ 0 & 0 & \cdots & e^{-d_N t} \end{pmatrix}. \tag{30}$$

Finally, substituting the above ansatz in equation (27), we come out with the following solution for problem (17),

$$Y(t)=X\begin{pmatrix} e^{-d_1 t} & 0 & \cdots & 0 \\ 0 & e^{-d_2 t} & \cdots & 0 \\ \vdots & \vdots & \ddots & \ddots \\ 0 & 0 & \cdots & e^{-d_N t} \end{pmatrix} X^{-1}Y_0. \tag{31}$$

To this point, it is relevant to underline that we are aware that problem (17) has a well-known solution. However, we must point out that the discussed solution is a robust algorithm, under computational point of view, to solve the problem with large N (N of order of 1, 500) and a small computational effort. Furthermore, this methodology can also be readily applied for the solution of equation (17) with boundary condition. This kind of a problem appears in the solution of discrete ordinates equation in a slab by the LTS_N approach. For more information, see [35].

Next, we extend this methodology to the solution of more general linear ordinary differential equation having the entries of the matrices E and F varying with time. To reach this goal, let us consider next the problem:

$$E(t)\frac{dY(t)}{dt}+F(t)Y(t)=0 \qquad t>0, \tag{32}$$

subject to the initial condition,

$$Y(0)=Y_0. \tag{33}$$

To solve this problem by the Laplace transform technique, we perform a stepwise approximation of the matrices E(t) and F(t). So far, bearing in mind the interest of finding a solution for the time ranging from zero to a prescribed time T, we split the interval (0, T) into sub-interval (t_{n-1}, t_n) for n=1:M, where M denotes the number of sub-intervals. For each sub-interval, we take the averaged values for the entries of matrix E(t) and F(t). It turns out that problem (28) simplifies to the recursive set of problems,

$$E^m\frac{dY^m(t)}{dt}+F^mY^m(t)=0, \tag{34}$$

for t in the sub-interval (t_{m-1}, t_m) and m ranging from 1 to M. Here, E_m and F_m are now constant matrices. Henceforth, the previously discussed solution can be applied in a straightforward manner. Indeed, the solution for problem (34) generically reads like,

$$Y^{(m)} = X^{(m)} \begin{pmatrix} e^{-d_1^{(m)}t} & 0 & \cdots & 0 \\ 0 & e^{-d_2^{(m)}t} & \cdots & 0 \\ \vdots & \vdots & \ddots & \ddots \\ 0 & 0 & \cdots & e^{-d_N^{(m)}t} \end{pmatrix} X^{(m)-1} Y^{(m-1)}, \tag{35}$$

for t in the interval (t_{m-1}, t_m) and the initial condition $Y^{(m-1)}$ given by the previously calculated solution at t_{m-1} for t in the interval (t_{m-2}, t_{m-1}).

Focusing our attention on the task of searching analytical solution for the GITT linear transformed equation, in the sequel, we report an analytical alternative approach to solve equation (19), skipping the stepwise approximation of matrices E(t) and F(t) entries. To hit this objective, we solve problem (33) by the decomposition method proposed by [36]. In order to apply the decomposition method, let us recast equation (34) as,

$$\bar{E}\frac{dY(t)}{dt} + \bar{F}Y(t) = A(t)\frac{dY(t)}{dt} + B(t)Y(t), \tag{36}$$

where the matrices A(t) and B(t) are, respectively, expressed as,

$$A(t) = E(t) + \bar{E} \tag{37}$$

and

$$B(t) = F(t) + \bar{F}. \tag{38}$$

Here, $\bar{E}$ and $\bar{F}$ are constant matrices properly chosen. Now, expanding the solution of equation (36) in the truncated series,

$$Y(t) = \sum_{k=0}^{L} u_k(t) \tag{39}$$

and replacing this ansatz in equation (36), we get,

$$\bar{E}\frac{d}{dt}\sum_{k=0}^{L} u_k(t) + \bar{F}\sum_{k=0}^{L} u_k(t) = A(t)\frac{d}{dt}\sum_{k=0}^{L} u_k(t) + B(t)\sum_{k=0}^{L} u_k(t). \tag{40}$$

From equation (40), we are in a position to construct the following recursive set of linear ordinary differential matrix equations with constant matrices,

$$\bar{E}\frac{du_0(t)}{dt}+\bar{F}u_0(t)=0, \tag{41}$$

$$\bar{E}\frac{du_1(t)}{dt}+\bar{F}u_1(t)=A(t)\frac{du_0(t)}{dt}+B(t)u_0(t), \tag{42}$$

$$\bar{E}\frac{du_2(t)}{dt}+\bar{F}u_2(t)=A(t)\frac{du_1(t)}{dt}+B(t)u_1(t) \tag{43}$$

and so forth. Generically, we may write,

$$\bar{E}\frac{du_n(t)}{dt}+\bar{F}u_n(t)=A(t)\frac{du_{n-1}(t)}{dt}+B(t)u_{n-1}(t), \tag{44}$$

for n = 1:L. Given a closer look at the above equations, we readily realize that the RHS of the recursive set of equations from (42) to (44) are known terms. Therefore, to obtain the solution of these equations, we must solve the following problem with constant matrices and source, namely,

$$\bar{E}\frac{dY(t)}{dt}+\bar{F}Y(t)=S(t) \qquad t>0, \tag{45}$$

which has the well-known solution,

$$Y(t)=Y_h(t)Y_0+Y_h(t)*S(t), \tag{46}$$

whereas the homogeneous solution $Y_h(t)$ is given by equation (35). Here, star denotes convolution. Therefore, the solution of the generic equation (44) has the form,

$$\begin{aligned} Y(t)=X&\begin{pmatrix} e^{-d_1 t} & 0 & \cdots & 0 \\ 0 & e^{-d_2 t} & \cdots & 0 \\ \vdots & \vdots & \ddots & \ddots \\ 0 & 0 & \cdots & e^{-d_N t} \end{pmatrix} X^{-1}Y_0 + \\ &+\int_0^t \left(X(t')\begin{pmatrix} e^{-d_1 t'} & 0 & \cdots & 0 \\ 0 & e^{-d_2 t'} & \cdots & 0 \\ \vdots & \vdots & \ddots & \ddots \\ 0 & 0 & \cdots & e^{-d_N t'} \end{pmatrix} X(t')^{-1} S(t-t') \right) dt' \end{aligned} \tag{47}$$

and consequently the solution of equation (32) is well determined by equations (39) and (41) to (44).

Finally, regarding the task of solving second-order ordinary differential transformed equation, we must recall that this equation can be transformed into a set of first-order matrix differential equation by changing the variable. Indeed, performing the substitution of,

$$Y(t) = Z_1(t) \tag{48}$$

and

$$\frac{dY(t)}{dt} = Z_2(t), \tag{49}$$

in the following transformed problem:

$$\frac{d^2Y(t)}{dt} + A(t)\frac{dY(t)}{dt} + B(t)\,Y(t) = S(t), \tag{50}$$

we come out with the ensuing result:

$$\frac{d}{dt}\begin{pmatrix} Z_1(t) \\ Z_2(t) \end{pmatrix} + \begin{pmatrix} 0 & -1 \\ B(t) & A(t) \end{pmatrix}\begin{pmatrix} Z_1(t) \\ Z_2(t) \end{pmatrix} = \begin{pmatrix} 0 \\ S(t) \end{pmatrix}, \tag{51}$$

which is a set of first-order ordinary matrix differential equation that can be promptly solved by the previous discussed methods for nondegenerated matrix. Finally, to handle transformed problems with degenerated matrix, we proceed by performing the inversion of the symbolic matrix by the Schur decomposition approach. Furthermore, we make the Laplace transform inversion applying the Heaviside expansion formulation for eigenvalues with multiplicity larger than one. Hoping that we have completed the mathematical analysis regarding the task of solving analytically the GITT transformed problems, in the next section, we illustrate the GILTT aptness to simulate pollutant dispersion in the atmosphere by solving the two-dimensional steady-state advection-diffusion equation.

3.1.1. An example: Solution of the 2D advection-diffusion equation

To solve the problem (7) by the GILTT method we recast equation (7) as ($S = 0$):

$$u(z)\frac{\partial \overline{c}}{\partial x} = K_z(z)\frac{\partial^2 \overline{c}}{\partial z^2} + \left(\frac{\partial K_z(z)}{\partial z}\right)\frac{\partial \overline{c}}{\partial z} \tag{52}$$

with the following boundary conditions:

$$K_z \frac{\partial \bar{c}}{\partial x} = 0 \text{ at } z = h \tag{53}$$

$$u\bar{c}(x,z) = Q\delta(z - H_s) \text{ at } x = 0 \tag{54}$$

and we construct the following associated Sturm-Liouville problem:

$$\Psi_i''(z) + \lambda_i^2 \Psi_i(z) = 0 \text{ at } 0 < z < h \tag{55}$$

$$\Psi_i'(z) = 0 \text{ at } z = 0, h, \tag{56}$$

which has the well-known solution:

$$\Psi_i(z) = \cos(\lambda_i z), \tag{57}$$

where the eigenvalues λ_i are given by $\lambda_i = \frac{i\pi}{h}$ for $i=0,\ 1,\ 2,\ 3,\ \ldots$.Furthermore, The eigenfunctions $\Psi_i(z)$ satisfy the ensuing orthonormality condition:

$$\frac{1}{N_m^{1/2} N_n^{1/2}} \int_V \Psi_m(z)\Psi_n(z)dv = \begin{cases} 0, & m \neq n \\ 1, & m = n \end{cases}, \tag{58}$$

where N_m is expressed by:

$$N_m = \int_V \Psi_m^2(z)dv. \tag{59}$$

Having solved the Sturm-Liouville problem, we are in a position to construct the GILTT transform formula, which has the form:

$$\bar{c}(x,z) = \sum_{i=0}^{\infty} \frac{\overline{c_i(x)\Psi_i(z)}}{N_i^{1/2}}. \tag{60}$$

Now, replacing the above ansatz in equation (52) we have,

$$u\sum_{i=0}^{\infty}\frac{\overline{c_i'(x)}\Psi_i(z)}{N_i^{1/2}}=K_z\sum_{i=0}^{\infty}\frac{\overline{c_i(x)}\Psi_i''(z)}{N_i^{1/2}}+\left(\frac{\partial K_z}{\partial z}\right)\sum_{i=0}^{\infty}\frac{\overline{c_i(x)}\Psi_i'(z)}{N_i^{1/2}}. \tag{61}$$

Here, we adopt the prime notation for the first derivative and double the prime notation for the second derivative. Multiplying equation (61) by $\frac{\Psi_j(z)}{N_j^{1/2}}$ and integrating from z equal zero to h, we read:

$$\begin{aligned}&-\sum_{i=0}^{\infty}\frac{\overline{c_i'(x)\lambda}}{N_i^{1/2}N_j^{1/2}}\int_0^h u\Psi_i\Psi_j\,dz-\sum_{i=0}^{\infty}\frac{\overline{c_i(x)}\,{}_i^2}{N_i^{1/2}N_j^{1/2}}\int_0^h K_z\Psi_i\Psi_j\,dz+\\&+\sum_{i=0}^{\infty}\frac{\overline{c_i(x)}}{N_i^{1/2}N_j^{1/2}}\int_0^h\left(\frac{\partial K_z}{\partial z}\right)\Psi_i'\,\Psi_j\,dz=0\end{aligned} \tag{62}$$

for $j=0,\ 1,\ 2,\ \ldots,\ N$. Rewriting equation (62) in matrix fashion, we obtain,

$$Y'(x)+FY(x)=0, \tag{63}$$

where Y(x) is the column vector whose components are $c_j(x)$ and the matrix F is defined as $F=B^{-1}E$. Here, the entries of matrices B and E are, respectively, given by:

$$b_{i,j}=\int_0^h u\Psi_i\Psi_j dz \tag{64}$$

and

$$e_{i,j}=\int_0^h\frac{\partial K_z}{\partial z}\Psi_i'\Psi_j dz-\lambda_i^2\int_0^h K_z\Psi_i\Psi_j dz. \tag{65}$$

Transforming the boundary condition (54) by a similar procedure, we mean, multiplying the boundary condition (54) by $\frac{\Psi_j(z)}{N_j^{1/2}}$ and integrating from z = 0 to h, we obtain:

$$\int_0^h\sum_{i=0}^{\infty}\frac{\overline{c_i(0)}\,u\Psi_i\Psi_j}{N_i^{1/2}N_j^{1/2}}dz=\int_0^h\frac{Q\,\delta\,(z-H_s)\Psi_j}{N_j^{1/2}}dz. \tag{66}$$

Using the orthonormality property of the eigenfunctions $\Psi_j(z)$, and the integration property of the generalized delta function, we get the following transformed boundary condition:

$$c_0(0)=\frac{Q\Psi_0(H_s)\sqrt{h}}{u(z)\Psi_0^2(z)dz} \tag{67}$$

for j = 0, and

$$c_i(0)=\frac{Q\Psi_j(H_s)\sqrt{h/2}}{\int_0^h u(z)\Psi_j^2(z)dz} \tag{68}$$

for j ranging from 1 to N. To this point, we are in a position to write down the solution of equation (39), subjected to the boundary condition given by equations (67) and (68), using the results discussed in the previous section. Indeed, the solution writes like,

$$\bar{c}(x,z)=\sum_{i=0}^{\infty}\frac{\overline{\overline{c_i(x)}}\Psi_i(z)}{\mathrm{N}_i^{1/2}}. \tag{69}$$

Here, $c_i(x)$ are the N components of the vector Y(x) given by,

$$Y(x)=Z\begin{pmatrix} e^{-f_1x} & 0 & \cdots & 0 \\ 0 & e^{-f_2x} & \cdots & 0 \\ \vdots & \vdots & \ddots & \vdots \\ 0 & 0 & \cdots & e^{-f_Nx} \end{pmatrix}G(x)Z^{-1}Y(0), \tag{70}$$

where f_k, for k ranging from 1 to N, are the eigenvalues of the matrix of eigenfunctions and Z^{-1} its inverse. For more details and recent developments, see [29, 30, 37, 38, 39, 40, 41].

4. Remarks

To deal with more realistic situations we need to change to numerical methods. However, it is helpful to check firstly possible analytical solutions in order to obtain a known framework and test solutions. The analytical solutions are efficient for many applications, for example: supply analysis of contaminant scenarios, leading sensitivity analyses to investigate the effects of parameters included in contaminant transport, extrapolation over large times and distances where numerical solutions may be impractical, as screening models for transport processes that cannot be solved exactly, and for validating numerical solutions.

Today, air pollution problems are not treated in the manner described in the present chapter. There are various air pollution situations that require the use of complex mesoscale models to properly describe the dispersion processes and properly represent the relevant chemistry and emission processes. Complex models, such as the CMAQ model (the Community Multiscale Air Quality model), have been designed to simulate air quality by including state of the art techniques for modeling multiple air quality issues. However, in complex models, increasingly more processes, such as sea breeze circulations, urban heat islands, and waves, are represented. Therefore, these models are often perceived as "closed" that cannot without difficulty or effort report the influence of individual processes on air quality. Finally, for many scientific applications, analytical solutions have utility in understanding air dispersion phenomena and some air chemistry phenomena, showing their usefulness in environmental management.

Author details

Davidson Martins Moreira[1*], Tiziano Tirabassi[2] and Taciana Albuquerque[3]

*Address all correspondence to: davidson.moreira@fieb.org.br

1 Center for Integrated Manufacturing and Technology (CIMATEC), Salvador, Brazil

2 Institute ISAC of CNR, Bologna, Italy

3 Universidade Federal de Minas Gerais, Belo Horizonte, Brazil

References

[1] Arya, S. 1999. Air pollution meteorology and dispersion. Oxford University Press, New York, 310 pp.

[2] Seinfeld, J.H., and Pandis, S.N. 1998. Atmospheric chemistry and physics. John Wiley & Sons, New York, 1326 pp.

[3] Tirabassi, T. 2003. Operational advanced air pollution modeling. PAGEOPH 160(1–2), 5–16.

[4] Moreira, D.M., Tirabassi, T., and Carvalho, J.C. 2005a. Plume dispersion simulation in low wind conditions in stable and convective boundary layers. Atmos. Environ. 39(20), 3643–3650.

[5] Roberts, O.F.T. 1923. The theoretical scattering of smoke in a turbulent atmosphere. Proc. Roy. Soc. 104, 640–648.

[6] Rounds, W. 1955. Solutions of the two-dimensional diffusion equation. Trans. Am. Geophys. Union 36, 395–405.

[7] Smith, F.B. 1957a. The diffusion of smoke from a continuous elevated point source into a turbulent atmosphere. J. Fluid Mech. 2, 49–76.

[8] Smith, F.B. 1957b. Convection-diffusion processes below a stable layer. Meteorological Research Committee, N. 1048 and 10739, London.

[9] Scriven, R.A., and Fisher, B.A. 1975. The long range transport of airborne material and its removal by deposition and washout-II. The effect of turbulent diffusion. Atmos. Environ. 9, 59–69.

[10] Yeh, G.T., and Huang, C.H. 1975. Three-dimensional air pollutant modelling in the lower atmosphere. Bound. Layer Meteor. 9, 381–390.

[11] Berlyand, M.Y. 1975. Contemporary problems of atmospheric diffusion and pollution of the atmosphere. Translated version by NERC, USEPA, Raleigh, NC.

[12] Demuth, C. 1978. A contribution to the analytical steady solution of the diffusion equation for line sources. Atmos. Environ. 12, 1255–1258.

[13] Tagliazucca, M., Nanni, T., and Tirabassi, T. 1985. An analytical dispersion model for sources in the surface layer, Nuovo Cimento 8C, 771–781.

[14] Tirabassi, T., Tagliazucca, M., and Zannetti, P. 1986. KAPPA-G. a non-Gaussian plume dispersion model: description and evaluation against tracer measurements. JAPCA 36, 592–596.

[15] Tirabassi, T. 1989. Analytical air pollution advection and diffusion models. Water, Air Soil Poll. 47, 19–24.

[16] Van Ulden, A.P. 1978. Simple estimates for vertical diffusion from sources near the ground. Atmos. Environ. 12, 2125–2129.

[17] Tirabassi, T., and Rizza, U. 1995. A practical model for the dispersion of skewed puffs. J. Appl. Meteor. 34, 989–993.

[18] Nieuwsadt, F.T.M. 1980. An analytical solution of the time-dependent, one-dimensional diffusion equation in the atmospheric boundary layer. Atmos. Environ. 14, 1361–1364.

[19] Nieuwstadt, F.T.M., and de Haan, B.J. 1981. An analytical solution of one-dimensional diffusion equation in a nonstationary boundary layer with an application to inversion rise fumigation. Atmos. Environ. 15, 845–851.

[20] Catalano G.D. 1982. An analytical solution to the turbulent diffusion equation with mean vertical wind. In: Proceedings of the 16th Southeastern Seminar on Thermal Sciences, Miami, 19–21 April 1982, 143–151.

[21] Chrysikopoulos, C.V., Hildemann, L.M., and Roberts, P.V. 1992. A three-dimensional atmospheric dispersion-deposition model for emissions from a ground-level area source. Atmos. Environ. 26, 747–757.

[22] Lin, J.S., and Hildemann, L.M. 1997. A generalised mathematical scheme to analytically solve the atmospheric diffusion equation with dry deposition. Atmos. Environ. 31, 59–71.

[23] Brown, M.J., Arya, S.P., and Snyder, W. 1997. Plume descriptors from a non-gaussian concentration model. Atmos. Environ. 31, 183–189.

[24] Sharan, M., and Yadav, A.K. 1998. Simulation of diffusion experiments under light wind, stable conditions by a variable K-theory model. Atmos. Environ. 32, 3481–3492.

[25] Sharan, M., and Modani, M. 2006. A two-dimensional analytical model for the dispersion of air-pollutants in the atmosphere with a capping inversion. Atmos. Environ. 40, 3479–3489.

[26] Essa, K.S.M., Etman, S.M., and Embaby, M. 2007. New analytical solution of the dispersion equation. Atmos. Res. 84, 337–344.

[27] Vilhena, M.T., Rizza, U., Degrazia, G.A., Mangia, C., Moreira, D.M., and Tirabassi, T. 1998. An analytical air pollution model: development and evaluation. Contr. Atmos. Phys. 71, 315–320.

[28] Moreira, D.M., Moraes, A.C., Goulart, A.G., and Albuquerque, T.T. 2014. A contribution to solve the atmospheric diffusion equation with eddy diffusivity depending on source distance. Atmos. Environ. 83, 254–259.

[29] Wortmann, S., Vilhena, M.T., Moreira, D.M., and Buske, D. 2005. A new analytical approach to simulate the pollutant dispersion in the PBL. Atmos. Environ. 39, 2171–2178.

[30] Moreira, D. M., Vilhena, M.T., Tirabassi, T., Buske, D., and Cotta, R.M. 2005b. Near source atmospheric pollutant dispersion using the new GILTT method. Atmos. Environ. 39, 6289–6294.

[31] Cotta, R.M. 1993. Integral transforms in computational heat and fluid flow. CRC Press, Boca Raton.

[32] Cotta, R.M., and Mikhaylov, M. 1997. Heat conduction lumped analysis, integral transforms, symbolic computation. John Wiley & Sons, Chichester.

[33] Cheroto, S., Mikhailov, M.D., Kakaç, S., and Cotta, R.M. 1999. periodic laminar forced convection: solution via symbolic computation and integral transforms. Int. J. Therm. Sci. 38, 613–621.

[34] Alves, L.S., Cotta, R.M., and Pontes, J. 2002. Stability analysis of natural convection in porous cavities through integral transforms. Int. J. Heat Mass Transfer 45, 1185–1195.

[35] Segatto, C.F., and Vilhena, M.T. 1999. The state-of-the-art of the LTS_N method, mathematics and computation, reactor physics and environmental analysis in nuclear applications – International Conference, Madrid, 2, 1618–1631.

[36] Adomian, G. 1988. A review of the decomposition method in applied mathematics. J. Math. Anal. Appl. 1(135), 501–544.

[37] Buske, D., Vilhena, M.T., Moreira, D.M., and Tirabassi, T. 2007a. Simulation of pollutant dispersion for low wind conditions in stable and convective planetary boundary layer. Atmos. Environ. 41, 5496–5501.

[38] Buske, D., Vilhena, M.T., Moreira, D.M., and Tirabassi T. 2007b. An analytical solution of the advection-diffusion equation considering non-local turbulence closure. Environ. Fluid Mech. 7, 43–54.

[39] Moreira, D.M., Vilhena, M.T., Buske, D., and Tirabassi, T. 2009. The state-of-art of the GLTT method to simulate pollutant dispersion in the atmosphere. Atmos. Res. 92, 1–17.

[40] Tirabassi, T., Buske, D., Moreira, D.M., and Vilhena, M.T. 2008. A two-dimensional solution of the advection-diffusion equation with dry deposition to the ground. JAMC 47, 2096–2104.

[41] Tirabassi, T., Tiesi, A., Buske, D., Vilhena, M.T., and Moreira, D.M. 2009. Some characteristics of a plume from a point source based on analytical solution of the two-dimensional advection-diffusion equation. Atmos. Environ. 43, 2223–2229.

Chapter 16

Issues in the Identification of Smoke in Hyperspectral Satellite Imagery — A Machine Learning Approach

Mark A. Wolters and C.B. Dean

Additional information is available at the end of the chapter

http://dx.doi.org/10.5772/60214

1. Introduction

Observations from earth-orbiting satellites play an important role in the study of various large-scale surface and atmospheric phenomena. In many cases the data collected by such satellites are used and communicated in the form of raster images—three-dimensional data arrays where the first two dimensions define pixels corresponding to spatial coordinates. The third dimension contains one or more image *planes*. A greyscale image, for example, has one image plane, while a color (RGB) image has three planes, one each for the brightness in the red, green, and blue parts of the visible spectrum.

The present work is related to *hyperspectral* images, where the number of image planes is much greater than three. In a hyperspectral image with r planes there is associated with each pixel a set of r data values, each measuring a different part of the electromagnetic spectrum.

The general task of analyzing geographic remote sensing imagery is aptly described by Richards [1] (p. 79):

With few exceptions the reason we record images of the earth in various wavebands is so that we can build up a picture of features on the surface. Sometimes we are interested in particular scientific goals but, even then, our objectives are largely satisfied if we can create a map of what is seen on the surface from the remotely sensed data available...

There are two broad approaches to image interpretation. One depends entirely on the skills of a human analyst—a so-called photointerpreter. The other involves computer assisted methods for analysis, in which various machine algorithms are used to automate what would otherwise be an impossibly tedious task.

Here, we will consider methods that are useful for the second approach: computer-assisted photointerpretation. Computer-aided analysis is particularly helpful for hyperspectral images, which contain too many planes to be visualized in a simple human-readable form.

The present work can be viewed as a case study in the application of machine learning approaches to a difficult task in remote sensing image segmentation. The remainder of this section introduces the problem we are addressing, the data we are using, and the modelling approach we will follow. In Section 2, important ideas from the field of classification are introduced in a tutorial format for researchers who might not be familiar with the topic. Those with prior experience in the area may wish to skip the section. Sections 3 and 4 describe the methods used and the results obtained. Sections 5 and 6 provide discussion and conclusions.

1.1. The problem

The application of interest is the automated identification of smoke from forest fires using hyperspectral satellite images. Smoke released from forest fires can be transported large distances and affect air quality over large areas, making it a matter of population health concern. Despite the importance of smoke events, their spatial scale makes them difficult to quantify through direct measurement. Satellite imagery is an alternative information source that could potentially fill a data gap, providing information about smoke over large areas at times of interest.

The work reported here is the first step in a research stream with the ultimate goal of developing a system that can quantify smoke using moderate- to high-resolution remote sensing images covering large geographic areas, and do so with minimal human intervention. If smoke can be quantified through remote sensing image analysis, the resulting data could be used as input to deterministic predictive models of forest fire smoke dispersal, as a validation check for such models, or as an input to retrospective studies of the health impacts of smoke.

Our present objective is twofold: first, to report our current results in developing a classifier for smoke detection, and second, to stimulate other researchers to consider applying similar methods for their own problems in remote sensing image analysis.

1.2. The data

The region of interest in this study covers parts of western Canada and the northwestern United States, and is centered close to the city of Kelowna, British Columbia. It extends from 46.5° to 53.5° latitude, and from -126.5° to -112.5° longitude. Data come from the moderate resolution imaging spectroradiometer (MODIS) aboard the Terra satellite, which provides images with 36 planes covering different spectral bands ranging from the blue end of the visible spectrum (400 nm) to well into the infrared (14 μ m). More information about MODIS can be found in [3, 4].

The Terra satellite follows a polar orbit that allows MODIS to image most of the globe each day, with images captured at mid-morning local time. All data are freely available from the LAADS web data portal [5]. There are numerous data products available, at different levels of processing for different purposes. We used the Level 1B data at 1km resolution, which provides the hyperspectral data in calibrated form corrected for instrumental effects, but without further manipulation. The data are available in chunks called *granules*. Each granule holds the instrument's observations as it passed over a certain portion of the earth's surface during a particular five-minute time interval. If a study region does not happen to be covered by a single granule, it is possible to stitch the data from adjacent granules to cover the region. If the region is large enough, it may be necessary to stitch granules from different orbital passes. In our case, we only used data from time-sequential granules, and not those from different passes, because we found that the smoke and clouds in the scene could change significantly between orbital passes. Because of this it was not always possible to collect complete data for the entire region of interest on every day.

A total of 143 images were collected, one for each day covering the peak dates of the fire season (July 15 to August 31) for the years 2009, 2010, and 2012. Each image is approximately 1.2 megapixels in size, and has spatial resolution of approximately one kilometer per pixel. Images are in plate carrée projection. Any pixel that had data quality concerns (as indicated by error codes in the downloaded data) was excluded from the analysis. The entirety of band 29 was also discarded because of a known hardware failure, leaving 35 spectral bands to be used for classification purposes.

To aid in visualization of the data, an RGB version of each image was produced. Following [6], the RGB images were created by letting bands 1, 4, and 3 fill the red, green, and blue image planes, respectively. First, each of these three bands was run through a saturating linear brightness re-mapping, letting 1 percent of the pixels be saturated at each end of the brightness range. Then, a piecewise linear brightness transformation was carried out on each band, as in the reference.

The resulting RGB images were used for the important task of manually assigning each pixel to either the smoke or nonsmoke class—that is, for specifying what the "true class" of each pixel was. To make this task easier, fire locations (found by comparing bands 22 and 31, as in [7]) were overlaid on the RGB images. While the smoke was sometimes easy to distinguish from the rest of the image, there were also many cases where the choice of true class was quite ambiguous: regions where smoke and cloud were mixed, or regions where the smoke was not highly concentrated, for example. Nevertheless, each pixel in all 143 images was assigned a true class label on a best-efforts basis. The approach to assigning true labels was to assign the smoke class whenever a pixel appeared to have any level of smoke, even a thin haze. The end result was a set of 143 black and white *mask* images corresponding to the hyperspectral ones, with white pixels indicating smoke and black indicating nonsmoke. The complete set of masks comprised 90% nonsmoke pixels and 10% smoke pixels.

As will be shown at the end of this chapter, the difficulty assigning true classes with high confidence is a potentially critical limitation of the analysis. The manual approach to labelling was used nonetheless, since no alternative method exists for identifying smoke pixels across entire images. We note in passing that we have previously obtained some "gold standard" images by request from NASA, and in this case smoke was also identified as hand-drawn regions.

1.3. Modelling approach

The observed images are the product of natural processes that are very complex. From a statistical standpoint, a sequence of remote sensing images covering a particular region of the earth is a spatiotemporal data set with statistical dependence both within and between images. Physically, the presence of smoke in a particular region at a particular time is surely dependent on the characteristics of a particular fire, as well as on meteorological and topographical variables that vary over the region of interest and over time. There is thus ample scope for mathematical complexity in a model used for classification. Some decisions must be made at the outset about which aspects of the problem to include in our classifiers, and which to ignore. As the research is still in its early stages, three simplifying decisions have been made.

First, classification will be conducted based only on the spectral information in the images themselves; no ancillary information (for example, about wind, fire locations, or topography) will be used to aid prediction. This decision was made partly to limit model complexity, but also to ensure that our methods are wholly independent of any physics-based deterministic models (which they might eventually be used to validate). Using only the hyperspectral data also maximizes the applicability of the methods to other image processing tasks.

Second, the focus is on detecting only the presence or absence of smoke. A successful system will be able to classify images on a pixel-by-pixel basis into one of two categories, "smoke" or "nonsmoke."

Third, all pixels and all images are assumed to be independent of one another. While ignoring temporal dependence from image to image does not throw away much information—with images collected at a frequency of once per day, there is little correlation between smoke locations from one image to the next—ignoring spatial dependence within images is clearly making a compromise. Smoke appears in spatially contiguous regions, so knowledge that a certain pixel contains smoke should influence adjacent pixels' probability of being smoke. Nevertheless, spatial association between the outcomes introduces many technical difficulties, so it was not included at this stage of our study.

With these decisions, the smoke detection task becomes a typical *binary classification* or *binary image segmentation* problem, using the data in the 35 spectral bands as predictors. Simplifying the problem in this way is justified in a preliminary analysis. Our goal is to evaluate whether the spectral data contain enough information to allow the smoke and nonsmoke pixels to be distinguished from one another with reasonably high probability. If they do not, there is little to be gained from the added complexity of more sophisticated models; if they do, the simple independent-pixel smoke/nonsmoke model can be extended in a variety of ways to obtain further improvements. Furthermore, it will be seen that despite retreating to a simple model for classification, the problem is still high dimensional, computationally intensive, and challenging.

With these considerations in mind, we use logistic regression for building our classifiers. Logistic regression has convenient extensions for accommodating spatial associations, for handling multiple levels of smoke abundance, and for including additional predictor variables. We anticipate that a final, useful future system will be based on such an extended model.

http://dx.doi.org/10.5772/60214

All analyses presented here were carried out using the free and open source statistical computing software R [2]. An R script demonstrating much of the analysis is available on the corresponding author's website (www.mwolters.com); readers interested in working with the full data set (which is large) can contact the authors by email.

2. Binary classification concepts

Classification is the process of assigning a category (a class label) to an item, using available information about the item. We are interested in binary classification, where there are only two class labels. In our case, the labels are nonsmoke (class 0) and smoke (class 1), the items to be classified are image pixels, and the available information is the content of the hyperspectral image. We say we have "built a classifier" when we have established a rule that tells us how any given pixel in a new image should be classified.

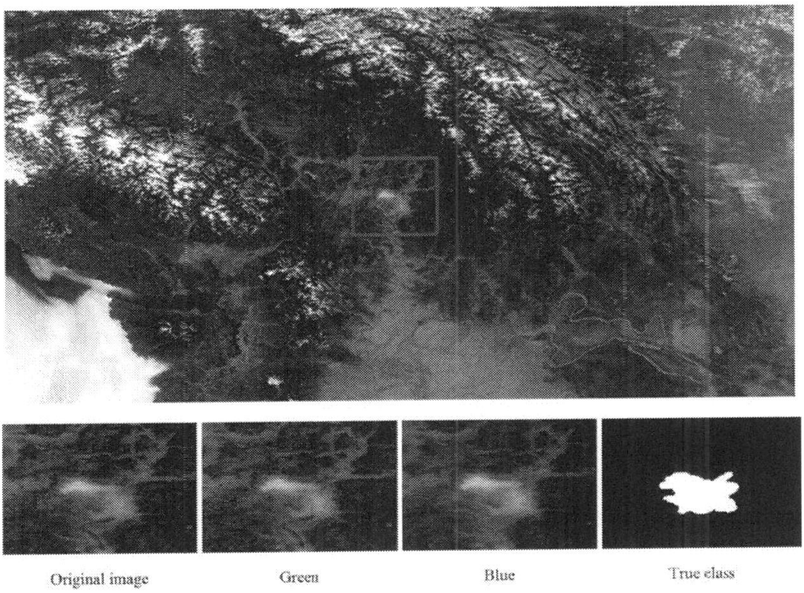

Figure 1. The data used for the example. Top: an RGB image of the study region, with regions of smoke outlined in red. The blue rectangle encloses the pixels that are used for the example. Bottom, from left to right: the RGB sub-image of the region of interest, the green channel, the blue channel, and the mask showing the true smoke (white) and true nonsmoke (black) regions.

Classifier building requires the availability of *training data*—a set of items where the true class labels are known. The reliance on training data is one reason classification is also known as *supervised learning*. One may think of an all-knowing supervisor who tells us the class membership of a subset of our items, but then goes home for the day, leaving us to learn for ourselves how to classify the remaining items. To prevent confusion, note that the alternative problem of *unsupervised learning* (where the wise supervisor never shows up, leaving all class labels unknown) is also known as *clustering*, and—although important in its own right—is not presently relevant.

Classification is a large topic. It is, in fact, the dominant activity in the field of machine learning. Consequently, no attempt is made here to provide a thorough review of the subject. Rather, a single classifier based on logistic regression will be discussed as a means of introducing common themes in classification. The logistic classifier is naturally suited to binary classification problems, and has a relatively simple form with strong connections to linear and nonlinear regression. This classifier will be used throughout the chapter.

Readers interested in further background on classification, and alternative classifiers, have many resources to turn to. The books [1, 8, 9, 10] provide accessible introductions to the topic, and [1] in particular discusses classification and many related topics in the context of remote sensing imagery. Note that while alternative classification methods may have better or worse performance in different situations, most of the important aspects of setting up and solving a classification problem remain the same regardless of the particular method chosen.

2.1. A small example

As an illustrative example, we restrict our attention to a small subset of the study data—a portion of a single image—and work with only the RGB image rather than the full hyperspectral data. The large image in Figure 1 shows the entire study region on the chosen date (and also provides an example of what the color images look like on a clear day). The picture contains two areas outlined in red. These are the areas that were deemed to contain smoke during the masking process. The blue rectangle in the image outlines the set of pixels used for this example. The four smaller images at the bottom of the figure show the example data in more detail: the RGB image, the information in the green channel, the information in the blue channel, and the corresponding mask showing the true classes.

The sub-image used for the example is 150 by 165 pixels (24750 pixels in all) and is centered on a smoke plume. To allow the problem to be visualized in two dimensions, we will consider only the green channel (G) and the blue channel (B) as predictors in our classifier.

2.1.1. Logistic classifier with two predictors

The logistic classifier is based on logistic regression, which is set up as follows. Let the true class (the response variable) of the i th pixel be Y_i, with $Y_i=1$ corresponding to smoke and $Y_i=0$ corresponding to nonsmoke. The true class is modelled as a Bernoulli random variable with $\pi_i = P(Y_i=1)$ being the probability of the smoke outcome. All pixels are assumed to be statistically independent.

http://dx.doi.org/10.5772/60214

Logistic regression models the log-odds of pixel i being smoke (the event $Y_i=1$) as a linear combination of predictor variables (the green and blue brightness values, in this case):

$$\log\left(\frac{\pi_i}{1-\pi_i}\right) = \beta_0 + \beta_1 G_i + \beta_2 B_i, \tag{1}$$

where G_i and B_i are the green and blue values of the i th pixel, and $\{\beta_0, \ \beta_1, \ \beta_2\}$ are the model coefficients. These three coefficients are to be estimated from a set of pixels for which both the responses and the predictors are known. Estimation is done using a weighted least squares or (equivalently) maximum likelihood approach. The process is called *model fitting* or *training*, and software for performing the estimation is readily available.

Once the parameters are estimated, the fitted model can be used to generate predictions for any given pixel, whether or not the response has been observed. Let x_j represent such a pixel, with predictor values G_j and B_j. Plugging G_j, B_j, and the fitted coefficients into the right hand side of (1), the equation can be solved for $\hat{\pi}_j$, the *fitted probability*. This quantity is the estimated probability that pixel j belongs to the smoke class.

The logistic regression model gives us fitted probabilities on a continuous scale from zero to one. To convert the model into a binary classifier, one need only specify a cutoff probability, c. If $\hat{\pi}_j$ is less than c, pixel j will be put into class 0 (nonsmoke), and if $\hat{\pi}_j$ is greater than c, it will be put into class 1 (smoke). We choose $c=0.5$, so that each pixel is put into the class that is more probable under the model.

Returning to the example data, the above procedure was followed using the 24750 chosen pixels and their true class labels as training data to fit model (1). The nature of the resulting fitted model is shown in Figure 2. The figure plots each pixel as a point in the (green, blue) plane. In machine learning, predictor variables are often called *features*, and so this plot considers each pixel in the model's *feature space*. We see that the smoke pixels generally occur at higher values of both blue and green, but that there is overlap between the two classes; the two classes are not completely separable. The fitted logistic regression model allows us to calculate a probability of being smoke for any point in the feature space. The thick line on the plot is the probability 0.5 contour of this probability surface; it is the decision boundary for our classifier with $c=0.5$. The model will classify any pixel above this line as smoke, and any pixel below the line as nonsmoke.

The inset image in the figure shows the classifier's predictions. White pixels in this image indicate pixels estimated to have greater than 50% chance of being smoke. The red outline indicates the boundary of the true smoke region. While most of the pixels are classified correctly, many are not.

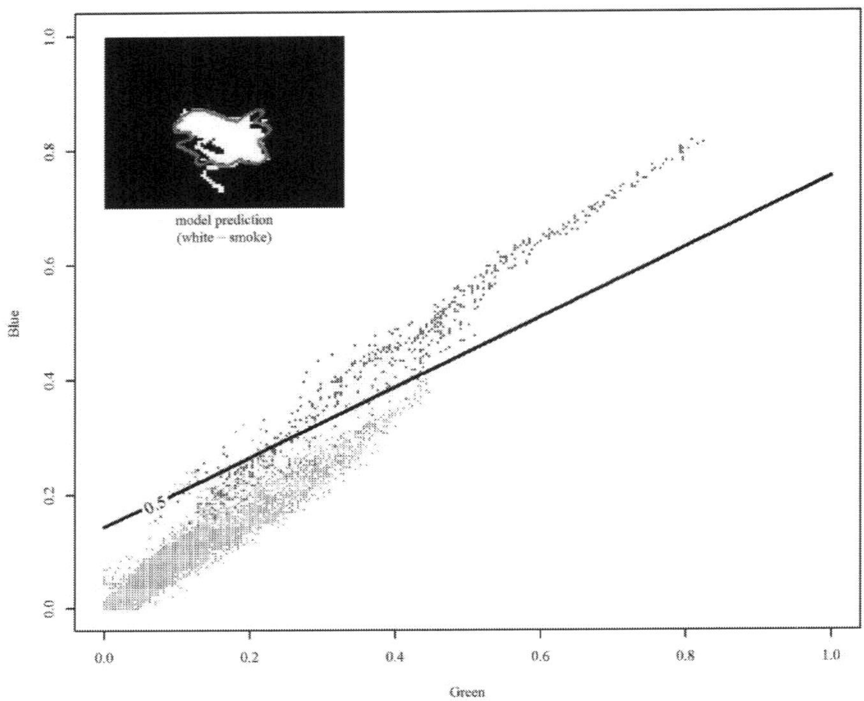

Figure 2. Results of fitting the two-predictor model (G, B) to the example image. Blue points are smoke pixels and red points are nonsmoke. The line on the plot gives the 50% probability line that can be used to discriminate one class from the other. The inset image shows the predicted classes using this model; the red outline in the inset is the boundary of the true smoke region.

2.1.2. Logistic classifier with expanded feature space

The mathematical structure of the previous model ensured that the decision boundary in Figure 1 had to be a straight line. This limited the ability of the classifier to discriminate between the two classes. To make the model more flexible, we can expand the size of the feature space by adding nonlinear functions of the original predictors G and B. For example, we can consider the model

$$\log\left(\frac{\pi_i}{1-\pi_i}\right) = \beta_0 + \beta_1 G_i + \beta_2 B_i, + \beta_3 \mathrm{G}_i^2 + \beta_4 \mathrm{B}_i^2 + \\ + \beta_5 \mathrm{G}_i \mathrm{B}_i + \beta_6 \mathrm{G}_i^3 + \beta_7 \mathrm{G}_i \mathrm{B}_i^2 + \beta_8 \mathrm{B}_i \mathrm{G}_i^2 + \beta_9 \mathrm{B}_i^3 + \beta_{10} \mathrm{G}_i^2 \mathrm{B}_i^2, \tag{2}$$

which includes the original variables G_i and B_i, along with squared and cubed terms (like G_i^2 and G_i^3) as well as products between the original variables taken to various powers (as in G_iB_i and $B_iG_i^2$). Borrowing terminology from industrial experimentation, we call the original variables *main effects* and any terms involving products of variables *interactions*.

The right hand side of model (2) is still a linear combination of various predictor variables, but we have expanded the feature space to ten dimensions. Considered as a function of G and B, the model is able to handle nonlinear relationships between these main effects. In Figure 3 we see the results of fitting this model to the example data. The figure shows the same scatter plot of the data, but now with the 50% contour line for this more flexible model. By adding extra features we can define a decision boundary with more complex shape. The additional shape flexibility of this boundary allows the classifier to correctly assign classes to a greater proportion of the pixels, as seen in the inset prediction image.

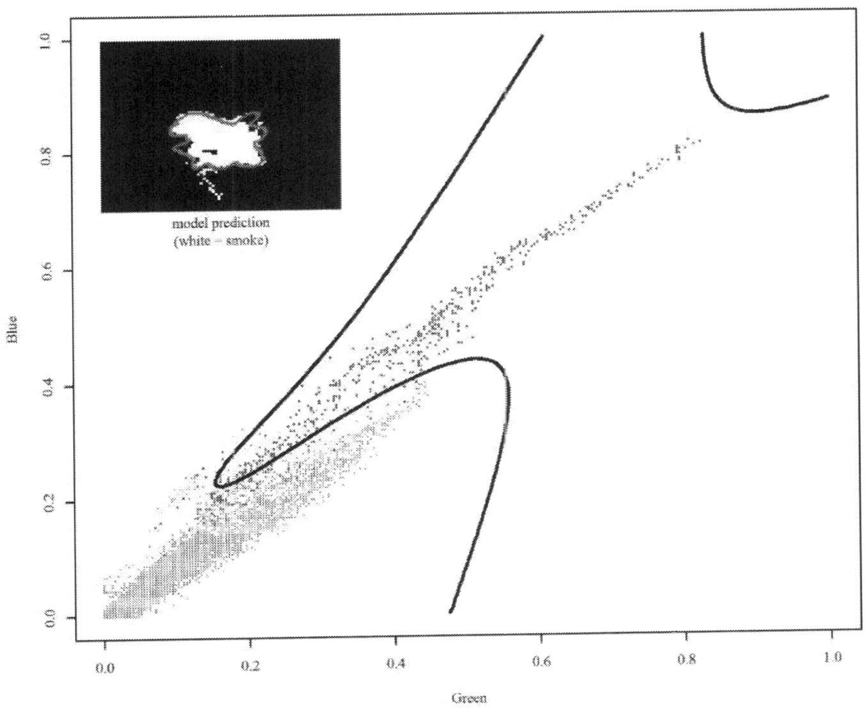

Figure 3. Results of fitting the example data to the 10-predictor model (G, B, G^2, B^2, GB, G^3, GB^2, BG^2, B^3, G^2B^2). The plot is constructed in the same way as the previous figure. In this case the class decision boundary can take a complex non-linear shape.

2.2. Other important concepts

The preceding example might tempt one to believe that simply adding more predictors to the model will always yield a better classifier. This is not true, however, for two reasons.

The first problem with arbitrarily growing the feature space is purely computational. In most problems (and certainly in the present study), the measured main effects are correlated with each other to varying degrees. When expanding the feature space, the variables in the model will increasingly suffer from a form of redundancy known *multicollinearity*: certain predictors can (almost) be written as linear combinations of the other predictors. When the degree of multicollinearity is mild, model fitting will still be possible, but the coefficient estimates can be grossly inaccurate (and can vary greatly from sample to sample). As the problem gets worse, fitting will fail due to the occurrence of numerically singular matrices in the estimation routine.

The multicollinearity problem does not preclude us from considering a large feature space, but it means we cannot include *all* variables from a large feature space in the model. This leads to the problem of *model (feature) selection*: when the number of potential predictors is large, we seek to choose a subset of them that produces a good classifier that is numerically tractable.

When selecting a model from a large collection of correlated predictors, it is important to remember that the coefficient estimate of a particular variable will vary depending on which other variables are included in the model. Further, the best-fitting models of two different sizes need not share their variables in common (the variables selected in the best five-variable model, for example, might not be present in the best ten-variable model). For these reasons it is best to consider the performance of a model as a whole, rather than paying undue attention to coefficient values, statistical significance tests, and the like.

The second problem is more fundamental, and can arise even when multicollinearity is not present. The predictions shown in the previous figures were predictions made on the training data itself; the same data were used both for model fitting and for evaluating performance. This circumstance leads to *overfitting* and poor *generalization* ability: the model fits the training data very well but, because the training data is only a sample from the population, the model's predictive power on new data suffers. When considering increasingly complex models, a point is reached at which additional complexity only detracts from out-of-sample prediction accuracy.

The remedy for overfitting again involves model selection. Because of overfitting, larger models are not necessarily better, so the challenge is to select a model of intermediate size that is best at what is really important, out-of-sample prediction. To do this, one must use different samples of the data for different parts of the procedure. Ideally, one portion of the data (a training set) is used for fitting, another portion (a *validation set*) for model selection, and a third portion (a *test set*) for final evaluation of predictive performance ([9], p. 222).

A final important consideration is the particular measure used for evaluating classifier performance. Any item processed by a binary classifier falls into one of four groups, defined by its true class (0 or 1) and its predicted class (0 or 1). The rates of these four outcomes can be displayed in a so-called confusion matrix, as shown in Table 1. The values $a, \ b, \ c, \ d$ in the table are the rates (relative frequencies) of the four possible outcomes. They must sum to 1. The values b and c (shown in bold) are the rates of the two types of errors: nonsmoke classified as smoke, and smoke classified as nonsmoke. The row sums f_0 and f_1 are the true proportions of items in each class.

Three error rates derived from the confusion matrix are considered subsequently. The *overall error rate* (OER$=b+c$) is simply the global proportion of pixels misclassified. The *classwise error rates* are the rates of misclassification in each class considered separately. We denote these by CER0$=b/f_0$ for the nonsmoke class, and CER1$=c/f_1$ for the smoke class.

Minimizing the OER will be taken as the primary goal of classifier construction. Note however, that our data set consists of 90% nonsmoke pixels (f_0=0.9), so focusing on overall prediction performance implicitly puts more weight on prediction accuracy in the nonsmoke class. Because the data are so unbalanced, even the naïve classification rule "assign all pixels to class 0" can achieve an error rate of only 10% (OER=0.1), but with the highly unsatisfactory classwise rates CER0=0 and CER1=1. More will be said about the trade-off between CER and CER in later discussion.

		Prediction		
		Class 0	**Class 1**	**Sum**
Truth	Class 0	a	$\mathbf{b}$	$a+b=f_0$
	Class 1	$\mathbf{c}$	d	$c+d=f_1$
	Sum	$a+c$	$b+d$	1

Table 1. A confusion matrix. Values in bold represent errors.

3. Experimental methods

The methods just described were applied to the full set of hyperspectral data. The logistic regression classifier was used, just as in the example. In the full-scale analysis, however, it was necessary to handle a much larger data set and a much larger pool of predictor variables. The following sections describe the methods used for preparing the data and searching for a suitable classifier.

3.1. Data splitting and sampling

This analysis took place in a data-rich context. Having a high volume of data is very advantageous, since the available pixels can be split into separate training, validation, and test groups with each group still having more than enough pixels to yield good estimates of the various quantities of interest. The data were randomly split into these three groups at the image level, with a roughly 50/25/25% split: 70 images (82×10^6 pixels) for training, 36 images (42×10^6 pixels) for validation, and 37 images (43×10^6 pixels) for testing.

The drawback of having this much data is the level of computational resources required to handle it. Fitting the logistic regression model requires matrix computations that are memory and computation intensive when the number of cases (pixels) or the number of predictors become large. To estimate a model with the 35 spectral bands as predictors using the full set of training images, for example, approximately 23 GB of RAM is be required just to hold the data in memory. Special techniques are required to perform regression computations on data sets this large. Furthermore, it is necessary to perform model fitting iteratively as part of a model search step, so simple feasibility is not sufficient. Computational run time is also an important factor.

A practical approach to working with such large data sets is to randomly sample a manageable subset of the data, and work with the sample instead. This approach will work well if the sample size can be chosen such that the computations are feasible and sufficiently fast, while still providing estimates of needed quantities (coefficient estimates, prediction error rates) that are sufficiently accurate.

To determine whether such a sample size could be found in the present case, a sequence of preliminary trials was carried out on the test and validation images. In these trials, the model with 35 main effects was fit to numerous independent training samples, and predictions were made on numerous independent validation samples. It was found that sampling 10^5 pixels was adequate for both the training and validation data. At this sample size, predicted probabilities from fitted models exhibited only minor variations (typically differing less than 0.02) when computed from different samples. Similarly, when the validation sample was this size, estimates of prediction error had variance low enough that it should be possible to estimate the prediction error rate on the full validation set to better than the nearest percentage point.

A working sample of 10^5 pixels was therefore drawn from the test images, and an equal-sized sample was drawn from the validation images. Subsequently all parameter estimation and model selection was done using these two samples, rather than the original images.

3.2. Model families considered

In an attempt to build a successful classifier, four groups of models were considered. Each group was defined by i) the set of candidate predictors that have the opportunity to be selected in the model, and ii) the methods used for model selection and model fitting. We attempted to find a single "best" classifier within each group, and carried forward those four best models for subsequent performance evaluations.

Scenario 1: *RGB model.* This model was the same as the first classifier shown in the earlier example, except with all three variables (R, G, B) used instead of only two. This model was included only as a reference point, since it was not expected to perform particularly well. There is only one possible model in this group, so no model selection step was necessary. Coefficients were estimated in the usual least-squares manner for logistic regression.

Scenario 2: *main effects model.* This model family used the 35 hyperspectral bands as candidate predictors. An optimal model with 35 or fewer variables was to be chosen by subset selection. Coefficients were estimated by least squares.

http://dx.doi.org/10.5772/60214

Scenario 3: *all effects model (subset selection)*. The third set of models included a greatly expanded set of predictors. The complete set of candidate variables for this case includes the following sets of variables:

- All 35 main effects.
- The 35 square-root terms.
- The 35 squared terms.
- The 595 interactions between different main effects.
- The 595 interactions between different square-root terms.
- The 595 interactions between different squared terms.
- The 1225 interactions between main effects and square-root terms.
- The 1225 interactions between main effects and squared terms.

In all, there are 4340 candidate variables in this collection. A best model consisting of a (relatively) small portion of these variables was found by subset selection, and coefficient estimation was done by least squares.

Scenario 4: *all effects model (LASSO selection)*. The fourth group of models used the same set of 4340 candidate predictors, but with model selection and parameter estimation carried out using the LASSO technique. Briefly, LASSO is a so-called *shrinkage* or *regularization* method, where parameter estimation and variable selection are done simultaneously. It works by introducing a penalty term into the least squares objective function used to fit the model. The nature of the penalty is such that certain coefficients are forced to take the value zero, effectively eliminating the corresponding variables from the model. The size of the penalty is controlled by a parameter; the larger this parameter, the more variables are removed from the model. The reader is referred to the literature for further details on LASSO and other shrinkage methods (for example, [11, 12, 9]). The LASSO-regularized logistic regression classifier was constructed using the R package glmnet [13].

3.3. Model selection

The main effects and all effects models required model selection by *best subsets*. For a given set of candidate predictors, this approach to model selection depends on two things: an objective function defining how "good" a particular model is, and a search procedure for finding the best model among all possibilities.

In the present case we were interested in out-of-sample prediction performance, so we used the validation sample of pixels to measure the quality of any proposed model. A straightforward measure of model quality is the prediction error rate on the validation data. While this measure could have been used, here a quantity known as *deviance* was used instead. The deviance is defined as −2 times the log-likelihood of the data under the model, and can be interpreted as a measure of lack of fit (smaller deviance indicates a better fit). For the logistic regression model with n pixels, the deviance is

$$-2\sum_{i=1}^{n} d_i, \quad \text{where} \quad d_i = \begin{cases} \log(\hat{\pi}_i) \text{ if pixel } i \text{ is smoke} \\ \log(1-\hat{\pi}_i) \text{ if pixel } i \text{ is nonsmoke,} \end{cases} \tag{3}$$

where $\hat{\pi}_i$ is the predicted probability of pixel i being in class 1. We can see from the equation that the i th pixel's deviance contribution, d_i, shrinks to zero when the predicted probability gets closer to the truth (i.e., when a smoke pixel's predicted probability approaches one, or when a nonsmoke pixel's predicted probability approaches zero). An advantage of the deviance is that it depends in a smooth and continuous way on the fitted probabilities, whereas the prediction error depends only on whether the $\hat{\pi}_i$ values are greater or less than the cutoff c.

In best subsets search, then, the objective function value for any proposed model was found by first estimating the model's coefficients using the training data, and then computing the deviance of the fitted model on the validation data.

Having defined an objective function, it was necessary to search through all possible models to find the best (i.e., minimum deviance) one. This task is challenging, because the combinatorial nature of subset selection causes the number of possible models to grow very quickly when the number of candidate predictors becomes large.

Let the *size* of a particular model be the number of predictors in the model, not including the intercept. Denote model size by k. For the main effects scenario with 35 predictors, there are a manageable 6454 possible models when $k=3$ (i.e., there are 6454 combinations of 3 taken from 35). When $k=5$, however, there are about 325 thousand models from which to choose; and when $k=15$, there are 3.2 billion models. For the all effects scenario with 4340 predictors, the situation is naturally much worse. Even for models of size 3, there are about 13.6 billion possible choices. For larger values of k, the number of possible models becomes truly astronomical, with approximately 10^{30} ten-variable models and about 10^{154} 70-variable models.

Clearly, it is not feasible to search exhaustively through all possible models for either the main effects or all effects scenario. Rather, a search heuristic is required to find a good solution in reasonable time. A traditional approach in such cases is to use sequential model-building procedures like forward, backward, or stepwise selection [14]. These methods have the advantage of convenience, but they lack a valid statistical basis and are generally outperformed by more modern alternatives.

An alternative option, that was pursued here, is to use a more advanced search heuristic to search the space of possible models. We used the function kofnGA, from the R package of the same name [15], to conduct model search using a genetic algorithm (GA). This function searches for best subsets of a specified size, using a user-specified objective function (which we chose to be the validation-set deviance). Instead of considering all possible model sizes, separate searches were run at a range of chosen k values. These were:

For the main effects model: $k=3, 5, 10, 15, 20, 25, 30$.

For the all effects model: $k=3, 10, 20, 30, 40, 50, 60, 70$.

By running the search at only these sizes, we expected to find a model close to the optimal size, without requiring excessive computation times. A discussion of GA methods is beyond the scope of this work, but references such as [16, 17, 18, 19] can be consulted for further information.

When using a search heuristic like GA on a large problem like this, we do not expect that the search will result in finding the single globally-optimal model in the candidate set. In fact if we were to run the search multiple times, it is likely that a variety of solutions will be returned. Nevertheless, the GA can be expected to find a good solution—that is, one with a validation-set deviance close to the minimum—in reasonable time. In practice we expect any model near the minimum deviance will have nearly equivalent predictive performance.

The model selection in the LASSO scenario was done quite differently. As mentioned previously, the LASSO solution depends on a regularization parameter that controls the complexity of the fitted model. For any given value of this parameter, a single model results, with some coefficients zero and some nonzero—the size of the model is implicit in the solution, and is not directly controlled. Model selection thus involves choosing only the value of the regularization parameter. Following the advice of [13], we used validation-set deviance as the measure of model quality for the LASSO fit, and chose the regularization parameter to minimize this quantity.

Note that the LASSO approach enjoys a computational efficiency advantage over the GA-based subset selection approach. For our large training and validation samples (10^5 pixels), fitting the LASSO at 100 values of the regularization parameter took approximately two hours on a contemporary desktop system, while a the longer GA runs (say, with all effects and $k = 50$) took an entire day. Given the overall timeframe of a study like this one, however, the run time difference is not viewed as especially important.

3.4. Performance evaluation

Predictive performance of the best models selected from each group was measured by the overall and classwise error rates OER, CER0, and CER1, as defined in Section 2.2. The probability cutoff c used to map the fitted probabilities onto the two classes was set to its default value of 0.5 for this performance comparison. There is no guarantee that 0.5 actually provides the best value, however. To investigate the impact of varying c, performance of the best model in group 3 was evaluated at a range of c values.

As an adjunct to quantitative assessment, qualitative analysis of model predictions was carried out by visual inspection of the predicted probability maps—greyscale images in which the intensity range [0, 1] represents the predicted probability of each pixel being smoke—from the best model in group 3. For all 37 test images, the probability maps were compared to the original RGB images, to learn more about which aspects of smoke detection were done well, and which were done poorly.

4. Results

The data splitting, sampling, and model selection procedures just described were carried out on the study data, with the net result of producing one best classifier from each of the four scenarios. These four best classifiers were subsequently used to generate predictions for every pixel in the 37 test images. The results of these tasks are presented below, beginning with model selection, and then moving on to the quantitative assessment of prediction performance. The qualitative assessment of performance is reviewed in Section 5.

4.1. Model selection results

The results of model selection are shown in Table 2 and Table 3. The first table lists all of the models considered, along with their deviance and their error rates on the validation data. The error rate estimates in the table are preliminary only, because they are measured on the same validation sample that was used to do variable selection. The final and most accurate measure of out-of-sample predictive performance (the error rates on the test images) are reported in the next section.

The four models selected as best in the four groups are shown in bold in Table 2. For model 1 (RGB), there was only one model, which was selected best by default. For models 2 and 3 (the main effects and all effects models), the best models had $k=20$ and $k=50$, respectively. For model 4 (the LASSO), the minimum-deviance approach chose a model with 109 variables.

Scenario/Model	Results on VALIDATION sample			
	Deviance	OER (%)	CER0 (%)	CER1 (%)
1. RGB	**58549**	**10.2**	**4.3**	**98.4**
2. Main effects, *k* variables				
$k=3$	57245	10.0	0.0	100.0
$k=5$	53162	9.8	0.4	94.3
$k=10$	50399	9.3	0.5	87.9
$k=15$	48521	8.7	0.5	83.0
***k*=20**	**48483**	**8.6**	**0.5**	**82.1**
$k=25$	48704	8.8	0.6	82.8
$k=30$	50144	8.8	0.7	81.7
3. All effects, *k* variables				
$k=3$	51262	9.6	0.4	92.9
$k=10$	42442	7.6	1.1	65.9
$k=20$	40180	7.2	1.1	62.2
$k=30$	39785	7.1	1.2	60.0
$k=40$	38600	6.8	1.3	55.7
***k*=50**	**38174**	**6.8**	**1.4**	**56.0**
$k=60$	38424	6.9	1.6	54.5
$k=70$	38475	6.8	1.6	53.7
4. All effects, LASSO*	**47711**	**8.1**	**1.6**	**66.6**

*The LASSO model shown is the minimum-deviance one, which had 109 nonzero coefficients.

Table 2. List of models considered, with results for the validation set.

Table 3 shows the particular combinations of variables that were chosen in the best models from each of the four groups. The main-effects-only model had 20 variables, the all-effects model had 50 variables, and the LASSO model had 109 variables (of which only 50 are shown). When regression models become this large, it is very difficult to glean any useful information from lists of included variables. Nevertheless, the table is presented for the sake of completeness.

1. RGB Image:
Red, Green, Blue (nonlinear transformations of bands 1, 4, and 3)
2. Main effects, *k*=20:
21, 31, 32, 24, 25, 36, 18, 7, 1, 23, 17, 6, 8, 30, 13, 11, 14, 16, 26, 15
3. All effects, *k*=50:
$\overline{19}:\overline{21}$, $\overline{26}:\overline{21}$, $24:\overline{26}$, $\overline{3}\ \overline{28}$, $\overline{5}:\overline{24}$, $23:\overline{1}$, $28:\overline{4}$, $\overline{2}:\overline{25}$, $33:\overline{21}$,
$\overline{30}:\overline{7}$, $\overline{8}:\overline{25}$, $27:\overline{31}$, $30:\underline{20}$, $22:\overline{2}$, $8:23$, $31:\overline{30}$,
$\overline{9}:\overline{18}$, $24:\overline{27}$, $\underline{23}:\underline{8}$, $30:\underline{34}$, $27:\overline{23}$, $8:\underline{32}$, $11:\overline{36}$,
$11:36$, $\overline{4}:\overline{9}$, $16:\underline{19}$, $23:\overline{19}$, $27:\overline{36}$, $7:\underline{23}$, $11:\underline{6}$,
$\overline{16}:17$, $5:\underline{1}$, $20:\underline{14}$, $\underline{19}:\underline{25}$, $\overline{36}:\overline{35}$, $\underline{7}$, $\overline{16}:\overline{17}$,
$20:\overline{34}$, $22:\underline{5}$, $\overline{23}:\overline{25}$, $\underline{12}\ \underline{26}$, $\underline{13}:33$, $\underline{7}:27$, $19:\underline{13}$,
$\overline{8}:\overline{6}$, $14:\underline{16}$, $35:\underline{8}$, $23:\overline{8}$, $11:\overline{16}$, $35:\overline{12}$
4. All effects, LASSO (109 variables, first 50 shown):
$\overline{24}:\overline{26}$, $\overline{20}:\overline{26}$, $\overline{18}:\overline{25}$, $\overline{7}:\overline{24}$, $6:\overline{24}$, $\overline{22}:\overline{26}$, $1:\overline{30}$, $1:\overline{25}$, $18:\overline{25}$,
$4:\overline{23}$, $1:\overline{32}$, $32:\underline{7}$, $8:\overline{32}$, $23:\underline{4}$, $13:\overline{31}$, $3:\overline{3}$,
$31:\underline{26}$, $2:\overline{30}$, $\overline{22}:\overline{36}$, $31:\underline{18}$, $\overline{10}:\overline{25}$, $\overline{24}:\overline{27}$, $26:\overline{22}$,
$26:\overline{20}$, $17:\overline{22}$, $20:\overline{36}$, $\overline{10}:\overline{32}$, $31:\underline{10}$, $\overline{16}:\overline{31}$, $\overline{27}:\overline{31}$
$30:\underline{10}$, $23:\underline{9}$, $4:\overline{36}$, $\overline{5}:\overline{27}$, $\overline{6}:\overline{18}$, $\overline{9}:\overline{32}$, $7:\overline{18}$,
$\overline{4}:\overline{36}$, $\overline{11}:\overline{20}$, $4:7$, $13:\overline{24}$, $21:\underline{27}$, $\overline{31}:\overline{32}$, $31:\overline{31}$,
$31:\underline{4}$, $\overline{27}:\overline{33}$, $\overline{3}:\overline{9}$, $16:\overline{31}$, $26:\overline{27}$, $\overline{20}:\overline{23}$

Table 3. Chosen variables for the best model in each category. Variables are listed in descending order of coefficient magnitude. See the text for a description of the notation.

A compact notation is used in the table to reduce the space consumed by long lists of variables. In this notation, each of the 35 spectral bands in the original images (the main effects) is represented by its band number. Squared terms are written with a bar over the band number, and square root terms are written with a bar underneath. Interactions between two terms are indicated by a colon. So, for example, the notation $\underline{9}$ refers to the square root of band 9, and $11:\overline{17}$ refers to the interaction between band 11 and the square of band 17.

4.2. Predictive performance

The final estimate of the performance of the four selected models is based on those models' predictions on the complete set 37 test images. Together these images contain over 43 million pixels that were not used in any way during the model fitting and variable selection processes.

Because they are previously unused, they provide a more accurate approximation of the predictive power of the models (better than the validation data, which was not used for parameter estimation, but was used repeatedly for variable selection). The results are shown in Table 4.

	OER (%)	CER0 (%)	CER1 (%)
Model 1: RGB image	10.4	0.5	98.6
Model 2: main effects, 20 variables	8.6	0.5	82.1
Model 3: all effects, 50 variables	8.1	1.9	63.5
Model 4: all effects, LASSO (109 variables)	7.8	1.2	66.0

Table 4. Summary of the selected models and their predictive performance on the test images.

Figure 4 illustrates the trade-off between the different error types as the cutoff c is varied, for the 50-variable all effects model. The plot shows OER, CER0, and CER1 as functions of the cutoff. We can see that the overall error rate is in fact minimized at the original cutoff of 0.5, so changing the cutoff to improve performance on the smoke class will unfortunately come at the cost of worse overall performance. This notwithstanding, both OER and CER0 are relatively flat over the cutoff range (0.3, 0.5). So, for example, setting the cutoff to 0.4 will reduce the classwise error rate of smoke pixels to 50%, while increasing the OER only slightly.

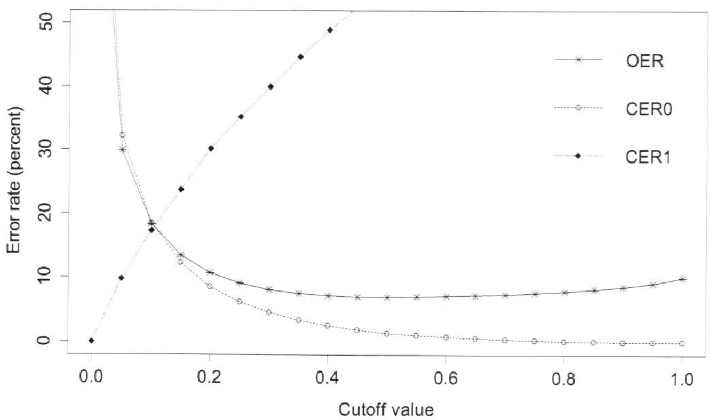

Figure 4. The effect of the decision cutoff on the overall and classwise error rates, for model 3.

5. Discussion

The experimental results are interpreted and discussed below, beginning with several remarks about model selection and performance evaluation, and followed by a qualitative evaluation of the classification results. Afterwards, a variety of suggestions for further improvement are provided.

5.1. Remarks on the selected models

The classification error rates were reported in Table 2 (for all models, on the validation set) and Table 4 (for the best models in each group, on the test set). Considering these tables, we see that our concern about the dominance of the smoke class (class 0) in the data set was justified. All of the models had overall error rates less than about 10%, which seems good at first glance. However in all cases this low error rate was achieved by having a very low error rate in the nonsmoke class (CER0) and a high error rate in the smoke class (CER1). This problem is particularly severe for smaller models and smaller sets of candidate variables, but even the best model in group 3 (the 50-variable model) had 56% misclassification of the smoke pixels.

Comparing the best models from each group, the only two models that can be considered even moderately successful are the two largest ones, the 50-variable all effects model (model 3) and the 109 variable LASSO model (model 4). There is little to separate these two classifiers: both have overall error rates of about 8% on the test set, with model 4 having a slight advantage; but model 3 has better performance on the smoke class.

Interestingly, these two models share only one variable in common (it happens to be 11:6). This is a consequence of the huge feature space and of the correlations among predictors. Two different models containing disjoint sets of variables can both have similar predictive power. This observation is related to the following two remarks.

Remark 1: physical interpretability of selected variables. It is desirable from a scientific and intellectual standpoint to be able to interpret the structure of a predictive model in terms of physical principles, but this is not always straightforward in a machine learning context. In the case of the spectral signature of smoke, a few general characteristics have been observed. Smoke scatters visible light [20], a component of it (organic carbon) is strongly absorbing below about 0.6 μm [21], and it is largely transparent in the middle infrared [22, 23]. We endeavored to interpret our models in light of these observations, but were unable to find any simple and unambiguous relationships based on the patterns of variables included in the models. This is often the price to pay for focusing on out-of-sample predictive accuracy: the classifier becomes a "black box" with internal structure that defies simple interpretation

Remark 2: interpretability of model coefficients. Noticeably absent from the discussion so far has been the actual values of the regression coefficients in the fitted models. This has been deliberate, because in a pure classification problem like this one the predictive performance of the model as a whole is the overriding concern. Interpretability of model coefficients is desirable, but is likely not achievable when we have models with dozens of predictors that are all interactions. Assessment of statistical significance of particular predictors also adds nothing to our understanding of the model as a classifier, and is best avoided.

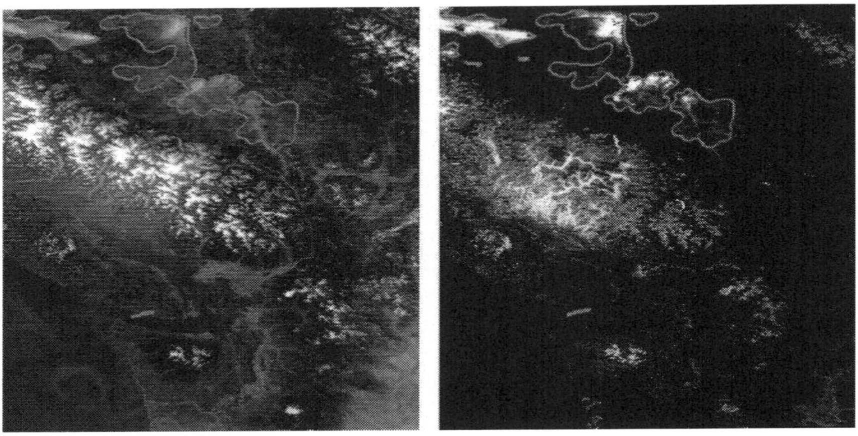

Figure 5. Results on a test image. Left: the RGB image. Right: the predicted probability map using the 50-variable model. The red contour delineates the true smoke region.

5.2. Qualitative performance analysis

Based purely on the observed numerical measures of prediction accuracy, it seems clear that none of the classifiers considered have performance good enough for real-world application, primarily because the majority of smoke pixels are misclassified in all cases. Visual inspection of the predictions on the test images can yield further insight into the nature of the problem, and possible causes of difficulty. Figure 5 and Figure 6 provide prototypical examples drawn from the test images. Our qualitative conclusions about predictive performance, based on the full set of 37 images, are listed below.

1. *Smoke-free images are generally classified well.* The classifier does have *some* ability to detect smoke, so it is still encouraging to observe that smoke-free images, or large regions that are smoke-free, are generally classified accurately. This can be observed in the bottom and left portions of Figure 6, which are assigned low probabilities throughout, despite the presence of clouds, water, and various types of terrain.

2. *Clouds and smoke can be distinguished well from one another.* It was observed that throughout the 37 test images, there were very few instances where cloud was erroneously identified as smoke. This provides at least some encouragement that the use of hyperspectral data holds benefits, because distinguishing clouds from smoke visually using the RGB images can be quite difficult.

3. *Snow and ice can be distinguished from smoke, but with greater difficulty.* A similar comment can be made about snow and ice, but less emphatically. The classifier generally performed well in separating smoke from snow and ice, but performance was less consistent. In certain images this task seemed to pose no problem, while in other images significant

numbers of snow or ice pixels were incorrectly labelled smoke. Both Figure 5 and Figure 6 provide some evidence of this, with moderate probabilities being mapped over the Coast Mountains in the upper left of either image.

4. *Co-located smoke and clouds present a problem.* The starting point for this problem is the assumption that smoke and clouds may both exist in the same pixel. Separation of smoke from clouds when both are in the same vicinity is a problem in two respects. First, when the masks were being prepared it was extremely difficult for the human interpreter to decide whether or not a given pixel in a cloudy region actually contains smoke. When clouds and smoke are mixed or adjacent, it is very difficult to distinguish one from the other using the RGB image alone. Second, because cloud is a significant constituent of the nonsmoke pixel class, the classifiers learned to assign low probability to pixels with the characteristics of clouds. An example of this problem can be seen in the upper right corner of Figure 6. In the RGB image, it is unclear if the bright feature in this corner is a cloud, and if so, whether there is also smoke present. From the probability map, it appears that there was indeed cloud in this region, which caused it to be assigned low probability.

5. *Prediction maps are unrealistically noisy.* Our mental model of the true scene in these images is of smoke regions being contiguous with relatively smooth boundaries. Because we are classifying pixels independently, however, this information is not incorporated into our procedures. The noisy nature of the probability maps is visible in both the smoke and nonsmoke regions ofFigure 5 andFigure 6.

Figure 6. Another example prediction. Top: RGB image. Bottom: predicted probability map.

6. *The quality of the training data is a major impediment to classifier construction.* Perhaps the most significant problem inherent in this study is uncertainty about the assigned classes in the original images themselves. Various portions of the images proved extremely difficult to assign to one class or the other with high confidence during the masking step. The aforementioned regions of mixed smoke and cloud provide one example. Regions where smoke becomes less concentrated provide another example (see Figure 5): where does the smoke end and the nonsmoke begin? In the same figure, we see a third example. A large number of pixels in a region over the mountains are "erroneously" assigned a high probability of being smoke. Is this a classification error, or an error in masking the original RGB image? The RGB image has a hazy appearance in this region, but it was not assigned to the smoke class due to the absence of a local fire and the general uncertainty about the nature of this hazy appearance. After the fact, it seems plausible that the classifier is detecting smoke that was erroneously labelled nonsmoke in the data set.

5.3. Opportunities for improvement

While the classification results were mixed, we feel there were enough positive elements to warrant further investigation, and that the overall approach can still be successful with appropriate modifications and extensions.

Probably the clearest opportunity for improvement is to alleviate the uncertainty in the true class labels that exists throughout the data set, and was illustrated in Figure 5 and Figure 6. The ambiguity in distinguishing smoke from nonsmoke at various places in the RGB images is a fundamental limitation. Simple approaches to solving this problem include considering only smoke plumes or "thick" smoke; excluding pixels that the photointerpreter finds ambiguous or that contain both cloud and smoke; or labelling images with more than two classes. More involved approaches include modelling each pixel as a mixture of different components, or modelling some continuous measure of smoke concentration rather than a binary presence/absence response. An unsupervised learning (clustering) approach or a semi-supervised method (where only some pixels are labelled) could also be considered, though such methods make quantitative performance assessment more difficult.

Another avenue for potential improvement of classification performance is to modify the feature space in the logistic model in the hopes of improving the separability of smoke and nonsmoke. While this could be done by adding even more factorial terms (cubic terms, higher-order interactions, and so on), it is unlikely that the benefit of doing so would outweigh the increase in computational burden. Instead, more focused modifications of the model could be considered. To reduce the effect of highly heterogeneous surface terrain in the nonsmoke class, for instance, a baseline spectrum (perhaps taken as an average of observations over recent clear-sky days) could be included as predictors in the model. Or each pixel could be assigned to a known ground-cover class at the outset, and these classes could be included in the model as categorical variables. Another option is to replace the fixed powers of reflectance we used (squared and square root terms) with spline functions, allowing data-adaptive nonlinear transformations of the variables to be used in the model. We anticipate exploring some of these alternatives in future work with these data.

http://dx.doi.org/10.5772/60214

Additional possibilities for improvement can be found by moving farther from the logistic regression framework. Under the assumption of independent pixels, for example, any of the many existing classification tools could be applied to the data. The support vector machine (see, e.g., [24], Ch. 11) in particular is a state-of-the-art method that has performed well across a variety of tasks and is worthy of consideration. If the independence assumption is dropped, the autologistic regression model [25], a model for spatially-correlated binary responses, is a natural fit for these data. This model would alleviate the problem of noise in the predicted probabilities, producing smoother and more accurate prediction maps. It is a natural extension of logistic regression to spatially-associated data. Finally, it may also be possible to incorporate relevant ancillary information (for example, prior knowledge of fire locations and wind directions) into a classification model to improve predictive power. Again, consideration of these alternatives and extensions are planned in future work.

6. Conclusion

The smoke identification problem provided a case study on the use of supervised learning to automate the process of recognizing features of interest in remote sensing images. The machine learning approach is especially attractive when working with hyperspectral images, because the high dimensionality of the data makes it very challenging for a human photointerpreter to consider all of the potential relationships in the data. Subject-matter knowledge can help to focus a human expert on certain models, relationships, or spectral bands, but automated procedures provide a valuable complementary approach. They can be used to search for more complex or previously unconsidered relationships, driven by the data itself If a machine learning procedure can be implemented successfully, another clear benefit is the ability to process data at a speed and scope not feasible by other means.

Our primary conclusion regarding the smoke identification goal is that the spectral information in the smoke and nonsmoke classes overlap to such a degree that it is not possible to construct a highly successful classifier—at least with the models and methods we employed. The results have some promising elements, however. Notably, it appears possible to distinguish smoke from cloud and snow when a) the smoke is not mixed with cloud, and b) the smoke is not too diffuse. Indeed, if the goal of the study were to find clear-sky smoke plumes only, the approach would be quite successful. Classification errors were largely attributable to the presence of cloud in a smoky region, to the smoke being too diffuse, or to inaccuracies introduced in the initial labelling of the data. Armed with this understanding, it should be possible to make considerable improvements to the results with adjustments to the methodology.

The problem used for this case study is a challenging image segmentation task, made more challenging by the loose definition of "smoke" used in the initial labelling of the data set. Reflecting this, the best classifiers we found were only partially successful. Still, the process of developing them has helped to provide insight into the problem and allows us to present both the advantages and challenges of the machine learning approach. With the dimensionality and throughput of remote sensing data ever on the rise, computer intensive techniques such as those explored here will be of increasing importance in the future.

Author details

Mark A. Wolters[1*] and C.B. Dean[2]

*Address all correspondence to: mwolters@fudan.edu.cn

1 Shanghai Center for Mathematical Sciences, Fudan University, Shanghai, China

2 Department of Statistical and Actuarial Sciences, Western University, London, Canada

References

[1] Richards JA. Remote Sensing Digital Image Analysis. 5th ed.: Springer-Verlag; 2013.

[2] R Core Team. R: A Language and Environment for Statistical Computing Vienna: R Foundation for Statistical Computing; 2014.

[3] National Aeronautics and Space Administration. MODIS Web. [Online].; 2014 [cited 2014 November 25. Available from: http://modis.gsfc.nasa.gov/about/specifications.php.

[4] MODIS Characterization and Support Team. MODIS Level 1B Product User's Guide. Greenbelt, MD: NASA/Goddard Space Flight Center; 2012. Report No.: MCST Document # PUB-01-U-0202-REV D.

[5] Goddard Space Flight Center. LAADS Web. [Online].; 2014 [cited 2014 November 25. Available from: http://ladsweb.nascom.nasa.gov/.

[6] Gumley L, Descloitres J, Schmaltz J. Creating Reprojected True Color MODIS Images: A Tutorial. Maison, WI: University of Wisconsin-Madison, Space Science and Engineering Center; 2010. Report No.: Version 1.0.2.

[7] Giglio L, Descloitres J, Justice CO, Kaufman YJ. An Enhanced Contextual Fire Detection Algorithm for MODIS. Remote Sensing of Environment. 2003; 87: p. 273-282.

[8] Bishop CM. Pattern recognition and machine learning New York: Springer; 2006.

[9] Hastie T, Tibshirani R, Friedman J. The Elements of Statistical Learning: Data Mining, Inference, and Prediction. 2nd ed.: Springer Science+Business Media; 2009.

[10] James G, Witten D, Hastie T, Tibshirani R. An Introduction to Statistical Learning with Applications in R: Springer Science+Business Media; 2013.

[11] Tibshirani R. Regression Shrinkage and Selection via the Lasso. Journal of the Royal Statistical Society, Series B. 1996; 58(1): p. 267-288.

[12] Efron B, Hastie T, Johnstone I, Tibshirani R. Least Angle Regression. The Annals of Statistics. 2004; 32(2): p. 407-451.

[13] Friedman J, Hastie T, Tibshirani R. Regularization Paths for Generalized Linear Models via Coordinate Descent. Journal of Statistical Software. 2010; 33(1).

[14] Miller A. Subset Selection in Regression. 2nd ed.: Chapman & Hall; 2002.

[15] Wolters MA. A Genetic Algorithm for Selection of Fixed-Size Subsets, with Application to Design Problems. Journal of Statistical Software. (to appear).

[16] Michalewicz Z, Fogel DB. How to Solve it: Modern Heuristics. 2nd ed.: Springer-Verlag; 2004.

[17] Rothlauf F. Design of Modern Heuristics: Principles and Application Heidelberg: Springer; 2011.

[18] Gendreau M, Potvin JY, editors. Handbook of Metaheuristics. 2nd ed.: Springer; 2010.

[19] Whitley D, Beveridge JR, salcedo CG, Graves C. Messy Genetic Algorithms for Subset Feature Selection. 1997..

[20] Li Y, Vodacek A, Zhu Y. An automatic statistical segmentation algorithm for extraction of fire and smoke regions. Remote sensing of environment. 2007; 108(2): p. 171-178.

[21] Jethva H, Torres O. Satellite-Based Evidence of Wavelength-Dependent Aerosol Absorption in Biomass Burning Smoke Inferred from Ozone Monitoring Instrument. Atmospheric Chemistry and Physics. 2011; 11: p. 10541-10551.

[22] Miura T, Huete AR, van Leeuwen WJD, Didan K. Vegetation Detection Through Smoke-Filled AVIRIS Images: An Assessment Using MODIS Band Passes. Journal of Geophysical Research. 1998; 103(D24): p. 32001-32011.

[23] Chu DA, Kaufman YJ, Remer LA, Holben BN. Remote Sensing of Smoke from MODIS Airborne Simulator During the SCAR-B Experiment. Journal of Geophysical Research. 1998; 103(D24): p. 31979-31987.

[24] Izenman AJ. Modern Multivariate Statistical Techniques: Springer Science+Business Media; 2008.

[25] Hughes J, Haran M, Caragea PC. Autologistic models for binary data on a lattice. Environmetrics. 2011; 22(7): p. 857-871.

Chapter 17

A Non-Homogeneous Markov Chain Model to Study Ozone Exceedances in Mexico City

Eliane R. Rodrigues, Mario H. Tarumoto and Guadalupe Tzintzun

Additional information is available at the end of the chapter

1. Introduction

In many cities around the world, air pollution is among the many environmental problems that affect their population. Among the many known facts about the impact of pollution on human health, we have that for ozone concentration levels above 0.11 parts per million (0.11ppm), the susceptible part of the population (e.g., the elderly, ill, and newborn) staying in that environment for a long period of time, may experience serious health deterioration (see, for example, [1–10]). Therefore, to understand the behaviour of ozone and/or pollutants in general, is a very important issue.

It is possible to find in the literature a vast amount of works that try to answer some of the many issues arising in the study of pollutants' behaviour. Depending on the type of questions that one is trying to answer, different methodologies may be used. Among the many works concentrating on the study of ozone behaviour are, [11–13] using extreme value theory to study the behaviour of the maximum ozone measurements; [14] using time series analysis; [15] using volatility models to study the variability of the weekly average ozone measurements; [13, 16] using homogeneous Poisson processes and [17, 18] using non-homogeneous Poisson models to analyse the probability of having a certain number of ozone exceedances in a time interval of interest; [19] using compound Poisson models to study the occurrence of clusters of ozone exceedances as well as their mean duration time; and [20] using queueing model to study the occurrence of cluster of ozone exceedances as well as their size distribution.

In the environmental area, it is also possible to find works using Markov chains models. Some of them are, [21, 22] where non-homogeneous Markov models are used to study the occurrence of precipitation. We also have [23] where those types of models are used to study tornado activity. In the case of ozone modelling we have, for instance, the works of [24–26] using time homogeneous Markov chains. In those works the interest was in estimating the probability that the ozone measurement would be above (below) a given threshold, conditioned on where it lays in the present and in the past days.

In [24], the order of the Markov chain was estimated using auto-correlation function. Its transition matrix was estimated using the maximum likelihood method (see, for instance, [27, 28], among others). In [25], the order of the chain was also considered an unknown quantity that needed to be estimated. The Bayesian approach (see, for example, [29]) was used to estimate the order as well as the transition probabilities of the chain. In particular, the maximum *à posteriori* method was used. In [26], the estimation of the order of the chain is performed using the Bayesian approach using the so-called trans-dimensional Markov chain Monte Carlo algorithm ([30, 31]). The transition matrix of the chain was obtained through the maximum *à posteriori* method. However, the common denominator of those works is that the Markov chain model used was a time homogeneous one. Since ozone data are not, in general, time homogeneous, the data had to be split into time homogeneous segments and the analysis was made for each segment separately.

Here, the interest also resides in estimating, for instance, the probability that the ozone measurement will be above a given threshold some days into the future, given where it stands today and in the past few days. Although in the present work we also use Markov chain models and the Bayesian approach, the novelty here is that the time-homogeneous assumption is dropped. Here, we consider a non-homogeneous Markov chain model. We assume that the order of the chain as well as its transition probabilities are unknown and need to be estimated. The chosen method of estimation is also the maximum *à posteriori*.

This work is presented as follows. In Section 2 the non-homogeneous Markov chain model is given. Section 3 presents the Bayesian formulation of the model. An application to ozone measurements from Mexico City is given in Section 4. In Section 5 some comments about the methodology and results are made. In an Appendix, before the list of references, we present the code of the programme used to estimate the order and the transition probabilities of the Markov chain.

2. A non-homogeneous Markov chain model

The mathematical model considered here may be described as follows. Let $N > 0$ be a natural number representing the number of years in which measurements were taken. Let T_i, $i = 1, 2, \ldots, N$, be natural numbers representing the amount of observations in each year. Hence, we have that for a given year i, either $T_i = 366$ or $T_i = 365$, depending on whether or not we have a leap year, $i = 1, 2, \ldots, N$.

Let $Z_t^{(i)}$ be the ozone concentration on the tth day of the ith year, $t = 1, 2, \ldots, T_i$, $i = 1, 2, \ldots, N$. Following [23], we will set $T_i = T = 366$, $i = 1, 2, \ldots, N$, with the convention that for non leap year, we assign $Z_T^{(i)} = 0$.

Remark. Since, we are taking all years of the same length we will drop the index i from the notation.

Denote by $L > 0$ the environmental threshold we are interested in knowing if the ozone concentration has surpassed or not. Define $\mathbf{Y} = \{Y_t : t \geq 0\}$ by,

$$Y_t = \begin{cases} 0, & \text{if } Z_t < L \\ 1, & \text{if } Z_t \geq L. \end{cases} \tag{1}$$

Hence, Y_t indicates whether or not in the tth day the threshold L was exceeded.

As in [25], we assume that $\mathbf{Y}$ is ruled by a Markov chain of order $K \geq 0$. In contrast with that work, in the present case the Markov chain is a non-homogeneous one. Hence, denote by $X^{(K)} = \{X_t^{(K)} : t = 1, 2, \ldots T\}$, the corresponding non-homogeneous Markov chain of order K. We assume that K has as state space a set $\mathcal{S} = \{0, 1, \ldots, M\}$, for some fixed integer $M \geq 0$, such that, $M \leq T$ with probability one.

Note that, $X^{(K)}$ has as state space the set $S_1^{(K)} = \{(x_1, x_2, \ldots, x_K) \in \{0, 1\}^K\}$, with $S_1^{(0)} = S_1^{(1)}$. Also, note that (see [25]), if the set of observed value is $(y_1, y_2, \ldots, y_T)$, then the transition probabilities of $X^{(K)}$ are such that

$$P(X_{t+1}^{(K)} = w \mid X_t^{(K)} = x_t = (y_{t+1}, y_{t+2}, \ldots, y_{t+K})),$$

is different of zero if, and only if, $w = (y_{t+2}, y_{t+3}, \ldots, y_{t+K+1}) \in S_1^{(K)}$, with $0 \leq t \leq T - K$. Therefore, w occurs, if and only if, the observation following $y_{t+1}, y_{t+2}, \ldots, y_{t+K}$, is y_{t+K+1}. This enables us to work with a more treatable state space for $X^{(K)}$, and therefore, to have a better form for the transition matrix.

Hence, as in [25, 32], we consider the transformed state space $S_2^{(K)} = \{0, 1, \ldots, 2^K - 1\}$, which is obtained from $S_1^{(K)}$ by using the transformation $f : S_1^{(K)} \rightarrow S_2^{(K)}$, given by, $f(w_1, w_2, \ldots, w_K) = \sum_{l=0}^{K-1} w_{l+1} 2^l$. Let $(x_1, x_2, \ldots, x_K) \leftrightarrow \overline{m}$ indicate that the state $(x_1, x_2, \ldots, x_K) \in S_1^{(K)}$ corresponds to the state $\overline{m} \in S_2^{(K)}$. Hence, the transition probabilities of $X^{(K)}$ may be written as (see, for instance, [25]),

$$P_{\overline{m}j}^{(K)}(t) = P(Y_{t+K+1} = j \mid X_t^{(K)} = (y_{t+1}, y_{t+2}, \ldots, y_{t+K}) \leftrightarrow \overline{m}), \tag{2}$$

where $\overline{m} \in S_2^{(K)}$, $j \in \{0, 1\}$, and $0 \leq t \leq T - K$.

Now, indicate by $Q_{\overline{m}}^{(K)}(t)$, $\overline{m} \in S_2^{(K)}$, $\overline{m} \in S_2^{(K)}$, the probability $P(X_t^{(K)} = \overline{m})$. Hence, when $t = 1$, we have that $Q_{\overline{m}}^{(K)}(1)$, $\overline{m} \in S_2^{(K)}$, is the initial distribution of $X^{(K)}$. When $K = 0$, we have that $P_{\overline{m}j}^{(0)}(t) = Q_j^{(0)}(t)$, $t = 1, 2, \ldots, T$, $j = 0, 1$, $\overline{m} \in S_2^{(0)} = \{0, 1\}$.

Remarks. 1. When $K = 1$, we have that $P_{\overline{m}j}^{(0)}(t)$, $j = 0, 1$, $t = 1, 2, \ldots, T - K$, are the usual one-step transition probabilities. When $K = 0$, the transition probabilities are just the probabilities $Q_{\overline{m}}^{(0)}(t)$, associated to each state $\overline{m} \in S_2^{(0)} = \{0, 1\}$, with $t = 1, 2, \ldots, T$.

2. Unless otherwise stated, from now on, we are going to use the state space $S_2^{(K)}$ and the corresponding transition probabilities.

3. $\mathbf{Y}$ is going to represent our observed data.

In addition to estimating the order K of the Markov chain, we will also estimate its transition probabilities $P_{\overline{m}j}^{(K)}(t)$ as well as the probabilities $Q_{\overline{m}}^{(K)}(t)$, $j \in \{0, 1\}$, $\overline{m} \in S_2^{(K)}$, for each t. We indicate by $P^{(K)}(t) = \left(P_{\overline{m}j}^{(K)}(t)\right)_{j\in\{0,1\},\overline{m}\in S_2^{(K)}}$, the transition matrix at time t. Note that, if $K = 0$, then $P_{\overline{m}j}^{(0)}(t) = Q_j^{(0)}(t)$, $j \in \{0,1\}$, $\overline{m} \in S_2^{(0)}$, $t = 1,2,\ldots,T$.

3. A Bayesian estimation of the parameters of the model

There are many ways of estimating the order and the transition matrix of a non-homogeneous Markov chain. One way of estimating the order is via the auto-correlation function associated to the chain throughout the years. Another way is to use the Bayesian approach. When it comes to estimating the transition probabilities we have, for instance, the maximum likelihood method ([33]) and the empirical estimator ([34]) which are essentially the same. In the present work, we will use the Bayesian approach (see, for instance, [29, 35]) to estimate the order and the transition probabilities. In particular, we are going to adopt the maximum *à posteriori* approach. Inference will be performed using the information provided by the so-called posterior distribution of the parameters. The posterior distribution of a vector of parameters $\boldsymbol{\theta}$ given the observed data $\mathbf{D}$, indicated by $P(\boldsymbol{\theta} \mid \mathbf{D})$, is such that $P(\boldsymbol{\theta} \mid \mathbf{D}) \propto L(\mathbf{D} \mid \boldsymbol{\theta})\, P(\boldsymbol{\theta})$, where $L(\mathbf{D} \mid \boldsymbol{\theta})$ is the likelihood function of the model, and $P(\boldsymbol{\theta})$ is the prior distribution of the vector $\boldsymbol{\theta}$.

In the present case, we have that the vector of parameter is $\boldsymbol{\theta} = (K, Q^{(K)}(1), P^{(K)}(t), t = 1,2,\ldots,T-K)$. If $K = 0$, the range of t is $\{1,2,\ldots,T\}$. The vector $\boldsymbol{\theta}$ belongs to the following sample space

$$\Theta = \bigcup_{K=0}^{M} \left(\{K\} \times \Delta_2^{2^K} \times \Delta_2^{(T-K)\,2^K}\right)$$

where $\Delta_2 = \{(x_1,x_2) \in \mathbb{R}^2 : x_i \geq 0,\, i = 1,2,\, x_1 + x_2 = 1\}$ is the one dimensional simplex. (Note that if we have $K = 0$, then the parametric space reduces to $\Theta = \Delta_2^T$.) In the present case we have $\mathbf{D} = \mathbf{Y}$

Let $(x_1, x_2, \ldots, x_K)$ be such that $Y_t = x_1$. Indicate by $n_{\overline{m}i}^{(K)}(t)$ the number of years in which the vector $(x_1, x_2, \ldots, x_K)$ corresponding to a state $\overline{m} \in S_2^{(K)}$ is followed by the observation i, $i = 0,1$. Also define $n_{\overline{m}}^{(0)}(t)$, $\overline{m} \in S_2^{(0)} = \{0,1\}$, as the number of years in which we have the observation $\overline{m}$ at time t, $t = 1,2,\ldots,T$. Additionally, let $n_{\overline{m}}^{(K)}$ indicate the number of years in which the state corresponding to the initial K days is equivalent to the value $\overline{m} \in S_2^{(K)}$, $K \geq 0$. In the case of $K = 0$, we have $n_{\overline{m}}^{(0)} = n_{\overline{m}}^{(1)}$, and $\overline{m} \in S_2^{(0)} = S_2^{(1)} = \{0,1\}$.

Therefore, since a Markovian model is assumed, the likelihood function is given by (see, for instance, [23, 33])

$$L(\mathbf{Y} \mid \boldsymbol{\theta}) \propto \left(\prod_{\overline{m}\in S_2^{(K)}} \left[Q_{\overline{m}}^{(K)}(1)\right]^{n_{\overline{m}}^{(K)}}\right) \left(\prod_{t=1}^{T-K} \left[P_{\overline{m}0}^{(K)}(t)\right]^{n_{\overline{m}0}^{(K)}(t)} \left[1 - P_{\overline{m}0}^{(K)}(t)\right]^{n_{\overline{m}1}^{(K)}(t)}\right). \quad (3)$$

Note that when $K = 0$, the expression (3) simplifies to

$$L(\mathbf{Y} \mid \boldsymbol{\theta}) \propto \prod_{t=1}^{T} \prod_{m=0}^{1} \left[Q_{\overline{m}}^{(0)}(t) \right]^{n_{\overline{m}}^{(0)}(t)},$$

where $Q_{\overline{m}}^{(0)}(t) = P(X_t^{(0)} = \overline{m}) = P(Y_t = \overline{m}),\ t = 1, 2, \ldots, T,\ \overline{m} \in S_2^{(0)} = \{0, 1\}$.

The prior distribution of the vector of parameters is given as follows. We assume a prior independence of $P^{(K)}(t)$ as functions of t. Also, since the forms of $P^{(K)}(t)$ and $Q^{(K)}(1)$ depend on the value of K, we have that for $\boldsymbol{\theta} = (K, Q^{(K)}(1), P^{(K)}(t), t = 1, 2, \ldots, T-K)$,

$$P(\boldsymbol{\theta}) = P(Q^{(K)}(1) \mid K) \left[\prod_{t=1}^{T-K} P\left(P^{(K)}(t) \mid K\right) \right] P(K),$$

where $P(Q^{(K)}(1) \mid K)$ and $P\left(P^{(K)}(t) \mid K\right)$ are the prior distributions of the initial distribution $Q^{(K)}(1)$ and of the transition matrix $P^{(K)}(t)$ given the order of the chain, respectively, and $P(K)$ the prior distribution of the order K.

Remark. When we have $K = 0$, the vector of parameters is $\boldsymbol{\theta}' = (Q_{\overline{m}}^{(0)}(t);\ t = 1, 2, \ldots, T)$, whose prior distribution is

$$P(\boldsymbol{\theta}') = \left[\prod_{t=1}^{T} P(Q^{(0)}(t)) \right],$$

where $P(Q^{(0)}(t))$ is the prior distribution of the probability vector $Q^{(0)}(t) = (Q_{\overline{m}}^{(0)}(t), \overline{m} = 0, 1),\ t = 1, 2, \ldots, T$.

Given the nature of transition matrices, we are going to assume that rows are independent. We also assume that, given the order K of the chain, each row of the transition matrix $P^{(K)}(t)$ will have as prior distribution a Dirichlet distribution with appropriate hyperparameters. Therefore, given that $K = k$, row $(P_{\overline{m}0}^{(k)}(t), P_{\overline{m}1}^{(k)}(t))$ has as prior distribution a Dirichlet$(\alpha_{\overline{m}0}^{(K)}(t), \alpha_{\overline{m}1}^{(K)}(t))$, $t = 1, 2, \ldots, T$; i.e.,

$$P\left(P^{(K)}(t) \mid K\right) = \prod_{\overline{m} \in S_2^{(K)}} \left(\frac{\Gamma(\alpha_{\overline{m}0}^{(K)}(t) + \alpha_{\overline{m}1}^{(K)}(t))}{\Gamma(\alpha_{\overline{m}0}^{(K)}(t))\, \Gamma(\alpha_{\overline{m}1}^{(K)}(t))} \left\{ \left[P_{\overline{m}0}^{(K)}(t) \right]^{\alpha_{\overline{m}0}^{(K)}(t)-1} \left[P_{\overline{m}1}^{(K)}(t) \right]^{\alpha_{\overline{m}1}^{(K)}(t)-1} \right\} \right)$$

for t in the appropriate range. In the case of initial distribution $Q^{(K)}(1)$, we also have a Dirichlet prior distribution, but now with hyperparameters $(\alpha_{\overline{m}}^{(K)}; \overline{m} \in S_2^{(K)})$ Therefore,

$$P\left(Q^{(K)}(1) \mid K\right) = \frac{\Gamma(\sum_{\overline{m} \in S_2^{(K)}} \alpha_{\overline{m}}^{(K)})}{\prod_{\overline{m} \in S_2^{(K)}} \Gamma(\alpha_{\overline{m}}^{(K)})} \prod_{\overline{m} \in S_2^{(K)}} \left[Q_{\overline{m}}^{(K)}(1) \right]^{\alpha_{\overline{m}}^{(K)}-1}.$$

If $K = 0$, then $Q^{(0)}(t)$ has as prior distribution a Dirichlet distribution with hyperparameters $(\alpha_{\overline{m}}^{(0)}(t); t = 1, 2, \ldots, T; \overline{m} \in S_2^{(0)} = \{0, 1\})$. Therefore, for any given t,

$$P\left(Q^{(0)}(t) \mid K = 0\right) = \frac{\Gamma(\sum_{\overline{m}=0}^{1} \alpha_{\overline{m}}^{(0)}(t))}{\prod_{\overline{m}=0}^{1} \Gamma(\alpha_{\overline{m}}^{(0)}(t))} \prod_{\overline{m}=0}^{1} \left[Q_{\overline{m}}^{(0)}(t)\right]^{\alpha_{\overline{m}}^{(0)}(t)-1}.$$

We assume that K has as prior distribution a truncated Poisson distribution defined on the set $\mathcal{S}$ with rate $\lambda > 0$; i.e.,

$$P(K) = \frac{\lambda^K}{K!} I_{\mathcal{S}}(K),$$

where $I_A(x) = 1$, if $x \in A$ and is zero otherwise.

Therefore, we have from [25, 32, 36], that the conditional posterior distribution of $P^{(K)}(t)$ given K, is

$$P\left(P^{(K)}(t) \mid K, \mathbf{Y}\right) \propto \prod_t \left\{ \prod_{\overline{m} \in S_2^{(K)}} \left(\left[P_{\overline{m}0}^{(K)}(t)\right]^{n_{\overline{m}0}^{(K)}(t) + \alpha_{\overline{m}0}^{(K)}(t) - 1} \left[P_{\overline{m}1}^{(K)}(t)\right]^{n_{\overline{m}1}^{(K)}(t) + \alpha_{\overline{m}1}^{(K)}(t) - 1} \right) \right\}.$$

Hence, $P\left(P^{(K)}(t) \mid K, \mathbf{Y}\right)$ is proportional to the product of Dirichlet distributions with hyperparameters $(n_{\overline{m}0}^{(K)}(t) + \alpha_{\overline{m}0}^{(K)}(t), n_{\overline{m}1}^{(K)}(t) + \alpha_{\overline{m}1}^{(K)}(t))$. The mode of each Dirichlet distribution is known and is given by (see [37]),

$$P_{\overline{m}i}^{(K)}(t) = \frac{n_{\overline{m}i}^{(K)}(t) + \alpha_{\overline{m}i}^{(K)}(t) - 1}{\sum_{j=0}^{1} \left[n_{\overline{m}j}^{(K)}(t) + \alpha_{\overline{m}j}^{(K)}(t) - 1\right]}, \quad i = 0, 1; \overline{m} \in S_2^{(K)}; K \in \mathcal{S}; t = 1, 2, \ldots, T - K. \quad (4)$$

Additionally, the posterior distribution of the initial distribution $Q^{(K)}(1)$ given K is

$$P(Q^{(K)}(1) \mid K, \mathbf{Y}) \propto \prod_{\overline{m} \in S_2^{(K)}} \left[Q_{\overline{m}}^{(K)}(1)\right]^{\alpha_{\overline{m}}^{(K)} + n_{\overline{m}}^{(K)} - 1}.$$

Therefore, $P(Q^{(K)}(1) \mid K, \mathbf{Y})$ is proportional to a Dirichlet distribution with hyperparameters $(\alpha_{\overline{m}}^{(K)} + n_{\overline{m}}^{(K)}; \overline{m} \in S_2^{(K)})$. Hence, as in the case of the posterior distribution of $P^{(K)}(t)$, the mode of $P(Q^{(K)}(1) \mid K, \mathbf{Y})$ is,

$$Q_{\overline{m}}^{(K)}(1) = \frac{n_{\overline{m}}^{(K)} + \alpha_{\overline{m}}^{(K)} - 1}{\sum_{\overline{m}' \in S_2^{(K)}} \left[n_{\overline{m}'}^{(K)} + \alpha_{\overline{m}'}^{(K)} - 1\right]}, \quad \overline{m} \in S_2^{(K)}; K \in \mathcal{S}. \quad (5)$$

When $K = 0$, we have

$$P(Q^{(0)}(t) \mid K = 0, \mathbf{Y}) \propto \left(\prod_{\overline{m}=0}^{1} \left[Q_{\overline{m}}^{(0)}(t) \right]^{n_{\overline{m}}^{(0)}(t) + \alpha_{\overline{m}}^{(0)}(t) - 1} \right), \quad t = 1, 2, \ldots, T.$$

Therefore, for each $t = 1, 2, \ldots, T$,

$$Q_{\overline{m}}^{(0)}(t) = \frac{n_{\overline{m}}^{(0)}(t) + \alpha_{\overline{m}}^{(0)}(t) - 1}{\sum_{\overline{m}'=0}^{1} \left[n_{\overline{m}'}^{(0)}(t) + \alpha_{\overline{m}'}^{(0)}(t) - 1 \right]}, \quad \overline{m} = 0, 1. \tag{6}$$

Furthermore, we also have, from [25], that

$$L(\mathbf{Y} \mid K) \propto \frac{\Gamma\left(\sum_{\overline{m} \in S_2^{(K)}} \alpha_{\overline{m}}^{(K)} \right)}{\Gamma\left(\sum_{\overline{m} \in S_2^{(K)}} \left[\alpha_{\overline{m}}^{(K)} + n_{\overline{m}}^{(K)} \right] \right)} \left(\prod_{\overline{m} \in S_2^{(K)}} \frac{\Gamma\left(n_{\overline{m}}^{(K)} + \alpha_{\overline{m}}^{(K)} \right)}{\Gamma\left(\alpha_{\overline{m}}^{(K)} \right)} \right)$$
$$\prod_{t=1}^{T-K} \left\{ \prod_{\overline{m} \in S_2^{(K)}} \left(\frac{\Gamma[\alpha_{\overline{m}0}^{(K)}(t) + \alpha_{\overline{m}1}^{(K)}(t)]}{\Gamma(\sum_{j=0}^{1} [n_{\overline{m}j}^{(K)}(t) + \alpha_{\overline{m}j}^{(K)}(t)])} \prod_{j=0}^{1} \frac{\Gamma(n_{\overline{m}j}^{(K)}(t) + \alpha_{\overline{m}j}^{(K)}(t))}{\Gamma(\alpha_{\overline{m}j}^{(K)}(t))} \right) \right\},$$

with the appropriate adaptation for the case of $K = 0$. Hence, the posterior distribution of the order K is

$$P(K \mid \mathbf{Y}) = \frac{1}{c} L(\mathbf{Y} \mid K) \frac{\lambda^K}{K!} \tag{7}$$

where $c = \sum_{k \in \mathcal{S}} L(\mathbf{Y} \mid K = k) \left(\lambda^k / k! \right)$ is the normalising constant.

Therefore, in order to obtain the probability of interest, we just have to use (7) to estimate the value of K that maximises that posterior probability, and then use (4), (5), and/or (6) (depending on the case), in order to calculate the corresponding transition matrix and initial distribution, respectively.

The hyperparameters appearing in the prior distribution will be considered known and will be specified later.

4. Application to ozone data from the monitoring network of Mexico City

In this section we apply the model to the Mexico City's ozone measurements. The data used consist of twenty two years of the daily maximum ozone measurements (from 01 January 1990 to 31 December 2011) provided by the monitoring network of the Metropolitan Area of Mexico City. The Metropolitan Area is divided into five regions, namely, Northeast (NE), Northwest (NW), Centre (CE), Southeast (SE), and Southwest (SW). The monitoring stations are placed throughout the city. Measurements in each monitoring station are obtained minute by minute and the averaged hourly result is reported at each station. The daily maximum measurement for a given region is the maximum over all the maximum averaged values recorded hourly during a 24-hour period by each station placed in the region. Since emergency alerts in Mexico City are declared regionally, we will analyse each region separately.

The Mexican ozone standard considers the threshold 0.11ppm (see [38]). Hence, we will take that value as one of our thresholds. Additionally, for comparison purpose, we will also take the threshold values 0.15ppm and 0.17ppm. One of the reasons for choosing these latter values is that we would like to know what would happen if the threshold for declaring emergency alerts in Mexico City was lowered to 0.17ppm. The reason for choosing the threshold 0.15ppm is because it is an intermediate value between the Mexican standard and 0.17ppm.

During the observational period considered here, we have that the mean of the daily observed measurements were 0.12, 0.098, 0.13, 0.12, and 0.14, in regions NE. NW, CE, SE, and SW, respectively, with corresponding standard deviations of 0.06, 0.04, 0.06, 0.05, and 0.06, for those same regions. The threshold 0.11ppm was either reached or exceeded in 4280, 3139, 4921, 4921, and 5711 days in regions NE, NW, CE, SE, and SW, respectively. In those same regions, the threshold 0.15ppm was reached or exceeded in 2460, 963, 2819, 2299, and 3594 days, and the numbers in the case of the threshold 0.17ppm are, 1769, 479, 1896, 1419, and 2660, respectively.

Even though it is a general belief that ozone measurements depend on the measurements of only a few days in the past, we are taking $M = 16$ when we consider the threshold values 0.15ppm and 0.17ppm. We have decided to do that because in previous works the order for homogeneous segments could have higher order. In the case of $L = 0.11$ppm, in some cases, larger values of M were needed. Hence, we also take $M = 16$, in the case of region NW, and we take $M = 18$ in the case of regions CE, NE, SE, and SW. In order to account also for the possibility of low order, we take $\lambda = 1$ in the prior distribution of K.

The hyperparameters of the Dirichlet prior distributions are assigned as in [25]. Therefore, the values of $\alpha_{\overline{m}i}^{(K)}(t)$, $\alpha_{\overline{m}}^{(0)}(t)$, and $\alpha_{\overline{m}}^{(K)}$ will belong to the set $\{3,4,5,6,7,8\}$. Hence, assign $\alpha_{\overline{m}i}^{(K)}(t) = 8$ for the coordinate corresponding to the $\max\{n_{\overline{m}0}^{(K)}(t), n_{\overline{m}1}^{(K)}(t)\}$. Depending on the difference $\max\{n_{\overline{m}0}^{(K)}(t), n_{\overline{m}1}^{(K)}(t)\} - \min\{n_{\overline{m}0}^{(K)}(t), n_{\overline{m}1}^{(K)}(t)\}$, an integer value in $\{3,4,5,6,7\}$ is assigned to the hyperparameter corresponding to $\min\{n_{\overline{m}0}^{(K)}(t), n_{\overline{m}1}^{(K)}(t)\}$. If we have $n_{\overline{m}i}^{(K)}(t) = 0$, then the value 3 is automatically assigned to the corresponding $\alpha_{\overline{m}i}^{(K)}(t)$. Similar procedure is applied in the cases of $\alpha_{\overline{m}}^{(0)}(t)$ and $\alpha_{\overline{m}}^{(K)}$.

Table 1 gives the values of $P(K \mid \mathbf{Y})$. Even though, $\mathcal{S}$ includes the values 0, 1, 2, and 3, since the posterior probabilities at those points are of order 10^{-8} and below, we have omitted those values of K. We use the symbol "-" to indicate that the specific value of K either was not considered in the corresponding region or the probability associated to it was small compared to the values shown.

Looking at Table 1 we may see that, if we consider the threshold $L = 0.11$ppm, then the selected order of the chain is K equal to 16 in the case of region NE, equal to 12 for region NW, and equal to 17 for regions CE and SE. When we consider region SW, the value of K is

	NE			NW			CE			SE			SW		
	0.11	0.15	0.17	0.11	0.15	0.17	0.11	0.15	0.17	0.11	0.15	0.17	0.11	0.15	0.17
$K = 4$	–	–	–	–	–	$< 10^{-7}$	–	–	–	–	–	–	–	–	–
$K = 5$	–	–	–	–	0.33	1	–	–	–	–	–	0.641	–	–	–
$K = 6$	–	–	0.02	–	0.67	$< 10^{-16}$	–	–	–	–	–	0.359	–	–	–
$K = 7$	–	–	0.93	–	–	–	–	–	0.33	–	–	–	–	–	–
$K = 8$	–	0.007	0.05	–	–	–	–	–	0.67	–	0.09	–	–	–	0.024
$K = 9$	–	0.173	–	–	–	–	–	–	–	–	0.9	–	–	–	0.635
$K = 10$	–	0.67	–	–	–	–	–	0.05	–	–	0.01	–	–	–	0.383
$K = 11$	–	0.15	–	0.47	–	–	–	0.68	–	–	–	–	–	–	0.003
$K = 12$	–	–	–	0.53	–	–	–	0.26	–	–	–	–	–	0.008	–
$K = 13$	–	–	–	–	–	–	–	–	–	–	–	–	–	0.095	–
$K = 14$	–	–	–	–	–	–	–	–	–	–	–	–	–	0.792	–
$K = 15$	–	–	–	–	–	–	–	–	–	–	–	0.012	–	0.105	–
$K = 16$	0. 66	–	–	–	–	–	0.01	–	–	0.004	–	–	–	–	–
$K = 17$	0.07	–	–	–	–	–	0.99	–	–	0.984	–	–	–	–	–
$K = 18$	0.27	–	–	–	–	–	–	–	–	–	–	–	–	–	–

Table 1. Posterior distribution of the order of the chain for all regions and threshold considered. The symbol "-" is used to indicate that the specific value of K either was not considered in the corresponding region or the probability associated to it was small compared to the values shown.

either larger than or equal to 18 with probability one. If we take into account the threshold $L = 0.15$ppm, then, also by looking at Table 1, we have that the chosen orders are 10, 6, 11, 9, and 14, in the cases of regions NE, NW, CE, SE, and SW, respectively. When we consider the threshold $L = 0.17$ppm, then the selected orders are 7, 8, and 9, for regions NE, CE, and SW, respectively. In the cases of regions NW and SE, the estimated order is 5. Therefore, using this information and (4), the corresponding transition and initial probabilities may then be calculated.

As an example, consider the case of region CE and the threshold 0.17ppm. In that case, we have that the order of the chain is $K = 8$. Therefore, $S_2^{(K)} = \{0, 1, \ldots, 255\}$. In Table 2, we have the approximated estimated values of the initial distribution $Q_{\overline{m}}^{(K)}(1)$, and of the transition probabilities $P_{\overline{m}0}^{(K)}(t)$, $t = 1, 2$. (We have truncated the values and the total sum is approximately one.) We use the notation $\overline{m}' - \overline{m}''$, to indicate that for all values of $\overline{m}$ in $\{\overline{m}', \overline{m}' + 1, \ldots, \overline{m}''\}$, the estimated probabilities are equal to the values shown.

Looking at Table 2, it is possible to see that the highest initial probability is that associated to the state $\overline{0}$, i.e., the first eight days of the year form a string of zeros, meaning that the concentration levels are below 0.17ppm. Additionally, once you have the information that the ozone concentration levels on the first eight days are below 0.17ppm, then the highest transition probability is also associated to the transition to zeros, i.e., the two days following the eight initial days with concentration below 0.17ppm are more likely to present lower concentration levels as well.

In order to illustrate the type of information that may be obtained using the methodology considered here, take the case of the year 2012 and region CE. Suppose we want to calculate the probability that during the first nine days of January we have that the ozone concentration is below 0.17ppm from the first eight days, and it is above it on the ninth. Therefore, we want to know the probability that (0, 0, 0, 0, 0, 0, 0, 0) is followed by one. Hence, we want the probability of having the following sequence of zeros and ones: 0, 0, 0, 0, 0, 0, 0, 0, 1. Therefore,

$$P((0,0,0,0,0,0,0,0,1)) = P_{\overline{0}1}^{(8)}(1) \times Q_{\overline{0}}^{(8)}(1) = 0.238 \times 0.0327 \approx 0.008.$$

$\overline{m}$	$Q^{(8)}_{\overline{m}}(1)$	$\overline{m}$	$P^{(8)}_{\overline{m}0}(1)$	$P^{(8)}_{\overline{m}0}(2)$
$\overline{0}$	0.0327	$\overline{0}$	0.762	0.889
$\overline{1}-\overline{15}$	0.0036	$\overline{1}-\overline{11}$	0.5	0.5
$\overline{16}$	0.0073	$\overline{12}$	0.5	0.286
$\overline{17}-\overline{23}$	0.0036	$\overline{13}-\overline{62}$	0.5	0.5
$\overline{24}$	0.0073	$\overline{63}$	0.286	0.8
$\overline{25}-\overline{27}$	0.0036	$\overline{64}-\overline{127}$	0.5	0.5
$\overline{28}$	0.0073	$\overline{128}$	0.5	0.444
$\overline{29}-\overline{31}$	0.0036	$\overline{129}-\overline{158}$	0.5	0.5
$\overline{32}$	0.0073	$\overline{159}$	0.5	0.286
$\overline{33}-\overline{62}$	0.0036	$\overline{160}-\overline{247}$	0.5	0.5
$\overline{63}$	0.0073	$\overline{248}$	0.286	0.5
$\overline{63}-\overline{125}$	0.0036	$\overline{249}$	0.5	0.5
$\overline{126}$	0.0073	$\overline{250}$	0.286	0.5
$\overline{127}-\overline{191}$	0.0036	$\overline{251}$	0.5	0.6
$\overline{192}$	0.0073	$\overline{252}-\overline{253}$	0.5	0.286
$\overline{193}-\overline{241}$	0.0036	$\overline{254}-\overline{255}$	0.5	0.5
$\overline{242}$	0.0073	–	–	–
$\overline{243}$	0.0036	–	–	–
$\overline{244}$	0.0073	–	–	–
$\overline{245}-\overline{247}$	0.0036	–	–	–
$\overline{248}$	0.0073	–	–	–
$\overline{249}$	0.0036	–	–	–
$\overline{250}$	0.0073	–	–	–
$\overline{251}-\overline{255}$	0.0036	–	–	–

Table 2. Transition probabilities $P^{(8)}_{\overline{m}0}(1)$ and $P^{(8)}_{\overline{m}0}(2)$ as well as the initial probabilities $Q^{(8)}_{\overline{m}}(1)$, for all values of $\overline{m} \in S_2^{(K)}$ in the case of region CE and threshold 0.17ppm. The notation $\overline{m}' - \overline{m}''$ is used to indicate that for all values of $\overline{m}$ in $\{\overline{m}', \overline{m}' + 1, \ldots, \overline{m}''\}$, the estimated probabilities are equal to the values shown.

In order to obtain the values of the probabilities of interest, recall that $P^{(K)}_{\overline{m}0}(t) = 1 - P^{(K)}_{\overline{m}1}(t)$, $\overline{m} \in S_2^{(K)}$, $t = 1, 2, \ldots, T-K$. Therefore, looking at Table 2, we have that $P^{(8)}_{\overline{0}1}(1)$ is one minus the value on the column corresponding to $P^{(K)}_{\overline{m}0}(1)$ with $\overline{m} = \overline{0}$. Similar comment is valid in the case of $Q^{(8)}_{\overline{m}}(1)$, $\overline{m} \in S_2^{(K)}$.

Suppose now that we want to know the probability of having (0, 0, 0, 0, 0, 0, 0, 0) followed by (1, 1). Hence, we want to know what the probability that (0, 0, 0, 0, 0, 0, 0, 0) is followed by one and that (0, 0, 0, 0, 0, 0, 0, 1) is followed by one. Therefore, we need to calculate

$$P((0,0,0,0,0,0,0,0,1,1)) = P^{(8)}_{\overline{255}\,1}(2) \times P^{(8)}_{\overline{0}1}(1) \times Q^{(8)}_{\overline{0}}(1) = 0.5 \times 0.238 \times 0.0327 \approx 0.0004.$$

Proceeding in this way we may calculate the probability of having any string of states at any time of the year.

If we compare to the actual measurements in the year 2012, then we have that in the first ten days, the sequence **Y**, in the case of region CE, has the configuration 0, 0, 0, 0, 0, 0, 0, 0, 0, 0. In fact, the estimated probability of that sequence of zeros and ones is $0.5 \times 0.762 \times 0.0327 \approx 0.0125$ which is three times higher than the probability of having (0, 0, 0, 0, 0, 0, 0, 0) followed by (1, 1). If we consider also the year 2013, the results are similar. Hence, the methodology used here can produce estimated values that may describe well the behaviour of the data.

5. Conclusion

In this work we have considered a non-homogeneous Markov chain model to study the ozone's behaviour in Mexico City. The interest resides in estimating the probability that the ozone level will be above (below) a certain threshold given that it is either above or below it in the present and in the recent past. Due to the nature of the questions asked here, a natural way of trying to answer them is to use Markov chain models. However, due to the non-homogeneity of the data, a non-homogeneous version of the chain is used.

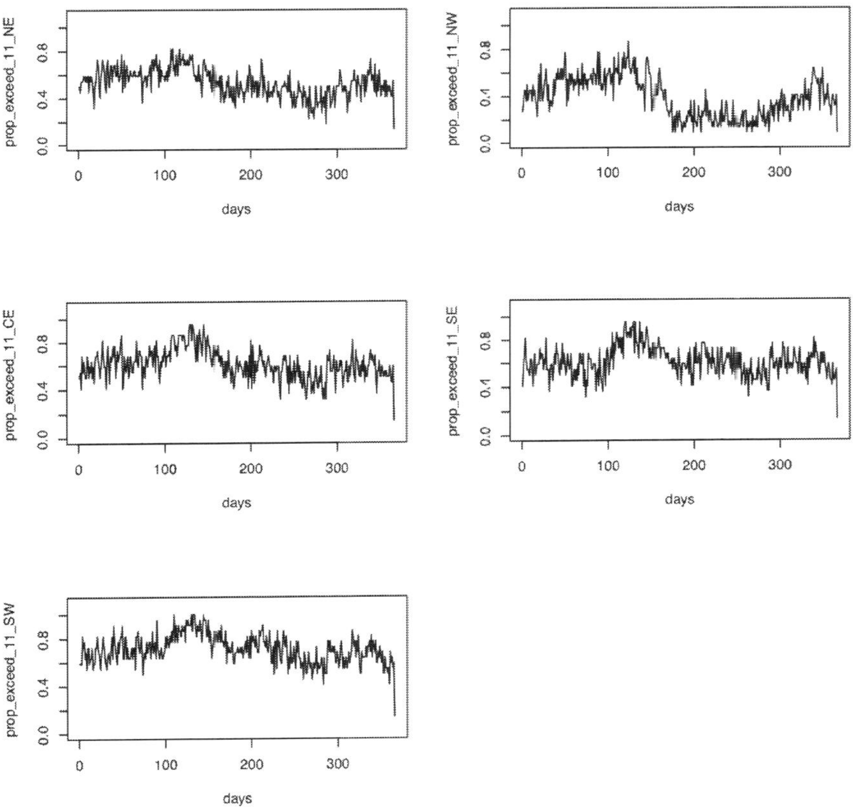

Figure 1. Proportion of years in which, for a given day, the threshold 0.11ppm was exceeded by the ozone concentration.

Using the Bayesian approach a maximum *à posteriori* estimation of the order of the matrix as well as its transition matrix and initial distribution was made. The results have shown that higher order should be considered for the chain. One explanation for that could be the way the empirical probability of having an exceedance in a given day behaves. That can be seen in Figure 1 where, as an illustration, the proportion of exceedances of the threshold 0.11ppm is presented for each region. The values correspond to the proportion of years in which in a fixed day the threshold was exceeded. As we vary the days, we have the behaviour throughout the 366 days. In that figure we have in the horizontal axis the days and in the vertical axis we have the values of $prop(t) = (1/N) \sum_{i=1}^{N} Y_t^{(i)}$, which represents the proportion of years with an exceedance in the tth day, $t = 1, 2, \ldots, T$. The notation $Y_t^{(i)}$ is used to indicate the variable Y_t defined in (1) on the ith year.

The plots in Figure 1 reflect well the fact that in region SW, in most of the days of the year, there are exceedances of the threshold 0.11. We may also see the influence of the seasons of the year. The hill between days 100 and 200 appearing in every plot, corresponds to measurements taken between April and June. Higher values occur during the days corresponding to approximately mid April to mid May. Those months are in the middle of Spring. During this season it does rain much in Mexico City. Additionally, there is a lot of sunlight. Hence, the ozone concentration is bound to be high, and as a consequence, the proportion of years in which exceedances occur at that period is large. The values decrease when the raining season starts (around the beginning of June).

If we consider the threshold values 0.15ppm and 0.17ppm, the behaviour of the proportion of years where in a given day exceedances occurred is similar to the case of 0.11ppm. The difference is that the values of the proportions are smaller. It is possible to see that the proportion of exceedances may vary according to the seasons of the year, and that, within a given season, changes are, in general, not drastic. Therefore, it is possible that measurements from more than a few days may have an influence on the behaviour of future measurements, and with that, make the estimation method considered here to produce high values for the order of the chain.

Acknowledgements

The authors thank the Editor for sending comments that helped to improve the presentation of the results. The authors also thank Peter Guttorp for providing a copy of his works related to applications of non-homogeneous Markov chains. ERR and MHT were partially funded by the project PAPIIT-IN102713-3 of the Dirección General de Apoyo al Personal Académico de la Universidad Nacional Autónoma de México (DGAPA-UNAM), Mexico. ERR also received funds from a sabbatical year grant from DGAPA-UNAM. ERR is grateful to the Departments of Statistics of the Universidade Estadual Paulista "Júlio de Mesquita Filho" - Campus Presidente Prudente, Brazil, and of the University of Oxford, UK, where parts of this work were developed, for all the support and hospitality received during her stay at those departments.

Appendix

In this Appendix we present the code of the programme in R used to estimate the order of the non-homogeneous Markov chain as well as its transition probabilities. The code it not optimal and can be highly improved, but in its present form it provides elements for estimating the necessary quantities.

```
# ESTIMATING THE ORDER OF THE CHAIN
ozonio_sw=read.table('data.txt', header=T)
attach(ozonio_sw)
anos=NCOL(ozonio_sw)
dias=NROW(ozonio_sw)
# assigning the value of M
M = 15 # for instance
# initialisation of the matrices to store the values of the likelihood
# for each order and day
term_init_like <- matrix(0, M+1, 1)
term_general_like <- matrix(0, dias, M+1)
#######
# Case K = 0
#######
# counting the numbers of ones and zeros in each row
count_init_0 <- matrix(0, dias, 2)
for(i in 1:dias){
for(j in 1:anos){
if(ozonio_sw[i,j] == 0){count_init_0[i,1] = count_init_0[i,1] + 1} }
count_init_0[i,2] = anos - count_init_0[i,1]}
# assigning the values of alpha
alpha_init_0 <- matrix(0,dias,2)
for(i in 1:dias){
for(j in 1:2){if(count_init_0[i,j] == 0){alpha_init_0[i,j] = 3} }
if ((count_init_0[i,1] == min(count_init_0[i,])) & (count_init_0[i,1] != 0)){
if(alpha_init_0[i,1] == 0){alpha_init_0[i,1] = 4
if(alpha_init_0[i,2] == 0){alpha_init_0[i,2] = 8}} }
if ((count_init_0[i,1] == max(count_init_0[i,])) & (count_init_0[i,2] != 0)){
if(alpha_init_0[i,1] == 0){alpha_init_0[i,1] = 8
if(alpha_init_0[i,2] == 0){alpha_init_0[i,2] = 4}} }
if ((count_init_0[i,1] == min(count_init_0[i,])) & (count_init_0[i,1] == 0)){
if(alpha_init_0[i,2] == 0){alpha_init_0[i,2] = 8} }
if ((count_init_0[i,1] == max(count_init_0[i,])) & (count_init_0[i,2] == 0)){
if(alpha_init_0[i,1] == 0){alpha_init_0[i,1] = 8} }
}
# calculating the value of the likelihood L(Y | K = 0)
prod_1_k0 <- matrix(0,dias,1)
prod_2_k0 <- matrix(0,dias,1)
prod_k0 <- matrix(0,dias,1)
for(i in 1:dias){
for(j in 1:2){
prod_1_k0[i] = (gamma(count_init_0[i,j]+alpha_init_0[i,j])/gamma(alpha_init_0[i,j])) }
prod_2_k0[i] = (gamma(sum(alpha_init_0[i,]))/gamma(sum(count_init_0[i,] +
 alpha_init_0[i,])))
prod_k0 [i] = prod_1_k0[i]*prod_2_k0[i]}
for(i in 1:dias){
term_general_like[i, 1] = prod_k0[i]}
#######
# Caso K = 1
#######
# the initial states in this case is the first row of the case K=0
# assigning the values of count_init_0[1,i] to count_init_1[i]
count_init_1 <- matrix(0,1,2)
for(j in 1:2){count_init_1[j]= count_init_0[1,j]}
# assigning the same values of alpha_init_0[1,i] to alpha_init_1[i]
alpha_init_1 <- matrix(0,1,2)
for(j in 1:2){alpha_init_1[j]= alpha_init_0[1,j]}
#############
# counting the number of transitions 0 -> 0, 0 -> 1, 1 -> 0 and 1 -> 1
# count_0_k1[i,1] counts the number of transitions from zero to zero in row i,
```

```
# count_0_k1[i,2] counts the number of transitions from zero to one in row i,
#count_1_k1[i,1] counts the number of transitions from one to zero in row i
# count_1_k1[i,2] counts the number of transitions from ne to one in row i
##############
count_0_k1 <- matrix(0, dias-1, 2)
count_1_k1 <- matrix(0, dias-1, 2)
for(i in 1:(dias-1)){
for(j in 1:anos){
if((ozonio_sw[i, j] == 0) & (ozonio_sw[i+1,j] == 0))
{count_0_k1[i,1] = count_0_k1[i,1] + 1}
if((ozonio_sw[i, j] == 0) & (ozonio_sw[i+1,j] == 1))
{count_0_k1[i,2] = count_0_k1[i,2] + 1}
if((ozonio_sw[i, j] == 1) & (ozonio_sw[i+1,j] == 0))
{count_1_k1[i,1] = count_1_k1[i,1] + 1}
if((ozonio_sw[i, j] == 1) & (ozonio_sw[i+1,j] == 1))
{count_1_k1[i,2] = count_1_k1[i,2] + 1}}}
# assigning the values of the values of alpha
# alpha_0_k1 is associated with the transitions 0 -> 0 and 0 -> 1
# alpha_1_k1 is associated with the transitions 1 -> 0 and 1 -> 1
alpha_0_k1 <- matrix(0, dias-1, 2)
alpha_1_k1 <- matrix(0, dias-1, 2)
for(i in 1:(dias-1)){
for(j in 1:2){
if(count_0_k1[i,j] == 0){alpha_0_k1[i,j] = 3}
if(count_1_k1[i,j] == 0){alpha_1_k1[i,j] = 3} }
if((count_0_k1[i,1] == count_0_k1[i,2]) & (count_0_k1[i,1] != 0) &
(count_0_k1[i,1] < 5) & (alpha_0_k1[i, 1] == 0)){alpha_0_k1[i,1] = 5
if(alpha_0_k1[i,2] == 0){alpha_0_k1[i,2] = 5} }
if((count_1_k1[i,1] == count_1_k1[i,2]) & (count_1_k1[i,1] != 0) &
(count_1_k1[i,1] < 5) & (alpha_1_k1[i, 1] == 0)){alpha_1_k1[i,1] = 5
if(alpha_1_k1[i,2] == 0){alpha_1_k1[i,2] = 5} }
if((count_0_k1[i,1] == count_0_k1[i,2]) & (count_0_k1[i,1] != 0) &
(count_0_k1[i,1] >= 5) & (alpha_0_k1[i, 1] == 0)){alpha_0_k1[i,1] = 7
if(alpha_0_k1[i,2] == 0){alpha_0_k1[i,2] = 7} }
if((count_1_k1[i,1] == count_1_k1[i,2]) & (count_1_k1[i,1] != 0) &
(count_1_k1[i,1] >= 5) & (alpha_1_k1[i, 1] == 0)){
alpha_1_k1[i,1] = 7
if(alpha_1_k1[i,2] == 0){alpha_1_k1[i,2] = 7} }
if((count_0_k1[i,1] == min(count_0_k1[i,])) & (count_0_k1[i,1] != 0)){
if(alpha_0_k1[i,1] == 0){alpha_0_k1[i,1] = 4}
if(alpha_0_k1[i,2] == 0){alpha_0_k1[i,2] = 8} }
if((count_1_k1[i,1] == min(count_1_k1[i,])) & (count_1_k1[i,1] != 0)){
if(alpha_1_k1[i,1] == 0){alpha_1_k1[i,1] = 4}
if(alpha_1_k1[i,2] == 0){alpha_1_k1[i,2] = 8} }
if((count_0_k1[i,1] == min(count_0_k1[i,])) & (count_0_k1[i,1] == 0)){
if(alpha_0_k1[i,2] == 0){alpha_0_k1[i,2] = 8} }
if((count_1_k1[i,1] == min(count_1_k1[i,])) & (count_1_k1[i,1] == 0)){
if(alpha_1_k1[i,2] == 0){alpha_1_k1[i,2] = 8} }
if((count_0_k1[i,1] == max(count_0_k1[i,])) & (count_0_k1[i,2] != 0)){
if(alpha_0_k1[i,1] == 0){alpha_0_k1[i,1] = 8}
if(alpha_0_k1[i,2] == 0){alpha_0_k1[i,2] = 4} }
if((count_1_k1[i,1] == max(count_1_k1[i,])) & (count_1_k1[i,2] != 0)){
if(alpha_1_k1[i,1] == 0){alpha_1_k1[i,1] = 8}
if(alpha_1_k1[i,2] == 0){alpha_1_k1[i,2] = 4} }
if((count_0_k1[i,1] == max(count_0_k1[i,])) & (count_0_k1[i,2] == 0)){
if(alpha_0_k1[i,1] == 0){alpha_0_k1[i,1] = 8} }
if((count_1_k1[i,1] == max(count_1_k1[i,])) & (count_1_k1[i,2] == 0)){
if(alpha_1_k1[i,1] == 0){alpha_1_k1[i,1] = 8} }
}
# calculating the value of the likelihood L(Y | K = 1)
# term corresponding to the initial distribution
```

```
prod_1_k1 = 1
for(j in 1:2){
prod_1_k1 = prod_1_k1*((gamma(count_init_1[j]+alpha_init_1[j])/
gamma(alpha_init_1[j])))*((gamma(sum(alpha_init_1))/
gamma(sum(count_init_1+ alpha_init_1))))}
term_init_like[2] = prod_1_k1
# term corresponding to the rest of the days (rows)
prod_2_k1 <- matrix(0,(dias - 1),1)
prod_3_k1 <- matrix(0,(dias - 1),1)
prod_k1 <- matrix(0,(dias - 1),1)
for(i in 1:(dias - 1)){
for(j in 1:2){
prod_2_k1[i] = (gamma(count_0_k1[i,j] + alpha_0_k1[i,j])/gamma(alpha_0_k1[i,j]))*
(gamma(count_1_k1[i,j] + alpha_1_k1[i,j])/gamma(alpha_1_k1[i,j]))}
for(j in 1:2){
prod_3_k1[i] = (gamma(sum(alpha_0_k1[i,])+1)*gamma(sum(alpha_1_k1[i,])))/
(gamma(sum(count_0_k1[i,]+alpha_0_k1[i,])+sum(count_1_k1[i,]+alpha_1_k1[i, )+1))}
prod_k1[i] = prod_2_k1[i] * prod_3_k1[i]}
for(i in 1:(dias-1)){
term_general_like[i, 2] = prod_k1[i]}
####
# Case K >=2
#####
# Transforming vector of zeros and ones in an element of S_{2}^{(K)}
# counting the initial values of the chain (K >= 2) initialisation of the
# matrices to store the values of the likelihood for each order and day
##########
prod_init <- matrix(0, 1, M)
for(K in 2:M){
count_init_k <- matrix(0, 1, (2^K))
mbase_init <- matrix(0, 1, anos)
alpha_init_k <- matrix(0, 1, 2^K)
# transforming the k-dimensional initial vector into an integer number
for(j in 1:anos){
for(l in 0:(K-1)){
mbase_init[j] = mbase_init[j] + ozonio_sw[l+1,j]*2^l}
# counting the number of m in the initial state
for(n in 0:(2^K-1)){if(mbase_init[j] == n){
count_init_k[mbase_init[j]+1] = count_init_k[mbase_init[j]+1] + 1}}}
# assigning the respective values of alpha_init_k
for(n in 0:(2^K - 1)){
if(count_init_k[n+1] == 0){
alpha_init_k[n+1] = 3}
if((count_init_k[n+1] == min(count_init_k)) & (alpha_init_k[n+1] == 0)){
alpha_init_k[n+1] = 3}
if((count_init_k[n+1] == max(count_init_k)) & (alpha_init_k[n+1] == 0)){
alpha_init_k[n+1] = 8}
if((abs(count_init_k[n+1]-max(count_init_k)) > 5) & (alpha_init_k[n+1] == 0)){
alpha_init_k[n+1] = 4}
if((abs(count_init_k[n+1]-max(count_init_k)) <= 5) & (alpha_init_k[n+1] == 0))
{alpha_init_k[n+1] = 6}}
# calculation of the first term in the product in the likelihood L(Y | K)
prod_1_init = 1
prod_2_init = 1
prod_init_k = 0
for(n in 1:(2^K)){
prod_1_init = prod_1_init*(gamma(count_init_k[n]+alpha_init_k[n])
/gamma(alpha_init_k[n]))}
prod_2_init = 1
sumalphacount = sum(alpha_init_k+count_init_k)
sumalpha = sum(alpha_init_k)
```

```
limsup = sum(alpha_init_k+count_init_k) - sum(alpha_init_k)
for(k in 0:(limsup-1)){prod_2_init = prod_2_init*((sumalphacount-1)-k)^(-1)}
prod_init_k = prod_1_init*prod_2_init
term_init_like[K+1] = prod_init_k
# initialisation of the matrices of interest in the case of i not the initial state
mbase <- matrix(0, dias - K, anos) # that is overline{m}
count_Km <- matrix(0, 2^K, 2) # matrix counting the transitions from m
s <- matrix(0, (dias-K), 2^K)
# transforming the vectors of length K into a number in the base 2 for
# each day for all years
for(i in 1:(dias-K)){
for(j in 1:anos){
for(l in 0:(K-1)){
mbase[i,j] = mbase[i,j] + ozonio_sw[i+l,j]*2^l}
# storing the number of mbase in row i in the vector s[day, mbase]
for(n in 0:(2^K-1)){
if(mbase[i,j] == n){s[i,n+1] = s[i, n+1] + 1} }
} # closes the j loop
} # closes the i loop
# counting the number of transitions for each day (day is kept fixed while
# counting goes through years)
# count_Km[m,1] counts the number of transitions m -> 0 in the 22 years for fixed i
# count_Km[m,2] counts the number of transitions m -> 1 in the 22 years for fixed i
# n_m0[dias,m] counts the number of transitions m -> 0 in the 22 years for each i
# n_m1[dias,m] counts the number of transitions m -> 1 in the 22 years for each i
n_m0 <- matrix(0, (dias - K), 2^K) # matrix counting m -> 0
n_m1 <- matrix(0, (dias - K), 2^K) # matrix counting m -> 1
for(i in 1:(dias-K)){
for(j in 1:anos){
if(ozonio_sw[i+K,j] == 0){
for(n in 0:(2^K-1)){
if(mbase[i,j] == n){
count_Km[n+1,1] = count_Km[n+1,1]+1}}}
if(ozonio_sw[i+K,j] == 1){
for(n in 0:(2^K-1)){if(mbase[i,j] == n){
count_Km[n+1,2] = count_Km[n+1,2]+1}}}  }
for(m in 1:(2^K)){
n_m0[i,m] = count_Km[m,1]
n_m1[i,m] = count_Km[m,2] }
count_Km <- matrix(0, 2^K, 2)
} # closes de i loop
# assignation of the values of the corresponding values of
# alpha the hyperparameter of the Dirichlet prior distribution
# alpha_m0 is associated to the transitions m -> 0 for each day and each m
# alpha_m1 is associated to the transitions m -> 1 for each day and each m
alpha_m0 <- matrix(0, (dias - K), 2^K)
alpha_m1 <- matrix(0, (dias - K), 2^K)
for(i in 1:(dias-K)){
for(m in 1:2^K){
if(n_m0[i,m] == 0){alpha_m0[i,m] = 3}
if(n_m1[i,m] == 0){alpha_m1[i,m] = 3}
} #closes the m loop
for(m in 1:2^K){
if((n_m0[i,m] == min(n_m0[i,m], n_m1[i,m])) & (alpha_m0[i,m] == 0)){
alpha_m0[i,m] = 4
if((abs(n_m0[i,m] - n_m1[i,m]) >= 5) & (alpha_m1[i,m] == 0)){alpha_m1[i,m] = 7}
if((abs(n_m0[i,m] - n_m1[i,m]) < 5) & (alpha_m1[i,m] == 0)){alpha_m1[i,m] = 5}}
if((n_m0[i,m] == min(n_m0[i,m], n_m1[i,m])) & (alpha_m0[i,m] != 0)){
if((abs(n_m0[i,m] - n_m1[i,m]) >= 5) & (alpha_m1[i,m] == 0)){alpha_m1[i,m] = 7}
if((abs(n_m0[i,m] - n_m1[i,m]) < 5) & (alpha_m1[i,m] == 0)){alpha_m1[i,m] = 5} }
if((n_m0[i,m] == max(n_m0[i,m], n_m1[i,m])) & (alpha_m0[i,m] == 0))
```

```
{ alpha_m0[i,m] = 8
if((abs(n_m0[i,m] - n_m1[i,m]) >= 5) & (alpha_m1[i,m] == 0)){alpha_m1[i,m] = 4}
if((abs(n_m0[i,m] - n_m1[i,m]) < 5) & (alpha_m1[i,m] == 0)){alpha_m1[i,m] = 7} }
if((n_m0[i,m] == min(n_m0[i,m], n_m1[i,m])) & (alpha_m0[i,m] != 0)){
if((abs(n_m0[i,m] - n_m1[i,m]) >= 5) & (alpha_m1[i,m] == 0)){alpha_m1[i,m] = 4}
if((abs(n_m0[i,m] - n_m1[i,m]) < 5) & (alpha_m1[i,m] == 0)){alpha_m1[i,m] = 7} }
} #closes the second m loop
} #closes the i loop
# calculation of the likelihood L(Y | K)
prod_1_km <- matrix(0, (dias - K), 1)
prod_2_km <- matrix(0, (dias - K), 1)
prod_km <- matrix(0, (dias-K), 1)
for(i in 1:(dias-K)){
for(n in 1:2^K){
prod_1_km[i] = (gamma(n_m0[i,m]+alpha_m0[i,m])*gamma(n_m1[i,m]+alpha_m1[i,m]))/
(gamma(alpha_m0[i,m])*gamma(alpha_m1[i,m]))
prod_2_km[i] = gamma(alpha_m0[i,m] + alpha_m1[i,m])/gamma(alpha_m0[i,m] +
n_m0[i,m] + alpha_m1[i,m] + n_m1[i,m])
} #closed the m loop
prod_km[i] = prod_1_km[i]*prod_2_km[i]
} # closes the i loop
for(i in 1:(dias-K)){
term_general_like[i,K+1] = prod_km[i]}
} #close the K loop
write.csv(term_general_like, file = "results-file-1.csv")
write.csv(term_init_like, file = "results-file-2.txt")
#
# TRANSITION PROBABILITIES - CASE OF REGION CE, THRESHOLD 0.17
#
K = 8 # may change for other regions and thresholds
# calculation of the normalising constant in the case of the initial distribution
somainit = 0
for(m in 1:2^K){
somainit = somainit + (alpha_init_k[m] + count_init_k[m] - 1)}
# calculation of the initial distribution
prob_init <- matrix(0, 2^K, 1)
for(m in 1: 2^K){
prob_init[m] = (alpha_init_k[m] + count_init_k[m] - 1)/somainit}
write.csv(prob_init, file = "prob_init_chain.txt")
# limiting the number of days
dd = 2 # because I need p_{mj}(1) and p_{mj}(2)
# alpha_m1[i,m] where i is day and m is the value of the vector
# the m1 indicates that m is followed by 1
p_m0 <- matrix(0, dd, 2^K)
p_m1 <- matrix(0, dd, 2^K)
# calculation of the transition probabilities m -> 0 and m -> 1
for(m in 1:2^K){
for(i in 1:dd){
p_m0[i,m] = (alpha_m0[i,m] + n_m0[i,m] - 1)/((alpha_m1[i,m] + n_m1[i,m] - 1)
+(alpha_m0[i,m] + n_m0[i,m] - 1))
p_m1[i,m] = (alpha_m1[i,m] + n_m1[i,m] - 1)/((alpha_m1[i,m] + n_m1[i,m] - 1)
+(alpha_m0[i,m] + n_m0[i,m] - 1)) }}
trans_mat <- matrix(0, 2^K, dd)
for(m in 1:2^K){
for(i in 1:dd){
trans_mat[m,i] = p_m0[i,m]}}
write.csv(trans_mat, file = "trans_mat.txt")
```

Author details

Eliane R. Rodrigues[1*], Mario H. Tarumoto[2] and Guadalupe Tzintzun[3]

1 Instituto de Matemáticas, Universidad Nacional Autónoma de México, Mexico

2 Faculdade de Ciênicas e Tecnologia, Universidade Estadual Paulista Júlio de Mesquita Filho, Brazil

3 Instituto Nacional de Ecología y Cambio Climático, Secretaría de Medio Ambiente y Recursos Naturales, Mexico

*Address all correspondence to: eliane@math.unam.mx

References

[1] Bell ML, McDermontt A, Zeger SL, Samet JM, Dominici F. Ozone and short-term mortality in 95 US urban communities, 1987-2000. Journal of the American Medical Society 2004; 292: 2372-2378.

[2] Bell ML, Peng R, Dominici F. The exposure-response curve for ozone and risk of mortality and the adequacy of current ozone regulations. Environmental Health Perspectives 2005; 114: 532-536.

[3] Cifuentes L, Borja-Arbuto VH, Gouveia N, Thurston G, Davis DL. Assessing the health benefits of urban air pollution reduction associated with climate change mitigation (2000-2020): Santiago, São Paulo, Mexico City and New York City. Environmental Health Perspectives 2001; 109: 419-425.

[4] Dockery DW, Schwartz J, Spengler JD. Air pollution and daily mortality: association with particulates and acid aerosols. Environmental Research 1992; 59: 362-373.

[5] Galizia A, Kinney PL. Long-term residence in areas of high ozone: association with respiratory health in a nationwide sample of nonsmoking adults. Environmental Health 1999; 99: 675-679.

[6] Gauderman WJ, Avol E, Gililand F, Vora H, Thomas D, Berhane K, McConnel R, Kuenzli N, Lurmman F, Rappaport E, Margolis H, Bates D, Peter J. The effects of air pollution on lung development from 10 to 18 years of age. The New England Journal of Medicine 2004; 351: 1057-1067.

[7] Gouveia N, Fletcher T. Time series analysis of air pollution and mortality: effects by cause, age and socio-economics status. Journal of Epidemiology and Community Health 2000; 54: 750-755.

[8] Loomis D, Borja-Arbuto VH, Bangdiwala SI, Shy CM. Ozone exposure and daily mortality in Mexico City: a time series analysis. Health Effects Institute Research Report 1996; 75: 1-46.

[9] Martins LC, de Oliveira Latorre MRD, Saldiva PHN, Braga ALF. Air pollution and emergency rooms visit due to chronic lower respiratory diseases in the elderly: an ecological time series study in São Paulo, Brazil. J. Occupational and Environmental Medicine 2002; 44: 622-627.

[10] WHO. Air Quality Guidelines-2005, Particulate Matter, Ozone, Ditrogen dioxide and Sulfur Dioxide. European Union: World Health Organization Regional Office for Europe; 2006.

[11] Horowitz J. Extreme values from a nonstationary stochastic process: an application to air quality analysis. Technometrics 1980; 22: 469-482.

[12] Smith RL. Extreme value analysis of environmental time series: an application to trend detection in ground-level ozone. Statistical Sciences 1989; 4: 367-393.

[13] Raftery AE. Are ozone exceedance rate decreasing?. Comment of the paper *Extreme value analysis of environmental time series: an application to trend detection in ground-level ozone by R. L. Smith*. Statistical Sciences 1989; 4: 378–381.

[14] Flaum JB, Rao ST, Zurbenko IG. Moderating Influence of Meteorological Conditions on Ambient Ozone Concentrations. Journal of the Air and Waste Management Association 1996; 46: 33-46.

[15] Achcar JA, Rodrigues ER, Tzintzun G. Using stochastic volatility models to analyse weekly ozone averages in Mexico City. Environmental and Ecological Statistics 2011; 18: 271-290.

[16] Javits JS. Statistical interdependencies in the ozone national ambient air quality standard. Journal of Air Pollution Control Association 1980; 30 58-59.

[17] Achcar JA, Fernández-Bremauntz AA, Rodrigues ER, Tzintzun G. Estimating the number of ozone peaks in Mexico City using a non-homogeneous Poisson model. Environmetrics 2008; 19: 469-485.

[18] Achcar JA, Rodrigues ER, Tzintzun G. Using non-homogeneous Poisson models with multiple change-points to estimate the number of ozone exceedances in Mexico City. Environmetrics 2011; 22: 1-12.

[19] Villaseñor-Alva JA, González-Estrada E. On modelling cluster maxima with applications to ozone data from Mexico City. Environmetrics 2010; 21: 528-540.

[20] Barrios JM, Rodrigues ER. A queueing model to study occurrence and duration of ozone exceedances in Mexico City. Journal of Applied Statistics 2014. dx.doi.org/101080/02664763.2014.939613

[21] Rajagopalan B, Upmanu L, Tarboton DG. Non-homogeneous Markov model for daily precipitation. Journal of Hydrology Engineering 1996; 1: 33-40.

[22] Hughes JP, Guttorp P, Charles SP. A non-homogeneous hidden Markov model for precipitation occurrence. Applied Statistics, Part 1 1999; 48: 16-30.

[23] Drton M, Marzban C, Guttorp P, Schafer JT. A Markov chain model for tornadic activity. American Meteorological Society 2003; 131: 2941-2953.

[24] Larsen LC, Bradley RA, Honcoop GL. A new method of characterizing the variability of air quality-related indicators. Air and Waste Management Association. International Specialty Conference, Tropospheric Ozone and the Environment. Pittsburgh, USA: California Air and Waste Management Series; 1990.

[25] Álvarez LJ, Fernández-Bremauntz AA, Rodrigues ER, Tzintzun G. Maximum a posteriori estimation of the daily ozone peaks in Mexico City. Journal of Agricultural, Biological, and Environmental Statistics 2005; 10: 276-290.

[26] Álvarez LJ, Rodrigues ER. A trans-dimensional MCMC algorithm to estimate the order of a Markov chain: an application to ozone peaks in Mexico City. International Journal of Pure and Applied Mathematics 2008; 48: 315-331.

[27] Cox DR, Lewis PA. Stochastic analysis of series of events. UK: Methuen; 1966.

[28] Rice JA. Mathematical statistics with data analysis. New York, USA: Wadsworth and Brook; 1988.

[29] Carlin BP, Louis TA. Bayes and Empirical Bayes Methods for Data Analysis. Second Edition. USA: Chapman and Hall/CRC; 2000.

[30] Green PJ. Reversible jump Markov chain Monte Carlo computation and Bayesian model determination. Biometrika 1995; 82: 711-732.

[31] Carlin BP, Chib S. Bayesian model choice via Markov chain Monte Carlo methods. Journal of the Royal Statistical Society Series B 1995; 57: 473-484.

[32] Boys RJ, Henderson DA. On determining the order of a Markov dependence of an observed process governed by a hidden Markov chain. Special Issue of Scientific Programming 2002; 10: 241-251.

[33] Fleming TR, Harrington DP. Estimation for discrete time non-homogeneous Markov chains. Stochastic Processes and their Applications 1978; 7: 131-139.

[34] Aalem OO, Johansen S. An empirical transition matrix for non-homogeneous Markov chains based on censored observation. Scandinavian Journal of Statistics 1978; 5: 141-150.

[35] Robert CP, Casella G. Monte Carlo statistical methods. New York, USA: Springer; 1999.

[36] Fan T-H, Tsai C-A. A Bayesian method in determining the order of a finite Markov chain. Communications in Statistics – Theory and Methods 1999; 28: 1711-1730.

[37] Evans M, Swartz T. Approximating Integrals via Monte Carlo and Deterministic Methods. Oxford Statistical Series 20. Oxford, UK: Oxford University Press; 2000.

[38] NOM. Modificación a la Norma Oficial Mexicana NOM-020-SSA1-1993. Diario Oficial de la Federación. 30 Octubre 2002. Mexico: 2002. (In Spanish.)

Section 3

Air Pollution Risk Assessment

Chapter 18

Radon Risk Analysis Through Geostatistical Tools Implemented in a WebGIS

Veronica Distefano, Sandra De Iaco,
Monica Palma and A. Spennato

Additional information is available at the end of the chapter

1. Introduction

Radon (Rn) is a colorless, odorless, tasteless, inert radioactive gas, and derives from the decay of uranium, which is a radioactive element that is found in small quantities in all sediments and rocks.

The International Agency for Research on Cancer (IARC) and the World Health Organization (WHO) classify Rn pollution as the second leading cause of lung cancer after smoking.

Since Rn is present, in the depths of the Earth, in gaseous phase, it reaches the surface because it interacts with other natural elements, such as uranium, thorium and radio (precursor elements); moreover other geo-lithological features, such as the mineralogical composition of the rocks, the underground permeability levels, the presence of faults, fractures and cavities, affect the transport of the Rn on the surface.

In this paper, the spatial distribution of the Rn concentrations in soil gas over a survey area located in the South of Apulian Region (Italy) and its prediction at unsampled points have been discussed. In particular, Ordinary Kriging (OK), Log-Normal Kriging (LK), Cokriging with indicator variable (ICK) and Kriging with Varying Means (KVM) [11, 21, 26, 27] have been used to predict Rn concentrations over the study area.

In this context, the integration of a Geographical Information System (GIS) and geostatistical tools can certainly support the evaluation of alternative scenarios, possible strategies for a sustainable development [1].

In addition, the geoprocessing tools (e.g., buffering, overlay, union/intersection, interpolation) can be used to combine and interpret data obtained from different sources. The construction of a GIS project facilitates the data sharing and the integration of environmental and demographic data, as well as the results of spatial analysis.

Thus, the usefulness of geostatistical techniques for monitoring Rn concentrations in soil gas and the integration of these results in an *ad hoc* WebGIS, named RnWebGIS, have been described. This innovative tool offers dynamic scenarios for monitoring and analyzing different environmental variables [7, 39]. The proposed RnWebGIS is based on an Open Source Software, which enables users to get online spatial and environmental information.

Hence, after a discussion on the integration of GIS and geostatistical tools (Section 2), a brief review of some geostatistical techniques for modeling and prediction purposes has been presented (Section 3). A thorough case study concerning the Rn concentration in soil gas measured in several sites over Lecce district has been developed (Section 4). Finally, the RnWebGIS, where all the available geo-lithological information of the area under study, as well as the Rn prediction results have been integrated, has been proposed and the steps for its implementation have been detailed.

2. GIS and geostatistical tools

Environmental risk management involves the integrated use of geostatistical techniques and several tools provided in a GIS.

A GIS is a repository of tools for capturing, storing, updating, manipulating, analyzing, and displaying all geographic information [9]. The data are stored as collections of relational tables logically associated with each other by shared attributes. On the other hand, Geostatistics provides advanced techniques to predict a variable of interest at unsampled points and to support decisions concerning monitoring, sampling, planning and re-qualification of a territory [10].

Fig. 1 shows a flowchart regarding the steps which characterize the connection between GIS and geostatistical tools.

The first step is based on the creation of a geodatabase, that is a relational database which stores a series of tables holding feature classes, raster datasets, attributes and geographic information represented by shapefiles (i.e. geospatial vector data format which can spatially describe vector features, such as polygon, line or point, with defined shape and geometric location). Then, the next step concerns the georeferencing, that is the process of assigning real-world coordinates to each pixel of the raster data. In this case, these coordinates are often obtained through field surveys, by collecting coordinates with a GPS device, or by using Google Earth software. On the other hand, the shapefile data do not need to be georeferenced, but simply loaded in a geodatabase by specifying the coordinate system associated to the datasets; hence, the vector features in the shapefile will be represented by using the given coordinate system.

The third step involves the implementation of the geostatistical analysis which is organized in various phases, such as the exploratory spatial data analysis (i.e. mapping the spatial distribution using thematic maps), the estimation and modeling the spatial correlation structure exhibited by the data, predicting the variable of interest at unsampled locations by using spatial interpolation techniques (such as Simple, Ordinary and Universal Kriging, Indicator Kriging, Cokriging) and computing the corresponding prediction errors. Finally, the last step consists in the visualization of geostatistical results by producing reliable prediction maps, probability maps, and other useful cartographic representations.

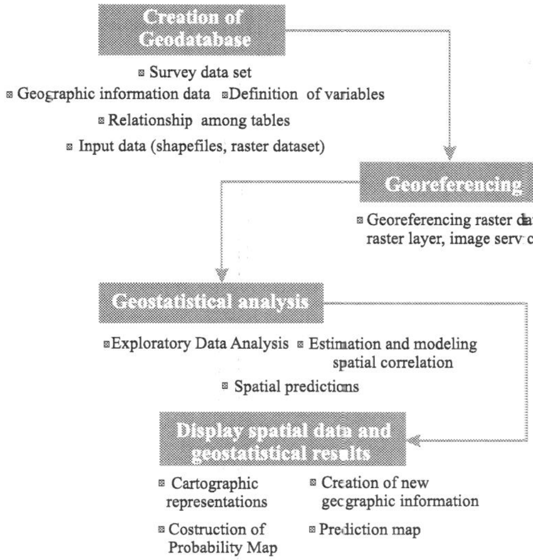

Figure 1. Integration between GIS and geostatistical tools.

The integration between GIS and Geostatistics has been of interest since the early 1990s, when [20] and [8] illustrated the potential benefits of a close link between information systems and Spatial Analysis. Many contributions, concerning the integration of spatial statistical methods and a GIS environment, have been proposed in the literature [2, 3, 6, 9, 14, 37] over the years.

In order to improve the integration between GIS and Geostatistics, different methods of spatial analysis have been implemented in several kinds of GIS softwares [1, 3]. For example, among the most widely used commercial GIS packages, it is worth citing *ArcGIS*, *Geomedia* and *Mapinfo*, which offer extensions for geostatistical analysis; on the other hand, among the *Open Source* packages it is worth mentioning *GRASS* and *Quantum GIS* which use R, Java and Python in order to implement the geostatistical tools. But, the collection of spatial data analysis softwares now available does not completely include all the computational aspects. Therefore, in order to satisfy the specific requirements of the user it is often necessary to develop *ad hoc routines* for sharing GIS data, implementing macros and customized libraries of geostatistical functions in GIS environment, creating suitable script [36].

3. Geostatistical basic notions

Geostatistical methods [15, 23, 28] are useful applied in many areas of environmental sciences (Geography, Geology, Environmental Sciences, Ecology, Meteorology, Agronomy), for studying phenomena which vary continuously over the survey domain. Usually, the final aim of geostatistical analysis consists of obtaining predictions of the variable of interest at unsampled points of the study area. In this case, the use of stochastic models to describe the spatial behaviour of the phenomenon under study, as well as its spatial uncertainty, is reasonable.

In the following sections, the most applied geostatistical techniques, developed for modeling and prediction purposes, both in the univariate and multivariate case, will be briefly described.

3.1. Modeling and prediction in the univariate case

In Geostatistics, the available spatial data concerning the measurements of a single variable, at n points located over the area of interest, are considered as a particular realization of the random variable $Z(\mathbf{u}_\alpha), \alpha = 1, 2, \ldots, n$, where $\mathbf{u} \in D \subseteq R^d$, $d \in N_+$. This random variable is called in [25] *regionalized variable*, in order to highlight the feature of the spatial dependency for the variable itself. The spatial distribution of the random variable Z is usually interpreted as the result of two overlapping structures: a systematic structure at the macroscopic level, which describes the global features of the same natural process, and another random structure at the microscopic level, which represents the local irregularity of Z.

In particular, by assuming the existence of the first and the second-order moments of the random variable Z, the basic stochastic model for Z is defined as follows:

$$Z(\mathbf{u}) = m(\mathbf{u}) + Y(\mathbf{u}), \qquad \mathbf{u} \in D \subseteq R^d,\ d \in N_+, \tag{1}$$

where $m(\mathbf{u})$, usually called *trend*, describes the large-scale variations and $Y(\mathbf{u})$ is a second-order stationary random field which represents the local fluctuations of Z.

Under the second-order stationary hypotheses, the expected value of Z exists and does not depend on $\mathbf{u} \in D$; moreover the variogram exists and can be defined as follows:

$$\gamma(\mathbf{h}) = 0.5\, E[Z(\mathbf{u}) - Z(\mathbf{u} + \mathbf{h})]^2, \tag{2}$$

where $\mathbf{h} = \mathbf{u} - \mathbf{u}'$ is the separation vector between two support points $\mathbf{u}$ and $\mathbf{u}'$ of the spatial domain D.

The variogram, by definition, is a measure of dissimilarity, in the sense that it increases when the difference between $Z(\mathbf{u})$ and $Z(\mathbf{u} + \mathbf{h})$ increases; thus, it is usually characterized by small values for short spatial distances.

After computing the sample variogram [32], as a measure of the spatial correlation of the variable of interest, a theoretical admissible model has to be fitted to the estimated variogram.

It is worth noting that the condition for a function to be a variogram is that it be conditionally negative definite [11]. However, given the difficulties to verify this condition, it is advisable for users to look for the best model among the wide parametric families whose members are known to be conditionally negative definite functions. This condition ensures a unique solution in the kriging system (system of linear equations based on the variogram matrix of observations in different spatial points of the domain, as it will be discussed hereafter) used to tackle the general problem of predicting the process of interest, starting from the observed values.

Various classes of variogram models have been proposed in the literature; however the selection of an appropriate class of models is in general based on its geometric features and theoretical properties [10, 12, 13].

Once the spatial correlation of the variable under study has been properly modelled, some well-known based-variogram prediction techniques can be used in order to estimate the analyzed variable at unsampled points over the domain of interest.

The most applied spatial predictor is the Ordinary Kriging (OK) [10, 23] which is based on the following estimator:

$$\widehat{Z}^{OK}(\mathbf{u}) = \sum_{\alpha=1}^{n} \lambda_{\alpha}^{OK}(\mathbf{u}) Z(\mathbf{u}_{\alpha}), \tag{3}$$

where the kriging weights λ_{α}^{OK}, $\alpha = 1, 2, \ldots, n$, depend on the variogram model and the sampling configuration

The predictor $\widehat{Z}^{OK}(\mathbf{u})$ satisfies both the unbiasedness and the efficiency condition.

In particular, the first property is satisfied if the weights λ_{α}^{OK}, $\alpha = 1, 2, \ldots, n$, of the linear combination are such that

$$\sum_{i=1}^{n} \lambda_{\alpha}^{OK}(\mathbf{u}) = 1. \tag{4}$$

On the other hand, $\widehat{Z}^{OK}(\mathbf{u})$ is efficient, if the weights are chosen in order to minimize the error variance $\sigma_E^2 = E[R(\mathbf{u})]^2 = E[\widehat{Z}(\mathbf{u}) - Z(\mathbf{u})]^2$, under the unbiasedness condition.

Finally, it is important to point out that, if the variable under study is characterized by a high coefficient of variation and/or by different levels of variability, traditional prediction methods, such as OK, might not be appropriate. In particular, when the experimental data shown a skewed distribution, it is possible to apply a non-linear transformation at the observed data, such that the distribution of the transformed data is symmetric. In many cases, the logarithmic transformation is useful for this aim. Hence, structural analysis and kriging are performed by considering the log-transformed data.

3.2. Modeling and prediction in the multivariate case

In many environmental applications, the phenomenon under study is the result of the simultaneous behavior of several variables, which are correlated each other and measured at some points of the area of interest.

In this case, a multivariate spatial structure [34] characterizes the observed data which can be considered as a realization of a multivariate random field (*MRF*) $\{\mathbf{Z}(\mathbf{u}),\ \mathbf{u} \in D \subseteq \mathbb{R}^d\}$, with

$$\mathbf{Z}(\mathbf{u}) = [Z_1(\mathbf{u}), Z_2(\mathbf{u}), \ldots, Z_p(\mathbf{u})]^T, \tag{5}$$

where $p \geq 2$.

Under second-order stationarity, the mean vector of $\mathbf{Z}$ exists and does not depend on $\mathbf{u}$, i.e.:

$$E[\mathbf{Z}(\mathbf{u})] = \mathbf{M} = [m_1, m_2, \ldots, m_p]^T, \qquad \mathbf{u} \in D, \tag{6}$$

and the $(p \times p)$ variogram matrix $\mathbf{\Gamma}$ exists and depends on the spatial separation vector $\mathbf{h}$ between two support points ($\mathbf{u}$ and $\mathbf{u}'$), i.e.:

$$\mathbf{\Gamma}[\mathbf{Z}(\mathbf{u}), \mathbf{Z}(\mathbf{u}')] = E\left\{[\mathbf{Z}(\mathbf{u}) - \mathbf{Z}(\mathbf{u}')][\mathbf{Z}(\mathbf{u}) - \mathbf{Z}(\mathbf{u}')]^T\right\} =$$

$$= \mathbf{\Gamma}(\mathbf{h}) = [\gamma_{ij}(\mathbf{h})], \tag{7}$$

where $\gamma_{ij}(\mathbf{h}),\ i,j = 1,2,\ldots,p,$ is the cross-variogram between $Z_i(\mathbf{u})$ and $Z_j(\mathbf{u}+\mathbf{h})$, when $i \neq j$, and the direct variogram of the $Z_i(\mathbf{u})$, when $i = j$.

In multivariate Geostatistics, *Ordinary Cokriging* (OCK) [34] is a well-known technique used to obtain prediction of one or all the variables under study, at unsampled points of the spatial domain. This technique considers the following linear predictor of the spatial random vector defined in (5):

$$\widehat{\mathbf{Z}}^{OC}(\mathbf{u}) = \sum_{\alpha=1}^{n} \mathbf{\Lambda}_{\alpha}^{OC}(\mathbf{u})\mathbf{Z}(\mathbf{u}_{\alpha}), \tag{8}$$

where $\mathbf{u} \in D$ is an unsampled point in the domain D, $\mathbf{u}_{\alpha} \in D,\ \alpha = 1,2,\ldots,n,$ are the sampled points in the domain D and $\mathbf{\Lambda}_{\alpha}^{OC}(\mathbf{u})$ are the $(p \times p)$ matrices of weights, which are determined by ensuring unbiasedness and efficiency conditions.

The cokriging predictor requires a model for the variogram matrix (7). The *Linear Model of Coregionalization* (*LMC*), widely discussed in [23] and [34], is the most commonly model used in the applications. In particular, the *LMC* can be written as

$$\mathbf{\Gamma}(\mathbf{h}) = \sum_{l=1}^{L} \mathbf{B}_l\, g_l(\mathbf{h}), \tag{9}$$

where $\mathbf{B}_l$, $l = 1,2,\ldots,L$, are $(p \times p)$ positive definite matrices, called *coregionalization matrices*, while $g_l(\mathbf{h}),\ l = 1,2,\ldots,L,$ are basic variograms which correspond to different scales of spatial variability.

A particular case of cokriging, which allows non-parametric, non-linear predictions [22], is the *Cokriging with Indicator Variable* (ICK), where the values observed for one and/or all the variables under study, are replaced by indicator values (that is 1 or 0, depending on a condition specified by the analyst which is of interest for the phenomenon under study).

Another very useful prediction technique, developed in the context of multivariate Geostatistics, is the *Kriging with Varying Means* (KVM), which corresponds to an extension of the *simple kriging* used in the univariate case. In the context of two or more variables, the stationary mean of the primary variable is replaced with a known varying mean according to the secondary information [21]. Hence, the KVM predictor is computed as follows:

$$\widehat{Z}^{KVM}(\mathbf{u}) = \sum_{\alpha=1}^{n} \lambda_{\alpha}^{KVM}(\mathbf{u})[Z(\mathbf{u}_{\alpha}) - m^{KVM}(\mathbf{u}_{\alpha})] \tag{10}$$

where $\mathbf{u} \in D$ is the unsampled point, while λ_{α}^{KVM}, $\alpha = 1, 2, \ldots, n$, are the weights of the predictor. Note that if the secondary variable is categorical, the primary local mean is the mean of the primary variable within a specific category of the secondary variable.

In the proposed case study, the above mentioned prediction techniques will be performed and the main prediction results will be discussed.

4. Case study: the survey area and the data set

The study area is located in the Southern part of Italy (Lecce district in the Apulian region), at 199 meters above the sea level (Fig. 2).

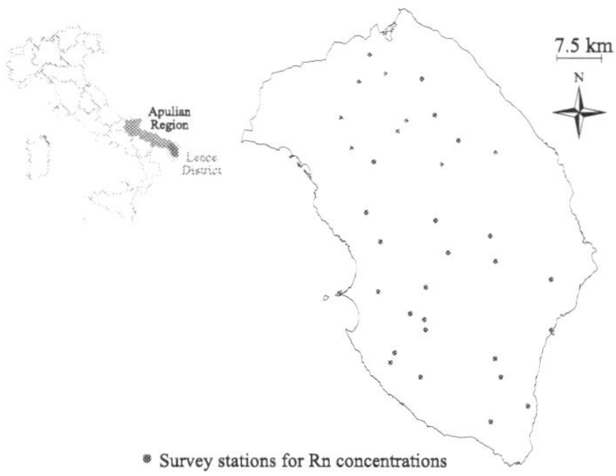

Figure 2. Location map of survey stations in Lecce district.

It is characterized by a limestone substrate and the presence of sink-holes, caves and underground drainages. Moreover, the area is characterized by 6 different lithotypes, with predominance of sandstone (45.7%), limestone and dolomitic limestone (29.4%), as well as siltstone and sandstone (23.9%). Because of this particular lithology, a medium-high level of underground permeability marks the area out. Fig. 3 illustrates the maps of geo-lithological features and permeability levels.

The available geo-lithological information about the survey area represents the *soft* information which can better explain the outdoor Rn concentrations and the corresponding exhalation rate from soil. In order to use the soft information in the modeling and prediction stages of the geostatistical analysis, a regular georeferenced grid covering the area under study (Fig. 4-a)) has been defined. The node separation distance (along the horizontal and vertical directions) has been set equal to 2 km. Thus, the soft data

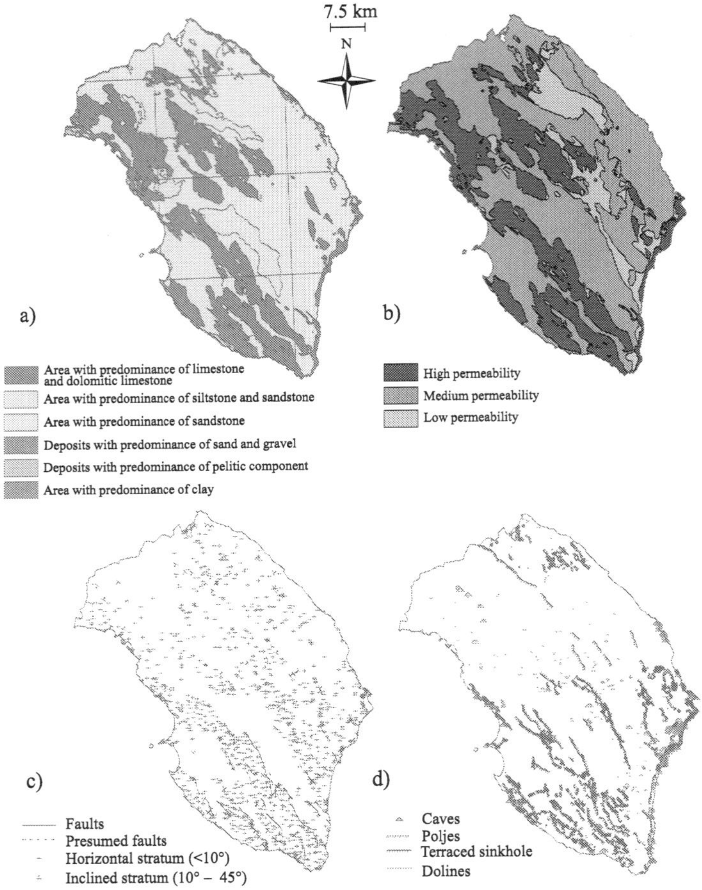

Figure 3. Maps of geo-lithological features in Lecce district. a) Map of geo-lithological features. b) Map of underground permeability levels. c) Map of tectonics features. d) Map of karst forms.

have been stored in a GIS project by using Arcmap of the ArcGIS software and the shapefile (data sources: http://www.sit.puglia.it) concerning the geo-lithological features, the underground permeability level, the tectonic (such as distance from tectonic elements) and karst features (such as distance from karst forms). In particular, by using the *spatial join* tool (a geoprocessing tool of Arcmap), all the soft data have been assigned to each grid point (for example, for each grid point it is possible to identify the type of geo-lithology composition, the distance grid point-faults, the distance grid point-karst forms).

On the other hand, the available Rn concentrations in soil gas have been considered as *hard* information. In particular, the in soil gas Rn levels have been measured at 32 locations over

Lecce district (Fig. 2), during a campaign conducted in June-July 2012 on the basis of a suitable spatial sampling.

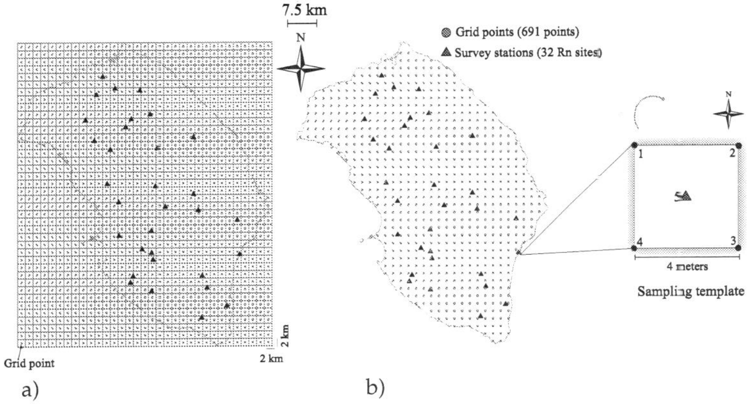

Figure 4. a) Regular grid of cell size 2 × 2 km and grid points with node separation distance equal to 2 km. b) Survey stations for Rn concentrations in soil gas, together with the sampling template.

For each selected site, the Rn concentrations have been detected by using an active sampling-based system, in which the Rn and its decay products are conveyed in proximity of the radiation detector through a mechanical pumping. A special instrument (AlphaGUARD radon monitor by Saphymo GmbH), equipped with an ionization chamber has been used. The chamber has been connected to a probe, introduced into the ground up to a maximum of 1 meter deep, and to a pump for the aspiration of Rn gas at fixed intervals of time. In particular, at each sample location, the measurements of the Rn levels in soil gas have been performed, every minute for 20 minutes, by aspirating the air through the probe introduced into the ground at 5 points according to a regular template (sampling template in Fig. 4). The template consists of a square with a side of 4 meters; thus the Rn measurements have been made both in the center and the corresponding vertices. Hence, both the center and the vertices of the square have been numbered, starting from the top of the North-West and proceeding clockwise, as considered in the sampling protocol.

Finally, the average of the observed Rn concentrations for each site has been assigned and georeferenced at the central point of the sampling template (number 5 in Fig. 4).

All the 32 sample values (hard data) have been georeferenced and stored in a GIS project; then they have been integrated with the soft information.

4.1. Geostatistical analysis: main results

In the following the spatial concentrations of the Rn in soil gas have been analyzed and predicted at unsampled points of the study area by applying different techniques, such as OK, LK, ICK and KVM

Note that, the above mentioned multivariate interpolation methods utilize both image-derived (soft) data which are more exhaustive and higher accurate than hard data, typically sparse over the domain. Using ICK and KVM is justified when the measurements of environmental variables can be characterized by a high coefficient of variation and/or by different level of variability; indeed in such cases traditional prediction methods, such as OK and LK, are not appropriate because they do not account for these features.

Ordinary kriging predictions

In order to predict Rn concentrations in soil gas by using OK method, the structural analysis has been performed on the measured values. After computing the sample variogram, the following Gaussian variogram model has been fitted:

$$\gamma(\mathbf{h})^{OK} = 80 + 263 \left[1 - \exp\left(-3||\mathbf{h}||/20\right)^2\right] \tag{11}$$

where nugget and sill correspond to 80 and 263, respectively, while range is equal to 20 km.

Fig. 5-b) shows the estimated Rn concentrations in soil gas obtained by using OK.

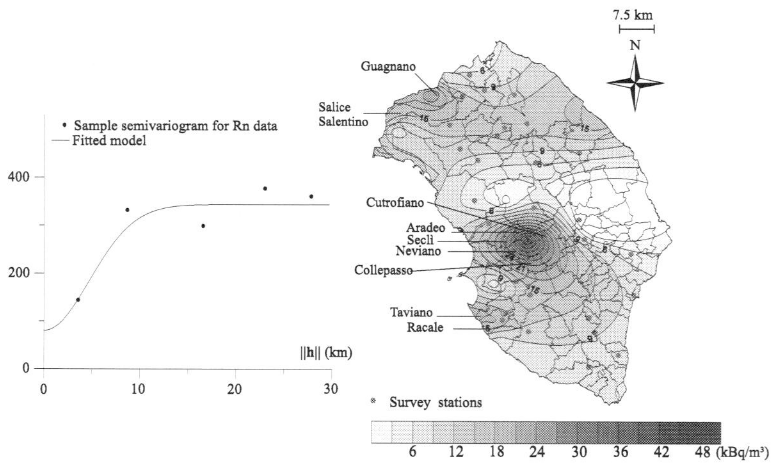

a) b)

Figure 5. a) Sample variogram and fitted model for Rn concentrations. b) Contour map of the estimated Rn concentrations using Ordinary kriging.

In particular, the prediction map in Fig. 5-b) highlights three Rn prone area, characterized by high Rn concentration values, in the North-Western, South-West and South-South-West of the domain of interest.

Log-Normal kriging predictions

In this case, the Rn measured values have been first processed by applying the logarithmic transformation. Hence, the structural analysis has been performed on the log-transformed data. In Fig. 6-a) the sample variogram computed for the log-transformed data is illustrated together with the fitted model. In particular, the variogram model considered for the LK prediction technique is given below:

$$\gamma(\mathbf{h})^{LK} = 0.1 + 1.2\left[1 - \exp\left(-3||\mathbf{h}||/20\right)^{2}\right] \qquad (12)$$

where nugget and sill correspond to 0.1 and 1.2, respectively, while range is equal to 20 km.

Fig. 6-b) shows the predictions for Rn concentrations obtaining by using LK, after a back-transformation on the data has been applied.

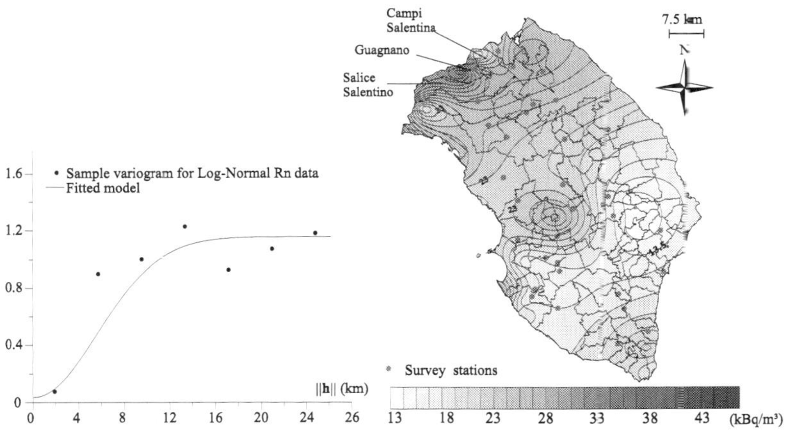

a) b)

Figure 6. a) Sample variogram and fitted model for Log-Normal Rn concentrations. b) Contour map of estimated Rn concentrations using LK, after a back-transformation on the data.

Despite the previous case (Fig. 5-b)), the prediction map illustrated in Fig. 6-b) highlights only one area with high Rn concentrations, located in the North-Western part of the domain of interest. From the prediction LK results, it is evident the significant *smoothing effect* produced by the interpolation process based on LK.

Cokriging predictions with indicator variable

This method has been applied by using a double source of information, i.e. the sampled Rn concentrations (primary variable) and the underground permeability levels (secondary or auxiliary variable). Evidently the auxiliary information is richer over the study area than the primary variable, hence it has been used to improve predictions of the Rn concentrations in soil gas. In particular, the auxiliary data, which represent the three permeability classes for any node of the grid, have been derived by the software Arcmap of ArcGIS and the shapefile concerning the underground permeability level in Lecce district. A value has been assigned to each point of the grid and to each survey station, according to the different permeability levels (the value 1 for low permeability, the value 2 for medium permeability and the value 3 for high level of permeability).

In this case, the following *LMC* [33, 34] has been fitted to the sample variogram matrix:

$$\mathbf{\Gamma} = \mathbf{B}_1\, g_1(\mathbf{h}) + \mathbf{B}_2\, g_2(\mathbf{h}) \tag{13}$$

where g_1 is a nugget model and g_2 is a gaussian model [23] with unit sill and range equal to 20 km; while $\mathbf{B}_l, l = 1,2$, are the following coregionalization matrices:

$$\mathbf{B}_1 = \begin{bmatrix} 80 & 2 \\ 2 & 0.125 \end{bmatrix}, \qquad \mathbf{B}_2 = \begin{bmatrix} 263 & 4.8 \\ 4.8 & 0.088 \end{bmatrix}. \tag{14}$$

The sample direct and cross variograms, shown in Fig. 7-a), have been modelled as follows:

$$\gamma_{11}(\mathbf{h}) = 80 g_1(\mathbf{h}) + 263 g_2(\mathbf{h}),$$

$$\gamma_{22}(\mathbf{h}) = 0.125 g_1(\mathbf{h}) + 0.088 g_2(\mathbf{h}),$$

$$\gamma_{12}(\mathbf{h}) = 2 g_1(\mathbf{h}) + 4.8 g_2(\mathbf{h}).$$

Finally, cokriging predictions of Rn concentrations in soil gas have been obtained as shown in Fig. 7-b).

In particular, the prediction map shown in Fig. 7-b) highlights three Rn prone areas, characterized by high Rn concentration values, in the North-Western, South-West and South-South-West of the domain of interest.

Kriging predictions with Varying Means

In order to apply this interpolation technique to the available data set, the Rn concentrations have been grouped by the three permeability classes, then the hypothesis of the presence of different mean levels of Rn concentrations in soil gas has been assumed.

After the structural analysis, the following variogram model (Fig. 8-a)) has been considered:

$$\gamma(\mathbf{h})^{KVM} = 170\ [1 - \exp(-3||\mathbf{h}||/15)] \tag{15}$$

where sill is equal to 170 and range equals to 15 km.

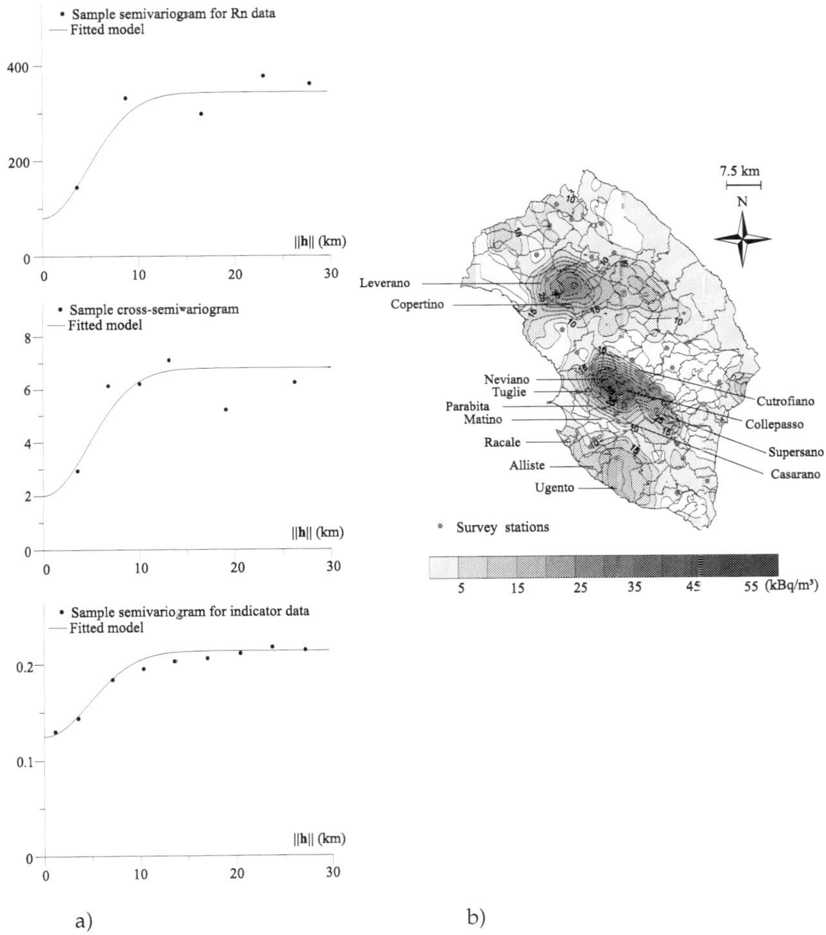

Figure 7. a) Sample direct and cross variograms and fitted model for Rn and indicator data. b) Contour map of estimated Rn concentrations using cokriging with indicator variable.

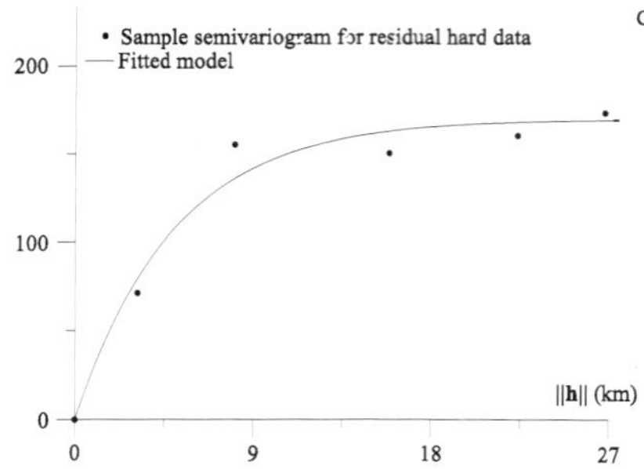

a)

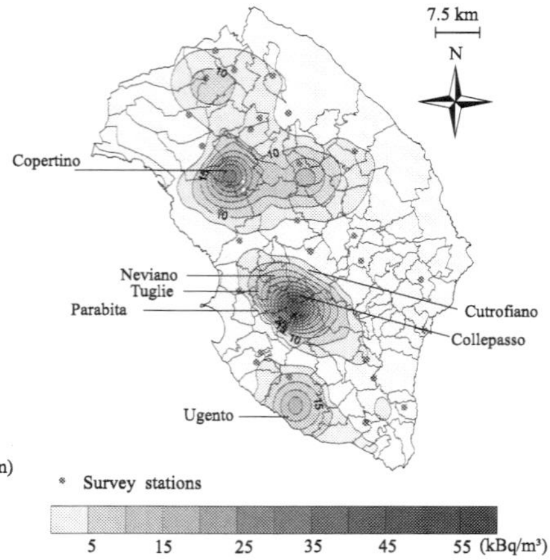

b)

Figure 8. a) Sample variogram and fitted model for residual hard data. b) Contour map of estimated Rn concentrations using KVM.

Hence, predictions for Rn concentrations in soil gas have been obtained for the study area (Fig. 8-b)) by using KVM.

In particular, the prediction map in Fig. 8-b) confirms that high Rn concentration values are located in the North-Western, South-West and South-South-West of the domain of interest.

4.2. A comparison among the prediction results

All the four contour maps illustrated in Figg. 5-b)-8-b) reproduce almost a similar behaviour, with high levels of Rn concentrations located in the Western part of the domain of interest. However, the predictions maps obtained from the KVM and ICK techniques, exploit the auxiliary information and identify the Rn prone areas more accurately than the ones computed by using the OK and LK techniques.

Moreover, the four interpolation techniques have highlighted the existence of a direction (135^0 from North to South, close to the Western coast of the study area), along with the Rn concentrations in soil gas present high levels.

The accuracy of the predictions has been measured through the *Root Mean Square Error* (RMSE) index [35, 38], defined as follows:

$$RMSE = \sqrt{\frac{1}{n}\sum_{\alpha=1}^{n}[z(\mathbf{u}_\alpha) - \widehat{z}(\mathbf{u}_\alpha)]^2}, \tag{16}$$

where $z(\mathbf{u}_\alpha)$ represent the measured values, $\hat{z}(\mathbf{u}_\alpha)$ correspond to the estimated values and n is the number of sample data.
By analyzing the RMSE shown in Fig. 9 it is evident that the accuracy of predictions is lower for OK and LK than the other two interpolation techniques.

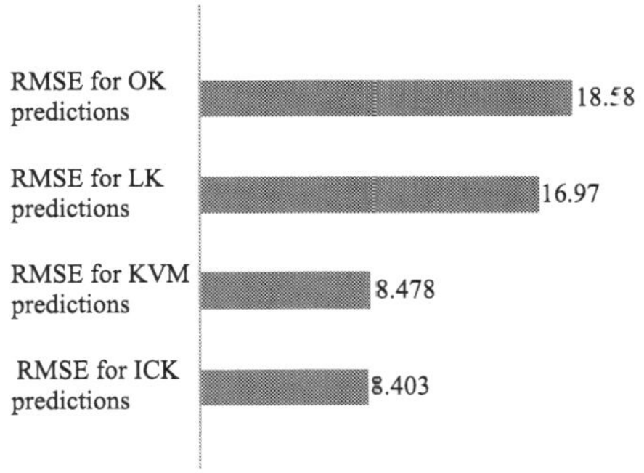

Figure 9. RMSE computed for the OK, LK, KVM and ICK predictions.

However, if the comparison is expressed in terms of the *relative percentage changes* of RMSE index:

- the use of LK method with respect to the OK, improves the predictions reliability of 9.5%;
- the use of ICK method with respect to the KVM, doesn't produce significant improvements for the predictions accuracy (just only 1%);
- the use of multivariate methods (such as KVM and ICK), improve the predictions reliability by about 51%, with respect to LK, and about 55% with respect to OK.

The better performance of the multivariate prediction techniques, compared with the univariate ones (OK and LK), is due to the appropriate use of the information given by the auxiliary variables (soft data). Moreover, among the univariate prediction techniques, the LK has shown a better performance since the variable under study has a skewed distribution.

5. Towards the implementation of a RnWebGIS

The evolution of information technology and GIS allow a potentially unlimited number of users to access and manage geospatial information online [18, 19].

A WebGIS offers an efficient way to provide geospatial information and Web Services to concurrent users without installing a GIS. For this reason a WebGIS system is defined as a web information system that provides geographic information on the web through the Hypertext Transfer Protocol (HTTP) and the HyperText Markup Language (HTML) [17].

Therefore, a WebGIS is different from other web maps, such as Google Earth or Google Maps, since it integrates Internet/web technologies with functionalities of a GIS structure [7].

Fig. 10 shows a schematic diagram regarding the WebGIS architecture, with the following essential steps:

- first step: a user sends a request to the *web browser*;
- second step: such a request is sent via Internet and processed by the server. Note that the WebGIS systems are equipped with a server mapping, such as *MapServer* [24] that obtains information from the geographic databases;
- third step: the request is displayed to the user through the *web browser*.

In the above-mentioned architecture the responsibility is on the server side, while the user (or client) needs only a browser in order to utilize the WebGIS and WebGIS applications. Therefore, the client sends requests to the GIS server which processes it and sends it to the client. Hence, the server performs the task of a *file server*, which contains the mapping data included in the WebGIS, while the client visualizes and consults the cartographic map. Therefore, the main elements of a WebGIS are the following: a) a web server, b) a scripting language, c) a database, d) a server mapping and e) an interface for the server mapping.

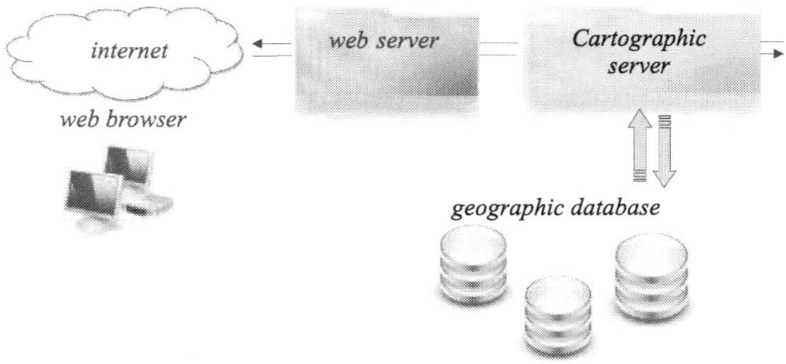

Figure 10. Schematic diagram of the WebGIS architecture.

In this paper, the RnWebGIS has been realized by using Open-source technologies, such as the *web server* Apache, a *MapServer* Project and the client interface *Pmapper*.

In the proposed RnWebGIS, the following databases have been included, such as:

- administrative boundaries:
  - the provincial borders;
  - the municipal boundaries;
  - the urban center boundaries;
- Rn concentrations and the related geographical information (Rn sites);
- the cartographic maps of the permeability and the faults in Lecce district;
- the Web Map Service (WMS), which allows to generate the satellite map of the Lecce district;
- the geostatistical results, which includes the prediction map of ICK and the level curves for the Rn concentrations.

It is important to highlight that the obtained RnWebGIS allows the visualization of alphanumeric information associated with geographic features of interest, through the use of appropriate tables, or to implement the connection to other *web* pages or *database*, by using suitable *hyperlinks* functions.

Through specific queries on thematic maps, the user can display, in tabular form, the required data. For instance, by using the *identify* tool in the navigation bar, the user can obtain detailed information about the "Rn sites " layer as shown in Fig. 11.

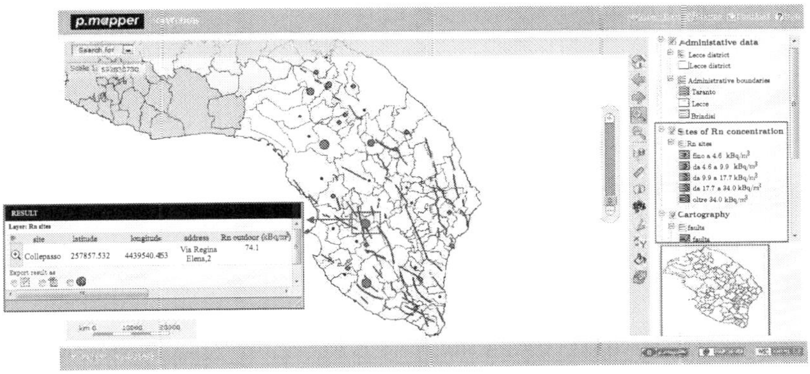

Figure 11. Map Overlay functions: combining layers referred to administrative data, Rn concentrations sites and tectonic faults map. The identify tool highlights an example of Rn site (Collepasso municipality) with high Rn concentrations. Note that the red points indicate the sample locations and their size is proportional to the measured Rn concentration.

Another example of query is obtained by using the *search for* tool, in top left of the *p.mapper* Framework (version 4.1.1.) as shown in Fig. 12.

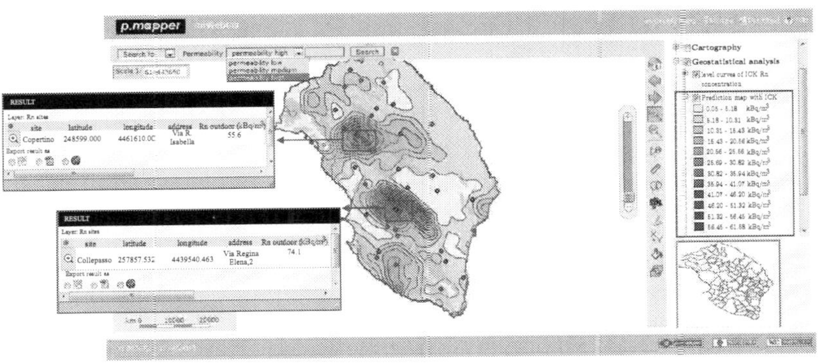

Figure 12. "Search for" tool with the visualization of the layer referred to "high" permeability overlapped on the "Rn site", "level curves of ICK" and "Prediction map of ICK" layers. The identify tool highlights two Rn sites (Collepasso and Copertino municipalities) with high Rn concentrations.

This function has been used to query the database associated with the layer "Permeability" and visualize, by using the thematic map, the study area with high level of permeability overlapped to the layers "level curves of ICK" and "Prediction map of ICK". In this case, the user can identify areas of high level of permeability close to area with high Rn concentrations.

Interactive navigation of maps is possible by using the tools implemented in the RnWebGIS, such as the use of *zooming/panning* available by selecting the object on the map.

On the other hand, in order to ensure the interoperability (i.e. exchange and/or sharing) of geographic data, it has been relevant to build a RnWebGIS according to the specifications of the services, defined by the *Open Geospatial Consortium* (OGC). Indeed, by using the features implemented in the RnWebGIS, the user can also integrate data located on different servers as well as data of different formats. This is possible by recalling the WMS given by the OGC that avoids duplication of data and, at the same time, provides updated geographical data which are certified by shared *standards*.

For example, it is worth highlighting that, by using the appropriate WMS, an orthophoto of Lecce district can be integrated with the layer "Prection map of OCK " as shown in Fig. 13.

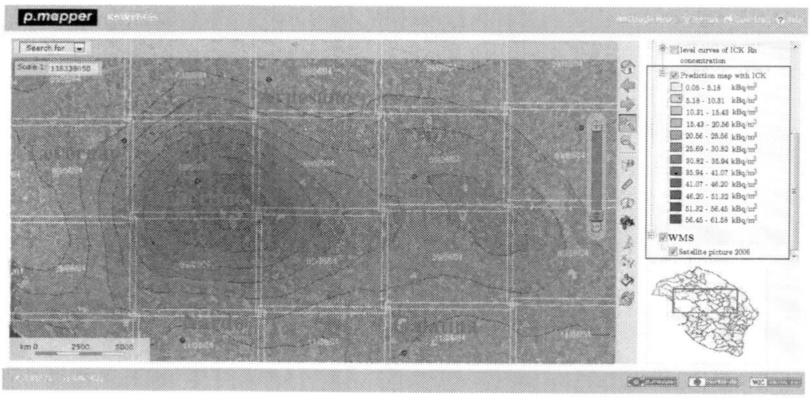

Figure 13. Orthophoto of Lecce district, obtained by the WMS and integrated with the "Rn sites", "level curves of ICK" and "Prediction map of ICK" layers.

6. Conclusions

In this paper, after introducing the usefulness of a GIS supported by geostatistical results, and reviewing some geostatistical modeling and predictions techniques used in the univariate and multivariate cases, a case study concerning a thorough geostatistical analysis of the Rn concentrations in soil gas over Lecce district has been discussed. In particular, four different spatial interpolation techniques, that is Ordinary Kriging, Log-normal Kriging, Cokriging with Indicator variable and Kriging with Varying Local Means, have been used in order to predict the Rn concentration levels over the study area. Then, an *ad hoc* WebGIS, called RnWebGIS, has been proposed. The map of the Rn predictions obtained by using Cokriging with Indicator variable, which performed more accurate predictions than the ones obtained by using the other methods, has been stored into the WebGIS, together with several

information about the geo-lithological features (such as permeability and tectonics maps), as well as administrative characteristics (such as demographic data and municipal boundaries) of the study area.
The proposed RnWebGIS highlights the usefulness of the integration between GIS and advanced geostatistical tools, in order to support the management of a sustainable development by taking into account even the risk of exposure to Rn pollution.

Author details

Veronica Distefano, Sandra De Iaco*, Monica Palma, Donato Posa, Alessandra Spennato

*Adress all corresspodence to: s.deiaco@economia.unile.it

University of Salento, Department of Management, Economics, Mathematics and Statistics, Italy

References

[1] Anselin, L. (1992). *Spatial Data Analysis with GIS*, National Center for Geographic Information and Analysis, University of California, USA.

[2] Anselin, L. (2000). Computing Environments for Spatial Data Analysis, *Journal of Geographical Systems* **2**: 201–220.

[3] Bivand, R.S. & Gebhardt, A. (2000). Implementing Functions for Spatial Statistical Analysis using the R Language, *Journal of Geographical Systems* **2**: 307–311.

[4] Beaubien, S.E., Ciotoli, G. & Lombardi, S. (2003). Carbon dioxide and radon gas hazard in the Alban Hills area (Central Italy), *J. Volcanol. Geotherm. Res.* **123**: 63–80.

[5] Beck, H.L. & Gogolak, C.V. (1979). Time-dependent calculations of the vertical distribution of 222Rn and its decay products in the atmosphere, *J. Geophys. Res.* **84**, doi: 10.1029/JC084iC06p03139.

[6] Boots, B. (2000). Using GIS to Promote Spatial Analysis, *Journal of Geographical Systems*, **2**: 17–21.

[7] Bottai, L., De Filippis, T., Mari, R. & Rocchi, L. (2010). *Servizi WebGIS sviluppo sostenibile: la condivisione dell'informazione meteorologica ed ambientale in Toscana*, Proceedings of the 14th National Conference ASITA, Brescia, Italy.

[8] Burrough, P.A. (2001). GIS and Geostatistics: Essential Partners for Spatial Analysis, *Environmental and Ecological Statistics* **8**: 361–377.

[9] Burrough, P.A. & McDonnell, R.A. (1998). *Principles of Geographical Information System*, Oxford University Press.

[10] Chilés, J. & Delfiner, P. (1999). *Geostatistics - Modeling spatial uncertainty*, Wiley.

[11] Cressie, N. (1993). *Statistics for Spatial Data*, Wiley, New York.

[12] De Iaco, S. & Posa, D. (2013). Positive and negative non-separability for space-time covariance models, *J. Stat. Plan. Infer.* **143**: 378–391.

[13] De Iaco, S., Posa, D. & Myers, D.E. (2013). Characteristics of some classes of space-time covariance functions, *J. Stat. Plan. Infer.* **143**: 2002–2015.

[14] De Iaco, S., Distefano, V. & Posa, D. (2013). Geostatistica e GIS, Tecniche e strumenti per il monitoraggio ambientale, in: *Lo sviluppo sostenibile: Ambiente, Risorse, Innovazione, Qualità*, Franco Angeli Editore, Italy.

[15] Delhomme, J.P. (1978). Kriging in the hydrosciences, *Adv. Water Resources* **1**: 251–266.

[16] Durrani, S.A. & Badr, I. (1995). Geostatistically controlled field study of radon levels and the analysis of their spatial variation, *Radiat. Meas.* **25**, 565–572.

[17] Gillavry, E.M. (2000). *Carthographic aspects of WebGIS*, Software Department of Cartography, Utrecht University.

[18] Goodchild, M.F. (1987). A spatial analytical perspective on geographical information systems, *Int. J. Geogr. Inf. Sci.* **1**: 327–334.

[19] Goodchild M.F. (1992). Geographical data modeling, *Comput. Geosci.* **18**: 401–408.

[20] Goodchild, M.F., Haining, R., Wise, S., Arbia, G., Anselin, L., Bossard, E.G., Brunsdon, C., Diggle, P., Flowerdew, R., Green, M., Griffith, D.A., Hepple, L., Krug, T., Martin, R.J. & Openshaw, S. (1992). Integrating GIS and Spatial Data Analysis: Problems and Possibilities, *International Journal of Geographical Information Science* **6**: 407–423.

[21] Goovaerts, P. (1997). *Geostatistics for natural resources evaluation*, Oxford University Press, New York.

[22] Journel, A.G. (1983). Non-parametric estimation of spatial distribution, *Mathematical Geology* **15**: 445–468.

[23] Journel, A.G. & Huijbregts, C.J. (1981). *Mining Geostatistics*, Academic Press, London.

[24] Kropla, B. (2006). *Beginning MapServer: Open Source GIS development*, Berkeley Apress Ed.

[25] Matheron, G. (1962). *Traité de Géostatistique appliquée*. Mémoires du Bureau de Recherches Geologiques et Minieres, Editions Technip, Paris, Tome I, n.14.

[26] Matheron, G. (1965). *Les variables régionalisées et leur estimation: une application de la théorie des fonctions aléatoires aux sciences de la nature*, Masson, Paris.

[27] Matheron, G. (1976). A simple substitute for conditional expectation: the disjunctive kriging. In: *Advanced Geostatistics in the mining industry*, Guarascio M. et al. (eds): 221:236.

[28] Monestiez, P., Courault, D., Allard, D. & Ruget, F. (2001). Spatial interpolation of air temperature using environmental context: Application to a crop model. *Environmental and Ecological Statistics* **8**: 297–309.

[29] O'Dea, J. & Dowdall, M. (1999). Spatial analysis of natural radionuclides in peat overlying a lithological contact in Co. Donegal Ireland, *J. Environ. Radioactiv* **44**: 107–117.

[30] Oliver, M.A. & Khayrat, A.L. (2001). A geostatistical investigation of the spatial variation of radon in soil, *Comput. Geosci.* doi:10.1016/S0098-3004(00)00133-3.

[31] Porstendorfer, J. (1994). Properties and behavior of radon and thoron and their decay products in the air, *J. Aerosol Sci.* **25**: 219–263

[32] Posa, D. & De Iaco, S. (2009). *Geostatistica. Teoria e Applicazioni*. Giappichelli Ed., Turin, Italy.

[33] Wackernagel, H. (1988). *Geostatistical techniques for interpreting multivariate spatial information*, in: Quantitative Analysis of Mineral and Energy Resources, Chung, C.F. et al. (Eds), 349–409.

[34] Wackernagel, H. (2003). *Multivariate Geostatistics: an introduction with applications*, Springer, Berlin.

[35] Wang, J., Krudy, G., Hou, T., Zhang, W., Holland, G., Xu, X., (2007). *Development of reliable aqueous solubility models and their application in druglike analysis*, Journal of chemical information and modeling. 47(4): 1395-1404.

[36] Wesseling, C.G., Karssenberg, D., Burrough, P.A. & Van Deursen, W.P.A. (1996). Integrating Dynamic Environmental Models in GIS, *Transactions in GIS* **1**: 40–48.

[37] Wise, S., Haining, R. & Ma, J. (2001). Providing Spatial Statistical Data Analysis Functionality for the GIS User: The SAGE Project, *International Journal of Geographical Information Science* **15** (3): 239–254.

[38] Yang, Y., Zhu, J., Tong, X., Wang, D., (2009). The spatial pattern characteristics of soil nutrients at the field scale, Computer and Computing Technologies in Agriculture II **Vol. 1**, Springer US: 125–134.

[39] Zhu, H.C., Charlet, J.M. & Poffijn, A. (2001). Radon risk mapping in southern Belgium: an application of geostatistical and GIS techniques, *Sci. Total Environ.* **272**: 203–210.

Chapter 19

Air Pollution in High-Risk Sites–Risk Analysis and Health Impact

M. Ferrante, M. Fiore, C. Copat, S. Morina, C. Ledda, C. Mauceri and G. Oliveri Conti

Additional information is available at the end of the chapter

http://dx.doi.org/10.5772/60345

1. Introduction

During the last century, at the top of industrialization phase, the excessive use of increasingly more complex inorganic and organic chemicals has caused serious environmental contamination all over the world with acute and chronic effects on population health.

The sites considered to be very contaminated and classified as dangerous are very numerous and often belong to industrial centres; in the past years, the contamination of soil and their contribution to air pollution have had an important role.

In countries characterized by a high socio-economic status, the industrial impact on air pollution is certainly less than the weight of traffic and urban heating, while in the developing and emerging countries, industrialization has still an important and unresolved role because of the lack of legislative management and control.

Epidemiological studies represent the scientific basis used to verify the existence of negative health effects caused by air pollution and to quantify the value of these effects, estimating the dose-response relationships. Many studies have demonstrated that air pollution is associated with a wide range of adverse health outcomes. Statistical analyses, conducted on monitoring and biomonitoring data, also show a relationship between the level of air pollution and the rate of mortality and morbidity. Unlike accidents, pollution is not considered a cause of immediate death. Actually, it is thought that it represents a cause of premature death. Air pollution is associated with a great number of illnesses, such as cancer, respiratory, cardiovascular and neurological diseases, and also diabetes and infertility problems.

The estimated cost of pollution becomes an economic evaluation of the risk of getting sick or dying prematurely. But the effects of pollution are particularly difficult to evaluate in terms of quantity and cost. In order to assess the related risk of air pollution, it is certainly very important to establish the effects of pollution on human health, but it is also needful to consider some health determinants such as genetic factors, age, sex, and lifestyle habits.

Considering all these aspects, our proposal includes a chapter about atmospheric and urban pollution describing their relative and assessed effects on human health, with a detailed regard on risk assessment. In the last decades, a large number of legislations have been issued accepting international subscribed agreement, such as the Kyoto protocol, to save public health and restrict the effects of pollution on climate change. We will also focus on indoor pollution, with its insufficient data and relative problems and difficulties in management, on methodologies of study, and on some case reports.

2. Description of air pollution and characterization of monitored atmospheric pollutants

Air pollution, both for indoor or outdoor environments, is caused by several causes like chemical, physical, or biological modification of the natural characteristics of the atmosphere. Most common household combustion devices, motor vehicles, and industrial process are anthropogenic sources of air pollution, but also natural sources are very important for air pollution assessment as the forest fires and volcanoes emissions. Particulate matter, carbon monoxide, ozone, nitrogen dioxide, and sulphur dioxide today are the most important pollutants for their serious health risks [1].

In Europe, emissions of many air pollutants have decreased significantly the last 30 years, and air quality has improved across the region. However, excess of air quality standards still occurs, especially in metropolitan areas, and air pollutants released in one country may be transported in the atmosphere, polluting or degrading the air quality in neighbouring countries. On the basis of this statement, air pollution can be considered as a local, pan-European, and hemispheric issue [2].

At present, particulate matter, nitrogen dioxide, and ground-level ozone are recognized as the three most dangerous pollutants that affect human health, ranging in severity of impact according to long-term and peak exposures. Moreover, benzo(a)pyrene is a carcinogenic substance of increasing concern, as its concentrations are above the threshold set to protect human health in several urban areas, especially in central and eastern Europe. Air pollution could also represent a danger for the environment, producing acidification, eutrophication, and damage to agriculture.

There are various sources of air pollution, both anthropogenic and of natural origin, in turn divided in stationary and mobile sources. The most common anthropogenic sources of pollutants are combustion processes used for electricity generation, transport, industry and households, industrial processes and solvent use, agricultural crops and livestock, waste incinerator and landfill. A stationary source of air pollution, also known as a point source, refers to a fixed emission source, represented by factories, power plants, dry cleaners, resi-

http://dx.doi.org/10.5772/60345

dential wood burners, dry lakebeds, and landfills. A mobile source of air pollution is a source capable of moving under its own power. "On-road" transportation is related to all vehicles; however, there is also a "non-road" or "off-road" transportation, including the gas-powered machines for garden maintenance or recreational vehicles.

People or their activities do not cause natural "air pollution," and emission sources generally comprise volcanic eruptions, windblown dust, sea-salt spray, wild animals in their natural habitat, and plants releasing volatile organic compounds [2].

In the evaluation of pollution effects, it must be kept in mind that dangerous substances can be emitted as primary or secondary pollutants. Primary pollutants are emitted directly into the air from pollution sources, while secondary pollutants are formed when primary pollutants undergo chemical changes in the atmosphere. For example, NO_X and SO_2 are directly emitted into the air following fuel combustion or industrial processes. In contrast, O3 and the major part of PM form in the atmosphere following emissions of various precursor species, and their concentrations depend strongly on meteorological conditions, such as high air temperatures and sunlight.

At present, the attention of scientist and legislator is focused on the major sources of dangerous substances, represented by urban and industrial activities.

In the evaluation of urban pollution, it is important to consider regional pollution, city pollution, real and *hot spot* events, determined by peaks of pollutants higher than average followed by gradual restoration of normal limits. Of these three components, the first two have direct effects on human and environmental health, as they expose population to contaminants for a longer time, causing chronic effects.

As regards industrial pollution, there are various unwanted substances and losses generated by industrial activities both for qualitative and quantitative aspects [3].

For this reason, to assess significant trends and to discern the effects of reduced anthropogenic precursor emissions, long time series of measurements and a continuous monitoring of air quality are needed [4].

The application of European Directives contained the emission of many pollutants, consequently reducing their levels in the atmosphere. However, while restrictions are imposed on industries, technologies are developed to reduce pollutants from road transport and the most harmful substances in the air are continuously monitored, air pollution is still a major problem and it is necessary to focus attention on its effects on human health and on methods to reduce these effects.

At present, based on *"The Clean Air Act"*, Quality Standards are set for six common air pollutants, also known as "criteria pollutants". They are particulate matter, ground-level ozone, carbon monoxide, sulphur oxides, nitrogen oxides, and lead. These pollutants can harm human health and the environment and cause property damage. The set of limits based on human health is called primary standards. Another set of limits intended to prevent environmental and property damage is called secondary standards [5].

The main pollutants measured to characterize and monitor the quality of the environment are shown below.

2.1. Particulate Matter (PM)

PM is one of the most important pollutants in terms of potential to harm human health, as it penetrates into low regions of the respiratory system. PM is a complex heterogeneous mixture of solid particles and liquid droplets found in the air, whose size and chemical composition change in time and space, depending on different emission sources and atmospheric and weather conditions. These particles come in many sizes and shapes and can be made up of hundreds of different chemicals. PM includes both primary and secondary PM; primary PM is the fraction of PM that is emitted directly into the atmosphere from construction sites, unpaved roads, fields, smokestacks, or fires, whereas secondary PM forms in the atmosphere following the oxidation and transformation of precursor gases (mainly SO_2, NO_X, NH3) and some volatile organic compounds (VOCs) that are emitted from power plants, industries and automobiles.

Particle pollution includes "inhalable coarse particles," with diameters larger than 2.5 μm and smaller than 10 μm, and "fine particles," with diameters that are 2.5 μm and smaller. Smaller sizes of PM such as $PM_{2.5}$, with a diameter up to 2.5 μm, are considered particularly harmful due to their greater ability to penetrate deep into the lungs.

2.2. Tropospheric or ground-level ozone (O_3)

Ozone (O_3) is a secondary pollutant formed in the troposphere from complex photochemical reactions following emissions of precursor gases such as NO_X and non-methane volatile organic compounds (NMVOCs), deriving from paint application, road transport, dry-cleaning, and other solvent uses. At the continental scale, methane (CH4) and carbon monoxide (CO) also play a role in O_3 formation. O_3 is a powerful and aggressive oxidizing agent, causing respiratory and cardiovascular problems and leading to premature mortality. High levels of O_3 can also damage plants, leading to reduced agricultural crop yields and decreased forest growth.

O_3 is found not only in the troposphere but also in the upper regions of the atmosphere. Three oxygen atoms constitute O_3, and it is a very reactive molecule. O_3 in the upper atmosphere protects the earth from the sun's harmful rays, whereas the ground level O_3 is considered as the main component of harmful smog.

In fact, the tropospheric O_3 is created by chemical reactions between oxides of nitrogen (NO_X) and volatile organic compounds (VOC) after its emission in the air from industrial facilities and electric utilities, motor vehicle exhaust, gasoline vapours, and chemical solvents. O_3 is likely to reach unhealthy levels on hot, sunny days in urban environments and can also be transported through long distances by wind so that even rural areas can experience high O_3 levels.

O_3 contributes to what we typically experience as "smog" or haze, which still occurs most frequently in the summertime but can occur throughout the year in some southern and

http://dx.doi.org/10.5772/60345

mountain regions. Ground-level O_3 that we breathe can harm human health, especially affecting people with lung disease, children, older adults, and people who are active outdoors.

Children are at greatest risk from exposure to O_3 because their lungs are still developing and they are more likely to be active outdoors when O_3 levels are high, which increases their exposure. O_3 also damages vegetation, in particular, trees and plants during the growing season, and ecosystems, including forests, parks, wildlife refuges, and wilderness areas.

2.3. Carbon monoxide (CO)

Carbon monoxide (CO) is a very dangerous gas because it is odorless and it is emitted generally from all combustion processes both natural and anthropogenic sources. Urban air pollution is characterized by high CO emissions to ambient air from vehicles movement. CO can cause several harmful health effects, all related to reduced oxygen delivery to the body's organs (especially the heart and brain) and peripheral tissues causing death at very high doses of exposure.

2.4. Sulphur dioxide (SO_2)

Sulphur dioxide (SO_2) is emitted by the combustion of fuels containing sulphur. It contributes to acid deposition and can cause adverse effects on aquatic ecosystems in rivers and lakes, and damage to forests.

Sulphur dioxide (SO_2) is a highly reactive gas. SO_2 is emitted mainly by fossil fuel combustion by power plants (73%) and other industrial process (20%). Smaller sources of SO_2 are released by other industrial processes such as mineral extraction and also by the burning of high-sulphur fuels. Many illnesses are related to SO_2; particularly, harmful effects are linked to the respiratory system.

2.5. Nitrogen oxide (NO_X)

Nitrogen oxide (NO_X) is emitted during fuel combustion, by industrial facilities and the road transport. NO_X contributes to the formation of secondary inorganic PM and tropospheric O_3, to acid deposition and to eutrophication. Of the chemical species that comprise NO_X, NO_2 is associated with adverse affects on health, such as inflammation of the airways and reduced lung function.

Nitrogen dioxide (NO_2) is a highly reactive gas. Other nitrogen oxides include the nitrous acid and nitric acid. EPA's National Ambient Air Quality Standard uses NO_2 as indicator for the class of nitrogen oxide. NO_2 is emitted from all vehicles and industrial plants, including power plants. In addition to the arising O_3 and PM, NO_2 is linked also to several adverse effects on the respiratory system. Both primary EPA standard (to protect health) and secondary EPA standard (to protect the public welfare) for NO_2 are established at 0.053 parts per million (53 ppb), averaged annually. EPA (January 2010) established an additional primary standard at 100 ppb. Primary standards are finalized to protect public health, including health of responsive populations, people with COPD, children, and the elderly. No European and extra-European countries have been found stranger to the problem.

2.6. Heavy metals

The main heavy metals in outdoor air are arsenic (As), cadmium (Cd), lead (Pb), mercury (Hg), and nickel (Ni); they are emitted mainly as a result of various combustion processes and industrial activities and generally can reside in or be attached to PM. Heavy metals are persistent in the environment and can be deposited on terrestrial or water surfaces, contaminating soils or sediments, and accumulating in food-chains.

2.7. Polycyclic Aromatic Hydrocarbons (PAHs) and benzo(a)pyrene (BaP)

PAHs are pollutants created during the incomplete burning of several organic substances like coal, oil, gas, wood, garbage, and also tobacco and smoked and grilled. Actually, there are more than 100 different PAHs identified. PAHs generally occur as complex mixtures of congeners. The primary sources of exposure to PAHs are inhalation of the compounds in tobacco smoke, wood smoke, and ambient air.

PAHs is a growing health concern in Europe, with particular regard to BaP, a polycyclic aromatic hydrocarbon, formed mainly from the incomplete burning of organic material such as wood and from car exhaust fumes especially from diesel vehicles. It is a known cancer-causing agent in humans. A main source of BaP in Europe is domestic home heating, in particular, wood burning, waste burning, coke and steel production, and mobile sources. Other sources include outdoor burning and rubber tyre wear. In Europe, BaP pollution is mainly a problem in certain areas such as western Poland, the Czech Republic, and Austria where domestic coal and wood burning is common.

Between the PAHs, the International Agency for Research on Cancer (IARC) considers BaP as 2A class, so carcinogen. However, the carcinogenicity of BaP was demonstrated through studies carried out with animals. Benzo(a)pyrene is a known mutagen, but it must be metabolized to be able to induce mutation. Exposure to BaP pollution is quite significant and ubiquitous. Central and Eastern European countries report air BaP concentrations above the target value. Approximately one quarter of the EU people were exposed to higher BaP concentrations respect to target value (1 ng/m^3) for the period 2008–2010. But almost the entire population was exposed to BaP concentrations above WHO reference level [6]. The increase of BaP emissions in Europe over the last years is therefore an important concern for public health. Regarding BaP, the Industrial Emissions Directive [7] regulates emissions from a large range of industrial sources.

2.8. BTEX

It is a group of chemical compounds including Benzene, Toluene, Ethylbenzene, and Xylenes. BTEXs are made up of naturally occurring chemicals that are found mainly in petroleum products such as gasoline. Everyone is exposed to small amounts of BTEX compounds in the outdoor air, at work, and in their homes. Besides the common daily exposures to BTEX, larger amounts can enter the environment from leaking underground storage tanks, overfills of storage tanks, fuel spills from auto accidents, and landfills.

Short-term exposure to gasoline and its components benzene, toluene, and xylenes has been associated with skin and sensory irritation, central nervous system (CNS) problems, and effects on the respiratory system. Chronic exposure to BTEX compounds can affect the kidney, liver, and blood systems. Long-term exposure to high levels of the benzene compound can lead to leukemia and cancers of the blood-forming organs [8].

2.9. Physical pollution

Physical pollution is pollution caused by colour (change), suspended solids, foaming, temperature conditions, or radioactivity, and it is characterized by its influence on environmental conditions caused by forces and operations of physics, such as noise, microwave radiation, vibration.

2.10. Microbiological factors

Microbiological factors in the environment represent an underestimated but insidious risk factor, which concern has increased with the introduction of advanced technologies in hospitals, industry, and agriculture. Microbial agents are transported and diffused in ambient through PM [9]. The main sources of microorganisms are Flügge droplets from human airways and conditioning systems, the last containing different bacteria such as *Legionella pneumophila*, virus, and mould.

To maintain a healthy environment, a strict monitoring must be undertaken for each of the contaminants mentioned above; legislation on environment sets a benchmark and provides a series of actions to hold the values of the contaminants within the limits.

As air pollutants can be transported many miles away from the area of their emission, monitoring these substances is more difficult than monitoring contaminants in water or soil, so it was decided to record ambient air quality data focusing on major pollutants. Continuously operating automatic analysers are the most important information source on air pollution levels [10]. According to current legislation, measurements are taken at permanent sites equipped with physico-chemical monitoring devices even if these stations provide only data on exposure levels which do not facilitate making conclusions on the effects or impacts to human health and the general environment. Moreover, spatial coverage and temporal definition of the carried measurements together with meteorological conditions influence the degree of representativeness of the data compared to the real state of pollution of the study area [11].

Recently, innovative technologies have been developed and direct damage of harmful effects can be demonstrated by monitoring living organisms, called biomonitors, which provide a measure of integrated exposure over a certain amount of time taking into account also climatic conditions and enrich the substance to be determined so that the analytical accessibility is improved and the measurement uncertainty reduced. Moreover, sampling is relatively simple and no expensive technical equipment is needed. The field of biomonitoring includes bioindication of effects and bioaccumulation of pollutants [12].

The information provided by mosses, lichens, and higher plants on the deposition on the effects of air pollutants is an important complement to the data acquired with automatic systems. Distribution of tree species on a national or supranational lets you draw cheaply and in a short time maps of diffusion and deposition of persistent pollutants or the effects of tropospheric O_3 and other phytotoxic pollutants [13].

3. Statistical methodologies to evaluate the risk assessment and air pollution correlated diseases

Extreme pollution episodes that took place in the period 1930–1960 have initiated the epidemiological studies for increase of health environmental concern. The association between air pollution and certain health variables was made clear by several studies [14, 15].

Thanks to the new approach, air pollution levels were reduced substantially, such that, for health effects assessment, now, longer monitoring plans are required. To this end, epidemiologists began to use some statistical models. In the 1970s firstly were used the dynamic regression models, models in which the relationship between the dependent and explanatory variables were distributed over time rather than being expected to occur simultaneously. However, the problem of these types of models is that they assume that the dependent variable is distributed normally, but this is a condition extremely rare to achieve [16].

In the 1990s were used for the first time the linear models based on Poisson regression because the event counts more typically have a Poisson distribution. These models use the "time" and its transforms as variable [16].

Also Poisson regression is particularly useful only when cases rather than the entire population can be enumerated.

Nevertheless, Poisson regression poses the problem that, if any of these unmeasured variables follows a cyclical component of varying frequency, the parametric functions of time cannot be easily adapted. So, these limitations led to the development also of nonparametric Poisson regression [17], which is well adapted to the irregular cyclic components of unmeasured variables reducing any potential confounding.

One difficulty with this method is that the researcher must specify the number of degrees of freedom, with several discrepancies arising as the way to calculate this. Because inappropriate determination of the number of degrees of freedom is frequently a bias in the estimates of nonparametric Poisson designs, epidemiologists focused on the case-crossover (CCO) [18] design that purported to control time trends. The CCO design is characterized by the fact that each subject serves as his or her own control. This design was initially used to assess the effect of exposures measured at an individual level and was not applicable to exposures with a time trend, such as air pollution. A variant was developed to bypass this bias [19], the bidirectional CCO, characterized by having control time periods before and after the event. This design was already appropriate for ecologic-type exposures, such as air pollution, water pollution, etc. In addition, pollution values are not affected by the presence of prior morbidity and mortality events. In fact, in the CCO design in air pollution, pre- and post-event exposure values are independent of the hazard-period exposure; those that are post-event referent can be appropriate. One advantage of CCO design respect to Poisson regression is its ability to assess potential effect modification at the individual level rather than at the group level [20]. As an alternative analytic methodology to Poisson regression, the CCO approach allows for direct modeling of interaction terms, rather than depending on multiple subgroup analyses [20].

Besides the individual characteristics during the statistical analysis of epidemiological data, other variables must be considered, taking into account that the chemical characteristics of air pollutants mixture and PM may change over time and depending on the geographical location, emission sources, atmospheric chemistry, and weather conditions [21].

Interest in health effects of air pollution became more intense after two US cohort studies suggested that exposure to fine PM in the air was associated with life shortening [22,23].

Exposure to pollutants such as airborne PM and O_3 has been associated with increases in mortality and hospital admissions mainly because of respiratory and cardiovascular disease, due to both acute and chronic exposure [24].

Health problems can include cancer, respiratory irritation, nervous system problems, and birth defects.

Air pollution, in the year 2012, was responsible for 3.7 million deaths, i.e., approximately 6.7% of the total deaths. In particular, air pollution is estimated to cause deaths from lung cancer (16%), COPD (11%), ischemic heart disease and stroke (>20%), and respiratory infection (13%). Furthermore, the IARC announced that it has classified outdoor air pollution as carcinogenic to humans (Group 1).

Typically, exposure to toxic components of air pollutant causes the well-established correlated diseases, respiratory and pulmonary ones, such as decreased lung function and increased incidence of chronic cough, bronchitis, chronic obstructive pulmonary disease, asthma, and conjunctivitis [25,26,27].

Recent clinical and epidemiological data suggest also that cardiovascular disease may be related to pollution [28,29] especially those associates with fine PM [30,31,32,33]. The PM's effects on the cardiovascular system seem to involve the activation of prothrombotic factors, leading to thrombosis, but destabilization of atherosclerotic plaques cannot also be excluded. In addition, there may be direct effects on the heart or indirectly on the nervous system.

Health effects have been seen at very low levels of exposure, and it is unclear whether a threshold concentration exists for PM and O_3 below which no effects on health are likely.

It could be expected that the impact caused by a preventable risk factor would decline if the exposure to that risk factor could be reduced or removed. According to this approach, the proportional reduction in the number of health problems or deaths as a result of reducing the risk factor is known as the attributable fraction (AF) [34].

Public health agencies concerned with air quality perform risk assessments to determine the increased risk of illness from a specific human exposure to a toxic air pollutant.

The risk assessment approach outlined by the WHO in the Environmental Burden of Disease (EBD) series [35,36] includes the following steps:

1. assessment of the air exposure of the population through data from Air model or monitoring networks. A target concentration is also needed to determine the attributable disease or the potential gains of a management plan;
2. sufficient number of persons exposed to air pollutants;
3. baseline data of incidence of the adverse health outcomes associated with air pollutants (like mortality rate); and
4. concentration-response functions (CRFs) related to the incidence of adverse health effects.

The US Environmental Protection Agency (US-EPA) divided the risk assessment process into the following four steps:

a. **Hazard Identification**: it allows determining the potential human health effects from exposure to a chemical. This is based on information provided by the scientific literature.
b. **Dose-Response Assessment:** it is the assessment of the relationship between a dose of chemical exposure and incidence or severity of the related adverse health effect. It takes into consideration the intensity and pattern of exposure and also age and lifestyle variables that may affect people's susceptibility. The dose-response relationship is evaluated differently for carcinogenic and non-carcinogenic pollutants. In fact, for carcinogens, it is assumed that there is a linear relationship between an increased dose of exposure and increase in cancer risk; this is expressed as slope factor (SF) and threshold is not accepted. For risks evaluation by inhalation of carcinogenic substances, US-EPA use the potency slopes to develop the unit risk factors (URFs). A URF is the upper-bound excess probability of contracting cancer as the result of a lifetime of exposure to a carcinogen at a concentration of 1 $\mu g/m^3$ in air. For inhalation effects from non-carcinogens, dose-response data are used to develop reference concentrations (RfCs) for both long-term and short-term exposures. Unlike carcinogens, non-carcinogens are assumed to have thresholds dose, so the injury does not occur until exposure has exceeded a threshold limit. An RfC is derived from a no-observed adverse effect level (NOAEL) or lowest observed adverse effect level (LOAEL) determined through human or animal exposure studies.
c. **Exposure Assessment**: it determines the extent (intensity, frequency, and duration, or dose) of human exposure to a chemical in the environment.
d. There are three components to exposure assessment:
   i. Estimation of the maximum quantity of each pollutant emitted from the source.
   ii. For each contaminant emitted, estimation of the resulting maximum annual average and (where applicable) maximum short-term average ambient air concentrations, using dispersion models, or air impact values based on dispersion models.
   iii. Estimation of the amount of contaminant taken in by a human receptor.

e. **Risk Characterization**: it is the final step in risk assessment where health risk is calculated and described thanks to data collected in the first three steps.

- **Carcinogens**

Human health risk estimates for inhalation of carcinogenic air pollutants are based on the following:

$$\text{Cancer Risk} = C \times URF$$

where

C = maximum annual average ambient air concentration of a pollutant ($\mu g/m^3$)

URF = pollutant-specific inhalation unit risk factor $(\mu g/m^3)^{-1}$

For routes of exposure other than inhalation, risk is calculated by multiplying the estimated chemical dose (in mg/kg/day) by the chemical-specific oral slope factor (in $(mg/kg/day)^{-1}$).

- **No carcinogens**

Human health risk estimates for inhalation of non-carcinogenic air pollutants are based on the following:

$$\text{Hazard Quotient} = C/RfC$$

where

C = maximum ambient air concentration, $\mu g/m^3$

RfC = pollutant-specific reference concentration, $\mu g/m^3$

The averaging time can be either annual, or a specific number of hours, depending on the basis of the reference dose [37]. For routes of exposure other than inhalation, the hazard quotient is calculated by ratio between the estimated chemical dose (in mg/kg/day) and the chemical-specific reference dose (in mg/kg/day).

Hazard quotients can be summed (separately for inhalation and oral exposures and for different averaging times) to give a hazard index.

4. Evaluation of indoor air pollution and comparison with the urban and high-risk sites

Clean air is a fundamental health determinant. The quality of indoor air (e.g., homes, offices, schools, day care centres, public buildings, where people spend a large part of their life). Hazardous substances emitted from buildings, construction materials, and indoor equipment or due to all human activities indoors lead to a broad range of health problems and may even be fatal for infants and children. Indoor environmental issues are still an open topic of public health, including health risks and the means by which human exposures can be reduced [38].

Mistakes of evaluation or unacknowledged by public about health risks associated with a variety of indoor environmental pollutants and sources of pollution (e.g., radon, mold and moisture, secondhand smoke, and indoor wood smoke) cause a minor efficacy of prevention and/or risk's low perception. In poorly ventilated dwellings, smoke in and around the house can exceed acceptable levels for fine particles 100-fold [39] with respect to outdoor levels already known. In fact, inadequate ventilation can increase indoor pollutant levels for incomplete dilution of these. High temperature and humidity levels can also increase concentrations of certain pollutants [40]. Outdoor air enters and leaves a house by infiltration, natural ventilation, and mechanical ventilation. If infiltration, natural ventilation, or mechanical ventilation is poor, the air exchange rate is low and pollutant levels can increase rapidly [41].

There are several sources of indoor air pollution including combustion of oil, gas, kerosene, coal, wood, and tobacco products; building materials and furnishings as diverse as deteriorated, asbestos-containing insulation, wet or damp carpet, cabinetry and objects made with certain pressed wood products (e.g., formaldehyde, pesticides, etc.); products for household cleaning and their maintenance (phtalates, bisphenol A, fragrances, pesticides, detergents, chlorine bleach, lye, ammonia, etc.), cosmetics for personal care (solvents, phtalates, bisphenol A, fragrances, mineral oil, ethoxylated surfactants and 1,4-dioxane, formaldehyde, lead, oxybenzone, parabens, toluene, triclosan, etc.), chemicals used for hobbies (solvents, lead, pesticides, mineral oil, cadmium, manganese dioxide, cobalt, formaldehyde, aromatic and chlorinated hydrocarbons, ethylene glycol monomethyl ether, ethylene glycol monobutyl ether, etc); central heating and cooling systems and humidification devices (mold, viruses, fungi, bacteria, mycotoxins, etc.); and outdoor sources such as radon, pesticides, and outdoor air pollution likely to come into the house where can concentrate. Some of these sources (building materials, outdoor sources, wood furnishings, etc.) release pollutants more or less continuously. Other sources, related to activities carried out in the home, release pollutants intermittently, and their concentrations can remain in the domestic air for long periods after these activities. The relative importance of any single source depends on how much of a given pollutant it emits and by the proven hazard of those emissions. In some cases, factors such as how old the source is and whether it is properly maintained are very significant (e.g., an improperly functioning gas stove can emit more CO than one that is properly adjusted).

Exposure to indoor air pollution is particularly high among women and young children who spend the most time in the domestic environment, and health effects may be shown soon after exposure or years later.

According to WHO, 4.3 million people a year die from the exposure to household air pollution [39]. In fact, especially cooking and heating with solid fuels (wood, charcoal, etc.) produce high levels of smoke both in and around the home and this contains a large variety of health-damaging pollutants [42].

Immediate effects such as irritation of the eyes, nose, and throat; headaches; dizziness; and fatigue may show up after a single acute exposure or repeated sub acute exposures. These are usually treatable and have short-term effects. Oftentimes, the simple elimination of a person's exposure to the identified source of pollution is sufficient for restoring health.

An additional problem is the recurrent multiple indoor pollution exposure of residents, complicating and amplifying the health effects [43].

Exposure to indoor air pollution leads to a wide range of respiratory diseases [39, 44] such as lung cancer [45, 46], ischemic heart disease [46], stroke [46], and cataract [46].

There is emerging evidence, although based on fewer studies, that suggests that household air pollution, especially in developing countries, may also increase the risk of other important child and adult illnesses such as the following:

- low birth weight and perinatal mortality (still births and deaths in the first week of life) [45, 46],
- asthma [39, 45],
- otitis media (middle ear infection) and other acute upper respiratory infections [39],
- tuberculosis [45],
- nasopharyngeal cancer [39],
- laryngeal cancer [39], and
- cervical cancer [39].

Also exposure to unhealthy concentrations of fine PM has been connected to increased respiratory/cardiovascular illnesses [39, 47]. In fact, the smaller air particles can penetrate into the deeper lung. PM is being linked to adverse birth outcomes [46], neurodevelopment, cognitive function [48], and diabetes [39].

Semple et al. [49] said that median $PM_{2.5}$ concentrations from 93 smoking homes were 31 μg/m^3 (ranged between 10 and 111 μg/m^3) and 3 μg/m^3 (ranged between 2 and 6.5) μg/m^3 for the 17 non-smoking homes and still showed that non-smokers living with smokers typically have average $PM_{2.5}$ exposure levels more than three times higher than the WHO guidance for annual exposure to $PM_{2.5}$ (10 μg/m^3). So, $PM_{2.5}$ pollution in indoor ambient where smokers live is approximately 10 times higher than that of non-smoking. A non-smoker living with a smoker is exposed to same $PM_{2.5}$ of a non-smoker living in a heavily polluted city such as Beijing. This condition is likely to be greatest and dangerous for the very young and for older members of the population because they typically spend more time at home. Zhou and colleagues [50] have demonstrated that in New York City, despite the ban on tobacco smoking, some hookah bars still serve tobacco-based hookahs, and in these particular indoor environments, the authors have found elevated concentrations of pollutants that may present a real health threat to visitors and employees. The mean real-time $PM_{2.5}$ level was 1179.9 μg/m^3, whereas the filter-based total PM mean was 691.3 μg/m^3. The mean real-time black carbon level was 4.1 μg/m^3, organic carbon was 237.9 μg/m^3, and CO was 32 ppm. Airborne nicotine was present in all studied hookah bars (4.2 μg/m^3).

Dorizas et al. [51] with their study showed that in nine naturally ventilated primary schools of Athens (Greece) during spring, PM concentrations were significantly affected by the ventilation rates and presence of students. Both PM_{10} and $PM_{2.5}$ were greater during teaching than the non-teaching hours, and, in many cases, the $PM_{2.5}$ concentrations exceeded their limit values. For most of the cases, the indoor to outdoor concentrations ratios of PM_{10} and $PM_{2.5}$ were much greater than one, indicating that the indoor environment was being mostly affected by indoor sources instead of the outdoor air. Furthermore, it was found that chalk and marker board usage significantly affects indoor pollutant concentrations.

No similar characteristics were found between indoor pollution and urban and high-risk sites outdoor pollution, because the poor ventilation of houses and buildings in general allows the concentration of chemical and biological pollutants that are not found in similar outdoor concentrations also in severe pollution event. Urban outdoor air pollution refers to the air pollution which the populations are exposed to, living in and around the urban area. Indoor air pollution refers to the pollutants found in indoors. An important difference is in the heavy metals concentration, generally most abundant and with major variability of species in urban and industrial air outdoor pollution compared with indoor pollution.

At the moment, in Italy, a reference rule has not been set. For this reason, until today, the main information concerning some guidelines or reference values in indoor air is obtained by the international scientific literature or by the few guidelines issued by other European countries or, for analogy, by other guidelines values regarding outdoor air. However, public health awareness on indoor air quality still lags significantly behind that of outdoor air quality.

The main areas at high risk of environmental crisis in Italy are described below.

5. Case reports: ILVA-Taranto, Melilli-Priolo-Augusta sites, Gela and Milazzo emerging situation, "Terra dei Fuochi" and Seveso

In order to cope with the huge problem of pollution, experts were mobilized at national and regional levels in the drafting of numerous scientific papers in order to initiate actions to protect health in the areas most affected by the emission of high quantities of contaminants. In many cases, the results of these studies support arguments about correlation between the exposition to pollutants and outbreak of neoplastic diseases; however, for some areas, it has been highlighted that there is a higher incidence in urban areas rather than in sites of national interest, leading to speculation about the etiologic implication of non-industrial contaminants.

At the moment in Italy, the industrial pole of Taranto, Melilli-Priolo-Augusta, Gela e Milazzo sites, and "Terra dei fuochi" and the outcome of the tragedy that occurred in 1976 in Seveso represent the most interesting areas for their extension and extent of involved population. An overview of current developments of studies and the status of remediation undertaken in these areas will be given below.

5.1. ILVA-Taranto

Founded in 1961, ILVA of Taranto is a steel plant at full cycle, where occur all the steps that lead from iron ore to steel [52–54]. The plant, which is partially impounded by order of the judiciary, is the largest steel maker in Europe. The area of Taranto is identified at high risk of environmental crisis because of an extensive industrial area developed close to the urban settlement. Industrial activities are responsible for environmental pollution, mostly due to polycyclic aromatic hydrocarbons (PAHs), heavy metals, organic solvents, polychlorinated biphenyls (PCBs), PM, and dioxin.

Table 1 shows the main pollutants found in soil, ground waters, and sediments of the area of Taranto.

Soil	antimony, arsenic, beryllium, cadmium, cobalt, chromium total, chromium VI, mercury, lead, nickel, zinc, copper, vanadium, cyanide, hydrocarbons C <12 and C"/> 12, PAHs, benzene, xylene, dioxins
Subsurface	antimony, arsenic, beryllium, cadmium, cobalt, chromium total, chromium VI, mercury, lead, nickel, zinc, copper, vanadium, cyanide, hydrocarbons C <12 and C"/> 12, PAHs, benzene, xylene, dioxins
Groundwater	arsenic, selenium, aluminium, iron, manganese, nickel, lead, cobalt, total chromium, chromium VI, cyanides, sulphates, nitrates, BTEXS, PAHs, hydrocarbons and MTBE
Sediments	arsenic, nickel, lead, total chromium, copper, mercury, zinc, total PAHs, PCBs

Table 1. Main pollutants found in soil, ground waters, and sediments of the area of Taranto

To define the extent of the health risk, rigorous epidemiological studies have been conducted to assess both the short-term effects that occur in acute as a result of rapid changes in the concentrations of pollutants and the long-term effects caused by prolonged exposures and which occur 10–15 years after the start of exposure.

In 2006, the Ministry of Health has funded a project called "SENTIERI" (Studio Epidemiologico Nazionale dei Territori e degli Insediamenti Esposti a Rischio da Inquinamento - National Epidemiologic Study of the Territories and Settlements Exposed to Risk from Pollution) with the purpose to analyse the mortality of populations residing in proximity of a number of industrial agglomerates which by their nature potentially have a high factor of hazardous health and/or environmental contamination such as to be classified as SIN (Sites of National Interest for the Remediation).

The population of Taranto has been the subject of several multi-centre epidemiological studies that have documented the role of air pollution on the increase in short-term and long-term effects [52,53].

Epidemiological analysis of the residents of the city revealed death rates from all causes, for lung, pleura, and bladder cancer and for non-Hodgkin lymphoma. A case-control study on incident cases of these diseases in Taranto suggested a possible link between risks and residence close to the sources of emissions.

A recent study called EpiAir in which subjects were recruited rigorously and it was taken into account the exact georeferencing of areas of residence and employment analysed the spatial variability of air pollutants in Taranto and showed that atmospheric pollution in this city is not distributed homogeneously, but it is mainly widespread in the areas adjacent to industrial pole. In the results of this study have been taken into account the socio-economic status and calculated indices of deprivation; these aspects have a great importance, as many individual habits, such as cigarette smoking, alcohol consumption, physical activity, and obesity, are often associated with social status so that the adjustment made for the socio-economic factor has also adjusted the individual variables not directly measured, providing very reliable data. Moreover, in this study, mortality rate and hospitalization rate were measured to evaluate both long- and short-term effects. Data about mortality are reliable, as 98% of the causes of death were recovered, thanks to the linkage of personal data with the database of the ASL and show an excess of mortality for cancer, cardiovascular, and respiratory diseases. As far as hospital admissions, evaluated from hospital discharge schedule, were concerned, it was difficult to get accurate data because the comparison was made only on the hospital area of Taranto, without taking into account the extra-regional mobility and the secondary diagnosis. Although these are limited, data show an increase of hospitalization for the abovementioned causes. The results of this study have strengthened the argument that there is a correlation between exposure to emissions from the steel plant of Taranto and an increase of cancer and cardiovascular and respiratory diseases [55].

The European Environment Agency, in the list that shows the 622 most polluting industrial plants in Europe, included more than 60 Italian companies; Ilva of Taranto is placed second. As a result, the situation with the city of Taranto and Ilva is currently the subject of discussion and great concern [53,54,56]. To control the situation in the area were enacted several regional laws aiming to carry out the assessment of environmental impact and damage health in order to take targeted action to protect the environment and public health. Moreover, in the Italian Act 6 of February 2014, strategies are defined to make the remediation of contaminated areas from emissions of ILVA.

5.2. Sicily air quality

In Sicily, four Sites of National Interest have been individuated and the industrial areas of Melili-Priolo-Augusta (Siracusa), Gela (Caltanissetta), and Milazzo (Messina) have been declared by national and regional legislation "areas high risk of environmental crisis." It is likely that the excesses of mortality and morbidity observed in areas of Melili-Priolo-Augusta, Gela, and Milazzo are attributable to occupational exposures and environmental concerns related to the number of plants and the consequent contamination of environmental matrices [57], although the last reports show how these rates do not exhibit a significant increase.

5.2.1. Melilli-Priolo-Augusta sites

The large industrialized coastal area of eastern Sicily within the territory of the municipalities of Augusta, Priolo Gargallo, and Melilli is defined "petrochemical pole of Siracusa." These territories started to be subject to industrialization in 1948 with the construction of a refinery (the RA.SI.O.M), and in the following years, other industrial facilities were born; currently, the petrochemical pole hosts five refineries of petroleum products, two centrals of ENEL, a gas plant and cogeneration, a factory of magnesite, a cement plant, a purifier of industrial and urban waste and a shipyard.

In these areas, the increase in the number of illness might be linked with PM, which are released into the atmosphere from industrial chimneys and have the ability to convey inside the body, through the lung, every kind of pollutants from the atmosphere. The residents, who have expressed their concerns to environment organizations and public heath institutions, have also noted this potential association.

Official data on emissions point out a long list of definitely carcinogenic and teratogenic chemicals (acrylonitrile, benzene, cadmium, hexavalent chromium, nickel, silicon, vanadium, dioxins, and furans), and many other potential hazards, according to the IARC (Agency International Agency for Research on Cancer).

About PM, the main substances contained in it are represented by heavy metals as shown by the results of a study conducted by the Faculty of Agriculture of the University of Palermo, which used lichen for biomonitoring as their action of bio-accumulators for heavy metals, highlighting a significant presence of these hazardous metals in the environment concerned [57, 58].

These pieces of evidence induced the Ministry of the Environment to recognize the Priolo-Augusta-Melilli site as "an area in environmental crisis for which it has become necessary a legislative and Financial able to address in a timely manner the dramatic emergency." The large amount of epidemiologic studies conducted in these areas highlights the role of atmospheric pollution in the development of some cancers although recent data published by AIRTUM (Associazione Italiana dei Registri Tumori – Italian Association of cancer registries) [59] show that the higher incidence of cancer in Sicily is found in the city of Catania and in other metropolitan areas, such as Palermo and Messina, as well as in big countries like Acireale and Gravina, leading to hypothesize the role of other risk factors probably linked to road traffic and unhealthy lifestyle, contributing in the outbreak of above mentioned diseases. According to the last published report, the industrial sites are not among the areas with the highest incidence of cancer, as supposed, and surprisingly, the SIN of San Filippo del Mela does not present excess in the incidence of tumors.

5.2.2. Gela and Milazzo emerging situation

In the area of Gela and the surrounding area rises one of the main industrial centres of the island. The area, declared "area at high risk of environmental crisis," includes the municipalities of Gela, Niscemi, and Butera. Within the bounded area subjected to remediation, there is a large industrial centre, consisting of plants for refining and extraction of crude oil and petrochemical plants. In particular, productions include polyethylene, molten sulphur, sulphuric acid and phosphoric acid, ammonia and fertilizer complexes [57].

The environmental impact of the refinery of Gela comes mainly from emissions into the atmosphere and the consequent presence of pollutants such as SO_2, NO_X, dust, H_2S, and CO. Conveyed emissions are mainly generated in the combustion process and are coming out of the chimneys.

Data provided by ARPA (Regional Agency for Environmental Protection) reveal exceedances in ambient town, some parameters such as benzene, methane, hydrocarbons, and PM_{10} containing heavy metals such as nickel. A scientific study has shown that the entire area of Gela is heavily influenced by emissions of metals and non-metals from both car traffic or vehicular pollution and industrial pollution. In particular, the study highlights pollution of the particulate present in the aerosol analysing the deposition of the particulates on pine needles. The analysis of pine needles is a fast method for monitoring the pollutants present in the air: the pine needles provide information for the long-term impact of even low levels of pollutants. As for neoplastic diseases, it has not been shown that there is an excess of incidence in this area, although some Sicilian studies highlight an increase of Relative Risk for these diseases compared to other areas of the Region not characterized by the presence of industrial agglomerates. Moreover, in this area, an important role is played by the current wind, responsible for the transport of substances emitted by plants in Gela for long distances, interestingly in particular to Niscemi where contamination of soils by the above mentioned pollutants was found.

The industrial area of Milazzo is characterized by the presence of a large industrial centre, including an oil refinery, a thermoelectric central, a co-generator, a steel mill, a plant for the recovery of lead from spent batteries, and different other smaller factories. In the vicinity of industrial installations, there are several common installations, among which are those belonging to "area at high risk of crisis environment."

A single study carried out in Milazzo area reveals that children are exposed to elevated concentrations of sulphur dioxide, nitric oxide, and $PM_{2.5}$. In more weeks, the average concentration evaluated for sulphur dioxide has exceeded 20 $\mu g/m^3$, and the $PM_{2.5}$ average in the study period was about 23 $\mu g/m^3$, i.e., twice the WHO air guideline limit. Also, epigenetic markers were associated with air pollutant concentrations – particularly with regard to the nitric oxide pathway [57, 58].

5.3. Terra dei fuochi

One of the greatest environmental and health criticality that covered Italy in the last two decades is linked to the illegal waste disposal, triggering serious repercussions both in the territories concerned and the health of the residents in these areas and leading the Institutions to take actions that stiffen confidence in themselves and protect the health of the population. Studies carried out in the areas affected by this phenomenon, although not definitive, confirmed an increase in health risks resulting from waste disposal, highlighting in particular an increase in mortality from all causes, the excesses of mortality from specific cancers (liver, lung, stomach, kidney, and bladder) as well as an increase in non-neoplastic diseases of the respiratory system. The limits of ecological studies mainly reside in the fact that it is not possible to assess the individual characteristics of involved subjects, data on the amount and types of substance emitted are not exact, and therefore, extent of population exposure is not certain and estimates arising from these studies do not provide reliable data on the correlation. Despite the necessity of deepening the studies above, on February 2014, the Italian Parliament approved a law for the protection of these specific areas, which has as main innovations the request for a mapping of polluted areas and crops of these territories, the allocation of funds to carry out a health screening, and the introduction of the crime of burning waste.

This law, if one side has paid attention on the problem, on the other hand has created problems for the application of screening, excellent in itself as a means of prevention, but totally inappropriate to the case in question. It should be noted that on such land occurred the burning of hazardous waste, even radioactive in nature, and that substances emanating from these processes have not been proven to be in correlation with cancers for which screening is scheduled (colorectal, breast, cervix). Moreover, alarmism generated by media has also created skepticism in consumers, even internationally, with a substantial drop in sales of the excellent products from Campania.

However, contrasting the illegality and protecting public health is a duty of the institutions, as well as civic, and to implement the cited actions, the Institutions with the collaboration of experts gathered in the Technical Committee for the Contaminated Territories (CTTC) that provide information on correct methodology to implement the definition of the criteria for the mapping of sites, for the identification of specific diseases to perform a targeted screening, and for implementing the operations of monitoring and remediation.

5.4. Seveso

On July 1976 in Meda (Italy) at the chemical plant ICMESA (Industrie Chimiche Meda Società Azionaria) occurred the Seveso disaster [60]. During the usual production of trichlorophenol, a fungicide, an uncontrolled reaction that sparked the safety valves of the tank reactor took place, releasing the deadly chemical vapour into the atmosphere. About 3000 kg of toxic cloud compound containing dioxin TCDD (2,3,7,8-tetrachlorodibenzo-p-dioxin) was released into the air. The wind immediately dispersed the toxic cloud eastwards: the dioxin could begin to generate its catastrophic effects, affecting the inhabitants of the area and also making uninhabitable the territory on which it was deposited.

The chemicals released into the air from the ICMESA were carried by the wind and caused the contamination of, mainly, four municipalities: Seveso, Meda, Desio, and Cesano Maderno.

Since no one was at the plant when it happened, the seriousness of the disaster was not immediately observed.

Dioxin has been considered to be the most toxic human-made substance; the Seveso disaster represents the highest known TCDD exposure to residential populations and probably the most studied dioxin contamination incident in history [60, 61].

The most common sign of human health problems was skin disorder. The immediate effects on the population were evident especially from a dermatological point of view: after two days already appeared the first cases of chloracne, a disease that is documented to be associated with dioxin.

In order to estimate health problems of population, a great number of laboratory tests were carried out on blood samples. In 1987, it became possible to measure low levels of dioxin in blood samples. Another important effect of the dioxin exposure concerned reproduction, in particular affecting the sex ratio (a higher proportion of females were born in the first 7 years after the accident). In addition, a large number of pregnancies ended as spontaneous abortions were reported.

The affected places still show signs of contamination. The new EU legislation on industrial equipment of chemical and their coordinated management was started in June of 1982, when the Council of Ministers approved the European Union's Council Directive 82/501/EEC, the Seveso Directive.

A central part of the directive was made by reference to the obligation of transparent public information about industrial accidents, as well as new security measures to be taken in case of an accident [60, 61].

The new concept was approved for industrial workers and the public, the right to know the substance and the form of the problems that could threaten them and the exact operation of the safety procedures.

6. Conclusions

From the analysis of situation in the sites at risk of environmental crisis on the Italian territory, the greatest criticalities are represented by the effects of emissions from industrial plants although increasing attention must be paid to the effects of vehicular traffic, as shown by recent data of RTI report, according to which the highest incidence of neoplastic diseases is recorded in areas characterized by the presence of large urban agglomerations.

As regards the comparison between the role of indoor and outdoor pollution in the onset of diseases related to environmental exposure, the second is still predominant and, in the case of sites of national interest, indoor air pollution can be regarded as a factor that may enhance the effects induced by outdoor pollutants.

Pollution situation is controlled through the application of numerous laws on environmental protection, which in our country are very stringent and thanks to which it was possible at least to contain both the emission of pollutants and their effects on human health.

The effect of environmental stressors on human health and policy-making today can be explained using the Driving Force-Pressure-State-Exposure-Effect-Action (DPSEEA) model [62]. Ideal environmental management is based on a reduction of environmental risk exposure, as well as the risks themselves, in all stages of the process. However, this policy-based intervention is not accurate or simple for several motifs (problematic approach for many variables as time lag between exposure reduction to environmental stressor and its negative effects on health or, e.g., as many other variables) that make the intervention very difficult to implement [63]. For DPSSEA model application, the national policy-based intervention, finalized to air pollutants reduction, is very closely to the air quality standards, as the setting-up of those standards can lead to management and reduction of sources of pollution so decreasing pollutant concentrations and subsequent relevant health risk factors to ensure healthy residents [62, 64]. Moreover, air quality standards setting may be considered to be a reliable method for reducing exposure to air pollutants.

Although Institutions have implemented several containment measures, environmental pollution neither has nor ceased to be a problem of Public Health. Currently, in Italy, there are about 15,000 sites to characterize, reclaim, and monitor that represent more than 2% of the entire territory and require the allocation of 25–30 billion euros.

To solve this problem, firstly, rigorous epidemiologic studies must be previewed and conducted, taking into account a careful evaluation phase to confirm the association between environmental exposure and health effects. Epidemiological studies should provide for a careful characterization of the sites, taking into consideration that the great variety of production processes gives rise to an equally wide variety of pollutants not known and for which are not yet known metabolism and mechanisms of interaction with other substances present or released into the environment.

Performing a thorough epidemiological investigation is essential to proceed to risk assessment and risk management. In the management of this route, greater emphasis must be attributed to activities of prevention departments and ARPA and technical committees consisting of experts who can give useful indications to legislators to improve and promulgate laws to protect collective health and must perform a central role. The aim of this cooperation is achieving a "sustainable risk," that is, the risk that remains after the interventions of control and prevention and that can be more easily managed by the competent bodies.

Nonetheless, it is necessary to consider that the bases to make the most of these actions are the training of the operators and the information, so that in this view, activities of continuous updating for public health professionals are promoted.

Author details

M. Ferrante*, M. Fiore, C. Copat, S. Morina, C. Ledda, C. Mauceri and G. Oliveri Conti

*Address all correspondence to: marfer@unict.it

Environmental and Food Hygiene Laboratories (LIAA), Department of Medical Science, Surgical and Advanced Technologies "GF Ingrassia", Hygiene and Public Health, University of Catania, Italy

References

[1] World Health Organization (WHO). http://www.who.int/en/.

[2] European Environmental Agency (EEA). http://www.eea.europa.eu/.

[3] Commission for Environmental Cooperation (CEC). The North American Mosaic: An Overview of Key Environmental Issues. www.cec.org.

[4] European Environmental Agency (EEA). Technical report No 9/2009.

[5] United States Environmental Protection Agency (US EPA). http://www.epa.gov/air/urbanair/.

[6] European Environmental Agency (EEA). Air Quality in Europe – 2013 Report. Report No 9/2013. ISSN 1725-9177.

[7] Europe Union (EU). Directive 2010/75/EU of the European Parliament and of the Council of 24 November 2010 on Industrial Emissions (integrated pollution prevention and control). 2010; OJ L 334: 17–119.

[8] Agency for Toxic Substances and Disease Registry (ATSDR). Interaction profile for Benzene, Toluene, Ethylbenzene, and Xylenes (BTEX). 2004.

[9] Pasquarella C, Pitzurra O, Savino A. The index of microbial air contamination. *Journal of Hospital Infection*. 2000; 46: 241–256.

[10] Zauli Sajani S, Scotto F, Lauriola P, Galassi F, Montanari A. Urban air pollution monitoring and correlation properties between fixed-site stations. *J. Air & Waste Manage. Assoc.* 2004; 54:1-6. ISSN 1047-3289.

[11] Allegrini I, De Santis F, Perrino C, Pietrodangelo A. Metodi innovativi per il monitoraggio dell'inquinamento atmosferico. *Silvae, Anno III*. 8: 45-64.

[12] International Atomic Energy Agency (IAEA). Biomonitoring of Atmospheric Pollution (with Emphasis on Trace Elements) – BioMAP II. IAEA-TECDOC 2003. 1338. ISBN 92–0–100803–1. ISSN 1011-4289.

[13] Bargagli R. Monitoraggio degli inquinanti atmosferici persistenti mediante i muschi e le piante superiori. *Biologia Ambientale*. 2006; 20 (2): 55-67.

[14] Firket J. The cause of symptoms found in the Meuse Valley during the fog of December 1930. *Bull Acad R Med Belg*. 1931; 11:683–741.

[15] Logan WPD. Mortality in the London fog incident, 1952. *Lancet*. 1953; 1:336–338.

[16] Saez M, Perez-Hoyos S, Tobias A, Saurina C, Barceló MA, Ballester F. Métodos de series temporales en los studies epidemiológicos sobre contaminación atmosférica [in Spanish]. *Rev Esp Salud Publica*. 1999; 73:133–143.

[17] Kelsall JE, Samet JM, Zeger SL, Xu J. Air pollution and mortality in Philadelphia, 1974–1988. *Am J Epidemiol.* 1997; 146:750–762.

[18] Maclure M. The case-crossover design: a method for studying transient effects on the risk of acute events. *Am J Epidemiol.* 1991; 133:144–153.

[19] Navidi W, Thomas D, Langholz B, Stram D. Statistical methods for epidemiologic studies of the health effects of air pollution. *Res Rep Health Eff Inst.* 1999; 86: 1–50.

[20] Figueiras A, Carracedo-Martinez E, Saez M, Taracido M. Analysis of case-crossover designs using longitudinal approaches a simulation study. *Epidemiology.* 2005; 16:239–246.

[21] WHO. Health relevance of particulate matter from various sources. Report on a WHO Workshop. Bonn, Germany, 26-27 March 2007.

[22] Dockery DW, Pope CA 3rd, Xu X, *et al.* An association between air pollution and mortality in six US cities. *N Engl J Med.* 1993; 329:1753–1759.

[23] Pope CA 3rd, Thun MJ, Namboodiri, MM *et al.* Particulate air pollution as a predictor of mortality in a prospective study of US adults. *Am J Respir Crit Care Med.* 1995; 151:669–674.

[24] Brunekreef B. Holgate ST. Air pollution and health. *The Lancet.* 2002; 360(9341):1233–1242.

[25] Dockery DW. Pope CA 3rd, Xu X, Spengler JD, Ware JH, Fay ME, Ferris BG Jr, Speizer FE. An association between air pollution and mortality in six U.S. cities. *N Engl J Med.* 1993; 329:1753-1759.

[26] Hoek G, Brunekreef B, Goldbohm S, Fischer P, Van den Brandt PA. Association between mortality and indicators of traffic-related air pollution in the Netherlands: a cohort study. *Lancet.* 2002; 360:1203-1209.

[27] Pope CA 3rd, Burnett RT, Thurston GD, Thun MJ, Calle EE, Krewski D, Godleski JJ. Cardiovascular mortality and long-term exposure to particulate air pollution: epidemiological evidence of general pathophysiological pathways of disease. Circulation 2004; 109:71-77.

[28] Brook RD, Franklin B, Cascio W, Hong Y, Howard G, Lipsett M, Luepker R, Mittleman M, Samet J, Smith SC Jr, Tager I. Air pollution and cardiovascular disease: a statement for healthcare professionals from the Expert Panel on Population and Prevention Science of the American Heart Association. *Circulation.* 2004; 109:2655-2671.

[29] Hoek G, Brunekreef B, Fischer P, van Wijnen J. The association between air pollution and heart failure, arrhythmia, embolism, thrombosis, and other cardiovascular causes of death in a time series study. *Epidemiology.* 2001; 12:355-357.

[30] Gold DR, Litonjua A, Schwartz J, Lovett E, Larson A, Nearing B, Allen G, Verrier M, Cherry R, Verrier R. Ambient pollution and heart rate variability. *Circulation.* 2000; 101:1267-1273.

[31] Wellenius GA, Schwartz J, Mittleman MA. Particulate air pollution and hospital admissions for congestive heart failure in seven United States cities. *Am J Cardiol.* 2006; 97:404-408.

[32] Zanobetti A, Schwartz J. The effect of particulate air pollution on emergency admissions for myocardial infarction: a multicity case-crossover analysis. *Environ Health Perspect.* 2005; 113:978-982.

[33] Pope CA 3rd, Muhlestein JB, May HT, Renlund DG, Anderson JL, Horne BD. Ischemic heart disease events triggered by short-term exposure to fine particulate air pollution. *Circulation.* 2006; 114:2443-2448.

[34] Prüss-Üstün A, Corvalán C. Preventing disease through healthy environments. Towards an estimate of the environmental burden of disease. WHO, 2006. http://www.who.int/quantifying_ehimpacts/ publications/preventingdisease/en/.

[35] Ostro B. Outdoor air pollution: assessing the environmental burden of disease at national and local levels. Geneva: World Health Organization. 2004. http://www.who.int/quantifying_ehimpacts/publications/ebd5/en/.

[36] Prüss-Üstün A, Mathers C, Corvalán C, et al. Introduction and methods: assessing the environmental burden of disease at national and local levels. Geneva: World Health Organization, 2003. http://www.who.int/quantifying_ehimpacts/publications/9241546204/en/.

[37] US-EPA. Guidelines for Carcinogen Risk Assessment, March 2005.

[38] Rivas I, Viana M, Moreno T, Pandolfi M, Amato F, Reche C, Bouso L, Àlvarez-Pedrerol M, Alastuey A, Sunyer J, Querol X. Child exposure to indoor and outdoor air pollutants in schools in Barcelona, Spain. *Environ Int.* 2014 Aug; 69:200-12. doi: 10.1016/j.envint.2014.04.009.

[39] WHO. Indoor Air Quality Guidelines: selected pollutants. 2010. ISBN 978 92 890 0213 4

[40] Pegas PN, Alves CA, Evtyugina MG, Nunes T, Cerqueira M, Franchi M, Pio CA, Almeida SM, Freitas MC. Indoor air quality in elementary schools of Lisbon in spring. *Environ Geochem Health.* 2011 Oct; 33(5):455-68.

[41] Kim C, Gao YT, Xiang YB, Barone-Adesi F, Zhang Y, Hosgood HD, Ma S, Shu XO, Ji BT, Chow WH, Seow WJ, Bassig B, Cai Q, Zheng W, Rothman N, Lan Q. Home kitchen ventilation, cooking fuels, and lung cancer risk in a prospective cohort of never smoking women in Shanghai, China. *Int J Cancer.* 2014 Jun 11. doi: 10.1002/ijc.29020.

[42] Gordon SB, Bruce NG, Grigg J, Hibberd PL, Kurmi OP, Lam KB, Mortimer K, Asante KP, Balakrishnan K, Balmes J, Bar-Zeev N, Bates MN, Breysse PN, Buist S, Chen Z,

Havens D, Jack D, Jindal S, Kan H, Mehta S, Moschovis P, Naeher L, Patel A, Perez-Padilla R, Pope D, Rylance J, Semple S, Martin WJ. Respiratory risks from household air pollution in low and middle income countries. *Lancet Respir Med*. 2014 Oct; 2(10): 823-860. doi: 10.1016/S2213-2600 (14) 70168-7.

[43] Colton MD, MacNaughton P, Vallarino J, Kane J, Bennett-Fripp M, Spengler JD, Adamkiewicz G. Indoor air quality in green vs conventional multifamily low-income housing. *Environ Sci Technol*. 2014 Jul 15; 48(14):7833-41.

[44] Alam DS, Chowdhury MAH, Siddiquee AT, Ahmed S, Hossain MD, Pervin S, Streatfield K, Cravioto A, Niessen LW. Adult cardiopulmonary mortality and indoor air pollution. a 10-year retrospective cohort study in a low-income rural setting. *GLOBAL HEART*. 2012; 7:(3):215–e221.

[45] Fullerton DG. Bruce N, Gordon SB. Indoor air pollution from biomass fuel smoke is a major health concern in the developing world. *Transactions of the Royal Society of Tropical Medicine and Hygiene*. 2008; 102:843-851.

[46] Ghosh R. Indoor smoke and prenatal and childhood growth: the role of (gestational) age. *World J Clin Pediatr*. 2013 Nov 8; 2(4):31-5. doi: 10.5409/wjcp.v2.i4.31.

[47] Challoner A, Gill L. Indoor/outdoor air pollution relationships in ten commercial buildings: $PM_{2.5}$ and NO_2. *Building and Environment*. 2014; 80:159e173.

[48] Tzivian L, Winkler A, Dlugaj M, Schikowski T, Vossoughi M, Fuks K, Weinmayr G, Hoffmann B. Effect of long-term outdoor air pollution and noise on cognitive and psychological functions in adults. *Int J Hyg Environ Health*. 2014 Sep 3. pii: S1438-4639 (14) 00077-7. doi: 10.1016/j.ijheh.2014.08.002.

[49] Semple S, Apsley A, Azmina Ibrahim T, Turner SW, Cherrie JW. Fine particulate matter concentrations in smoking households: just how much secondhand smoke do you breathe in if you live with a smoker who smokes indoors? *Tob Control*. 2014 Oct 20. pii: tobaccocontrol-2014-051635. doi: 10.1136/tobaccocontrol-2014-051635.

[50] Zhou S, Weitzman M, Vilcassim R, Wilson J, Legrand N, Saunders E, Travers M, Chen LC, Peltier R, Gordon T. Air quality in New York City hookah bars. *Tob Control*. 2014 Sep 16. pii: tobaccocontrol-2014-051763. doi: 10.1136/tobaccocontrol-2014-051763.

[51] Dorizas PV, Assimakopoulos MN, Helmis C, Santamouris M. An integrated evaluation study of the ventilation rate, the exposure and the indoor air quality in naturally ventilated classrooms in the Mediterranean region during spring. *Sci Total Environ*. 2014 Oct 6; 502C: 557-570. doi: 10.1016/j.scitotenv.2014.09.060.

[52] Tonelli F, Short S.W., Taticchi P. Case study of ILVA, Italy: The impact of failing to consider sustainability as a driver of business model evolution. www.isss.it, Istituto Superiore Sanità, accessed 10/09/2014.

[53] Boffetta P, La Vecchia C, Lotti M, Moretto A. Commenti al documento "Valutazione del danno sanitario stabilimento ILVA di Taranto ai sensi della LR21/2012. Scenari

emissivi pre-AIA (anno 2010) e post-AIA (anno 2016)" dell'ARPA Regione Puglia. pp.1-44. http://www.gruppoilva.com/items/23/allegati/1/ILVA_SpA_Allegato_Nota_Dir_205_2013.pdf accessed 21/10/2014.

[54] Banini T, Palagiano C. Environment and Health in Italian Cities: the case of Taranto.A. Malik et al. (eds.), *Environmental Deterioration and Human Health*. Springer Science+Business Media: Dordrecht. 2014. DOI 10.1007/978-94-007-7890-0_2.

[55] Mataloni F, Stafoggia M, Alessandrini E, Triassi M, Biggeri A, Forastiere F. Studio di coorte sulla mortalità e morbosità nell'area di Taranto. A color study on mortality and morbidity in the area of Taranto, Southern Italy. *Epidemiol Prev*. 2012; 36(5): 237-252.

[56] Comba P, Conti S, Iavarone I, Marsili G, Musmeci L, Pirastu R. Ambiente e salute a Taranto: evidenze disponibili e indicazioni di sanità pubblica.

[57] Musmeci L, Bianchi F, Carere M, Cori L. Environment and healthin Gela (Sicily): present knowledge and prospects for future studies.

[58] Cernigliaro A, Pollina Addario S, Nicolotti N. Research on the health state of the Sicilian population living in areas at environmental or natural risk: the experience of the Regional Department of Epidemiology.

[59] Pesce Paola, Castaing Marine, Leone Anna, Varvarà Massimo, Dinaro Ylenia, Madeddu Anselmo, Tisano Francesco, Contrino Lia, Tumino Rosario, Sciacca Salvatore. Differenze di incidenza oncologica e mortalità per sede nelle aree metropolitane e nei SIN in Sicilia: dati 2003-2011.http://www.registro-tumori-integrato.it/

[60] Hakkinen PJ. Seveso Disaster, and the Seveso and Seveso II Directives. *Elsevier*. 2005.

[61] Pesatori AC, Consonni D, Bachetti S, Zocchetti C, Bonzini M, Baccarelli A, Bertazzi PA. Short-and long-term morbidity and mortality in the population exposed to dioxin after the "Seveso accident". Industrial Health 2003; 41: 127–138

[62] Corvalán C, Briggs DJ, Zielhuis G; World Health Organization. Decision-making in environmental health: from evidence to action [cited 2014 Sep 12]. accessed 5/10/2014 from: http://apps.who.int/iris/bitstream/10665/42304/2/0419259406_part1.pdf?ua=1.

[63] Ha J. Applying policy and health effects of air pollution in South Korea: focus on ambient air quality standards. Environmental Health and Toxicology. Volume: 29:1-7, Article ID: e2014011, http://dx.doi.org/10.5620/eht.2014.29.e2014011

[64] WHO. Air quality guidelines for particulate matter, ozone, nitrogen dioxide and sulfur dioxide. Global update 2005. http://www.who.int/phe/health_topics/outdoorair/outdoorair_aqg/en/ accessed 11/10/2014.

Chapter 20

Treatment of Post-consumer Vegetable Oils for Biodiesel Production

Elaine Patrícia Araújo, Divânia Ferreira da Silva,
Shirley Nobrega Cavalcanti,
Márbara Vilar de Araujo Almeida,
Edcleide Maria Araújo and Marcus Vinicius Lia Fook

Additional information is available at the end of the chapter

http://dx.doi.org/10.5772/59874

1. Introduction

The current energy model based on petroleum shows signs of exhaustion, which is aggravating, as besides energy source petroleum is used extensively for the production of plastics, clothing, fertilizers and medicine, moving a true "Petroleum Civilization" [1]. Ally the question of exhaustion of petroleum reserves and its derivatives and the search for renewable energy sources, is also highlighted the issue of waste, which daily becomes one of major problems for humanity. Worldwide, approximately 60 million tons of edible vegetable oils - which, in most cases, are used for frying various types of food - are produced, according to data from the United States Department of Agriculture Food, published in 2000. A significant number of these oils are eliminated directly into the environment, harming these aquatic and terrestrial environments [2].

Almost all energy consumed in the world comes from non-renewable sources of fossil fuels, which cause great environmental impact. Alternative fuels for diesel engines are becoming increasingly significant due to the decrease of petroleum reserves and thus, increasing it's price, that reaches levels high enough to prevent it's use. Also the environmental impact caused by emissions of gases generated from burning of fossil fuels have been reason for research on alternative energy sources [3,4].

Due to emission of toxic gases by discharges from diesel vehicles, hundreds of researches warn that different pollutants emanating from the exhausts lie in the main causes of degradation of air quality in large urban centers.

The recycling of post-processed oil is minimal and has restricted applications, one being the use in the detergent industry and most recently, as biodiesel. This can be defined as the mono-alkyl ester derived from long chain fatty acids, from renewable sources such as vegetable oils or animal fats obtained by transesterification process, use of which is associated with replacement of fossil fuels in engines compression ignition [5]. It can also be defined as a biodegradable fuel derived from renewable resources obtained from the reaction of vegetable oil and animal fats which, stimulated by a catalyst, react chemically with methanol or ethanol. This can be done with any fresh or post-consumer vegetable oil, waste or sludge. Several studies have shown that the obtaining of methyl and ethyl esters from soybean oil, canola, sunflower, palm, castor, and also post-consumer frying oil, is recommended, since it has lower incomplete combustion of hydrocarbons and lower emissions of carbon monoxide, particulate matter, nitrogen oxides and soot [3, 6].

The biodiesel obtained from post-consumer frying oil, according to studies, decreased smoke, demonstrating that has effective benefit in reusing this oil for biofuel production, featuring a more suitable destination to this agro-industrial waste that, in Brazil, is commonly discarded and/or partially reused, but often in inadequate ways [7].

Vegetable oils have many advantages as alternative fuels when compared to diesel: they are natural liquids, renewable, with high energy value, low sulfur content, low aromatic content and biodegradable. However, despite the use of these oils being favorable from the point of view of energy, its direct use in diesel engines is very problematic. Studies performed with various vegetable oils showed that its direct combustion leads to a series of problems: carbonisation in the injection chamber, contamination of the lubricating oil, among others [8].

The emission of toxic gases by motor vehicles is a major source of air pollution. In cities, these vehicles are responsible for the emission of harmful gases such as carbon monoxide (CO), carbon dioxide (CO_2), nitrogen oxides (NO_x), sulfur dioxide (SO_2), hydrocarbons (HC), lead, smoke and particulates. Studies have been conducted in order to quantify and estimate the use of various energy sources on the increase of CO_2. The main sources of energy considered more polluting in terms of CO_2 emissions are: Liquefied Petroleum Gas (LPG), natural gas, fuel oil and diesel oil [9].

In general, air pollution affects health, generating both acute effects such as eye irritation and coughing, which are temporary and reversible, and chronic effects, which are permanent and cumulative with demonstrations in the long run, of causing severe respiratory diseases. There may also be structural corrosion and degradation of buildings and work of art. In heavily polluted cities, these disturbances are exacerbated in winter with temperature inversion, when a layer of cold air forms a bell high in the atmosphere, trapping hot air and preventing dispersion of pollutants. Compared to different sources of emissions, the diesel has the highest emission of toxic gases, contributing to the rise of the various environmental scenarios, social and economic [10]. Due to this problem, various studies are being conducted with post-consumption vegetable oils for biodiesel production, therefore, is an alternative renewable fuel that releases less harmful gas emissions compared to conventional fossil fuel (diesel). The most common method of making biodiesel is by transesterification reaction of vegetable oils or animal fats, with a short chain alcohol [11].

Biodiesel in its pure form (B100) can allow the net emission of carbon dioxide (CO_2) - the main from the Greenhouse Gas (GHG) emissions be reduced by 80%. This has a positive impact on the environment because it decreases air pollution in large urban centers (the B100 blend provides a 90% reduction of smoke and eliminate the sulfur oxide, responsible for acid rain), thus improving the quality of life and reduced spending in the health system population [12].

The recycling of post-consume discarded vegetable oils contributes to reduce the uncontrolled and harmful environmental disposal, and may have competitive price on fossil fuels. However, the use of this oil in biodiesel production requires treatment prior to transesterification reaction, which comprises the removal of contaminants solid particles and the appropriateness of color and odor. For this reason, the main objective in this research was to use clays of the northeastern semi-arid region of Paraíba/Brazil to evaluate its potential in the treatment of post-consumer vegetable oils for biodiesel production.

1.1. Post-consumer vegetable oils

Among renewable energy resources, the use of biomass, in its different forms (solid, liquid and gas), was intensively researched in recent years as an alternative to minimize adverse environmental impact and the uncertainty in future supply of fossil fuels. Despite the possible environmental benefits in the use of vegetable oils as a substitute for diesel, barriers economically and ethically motivated the search for alternative raw materials for biofuel production [13, 14].

Among the alternatives studied, the reuse of Waste Vegetable Oils (WVO) and fats of processes of frying various foods has been shown to be attractive because the advantage in that the vegetable oil as fuel after its use in the food chain, thus resulting in a second use, or even an alternative destination to a residue of food production destination [15]. Among the several aspects that motivated the study of vegetable oils as fuel potential are:

- It's (liquid) physical condition and its high specific energy content (MJ/kg fuel) when compared with other fuels derived from biomass;
- The fact that they are produced from different oilseeds (soybean, rapeseed, palm, etc.) under different climatic conditions;
- High energy productivity of some oilseeds (above 150 GJ/ha for palm oil);
- The possibility of using oil and its derivatives in high efficiency energy conversion engines, such as diesel engines.

The oils used in frying have important nutritional aspects, involving the transport of fat soluble vitamins, supply of essential fatty acids of the ω3 and ω6 series, precursors of eicosanoids, the energetic power and present a wide acceptance by the various social groups [16].

Studies conducted in the city of Valencia/Spain, concludes that it is attractive, the environmental point of view, obtaining biofuel from WVO. A selective collection system, established by the city council, supported the project to produce biofuel to supply 480 city buses, with a demand of approximately 42,000 liters/day. The ultimate goal of the project was the elimination, large scale, the WVO plumbing the sanitary sewer system of the city, about 10,000 t/a [13].

In Brazil, it is common the use of soybean oil (nationally) and rice oil (in the south) to processes of frying food in shops. Soybean oil contains 15% of saturated fatty acids, 22% of oleic acid, 54% of linoleic acid and 7.5% linolenic acid. The rice oil contains about 20% of saturated fatty acids, 42% of oleic acid, 36% of linoleic acid and 1.8% of linolenic acid. Soybean oil, by presenting a lower composition of saturated fatty acids and higher in polyunsaturated fatty acids, is more susceptible to degradative processes [17].

The physical changes which occur in the oil or fat during the frying process include dimming, increase in viscosity, decrease in smoke point and foaming. Chemical changes can occur by three different types of reactions: the oils and fats can hydrolyze to form free fatty acids, monoacylglycerol and diacylglycerol; can oxidize to form peroxides, hydroperoxides, conjugated dienes, epoxides, ketones and hydroxides; and may decompose into small fragments or remains in the triacylglycerol and to associate, leading to dimeric and polymeric triglycerides [16,18].

During the process of frying oils and fats are exposed to the action of three agents that contribute to compromise their quality and modify its structure: the moisture from foods, which is the cause of hydrolytic alteration; oxygen, which in contact with the oil, for prolonged periods, causes oxidative modification and the high temperature in the operation, 180°C, causing thermal alteration [19].

The usage time of the oil varies from one establishment to another, mainly due to the lack of legislation to determine the exchange of post-consumer oil [7]. There is no single method by which it is possible to detect all situations involving the deterioration of oils in the frying process. The determination of the optimal point for disposal has significant economic impact resulting in a higher cost when oil is discarded before its effective degradation, and loss of quality of food, when discarded later. Some indicators used by restaurants and cafeterias, to determine the point of discharge of oil or fat are: color change, formation of smoke and foam during the frying process and changes in aroma and taste [20].

In frying temperature (170 to 180°C) occurs in reactions with air, water and food components. The oil and the vegetable fat used in frying process by immersion represent a major risk of environmental pollution since most commercial establishments (pubs, restaurants, coffee shops, etc.) and residential discard the residual oil into the sewer system difficult to treat these. However, this material can be used as raw material for biodiesel production [21, 10].

The transformation of the used cooking oil into biodiesel brings significant environmental improvements. Initially, the byproduct that would be discarded in the environment receives a new use, no longer willing improperly. Thus, reducing the consumption of fossil fuels (diesel oil) occurs, in addition to encouraging the use of renewable fuels.

For the manufacture of biodiesel, it is necessary to invest in the industry of purification and transformation. Biodiesel is biodegradable fuel derived from renewable sources (vegetable oil or animal fat) which can be obtained by different processes, such as cracking, esterification and transesterification [22].

The lipids are oils and fats insoluble in water, animal or vegetable source, and consist of triglycerides or triglyceride esters formed from glycerol and fatty acids. The present fatty acids are generally saturated carboxylic acids with 4 to 24 carbon units in the chain and unsaturated carboxylic acids with 10 to 30 carbons and 1 to 6 double bonds in the chain [23].

The vegetable oils are natural products consist of a mixture of esters of glycerol derivatives, which contain fatty acid chains from 8 to 24 carbon atoms having different degrees of unsaturation (Figure 1). Different species show variations in oil molar ratio between the different fatty acids present in the structure [7].

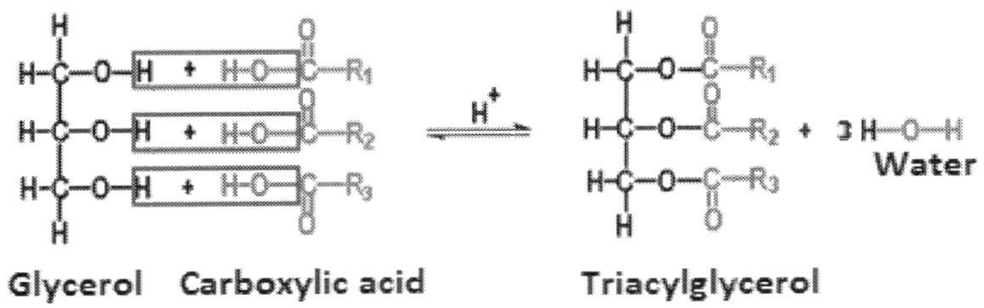

Source: [24].

Figure 1. Formating of Trigliceride: a molecul of glicerol and a molecul of fatty acid.

Firestone et al. [25] comment that in some countries such as Belgium, France, Germany, Switzerland, Netherlands, United States and Chile, there are rules on the conditions under which a vegetable oil used for frying should be discarded. But in Brazil, as in many other countries there are no laws and regulations establishing limits to the changes in these oils. An estimated damaged by oil frying process must be discarded when their content of polar compounds meet above 25%. Another aspect that must be considered is the percentage of free fatty acids, for which the laws set, limits around 1 and 2.5%.

One of the main causes of the degradation of oils and fats is rancidity, which is associated with the formation of organoleptically, creates unacceptable product due to occurrence of foreign odors and flavors, and the loss of product color, and inactivation of vitamins polymerization [26]. The rancidity can be classified as:

- Hydrolytic rancidity - occurs in the presence of moisture due to the action of lipases that catalyze the hydrolysis enzymes, releasing fatty acids, and
- Oxidative rancidity or oxidation occurs due to non-enzymatic action of lipoxygenase enzymes or by action, such as autoxidation and photooxidation.

According [17] consumption of fried foods and frozen pre-fried, induces higher intake of oil through the frying process. During these processes, there are several forms of lipid deterioration that compromises the quality of the oil, they are:

- Hydrolysis involves the cleavage of the glyceride ester with formation of free fatty acids, mono glycerides, diglycerides and glycerol. It is a reaction that occurs due to the presence of water at high temperatures, which can result in products with high volatility and high chemical reactivity;
- Consisting of degradative oxidation process in which atmospheric oxygen dissolved in the oil or reacts with unsaturated fatty acids, producing sensory unacceptable products with unpleasant smells and flavors for human consumption (Figure 2);
- Polymerization that occurs when two or more fatty acid molecules combine as a consequence of changes in the oxidation process and high temperatures.

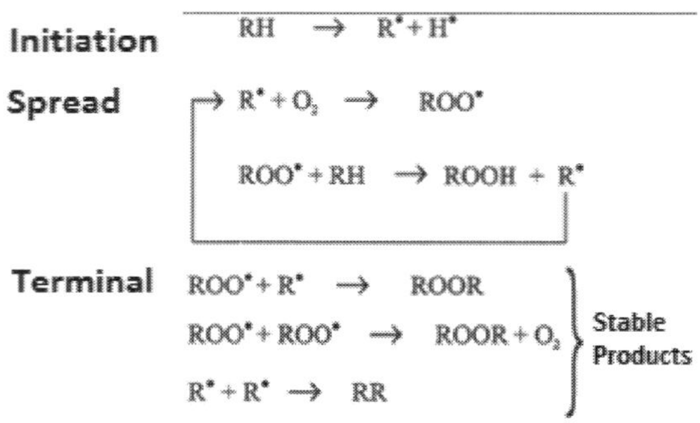

Source: [27].

Figure 2. General scheme of the lipid oxidation mechanism.

Several studies with oil heated for long periods at high temperatures showed that the resulting product contains more than 50% of polar compounds which are degradation products of triglycerides (polymers, dimers, oxidized fatty acids, diglycerides and free fatty acids). These oils with high contents of polar compounds can cause severe irritation of the gastrointestinal tract, diarrhea, reduction in growth, and in some cases death of laboratory animals [28]

When oils are used at high temperatures or are reused, they release a toxic substance, acrolein, which interferes with the functioning of the digestive and respiratory system, mucous membranes and skin, and can even cause cancer [29].

The resulting polymers increase the viscosity of the oil. The frying process characteristics such as browning develops, an increase in viscosity, decrease in smoke point and foaming affecting the quality of the oil [16].

Brazil does not have any regulation that legally defines the monitoring of disposal for oil and frying fats. There are regulations governing the suitability of oil for consumption in Brazil, the NTA 50, citing some physicochemical items to control the suitability of this oil: iodine value, peroxide value and acid value, however not refer to oils and cooking fats [16].

http://dx.doi.org/10.5772/59874

1.2. Clays

According to Santos [30], clays are natural earth materials which exhibit fine-grained (typically with a diameter of less than 2 μm particles) and are formed by chemically hydrated silicates of aluminum, iron and magnesium. These are composed of small crystalline particles of a limited number of minerals, clay minerals. In addition to these clay minerals, clays may also contain organic matter, soluble salts, particles of quartz, pyrite, calcite, and other residual amorphous mineral reserves. The main factors that control the properties of clays are the mineralogical and chemical composition of clay minerals of non-clay minerals and their particle size distributions; electrolyte content of exchangeable cations and soluble salts; nature and content of organic components and textural characteristics of the clay.

Brazil has industries that utilize different types of clays for several purposes: fabrications red ceramic, white ceramic, refractory materials; the manufacturers of rubber and plastics used as the active and inert fillers; metallurgical industry uses clays as binders for molding sands for the casting of metals and for pelletizing iron ores; industries of edible oils and petroleum use them as bleaching agents of vegetable and mineral oils; can also be used as thixotropic agents in mud for drilling for oil drilling and water; There are special clays are used as catalysts in cracking of oil to produce gasoline and are used for special purposes being used as filler for soap and tissues, as pigments for paints, in the manufacture of pharmaceutical products [30].

Determining the result of the technological properties of these properties whose function is to complement function test results traditional characterization as: X-ray diffraction, X-ray fluorescence, particle size analysis. With these results together with the results of the technical properties (physical and mechanical properties) can indicate the proper use of a clay and establish accurate or necessary for better performance properties to which the clay is subjected [31].

The importance and diversity of use of clays is a result of its particular characteristics. This difference makes the clays of the most used materials, either on his great geological variety or offers a set of essential and indispensable factors in numerous industrial processes [31].

The bentonite is a layered clay mineral composed montmorillonite that is an aluminosilicate trifórmico the type crystalline structure appearing as a layer of alumina octahedrons between two layers of silica tetrahedra with adjacent margins primarily. Their composition is variable due to ease of isomorphic substitutions (may contain FeO, CaO, Na_2O and K_2O), which causes a negative charge density on the surface of the smectite clay and require cations to compensate for these loads, the exchangeable cations [30].

In Brazil, the terms bentonite clay materials are used to montmorillonite without any information about the geological origin or mineralogical composition. The chemical composition and method of the unit cell of the "theoretical" montmorillonite or end of the series is $(Al_{3,33}Mg_{0,67})$ Si_8O_{20} (OH) $4.M^{+}_{10.67}$, where M^{+1} is a monovalent cation. This formula shows that the unit cell has a negative electrical charge due to isomorphic substitution of Al^{3+} by Mg^{2+}. The cation M^+ which balances the negative charge is called exchangeable cation since it can be changed in a reversible way, by other cations. The content of exchangeable cation, expressed in milliequivalents of cation per 100g of clay is called CEC - cation exchange capacity. The cation M^+ interplanar occupies the space of the two layers 1 and may be anhydrous or hydrated. As the size of the dry cation and the number of layers of water molecules coordinated to the cation, it may have different values of basal interplanar distance [32].

According Centre for Mineral Technology – CMT [33], is the term given to a smectite group of minerals consisting of: montmorillonite, beidellite, nontronite, hectorite and saponite, in which each of these minerals form a similar structure, but each is chemically different. The most common mineral in the economic deposits of smectite is montmorillonite group. The calcic and sodic varieties, based on exchangeable cation, are the most abundant.

Amorim et al. [34] commented that according to geologists, the bentonite is formed by devitrification and chemical alteration of volcanic ash. For many years, scholars have used the origin of these clays as part of its definition, but in some countries as their deposits were not originated by volcanic action, other definition came to be used: bentonite clay is composed of any clay mineral montmorillonite, smectite group the and whose properties are established by this clay mineral.

The bentonite clays have distinct and peculiar to increase to several times its original volume when wetted with water and form thixotropic gels in aqueous media at low concentrations, interplanar spaces reaching up to 100 Å, high surface area and cation exchange capacity. These are characteristics that make the bentonite a wide range of applications in various technological sectors from the preparation of nanocomposites by the use as decolorizing agent [35].

Deposits of bentonite clays of Paraíba form the largest, and the most important deposit is located in Brazil. Their occurrences are located in the city of Boa Vista, and its deposits are mines Lages, Bravo, Jua and Canudos. In 2004, Paraíba State was the main producer of crude bentonite with 88% of national production, followed by São Paulo (7.3%), Rio de Janeiro (4.4%) and Paraná (0.2%). The production of bentonite in Brazil, which focuses on two products, activated bentonite clay and dry ground, grew by 14% [34, 36].

According to data released by the National Department of Mineral Production – NDMP [37], the state of Paraíba is currently the most significant source of bentonite clay, bentonite deposits it's being located mainly in the city of Boa Vista. Its reserves amount to about 70% of bentonite clays throughout Brazil.

The national reserves of bentonite represent about 3% of world reserves. Brazilian production is around 300 000 t/a which represents 3% of world consumption. The average price of bentonite is about $107/t, while the activated bentonite can reach $1,800/t. Also according to [38], the market for bentonite is very concentrated in the United States, the world's largest producer and has high investments made in this industry, which has provided diversification in its use and application [38].

Clay minerals of the smectite (montmorillonite) group are composed of two layers of a tetrahedral silicate with an octahedral core sheet joined together by common oxygen atoms to the leaves and in the space between the sheets are adsorbed water molecules and exchangeable cations, which may be Ca^{2+}, Mg^{2+} or Na^+ or both (Figure 3) [30].

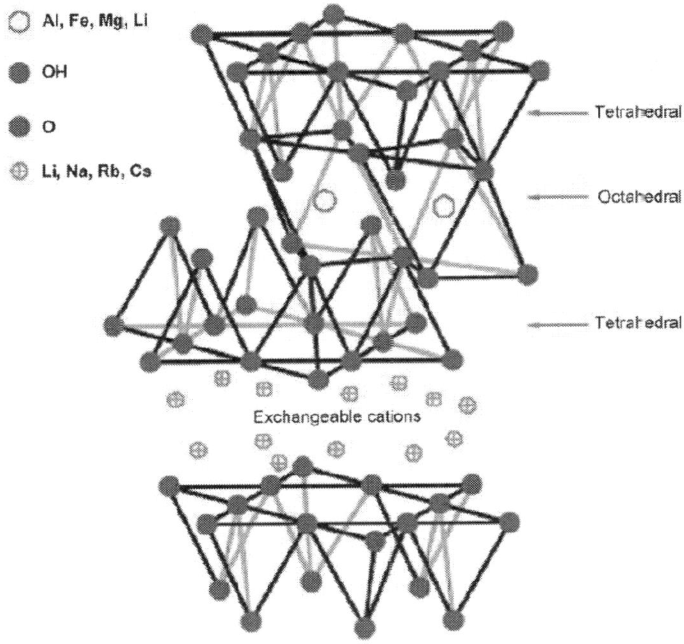

Source: [39, 40].

Figure 3. Crystal structure of the clay mineral montmorilonitico.

The bentonite clay is classified according to their exchangeable cations present in [39]:

- Homocationica: when there is a predominance of a type of exchangeable cation such as sodium or calcium, so called sodic or calcic bentonite.
- Polycationic: when there is no predominance of one type of exchangeable cation. Cations such as sodium, calcium, potassium and others can be present in similar concentrations.

Treatment with acid serves to Dissolve some impurity of bentonite; replace calcium and other cations intercalated by ions H_3O^+ hidroxônio and dissolve in the octahedral layers of two layers: one, some cations Mg^{2+}, Al, Fe^{3+} or Fe^{2+}. The acid treatment causes significant morphological changes in the crystal structure of montmorillonite during and after acid activation. The montmorillonites activated by acids, are commonly used for bleaching of edible oils and fats [40].

1.3. Clays for treatment of post-consumer vegetable oils

Clays have been used by mankind since antiquity for manufacturing ceramic objects, such as bricks and tiles, and more recently, in several technological applications. These are used as adsorbents in bleaching processes in the textile and food industry, in processes of soil remediation and landfill. The interest in its use has been gaining momentum mainly due to the search for materials that do not harm the environment when discarded, the abundance of world reserves and its low price. In the oil industry, the clays that are used for bleaching these oils are called "bleaching earth", "soil bleach", "clarifying clay" or "adsorbent clay" to indicate that clays in the natural state or after chemical or thermal activation, have the property of coloring materials present in adsorbing mineral oils, animal and vegetable [32].

The bentonite clays according to Santos [32] can be classified according to their adsorptive properties: montmoriloniticas bentonite-type clays, which are virtually inactive and inativaveis ; montmoriloniticas inactive clays, but highly activatable by acid treatment; extremely active and activatable clays by acid treatment; active clays and whose activity is little affected by acid treatment; active clays whose activity is decreased by acid treatment.

According to the adsorptive and catalytic properties, the activated bentonite clays are used industrially as catalysts, adsorbents and catalyst supports. However, in terms of consumption, the most important use of this material and purification, bleaching and stabilization of vegetable oils. The adsorptive capacity of these materials increases with treatment with strong acid, typically sulfuric or hydrochloric acid are used. The presence of these acids modifies the structure of clays [41].

The adsorptive capacity of clay bleaching increases with the increase of the specific area. The bleaching earth adsorbs some better connection than others or even ceases to adsorb some. Polar or polarizable molecules are well adsorbed by bleaching earth. However, the adsorptive ability of the bleaching earth is reduced if the oil contains soaps or gums in excess to neutralize the acid sites of the same is true when there are many free fatty acids, which, as highly polar compounds occupy part of the surface of the clay mineral [26].

The power of bleaching clay may be due alone or in combination, the following factors: simple filtration, which corresponds to the retention of colored particles dispersed in oil in the capillaries clay; the selective adsorption of dissolved dyes and catalytic activity of the clay [32].

The time of bleaching oils suffer limitations due to the bleaching temperature. For this, we used 0.75% of smectite clays activated bleaching processes structured in three different temperature levels (82°C, 104°C and 138°C) and five levels of time (5 min., 10min., 15 min., 35min., and 55min.). It was observed that the red color of the oil fell to the lowest level when the highest temperature is used. At this temperature, however, the color began to darken oil from the time of bleaching, coming, at the end of the bleaching, become darker than the other two processes [26].

For an acid activated bentonite clay may be used as a decolorizing agent is necessary to have the following requirements: the pH is between 6.0 and 7.5; porosity between 60 and 70%; no catalytic activity in the case of edible oils and fats to prevent the generation of undesirable odors and tastes after bleaching; low oil retention in filtration and good filterability [40].

2. Materials and methods

The calcic clays used for the treatment of post-consumer vegetable oils were the bentonite clay (Figure 4) trade name Tonsil and Aporofo with a particle size of 200 mesh (0.074mm) mesh provided and identified by the company BENTONISA - The Bentonite Nordeste S/A, located in João Pessoa-Paraíba.

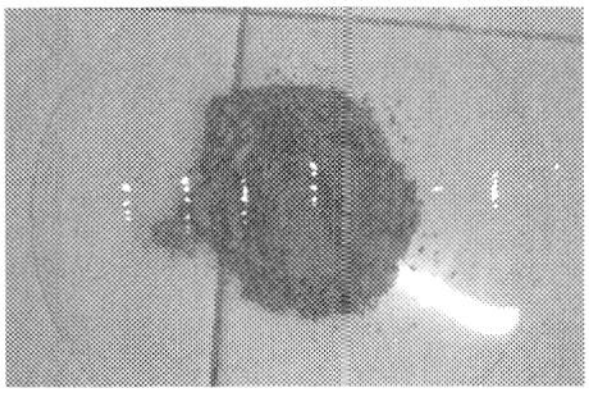

Source: Research data.

Figure 4. Calcium bentonite clay used for the treatment of post-consumer vegetable oils.

2.1. Post-consumption vegetable oil

Residual soybean oil, mixture of soybean oil and hydrogenated fat residual: samples of raw materials found in some homes in the city of Campina Grande-Paraíba were collected. These oils had a dark color and unpleasant odor. A sample of fresh vegetable oil from soybeans, Figure 5 was acquired in a business in order to make a comparison with the samples of vegetable oils untreated post-consumer and post-consumer-treated clays under study. Soybean oil was chosen because it is the most widely used in the domestic market and for having little commercial value in relation to other edible vegetable oils, such as olive oil, sunflower oil and corn oil.

Figure 5. Samples of fresh vegetable oils (a) and post-consumer without treatment (b).

2.2. Methodology

Figure 6 shows the flowchart of the methodology used for the bleaching of treaties with calcic bentonite clay, Tonsil and Aporofo, Paraíba region of post-consumer vegetable oils. This method was adapted from de Santos [32] literature.

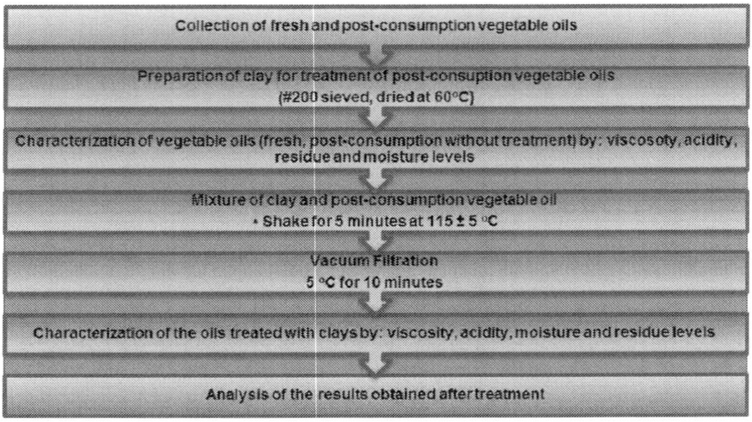

Source: Personal archive.

Figure 6. Steps in the treatment of post-consumer vegetable oils process.

Source: Research data.

Figure 7. Steps used to treat post-consumer vegetable oils: (a) mixing oil with clay, (b) filtration processes the oil with clay, (c) fresh oils and vegetable consumption without post-treatment and (d) post-consumer oil treated and fresh.

2.2.1. Characterization of post-consumer vegetable oils

2.2.1.1. Kinematic viscosity

The kinematic viscosities of fresh oils, post-consumer and post-consumer treated with Tonsil and Aporofo clays were measured by a CANNON-FENSKE viscometer thermostat of the brand Quimis according to ASTM 445, 220V, 40°C. For the determination of kinematic viscosity, was used an oil standard viscosity suitable for the viscometer/viscosity range. The standard viscometer is filled with oil by immersing the tube containing the oil bath, waiting 5 to 10 minutes for thermal equilibrium to occur. A reading is held on the 1st and second bulb, noting the result. This procedure is repeated eight times. Then makes an average of measurements and calculates the calibration factor of the tube:

$$F=\frac{V}{t} \tag{1}$$

F = Calibration Factor

V = Standard viscosity

t = Time spent in seconds

The stirred sample is transferred to a beaker and then the viscometer reservoir is filled with this sample, adapting a stopper at the end, in order to promote complete sealing, thus avoiding leakage of oil. The viscometer is then transferred to a thermostatic bath at a test temperature of 40°C. Then the stopper is removed allowing the flow of oil. It is noted that the time was spent for the oil to drain from the first to second and second to third meniscus viscometer. The kinematic viscosity is calculated by the formula described below, and the result is presented in mm^2/s:

$$V_{(cin)}=T\times F \tag{2}$$

where:

V = Kinematic viscosity at the test temperature, in seconds.

T = Time in seconds obtained by the sample flow.

F = calibration factor.

2.2.1.2. Acidity

The acid content of the oils was determined by titration according to ASTM standard D664 24A. For this test, the following reagents are used: isopropyl alcohol ($CH_3CH(OH)CH_3$);

phenolphthalein indicator ($C_2OH_{14}O_4$); barium hydroxide ($Ba(OH)2.8H_2O$); potassium hydroxide (KOH); potassium hydrogen phthalate ($C_8H_5KO_4$). To prepare the solution of 0.1 NKOH are weighed 5.6g of potassium hydroxide. The solution is transferred to a 1000mL volumetric flask, where it is allowed to stand for 24 hours.

After this time is added 2mL of barium hydroxide to this solution is allowed to stand for 24 hours. Is added 2mL of a solution of barium hydroxide precipitation and if the solution is left standing for 24 hours. If there is no precipitation, the solution is filtered with Millipore filter assembly. Then collects the filtered solution to calculate the factorization of 0.1 N KOH, weighting 0, 3500g of potassium. Hidrogenphthalate is added to the flask 50mL of distilled water and six drops of phenolphthalein indicator. Titrate with 0.1 NKOH It is a white 50mL of distilled water added six drops of phenolphthalein indicator. Titrate again with 0.1 NKOH, recording the volume required. The calculation of the factorization is performed by the following formula:

$$N=\frac{\frac{P\times 9,99}{100}}{0,2042\times(A-B)}\times 56,1= \tag{3}$$

Where:

P = weight of potassium hydrogen phthalate (grams)

A = Volume of spent KOH titration of potassium hydrogen

B = Volume of spent KOH titration white

V = Normal Concentration

M. Eq. 56.1 KOH =

Purity = 99.9 hidrogenphthalate

Constant = 0, 2042

Constant = 100

In the titration, solvent is titrated in the absence of oil. Initially weighed in an Erlenmeyer ± 2.5g of oil. Added 50mL Erlenmeyer flask with the solvent in the oil, the measuring cylinder and 4 to five drops of phenolphthalein. It is a plug inserted into the Erlenmeyer flask (magnet) for mechanical agitation. Drops of KOH solution are added to the Erlenmeyer flask until the appearance of a slight pink tint. It is noted the amount of KOH. The calculation is done by the NHS expressed below formula and the result is displayed in mg KOH/g.

$$\frac{\text{Factor (KOH Volume - Volume of White)}}{\text{Weight of the sample}} \tag{4}$$

2.2.1.3. Residue

The residue content of the oils was examined in a model centrifuge. 215, brand FANEM, voltage of 220 V. Two tubes are filled with 100 mL sample and then are placed in a centrifuge. The process is centrifuged for 30 minutes at 1,500 rpm.

The result of the residue content is presented in percentage (%).

2.2.1.4. Moisture content

The moisture content of fresh oils, untreated post-consumer and post-consumer treated were analyzed by means of a water condenser with the heating mantle, make Quimis, 220V, Q.321.24 model. Initially, it is checked whether the oil for contamination by water, through the test on a hot plate apparatus. Oil drops are dripped with a glass rod and verifies whether precipitation occurred this oil, i.e., if is detected the presence of water. After this procedure, other tests are initiated to know the amount of oil contamination by water. 0.01mL of sample and 100mL of Xilou into a 500mL flask are added. The water condenser is turned on and starts heating to a temperature of 150°C and adjusted so as to provide reflux for 2 to five drops/second. The process of distillation continued/continues until no more water appears nowhere in the unit, except in the collector. After distillation, the collector is cooled to room temperature. After this process, the reading of the volume of water in the sink is performed. The moisture content is calculated by the following formula:

$$\frac{\text{Moisture content }(\%) = \text{Volume of water in the collector }(\text{in mL}) \times 100}{\text{Sample Volume }(\text{mL})} \tag{5}$$

The results of moisture content are presented in %

Tests of kinematic viscosity, acidity levels, residue and moisture were conducted in the laboratory LUBECLEAN- Distributor Cleansing and Lubricants LTD, located in João Pessoa – Paraíba.

3. Results and discussion

3.1. Analysis of kinematic viscosity, acidity, residue and moisture content

The procedures adopted in this study allowed analyzing comparatively the results Tonsil and Aporofo bentonite clay in viscosity, the levels of acidity, residue and moisture from fresh vegetable oils, untreated post-consumer and post-consumer treaty Table 1 illustrates these results.

As can be seen in Table 1, the oil treated with Tonsil clay had a lower viscosity compared to the crude oil and indicating the possibility of use as a biofuel is also a better efficiency than the Aporofo clay. The viscosity is a measure of the internal resistance to flow of a liquid, is an important property of vegetable oils because its control feature is intended to maintain its lubricating the engines, as well as proper operation of systems and injection pumps fuel [7]. Values above or below the viscosity specified by the NAP (National Agency of Petroleum, Natural Gas and Biofuels) range can lead to excessive wear on parts self - lubricating the injection system, with an increased work and leak in the fuel pump, as well as providing inadequate fuel atomization, incomplete and with consequent increase in the emission of smoke and particulate matter combustion [42].

	Fresh vegetable oil	Un-treated post consumption vegetable oil	Post consumption vegetable oil treated with Tonsil clay	Post consumption vegetable oil treated with Aporofo clay
Viscosity (mm^2/s)	34,57	35,45	32,47	35,33
Acidity (mg KOH/g)	0,64	2,90	2,44	2,27
Residue (%)	0,0	0,1	0,0	0,0
Moisture content (%)	0,0	0,0	0,2	0,3

Table 1. Viscosity values, levels of acidity, and the residue moisture of fresh vegetable oils, untreated post-consumer and post-consumer and treated with Tonsil and Aporofo clay.

Viscosity is a measure of the resistance offered to the flow of diesel engines. The key is to provide a proper atomization of the oil and preserve its lubricating characteristics. Kinematic viscosity of the biodiesel increases with increasing carbon chain and is inversely proportional to the number of unsaturation present thereon. For the same unsaturated compound has a higher viscosity dependence on the configuration of the double bond (cis or trans) than the position of the same [43].

The viscosity directly influences the atomization; that is, the higher the viscosity, the greater the average size of the droplets of the fuel sprays in the combustion chamber. Accordingly, larger droplets resulting in poorer and slower burning mixtures damage the ignition and combustion efficiency. Therefore, an increase in viscosity increases the time delay in the ignition cycle diesel engines.

The viscosity of the biodiesel with increasing carbon chain length and degree of saturation and influences the combustion process in the combustion chamber of the engine. High viscosity causes heterogeneity in the combustion of biodiesel, due to decreased efficiency of atomization in the combustion chamber, causing the deposition of waste in the internal parts of the engine [45].

In Brazil, there was an attempt to make a single specification for biodiesel (B100) similar to those already existing in some countries. However, there are characteristics that differentiate Brazil, a tropical country, countries like those in Europe: the temperature, which is high during most of the year and another important factor, is that in Europe reproduces biodiesel from oils of a single species vegetable, rapeseed. As in Brazil a wide diversity of species of potential use for the production of oils in many cases it is impossible to attain the viscosity values as specified. The specifications for conventional diesel and biodiesel in Brazil are: viscosity at 40°C - NAP 310/01 (diesel) and NAP 255/03 (biodiesel): 2.5-5.5 mm^2/s. The viscosity of the biodiesel is considerably reduced compared to when the source oil is passed to a subsequent process which is the transesterification reaction (which is the step of conversion of the oil or fat to methyl or ethyl esters of fatty acids, which is the biodiesel) [46].

With respect to acid content found in the samples studied after consumption of processed vegetable oils, can be seen a decreased value compared to vegetable oil consumption without post-treatment. NAP Ordinance 42/2004 establishes a value of ≤ 0.80 mgKOH/g for biodiesel (B100) [42]. However, even if the treated vegetable oils do not present results of acidity within the values established by the NAP, these oils must undergo a subsequent treatment process that is the transesterification reaction. The acidity can partly reveal the condition of the oils and fats because the higher the number, the greater the hydrolysis of frying oil, with a consequent increase in fatty acid content. The ideal is that the oil is index less than 2mgKOH/g acidity, in order to have a good reaction yield in the production of biodiesel and also to avoid problems in diesel engines [5].

Argue that the acidity can be defined as the mass (in mg) of potassium hydroxide required to neutralize the free fatty acids not esterified. He also reveals the conservation status of biodiesel, because the hydrolysis of esters occurs with consequent lowering of pH due to the increased content of fatty acids over time [47].

The condition of the oil is closely related to the nature and quality of the raw material, with the quality and purity of the oil with the processing, and especially with the storage conditions. The breakdown of triglycerides is accelerated by heat and light as rancidity is almost always accompanied by the formation of free fatty acid.

The monitoring of acidity in the biodiesel is of great importance during storage, in which the change of values in this period can mean the presence of water. All the rules described above established maximum acidity of 0.5 mg KOHg [45]. The acidity of biodiesel produced, measured in mg KOH per gram of sample, must be strictly within fixed parameters. The National Petroleum Agency recommends an acid, for any sample of biodiesel produced, less than 0.80 mgKOH/g.

A measure of acidity is a variable directly related to the quality of the oils, processing and storage conditions. According to Ordinance No.482 of National Health Surveillance Agency (NHSA 2), the acidity is one of the quality characteristics of various vegetable oils. The lipid materials undergo chemical changes during still in use as heat transfer medium. The acidity is determined by the amount of base required to neutralize the free fatty acids, the acidity increases with the deterioration of the oil during the exchange of heat, such as the frying process, for example [48].

High levels of acidity have a very adverse effect on the oil quality, as to make it unfit for human consumption or even for fuel purposes. Furthermore, the pronounced acidity oils can catalyze intermolecular reactions of triglycerides while affecting the thermal stability of the fuel in the combustion chamber. Also, in the case of use of the fuel oil has a significant free acidity corrosive action on the metallic components of the engine. The acid is a crucial examination for oil and biodiesel since high acidity reaction makes it difficult to produce biodiesel, biodiesel while an acid may cause corrosion of the engine or deterioration of biofuel [8].

An analysis of the results, it is found that soybean oil, despite the relatively high level of acidity, which is a limiting factor of process yield, was higher in the transesterification reaction of the mixture of soybean oil/hydrogenated fat. The yields obtained with either soybean oil as with the mixture, indicate that these materials have significant potential for the production of methyl esters of fatty acids. The positive aspect of biodiesel can be explained by the fact of not having nitrogen or sulfur in structures. Thus, they do not contribute to the acidification of precipitation [13].

The contents of waste clay substantially removed the impurities that were present in the oils without treatment, since the oil recovery achieved by means of a first step which is the bleaching process, is one of the most important steps in refining vegetable oils and has a fundamental role to eliminate substances which color instability and oil (residue). In this step, the decolorization of the oil occurs through the adsorption of pigments, which can be done using clays as adsorbent material [24].

The method of treatment is aimed at reducing the amount of impurities and substances which color the oil. Many of these substances act as catalysts of undesirable reactions, such as oxidation, interfering negatively on the physicochemical characteristics of the oil [24].

With respect to moisture content, all results presented are within the specifications of the NAP. Claim that the oil quality influences the transesterification reaction. So is that the oil is ideal index less than 2.0mgKOH/g of oil and moisture content below 0.5% for the purpose of a good reaction yield in obtaining biodiesel acidity and avoids problems microemulsion, corrosion, among others [49].

With respect to the final destination of clays retained on the filter, these can be applied in layers of compression earthworks or can be converted to non-polluting clusters applicable in the formation of landfills [50].

4. Conclusions

- Tonsil bentonite clay had a potential post-treatment in vegetable oil consumption more efficiently compared with the Aporofo bentonite clay, since clay removed many impurities from the oil.

- The post-consumer oil treated with Tonsil bentonite clay showed a viscosity less than the fresh oil and post-oil consumption without treatment, indicating an excellent adsorptive activity of the clay and the possibility of potential use oil as biofuel.

- Vegetable oils treated clay had a water content lower than after consumption untreated oil acidity.

http://dx.doi.org/10.5772/59874

- The oils treated and characterized for the levels of moisture and residue showed values within the standards established by the NAP.
- The post-consumer vegetable oils showed similar sensory and physicochemical characteristics, but differed substantially from fresh oil.
- After a period of storage, vegetable oils treated post-consumer showed stabilization of properties according to the requirements for processed oil.
- Tonsil bentonite clay may be considered more efficient for the post-consumer treatment vegetable oils since they showed better results than those treated with Apororofo clay.

Acknowledgements

The authors thank the Federal University of Campina Grande, the Graduate Program in Science and Engineering of Materials, the Company LUBECLEAN - Distributor Cleansing and Lubricants LTD, the BENTONISA - Bentonite Northeast S/A and CNPq for fellowships and financial support.

Author details

Elaine Patrícia Araújo[1*], Divânia Ferreira da Silva[1], Shirley Nobrega Cavalcanti[1], Márbara Vilar de Araujo Almeida[2], Edcleide Maria Araújo[1] and Marcus Vinicius Lia Fook[1]

*Address all correspondence to: elainepatriciaaraujo@yahoo.com.br

1 Department of Materials Engineering, Federal University of Campina Grande, Brazil

2 Department of Civil Engineering, Federal University of Campina Grande, Brazil

References

[1] Hocevar, L. Biocombustível de óleos e gorduras residuais - A Realidade do sonho. In: II Congresso Brasileiro de plantas oleaginosas, óleos, gorduras e biodiesel, Varginha, São Paulo, p. 953-957, 2006.

[2] Bhattacharya, A. B., Sajilata, M. G., Tiwari, S. R., Singhal, R. S. Regeneration of thermally polymerized frying oils with adsorbents. Food Chemistry. v.110, p. 562-570, 2008.

[3] Torres, E. A., Chirinos, H. D., Alves, C. T., Santos, D. C., Camelier, L. A. Biodiesel: o combustível para o novo século. Bahia Análise & Dados. Salvador, v. 16, p.89-95, 2006.

[4] Lapuerta, M.; Armas, O.; Ballesteros, R.; Fernández, J.; Diesel emissions from biofuels derived from Sapinsh potential vegetable oils. Fuel, China, v. 84, p.773-780, 2005.

[5] Dantas, H. J., Candeia, R. A., Conceição, M. M., Silva, M. C. D., Santos, I. M. G., Sousa, A. G. Caracterização físico-química e estudo térmico de biodiesel etílico de algodão. http://www.biodiesel.gov.br (accessed 28 August 2008).

[6] Murillo, S., Míguez, J. L., Porteiro, J., Granada, E., Morán, J. C. Performance and exhaust emissions in the use of biodiesel in outboard diesel engines. Fuel, v.86, p. 1765-1771, 2007.

[7] Neto, P.R.C., Rossi, L. F. S., Zagonel, G. F., Ramos, L. P. Produção de biocombustível alternativo ao óleo diesel através da transesterificação de óleo de soja usado em frituras. Química Nova. São Paulo, v.23, 19p. 2000.

[8] Dantas, H. J. Estudo Termoanalítico, Cinético e Reológico de Biodiesel derivado do Óleo de Algodão (*Gossypium hisutum*). 122f. (MS in Química/Química analítica), Centro de Ciências Exaras e da Natureza. Federal University of Paraíba, João Pessoa, 2006.

[9] Graboski, M. S., Mccormick, R. L. Combustion of fat and vegetable oil derived fuels in diesel engines. Prog. Ennerg. Combust. Sci. Colorado, v.24, n.1, p. 57-64, 1998.

[10] Castellanelli, C.A. Study the feasibility of producing biodiesel, obtained through the used frying oil in the town of Santa Maria - RS. p. 1-30. Dissertation (MS in Production Engineering), the Engineering Institute of Paraná. Federal University of Santa Maria, Santa Maria, 2008.

[11] Issariyakul, T., Kulkarni, M. G., Meher, L. C., Dalai, A. K., Bakhshi, N. N. Mixtures of Biodiesel production from canola oil and used cooking oil. Chemical Engineering Journal. Canada, V.140, p.77-85, 2008.

[12] Alberice, R. M.; BRIDGES, F.F.F. Recycling used cooking oil through the manufacture of soap. Environmental Engineering. Espírito Santo do Pinhal, v. 1, n.1, p. 073-076, 2004.

[13] Dorado, M. P., Cross, F., Palomar, J. M., Lopez, F. J. In approach to the economics of two vegetable oil based biofuels in Spain. Renewable Energy. v. 31, p. 1231 - 1237, 2006.

[14] Vecchi, C. C. C., Tarozo, R., Pinto, J. P., Faccione, M., Guedes, C. L. B. Thermal and photochemical degradation process of biodiesel from soybean oil. In: international Congress of r&d in Oil and Gas, 3rd, 2005, Salvador. Proceedings of the 3rd Brazilian Congress of R & D in Oil and Gas, Brazil, 6p, 2005.

[15] Texeira-Neto, E., Teixeira-Neto, A. Chemical modification of clays: scientific and technological challenges for obtaining new products with higher added value. Quim. Nova. São Paulo, v. 32, n. 3, p. 809-817, 2009.

[16] Sanibal, E. A. A., Filho, J. M. Physical, chemical and nutritional oils subjected to frying process changes. Notebook Technology Food & Beverage. Sao Paulo, p. 48-54, 2008.

[17] Vergara, P., Wally, A. P., Pestana, V. R., Bastos, C.; Zambiazi, R. C. Study of the behavior of soyabean oil and rice reused in successive fried potatoes. B. CEPPA, v. 24, p. 207-220, 2006.

[18] AL-Kahtani, H. A. Survey of quality of used frying oils from restaurants. J. Am. Oil Chem. Corporation. Champaign, v.68, n. 11, p. 857-862, 1991.

[19] Corsini, M. S., Jorge, N. Alterações Oxidativas em óleos de algodão, girassol e palma utilizados em frituras de mandioca palito congelada. Alim. Nutr. Araraquara, v.17, p. 25-34, 2006.

[20] O'Brien, R. D. Fats and oils: formulating and processing for applications. Technomic Publishing. Lancaster, p. 385 – 410, 1998.

[21] Tiritan, M. G., Ferreira, E. S. Produção de biocombustível de oleo de fritura com etanol hidratado utilizando planejamento fatorial. In: Anais II Congresso da Rede Brasileira de Tecnologia de Biodiesel, Brasília, 2007.

[22] Silva, C. V. Reaproveitamento do óleo de cozinha como tema nas aulas de educação ambiental. 2010. 48f. (Graduate in Química), Centro de Ciências e Tecnologia, Universidade Estadual da Paraíba, PB, Brasil.

[23] Mothé, C. G., Correia, D. Z. In: Anais do 2° Congresso Brasileiro de Plantas Oleaginosas, Óleos, Gorduras e biodiesel, Varginha, p. 547-551, 2005.

[24] Oliveira, C. G. Proposta de modelagem transiente para a clarificação de óleos vegetais- experimentos cinéticos e simulação do processo industrial. 164p. Dissertação (MS in Engenharia Química), Centro Tecnológico. Universidade Federal de Santa Catarina, Florianópolis, 2001.

[25] Firestone, D., Stier, R. F., Blumenthal, M. M. Journal of Food Technology. p. 90-94, 1991.

[26] Baraúna, O. S. Pigment adsorption process with vegetable oil acid-activated smectite clays.173P. Thesis (Ph.D. in Chemical Engineering), School of Chemical Engineering. State University of Campinas, Campinas, 2006.

[27] Ramalho,V. C., Jorge, N. Ação antioxidante de a-tocoderol e de extrato de alecrim em óleo de soja submetido à termooxidação. Química Nova. São Paulo, v 29, p. 755-760, 2006.

[28] Cella, R. C. F., Reginato, D. A. B.; Spoto, M. H. F. Ciência e Tecnologia de Alimentos. São Paulo, p.1001-1007, 2005.

[29] Daily Borborema. Biodiesel from frying oil. Campina Grande, December 6, 2008.

[30] Santos, P. S., Science and Technology of Clays. 2 ed. São Paulo: Editora Edgard Blücher Ltda, 1989. 408p.

[31] Dutra, R. P. S., Varela, M. L., Birth, R. M., Gomes, U. U., Paskocimas, C., Melo, P. T. Evaluation of the potential of clays of Rio Grande do Norte - Brazil. Industrial ceramics. São Paulo, v. 11, p. 42-46, 2006.

[32] Santos, P. S. Ciência e Tecnologia de Argilas. 2 ed. São Paulo: Editora Edgard Blücher Ltda., 1992. p. 650-673.

[33] CETEM. Centro de Tecnologia Mineral. Bentonita. Rio de Janeiro, p. 217-230, 2005. (Boletim Técnico,115-00).

[34] Amorim, L. V., Viana, J. D., Farias, K. V., Barbosa, M. I. R., Ferreira, H. C. Comparative study between varieties of bentonite clays of Boa Vista, Paraíba. Matter Magazine. Rio de Janeiro, v. 11, p. 30-40, 2006.

[35] Leite, I. F. Preparação de nanocompósitos do poli (tereftalato de etileno)/Bentonita. 86f. Dissertação (MS em Ciência e Engenharia de Materiais), Centro de Ciências e Tecnologia. Universidade Federal de Campina Grande, Campina Grande, 2006.

[36] Barbosa, R; Araújo, E. M, Melo, T.J.A., Ito, E.N. Preparação de argilas organofílicas e desenvolvimento de nanocompósitos de polietileno. Parte2: Comportamento de Inflamabilidade. Polímeros: Ciência e Tecnologia. São Carlos, v.17, p.104-112, 2007.

[37] DNPM, Departamento Nacional de Produção Mineral, Bentonita, Sumário Mineral Brasileiro, 2007.

[38] Silva, A. R. V., Ferreira, H. C. Argilas bentoníticas: conceitos, estruturas, propriedades, usos industriais, reservas, produção e produtores/fornecedores nacionais e internacionais. REMAP-Revista Eletrônica de Materiais e Processos. Campina Grande, v. 3.2, p. 26-35, 2008.

[39] Barbosa, R. Effect of quaternary ammonium salts on organophilization of national bentonite clay nanocomposites for the development of high density polyethylene (HDPE). p. 7-12. (MS in Chemical Engineering), Centre for Science and Technology. Federal University of Campina Grande, Campina Grande, 2005.

[40] Coelho, A. C. V., Santos, P. S., Santos, H. S. Argilas especiais: argilas quimicamente modificadas - uma revisão. Química Nova. São Paulo, v. 30, p.1282-1294, 2007.

[41] Foletto, E. L., Volzone, C., Morgado, A. F., Porto, L. M. Influência do tipo de ácido usado e da sua concentração na ativação de uma argila bentonítica. Cerâmica. São Paulo, p. 208-211, 2001.

[42] Melo, J. C., Teixeira, J.C., Brito, J. Z, Pacheco, J. G. A., Stragevitch, L. Production of biodiesel from myrtle oil. Biodiesel. Available at: www.periodicosdacapes.com.br. (accessed 10 September 2012).

[43] Knothe, G., Steidley, K.R. Fuel 84 (2005) 1059.

[44] Teixeira, C. V., Colaco, J. M. Viscosity and performance of diesel/biodiesel blends on a single cylinder engine. http://rmct.ime.eb.br/arquivos/RMCT_1_tri_2013/RMCT_007_E4A_11.pdf. (accessed 10 April 2014).

[45] Lobo, I. Ferreira, S. P. Biodiesel: parameters of quality and analytical methods. Quim. Nova, Vol 32, Nc. 6, 1596-1608, 2009.

[46] Gomes, L. F. S. Potencial de produção de biodiesel a partir do óleo de frango nas cooperativas do Oeste do Paraná. 81f. (MS in Engineering Agrícola), Centro de Ciências Exatas e Tecnológicas. Universidade Estadual do Oeste do Paraná, Cascavel, 2005.

[47] Vasconcelos, A. F. F.; Dantas, M. B.; Lima, A. E. A.; Silva, F. C.; Conceição, M. M.; Santos, I. M. G.; Souza, A. G. Compatibilidade de misturas de biodiesel de diferentes oleaginosas. http://www.biodiesel.gov.br. (accessed 10 March 2013).

[48] Leung, D.Y.C., Guo, Y.Transesterification of used frying neat and oil: optimization for biodiesel production, Fuel Processing Technology, V.87, p. 883-890, 2006.

[49] Santos, R. B. dos, Serrate, J. W., Caliman, L. B., Lacerda, J. R, V.; Castro, E. V. R. Avaliação do uso de óleo residual usado em fritura para a produção de biodiesel e estudo da transesterificação de óleo de soja com álcoois de cadeia de até quatro carbonos. http://www.periodicoscapes.gov.br. (accessed 03 August 2011).

[50] Ramos, A. A. P Study on the recycling process of industrial lubricant level. 67p. (Graduate Mechanical Engineering), Faculty of Engineering. Pontifical Catholic University of Rio Grande do Sul, Porto Alegre, 2008.

Section 4

Indoor Air Pollution

Chapter 21

Indoor air Quality Improvement Using Atmospheric Plasma

Kazuo Shimizu

Additional information is available at the end of the chapter

http://dx.doi.org/10.5772/60138

1. Introduction

Air pollution is serious problem for human health and for the environment including causing global warming. An interdisciplinary collection of new studies and findings regarding air pollution was previously published [1]. Governments around the world are managing air quality in their countries for the health of their citizens. The management of air pollution involves understanding sources of air pollution, monitoring contaminants, modeling air quality, performing laboratory experiments, controlling indoor air pollution, and eliminating contaminants through various methods. Research activities are carried out on every aspect of air pollution and its control throughout the world to respond to public concerns.

Another book raised concerns about environmental air pollution including indoor air pollution as well. In recent years, the deterioration of indoor air quality (IAQ) has become a concerning issue [2]. Modern construction methods make the spaces inside buildings more air tight. And, various harmful chemicals are used in building materials. Among the pollutants are volatile organic compounds contained in coated materials and adhesives. While activated charcoal filters used in homes and buildings can remove organic molecules, they have a limited life-time and they must be replaced regularly to maintain good performance.

Another pollutant is particulate matter ranging from sub-micron to few micron meters including dust pollution caused by cigarette smoke, house dusts, and various fungi caused by hot and humid air. Air filters have been used to improve indoor air quality for many years. However, most filters used in homes have poor collection efficiencies for smaller particles (less than 10 micrometer in diameter). This low collection efficiency is a problem, especially for 2nd hand smoke, dust and pollen particles [3-5].

Living microorganisms such as bacteria and viruses also contaminate the indoor air. They may bring on respiratory disorders, asthma, or influenza. And, activated charcoal filters can incubate bacteria. So the other sterilization method is required for living microorganisms or virus.

Finally, while sterilization devices for indoor air may be commonly used in medical facilities (hospitals, clinics, etc.), these devices are uncommon for use in homes and general office buildings. While these issues could be eliminated by natural air flow, this method is energy inefficient and many houses cannot use this method. Because we spend most of our time indoors, indoor air quality is an important factor for healthy and comfortable lives.

From these points of view, there is a need for reliable devices for the home that can remove organic pollution and that can sterilize indoor air. Devices using atmospheric pressure plasma technologies, especially microplasmas, are very promising.

This chapter deals with improving indoor air quality using atmospheric plasma treatment [6]. Discussed are the results of a series of experiments on particulate matter (PM) precipitation and removal, odor control targeting ammonia and sterilization of *E. coli.* While such experiments are often performed in a small experimental chamber, these experiments were carried out in the relatively large space (23.4 m^3) shown in Fig. 1. These results reveal the performance of commercially available air treatment devices.

Figure 1. The experimental room for measuring indoor air quality improved by atmospheric plasma.

2. Experimental set up for PM precipitation

2.1. About particulate matter in the indoor air environment

Indoor particulate matter is a mixture of substances including carbon (soot) emitted by combustion sources, tiny liquid or solid particles in aerosols, fungal spores, pollen, and toxins present in bacteria (endotoxins). In many homes, most of the airborne particulate matter comes from the outside. However, some homes do have significant sources of indoor particulate matter that comes from various sources: cigarette smoking is the greatest single source of particulate matter in homes and buildings where people smoke; cooking: especially frying and

http://dx.doi.org/10.5772/60138

sautéing; combustion appliances: for example, furnaces without a proper air filter; non-vented combustion appliances like gas stoves; wood-burning appliances like wood stoves and fireplaces: especially if the smoke leaks or *backdrafts* into the home; and mold growth [7].

Since indoor air contains various types of particulate matter, improving air quality requires a combination of high performance filters and active treating devices such as electrostatic precipitators and atmospheric plasma devices [8, 9].

2.2. Electrodes for particulate matter collection

The reactor shown in Fig. 2 consisted of a corona discharge needle type electrode system, a ground electrode, and a mesh filter electrode for collecting particulate matter. The needle type electrode system shown in Fig. 2 consisted of 5 parallel wires to which 4 needle type electrodes were attached to each wire. The diameter of the electrode assembly was 55 mm. The ground electrode consisted of parallel wires positioned at a discharge gap of about 5mm from needle type electrodes. At the outlet, a mesh filter placed 5mm from grounded electrode captured particles.

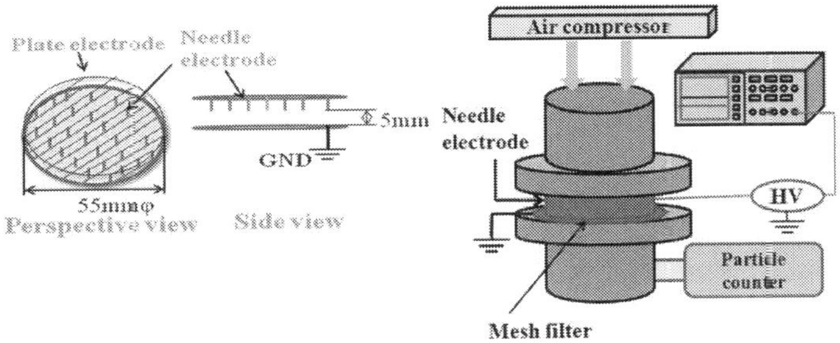

(a) A structure of the needle electrodes.

(b) Reactors consisted by the needle electrodes and mesh filters.

Figure 2. Schematic diagram of the needle electrodes and the reactor system with the mesh filters.

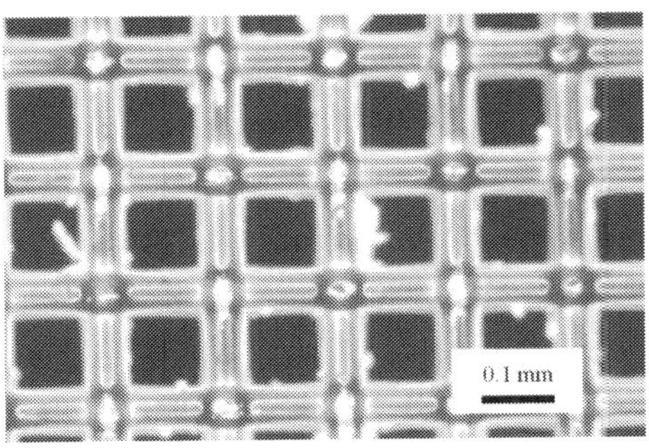

Figure 3. Mesh filter used for collecting dust. (x100)

Mesh filters like the one in Fig. 3 were used in this study. One side of the filters were coated by metal. The characteristics of the mesh filters are given in Table 1.

Aperture size [mm]	0.1
Wire diameter [mm]	0.06
Mesh count [mesh /inch]	150

Table 1. Characteristic of mesh filters.

We used the large, experimental room in Fig. 1 to carry out our indoor air quality experiments using the reactor in Fig. 2. We measured particulate matter collection, and ammonia removal from the indoor air. The fan in the corner of the experimental room shown in Fig. 4 circulated indoor air at a flow rate of 4.6 m^3 / min.

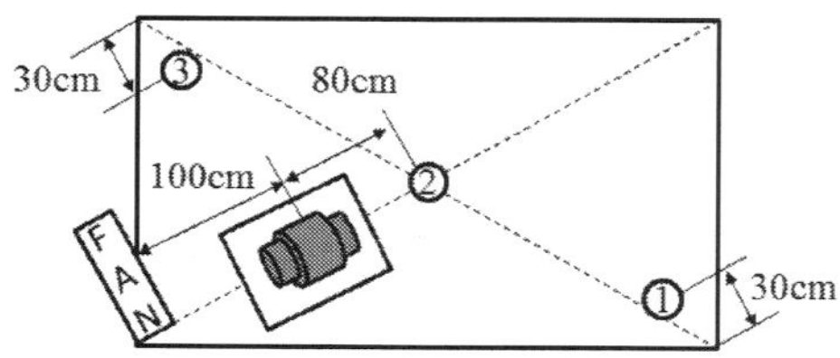

Figure 4. A reactor and a fan arrangement in the experimental room.

The reactor was placed 30 cm above the floor and 100 cm from fan. The gas flow through the reactor was about 112 L / min. The reactor was driven by a negative DC voltage of 5 kV.

To evaluate the collection of particulate matter, we took particle measurements at three locations #1, #2 and #3 in Fig. 4. We measured the number of particles using an optical particle counter (SHIMADZU, MODEL 3886). The diameter of the measured particles was 0.5 μm. During the measurement time of 2 hours, the relative humidity was 67% and the temperature was 23.7 C.

2.3. Microplasma generation for indoor air treatment

The microplasma electrode has been described in previous publications [6, 10]. In this study, the volume of the experimental room shown in Fig. 1 (23.4 m^3) was much larger that the experimental chambers used in previous studies. So the microplasma electrode in Fig. 5 was enlarged corresponding to the area.

The microplasma electrodes consisted of a high voltage electrode and a grounded electrode separated by a discharge gap in the range 30 to 100 m. Both electrodes were coated with dielectric materials. When an AC voltage in the range from several hundred volts to 1 kV was

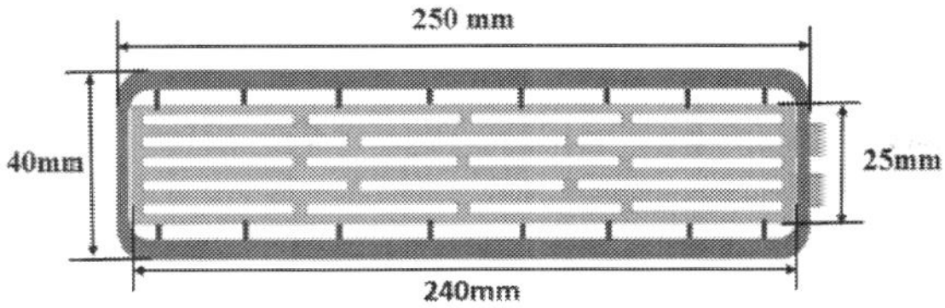

Figure 5. Microplasma electrodes utilized in this study.

applied to the high voltage electrode, numerous streamers were generated between the electrodes as shown in fig. 6 [11].

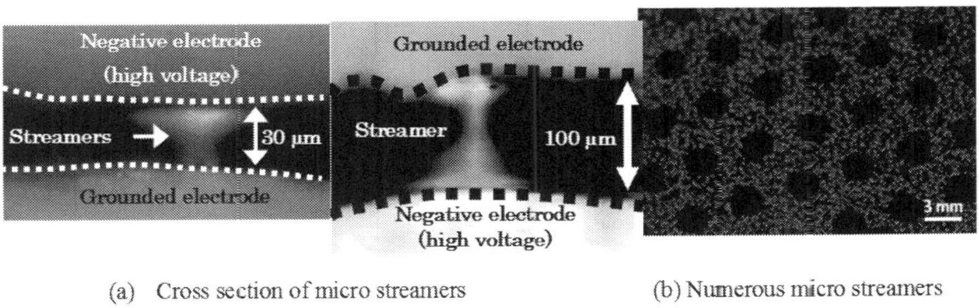

Figure 6. Generation of micro streamers (a); cross section, (b) front view.

The basic characteristics of a microplasma are similar to that of a typical dielectric barrier discharge (DBD) that typically has a discharge gap of 1mm or more. However, the diameters of streamers in microplasmas were observed to be in the range 10-20 um, which is narrower than that of DBD streamers [11]. The numerous microplasma streamers generate various active species, such as radicals, ions, and ozone that enhance the chemical reactions in the gas phase describing the next section.

3. Results and discussion

3.1. PM collection by the needle electrodes with the mesh filter

The particle concentrations in Fig. 7 were measured in the experimental room while applying a negative DC voltage of 5 kV.

The measured particle concentrations at the 3 measurement points shown in Fig. 4 were consistent indicating that the air in the experimental room was well mixed by the fan.

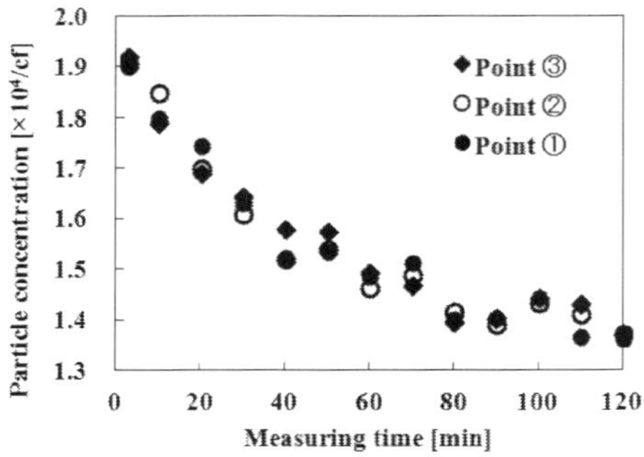

Figure 7. Particle concentration versus treatment time at 3 measurement points in the experimental room.

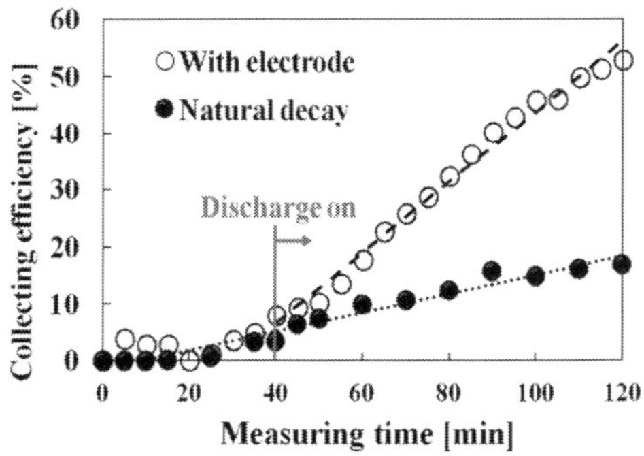

Figure 8. Decrease of the particles in the experimental room measured at position #2.

Airborne particles will deposit naturally on surfaces resulting in the natural decay of the particle concentration shown in Fig.8. When the power supply of the reactor was turned on for 40 minutes, the deposition of particles on the mesh filter within the reactor increased the decay rate compared to the natural decay (without any discharge). The collection efficiency of particle collection is calculated using equation (1) [12, 13].

$$\eta = \left(1 - \frac{N_A}{N_B}\right) \times 100(\%) \tag{1}$$

Particulate matter (PM) charged by the needle electrode was then trapped by the mesh filters. The effect of mesh filters and the discharge voltage, the number of filters and the flow rate have been previously published [14].

3.2. Odor removal for improving the indoor air quality

This section presents the removal process results from the experimental room (23.4 m^3) shown in Fig. 1 of chemical substances that deteriorate the indoor air quality. The removal processes target various chemical substances such as volatile organic compounds (VOCs) derived from the building materials, ammonia and others derived from the human being and animals as well. Treatment of VOCs by microplasmas has been published previously [6, 15].

We used the experimental setup shown in Fig. 9 to measure the removal of ammonia from indoor room air. The evaporating dish with 3.5ml of liquid phase ammonia (25%) diffused into the air forming a relatively high initial concentration of ammonia. After confirming the initial concentration of the gas phase, ammonia reached a constant value, the experiment room was ventilated, to evaporate the ammonia again for 10 minutes. Using this procedure, the initial concentration of ammonia was set to 25ppm, which is an acceptable concentration according to the guideline values. Ammonia concentration changes in the experimental room were measured for 2 hours with a negative voltage of 5 kV applied to the needle. The needle electrode was placed in front the fan shown in Fig. 4 at a distance of 100 cm. To enhance the ammonia removal reaction, a fine water mist (3 μm) was added by an ultrasonic wave humidifier [16, 17].

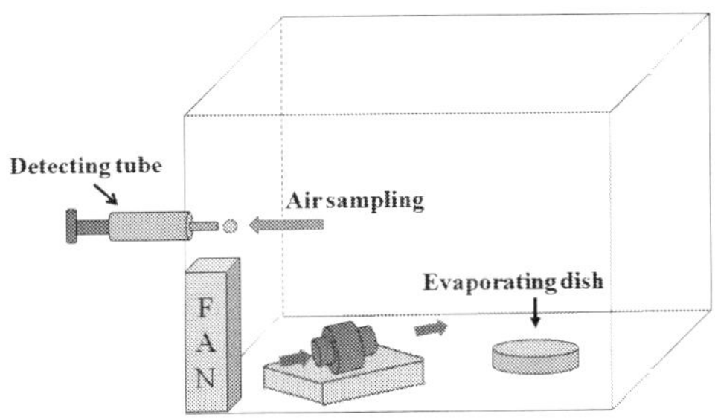

Figure 9. Experimental setup of ammonia removal in the experimental room (23.4 m^3)

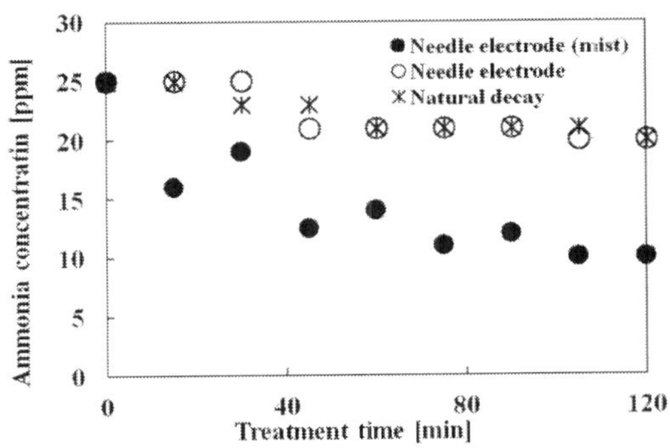

Figure 10. Ammonia concentration versus treatment time by the needle electrode.

During the 2 hours process, the ammonia concentration variation while energizing the needle electrode was the same as the concentration variation due to natural decay. This indicates that the needle electrode removed little ammonia. Note that the ozone concentration in the experimental room after two hours was 0.6 ppm. This relatively low ozone concentration resulted in no measureable reduction in the ammonia concentration. However, when the test was repeated with a fine water mist introduced into the needle electrode, ammonia removal reached 50%.

Well known gas phase reactions of ammonia with OH radicals and nitrogen oxides generate ammonium nitrate [18, 19].

$$NO{+}OH{+}N_2 \rightarrow O{+}O_2{+}N_2 \tag{2}$$

$$NO_2{+}OH{+}N_2 \rightarrow OH{+}_3{+}N_2 \tag{3}$$

$$NH_3{+}HNO_2 \rightarrow HN_4NO_2 \tag{4}$$

$$NH_3{+}HNO_3 \rightarrow HN_4NO_3 \tag{5}$$

These reactions suggest why introducing a fine water mist into the needle electrode greatly increased rate of ammonia removal.

3.3. Sterilization of *E. Coli*

E. coli may be sterilized by the various active species formed by a microplasma discharge such as ozone, ions, and radicals [20]. Sterilization of *E. coli* (Migula 1895) was carried out using microplasma electrodes in the experimental room shown in Fig. 1. *E. coli* deposited on stamp media agar "Tricolor (El Mex, Inc.)" were placed at the seven positions shown in Fig. 11 in the experimental room.

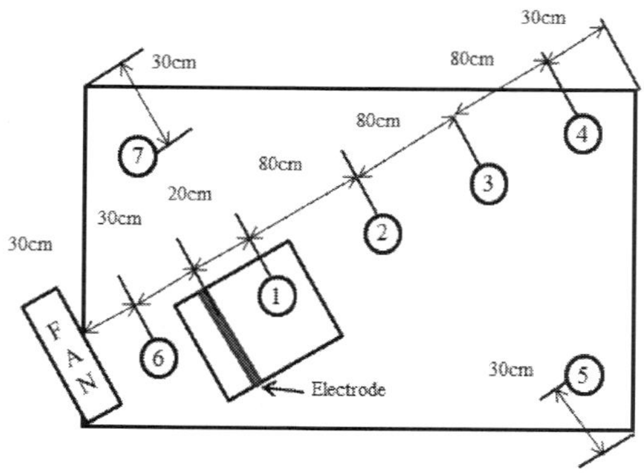

Figure 11. Position of the agar media placed in the experimental room.

The target bacteria were placed and treated with the microplasma for 2 hours. An 8th control medium with target bacteria was kept outside of the experimental room. Within the room, the microplasma electrodes were driven by a high-frequency alternating voltage of 27 kHz at voltages of 0.8, 0.9 and 1.0 kVp-p by an inverter and a neon transformer. After treatment, the 7 agar media and the 8th control medium were cultured for 24 hours in an incubator set to 37 $^{\circ}$C. The sterilizing rate was obtained based on the control colonies number compared with the colonies number placed in the each position 1 to 7.

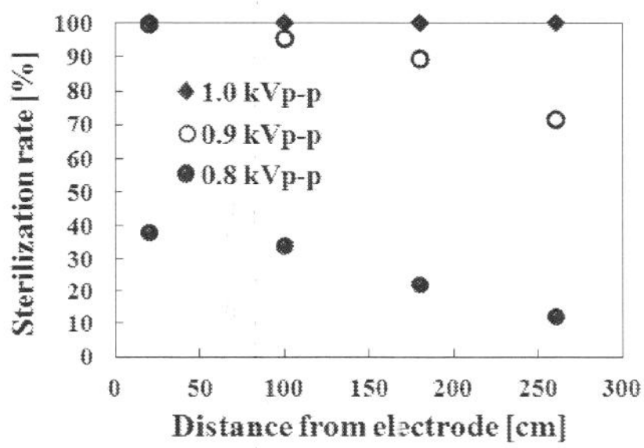

Figure 12. Sterilization rate for various discharge voltage at each position in the experimental room.

The sterilization rate increased with increasing voltage and decreased with distance from the electrode with an applied voltage of 0.8 and 0.9 kVp-p. There was no decrease of the sterilization rate with distance from the electrodes when applied voltage was 1.0 kVp-p. Increasing the voltage from 0.8 to 0.9 kVp-p resulted in a considerable increase in the sterilization rate. In the large experimental room, the effect of radicals is likely small since life time of radicals is short compared with the transit time of air moving from the reactor to the agar media test location. Consequently, radicals could not reach the target to sterilize [21, 22]. Figure 13 also shows the plasma treated *E. coli* for various distances from the electrode at applied voltage of 0.9 kVp-p.

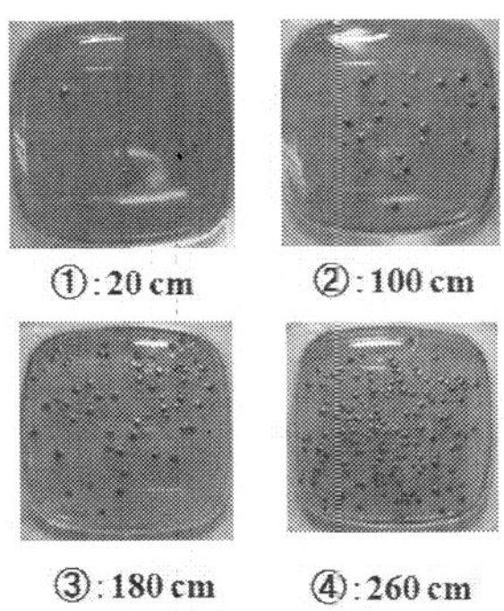

Figure 13. Plasma treated *E. coli* samples (Applied voltage 0.9 kVp-p).

We also measured the ozone and ions concentration at each position in Fig. 11 in the experimental room. Figure 14 shows the ozone and ions concentration of each position from the electrode, respectively. While the ozone concentration was very high at the outlet of the microplasma electrode, it decreased within a very short distance to a relatively uniform level that varied with voltage. On the contrary, ion concentrations varied strongly with position. This variation of ion concentrations with position was considerably different than the small chamber results. Ion concentrations depended on the distance from the microplasma electrode because ions have various reaction rate, and high reaction rate ions decayed with the distance by reacting with neutral molecules [23, 24].

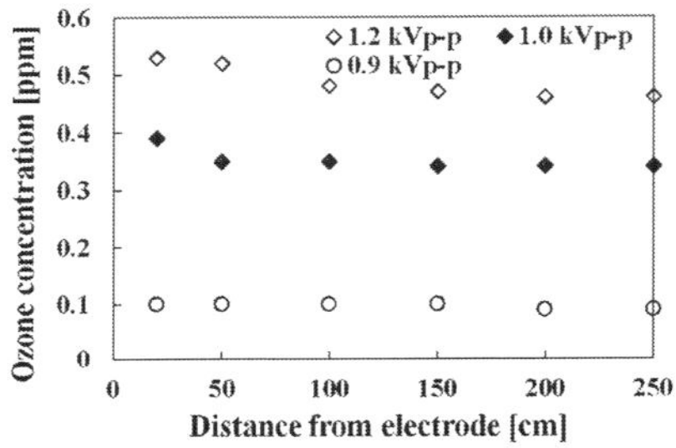

(a) Ozone concentration (Tr= 30 minutes)

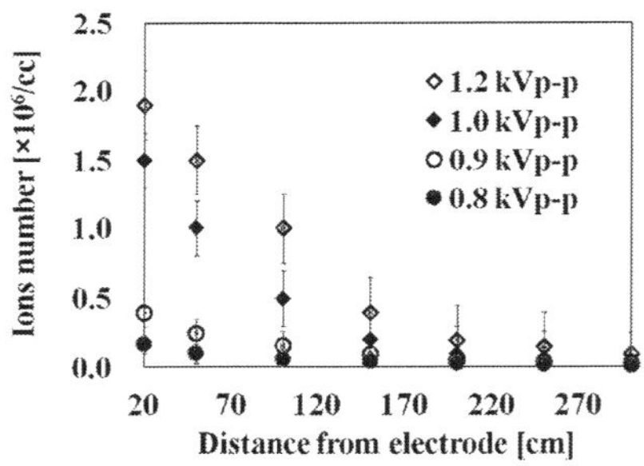

(b) Ion concentration (Tr= 30 minutes)

Figure 14. Ozone and ion concentration for various positions in the experimental room.

Even low concentrations of ozone in actual rooms could arise health issues [25, 26], so we investigated both ozone and ion concentration in the large experimental room. The variations of both ozone and ion concentrations are shown in Fig. 15. Ozone concentration increased with the process time, and exceeded the EPA regulatory [27] when the applied voltage was 0.9 kVp-p or higher. The concentrations of ions stabilized as shown in Fig. 15 during the process time (2 hours) at a level that depended on the applied voltage level.

The concentrations of ozone and ions shown in figures 14 and 15 suggest that the sterilization rates shown in Fig. 12 were related to the ozone concentration. However, relatively low

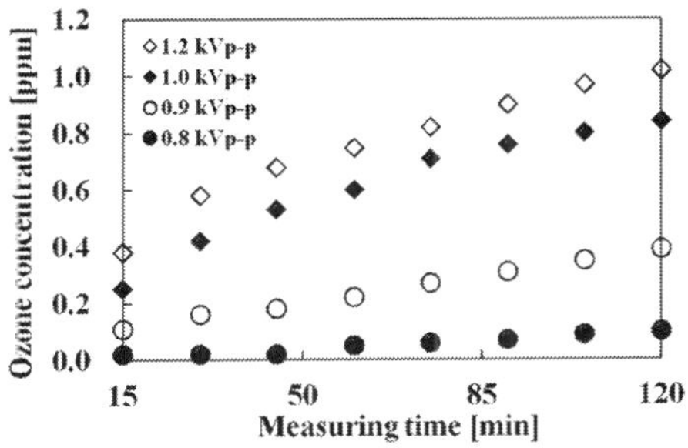

(a) Ozone concentration change (Position; 100cm away from electrodes)

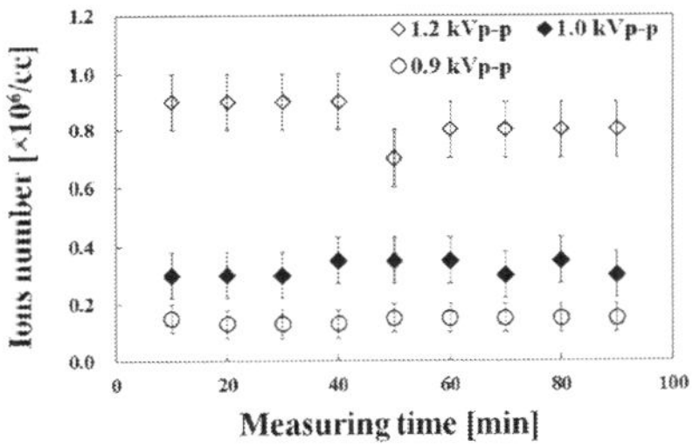

(b) Ions concentration change (Position; 100cm away from electrodes)

Figure 15. Variations in ozone and ion concentrations with microplasma treatment time in the experimental room.

concentrations of ozone are generally ineffective in sterilizing bacteria [28]. Similarly, the relatively low concentrations of ions diluted in the experimental room are also generally ineffective in sterilizing bacteria. Note that ions are only weakly reactive compared to highly reactive radicals. Consequently, there should be some reactions with reactive radicals that usually have short life time, such as OH radicals [21].

The ESR analysis of the agar medium in Fig. 16 with spin trap agent (5-dimethyl-1-pyrroline-N-oxide (DMPO)) indicate that the sterilization process may be the OH regeneration process shown in Fig. 17. Low concentration of ozone that reach the agar surface react as in equations (6) – (9) to generate OH radicals that sterilize the bacteria [29, 30].

$$e + O_2 + O y O(^3P) + O(^3P, ^1D) \tag{6}$$

$$O(^3P) + O_2 + M\ O)r_3 + M \tag{7}$$

$$O(^1D) + H_2O \rightarrow 2\ OH \tag{8}$$

$$O_3 + H_2OH)_2 + OH \tag{9}$$

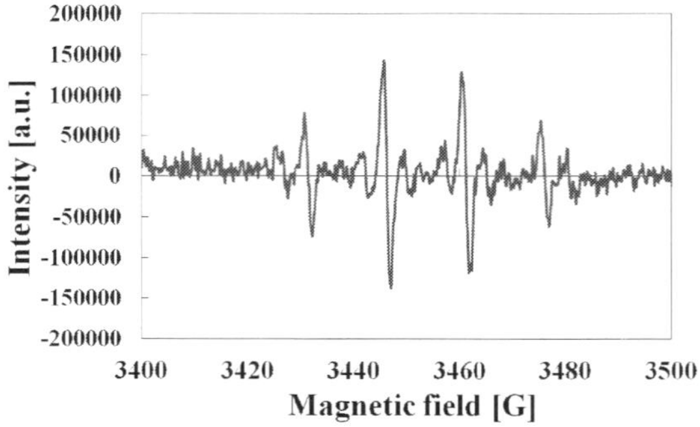

Figure 16. ESR analysis of the medium by plasma exposure.

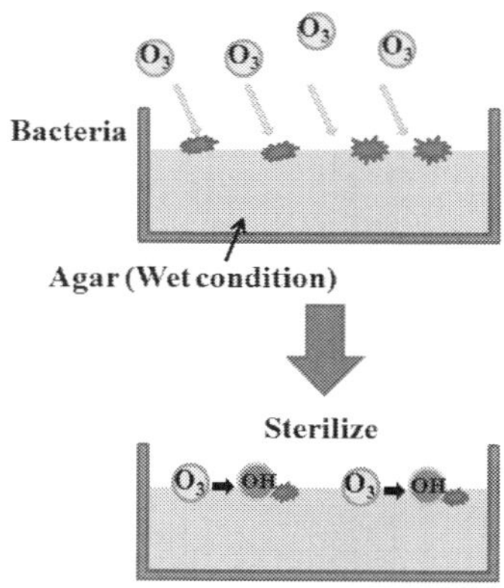

Figure 17. OH generation and sterilization process on the surface of the agar medium.

3.4. Sterilization of *E. coli* with ethanol and microplasma

In the large experimental room shown in Fig. 1, *E. coli* was sterilized by O_3 generated by a microplasma rather than by OH radicals whose short lifetime made them ineffective as a sterilization agent. We have enhanced the sterilization effect of bacteria in the small space experimentally as shown in Fig. 18.

A circular microplasma electrode (Diameter 50 mm) was installed in the experimental chamber having an inner diameter of 65 mm. *E. coli* was cultivated on the agar medium in the Petri dish. Air from the cylinder flowed at 10 L/min through a chamber containing liquid ethanol (70 vol %) that evaporated to generate 1.3% gas phase ethanol. The distance between the microplasma electrode and the Petri dish shown in Fig. 18 was 23 mm.

http://dx.doi.org/10.5772/60138

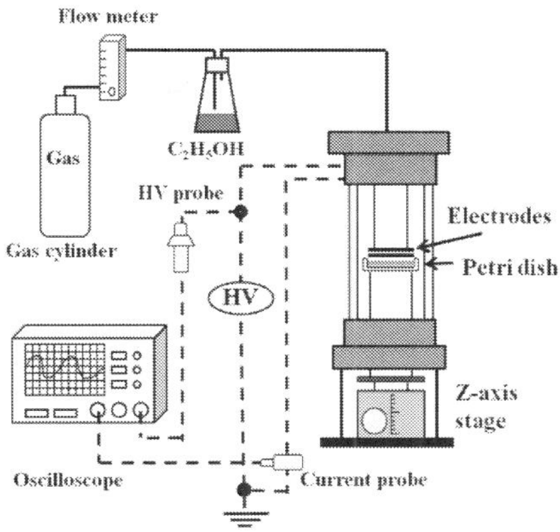

Figure 18. Experimental setup for sterilization of *E. coli* by microplasma with addition of ethanol.

With 1.3% gas phase ethanol, the microplasma sterilization process is greatly improved. Without microplasma treatment and only using ethanol, the decrease of bacteria colonies shown in Fig. 19 was minimal. With air microplasma treatment having no ethanol, the sterilization rate reached to 2digits (about 99%) as previously reported [6]. However, the microplasma treatment with 1.3% ethanol as an additive achieved a sterilization rate to 6digits with a 60 seconds treatment time.

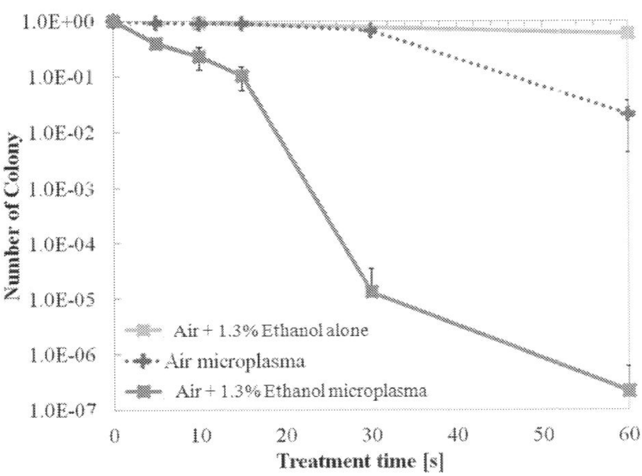

Figure 19. Sterilization of *E. coli* by microplasma and addition of ethanol.

For sterilizing E. coli in a large room such as the experimental room shown in Fig. 1, the life time of highly reactive radicals is too short for effective performance. Thus, experimental work was done to extend the distance between the microplasma electrode and the Petri dish in the small experimental chamber shown in Fig. 18. We found as shown in Fig. 20 that the sterilization performance of the microplasma with ethanol was effective at more than 20 cm away from the Petri dish. Relatively long life species must be generated by the microplasma with ethanol.

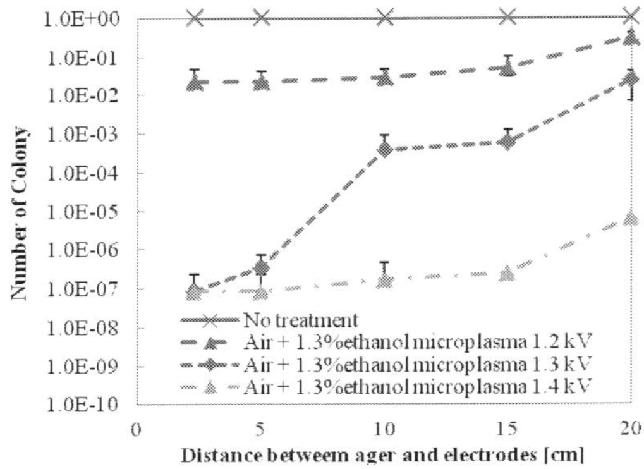

Figure 20. Sterilization rate dependencies with the distance between microplasma electrode and Petri dish.

Byproduct analysis of the process gas was carried out using FT-IR to identify the relatively long life species in the sterilization process. We found the undesirable byproducts shown in Fig. 21 with the ethanol additive process such as CO, and EOG (ethylene oxide gas). EOG is undesirable because it is well known to cause cancer. However, it should be noted that this process would be suitable for applications having a controlled room without any human [31].

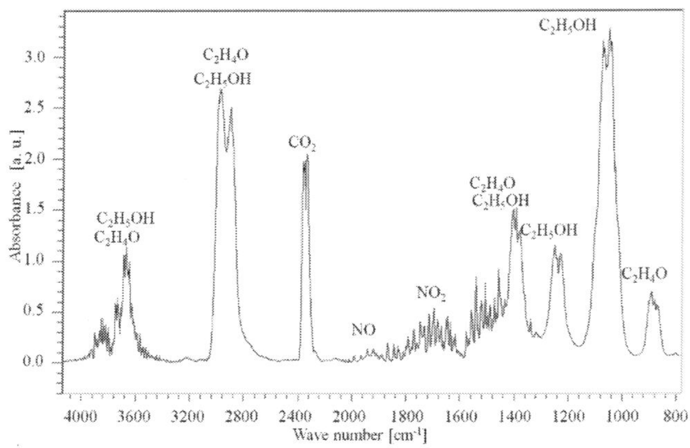

Figure 21. Byproduct analysis by FTIR for the microplasma treatment with addition of ethanol.

The shape of E. coli in the SEM images shown in Fig. 22 was completely changed by the microplasma process with ethanol addition. Note that this microplasma process is a remote plasma process. The electric fields and forces acting within the microplasma discharges are not acting on sample. Only various active species or chemical substances generated by the microplasma play a role in sterilizing the *E. coli.*

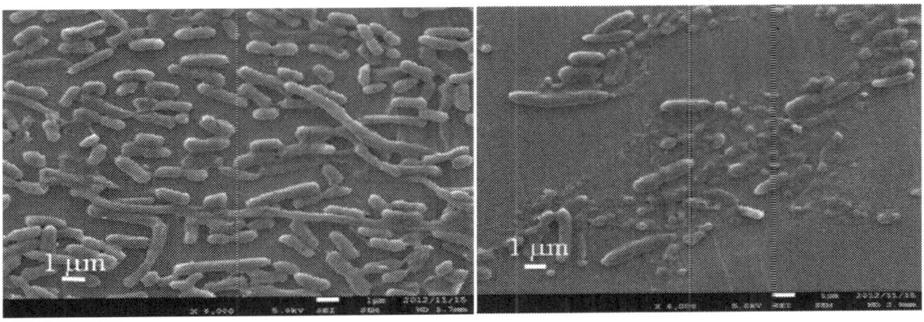

(a) Before treatment (b) after 60 s of microplasma exposure

Figure 22. SEM images of microplasma treatment with ethanol addition (Vp=1.4 kV, treatment time= 60 s., ethanol concentration 1.3%, x 6, 000).

4. Conclusions

Since indoor air quality (IAQ) is so important for our health, indoor air purifiers are commercially available in Japan and in countries around the world. We have investigated the effect of atmospheric plasma including microplasma based on DBD technology both in a small experimental chamber and in a large experimental chamber that is the size of a room. Our results show that a low concentration of ozone played a role in the sterilization process and in decomposing harmful chemical substances such as ammonia. Conducting mesh and electrostatic filters also played a role in collecting particulate material (PM), which are fine particles.

Plasma treatment technologies are relatively new and their performance remains uncertain. Atmospheric plasma can generate ozone various kinds of radicals, ions, and other active species. We propose various plasma treatments for applications to improve the public health. Commercially available plasma devices based on the corona discharge technology and electrostatic precipitation generate low concentrations of ozone. Our results show that ozone at low concentrations is a surprising effective sterilization agent. However, the academic sector has an important responsibility to public health to reveal that plasma processes can also generate harmful byproducts. Our results with investigating plasma technologies to improve the indoor air quality show great promise. It is our pleasure to understand the usefulness of atmospheric microplasmas and their applications

Acknowledgements

The author thanks to the Dr. Marius Blajan, and Mr. Yusuke Kurokawa of Shizuoka University for fruitful discussions. The author also thanks to Mr. Masataka Sano of the Ceraft Co. Ltd. for their technical support.

Author details

Kazuo Shimizu*

Address all correspondence to: shimizu@cjr.shizuoka.ac.jp

Shizuoka University, Japan

References

[1] Vanda Villanyi, Air Pollution, Intech; 2010.

[2] Michael Theophanides, Jane Anastassopoulou and Theophile Theophanides. Air Polluted Environment and Health Effects. Prof. JosÃ© Orosa (Ed.) Indoor and Outdoor Air Pollution. ISBN: 978-953-307-310-1, InTech; 2011

[3] Silvia Vilcekova. Indoor Nitrogen Oxides. Dr. Farhad Nejadkoorki (Ed.) Advanced Air Pollution. ISBN: 978-953-307-511-2, InTech; 2011.

[4] Ricardo Araujo and João P. Cabral. Fungal Air Quality in Medical Protected Environments. Ashok Kumar (Ed.) Air Quality. ISBN: 978-953-307-131-2, InTech; 2010.

[5] Risto Kostiainen. Volatile Organic Compounds in the Indoor Air of Normal and Sick Houses. Ahnospheric Enuironment 1995; 29 (6) 693702.

[6] Kazuo Shimizu. Indoor Air Control by Microplasma. Dr. Farhad Nejadkoorki(Ed.) Advanced Air Pollution, ISBN: 978-953-307-511-2, InTech; 2011.

[7] Heidi Ormstad. Suspended particulate matter in indoor air: adjuvants and allergen carriers. Toxicology 2000; 152 53–68.

[8] Markus Sillanpää, Michael D. Geller, Harish C. Phuleria, Constantinos Sioutas. High collection efficiency electrostatic precipitator for in vitro cell exposure to concentrated ambient particulate matter (PM). Aerosol Science 2008; 39 335 – 347.

[9] Shoji Koide, Akira Nakagawa, Katsuhiko Omoe, Koichi Takaki, Toshitaka Uchino. Physical and microbial collection efficiencies of an electrostatic precipitator for abat-

ing airborne particulates in postharvest agricultural processing. Journal of Electrostatics 2013; 71 734-738.

[10] M. Blajan, and K. Shimizu. Phenomena of Microdischarges in Microplasma. IEEE Trans. on PS 2012; 40 (6) 1730-1732.

[11] M. Blajan, and K. Shimizu. Temporal evolution of dielectric barrier discharge microplasma. Appl. Phys. Lett 2012; 101 104101.

[12] L. F. Gaunt, J. F. Hughes, N. M Harrison. Removal of domestic airborne dust particles by naturallycharged liquid sprays. J. Electrostat 2003; 58 159-169.

[13] G. D. Conanan, F. C. Lai. Performance Enhancement of Two- Stage Corona Wind Generator in a Circular Pipe. Proc. 2012 Electrostatics Joint Conference, Cambridge, K3, June 12-14, 2012.

[14] K. Shimizu, Y. Kurokawa, M. Blajan. Basic Study on Indoor air Quality improvement by Atmospheric Plasma. IEEE IAS Annual Meeting 2014; 2014-EPC-0308, 2014.

[15] K. Shimizu, T. Kuwabara, and M. Blajan. Study on Decomposition of Indoor Air Contaminants by Pulsed Atmospheric Microplasma. Sensors 2012. 12. 14525-14536.

[16] Jan Mertens, C. Anderlohr, P. Rogiers, L. Brachert, P. Khakhariac, E. Goetheer, K. Schaber. A wet electrostatic precipitator (WESP) as countermeasure to mist formation in amine based carbon capture. International Journal of Greenhouse Gas Control 2014; 31 175–181.

[17] Yih-Juh Shiau, Chien-Young Chu. Comparative effects of ultrasonic transducers on medium chemical content in a nutrient mist plant bioreactor. Scientia Horticulturae 2010; 123 514–520.

[18] L. Komunjer, C. Affolter. Absorption–evaporation kinetics of water vapour on highly hygroscopic powder: Case of ammonium nitrate. Powder Technology 2005; 157 67 – 71.

[19] L. Xia, L. Huang, X. Shu, R. Zhang, W. Dong. Removal of ammonia from gas streams with dielectric barrier discharge plasmas. Journal of Hazardous Materials 2008; 152 (1) 113-119.

[20] K. Shimizu, M. Blajan, and S. Tatematsu. Basic Study of Remote Disinfection and Sterilization Effect by Using Atmospheric Microplasma. IEEE Trans. on Ind. Appl. 2012; 48 (4) 1182-1188.

[21] Ryo Ono, Tetsuji Oda. Measurement of hydroxyl radicals in pulsed corona discharge. Journal of Electrostatics 2002; 55 333–342.

[22] S. Kanazawa, H. Tanaka, A. Kajiwara, T. Ohkubo, Y. Nomoto, M. Kocik, J. Mizeraczyk, Jen-Shih Chang. LIF imaging of OH radicals in DC positive streamer coronas. Thin Solid Films 2007; 515 4266–4271.

[23] K. Nagato, Y. Matui, T. Miyata, T. Yamauchi. An analysis of the evolution of negative ions produced by a corona ionizer in air. Int. J. Mass Spectrom 2006; 248 (3) 142-147.

[24] K. Sekimoto. Influence of needle voltage on the formation of negative core ions using atmospheric pressure corona discharge in air. Int. J. Mass Spectrom 2007; 261 (1) 38-44.

[25] Charles J. Weschler. New Directions: Ozone-initiated reaction products indoors may be more harmful than ozone itself. Atmospheric Environment 2004; 38 5715–5716.

[26] William W. Nazaroff, Charles J. Weschler. Cleaning products and air fresheners: exposure to primary and secondary air pollutants. Atmospheric Environment 2004; 38 2841–2865.

[27] U.S. Environmental Protection Agency, Office of Air and Radiation, Office of Air Quality Planning and Standards. Regulatory Impact Analysis -Final National Ambient Air Quality Standard for Ozone-. 2011.

[28] F. M. Gabler, J. L. Smilanick, M. F. Mansour, H. Karaca. Influence of fumigation with high concentrations of ozone gas on postharvest gray mold and fungicide residues on table grapes. Postharvest Biology and Technology 2010; 55 85-90.

[29] B. Eliasson, M Hirth and U. Kogelschatz. Ozone synthesis from oxygen in dielectric barrier discharges. J. Phys. D: Appl. Phys. 1987; 20 (11) 1421-1437.

[30] H. S. Ahn, N. Hayashi, S. Ihara and C. Yamabe. Ozone Generation Characteristics by Superimposed Discharge in Oxygen-Fed Ozonizer. J. Appl. Phys 2003; 6578- 6583.

[31] P.A. Schulte, M. Boeniger, J.T. Walker, S.E. Schober, M.A. Pereira, D.K. Gulati, J.P. Wojciechowski, A. Garza, R. Froelich, G. Strauss, W.E. Halperin, R. Herrick and J. Griffith. Biologic markers in hospital workers exposed to low levels of ethylene oxide. Mutation Research/Genetic Toxicology 1992; 278 237-251.

Chapter 22

Biological Contamination of Air in Indoor Spaces

Anca Maria Moldoveanu

Additional information is available at the end of the chapter

http://dx.doi.org/10.5772/59727

1. Introduction

The indoor environment is a very complex environment. Pollutants are numerous in both outdoor and indoor air. While the same pollutants (biological, chemical, or radioactive) can be identified in indoor and outdoor spaces, their concentrations can be very different. More specifically, extremely high concentrations can accumulate indoors. Many bacteria that belong to the normal microflora of the human skin are continuously emitted into the immediate surroundings of humans, and accumulate indoors.

The pollution of the indoor spaces is an important element that can affect human health. People spend around 90% or more of their time of a day indoors. Indoor spaces include the inside of houses, schools, university rooms, social buildings like shops, cars, planes, and workplaces. The outdoor and indoor environments are linked. The two elements cannot be separated. Biological agents are defined as agents or microfragments from plant or animal matter or from microorganisms.

Elements identified in the air include bacteria, either in vegetative status or spores, fungi, yeasts, microbial toxins and secondary metabolites like bacterial endotoxin, peptidoglycans or fungal β(1,3)-glucans, volatile organic compounds, pollens, pet and insect allergens, other allergens, viruses, protozoa, etc. [1-3].

Many bacteria, fungi, viruses, and protozoa may be infectious to humans and they may cause specific infectious diseases. Infectious agents are usually received from other humans, from animals (these infections are zoonoses), or from environmental sources such as contaminated water, soil, etc.

Biological contamination in indoor air may have sources like outdoor air, human body, bacteria that are growing indoors, and pets.

2. Biological contamination of indoor air

2.1. Bacteria in the indoor environments

The most common building-associated bacteria are saprophytic bacteria of the normal human skin, mouth, and nose that are emitted into the indoor air and bacteria originating from outdoors air [1-3]. Other common bacteria of indoor environments include heterotrophic bacteria that grow in the water reservoirs or moist sites of the building, such as bathroom sinks. Specifically, legionellae and nontuberculous environmental mycobacteria develop in biofilms of water pipelines or in water reservoirs of cooling systems. Actinobacteria (especially Streptomycetes), *Bacillus* species and various other bacteria grow in moist building materials together with fungi. Elements from bacterial structures released in air include bacterial cells, bacterial spores, peptidoglycans, microbial volatile organic compounds, exotoxins, and other bacteria growing metabolites [1].

Humans are an important source of indoor bacteria. The upmost layer of the normal human skin is continuously renewed, and skin scales containing bacteria are shed into the environment [1]. Bacteria in the respiratory airways are eliminated through Pflügge droplets while talking, coughing, or sneezing. The level of air contamination is dependent on the number of persons inside a room and the efficiency of the ventilation system (natural or artificial ventilation). Bacteria that can be identified in indoor air are micrococci, staphylococci, streptococci, and corynebacteria [1].

The presence of pets in indoor spaces can represent an important source of air contamination and can be linked with the level of indoor endotoxins. The presence of dogs and cats can be the main predictors of endotoxin levels in house dust [1, 4-7]. Other predictors are the presence of vermin, such as mice, and infrequent cleaning, which indicates poor hygienic conditions in the home [1]. Storage of organic household waste indoors also increases bacterial contamination in the indoor environment [1].

Bacteria actively growing or accumulating in the indoor environment may cause health effects and may require specific actions to control growth and prevent the exposures.

Bacterial growth can be found wherever standing water is present and Gram-negative bacteria can be identified in this environment. The presence of bacteria, either as viable bacteria or bacterial spores, mycotoxins, chemical markers like β-glucans and volatile organic compounds, and endotoxins, represents an indication of high humidity in the indoor environment [1].

2.2. Fungi in indoor environments

Fungi are present in both outdoor and indoor air. The levels found in the outdoor air can range from less than 100 spores/m^3 up to100 000 spores/m^3. Normally, the fungi levels are lower in indoor spaces compared with outdoor spaces. If the conditions that permit the fungal growth are present, for example, in damp houses, the levels of fungi in the indoor air can reach very high concentrations. Fungi level measurements are expressed in colony-forming units (CFUs) per cubic meter of air

The levels of fungi and the type of fungi identified in indoor air depend on the season, construction features, age, and use of the building as well as the ventilation rate.

The most common fungal genera occurring in indoor environments are *Penicillium, Aspergillus, Cladosporium, Alternaria,* and yeasts [1-3]. These genera are also the most frequently occurring fungi in outdoor air.

Spores of fungi are present everywhere, and they are able to germinate wherever there is water available and the ranges between 0.80 and 0.98. Fungi need carbohydrates, proteins, and lipid to develop. They can find all these elements in house dust, construction materials like wallpaper or textiles, paint, glue, wood, paper and books, stored food, or deposit of cooking oil. Fungi can grow on inert materials like ceramic tiles [2].

2.3. Allergens in indoor environment

Allergens are a group of agents that may cause a specific IgE-mediated reaction in humans. Fungal allergens can trigger type I allergic reactions and IgE sensitization to fungal species like *Alternaria, Penicillium, Aspergillus,* and *Cladosporium* spp. All these types of fungi species can induce allergic respiratory diseases, especially acute asthma attacks. Fungal allergens can also trigger type III-IgG-mediated allergic reactions (*Penicillium* and *Aspergillus)* and even type IV allergic reactions. This is the premise of the onset of hypersensitivity pneumonitis [2].

Allergens may also be present in indoor air due to house dust mites. The mite species that produce allergens are *Dermatophagoides pteronyssinus* (major allergens—Der p I and Der p II) and *Dermatophagoides farina* (major allergen—Der f I).

2.4. Endotoxins

These are toxins integrated in the outer membrane of Gram-negative bacteria, which are composed of proteins, lipids, and lipopolysaccharides [2, 4-8]. The bacterial cell lysis determines the release of endotoxins. In the air, endotoxins are bound to coarse particles. Exposure to different levels in air has different results: if the exposure is to high levels, respiratory symptoms or nonallergic asthma can develop, while if the exposure is at low and moderate levels, it is possible to have a protective action against allergies and asthma [2, 4-8].

2.5. (1 → 3)- β -D-glucans

It is a structural element of the cell wall of fungi and some bacteria, and it can be found in house dust. (1 → 3)-β-D-glucans do not have allergic properties. However, they do have immune-modulating properties. Exposure to (1 → 3)-β-D-glucans in home environment can induce respiratory diseases [2, 9, 10] and may increase peak flow variability in asthmatic children.

2.6. Mycotoxins

These are toxins (secondary metabolites) produced by fungi that interfere with RNA synthesis and can cause DNA damage. Some mycotoxins like aflatoxins produced by *Aspergillus flavus* and *Aspergillus parasiticus* induce carcinogenic effects [2, 11]. Indoor exposure to mycotoxins produced by *Stachybotrys chartarum* can be associated with acute pulmonary hemorrhage.

2.7. Fungi and Volatile Organic Compounds (VOC) in indoor air

Fungi can produce around 200 volatile organic compounds. The identification of the VOCs in indoor air shows that there is a fungal growth present, even if the quantitative measurements are negative [11]. One of the most important substances that can be identified in indoor air is formaldehyde that can induce cancer.

2.8. Viruses

Indoor air humidity can induce a longer survival time of respiratory viruses and increased risk of respiratory infection and allergic diseases.

3. Exposure evaluation of biological contamination of indoor air

Indoor air levels of bacterial concentrations of <1, 000 CFU/m^3 can be regarded as "low, " and concentrations of >5, 000 CFU/m^3 can be regarded as "high" [1]. These numbers reflect mainly the crowdedness of the room and the efficiency of the ventilation and thus are a measure of the hygienic quality of the indoor air.

Characterizing the bacterial flora in the sample may give additional information about the sources of the bacteria present. If Gram-positive cocci and corynebacteria, the common bacteria of the human skin, are the dominating groups, this indicates human sources [1].

The presence of Gram-negative bacteria shows an active bacterial growth due to the presence of water indoors.

Gram-positive bacteria with exo- and endospores like *Streptomyces* and *Bacillus* can grow on moist building materials. Their spores are very resistant and can survive even if the air humidity is low.

The presence of fecal bacteria like the group of *Enterobacteriaceae* is considered heavy contamination of indoor air, coming probably from a sewage leakage. The presence of actinomycetes shows the growth of microbes in the building material.

The presence of two groups of bacteria, that is, *Legionella* and actinomycetes, has a special importance in indoor air.

Legionella spp. are Gram-negative environmental bacteria that have found a remarkable ecological niche in man-made water systems, especially in those serving air conditioning systems for heat transfer [1].

http://dx.doi.org/10.5772/59727

4. Health effects induced by indoor exposure to biological contaminants

Indoor air contamination can generate a multitude of effects on human health. The effects are especially on the respiratory system. Allergic reactions can be induced by the exposure to different biological agents that are identified in indoor air. Also, there is evidence to suggest that exposure to biological pollutants in indoor air can be associated with symptoms with neurological basis like headaches, fatigue, and forgetfulness.

Some of the effects on the respiratory system related to the biological contamination of indoor air are due to the inflammation of the airways. This can appear in both allergic (with the implication of IgE and IgG antibodies) and nonallergic respiratory diseases The separation between those two entities is important in the case of asthma and rhinitis.

4.1. Rhinitis

Rhinitis is induced by the inflammation of the mucosa of the upper respiratory tract. Symptoms of rhinitis are nasal congestion, rhinorrhea, sneezing, and conjunctivitis. Rhinitis can be of allergic or nonallergic cause [1, 11].

4.2. Asthma

Allergic asthma involves IgE-mediated sensitivity and eosinophils in presence of an atopic underlying reaction of the respiratory airways [1, 12-18]. Among adults, there is also an increasing prevalence of asthma. People with moderate to severe asthma will react adversely toward biological and other contaminants in the indoor environment due to a nonspecific bronchial hyperresponsiveness. That is an exaggerated bronhoconstrictor response to numerous numbers of stimuli. The severity of asthma and the need of medication in house dust mite allergic patients is closely related to the amount of endotoxin in house dust in a dose-dependent manner [1].

Pollen allergens interact with other sources of fine particles in polluted air, which concentrate the allergens and trigger asthma attacks. As for the interaction between biological and nonbiological agents, airborne particles such as diesel exhaust particles have an adjuvant activity for IgE antibody and cytokine production [1]. The intranasal instillation of ragweed allergens with diesel exhaust particles in allergic patients significantly increased allergen-specific IgE production and enhanced Th2-type cytokine mRNA expression compared to allergic patients instilled with ragweed alone [1].

Both allergens and viruses have been known to worsen asthma symptoms. A study in the UK demonstrated that allergen exposure and viral infections act synergistically to trigger exacerbations in adults [1].

4.3. Chronic bronchitis

Chronic bronchitis is defined as cough with phlegm (productive cough) for at least 3 months per year for 2 years or more [1, 17-19]. Chronic bronchitis may be present with or without airways obstruction. The underlying mechanism is an increase in mucus secreting glands and alterations in the characteristics of the mucus itself [1]. Airway inflammation is present, and an increased number of neutrophils can be observed [1].

Subjects with chronic bronchitis may also have symptoms such as dyspnea, chest tightness, and wheezing, and there is thus a considerable overlap in symptoms between allergic asthma, nonallergic asthma, and bronchitis [1]. Biological contamination through the presence of endotoxin and β-glucan can be incriminated in the pathogenesis of bronchitis [1].

4.4. Chronic airflow obstruction

Several studies have shown associations between chronic airflow obstructions as measured by spirometry (particularly FEV1) and organic dust exposures in industrial and agricultural occupational populations [1, 19]. Endotoxin exposure may play an important role.

4.5. Hypersensitivity Pneumonitis (HP) or Extrinsic Allergic Alveolitis (EAA) [1, 2, 20]

The disease, also known as "farmers lung" or "pigeon breeder's lung, " is one of the most severe diseases induced by fungi and thermophilic actinomycetes (spore-forming bacteria) exposure. The clinical aspect of HP is represented by severe pulmonary disease with pneumonia-like fever, cough, tightness of chest, shortness of breath, inflammation of pulmonary parenchyma, and development of pulmonary granulomas. HP is very similar to organic dust toxic syndrome (ODTS).

4.6. Organic Dust Toxic Syndrome (ODTS)

The disease is similar to influenza and is characterized by symptoms like acute fever, dry cough, dyspnea, chest tightness, headache, shivering, malaise, fatigue, and muscle and joint pains. ODTS is a nonallergic neutrophil-mediated airway inflammation [1].

4.7. Atopic allergic dermatitis

This disease can be induced by exposure to the biological indoor air pollutants.

4.8. Sick building syndrome

In the 1988s, the concept of "sick building syndrome" (SBS) was established by the WHO [1, 3, 21, 22]. The label sick building syndrome (SBS) can be applied when the inhabitants or occupants of different indoor spaces develop acute health problems or discomfort inside the rooms. The exposed people have the following symptoms: eye, nose, and throat irritation and dryness; itching and dryness of skin and eyes; dry cough; nausea; vomiting; headache; difficulty in concentration; fatigue; etc. The symptoms disappear after cessation of the exposure.

5. Aim and purpose of the study

The aim of the study was to assess indoor air contamination from the point of view of bacteriological and fungal contamination of different indoor spaces like houses, schools, offices, archives, and spaces with commercial destinations [23-25].

6. Materials and methodology of investigation

The methodology of the study implies the investigation of the levels of indoor air contamination with fungi and bacteria. The destination of the investigated indoor spaces was very diverse. The levels of bacteriological contamination were assessed in indoor spaces designated as living spaces, social use spaces like shops, and work and education spaces like schools, offices, and archives.

Quantitative and qualitative assessments were made to evaluate the levels of air contamination. Passive and active methods were used to sample the air from the investigated sites. The results were compared to the enforced recommendations in Romania, at this moment.

6.1. Investigated places

6.1.1. Houses in Bucharest City

There were 10 investigated apartments, with different numbers of room situated at different floors and with different geographical orientations and different characteristics.

Type of investigated apartments:

a. One apartment with one room, one inhabitant, nonsmoker

b. One apartment with one room, two inhabitants, one smoking inhabitant

c. Apartment with two separates rooms, two nonsmoking inhabitants

d. Apartment with two separates rooms, three smoking inhabitants

e. Apartment with two rooms, with five nonsmoking inhabitants

f. Apartment with two separates rooms, four inhabitants, one smoking person

g. Apartment with three separates rooms, three nonsmoking inhabitants

h. Apartment with three separates rooms, five inhabitants, one smoking person

i. Apartment with three rooms, five nonsmoking inhabitants

j. Apartment with three rooms, with four inhabitants, one smoking person

For an accurate assessment of the biological contamination of indoor air, the measurements were made three times in a year, in the winter, in the summer, and in the autumn, 3 days consecutively, in the kitchen and in the living room of each investigated apartment.

6.1.2. Schools

There were two investigated schools, one in Bucharest and one in Giurgiu City.

In the school in Bucharest, the investigations were made

- in an empty classroom, before 8:00 a.m.
- during the school day (at 9.30 a.m.), in a classroom with 23 children
- in a classroom with 36 children and 1 teacher
- in a classroom with 23 children and 1 teacher
- in a classroom (at 11:00 a.m.) with 24 children and 1 teacher
- in the school library (at 10:00 a.m.) with 3 people inside
- in study room number 1, next to the library with 15 children inside
- in study room number 2, next to the library, room that is connected to the hall in front of the school library
- in the hall of the school
- in the medical ward, with 8 people inside the room.

In the school in Giurgiu, investigations were made

- in a classroom at 8:00 a.m.
- in a classroom at 11:00 a.m.
- in a classroom at 1:00 p.m.
- in the school hall.

6.1.3. Offices

Fungi were investigated in eight offices.

6.1.4. Archives

Fungi were investigated in 10 archive rooms, where the possibility of contamination was presumed.

6.1.5. Shops

The levels of bacteria in air were investigated (a) in a supermarket in Bucharest City with ground floor and 4 floors and (b) in a shop with only a ground floor.

The investigations were made in the morning when the shop opens and in the evening when the shops were nearly to be closed.

6.2. Sampling methods

For the bacteriological and fungal evaluation, the sampling methods used were a passive method and an active one [26, 27].

6.2.1. Passive sampling method

The total number of mesophilic aerobic bacteria, yeast, and molds in the air of selected rooms was determined using the Koch sedimentation method. Bacteria from air were harvested using the gravitational method on the Petri plates with specific growth media. The measurements of levels of bacteria in air were made in several points on each location. The results were expressed as CFU/m^3. The equation used for calculating the results was as follows:

$$CFU/m^3 = n \times 10{,}000/s \cdot t/5$$

where n is the number of colonies on the Petri plate, s is the surface of the Petri plate, and t is the time of Petri plate exposure. Petri dishes with growth media were exposed for 10 min.

Results obtained by the Koch sedimentation method are less accurate than those from impaction methods with the use of an air sampler. The sedimentation method does not require expensive instrumentation and is cheap and simple. The sedimentation method does not permit exact quantitative determination. Still, this method permit an assessment of the types of microorganisms present in the air and can give an approximation of bacterial and fungal concentration in the air of investigated indoor spaces.

6.2.2. Active sampling method

This method used an impactor sampler M.A.S.Q.-90, with 2 l/min airflow and a total air volume equal to 200 l.

The following growth media were used [28]:

- tryptic soy agar (TSA) to determine the total number of bacteria
- blood agar to determine the presence of *Staphylococcus* and *Streptococcus*
- Czapek Dox Agar (Oxoid) for fungi identification

To obtain the total number of bacteria in the air of investigated rooms, the Petri dishes with growth media were incubated at 37°C for 24 h and then left for another 24 h at room temperature. To obtain the number of fungi colonies, the Petri dishes were exposed for 10 days at 25°C.

The identification of bacteria types was made by macroscopic estimation (description of colony) and by microscopic estimation.

Diagnosis of filamentous fungi was based on estimation of morphological features of growth on Czapek medium as well as on microscopic observation.

6.3. Indicators used to assessed the air contamination of indoor air

The following were assessed and counted on a Petri dish:

a. The total number of bacteria, expressed in CFU/m^3 (colony-forming units/cubic meter, CFU/m^3)

b. The presence of *Streptococcus* alpha and beta hemolytic

c. The presence of *Staphylococcus aureus* and non-*aureus* with and without hemolytic proprieties

d. The presence of coliform bacteria as an indicator of enteric contamination of air

e. The total number of fungi expressed in CFU/m^3 as quantitative assessment of contamination

f. The types of fungi as a qualitative assessment of indoor air contamination with fungi

6.4. References values

To have a good assessment of the results that were obtained, it was useful to have as references standardized values for the biological contamination of indoor air.

At present, there are no standards or guidelines for microbiological quality of indoor air. Each country has individual regulations and recommendations.

The Commission of the European Communities (CEC) defines the following level of bacteriological air contamination:0 undetectable, 1-499 CFU/m^3 low, 500-999 CFU/m^3 medium, >1, 000 CFU/m^3 high, and 2, 000 CFU/m^3 very high. A value of 10, 000 CFU/m^3 of nondescript airborne microbes could therefore be considered a hazardous level for indoor environment [12].

In 2001, the American Industrial Hygiene Association (AIHA) published a proposition of guidelines that stipulate 1, 000 of viable CFU/m^3 in air as an upper limit for concentration in indoor environment and for the amount of fungal spores in different indoor environments. It is proposed that the indoor air levels of microorganisms should be less than 500 CFU/m^3 for residential buildings and less than 250 CFU/m^3 for commercial buildings [29, 30].

According to current Swedish requirements, the number of 500 colony-forming units (CFU) of bacteria and 300 CFU of fungal spores in 1 m^3 can be accepted in an indoor environment [31].

The regulation in Brazil stipulates that the upper limit for total amount of airborne microorganisms (especially fungi) should be 750 CFU/m^3 [32, 33].

In Hong Kong, the indoor air levels of microorganism should be less than 1, 000 CFU/m^3 of bacteria. The levels are considered excellent if they are less than 500 CFU/m^3 [34, 35].

In Taiwan, the indoor air levels of bacteria should not exceeded 500 CFU/m^3 and 1, 000 CFU/m^3 for fungi [36, 37].

In India, there are no guidelines for the levels of bacteriological contamination of air [38].

In Singapore, the regulation for indoor air quality imposes an upper limit for the concentration of bacteria of 500 CFU/m^3 [39].

http://dx.doi.org/10.5772/59727

The American Conference of Governmental Industrial Hygienists (ACGIH) stipulates that the counts for total bacteria should not exceed 500 CFU/m^3[40].

WHO (1988) has set a limitation of fungi counts at 500 CFU/m^3 [41].

The Korean Indoor Air Quality Control in Public Use Facilities Act prescribes a total airborne bacterial concentration below 800 CFU/m^3 for hospitals, nurseries, public nurseries, and hospitals for the elderly and post partum care centers [42, 43].

In Romania, the reference value for indoor air contamination in a house is equal to 2, 500 CFU/m^3 and 550 CFU/m^3 for fungi. In schools, the level of bacteriological contamination of air should not exceed 1, 500 CFU/m^3 for bacteria at 37°. The presence of coliform bacteria and hemolytic bacteria are not allowed in air inside education units.

7. Results

7.1. Houses in Bucharest City

The highest value for air contamination with germs at 37°C was equal to 21, 466 CFU/m^3 in one of the apartments with one room (apartment B).

In the kitchens, the highest value of contamination identified was equal to 12, 701 CFU/m^3, the minimum value was equal to 201 CFU/m^3, and the highest average value was equal 3, 200 CFU/m^3 (Fig. 1).

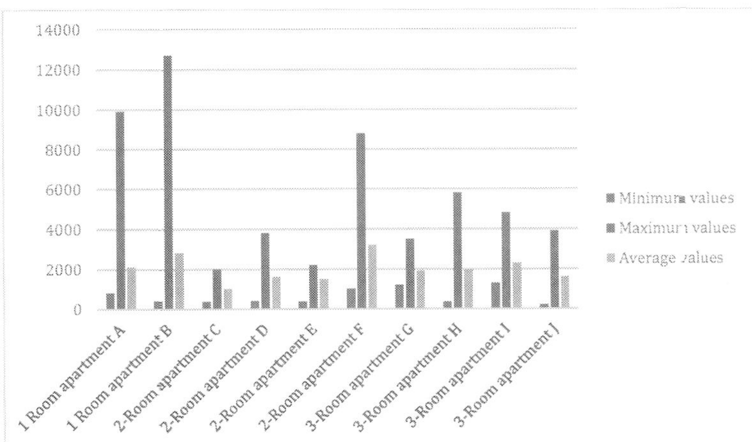

Figure 1. Bacteriological contamination of kitchen air (CFU/m^3 at 37°C).

In the living rooms, the highest identified value of contamination was equal to 21, 466 CFU/m^3, the minimum value was equal to 210 CFU/m^3, and the highest average value was equal 4, 731 CFU/m^3 (Fig. 2).

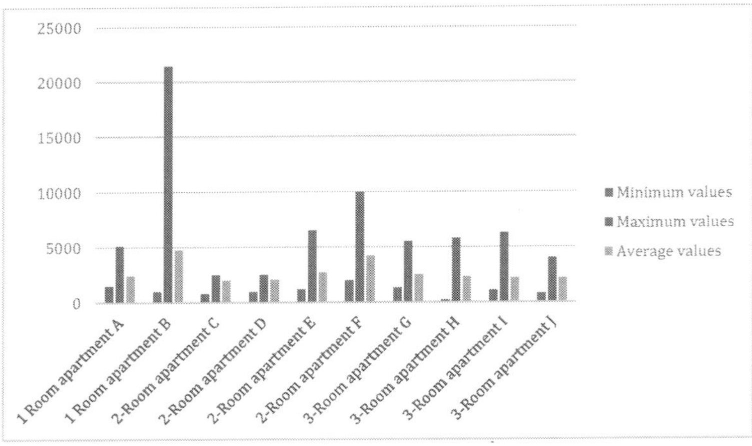

Figure 2. Levels of bacteriological air contamination in living rooms (CFU/m^3 at 37°C).

Taking into account the moment of the day, the highest bacteriological contamination (average values) was found in the living room, in the evening, and was equal to 6, 665 CFU/m^3 in the one room apartment—apartment B (Fig. 3).

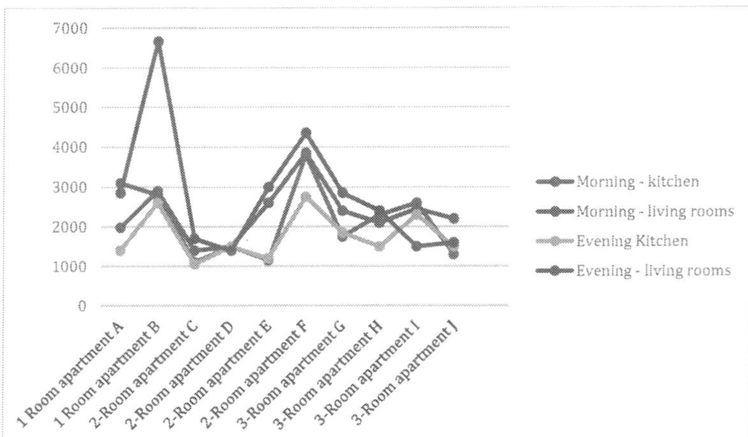

Figure 3. Average numbers of germs at 37°C in the morning and in the evening in investigated apartments.

The indoor air contamination is higher in the fall and in the winter, reaching a level equal to 7, 853 CFU/m³ in one room apartment (average values)—apartment B (Fig. 4).

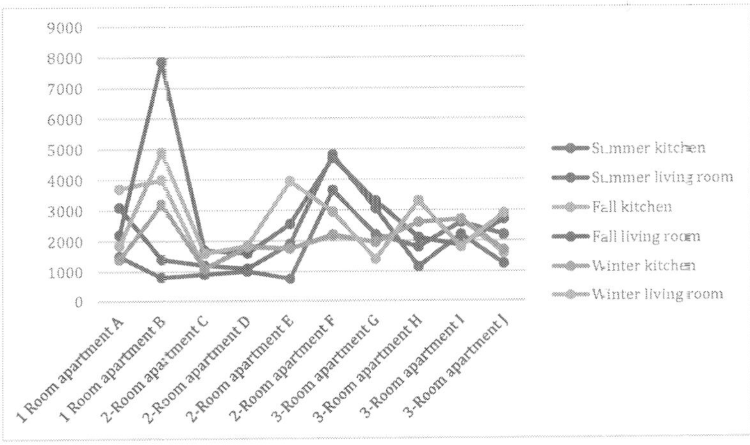

Figure 4. Average numbers of germs at 37°C according to the season in investigated apartments.

In connection to the smoking habit of the inhabitants, apartments with one and two rooms had a higher bacteriological contamination of air (Fig. 5).

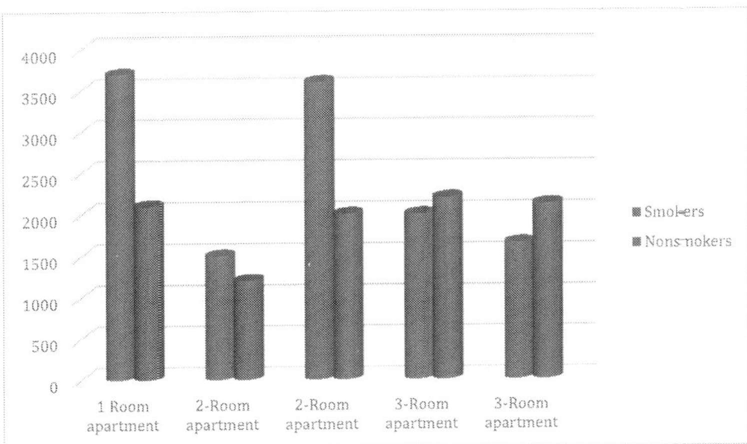

Figure 5. Number of germs at 37°C in houses according to the smoking habit.

In connection to the type of apartment, bacteriological contamination is higher in the apartments with two linked rooms. Bacteriological contamination is also presented in the three-room apartment (Figs. 6 and 7).

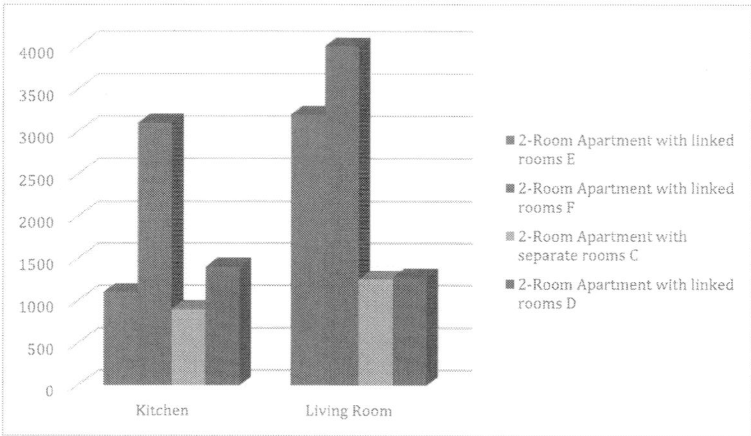

Figure 6. Number of germs at 37°C and the type of apartment (apartments with two rooms).

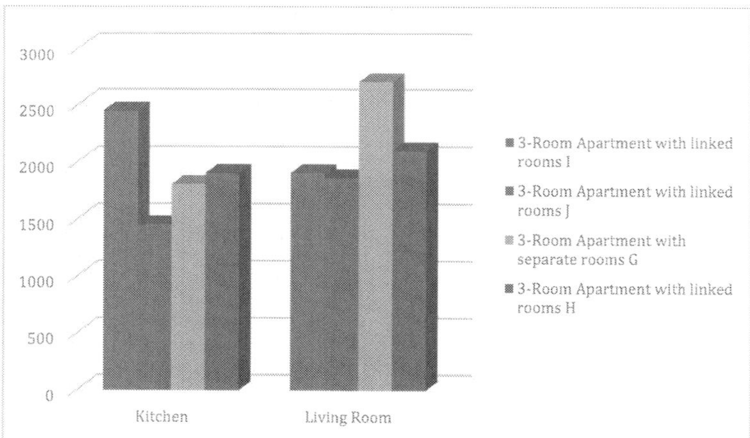

Figure 7. Number of germs at 37°C and the type of apartment (apartments with three rooms).

Contamination of air with staphylococci reached the highest value equal to 7, 988 CFU/m^3 in the kitchen and 10, 157 CFU/m^3 in the living room (Figs. 8 and 9).

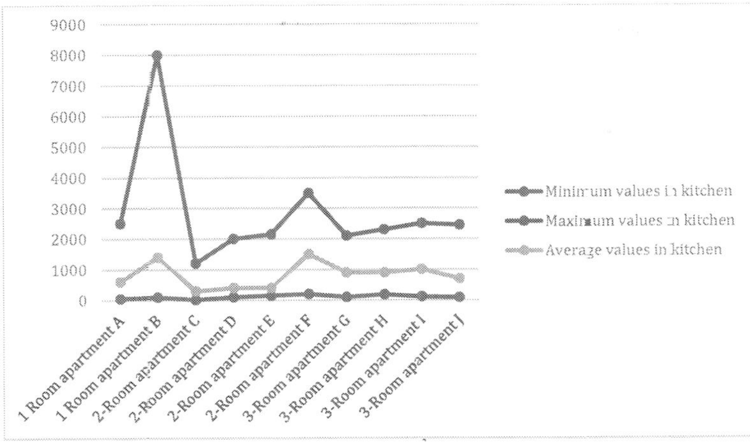

Figure 8. Contamination with staphylococci in kitchen air (CFU/m^3).

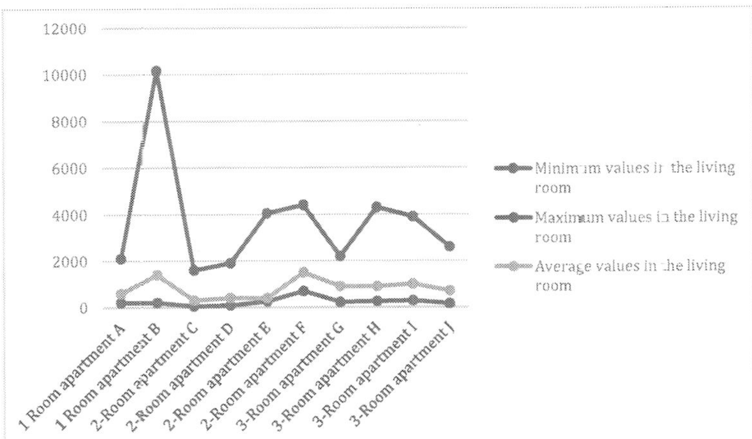

Figure 9. Contamination with staphylococci in living rooms air (CFU/m).

The ratio between staphylococci/germs at 37°C is around 40% in the kitchen and over 50% in the living rooms (Fig. 10).

The assessment of the fungal contamination showed that in many homes, the level of fungal contamination was very high in all investigated rooms. The levels of fungi were between 635 and 10, 859 CFU/m^3. The highest values measured were 2, 802 CFU/m^3 in the living room, 2,

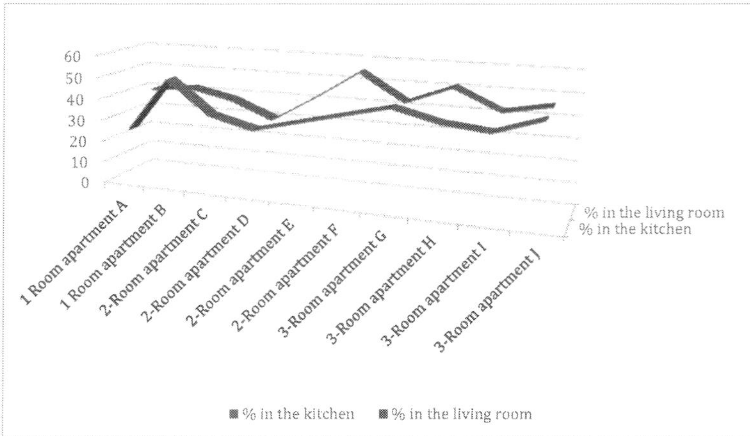

Figure 10. Ratio between staphylococci/germs at 37° in investigated houses.

302 CFU/m³ in the kitchen, 7, 675 CFU/m³ in the bathroom, and 10, 859 CFU/m³ in the bedroom (highest overall). The species of fungi identified were *Penicillium, Aspergillus niger, Rhizopus, Mucor, Cladosporium herbarum,* and *Fusarium.*

7.2. Schools

In a school in Bucharest, the following results were obtained: in an empty classroom, before 8:00 a.m., the value identified was 25 CFU/m³; during the school day (at 9.30:00 a.m.) in a classroom with 23 children, the value identified was 67 CFU/m³; in a classroom with 36 children and 1 teacher, the value identified was 229 CFU/m³; in a classroom with 23 children and 1 teacher, the value identified was 140 CFU/m³; in a classroom (at 11:00 a.m.) with 24 children and 1 teacher, the value identified was 1, 465 CFU/m³ with 38 CFU/m³ *Streptococcus* alpha-hemolytic present; in the school library (at 10:00 a.m.) with 3 people inside, the biological contamination of air was equal to 45 CFU/m³; in study room number 1, next to the library with 15 children inside, the level of air contamination was equal to 89 CFU/m³; in study room number 2, next to the library, the room connected to the hall in front of the school library, the level of biological air contamination was equal to 427 CFU/m³; the level of air contamination in the school hall was equal to 180 CFU/m³; in the medical ward, with 8 people inside the room, the level of biological air contamination was equal to 60 CFU/m³, 1 CFU/m³ *Streptococcus* alpha hemolytic, and 1 CFU/m³ *Streptococcus* beta hemolytic present (Fig.11).

In a school in Giurgiu City, the following results were obtained: at 8:00 a.m., a number of 1, 259 CFU/m³ was identified; at 11:00 a.m., a number of 1, 732 CFU/m³ was identified; at 1:00 p.m., a level of biological air contamination equal to 472 CFU/m³ was identified. In the school hall, the level of air contamination was equal to 2, 202 CFU/m³, the level of contamination of the air with *S.* non-*aureus* was equal to 1, 259 CFU/m³ at 8:00 a.m. and 11:00 a.m., and 472 CFU/m³ at 1:00 p.m.; the level of contamination with *Streptococcus* alpha-hemolytic was equal to 157

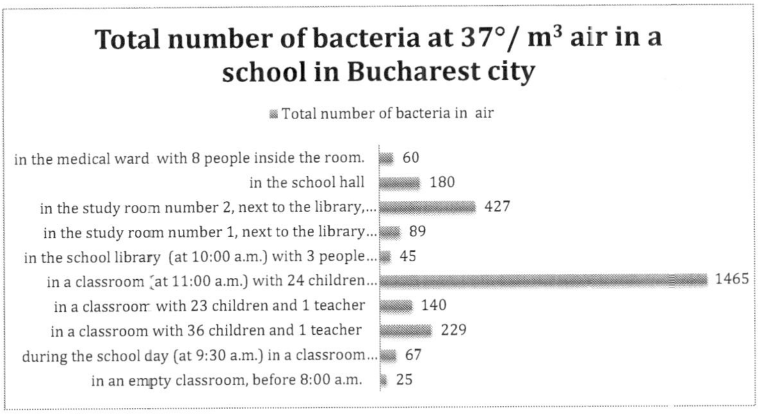

Figure 11. Level of bacteriological contamination of air in a school in Bucharest City.

CFU/m^3 at 11:00 a.m. In a classroom and in the school hall, at 1:00 p.m., coliform bacteria were identified as an indicator of enteric contamination of air. The level of contamination with coliform bacteria was equal to 315 CFU/m^3. The detected levels of coliform bacteria can be dangerous for children health (Fig.12).

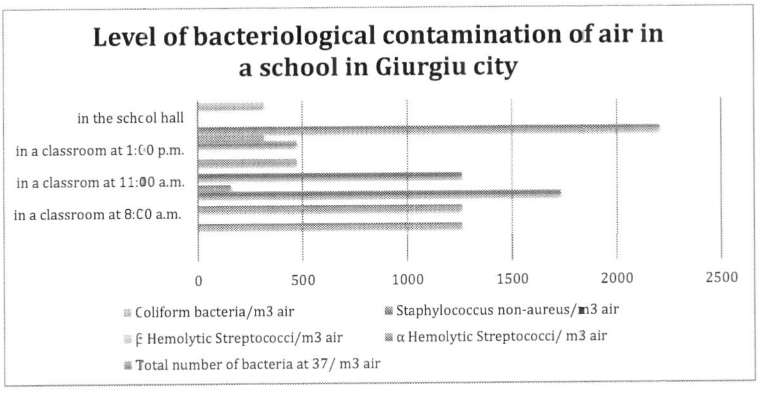

Figure 12. Levels of bacteriological air contamination in a school in Giurgiu City.

7.3. Offices

The levels of fungal contamination were between 734 and 4, 034 CFU/m^3. The fungi species identified were *Penicillium* spp., *A. niger*, *A. flavus*, *Aspergillus versicolor*, *Mucor*, and *Cladosporium*.

7.4. Archives

The levels of fungal contamination were between 655 and 1, 490 CFU/m^3. By a qualitative assessment of air pollution with fungi, the following types of fungi were identified: *Penicillium* spp., *A. niger, A. flavus, A. versicolor, Mucor, Cladosporium.*

7.5. Shops

In a supermarket with ground floor and 4 floors, the levels of bacteriological contamination are shown in Table 1.

Average values		Total number of bacteria at 37°/m^3 air	Total number of staphylococci/m^3 air	Hemolytic staphylococci/m^3 air	*Staphylococcus aureus/m^3 air*	α Hemolytic streptococci/m^3 air	β Hemolytic streptococci/m^3 air
Ground floor	M	5, 140.75	3, 296	161.25	197.72	328.75	3.24
	E	13, 836.00	10, 397.25	404.5	240.75	964.25	0
First floor	M	3, 919.25	2, 015.5	137.5	170.55	213.66	0
	E	9, 258.75	7, 316	372	232.5	569.33	0
Second floor	M	6, 155.45	4, 276.33	299.75	226.25	182.5	0
	E	10, 916.75	8, 705.5	339	286.25	549.16	0
Third floor	M	6, 946	4, 914.25	280.25	141	273.5	0
	E	13, 509.25	10, 800.5	419.5	316	775.5	0
Fourth floor	M	6, 473.75	4, 292.75	301.25	189.97	228	0
	E	10, 352.25	7, 759.75	405.25	267.78	560	0

Table 1. Levels of bacteriological air contamination in a supermarket in Bucharest City (M = morning, E = evening)

The highest levels of contamination were registered in the evening and were equal to 13, 836 bacteria at 37°/m^3 air at the ground floor and 13, 509 bacteria at 37°/m^3 air at the third floor. In the second and fourth floors, in the evening, the level of bacteria in air were equal to 10, 917 bacteria at 37°/m^3 air and 10, 352 bacteria at 37°/m^3 air. In the first floor, in the evening, levels of bacterial contamination equal to 9, 259 bacteria at 37°/m^3 air were registered. The levels of air contamination were the lowest in the first floor (Fig. 13).

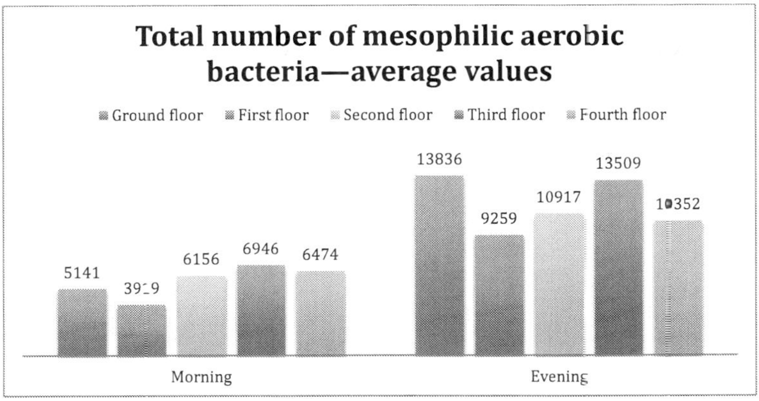

Figure 13. Levels of bacteriological contamination of air in a supermarket—average values.

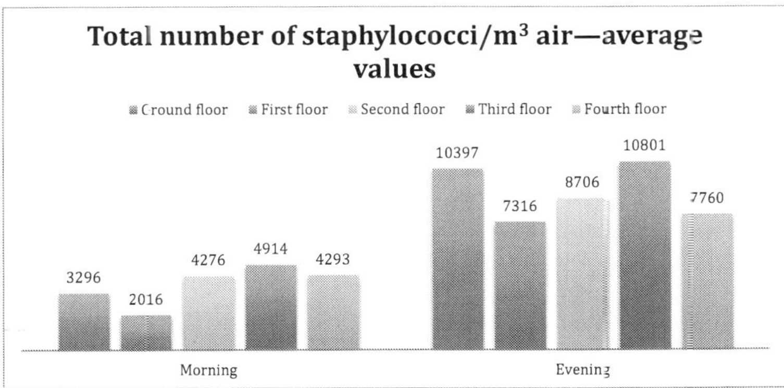

Figure 14. Total number of staphylococci/m^3 air—average values.

The contamination of air, in the case of the supermarket with a ground floor and other four floors, with staphylococci was high in the evening. The highest levels were registered in the third floor with a contamination burden equal to 10, 801/m^3 air, followed by the ground floor with 10, 397/m^3 air, the second floor with 8, 706/m^3 air, the fourth floor with 7, 760/m^3 air, and the first floor with 7, 316/m^3 air. It can be seen that the levels of air contamination were the lowest in the first floor (Fig.14).

Hemolytic staphylococci/m^3 air and *S. aureus*/m^3 air were identified in the same amount in the evening at al floors.

α Hemolytic streptococci/m³ air were identified in the highest levels in the evening at the ground floor. In the evening, at the other floors, the levels were almost equal.

β Hemolytic streptococci/m³ air were identified at the ground floor only.

In a shop with only a ground floor, the levels of bacteriological contamination are presented in Table 2.

Average values		Total number of bacteria at 37°/m³ air	Total number of staphylococci/m³ air	Hemolytic staphylococci/m³ air	*Staphylococcus aureus/m³ air*	α Hemolytic streptococci/m³ air	β Hemolytic streptococci/m³ air
Ground floor	M	2, 398.16	1, 323.2	78	47.3	86.5	0
	E	3, 107.5	2, 217.85	90.08	164.76	75.65	0

Table 2. Levels of bacteriological air contamination in a shop with only a ground floor in Bucharest City (M = morning, E = evening)

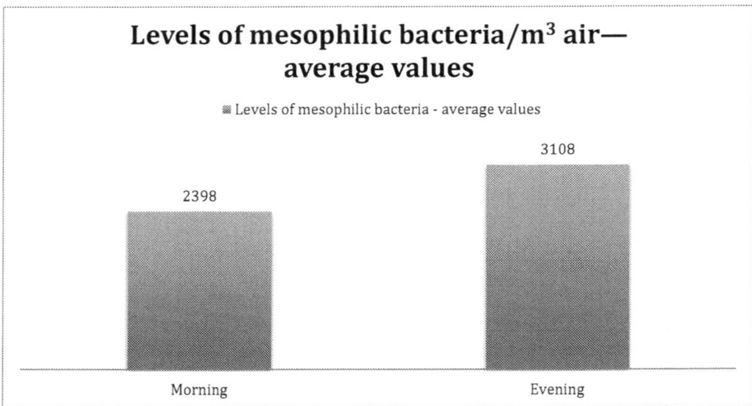

Figure 15. Levels of bacteriological contamination of air in shop with only a ground floor.

In the shop with only a ground floor, the levels of air contamination registered higher values in the evening than in the morning measurements for mesophilic bacteria, total number of staphylococci, hemolytic staphylococci, *S. aureus*, and α hemolytic streptococci (Fig.15).

8. Discussions

In the investigated houses, the highest value for air contamination with germs, at 37°C, was found in one of the apartments with one room (apartment B). Regarding the time of the day, the highest level of bacteriological contamination was found in the evening, in the one room apartment—apartment B. Taking into account the time of the year, the indoor air contaminations were higher in the fall and winter, especially in the one room apartment—apartment B.

Considering the smoking habit of the inhabitants, apartments with one and two rooms had a higher bacteriological contamination of air where there were smokers present.

Considering the type of the apartments, in apartments with more than one room, bacteriological contamination is at the highest level in the apartment with two linked rooms.

The ratio between staphylococci/germs at 37°C is around 40% in the kitchen and over 50% in the living rooms.

In a previous study [44], the levels of bacteriological and fungal pollution of indoor air were as follows:

In summer, the maximum levels of bacteria were equal to 57, 500 CFU/m^3 in the kitchen and bathroom and 32, 500 CFU/m^3 in the living room with average values between 6, 090 and 13, 241 CFU/m^3.

In winter, the maximum levels of bacteria were equal to 12, 600 CFU/m^3 in the kitchen, 50, 400 CFU/m^3 in the bathroom, and 6, 300 CFU/m^3 in the living room with average values between 2, 406 and 7, 903 CFU/m^3.

The highest levels of staphylococci in air were, in the summer, between 26, 125 CFU/m^3 in the living room and 55, 000 CFU/m^3 in the bathroom with average values between 4, 308 CFU/m^3 and 7, 200 CFU/m^3. In winter, the maximum values were between 10, 250 CFU/m^3 in the kitchen and 55, 000 CFU/m^3 in the bathroom, with the average values between 3, 219 and 6, 421 CFU/m^3.

The air pollution with fungi was visualized as a layer on the Petri dish in the kitchen and bathroom, and in the living room the maximum level registered was equal to 1, 820 CFU/m^3. The fungi species identified were the same like in the present study.

Another study performed in Torun University in Poland [45] had the following results:

- The levels of air contamination with bacteria were between 7 and 3, 313 CFU/m^3 as the background and between 53 and 5, 673 CFU/m^3 air at different sample size. The amount of bacteria found in the outdoor air ranged from 10 to 310 CFU/m^3.
- The air pollution with fungi ranged between 0 and 893 CFU/m^3 in the morning and 0 and 1, 373 CFU/m^3 in the afternoon. The outdoor fungal background ranged between 217 and 3, 750 CFU/m^3.

Another study that investigated the levels of biological contamination of indoor air in eight libraries of the Jimma University in Addis Ababa [46] had the following results:

- The levels of bacteria were between 367 and 2, 595 CFU/m^3 with an average value equal to 1, 476. CFU/m^3.
- The levels of fungi were between 524 and 1, 992 CFU/m^3 with an average value equal to 1, 087 CFU/m^3.

- The isolated bacteria were Micrococcus sp., S. aureus, Streptococcus pyogenes, Bacillus sp., and Neisseria sp.
- The isolated fungi were *Cladosporium* sp., *Alternaria* sp., *Penicillium* sp., and *Aspergillus* sp.

The microbiological quality of air in the investigated rooms in the Bucharest school as well as in the Giurgiu school changed significantly during the school day. The highest levels were found around 11:00 a.m. in classrooms with large numbers of persons inside. Most of the levels identified in the study were under the limit of 1, 000 CFU/m^3, with the exceptions of one measurement in the school in Bucharest and four measurements in the school in Giurgiu City. Also, the presence of *Streptococcus* beta hemolytic and coliform bacteria in classrooms air represents a threat to the health of children. This threat appears to be greater in the school in Giurgiu City.

The fungi levels identified in offices and archives were over the limit accepted in Romania. Isolated fungi species include *Cladosporium, A. flavus, A. niger*, and *Penicillium* sp.

In the case of the shop with only a ground floor and the supermarket, it is obvious that the size of the shop, the numbers of floors, and the numbers of departments in the shop can induce a high bacteriological contamination of air due to the high number of people visiting the shops. The air contamination in the supermarket was found to be higher at the ground floor and at the third and fourth floors. The high level of bacteriological contamination at the ground floor is probably due to the large number of visitors, while at the third and fourth floor it is probably due to the accumulation of bacteria coming from the first and second floors. In the shop with only a ground floor, the levels of air contamination in the evening were lower than in the morning. It is probable that this type of shop has better ventilation. High levels of staphylococci and the presence of *Streptococcus* in air can be dangerous for human health, and it is necessary to have a very strong disinfection action.

9. Conclusions

From this study, several conclusions can be drawn:

a. Biological contamination (bacteriological and fungal contamination) represents a constant element that was identified in all Romanian houses.

b. All results showed that the levels of bacteria in indoor air are higher than the allowed limit (2, 500 CFU/m^3 air for bacteria at 37°—limits admitted in Romania for rooms in a house). In schools, the levels of bacteriological contamination of air must not exceed 1, 500 CFU/m^3 for bacteria at 37°. The presence of coliform bacteria and hemolytic bacteria are not allowed in the air inside education units).

c. Concerning the indoor air pollution with fungi, the results showed a high level of contamination exceeding the allowed limit in Romania of 550 CFU/m^3 for fungi in air.

d. Biological contamination of air in schools can represent a threat for children health. This situation is obvious especially in schools outside Bucharest.

http://dx.doi.org/10.5772/59727

e. In work environment like offices or archives, the presence of fungi may represent a health risk for those who are exposed during the work process.

f. In the shops, a high biological contamination can be identified not only due to the numerous visitors but also by the numerous vendors present inside. The employees can be exposed during the working days to an important bacteriological load.

g. The hygienic quality of indoor air is revealed by the levels of biological air contaminants (bacteria and fungi).

h. A poor biological quality of air in all investigated environments was identified in the present study.

i. A constant exposure to biological contaminants (bacteria and fungi) from air can induce specific pathologies like asthma, allergic diseases, digestive diseases, and other infectious diseases. Among the isolated fungi, there were strongly allergenic and toxic species. The increased level of fungal flora in insufficiently ventilated rooms could be a reason for serious health problems of people occupying and working in those rooms.

j. All the investigated places require measurement to ensure good ventilation and proper measures of hygiene.

Author details

Anca Maria Moldoveanu[1,2]

Address all correspondence to: anca.moldoveanu@gmail.com

1 Carol Davila University of Medicine and Pharmacy, Bucharest, Romania

2 National Institute of Public Health, Bucharest, Romania

References

[1] Nevalainen A, Morawska Lidia, Biological Agents in Indoor Environments Assessment of Health Risks, Work conducted by a WHO Expert Group between 2000-2003. Edited by Aino Nevalainen and Lidia Morawska (2009).

[2] WHO Guidelines for Indoor Air Quality: Dampness and Mould, World Health Organization, Copenhagen, Denmark (2009).

[3] World Health Organization. Indoor Air Pollutants: Exposure and Health Effects, Copenhagen, WHO Regional office for Europe (EURO Reports and Studies No 78) (1983).

[4] Park JH, Spiegelman DL, Gold DR, Burge HA, Milton DK. Predictors of airborne endotoxin in the home. Environ Health Persp (2001), 109:859-64.

[5] Bischof W, Koch A, Gehring U, Fahlbusch B, Wichmann HE, Heinrich J. Predictors of high endotoxin concentrations in the settled dust of German homes. Indoor Air (2002), 12(1):2-9.

[6] Baun-Fahrlander C, Riedler J, Herz U, Eder W, Waser M, Grize L, Maisch S, Carr D, Gerlach F, Bufe A, Lauener RP, Schierl R, Renz H, Nowak D, Von Mutius Erika. Environmental exposure to endotoxin and its relation to asthma in school-age children. N Engl J Med (2002), 347(12):869-77.

[7] Waser M, Schierl R, Erika Von Mutius, Maisch S, Carr D, Riedler J, Eder W. Schreuer M, Nowak D, Braun-Fahrländer C. Determinants of endotoxin levels in living environments of farmers' children and their peers from rural areas. Clin Exp Allergy (2004), 34:389-397.

[8] Park J-H, Spiegelman DL, Burge HA, Gold DR, Chew GL, Milton DK. Longitudinal study of dust and airborne, endotoxin. Environ Health Perspect (2000), 108(11): 1023-1028.

[9] Douwes J. (1-3)-β-D-glucans and respiratory health: a review of the scientific evidence. Indoor Air (2005), 15: 160-169.

[10] Douwes J, Zuidhof A, Doekes G, van der Zee S, Wouters I, Boezen HM, Brunekreef B. (1 → 3)-β-D-glucans and endotoxin in house dust and peak flow variability in children. Am J Respir Crit Care Med (2000), 162: 1348-1354.

[11] Haalen Khan AA, Karuppayil SM. Fungal pollution of indoor environments and its management. Saudi J Biol Sci (Oct 2012), 19(4): 405-426.

[12] European Collaborative Center—Indoor Air Quality & Its Impact On Man, Report No. 12, Biological Particles in Indoor Environment, Luxembourg (1993).

[13] Institute of Medicine (IOM). National Academy of Sciences, Committee on the Assessment of Asthma and Indoor Air. Clearing the Air: Asthma and Indoor Air Exposures. National Academy Press, Washington DC (2000).

[14] Simpson JC, Niven RM, Pickering CA, Fletcher AM, Oldham LA, Francis HM. Prevalence and predictors of work related respiratory symptoms in workers exposed to organic dusts. Occup Environ Med (1998), 55(10):668-672.

[15] Flamant-Hulin M, Annesi-Maesano I, Caillaud D. Relationships between molds and asthma suggesting non-allergic mechanisms. A rural-urban comparison, Pediatr Allergy Immunol (2013), 24: 345-351.

[16] Watcharoot K, Mendell MJ, Gaffin JM, Wang G, Phipatanakul W. Indoor environmental exposures and exacerbation of asthma: an update to the 2000 review by the Institute of Medicine. Environ Health Perspect (January 2015), 123(1): 6-20.

[17] Mendell MJ, Mirer Anna G, Kerry Cheung Kerry, Tong My, Douwes J. Respiratory and allergic health effects of dampness, mold, and dampness-related agents: a review of the epidemiologic evidence. Environ Health Perspect (2011), 119:748-756.

[18] Annesi-Maesano I. Indoor exposition to molds and health outcome. Rev Allergy Clin Immunol (2013), 23; 21-25.

[19] Simpson JC, Niven RM, Pickering CA, Fletcher AM, Oldham LA, Francis HM. Prevalence and predictors of work related respiratory symptoms in workers exposed to organic dusts. Occup Environ Med (1998), 55(10):668-672.

[20] Fung F, Hughson WG. Health effects of indoor fungal bioaerosol exposure. Proc Indoor Air 2002, 46-51.

[21] Joshi SM. The sick building syndrome. Indian J Occup Environ Med (Aug 2008), 12(2): 61-64.

[22] Shoemaker RC, House DE. Sick building syndrome (SBS) and exposure to water-damaged buildings: time series study, clinical trial and mechanisms, Neurotoxicol Teratol (2006), 28(5):573-588, Epub 2006.

[23] Fraczek K, Górny RL. Microbial air quality at Szczawnica sanatorium, Poland. Ann Agric Environ Med (2011), 18, 63-71.

[24] Madureira J, Paciência I, Rufo JC, Pereira C, Teixeira JP, de Oliveira FE. Assessment and determinants of airborne bacterial and fungal concentrations in different indoor environments: homes, child day-care centres, primary schools and elderly care centres. Atmos. Environ (2015), 109: 139e146.

[25] Solomon J. Leviston-Porter School District—Annual Microbiological Assessment. Leviston-Porter Elementary School, New York (2005).

[26] Stryjakowska-Sekulska M, Piotraszewska-Pajak A, Szyszka A, Nowicki M, Filipiak M. Microbiological quality of indoor air in university rooms. Polish J Environ Stud (2007), 16(4): 623-632.

[27] Méheust D, Le Cann P, Reboux G, Millon L, Gangneux J-P. Indoor fungal contamination: health risks and measurement methods in hospitals, homes and workplaces, http://informahealthcare.com/mby. ISSN: 1040-841X (print), 1549-7828 (electronic). Crit Rev Microbiol. Online: 2013 Informa Healthcare USA.

[28] Ulla L. Characterization of microorganisms in indoor environments. Publications of the National Public Health Institute—Academic Dissertation, Kuopio (2008).

[29] IRSST, Occupational Health and Safety Research Institute Robert Sauvé: Bioaerosol in Workplace: Evaluation, Control and Prevention Guide. Technical guide, 2001, www.irsst.qc.ca.

[30] AIHA. Field Guide for the Determination of Biological Contaminates in Environmental Samples. Fairfax, VA: American Industrial Hygiene Association (1996).

[31] Abel E, Andersson JV, Dawidowicz N, Christophersen E, Hanssen SO, Lindén A-L, Lindvall T, Pasanen AL. The Swedish key action 'The Healthy Building'—research results achieved during the first three years period 1998-2000, Indoor Air 2002. Proceedings of the 9th International Conference on Indoor Air Quality and Climate (2002), 2:996-1001.

[32] de Aquino Neto FR, de GóesSiqueira LF. Guidelines for indoor air quality in offices in Brazil. Proceedings of Healthy Buildings (2000), 4:549-554.

[33] Nunes ZG, Martins AS, Altoe ALF, Nishikawa MM, Leite MO, Aguiar PF, Fracalanzza SEL. Indoor air microbiological evaluation of offices, hospitals, industries, and shopping centers. Mem Inst Oswaldo Cruz (2005), Vol. 100(4): 351-357.

[34] Lee SC, Li W-M, Ao C-H. Investigation of indoor air quality at residential homes in Hong Kong—case study. Atmos Environ (2002), 36:225-237.

[35] Li WM, Lee SC, Chan LY. Indoor air quality at nine shopping malls in Hong Kong. Sci Total Environ. (Jun 2001), 12;273(1-3):27-40.

[36] Huang P-Y, Shi Z-Y, Chen C-H, Den W, Huang H-M, Tsai J-J. Airborne and surface-bound microbial contamination in two intensive care units of a medical center in Central Taiwan. Aerosol Air Qual Res (2013), 13: 1060-1069.

[37] Hsu C-H, Lu M-C, Huang D-J. Application of chlorine dioxide for disinfection of student health centers. Environ Monit Assess (2012) 184:741-747.

[38] Katiyar V. Assessment of indoor air micro-flora in selected schools. Adv Environ Res (2013), 2(1): 61-80.

[39] Obbard JP, Fang LS. Airborne concentrations of bacteria in a hospital environment in Singapore. Water Air Soil Pollut (2003), 144 (1): 333-341.

[40] American Conference of Governmental Industrial Hygienists (ACGIH). Guidelines for the Assessment of Bioaerosols in the Indoor Environment. ISBN-13: 978-0936712833 (1989).

[41] World Health Organization (WHO). Indoor Air quality: Biological Contaminants: European Series, Number No. 31. WHO Regional Publication, Copenhagen (1988).

[42] Ministry of Environment (MOE). Korean Indoor Air Quality Control in Public Use Facilities, etc. Act. Republic of Korea (2011).

[43] Lee JH, Ahn KH., Yu IJ. Outbreak of bioaerosols with continuous use of humidifier in apartment room. Toxicol Res (June 2012), 28(2): 103-106.

[44] Moldoveanu AM. Indoor air pollution in Romanian homes. In: Environmental Technologies—New Developments. ISBN: 978-3-902613-10-3. I-Tech Education and Publishing, Wien (2008).

[45] Kalwasińska A, Burkowska A, Wilk I. Microbial air contamination in indoor environment of the university library. Ann Agric Environ Med (2012); 19(1): 25-29.

[46] Hayleeyesus SF, Manaye AM. Microbiological quality of indoor air in university libraries. Asian Pac J Trop Biomed (May 2014), 4(Suppl 1): S312-S317.

Chapter 23

Microclimate and Indoor Air Quality in an Operating Theatre under real use Conditions — An Experimental and Numerical Investigation

Carla Balocco and Giuseppe Petrone

Additional information is available at the end of the chapter

http://dx.doi.org/10.5772/59671

1. Introduction

The study of OT ventilating and IAQ conditions is particularly important. Campaigns of experimental measurements in OTs are usually expensive, invasive and, in many countries, require complex procedures for legal authorization. CFD simulation has opened up low cost methods to provide useful indications on proper indoor microclimate conditions and IAQ. [1, 2,3]. There is a great dealing the literature on airflow patterns, air velocity and temperature distribution in unidirectional (vertical downward) airflows, as well as other applied ventilating layouts, such as air curtains use to achieve a chosen air quality level inside OTs using CFD transient simulations [4.5,6,7,8]. Other authors have studied the airflow and temperature distribution modifications due to different surgical lamp shapes and locations [9]. In particular [10], the effectiveness of the ventilation system and airborne bacteria removal due to the HVAC plant have been analysed by means of CFD simulation supported by experimental data obtained in a test room A comprehensive experimental and numerical analysis has been proposed by Kameel and Khalil [11] concerning both the airflow regimes and heat transfer inside an OT considering the correct operational use conditions. Few analyses have been made concerning the effect of moving objects on OT airflow: the "computationally expensive" moving mesh approach has been applied in order to manage the fluid-solid interface during transient simulations [12,13], but similar investigations were proposed for a hospital isolation room [14,15] and an industrial cleanroom [16]. An impressive study concerning the influence of movements on contaminant transport has been proposed by Brohus et al. [17] using the indirect approach, based on distributed momentum and turbulent kinetic energy sources. Recently, the present authors proposed an alternative indirect numerical approach in order to simulate solid objects moving on a fixed mesh in forced [18] and natural convection flows [19].

Definition and quantification of the contaminant sources under OT effective use conditions is a complex task. Existing databases refer to constant particle mass emission of a person as a function of his/her activity. Several studies analyze different sources and human activities causing high levels of indoor particle concentration, diffusion and re-suspension processes. Some authors [20] present studies on the influence of periodic bending movement of the surgeon on the air flow field and bacteria carrying particle distribution using a combined approach based on an Eulerian RANS model, modified drift-flux and moving mesh. In the study by Brohus et al. [17], bacteria attributed to the skin flake emission of staff and patient are simulated as a gaseous contaminant. The most widely used experimental methods allow particle size calculation for discrete intervals, so as to check whether microclimatic and IAQ conditions are in compliance with standards and specific rules. Some authors have developed a useful method for determining individual emission rates from ambient air contaminant concentrations caused by multiple indoor sources [21]. Other studies provide a sampling method integrated with statistical analysis. [22]. In a recent paper bacteria sources are investigated [23] in a university classroom both occupied and vacant as concentration differences due to the two conditions. Besides the influence of the airflow pattern, it has been demonstrated that particle dynamic behaviour is strongly dependent on particle size and size-related forces [24]. Chih-Shan Li et al. [25] carried out a series of field tests by active and passive sampling of air and surface measuring bacterial and fungal concentrations without referring to particle size and diameter. Another paper [26] based on particle counting combined with statistical analysis, evaluates aerobic bacterial sedimentation and connected index of microbial air contamination. There, results refer to different particle diameters but not emission sources. Some authors [1] have carried out transient simulations using the Renormalization Group (RNG) k-ε turbulence model, assuming a released rate of bacteria-carrying airborne particulates from surgical staff of 100 BCP/min per person and for patients of 400 BCP/min per wound. In general, CFD simulation concerning aerosol particle transport and diffusion processes can be solved by three main different numerical approaches: the first method consists in solving the particle concentration field by using diffusion-transport equations based on "passive scalar transport" in which the vector of particle transport is the motion field of the fluid but gravitational effects and frictional forces on particle motion are disregarded; the second is the Eulerian-Eulerian method and the third is the Eulerian-Lagrangian one. These last two methods solve the airflow field based on the Eulerian approach, but the particle phases are treated differently: in the first, particle concentration is directly solved using gravitational effect in the transport term, in the second one an ordinary differential equation is solved for any particle path. A modified drift-flux model based on the second approach has been applied to modelling particle transport [27,28,20]. At present, there are not many studies in the literature concerning numerical modelling in which the contaminant concentration for the IAQ analysis has been performed starting from sources of particle emission with a distribution dependent on particle diameter [29,30]. However, Italian and international standards require, for environment classification, a particle count for diameter dimension. Actually, there is a lack of data concerning specific emission sources in terms of particles issued in time unit by people. A recent study provides the quantification of size-resolved particle concentrations in indoor air of a classroom under occupied and vacant conditions [29]. Another important

http://dx.doi.org/10.5772/59671

numerical study provides information on airflow, particle deposition and movement in two different spaces equipped with displacement and mixing ventilation modes [30]. In our present research airflow and climate in an existing orthopaedics OT under both the "at rest" and real use conditions were investigated by experimental measurements and numerical simulations. The "operational" conditions should be understood with faked surgery, split into two conditions respectively "correct operational use" and "incorrect operational use". Two main purposes were pursued in our investigation: the first one consists in exploiting the experimental acquisitions in order to check whether the environmental parameters respect standard requirements. Numerical results are then correlated with the microclimatic parameters suggested by the standards to verify the model suitability for assessing the OT indoor climate according to suggested standard limits, starting from the design variables of the room and HVAC system. The second purpose focuses on the airflow and microclimatic parameter variations induced by the operative use conditions. Our study develops an investigation and applied research to obtain a distribution of emitted particles as a function of their diameter, assuming that the amount of particle mass emitted by a person in a time unit does not change with their size. Starting from this assumption, the distribution of emitted particles will consist of a larger number of a smaller size, and vice versa. Our study investigates the thermo-fluid dynamics combined with the analysis of the concentration of diluted gas phase and particle aerosol dispersion in the air. Some IAQ indexes, widely used in the literature, were calculated as global and local values referring to the total volume of the room but also to the critical zones defined by ANSI/ASHRAE [31]. The present investigation is organized, referring to air flow and climate, ventilation efficiency and IAQ in the investigated real OT, as follows: the second section presents the experimental measures; the third one concerns the numerical tools and model validation; results are reported in the fourth section; then, conclusions, basic findings and possible developments due to the application of our proposed approach to similar cases, follow.

2. Materials and methods

2.1. Experimental measurements and data analysis

Experimental measurements were conducted in a real orthopaedic OT of the University Hospital of Parma (Italy) referring to [32,33] and during the different conditions as given: **at rest**, i.e. room with all services functioning and equipment installed and operable/operating, but without surgical/healthcare staff and the patient; **operational "correct use condition"** reproducing the effective use conditions of the OT, with all services and equipment functioning, surgical and healthcare presence. Experimental data were collected after installation of all acquisition equipment in the room and during simulated hip surgery lasting 20-25 minutes and involving some PhD students. The surgical staff and healthcare assistants were in upright stationary positions and surrounding the patient lying on the operating table. The sliding door was closed; **operational "incorrect use condition"**, i.e. the same conditions described for operational "correct use", but considering the sliding door opening and closing and one healthcare assistant walking from outside through the door and up to the patient's head on the operating table, and then turning round and moving back to the corridor. The HVAC system was working during all tests, supplying a constant incoming airflow rate at controlled temperature. The measurements campaign was carried out over a two day period, from 8 a.m. to 2 p.m. Microclimate parameters were measured over time at discrete points PT01, PT02, PT03 that are shown in Figure 1. In particular: 1 hot wire anemometer, 1 air-temperature and humidity sensor and 1 globe thermometer were set below the central air inlet diffuser on the plane of the surgical table (PT01); 1 air temperature and humidity sensor and 1 differential pressure sensor near the sliding door on a tripod (PT02); 1 differential pressure sensor and 1 air temperature and humidity sensor were set near the air outlet diffuser on a utility pole (PT03). Differential air pressure, between monitored room and the adjoining ones, was evaluated in proximity to the sliding door, at 1 m with respect to the floor. Some telescopic tripod poles were used for instrument arrangement during tests. Instruments were connected to a radio master R-Log data logger system. Measurements were made every two seconds. In Table 1 the average values of experimental measures upon N samples ($EXP_{AV} = 1/N\sum_{i=1}^{N} EXP_i$) are shown for the different room conditions. Data dispersion with respect to the average values was quantified by means of the sample standard deviation $\sigma_{(N-1)}$, computed using Bessel's correction and the percentage deviation with respect to the average value is also shown in Table 1. For air active sampling, a particle counter (Climet CL 754) was used. The equipment can sample a volume of 75 l/min of air and gives the number of suspended particles, divided according to the diameter (≥0.3; ≥0.5, ≥1.0; ≥5.0 μm), allowing one to deduce particle concentration in the diameter ranges (0.3-0.5 μm); (0.5-1 μm); (1-5 μm). Experimental data of air temperature and relative humidity (RH) are globally consistent: standard deviations are narrow, much more for temperature (less than 1 °C) than for RH (less than 3 percentage points); normalized deviation does not exceed a 6% threshold in any cases. At point PT01 a detected dispersion of experimental data appears, that becomes very significant for air velocity measurements. The higher variation of the measured velocity in PT01 during operational conditions, due to medical staff movement, produces an increase in RH levels in the central zone (PT01). Temperature values, recorded during operational conditions, are slightly lower than values in at rest conditions, because the local ventilation system operates in the full working phase. Analysing particle concentration measurements, excluding PT1 at operational conditions for particles of 0.3 μm diameter and PT05 at rest conditions for particles of 5 μm diameter, particle distribution for each diameter is similar for the two OT use conditions at the different measurement points.

As expected, the highest particle concentration values are during operational conditions. Referring to the absolute value, the higher differences between the two OT use conditions were checked at PT05 for each particle diameter, while the lower differences were checked at PT01. Particle concentration measured at PT01 for the two use conditions of OT and for each diameter, respects the standard limit imposed by the ISO5 at rest classification [34,35] and Grade B EU-GMP clean-room classification (Annex1), in accordance with EN ISO 14644, for both at rest and operational conditions. Referring to the GMP at operational conditions, at PT01 the conditions imposed by Grade B are respected by all the various particles (for all different

	T [°C]			T_{mr} [°C]	U [m/s]	ϕ [%]			
	PT01	PT02	PT03	PT01	PT01	PT01	PT02	PT03	
EXP_{AV}	**21.2**	**21.4**	**21.3**	**21.6**	**0.29**	**63.2**	**61.5**	**61.9**	At rest
$\sigma_{(N-1)}$	0.76	0.33	0.42	0.43	0.03	2.70	1.08	1.45	
$\iota = \sigma_{(N-1)}/EXP_{av}$	*3.6%*	*1.5%*	*2.0%*	*2.0%*	*10.9%*	*4.3%*	*1.8%*	*2.3%*	
EXP_{AV}	**20.1**	**21.1**	**21.1**	**20.9**	**0.23**	**65.6**	**61.8**	**62.1**	Operational "correct use"
$\sigma_{(N-1)}$	1.03	0.41	0.44	0.59	0.07	2.91	1.32	1.10	
$\iota = \sigma_{(N-1)}/EXP_{av}$	*5.1%*	*1.9%*	*2.1%*	*2.8%*	*29.3%*	*4.4%*	*2.1%*	*1.8%*	
EXP_{AV}	**20.1**	**21.3**	**21.2**	**20.9**	**0.28**	**67.1**	**61.8**	**62.1**	Operational "incorrect use"
$\sigma_{(N-1)}$	0.71	0.30	0.35	0.74	0.08	2.71	0.97	1.45	
$\iota = \sigma_{(N-1)}/EXP_{av}$	*3.6%*	*1.4%*	*1.6%*	*3.6%*	*30.0%*	*4.0%*	*1.6%*	*2.3%*	

Table 1. Microclimatic parameters measured at the discrete points for different conditions: average values (EXP_{av}), sample standard deviation ($\sigma_{(N-1)}$) and percentage error (ι).

diameters) at operational conditions, and the concentration of 0.5 μm diameter particles is within the standard limits for operational conditions, but in particular at PT01, PT03, PT04 and PT05, it is higher.

2.2. Numerical modelling

A solid model of the OT was made: the OT geometry is outlined by a rectangular-shaped room with smoothed corners, a 43 m^2 base area and 120 m^3 volume (Figure 1).

A sliding door 1.4 m wide and 2.2 m high connects the room to an entry corridor. The locations of the staff members, surgical lamps and equipment for the operating table, defined in the model are shown in the same Figure 1. The surgical staff and patient bodies are outlined as rectangular solid boxes. The room is equipped with an operating bed and lighting system made up of three joined arms, each one holding three lamps. Two further solid boxes represent an operating trolley and an electro-medical device. One of the walls of the room has a window, facing an internal space. There are two rectangular supply ceiling diffusers each of 0.56 m^2 surface, located in the central zone of the ceiling, that strengthen the unidirectional flow. Each diffuser provides a constant flow rate of 3969 m^3h^{-1} of fresh air, so that there are 66 total air changes per hour in the room. Although the real HVAC system is temperature-controlled by a remote ambient thermostat located inside the room, in the numerical models a constant inlet temperature value is set for the incoming fresh air. Two groups of 14 conical outgoing grilles (cross-section of 0.0128 m^2) are arranged over two of the opposite four walls corresponding to the OT smoothed corners. Following the indications given in the standards [36,37], the global OT air-volume (TV) is divided into 3 zones, labelled from now on as Breathing Zone (BZ, pink

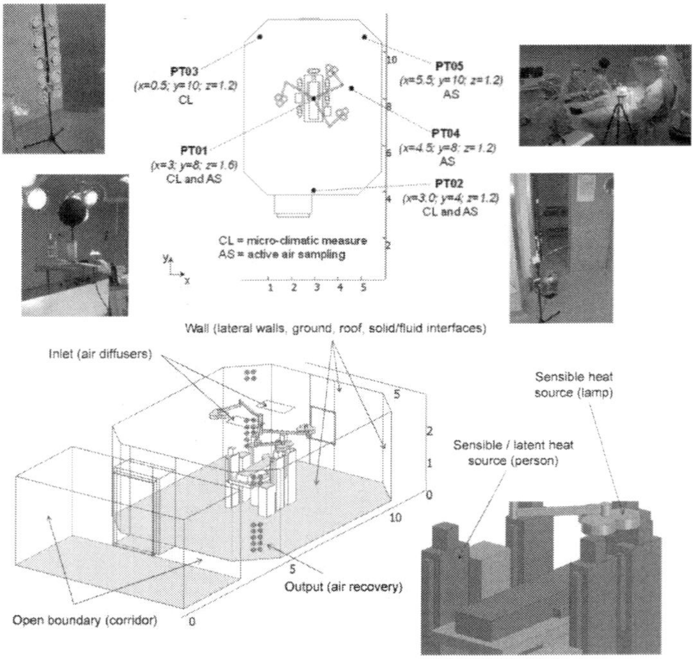

Figure 1. System description: in the map (upper portion of the figure) positions (labelled as PT01, PT02, PT03, PT04 and PT05) where instruments were located are shown; some photos taken during the experimental campaign are provided. In the bottom portion, the geometry of the numerical model: solid elements used for simulating people, room furnishing (operating table, trolleys) and equipment (ceiling diffusers, lighting system) are shown as well as some indications about physical constraints for the system.

in Figure 2), Occupied Zone (OZ, blue in Figure 2) and Peripheral Zone (PZ, corresponding to TV-OZ). This allowed the computation of air quality indexes for each zone. Numerical models were built-up to simulate airflows, climate and air quality conditions in the real orthopaedic OT used for experimental measurements.

Commercial software, that allows multi-physical analyses through solution of the related governing equations by a Finite Element approach [38], was used. Fluid-dynamics and thermal analysis were firstly solved under the assumption of Newtonian fluid and incompressible flow. A standard k-ε closure scheme [39] was used for solving velocity and pressure field by an eddy viscosity approach. Then, further transport-diffusion equations were considered for solving RH, CO_2 concentration, mean age of air and particulate concentration for diameters 0.4, 0.75, 3 μm. These dimensions represent the average particle diameter referring to the particle diameter range that were experimentally measured ((0.3-0.5 μm); (0.5-1 μm); (1-5 μm)). The basic formulation of the PDE used for computations is:

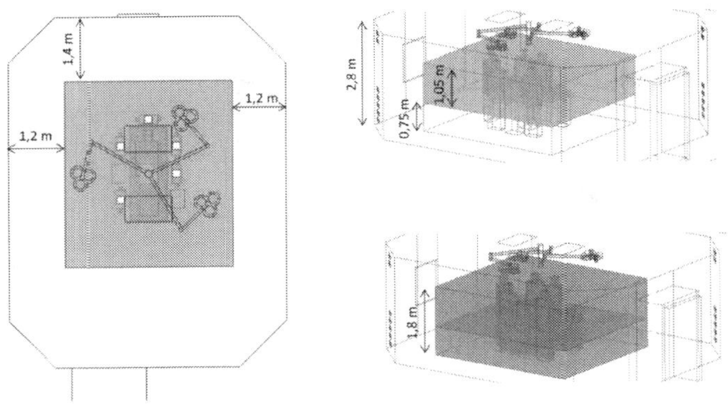

Figure 2. OT zones description: BZ (pink) and OZ (blue).

$$\frac{\partial(\rho\phi)}{\partial t}+\nabla\cdot(\rho\phi\mathbf{U})=\nabla\cdot(\Gamma\nabla\phi)+\Lambda \tag{1}$$

where ρ is the fluid density and $\mathbf{U}(u, v, w)$ is the air velocity vector. Table 2 shows the analytical formulation of different terms in eq. (1), depending on the specific physics referred to. From a physical point of view, source term F corresponds to the buoyancy force, Q_S is the metabolic sensible heat and Q_L the latent one. For comparing room ventilation performance in different conditions, the mean age of air (τ) was computed: it quantifies the average lifetime of air at a particular location of the room [41,42,43] once a steady state is achieved in terms of airflow patterns. In solving the particulate concentration, because we considered diameters higher than 0.01 μm, the Brownian diffusivity was disregarded with respect to turbulent diffusivity, that was assumed to be equal to the air kinematic turbulent viscosity [44]. Adopting an Euler approach, the following settling velocity was added to the transport vector vertical component:

$$\mathbf{w}_S=\frac{C_c\rho_p d_p^2}{18\mu}\cdot\mathbf{g} \tag{2}$$

with C_c the Cunningham coefficient, ϱ_p and d_p particulate density and diameter, respectively. Source terms and physical properties values used in governing equations are listed in Table 3.

Equation	ϕ	Γ	Λ
Continuity	1	0	0
Momentum	$\mathbf{U}$	$\mu + \mu_T$	$-\nabla p + \mathbf{F}$
Turbulent kinetic energy	k	$\mu + \frac{\mu_T}{\sigma_k}$	$\frac{1}{2}\mu_T\left[\nabla\mathbf{U} + (\nabla\mathbf{U})^{\mathrm{T}}\right]^2 - \rho\varepsilon$
Dissipation rate of kinetic energy	ε	$\mu + \frac{\mu_T}{\sigma_\varepsilon}$	$\frac{1}{2}C_{\varepsilon 1}\frac{\varepsilon}{k}\mu_T\left[\nabla\mathbf{U} + (\nabla\mathbf{U})^{\mathrm{T}}\right]^2 - \rho C_{\varepsilon 2}\frac{\varepsilon^2}{k}$
Energy	T	$\frac{\lambda}{C_p}$	$\frac{Q_S}{C_p}$
Relative humidity	RH	$\frac{\delta_p p_{sat}}{\xi} + D_w$	$\frac{Q_L}{\xi}$
Mean age of air	τ	Υ	ρ
CO_2 concentration	CO_2	$\frac{D_{CO_2}}{\rho}$	0
Particulate concentration	C_{d_i}	$\frac{D + \varepsilon_p}{\rho}$	0

Table 2. Analytical formulation of terms in eq. (1), depending on the specific physics referred to.

	ρ	η	λ	C_p	F	Q_S	Q_L
Material	[kg/m^3]	[Pa s]	[W/(m K)]	[J/(kg K)]	[N/m^3]	[W]	[W]
Glass (window, lamp)	2500	-	1.00	800	-	4025 (lamps)	-
Wood (door)	500	-	0.50	1000	-	-	-
Aluminium (operating table, electro-medical case, lamp arms)	2700	-	160	900	-	-	-
Medical staff	950	-	0.62	4180	-	73.3	58.6
Patient	950	-	0.62	4180	-	65.9	30.7
Air	$p/(RT)$	2E-5	0.04	1004	$\rho g\beta\Delta T$	-	-

Table 3. Physical properties of materials and source term values.

Movements of the sliding door and of one healthcare assistant were numerically simulated to study the incorrect operational use conditions. In transient analyses, once the sliding door is open, one assistant is expected to walk from the corridor space inside the room, moving along the surgical table up to the patient's head, then turning around and leaving the room. The mean velocity values applied for a person walking and door opening/closing are respectively 0.7 m s^{-1} s and 0.28 m s^{-1}. The procedure adopted for simulating the "moving objects" inside the room was presented in detail in recent studies by the authors [18,19]. It is mainly based on the definition of specific source terms in the governing equations, assuming assigned values in the portions of the computational domains where the solid objects are located at a chosen time. Solid objects were not explicitly designed in the geometrical model because their movement is driven by assumed values 0 or 1 in specific logical functions during time. Solid object time-displacements and paths considered during transient analyses are shown in Figure 3. The black line represents displacement of the "moving object" sliding door (x direction into the model reference system), while the red line indicates movement belonging to the "moving object" healthcare assistant (y direction). The grey dashed lines schematically separate the different steps characterizing the simulated dynamics, that are highlighted by progressive numbers from 1 to 7, as shown in the grey horizontal bars in the upper part of Figure 3. Assuming the initial instant with the door closed and medical assistant standing still in front of it, in the corridor, 7 consecutive steps were simulated: step1 (0-5 seconds): sliding door opens, and medical assistant is standing still, waiting for the door to open; step 2 (5-11 seconds): the door is completely open and medical assistant is walking through it. After 1 second, the door starts to close, while the person is walking in the room; step 3 (11-15 seconds): the door is closed and medical assistant walks through the room until he reaches the operating table top; step 4 (15-27 seconds): medical assistant stops for 3 seconds, then moves in the opposite direction towards the sliding door and finally stops inside the room in front of the door waiting for it to open; step 5 (27-32 seconds): medical assistant is standing still until the door is completely open; step 6 (32-33 seconds): the door is completely open and the person is walking through it and then stopping in the corridor; step 7 (33-38 seconds): sliding door closes.

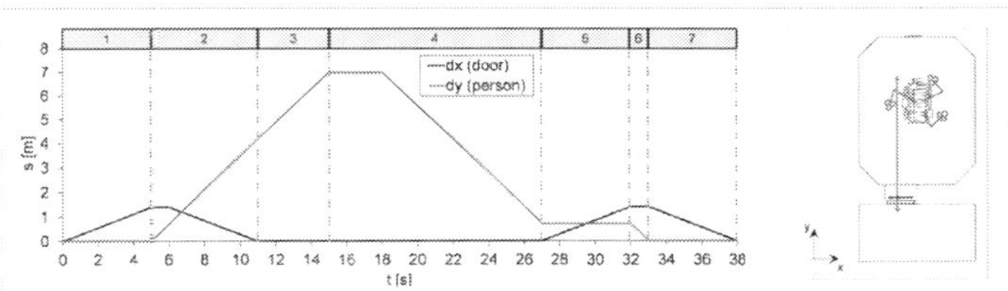

Figure 3. Path on the room map (right side) and time-displacement (left side) of the "moving objects": the black line identifies the transient position of the sliding door (moving along the x-axis in the chosen system of coordinates), the red one refers to the healthcare assistant (moving along the y-direction). The grey dashed lines separate the different steps characterizing the simulated dynamics (numbers in the grey horizontal bars).

The computational domain, representing the corridor, was considered only for simulations involving the sliding door opening/closing (incorrect use conditions). In this case, open boundary conditions were considered at the corridor transversal section. The following first-type boundary conditions for supplied air at the ceiling diffusers were assumed: velocity magnitude (1.97 m/s), turbulence intensity (5%), air temperature (18 °C), RH (60%), CO_2 concentration (350 ppm) and mean age of air (0 s). For modelling persons breathing, a RMS value, corresponding to the sinusoidal trend for inhaled/expired air, and for CO_2 emission rate into the room were assumed. Adopted values in simulations were 2.0 m^3/h (medical staff) and 0.3 m^3/h (patient) for breathed airflow and 0.080 m^3/h (medical staff) and 0.012 m^3/h (patient) for CO_2 emission rate. Outflow conditions were considered for all dependent variables at the recovery grids. At each solid/fluid interface, logarithmic wall functions were applied in the near wall airflow, that was considered parallel to the wall and being in a wall offset equals one hundred viscous units. Turbulent production was established equal to dissipation at walls.

For the remaining dependent variables impermeable/insulation conditions at walls were imposed. The occupants contribution for particle emission rate was applied as a boundary flux [particle/(m^2 s)] at the occupants/surrounding air interfaces. The procedure adopted to assess the particulate flux depending on particle dimension is explained in the next section of the present chapter. In thermal analysis, a convective thermal flux was applied to the walls, considering a heat exchange coefficient of 7.7 (W/(m^2 K)) and a constant temperature of 20 °C in the adjacent rooms. Insulation conditions were applied to the solid/fluid interfaces for all other dependent variables that were solved. For the "at rest" conditions, the lamps were considered the only internal sources of sensible heat. Otherwise, in operational "correct use" conditions, internal heat sources were related to medical staff and patient presence, as sensible and latent heat loads. Other boundary conditions were not considered to change from "correct use" to "incorrect use" operational conditions: heat and vapour source values were unchanged, except for the additional load due to the walking healthcare assistant, whose location was variable in accordance with the moving object position during time. The governing equations together with their boundary conditions were spatially discretized on non-structured grids, made of second order tetrahedral elements. Steady state solutions of discrete equations were carried-out by applying an iterative dumped Newton-Raphson scheme [45] based on the discretized PDE linearization by a first-order Taylor expansion. Algebraic systems of equations coming from differential operator discretization were solved by a PARDISO package, a direct solver particularly efficient to solve unsymmetrical sparse matrixes by a LU decomposition method. The convergence criterion was set to 1E-5. Time integration of governing equations for transient simulations was performed applying an Implicit Differential-Algebraic (IDA) solver [46], which uses variable-order and variable-step-size Backward Differentiation Formulas (BDF). Because the time-matching scheme is implicit, a nonlinear system of equations was solved at each time step. All computations were carried out on a workstation with two 64-bit 6-core/12-thread processors speeding up to 2.3 GHz of frequency and handling 128 GB of RAM.

3. Use of microclimatic measurements for numerical model validation

The influence of mesh refinement on results deviation was investigated to assure mesh independent results and assess the model reliability. Simulations were carried out in conditions labelled as "at rest". The accuracy and independence of results from mesh refinement were estimated by evaluating result variation (magnitude of air velocity and temperature) at point PT01 (see Figure 1), by increasing the number of elements of the computational grid. Table 4 provides the absolute deviation and normalized variance of the values compared to those obtained using the finest mesh (M4). It can be gathered that over a certain value of mesh refinement, the relative variance is lower than 2.5% for the velocity magnitude, i.e. the most sensitive variable with respect to mesh size. Numerical results were also compared with experimental data to check and validate the numerical model. Comparisons were made in points where probes were located (PT01, PT02, PT03). An extract of comparison is given in Figure 4a-c, where experimental and numerical results obtained for operational "correct use conditions" are compared with each other. In this figure the average value of the acquired data

($EX\,P_{AV}$) and the numerical result obtained in steady state conditions (NUM), were plotted together. The experimentally detected absolute deviation above ($EX\,P_{MAX} - EX\,P_{AV}$) and below ($EX\,P_{AV} - EX\,P_{MIN}$) the average value is presented by means of error bars. In the same diagrams, circled symbols to indicate the maximum difference between acquired data and simulations ($\max_i\{\mid EX\,P_i - NUM \mid\}$), and squared symbols to present the difference between experimental average data and numerical results ($\mid EX\,P_{AV} - NUM \mid$) were used. Deviations were normalized using experimental average values as reference for temperature and velocity (Figure 4a and 4c), while the absolute values of deviation for RH (Figure 4b) was retained.

Mesh #		M1	M2	M3	M4
Number of elements		517633	774 220	1092784	1714306
Mesh refinement		-	33%	29%	36%
Velocity magnitude in (3; 8; 1.4)	[m/s]	0.303	0.305	0.309	0.316
Absolute gap	[m/s]	0.013	0.012	0.007	0.000
Relative gap	[-]	4.2%	3.74%	2.20%	0.00%
Temperature in (3; 8; 1.4)	[°C]	20.145	20.154	20.156	20.156
Absolute gap	[°C]	0.011	0.002	0.000	0.000
Relative gap	[-]	0.05%	0.01%	0.00%	0.00%

Table 4. Mesh accuracy tests.

For temperature (Figure 4a), the maximum normalized difference between numerical and experimental values is 12.3% in PT01, 4.6% in PT02 and 5.6% in PT03. Referring to the average experimental values, the relative gaps decrease to 6.3% (PT01), 1.7% (PT02) and 1.8% (PT03). Comparison of numerical temperature in PT01 with the experimental T_{mr} highlights a maximum normalized difference of 5.6% and a normalized deviation to the average value of 2.3%. Referring to RH (Figure 4b), a maximum absolute deviation of 6.3 (PT01), 4.8 (PT02) and 4.5 (PT03) percentage points, and an absolute deviation to the average value of 1.9, 2.9 and 3.4 percentage points, respectively can be found. Analysing the air velocity value, obtained under the plenum (Figure 4c), the detected range of variation is quite high with respect to the mean value (0.23 m/s), and the simulated velocity magnitude (0.27 m/s) highlights a consistent gap. However, it should be noticed that numerical values globally stand within the range of variation of the measurements, foregrounding a satisfactory agreement with the experimental data. A good agreement can be noted between microclimatic experimental with numerical results with relative differences lower than 10%. Comparison between microclimate experimental data with average parameter values, suggested by Italian and International standards, show that [47, 48, 49, 50,51, 32, 52] the average temperature values at probe locations are within the suggested limits. This is always confirmed for the "at rest" condition, but for the "operational" one, the acquired values exceed the recommended thresholds due to medical staff presence and movements that produce local temperature and airflow modifications. The air velocity values globally fit the standard limits, suggested for unidirectional flow.

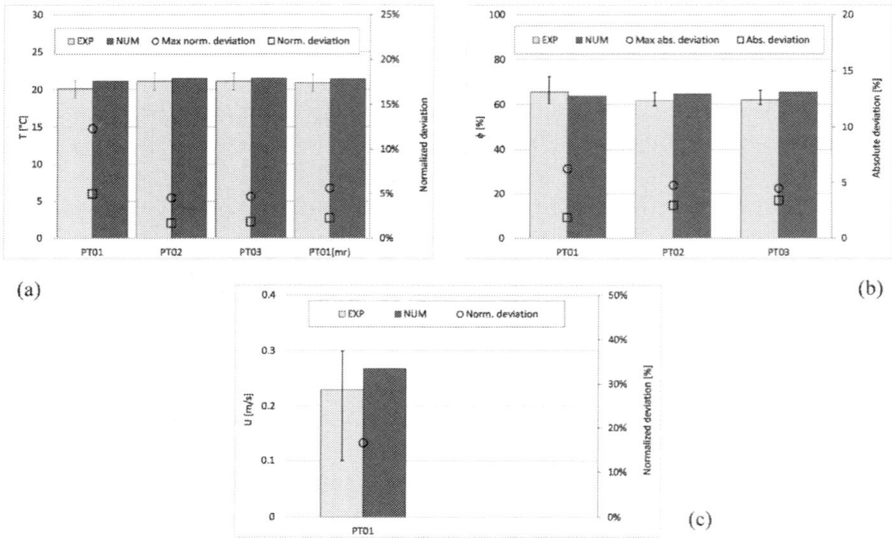

Figure 4. Comparison between numerical results (NUM, black grey) and mean-average experimental data (EXP, light grey). Error bars indicate the detected experimental deviation (minimum and maximum value). Circled and squared symbols (referring to the second y-axis) indicate the maximum deviation of numerical values from experimental data, and deviation of numerical results from experimental time averaged data, respectively. Temperature (a), RH (b) and velocity (c) values are reported at locations PT01, PT02 and PT03 (as shown in Figure 1).

4. Use of microbiological measurements for numerical model implementation and validation

Microbiological measurements of particle concentration were used to assess particle emission rates imposed as boundary conditions in the numerical models. Particle emission rate per person, distinguished for dimension, was not known a priori during our research. Zhao et al. [29] considered a constant generation rate of 0.0916 μg/s per person for each particle size group. For each particle dimension, referring to an assumed particle density of 1.05E-6 μg/μm^3, scaling the emission rate by the particle volume, it is possible to obtain the related number of particles emitted by a person in the time unit. Quian et al. [30] used a different method to assess particle emission rate, divided for different diameters, caused by the occupants in a classroom. They experimentally measured the particulate concentration at specific locations in the classroom for similar environmental conditions, but with and without people presence. They argued that

http://dx.doi.org/10.5772/59671

difference in acquisitions could be attributed to emissions due to occupant contribution, and then proposed a particulate emission distribution as a function of particle diameter. The emission rate refers to how many particles of a specific dimension are supplied by one person in the time unit to the surrounding environment. The two above mentioned approaches provide different results. In our investigation, we evaluated particle emission due to each occupant, differentiated for diameter dimension, using the experimental acquisitions of particle concentration in the real OT with and without patient and medical staff presence. In particular, we applied a similar method used in [30] to assess the particle emission rate by occupants to be used in the numerical models. Otherwise, in our case we combined the available experimental data with results coming out from numerical simulations to assess the emission rate of particles per person depending on particle dimension The procedure used is given in the flow-chart diagram of Figure 5: the caption "sampling points" refers to points PT01, PT02, PT04 and PT05, "particle diameter range" means (0.3-0.5 μm); (0.5-1 μm); (1-5 μm), and "average particle diameter" refers to the average values of the particle diameter (0.4; 0.75; 3 μm) computed for each of the previous ranges.

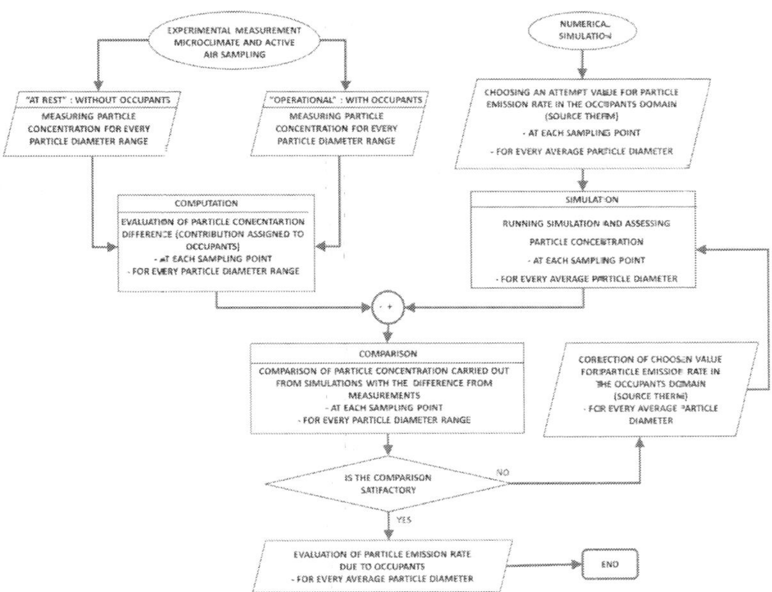

Figure 5. Flowchart of method used for estimating the occupant particle emission rate per diameter by an iterative cross-comparison between numerical and experimental data.

Comparison between experimental and numerical (NUM) particle concentration due to occupants in the OT is shown in Figure 6. In this figure, experimental value (EXP) for each diameter (d_i) represents the difference between average data (Av) acquired for the at rest (rest) and operational (oper) conditions, computed as follows:

$$EXP_{d_i} = Av\left\{EXP_{d_i}\right\}_{oper} - Av\left\{EXP_{d_i}\right\}_{rest} \tag{3}$$

while the error bars are plotted by assuming the following:

$$Err_{d_i}^{+} = \max\left\{EXP_{d_i}\right\}_{oper} - \min\left\{EXP_{d_i}\right\}_{rest} \tag{4}$$

$$Err_{d_i}^{-} = \min\left\{EXP_{d_i}\right\}_{oper} - \max\left\{EXP_{d_i}\right\}_{rest} \tag{5}$$

Numerical values plotted, refer to the "final" particle emission rate per average diameter applied to the exposed person surfaces carried out from application of the computing procedure shown in Figure 5. Values of emission rate for diameter range are provided in Table 5, where values suggested in the previous literature are also reported.

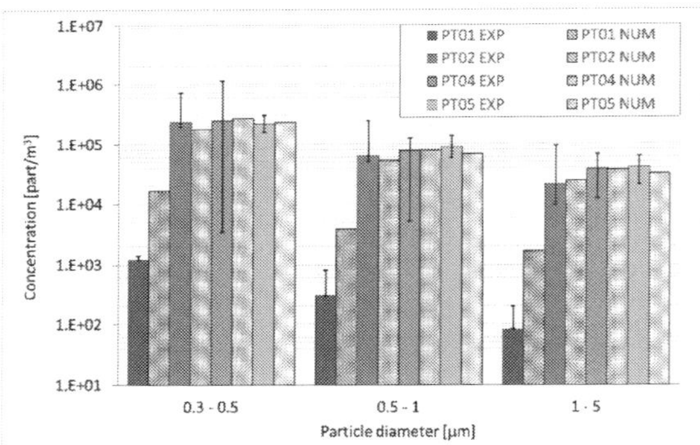

Figure 6. Particles concentration per diameter in different locations: comparison between experimental and numerical values.

Despite an iterative application of our proposed "guess and check" procedure, experimental/ numerical difference in particle concentration at point PT01 remained quite high: the numerical model overestimates particle contents in the air at this location. Otherwise, a good agreement can be pointed out from comparison of particle concentration at the other different locations. Obtained results of particle emission rate per diameter and per person (Table 5), provide lower values with respect to those proposed by Quian et al. However, this is an expected result,

		Particle diameter [µm]							
		0.4	0.5	0.75	1	2	2.5	3	5
Present study	[part/s]	1.01E+05		2.96E+04				1.36E+04	
Zhao					1.67E+05		1.07E−04		1.33E+03
Quian		3.95E+06			7.58E+05		3.16E−05		2.43E+05

Table 5. Particle emission rate (particles/s) from persons differentiated by diameter.

because measurements by Quian [30] refer to a different indoor environment (university classroom) with different use conditions, occupied by students.

5. Numerical results

5.1. Microclimate and ventilation assessment

This chapter section is devoted to numerical results analysis and discussion. Figure 7-a-c provides the air velocity fields in a horizontal section (z=1.5) for the different room conditions studied. Transient simulation results (for the "incorrect use conditions", Figure 7c) refer to step 3, when the door is completely closed and the medical assistant walks through the room, until he reaches the top of the operating table. In particular, Figure 7c refers to the first 13 seconds, when the moving person has reached the table midpoint. Distribution of velocity magnitude in the operational zone is significantly modified from the "at rest" towards both the "operational conditions". In Figure 8a-c velocity profiles along x (y=8; z=1.5), y (x=3; z=1.5) and z (x=3; y=8) direction are given for the different room conditions. The "incorrect use conditions" still refer to the time instant t=13 s.

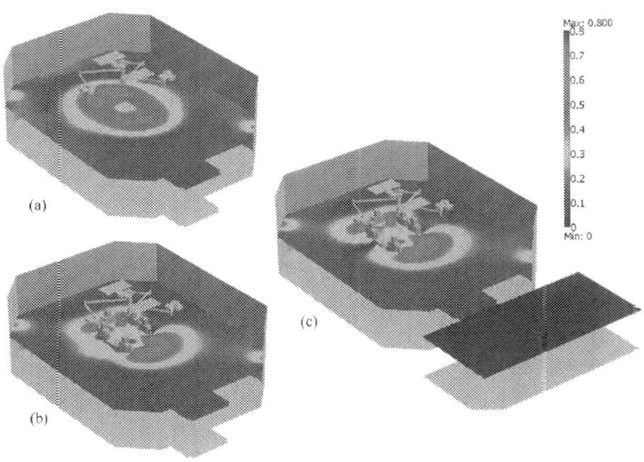

Figure 7. Air velocity field [m/s] in a horizontal plane (z=1.5) for "at rest" (a), "correct operational use" (b) and "incorrect operational use" (c).

In these diagrams zero values represent the "imprint" of a fixed object/person or moving person standing in that location for the considered instant. Velocity profiles underline the modifications of the air flow patterns due to the operational conditions. In particular, air flow in the surgical zone is strongly modified by medical staff presence: important differences, produced by this effect, are shown with the gap between dashed lines and continuous/dashed-dotted lines in Figs. 8a-8c.

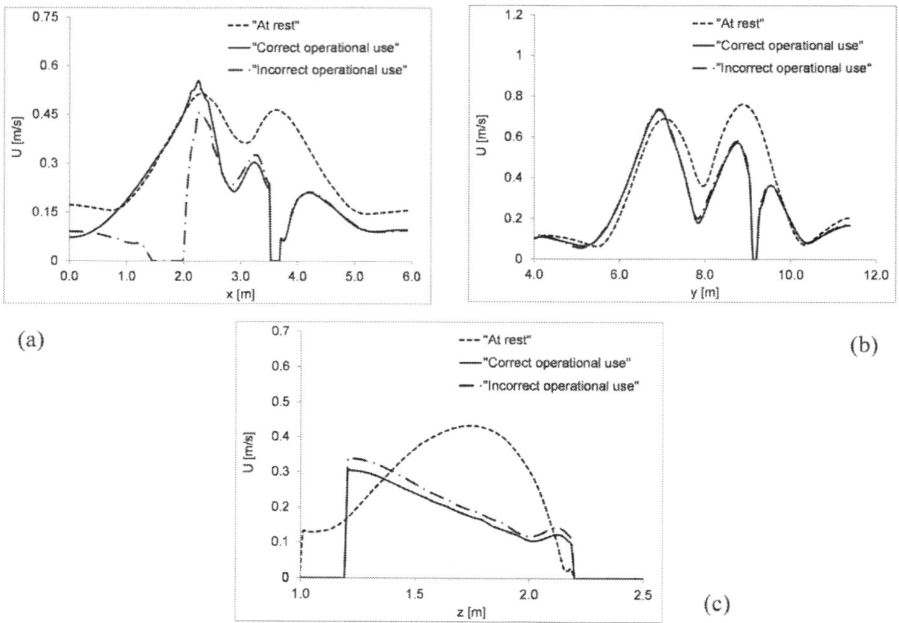

Figure 8. Velocity profiles for "at rest" (dashed line), "operational correct use (continuous line) and "operational incorrect use" (dotted-dashed line) along x (a) (y=8; z=1.5), y (b) (x=3; z=1.5) and z (c) (x=3; y=8).

The effect due to the person movements on the airflow patterns in the surgical zone, appears to be less important. In Figs.8b,8c continuous and dashed-dotted lines are almost overlapped, while in Figure 8a they are remarkably distant only for x<2.5. Velocity profiles show variable trends and high curve slopes in each case: this is really very important when the efficacy of "unidirectional" or "laminar" airflow is discussed for similar applications. Now the influence of OT use conditions on indoor thermal field variations, is discussed. Figure 9 shows the air temperature distribution by means of contour plots and horizontal slice (z=1.5), obtained for "incorrect use conditions", step 3, time 13 s. Thermal "imprint" of a walking person is clear. On the left side of Figure 9, thermal profiles obtained along the x axis (y=8; z=1.5, see the line sketch in the coloured map) for the different room conditions, is provided. Temperature variation is evident in the "correct operational use" compared with the empty room characterized by a predominant isotherm profile. The additional thermal load, due to the walking person, produces a further local temperature increase. In the operating zone thermal levels remains very close to the design value despite the different use conditions. The mean air temperature value, computed all over the OT, is within the limits suggested by the Italian and International standards for the correct use conditions (23.5 °C) and slightly outside the limits for the incorrect use conditions (24.4 °C). Taking into account results obtained on the air RH, it can be detected that vapour production, due to persons presence, determines an air moisture content that is not well balanced by the incoming air at the considered hygrometric conditions. Figure10 shows (left side) RH distribution in a horizontal slice (z=1.5) for "correct operational conditions".

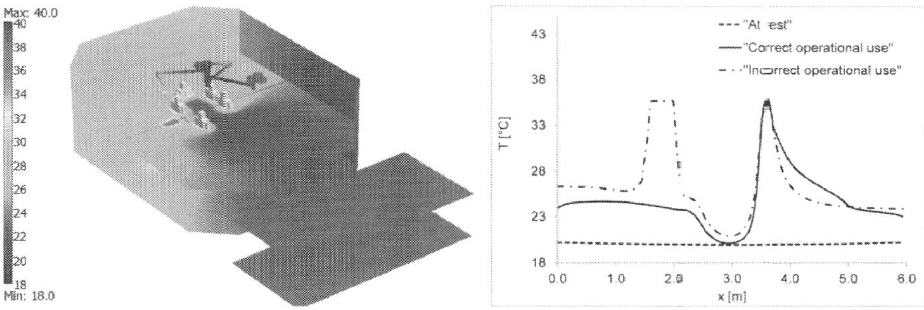

Figure 9. On the left side: temperature [°C] map plotted on contours and in a horizontal slice (z=1.5) for the "incorrect operational use" (step 3, time 13 s). On the right side: temperature profiles along the x-direction (y=8; z=1.5, see the line sketched on the slice) for "at rest" (dashed line), "operational correct use" (continuous line) and "operational incorrect use" (dotted-dashed line) conditions.

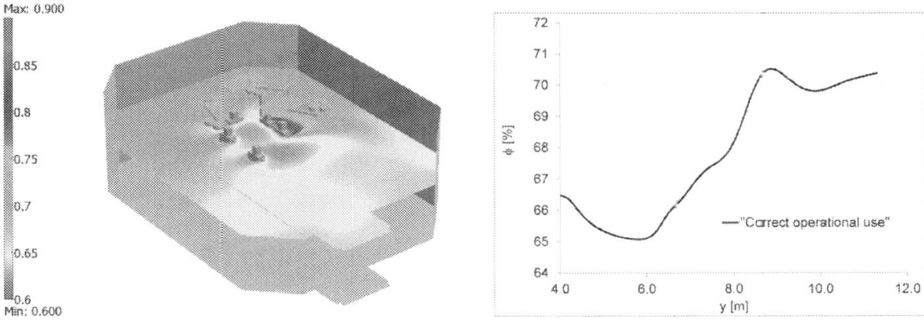

Figure 10. On the left side: RH (%) field in a horizontal plane (z=1.5) for the "correct operational use". On the right side: RH profile along the y-direction (x=1.1; z=1.5) for "operational correct use" (continuous line) conditions.

The mean value computed all over the room in this condition is 67.8%, which is exceeding any maximum threshold suggested by all the standards. The RH profile lying on a horizontal line along the y axis (x=1.1; z=1.5, sketched in the horizontal slice) is also reported in the right portion of Figure10. An increasing level of the moisture content along the y-direction can be noticed: RH at the back of the operating zone presents higher level. This could be due to the lack of ventilation all over the room that can determine stagnation zones. The mean age of air (τ) was also evaluated, using the steady airflow achieved for "at rest" and "correct operational conditions" as transport field for τ computations. The lower τ value corresponds to the higher air washing effect of the ventilation system for the considered zone. Results are plotted in Figure 11, where τ distribution is reported for both the analyzed conditions. In the same, the τ profiles along a horizontal line lying on the x direction (y=8; z=1.5, sketched on the horizontal slices) are provided. The τ values are not critical in both analyzed conditions. A much more uniform distribution is found for the empty room, and in this condition the medical staff act as a "constraint" for the local airflow, allowing a slightly lower air age value. It should be noticed that a lower air age does not directly mean better air quality. We were finally interested in assessing the effect of the sliding door opening/closing during the simulated "incorrect operational conditions". Figure12 provides the air velocity field on a horizontal plane (z=1.5) for step 2, at time 7 seconds: the medical assistant is walking in the room and the sliding door is shutting behind his back. In the same figure, as an enlargement, air velocity vector distribution is shown for the zone close to the sliding door. The important velocity field variation, in the zone of the sliding door, and also the one due to the medical assistant's movement through the room, is evident. As a consequence, an important rate of air outflows from the room. Then a total amount of 16.3 m^3 of air outflows toward the corridor during steps 5-6-7 (door opening/person crossing/door closing) was estimated.

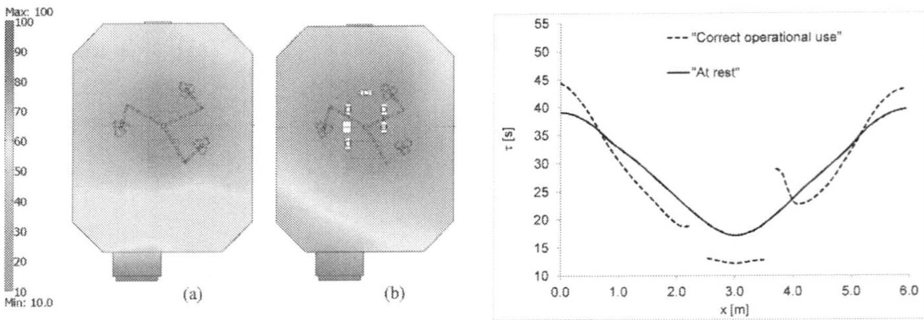

Figure 11. On the left side: mean age of air [s] in a horizontal slice (z=1.5) for the "at rest" (a) and "correct operational use" (b). On the right side: mean age of air profiles along the x-direction (y=8; z=1.5, see the line sketched on the slices) for "at rest" (continuous line) and "operational correct use" (dashed line) conditions.

Figure13 provides a representation of the airflow rate (continuous black line) and the total volume of air (grey-filled surface) outgoing the OT during the door opening. Indeed, the effect of the door opening on the average pressure level inside the OT was also investigated. Figure13 also shows the average pressure trend as a function of time during step 1 (door opening): a significant pressure decrease can be seen from the plotted data. Starting from its initial value, i.e. 32.6 Pa, the OT average overpressure with respect to the corridor level becomes very low, i.e. 1.2 Pa. The overpressure variation due to an "unforeseen event" occurring during an "incorrect use condition" of the OT, can determine a temporary non compliance of the pressure scheme with limits suggested by all the considered standards giving specific indications for it.

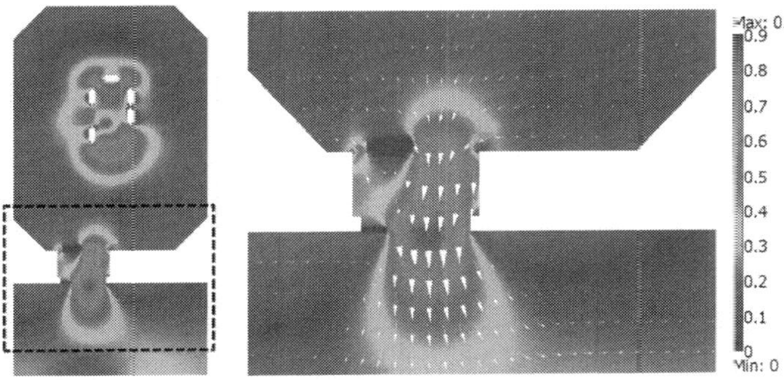

Figure 12. Air velocity field [m/s] in a horizontal slice (z=1.5) for step 2 and time 7 s (left side) and an enlargement with velocity vectors representation, in the proximity of crossing zone of the door towards the corridor (right side) during "incorrect operational use" conditions.

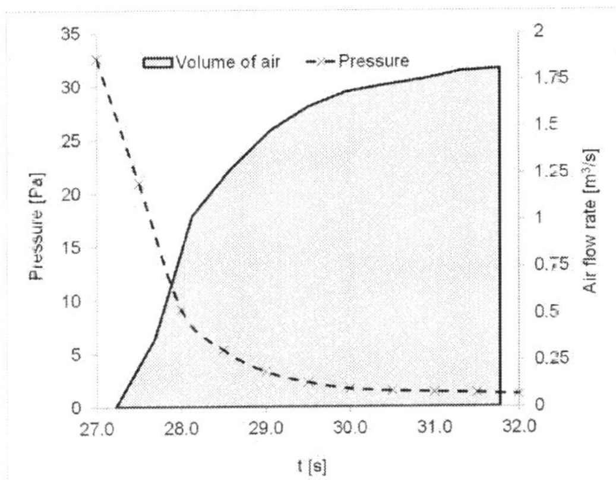

Figure 13. Time evolution of average OT pressure (dashed black line), airflow rate (continuous black line) and total volume of air out-coming the OT (grey-filled surface) during the sliding door opening (Step 5 of the "incorrect operational use" conditions).

5.2. IAQ indexes evaluation

Referring to the air freshness concept, the calculated CO_2 concentration, was found to be considerably below the critical limit of 3000 parts per million (ppm) in the global room volume and 1000 ppm in the operating table zone. CO_2 concentration is high only in the breathing zone where staff are exhaling. The driving effect due to the velocity field prevails over its spatial distribution. The simulation results obtained for CO_2 concentration and its distribution in the BZ and OZ, but also in the PZ and in the total volume of the OT, are in agreement with those provided by recent studies [15, 53]. Anyway, results showed that there is a significant increase in the CO_2 concentration from the fundamental zone BZ, to the OZ but progressively to the PZ and TV. Moreover, some IAQ indexes were computed using simulation results and discussed. The proposed indexes are usually applied for IAQ assessment and a quantitative evaluation of ventilation system performance with regard to contaminant removal and infection risk control [54, 55, 56]. Once the distribution of the dependent variables inside the OT, i.e. air velocity, CO_2 and particle concentration, were evaluated by simulations, IAQ indexes were calculated. Some of them were expressed in the form of a continuous distribution (local indexes), others were referred to the average values of dependent variables, in the BZ, OZ, PZ and TV. The first investigated IAQ parameter was the mean age of air (τ) [41, 42, 43]. The air age concept is generally defined as the average time for air to travel from a supply inlet area to any location in a forced ventilated room [57, 58, 59, 8]. The mean age of air was calculated as a dependent variable, as explained in the modeling section. It provides a measure of air freshness, so its lower values are more favourable. The τ parameter trend was controlled during a transient simulation of 1800 seconds, starting from an initial state corresponding to the steady state condition discussed in the above section. Figure 14 shows a spatial distribution of τ in a horizontal and in a vertical slice of the room, once the transient time to achieve the steady state was expired. Results show that lifetime of air located in the central portion of the OT is much lower than that concerning the peripheral portion of the room (Figure 14).

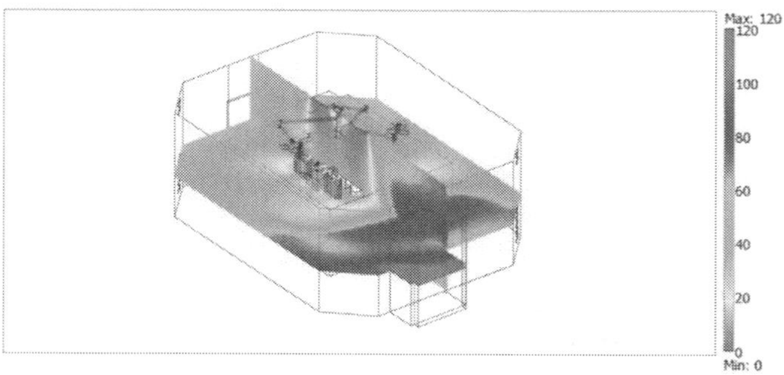

Figure 14. Mean age of air in horizontal (z=1) and vertical (x=3.0) slices.

To quantify this result, the average value of τ was computed in the OT different zones (i.e. BZ, OZ, PZ, TV), and called as τ_{Zj}, where Zj means the generic j-zone (Table 6). Generally, these

http://dx.doi.org/10.5772/59671

values are quite low. Comparison of these values with the theoretical residence time of air inside the OT (defined as the ratio between the total volume of the room (V_{TV}, m³) and the mass flow rate of incoming ventilating air (V_{vent}, m³/s)), shows that the ratio is always higher than 1.

	BZ	OZ	TV	PZ
τ_{ZJ} [s]	25.6	27.7	46.3	51.1
ACE [-]	210%	194%	116%	105%

Table 6. Values of mean age of air and Air Change Efficiency in the different zones.

Indeed, this comparison consists in computing the following Air Change Efficiency (ACE) index:

$$ACE = \frac{V_{TV} / \dot{V}_{vent}}{\tau_{Z_j}} \cdot 100 \tag{6}$$

whose values in the different zones are given in Table 6. The ACE index measures how effectively ventilation systems replace the air in a room with fresh air. In the BZ the average lifetime of air is more than 2 times lower than the theoretical residence time (about 53.7 seconds), that can be deduced analytically. The Local Air Change Efficiency (LACE) is expressed by the following expression:

$$LACE = \frac{V_{TV} / \dot{V}_{vent}}{\tau} \cdot 100 \tag{7}$$

The LACE index characterizes the conditions at a specific point (defined as the ratio between the minimum replacement time, as previously defined, and the local mean age of the air). It is possible to observe (Figure15) that the zone corresponding to the operating table is more favourable from this point of view, reaching LACE values of up to 500-600%.

Knowing the concentration field computed for CO_2 and particles, the Ventilation Effectiveness (VE) index was also assessed. The VE index measures how quickly a contaminant is removed from an air volume by quantifying the efficiency with which the internal pollutant is diluted or removed. It depends on the airflow patterns, and is expressed as follows:

$$VE = \frac{C_E - C_S}{C_{Z_j} - C_S} \tag{8}$$

where C_E is the mean value of contaminant concentration (i.e. CO_2 and particles) calculated at the air-recovery grilles (Exhaust), C_S is the contaminant concentration at the air inlet diffusers (Supply) and C_{Zj} is the mean value of the contaminant concentration in a specific OT zone (i.e. BZ, OZ, PZ, TV).

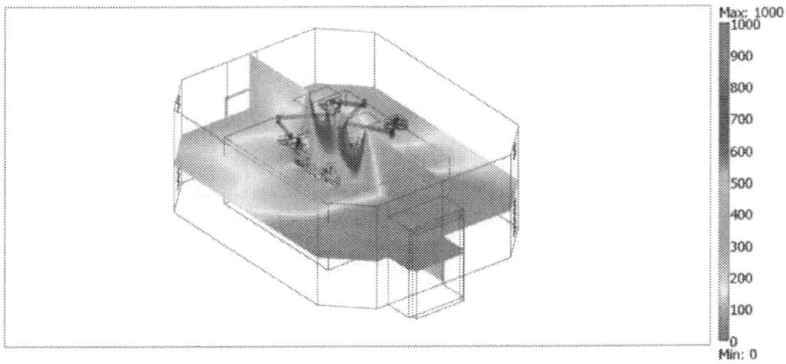

Figure 15. LACE index in horizontal (z=1.2) and vertical slices (x=3.0).

Similarly, the Contaminant Removal Effectiveness (CRE) index, expressed by the ratio between the concentration of contaminants at the exhaust point and the mean value of contaminant concentration within a specific zone was evaluated:

$$CRE = \frac{C_E}{C_{Z_j}} \tag{9}$$

Figure 16 shows the VE and CRE indexes, computed using CO_2 and particle concentrations. Because the particle concentration value was assumed to be zero at the inlet air diffusers (C_S=0), VE and CRE expressions correspond to each other. Due to the very low effect of the settling velocity on particle concentration distribution, a very low quantitative difference was found in computing the CRE (or VE) index as a function of the different particle diameter ranges studied. Therefore, we referred to a single CRE index representative for any particle diameter range studied. Values of the computed global indexes are shown in Figures 17 (CO_2) and 18 (particles). The CRE distribution at a specific point, that is known as Local Contaminant Removal Effectiveness (LCRE) was also calculated by:

$$LCRE = \frac{C_E}{C} \tag{10}$$

Local index distributions are shown in Figure 17 for CO_2 and in Figure 18 for the particles concentration. Comparison between the LCRE indexes, computed for CO_2 and particles, highlights the combined effect of the two sources (nose for CO_2 and body surface for particles) which provides very different trends. The mass transport effect is particularly evident in both cases.

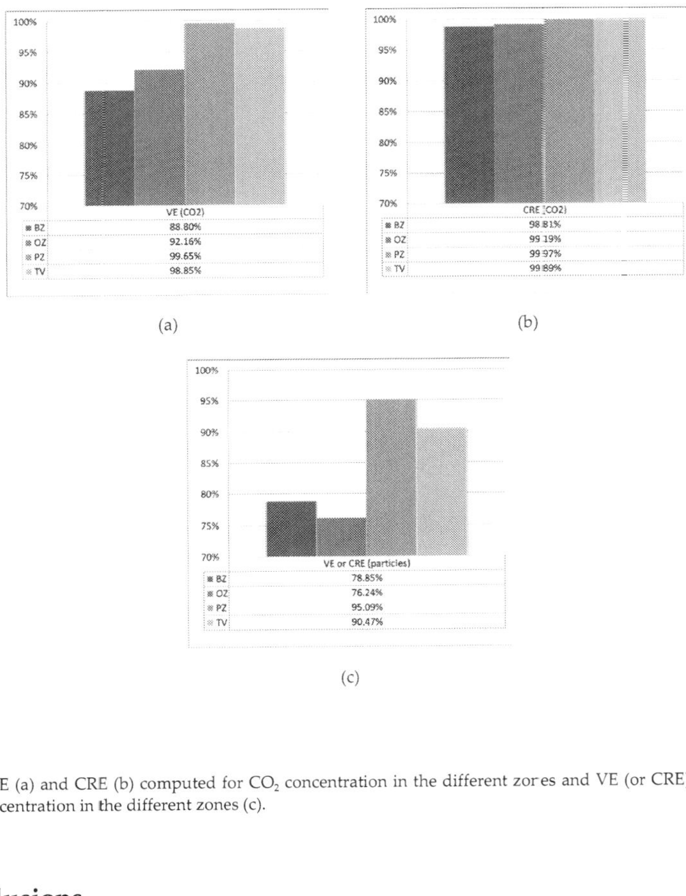

Figure 16. VE (a) and CRE (b) computed for CO_2 concentration in the different zones and VE (or CRE) computed for particles concentration in the different zones (c).

6. Conclusions

An experimental and numerical investigation on the airflow patterns and thermal field in a real OT is presented in this chapter. Different scenarios were considered, then measured and simulated, representative of "at rest" and "operational/effective use" conditions of the OT.

Numerical models were successfully validated against experimental data. In this chapter crucial results of our investigation, obtained through comparison and discussion of quantitative microclimatic parameters (air velocity, temperature, RH and pressure) the influence of unforeseen movements of medical staff, sensible and latent heat, but also CO_2 and particles, released by persons in the ambient and sliding door opening/closing on the OT climate and IAQ, are provided. It can be noticed that, variations in use (resulting in different internal sensible and latent heat loads, moving objects and room confinement) can play an important role in terms of microclimatic system performance against standard suggestions, even those lacking compliance with standard limits are found to be mainly local or temporary.

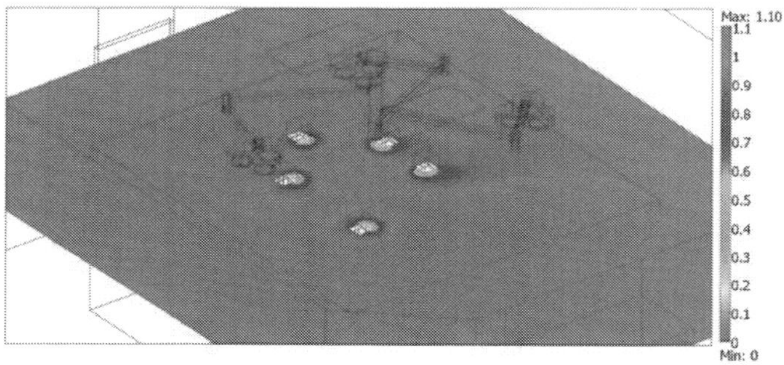

Figure 17. LCRE for CO_2 in a horizontal slice (z=1.6).

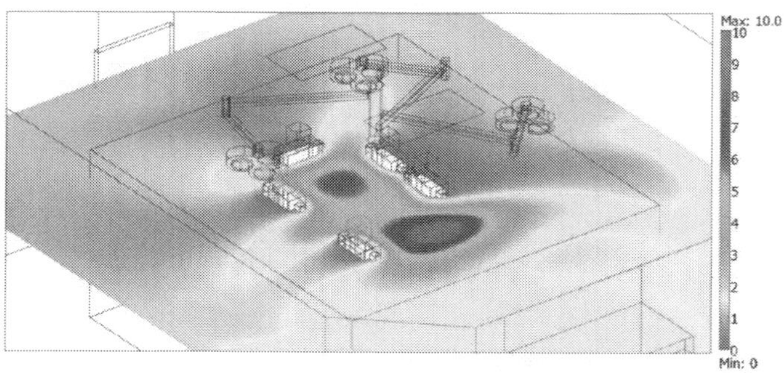

Figure 18. LCRE for particles concentration in a horizontal slice (z=1.4).

From the case study presented here, the efficiency of the AHU-HVAC system for providing the right indoor microclimatic conditions in compliance with standard thresholds is seen not only related to a good plant design, but also to medical staff and assistant behaviour and room use. Measurements of particle concentration, with and without persons, are used in combination with iteratively computed numerical results in order to assess particle emission rate by the occupants for given particle dimensions. Particle concentration fields are also numerically solved for several particle diameters, by using an Euler approach based on the Cunningham formulation of settling velocity. Numerical results are successfully checked against the experimental evidence. CO_2 concentration levels numerically computed in the OT agree well with data reported in the literature for similar applications. Some consolidated indexes, adopted for monitoring the IAQ, are computed for gaseous contaminant (CO_2) and particle concentration, both in terms of spatial distribution and overall values referring to specific OT zones. Results obtained, by applying our proposed methodology for estimating the particle emission rate, highlight a good agreement with the small number of contributions in the literature concerning particle emission for different diameters. Our study contributes to better understanding the additional environmental "load" induced by the medical staff in an OT, based on an innovative strategy proposed and applied to quantify the particle emission rate released by occupants, for given ranges of particle diameters. Our integrated experimental and numerical approach, is in accordance with some recent surgical infection control guidelines, that highlight the importance of surgical staff behaviour control in order to decrease air contamination and wound colonisation. These recommendations include restricting the movements and the number of persons in the OT, but they are often general and based on expert advice. An integrated approach, such as that proposed here, based on CFD application and periodic experimental monitoring campaigns of OT, can contribute to providing valuable suggestions for medical staff information and education concerning the analysed topics, and in general can hopefully stimulate careful considerations on specific procedures for OT design and use.

Nomencalture

Symbol		S.I. Unit
C_p	Specific heat at constant pressure	[J/(kg K)]
D_w	Vapour diffusivity	[m^2/s]
F	Magnitude of buoyancy force	[N/m^3]
h	Coefficient of heat transfer	[W/(m^2 K)]
$\mathbf{I}$	Identity tensor	
k	Turbulent kinetic energy	
p	Pressure	[Pa]
$\mathbf{n}$	Normal unit vector	
M	Vapour production	[kg/(s m^3)]
Q	Sensible heat source	[W/m^3]
$\mathbf{U} \equiv (u, v, w)$	Velocity vector	[m/s]
U	Magnitude of velocity vector	[m/s]
t	Time	[s]
T	Temperature	[K]

Greek symbols

δ_p	Vapour permeability in the air	[kg/(m s Pa)]
δ_w	Wall offset	[m]
ε	Dissipation rate of turbulent kinetic energy	[-]
ϕ	Relative humidity	[%]
ι	Percentage error	[-]
κ	Karman's constant	[-]
ξ	Moisture capacity	[kg/m^3]
λ	Thermal conductivity	[W/(m K)]
μ	Dynamic viscosity	[Pa s]
ρ	Density	[kg/m^3]
σ	Standard deviation	
τ	Mean age of air	[s]

Subscript

AV	Average
in	Incoming
MAX	Maximum
MIN	Minimum
$neigh$	Neighbouring
mr	Mean radiant
out	Outgoing
sat	Saturation
T	Turbulent

Acknowledgements

The Authors thank Prof. Pietro Vitali (Hospital Hygiene Unit, University Hospital of Parma), Prof. Cesira Pasquarella (Department of Biomedical, Biotechnological and Translational Sciences, University of Parma), Prof. Roberto Albertini (Department of Clinical and Experimental Medicine, Unit of Medical Immunology, University Hospital of Parma), Prof. Giuliano Cammarata (Department of Industrial Engineering, University of Catania) for their crucial cooperation and research support. The authors want also to thank the PhD students engaged for the hip surgery simulation and indoor microclimatic measurements, active and passive air samplings.

http://dx.doi.org/10.5772/59671

Author details

Carla Balocco[1] and Giuseppe Petrone[2]

1 Department of Industrial Engineering, University of Firenze, Firenze, Italy

2 Department of Industrial Engineering, University of Catania, Catania, Italy

References

[1] Chow TT, Yang XY. Performance of ventilation system in a non-standard operating room. Building and Environment 2003; 38(12) 1401-1411.

[2] Chow, T.T., Fog, K.F., Tsang, C.F., Qiuwang, W. Comparison of performances of displacement and mixing ventilations. Part II: indoor air quality. Int.J. of Refrigeration 2005; 28 288-305.

[3] Karthikeyan CP, Samuel Anand A. CO_2 dispersion studies in an operation theatre under transient conditions. Energy and Buildings 2008; 40 231-239.

[4] Pfost JF. A re-evaluation of laminar air flow in hospital operating rooms. ASHRAE Transactions 1981; 87(2) 729-739.

[5] Ho SH, Rosario L, Rahman MR. Three-dimensional analysis for hospital operating room thermal comfort and contaminant removal. Applied Thermal Engineering 2009; 29 2080-2092.

[6] Memarzadeh F, Manning A. Comparison of operation room ventilation systems in the protection of surgical site. ASHRAE Transactions 2002; 108(22) 3-15.

[7] Memarzadeh F, Manning A. Reducing risks of surgery. ASHRAE Journal 2003 ; 45(2) 28-33.

[8] Mendez C, San Jose′ JF, Villafruela JM, Castro F. Optimization of a hospital room by means of CFD for more efficient ventilation. Energy and Buildings, 2008; 40 849–854.

[9] Chow TT, Zhang L, Wei B. The integrated effect of medical lamp position and diffuser discharge velocity on ultraclean ventilation performance in an operating theatre. Indoor Built Environment 2006; 15(4) 315-331.

[10] Zoon WAC, Van der Heijden HGM, Loomans MGLC, Hensen JLM. On the applicability of the laminar flow index when selecting surgical lighting. Building and Environment 2010; 45(9) 1976-1983.

[11] Zoon WAC, Loomans MGLC, Hensen JLM. Testing the effectiveness of operating room ventilation with regard to removal of airborne bacteria. Building and Environment 2011; 46 2570-2577.

[12] Kameel R, Khalil EE. Experimental and numerical investigations of airflow regimes in air-conditioned operation theatres. Proc. 14th Usnctam, Symposium On Recent Advances In Three Dimensional Flow Modeling 2003; Paper 52.

[13] Dong S, TU G, CAO R, YU Z. Numerical study on effects of door-opening on airflow patterns and dynamic cross-contamination in an ISO class 5 operating room. Trans. Tianjin Univ. 2009; 15 210-215.

[14] Shih YC, Chiu CC, Wang O. Dynamic airflow simulation within an isolation room. Building and Environment 2007; 42 3194-3209.

[15] Santa Cruz A, Guillou S, Makhloufi R. Effect of moving person on airflow patterns in a ventilated enclosure (operating theatre applications). ITBM RBM 2007; 28 49-55. (In French).

[16] Yang SJ, Fu WS. A numerical investigation of effects of a moving operator on airflow patterns in a cleanroom. Building and Environment 2002; 37 705-712.

[17] Brohus H, Balling KD, Jeppesen D. Influence of movements on contaminant transport in an operating room. Indoor Air 2006; 16 356–372.

[18] Balocco C, Petrone G, Cammarata G. Assessing the effects of sliding doors on an operating theatre climate. Building Simulation: An International Journal 2012; 5(1) 73-83.

[19] Balocco C, Petrone G, Cammarata G. Numerical multi-physical approach for the assessment of coupled heat and moisture transfer combined with people movements in historical buildings. Building Simulation: An International Journal 2013; 5(1) 73-83

[20] Chow, TT, Wang, J. Dynamic simulation on impact of surgeon bending movement on bacteria-carrying particles distribution in operating theatre. Building and Environment 2012; 57 68-80

[21] Franke, JE, Wadden RA. Indoor Contaminant Emission Rates Characterized by Source Activity Factors, Environmental Science and Technology 1987; 21 45-51

[22] Friberg B, Friberg S, Östensson R, Burman LG. Surgical area contamination-comparable bacterial counts using disposable head and mask and helmet aspirator system, but dramatic increase upon omission of head-gear: an experimental study in horizontal laminar air-flow, Journal of Hospital Infection 2001; 47 110-115.

[23] Hospodsky D, Qian J, Nazaroff WW, Yamamoto N, Bibby K, Yazdi HR, Peccia J. Human Occupancy as a Source of Indoor Airborne Bacteria, Plos One Sources of Indoor Airborne Bacteria 2012; 7(4) 1-10.

[24] Nazaroff, WW. Indoor particle dynamics. Indoor Air 2004; 7 175-183.

[25] Li Chih-Shan, Hou Po-An. Bioaerosol characteristics in hospital clean rooms. The Science of the Total Environment 2003; 305 169–76.

[26] Yu BF, Hu ZB, Liu M, Yang HL, Kong QX, Liu YH. Review of research on air-conditioning systems and indoor air quality control for human health, International Journal of Refrigeration 2009; 32 3-20.

[27] Chen FZ, Yu Simon CM, Lai Alvin CK. Modeling particle distribution and deposition in indoor environments with a new drift-flux model. Atmospheric Environment 2006; 40 357-367.

[28] Gao NP, Niu JL. Modeling particle dispersion and deposition in indoor environments, Atmospheric Environment 2007; 41 3862-3876.

[29] Zhao B, Zhang Y, Lia X, Yang X, Huang D. Comparison of indoor aerosol particle concentration and deposition in different ventilated rooms by numerical method, Building and Environment 2004; 39 1-8.

[30] Qian J, Hospodsky D, Yamamoto N, Nazaroff WW, Peccia J. Size-resolved emission rates of airborne bacteria and fungi in an occupied classroom, Indoor Air 2012; 22 339–351.

[31] ANSI/ASHRAE Standard 62.1-2007, Ventilation for Acceptable Indoor Air Quality

[32] HTM 03-01-2007. Heating and ventilation systems Health Technical Memorandum 03-01: Specialised ventilation for healthcare premises Part A: Design and validation. Edinburgh, UK: TSO.

[33] UNI EN ISO 14644-3-2006. Cleanrooms and associated controlled environments - Part 3: Test Methods.

[34] UNI EN ISO 14644-1-1999. Cleanrooms and associated controlled environments - Part1: Classification of air cleanliness (ISO 14644-1:1999)

[35] UNI 11425-2011. Surgery operating theatre, ventilation and air-conditioning system for contamination control (VCCC) - Design, construction, commissioning, qualification, management and maintenance. (In Italian).

[36] ANSI/ASHRAE Standard 62.1-2004 Ventilation for Acceptable Indoor Air Quality

[37] ANSI/ASHRAE Standard 55-2005 Thermal Environmental Conditions for Human Occupancy.

[38] Comsol. Introduction to COMSOL Multiphysics. 2012; v. 4.3. http://www.comsol.com/.

[39] Ignat L, Pelletier D, Ilinca F. A universal formulation of two-equation models for adaptive computation of turbulent flows. Computer Methods in Applied Mechanics and Engineering 2000; 189 1119–1139.

[40] Launder BE, Spalding DB. The Numerical Computation of Turbulent Flows. Computer Methods in Applied Mechanics and Engineering 1974; 3 269-289.

[41] Abanto J, Rarrero D, Reggio M, Ozell B. Air flow modelling in a computer room. Building and Environment 2004; 391393-1402.

[42] Zhang L, Chow TT, Fog KF, Tsang CF, Qiuwang W. Comparison of performances of displacement and mixing ventilations. Part II: indoor air quality. International Journal of Refrigeration 2005; 28 288-305.

[43] Petrone G, Cammarata L, Cammarata G. A multi-physical simulation on the IAQ in a movie theatre equipped by different ventilating systems. Building Simulation: An International Journal 2011; 4 21-31.

[44] Hinze JO. Turbulence. 2nd ed. New York: McGraw Hill; 1975.

[45] Deuflhard P. A modified Newton method for the solution of ill-conditioned systems of nonlinear equations with application to multiple shooting. Numerical Mathematics 1974; 22 289-315.

[46] Hindmarsh H, Brown AC, Grant PN, Lee KE, Serban SL, Shumaker R, Woodward DE, Woodward CS. SUNDIALS: Suite of Nonlinear and Differential/Algebraic Equation Solvers. ACM Trans. Math. Software 2005; 31 363-396.

[47] ONORM H 6020-2007. Ventilation and air conditioning plants for locations for medical use - Design, construction, operation, maintenance, technical and hygienic inspections.

[48] VDI 2167-2004. Part 1. Building services in hospitals – Heating, ventilation and air-conditioning

[49] DPR 14-01-1997. Approvazione dell'atto di indirizzo e coordinamento alle regioni e alle province autonome di Trento e di Bolzano, in materia di requisiti strutturali, tecnologici ed organizzativi minimi per l'esercizio delle attività sanitaria da parte delle strutture pubbliche e private. In: G.U.R.I. 20/02/1997 n.42. (In Italian).

[50] ISPESL-2009. Dipartimento Igiene del Lavoro, Linee guida per la definizione degli standard di sicurezza e di igiene ambientale dei reparti operatori. http://www.ispesl.it/. (In Italian).

[51] NBHF-2004. Building guidelines for a surgical department.

[52] ANSI/ASHRAE 170-2008. Ventilation of Health Care Facilities.

[53] Al-Waked R. Effect of ventilation strategies on infection control inside operating theatres. Engineering Application of Computational Fluid Mechanics 2010; 4(1) 1-16.

[54] Villafruela JM, Castro F, San José JF, Saint-Martin J. Comparison of air change efficiency, contaminant removal effectiveness and infection risk as IAQ indices in isolation rooms. Energy and Buildings 2013; 57 210-219.

[55] Kwon KS, Lee IB, Han HT, Shin CY, Hwang HS, Hong SW, Bitog J P, Seo IH, Han CP. Analysing ventilation efficiency in a test chamber using age-of-air concept and CFD technology. Biosystems Engineering 2011; 110 421-433.

[56] Meiss A, Feijo-Mu˜noz J, Garcia-Fuentes MA. Age-of-the-air in rooms according to the environmental condition of temperature: A case study. Energy and Buildings 2013; 67 88-95.

[57] Sandberg M, Sjöberg M. The use of moments for assessing air quality in ventilated rooms. Building and Environment 1983; 18(4)181–197.

[58] Federspiel CC. Air-change effectiveness: theory and calculation methods. Indoor Air 1999; 9 47–56.

[59] Xianting Li, Dongning Li, Xudong Yang, Jianrong Yang. Total air age: an extension of the air age concept. Building and Environment 2003; 38 1263 – 1269.

Chapter 24

Indoor Air Quality and Risk Factors Associated with Respiratory Conditions in Nigeria

Godson Rowlanc Ana, Oyewale Mayowa Morakinyo and Gregory Adekunle Fakunle

Additional information is available at the end of the chapter

http://dx.doi.org/10.5772/59864

1. Introduction

Clean air is a basic requirement for life and healthy living. The quality of air in homes, offices, schools, day care centres, public buildings, health care facilities and other private and public buildings where people spend over 80% (3/5) of their time daily is crucial for healthy living and people's well-being. The National Health and Medical Research Council (NHMRC) defines indoor air as 'air within a building occupied by people of varying states for a period of at least one hour' (NHMRC, 1996). Buildings covered by this definition include homes, schools, restaurants, public buildings, residential institutions, offices, etc. (www.arf.org.au)

Indoor Air Quality (IAQ) is an increasing concern in the world today. In fact, the mere presence of people in a building can significantly alter indoor air quality. Indoor air pollution is present in virtually every indoor space, with the exception of strictly controlled and sterile spaces in pharmaceutical, medical and research facilities. Indoor pollutants may originate from human activities, building materials and carpets; they may also penetrate from outdoor environments by forced ventilation, diffusion or infiltration [8]. Humans are inevitably exposed to these pollutants, considering the amount of time spent indoors, but the influence of the pollution on human health may vary, depending on age, sex, nutritional status, physiological conditions, and individual predisposition.

There is increasing evidence linking indoor air pollution to increased risk of respiratory tract infections, exacerbations of inflammatory lung conditions, development of chronic obstructive lung disease, cardiac events, stroke, eye disease, tuberculosis, cancer and hospital admissions especially in women and children who are the most exposed [40, 54, 59, 63].Acute Respiratory Infections (ARIs) are responsible for one-third of hospital admissions and 20%–30% of deaths among children under 5 years [19, 48] in developing countries.

However, information regarding IAQ in Nigeria is scanty. Aside from the data collected by a few individuals and corporate organisations at scattered locations, there is no comprehensive and empirical database on the magnitude of the health risks and deleterious effects associated with exposure to poor IAQ by occupants in different built environments and the general population in Nigeria.

2. Brief profile of Nigeria

Nigeria is located in western Africa on the Gulf of Guinea and has a total area of 923,768 km^2. It shares a 4,047 km border with Benin (773 km), Niger (1497 km), Chad (87 km) and Cameroon (1690 km), with a coastline of at least 853 km. Nigeria lies between latitudes 4° and 14°N, and longitudes 2° and 15°E. Nigeria has about174 million inhabitants and is the most populous country in Africa and the seventh most populous country in the world. The country has over 500 ethnic groups, of which the three largest are the Hausa, Yoruba and Igbo. Nigeria is the 12th largest producer of petroleum in the world and the 8th largest exporter, and has the 10th largest proven reserves. Petroleum plays a large role in the Nigerian economy, accounting for 40% of GDP and 80% of Government earnings.

3. Indoor air pollution

Indoor air pollution refers to the amount of chemical, biological and physical contaminants in the air inside a building. There is a wide range of indoor air pollutant sources in houses including building materials, furniture, central heating and cooling systems, household cleaning products, paints, tobacco smoke, office machines and a variety of other products used in daily activities (EPA, 2009).

Nigeria, unlike the developed countries is faced with several environmental, social and economic challenges such as inadequate electric power supply, poor waste disposal system, air pollution, water pollution, noise pollution, unemployment, inadequate of water supply, etc. [67]. Most households in Nigerian cities operate small capacity fossil fuel electric power generators for electricity supply (ECN, 2009).

A study carried out by [67], revealed that small household generators in Nigeria operate an average of 6 h daily, while the average distance of generator away from building was 5.6 m. These alongside poor ventilation have influenced the quality of indoor air in the households [46].

Another major source of pollution in the indoor environment is household combustion of coal or biomass for cooking and heating. It is estimated that more than 50% of the world's population depends on animal dung, wood, crop waste or coal to meet their most basic energy needs (WHO, 2005). Firewood combustion for cooking is a common practice in most rural communities of developing nations including Nigeria. Its use as an energy source is widely reported to impinge on the environment, particularly the quality of air and the health of populations especially women [5].

Based on a study designed to determine exposure to emissions from firewood cooking stove and the pulmonary health of women in Olorunda community in Ibadan, Nigeria, [5] found that the mean CO levels for fire wood stove emissions at six kitchen locations combined were 20.0±0.8 ppm (morning) and 24.2 ± 0.5 ppm (evening). These values were 2 and 2.4-folds higher, respectively, than the World Health Organization (WHO) limit of 10 ppm. The mean SO_2 levels for firewood stove emissions at the six locations were 0.505 ppm (morning) and 0.683 ppm (evening). These values were 6.3- and 8.5-folds higher, respectively, than the WHO limit of 0.08 ppm. The mean NO_2 levels for firewood stove emissions at the six locations were 0.517 ppm (morning) and 0.81 7 ppm (evening). These values were 25- and 41-fold higher, respectively, than the WHO limit of 0.02 ppm(Fig 1).

In this study, although the levels of NO_2 were low, there was a significant negative correlation between NO_2 and lung function (r = – 0.488), which indicated that chronic exposure to NO_2 is likely to impair lung functions (Ana *et al.*, 2012).

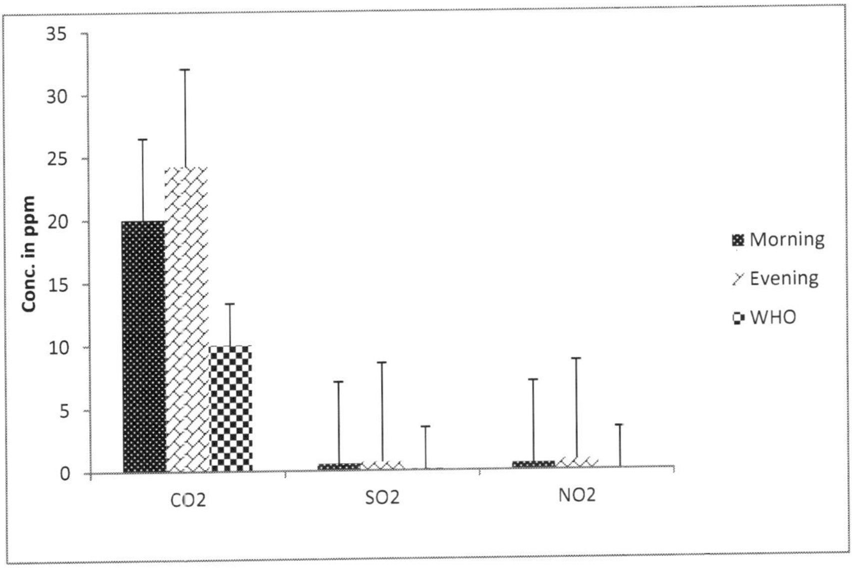

Source : Ana *et al.*, 2012

Figure 1. Mean morning and evening gaseous concentrations at six kitchen locations in comparison with WHO limits

4. Housing and indoor air quality

Housing is a place people spend most of their time and one of 'the main settings that affect human health' [12]. It is well established that housing is a 'key determinant of health' [62]. Good housing and good health go together [62]. Housing, as a neglected site for public health action, has been identified in a number of recent publications (Breysse *et al.*, 2005). Housing, however, encompasses a large range of factors, including biological (mould, cockroaches, dust mites, etc.), chemical (tobacco smoke, paints, etc.) and structural (moisture, heat ventilation, etc.).

A large number of the population in Nigeria resides in houses with poor housing conditions (as shown in Figure 2.). Provision of adequate housing has been a problem in developing countries like Nigeria. Shortage of decent accommodation, particularly in cities, has resulted in individuals occupying bungalows or rooms above specified capacity. More so, in the quest to provide themselves with urgent accommodation, people have resorted to building houses outside required specifications and/or at indecent locations. Because of this problem, there is no available information on the indoor air quality of living homes in the country.

5. Indoor air environment

Indoor environments are fundamental environmental factors capable of impacting health [16, 32]. Air quality of indoor environments is one of the main factors affecting the health, well-being and productivity of people. The effect on health rises as exposure to and density of air pollution increases [35, 42].

In indoor air environments (non-industrial), the most important source of airborne bacteria is the presence of humans (Kim and Kim, 2007). Major indoor activities such as talking, sneezing, coughing, etc. can concentrate airborne biological particulate matter in the indoor environment. House dust, pets and their bedding material, wood materials and furniture stuffing could serve as reservoir from which spores of *Alternaria, Aspergillus, Botrytis, Cladosporium, Penicillium, Scopulariopsis* spp. and yeast cells are occasionally released into the indoor air [17].

6. Bioaerosols in the indoor air environment

Bioaerosols consist of aerosols containing microorganisms (bacteria, fungi, viruses) or organic compounds derived from microorganisms (endotoxins, metabolites, toxins and other microbial fragments) (Heikkienenet *al.*, 2005). Aerosols of biological origin form a significant portion of the total atmospheric aerosols, sometimes reaching up to 50% of all aerosol particles [37].

Bioaerosols vary in size (20 nm to >100 nm) and composition depending on the source, aerosolisation mechanisms and environmental conditions prevailing at the site [53]. The

http://dx.doi.org/10.5772/59864

inhalable fraction ($PM_{2.5}$) is of primary concern because it is the most susceptible portion of the bioaerosols to reach the deeper parts of the respiratory system (Pope *et al.*, 1995).

(a) (b) (c) (d)

Plate 1. Typical housing structure in urban (a and b) and rural (c and d) settings in Nigeria

Indoor air contains a complex mixture of bioaerosols such as fungi, bacteria and allergens along with non-biological particles (e.g., dust, smoke, particles generated by cooking, organic and inorganic gases) [33]. Airborne microorganisms might pose an environmental hazard when present in high concentrations in indoor environments resulting in health problems [68].

Bioaersosols contribute to about 5%--34% of indoor air pollution [66]. The source of bioaerosols in indoor air includes furnishing and building materials, microbiological contamination within the walls and ceilings and floor activities. The outdoor environments are oftentimes major sources of indoor bioaerosols as particles are transferred indoor through openings of the building envelope (windows, doors). However, one of the most important factors affecting indoor air quality is how the building is heated, ventilated, air-conditioned and its occupancy [9].

6.1. Bacteria in indoor air environment

Airborne bacteria are major issues in indoor environments. The potential health risks resulting from exposure to airborne bacteria can occur at workplaces and residential locations [28]. A study conducted in Nigeria by [7] in day care centres, reported that the genera of bacteria isolated from the indoor environment included *Pseudomonas, Staphylococcus* and *Bacillus* spp., in both wet and dry seasons.

Elsewhere, the most important bacterial strains found in an indoor atmosphere are representatives of the genera *Bacillus, Micrococcus, Kocuria and Staphylococcus, Streptomyces albus, Pantoea agglomerans, Pseudomonas chlororaphis, Arthrobacter globiformis, Thermoactinomyces vulgaris and Corynebacterium* spp.[23].

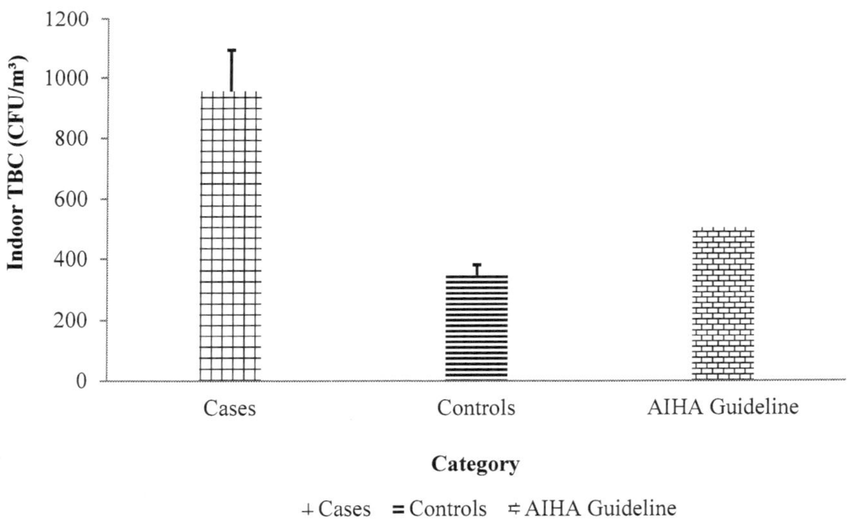

Source: Ana et al., 2013

Figure 2. Mean Indoor TBC among cases and controls as compared with AIHA Guideline

Also, in a study designed to determine the burden of airborne microbes in indoor environment that predispose children under the age of 5 to acute respiratory infections, the indoor airborne bacterial load among cases ($9.6x10^2 cfu/m^3$) was higher than the acceptable limit

http://dx.doi.org/10.5772/59864

(≤5.0x10^{2}cfu/m^3) proposed by the American Industrial Hygiene Association (AIHA, 2001) for residential locations compared to controls (3.5x10^{2}cfu/m^3) [7] (Fig. 2).

Like fungi, bacterial growth takes place in the presence of sufficient moisture. In the indoor environments, sources of required moisture for microbial growth include drip pans of air-cooling devices, freezers or refrigerator, humidifiers and wet surfaces of bathrooms and kitchen [45]. Water may also originate from leaks in plumbing or roof, or be a result of condensation or harsh rain or snowing [65]. Standing water is a potential source of bacteria, and once disturbed, bacteria may become airborne [65].

6.2. Fungi in indoor air environments

The building-associated fungi consist of filamentous microfungi (moulds) and yeasts. The most common fungal genera occurring in indoor environments are *Penicillium, Aspergillus, Cladosporium, Alternaria* and yeasts [36].

In the same study carried out by [6] they showed that genera of fungi isolated in indoor environment of day care centres included *Aspergillus, Penicillum, Geotricum, Fusarium* and *Candida* spp., and in the dry season *Aspergillus, Penicillum, Geotricum* and *Fusarium* spp.,

Their presence in indoor air is as a result of transportation from outside environment via building materials, carpets, furniture, wallpapers, etc. Ventilation and air-conditioning systems are another common way of penetraton of fungi into the buildings. The rate of further growth, spreading and multiplication depends exclusively on moisture content in indoor air, regardless of the type of surface [30].

Natural food source for fungi vary from plant, animal and human particles in house dust, to fragments of construction materials such as floor and wall textile coverings, furniture, residue of cooking traces, food storage and paper materials. Since these materials occur in good quantity in every building, and considering that optimal temperature for fungi growth ranges from 10–35 ºC, the only limiting factor for development of fungi and mould contamination is dampness [31].

Fungi may be extremely harmful for human health, but may also destroy the building itself (Figure 4.), particularly wooden parts, such as roofs and other materials. Some fungi species produce strong allergens, which initiate immune reaction type I (IgE mediated). For example, the indoor contamination with *Alternaria, Penicillium, Aspergillus* and *Cladosporium* spp., is related to asthma and other allergic respiratory diseases. Some of these species, such as *Penicillium* and *Aspergillus* can also induce type III allergy (IgG mediated), while at high concentrations, may also initiate combined type III and IV reaction manifested as hypersensitivity pneumonitis. Major fungal allergens are isolated and identified (such as Cla h I from *Cladosporium herbarum,* Alt a I and Alt a II from *Alternaria alternata* and Asp f I and Asp f III from *Aspergillus fumigatus*) [22].

Fungi mycotoxins have strong genotoxic, carcinogenic and immunotoxic potential. The carcinogenic effects of aflatoxin (mycotoxin produced by *Aspergillus flavus* and *Aspergillus parasiticus*) are well known. The most important mycotoxins related to indoor air contamination are trichothecenes, generated by fungi *Stachybotrys chartarum* (macrocyclic trichothecenes, trichodermin, sterigmatocystin and satratoxin G) [10].

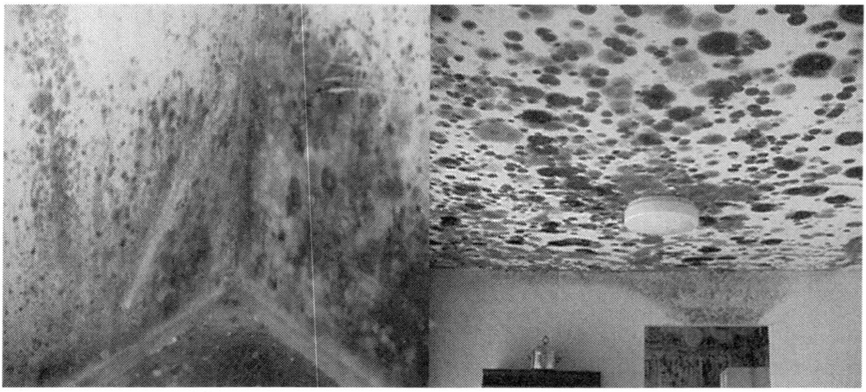

Plate 2. Fungi on the wall and ceiling of a typical indoor environment

6.3. Viruses in indoor environment

Viruses are small (20-400 nm), obligate intracellular parasites. They represent a common cause of infectious disease acquired indoors, as they are easily transmitted especially in crowded, poorly ventilated environments [72].

Airborne transmission of viral agents occurs when viruses travel on relatively large respiratory droplets (> 10 μm) that people sneeze, cough or exhale during conversation or breathing (primary aerosolisation). A single cough can release hundreds of droplets while a single sneeze can release about 40 000 droplets at speeds of up to 50--200 miles per hour with each droplet containing millions of viral particles. Aerosol droplets travel only short distances (1-2 meters) before settlings on surfaces, where viruses can remain infectious for hours or days.

Poorly ventilated and crowded indoor environments are favourable conditions for viral transmission. Hospitals, in particular, are environments where viral aerosols can be particularly hazardous, since patients tend to be especially prone to infection due to pre-existing illness. Elderly patients, children, cancer patients, immunocompromised or immunosuppressed patients are most at risk.

6.4. Protozoa in indoor environment

Protozoa may also be present in indoor air in damp buildings. [77] detected amoebae in 22% of the 124 samples of various materials collected from buildings with evident moisture damage. Field studies on the presence and concentrations of protozoa in indoor air, as well as health aspects of these microorganisms in given conditions, are still lacking, with the exception of one in vitro study conducted by the same authors, who suggested that amoebae act synergistically with certain bacteria, enhancing their cytotoxic and pro-inflammatory potential [78].

7. Prevalence of respiratory problems in Nigeria

Acute respiratory infections are a major cause of morbidity and mortality among Nigerian children. It was estimated that pneumonia accounted for 20% of deaths in children under the age of 5 years in Nigeria between the year 2000 and 2003 (WHO, 2007).

A prospective cohort study in Ilorin, Nigeria, reported that the rate of acute respiratory infection was three episodes per child per year with pneumonia being responsible for 1.3 episodes per child per year [27]. In another hospital-based study in Ibadan, 28.4% of children admitted to the hospital with acute lower respiratory tract infection had acute bronchiolitis with respiratory syncytial virus being the most common viral aetiologic agent [39].

There is scanty data on the bacterial aetiology of pneumonia in Nigerian children.According to WHO, (2002), about 20% of all deaths in children under 5 years are due to Acute Lower Respiratory Infections (ALRIs - pneumonia, bronchiolitis and bronchitis); 90% of these deaths are due to pneumonia. There is a seasonal variation in acute respiratory infections in Nigerian children with more episodes occurring during the rainy season [5, 27, 39].

8. Indoor environmental risk factors for respiratory conditions

8.1. Overcrowding

Many children are exposed to very crowded conditions at home, schools (Figure 5.) etc., and this increases risk of transmission of illness. Most studies in developing countries have found that the average area of habitable space per person is well below the WHO recommendation of $12m^2$ [15].

According to a study by [7], the mean number of occupancy among children under the age of 5 admitted in a tertiary health facility for ARIs was 6.0±1.5 as compared to 4.0±1.0 among controls. A positive association was found between the level of occupancy and indoor total bacterial count. A similar study by [70] recorded the highest bacterial burden in an over-crowded environment. This suggests that the number of persons in the household is directly proportional to the level of bacteria build-up in the indoor environment.

8.2. Environmental tobacco smoke among parents

More than 150 published studies have shown a significant relationship between Environmental Tobacco Smoke (ETS) and respiratory illness in children. Meta-analyses revealed a strong evidence for associations between both prenatal maternal smoking and postnatal ETS exposure and risk of ARI in children [21].

[6] in their study indicated that parental smoking or any other smoker in the house as a risk factor for ARIs in children under the age of 5 (OR = 4.7; CI = 0.9-2.17, $p<0.05$). This could be due to the accumulation of emissions from cigarette smoking in the indoor environment as a result of inadequate ventilation [13].

8.3. Household biomass fuel

Biomass fuel is any material derived from plants or animals which is deliberately burnt by humans. Firewood combustion for cooking is a common practice in most rural communities of developing nations including Nigeria [4]. Majority of the households burn biomass fuels in open fireplaces, consisting of simple arrangements as three rocks (Figure 7.), a U-shaped hole in a block of clay, a pit in the ground or in poorly functioning earth or metal stoves. The process of combustion in most of these stoves is incomplete, resulting in substantial emissions which, in the presence of poor ventilation, produce very high levels of indoor pollution (WHO, 2000).

Plate 3. Overcrowding condition in a typical school setting

Many of the substances in biomass smoke have adverse health effects. The most important ones are carbon monoxide, nitrous oxides, sulphur oxides, formaldehyde and polycyclic organic matter, including carcinogens such as benzo[a]pyrene. Particles with diameters <10 microns (PM_{10}), and particularly those <2.5 microns in diameter ($PM_{2.5}$), can penetrate deeply into the lungs (USEPA, 1997).

http://dx.doi.org/10.5772/59864

A Nigerian study on the concentration of indoor air pollutants due to the use of firewood for cooking and its effects on the lung function of women living in the selected homes found that the concentration of PM_{10} and gaseous emissions such as CO and NO_2 significantly exceeded the WHO limits by several folds. Chronic exposure to high level of the indoor air pollutants particularly the respirable particulate matter could possibly compromise the lung function status of women [4].

[47] in their study of the effects of stove intervention on household air pollution and the respiratory health of women and children in rural Nigeria, demonstrated an association between cooking with biomass fuel, increased prevalence of respiratory symptoms and presence of obstructive lung disease. The high frequency of respirator y and other exposure-related symptoms seen in mothers and children at baseline (pre-intervention) were substantially reduced a year later following replacement of the traditional stoves with low-emission stoves (Fig.3). This observation is consistent with the report of higher frequency of respiratory symptoms, especially cough in women exposed to biomass smoke in Mozambique [24], Nepal [60], Pakistan [61], Mexico [56] and Guatemala [20, 64].

Although the incidence of some respiratory symptoms may be lower in children compared to mothers, it is evident that the children are also exposed to the same $PM_{2.5}$ and CO levels, as 94.1% of the mothers always have their children with them during cooking [47]. Studies have also suggested that children exposed to household air pollution early in life may have impaired lung development, indicating that the impact of exposure to biomass smoke may continue into adulthood [2].

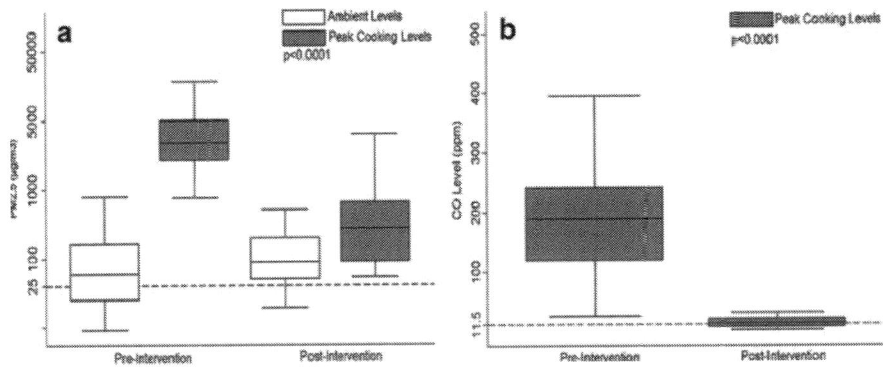

Source: Oluwole et al., 2013

Figure 3. Comparison of household particulate matter (a) and carbon monoxide (b) concentration levels before and 1 year after distribution and monitored use of low-emission stoves. The dotted lines represent WHO recommended levels. It has been highlighted to show that cooking time concentrations of $PM_{2.5}$ and CO in all households sampled were significantly above the recommended levels before distribution and monitored use of low-emission cooking stoves

Over 40 % of the participating mothers and children in our study had mild to moderate obstructive defects. This impaired lung function is likely caused by particulate matter (PM) in firewood smoke, which is believed to induce oxidative injury through its ability to carry adherent metals into the lungs and cause inflammation [57].

Source: Oluwole et al., 2013

Plate 4. Simple arrangement of three stones where biomass burns in an indoor environment

Exposure to these pollutants is reported to be higher in women and children (Figure 8.) [1, 11]. It has been shown that exposure to biomass fuel smoke is responsible for a number of respiratory diseases such as Acute Respiratory Infections (ARI), Chronic Obstructive Pulmonary Disease (COPD), Tuberculosis and Asthma; Low Birth Weight; Cataract and Blindness [14]. The adverse effects on respiratory health of products of incomplete solid-fuel combustion are summarized in Table 1.

Compound	Potential Health Effect
Breathable particulate matter	Wheezing, exacerbation of asthma, respiratory infections, chronic bronchitis and chronic obstructive pulmonary disease.
Carbon monoxide	Low birth weight, increase in perinatal deaths.
Polycyclic aromatic hydrocarbons	Lung cancer, cancer of mouth, nasopharynx and larynx.
Nitrogen oxides	Wheezing, exacerbation of asthma, respiratory infections, reduced lung function in children.
Sulphur oxides	Wheezing, exacerbation of asthma, exacerbation of chronic obstructive pulmonary disease, cardiovascular disease.

Source: Perez et al., 2011

Table 1. Adverse effects on respiratory health associated with indoor air pollution

http://dx.doi.org/10.5772/59864

Source: Oluwole et al., 2013

Plate 5. Mother and child exposed to biomass emissions in an indoor environment

8.4. Dampness in indoor environments

Numerous studies have examined the potential association between damp housing conditions and respiratory ailments among occupants. It was discovered that increase humidity leads to increase mould growth and exposure, which could result in asthma and other respiratory conditions. The presence of home dampness has been reported to affect about 38% of Canadian homes [18]. A Canadian study found that children living in damp or mouldy homes were 32% more likely to have bronchitis [18] (secure.cihi.ca). Substantial problems have been identified in some First Nations communities due to a combination of inappropriate housing design, poor construction, inadequate maintenance and poor ventilation [41].

8.5. House dust mites and cockroaches in indoor environments

House Dust Mites (HDMs) thrive in the dust of homes, particularly in the presence of high indoor relative humidity (secure.cihi.ca). In order to survive and multiply they require a relative humidity in excess of 45%–50%, but their activity, including feeding and maturation is more rapid at higher rates of relative humidity, which was confirmed in field studies [79].

Cockroaches are implicated as a major cause of asthma among inner city children in the U.S., resulting in increased hospital admissions, school absenteeism, and unscheduled medical visits for asthma [58].

8.6. Sanitation and housing quality

[15] found that children with respiratory illness come from houses with poorer sanitation than controls, while in developed countries promotion of hand washing has been associated with reduced incidence of respiratory illness.

8.7. Indoor air meteorological conditions

A case-control study carried out in Ibadan among houses of children under the age of 5 revealed that a higher proportion of houses visited among cases and controls recorded indoor relative humidity (RH) values above the comfort level (30%–60%)(Fig. 4). The high RH (above comfort level) observed among a large proportion of houses among cases could be as a result of high moisture content. The mean indoor and outdoor air RH among cases (69.6 ± 4.7; 67.2 ± 5.0) was found to be higher than among controls (63.1 ± 6.5; 66.1 ± 7.1) [7]. With such high relative humidity levels, microorganisms such as fungi and bacteria can survive on non-living materials including dusts (Choa *et al.*, 2002). High relative humidity above 70% also tends to favour the survival of viruses that infect the membrane of the respiratory tract.

8.8. Inhalable particulate matter burden

Particulate matter (PM) has been recognized in recent years as the most dangerous and widely spread air pollutant [43] affecting more people than any other pollutant and contributing to the risk of developing cardiovascular and respiratory diseases, as well as lung cancer. PM_{10} is associated with increased mortality and morbidity in many cities worldwide and the risk is greatly pronounced among susceptible populations, including the elderly, children and people with pre-existing health conditions [55].

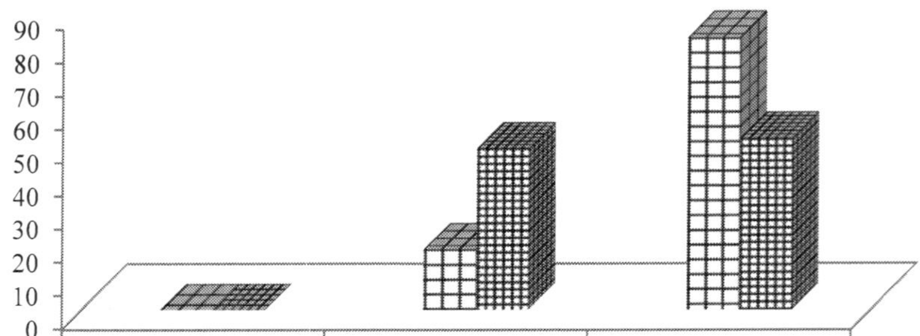

	Low	Comfort	High
+ Case	0	18.2	81.8
⊞ Control	0	48.5	51.5

Key: Low = <30%, Comfort = 30 – 60%, High = >60%
Source: Ana et al, 2013

Figure 4. Indoor relative humidity for houses among cases and controls

[6] in their study of inhalable particulate matter burden in selected day care centres in Ibadan, Nigeria, observed that the mean indoor PM_{10} readings for wet season (73.4 ± 54.4μg/m³) and dry season (296.3 ± 61.6μg/m³) significantly exceeded the WHO guideline limit of 50μg/m³ (Table 2).The indoor/outdoor PM_{10} ratio was 1.38 ± 0.97 and being greater than 1 indicated a possible indoor source of pollution. High particulate matter concentration observed was indicative of the poor indoor air quality condition in the day care centres.

Parameter	Season	Mean ± Standard deviation	p-value
Indoor condition			
Particulate Matter (PM_{10})	Wet	88.6 ± 55.6	<0.001
	Dry	299.8 ± 62.9	
Source: Ana and Umar, 2013			

Table 2. Indoor particulate matter concentration in the wet and dry seasons

9. Conclusion

Everyone has the right to a good standard of living, adequate for the health and well-being of his family. Over the years, man has built houses to protect himself from environmental hazards. Buildings do not always protect their occupants from pollution. The quality of housing conditions plays a decisive role in the health status of the residents. The bioaerosols, dust and gaseous pollutants trapped or growing on the inside may well exceed those outdoors and therefore exacerbate health problems.

A polluted indoor air is a risk factor for human health globally. The magnitude of the health risks and risks factors associated with exposure to poor indoor air quality especially among children under the age of 5 are evident in Nigeria. Overcrowding, the use of generators and biomass fuel have been identified as the critical environmental problems which increases the likelihood of respiratory symptoms and/or obstruction in those exposed on a regular basis in Nigeria.

Health policy changes regarding construction of homes with better ventilated kitchens and use of environmentally friendly, low-emission and energy-efficient cooking stoves, aimed at mitigating the deleterious effects of exposure to indoor air pollutants in mothers and children will significantly improve indoor air quality and reduce the risk of respiratory symptoms.

Indoor Air Quality will continue to be an essential field in environmental health studies and interventions because exposure to varied indoor substances will likely increase in the coming years. Research on the molecular characterisation of indoor bioaerosols, genetic susceptibility to indoor pollutants and their carcinogenic effect are required.

Author details

Godson Rowland Ana[1,2*], Oyewale Mayowa Morakinyo[1] and Gregory Adekunle Fakunle[1,3]

*Address all correspondence to: anagrow@yahoo.com

1 Department of Environmental Health Sciences, Faculty of Public Health, University of Ibadan, Nigeria

2 Climate and Society Programme Centre for Sustainable Development, University of Ibadan, Nigeria

3 Department of Medicine, University College Hospital, Ibadan, Nigeria

References

[1] Albalak, R., Frisancho, A.R. and Keeler, G.J. (1999). Domestic biomass fuel combustion and chronic bronchitis in two rural Bolivian villages. *Thorax;* 54 (11): 1004-1008.

[2] Almond, D. (2006). Is the 1918 influenza pandemic over? Long-term effects of in-utero influenza exposure in the post 1940 USA population? *Journal of Political Economy;* 114(4):672 -712

[3] American Industrial Hygiene Association (AIHA) (2001). Report of microbial growth task force. From http://www.hc-sc.gc.ca/ewh-semt/pubs/air/fungal-fongique/references-eng.php (retrieved 23 January 2009). Fairfax, VA1.

[4] Ana, M.G., Dairo, D., Adebowale, S. and Ana, G. (2012). A hospital-based pilot survey of acute infections among under-Five Children in Ibadan, Nigeria. Proceedings of the Ibadan Sustainable Development Summit; 185 – 191.

[5] Ana, G., Adeniji, B., Ige, O., Oluwole, O., Olopade, C.O. (2012). Exposure to emissions from firewood cooking stove and the pulmonary health of women in Olrunda community, Ibadan, Nigeria. Air QualAtmos Health (online version). doi:10.1007/s11869-012-0183-6

[6] Ana, G.R., Fakunle, G.A. and Ogunjobi, A.A. (2013). Indoor airborne microbial burden and risk of acute respiratory infections among children under five in Ibadan, Nigeria. *Indoor and Built Environment;* 0(0): 1-7.

[7] Ana, G.R.E.E. and Umar, Z.O. (2013). Indoor Air Quality and Respiratory Symptoms experienced among under-five children in Daycare centres in Ibadan North Local Government Area, Oyo State, Nigeria. An unpublished Master's dissertation in the Department of Environmental Health Sciences, University of Ibadan, Nigeria.

[8] Anyanwu, E.C. (2011). Advances in Environmental Health Effects of Toxigenic Mold and Mycotoxins. New York: Nova Science Publishers, Inc.

[9] Bartlett, K.H., Kennedy, S.M., Brauer, M., Van Netten, C. and Dill, B. (2004). Evaluation and determinants of airborne bacterial concentrations in school classrooms. *J Occup Environ Hyg 2004*; 1: 639-647.

[10] Bloom, E., Bal, K., Nyman, E., Must, A. and Larsson, L. (2007). Mass spectrometry-based strategy for direct detection and quantification of some mycotoxins produced by *Stachybotrys* and *Aspergillus spp.* in indoor environments. *Appl Environ Microbiol*; 73(13): 4211-4217.

[11] Boadi, K.O and Kuitunen, M. (2006). Factors affecting the choice of cooking fuel, cooking place and respiratory health in the Accra metropolitan area, Ghana. *Journal of biosocial Science*;38: 403-412.

[12] Bonnefoy, X., Braubach, M., Krapavickaite, D., Ormandy, D. and Zurlyte, I. (2003). Housing conditions and self-reported health status: a study in panel block buildings in three cities of Eastern Europe, *Journal of Housing and the Built Environment*; 18: 329-352.

[13] Brims, F. and Chauhan, A.J. (2005). Air quality, tobacco smoke, urban crowding and daycare: modern menaces and their effects on health. *Journal of Pediatrics Infectious Disease;* 11:152-156.

[14] Bruce, N., Perez-Padilla, R. and Albalak, R. (2000). Indoor air pollution in developing countries: a major environmental and public health challenge. *Bull WHO*; 78: 1080-1092.

[15] Cardoso, M.R., Cousens, S.N., De Goes Siqueira, L.F., Alves, F.M. and D'Angelo, L.A. (2004). Crowding: risk factor or protective factor for lower respiratory disease in developing countries. *Biomedical Journal of Public Health*;3(4):19.

[16] Cobanglu, N. and Kiper, N. (2006). Binaicisolunanhavadatehlikeler. CocukSagligive-HastaliklariDergisi; 49: 71-75.

[17] Cox, C.S. and Wathes, C.M. (1995). Bioaerosols handbook. New York. Retrieved September 12, 2011, from www.alibris.com/search/books/./ Bioaerosols%20Handbook.

[18] Dales, R.E., Zwanenburg, H., Burnett, R. and Franklin, C.A. (1991). Respiratory health effects of home dampness and molds among Canadian children. *American Journal of Epidemiology*;134.2: 196-200.

[19] Demers, A.M., Morency, P., Mberyo-Yaah, F., Jaffer, S., Blais, C. and Somse, P. (2000). Risk factor for mortality among children hospitalized because of acute respiratory infections in Bangui, Central African Republic. *J Pediatr Infect Dis.*; 19: 424-432.

[20] Diaz, E., Bruce, N., Pope, D., Lie, RT., Diaz, A., Arana, B. *et al.* (2007). Lung function and symptoms among indigenous Mayan women exposed to high levels of indoor air pollution. *Int J Tuberc Lung Dis.;* 11(1 2): 1372-1379.

[21] DiFranza, J.R., Aligne, C.A. and Weitzman, M. (2004). Prenatal and postnatal environmental tobacco smoke exposure in children's. *Pediatrics;* 113(4): 1007-1015.

[22] Douwes J. (2005). (1: >3)-Beta-D-glucans and respiratory health: a re-view of the scientific evidence. *Indoor Air;* 15(3): 160-169.

[23] Dutkiewicz, J., Krysinska-Traczyk, E., Skorska, C. *et al* (2002). Exposure to airborne microorganisms and endotoxin in a potato processing plant. *Ann Agric EnvironMed;* 9: 225-235.

[24] Ellegard, A. (1996). Cooking fuel smoke and respiratory symptoms among women in low income areas in Maputo. *Environ Health Perspect;* 104(9):980-985

[25] Energy Commission of Nigeria (ECN) (2009). 60 Nigerians Now Own Power Generators. Adopted from Vanguard Newspaper, 26th January, 2009. [Online] Available at www.energy.gov.ng.

[26] Environmental Protection Agency (EPA) (2009). Residential Air Cleaners. Indoor Air Quality, Available from http://www.epa.gov/iaq/pdfs/residential_air_cleaners.pdf.

[27] Fabule, D., Parakoyi, D.B. and Spiegel, R (1994). Acute respiratory infections in Nigerian children: prospective cohort study of incidence and case management. *J Trop Pediatr;* 40: 279-284.

[28] Halide, A., Ahmet, A., Muserref, T.O. (2009). Indoor and outdoor airborne bacteria in child day-care centers in Edirne City (Turkey), seasonal distribution and influence of meteorological factors", Environmental Monitoring and Assessment; 164: 53-66

[29] Johnson, A.W. and Aderele, W. I. (1996). The association of household pollutants and socio-economic risk factors with the short-term outcome of acute lower respiratory infections in hospitalized pre-school Nigerian children. *Annal Tropical Paediatric;* 12(4): 421-432.

[30] Grant, C., Hunter, C.A., Flannigan, B. and Bravery, A.F. (1989).The moisture requirements of moulds isolated from domestic dwellings. *IntBiodeterior;* 25(4): 259-284.

[31] Green, B.J., Tovey, E.R., Serkombe, J.K., Blachere, F.M., Beezhold, D.H., Schmechel, D. et al. (2006). Airborne fungal fragments and allergenicity. *Med Mycol;* 44(1): 245-255.

[32] Gocgeldi, E., Berdan, M.E., Ucar, M., Turker, T., Istanbulluoglu, H., Gulec, M. and Hasde, M. (2011). Analysis of children's rooms in terms of microbiological air quality. *J. Exp. Integr. Med.;* 1: 51-58.

[33] Hargreaves, M., Parappukkaran, S., Morawska, L., Hitchins, J., Congrong, H. and Gilbert, D. (2003). A pilot investigation into associations between indoor airborne

fungal and non-biological particle concentrations in residential houses in Brisbane. *Sci Total Environ*; 312: 89-101.

[34] Heikkienen, M.S.A., Hjelmroos-Koski, M.K., Haggblom, M.M., Macher, J.M. (2005). Bioaerosols. In: Ruzer, L.S., Harley, N.H., Eds. *Aerosols Hand-book.* Boca Raton: CRC Press; pp. 291-342.

[35] Hoskins, J.A., (2007). Health effects due to indoor air pollution. *Indoor Built Env.;* 12: 427-433.

[36] Hyvärinen, A., Meklin, T., Vepsäläinen, A. and Nevalainen, A. (2002). Fungi and actinobacteria in moisture-damaged building materials- concentrations and diversity. *International Biodeterioration and Biodegradation;* 49: 27-37.

[37] Jaenicke, R. (2005). Abundance of cellular material and proteins in the atmosphere. *Science*; 308:73.

[38] Johnson, A.B.R. and Aderele, W.I. (1992). The association of household pollutants and socio-economic risk factors with the short-term outcome of acute lower respiratory infections in Nigerian children. *Annals of Tropical Pediatrics;* 12: 421-432.

[39] Johnson, A.W, Aderele, W.I, Osinusi, K. et al (1996). Acute bronchiolitis in Tropical Africa: a hospital based perspective in Ibadan, Nigeria. *PediatrPulmonol;* 24: 236-247.

[40] Lin, H.H., Murray, M. *et al.* (2008). Effects of smoking and solid-fuel use on COPD, lung cancer, and tuberculosis in China: a time-based, multiple risk factor, modelling study. *Lancet;* 372(9648): 1473-1483.

[41] Lawrence, R. and Martin, D. (2001). Moulds, moisture and microbial contamination of first nations housing in British Columbia, Canada. *International Journal of Circumpolar Health;* 60(2): 150-156.

[42] Li, Y., Leung, G.M., Tang, J.W. and Yang, X. *et al.*, (2007). Role of ventilation in airborne transmission of infectious agents in the built environment-a multidisciplinary systematic review. *Indoor Air;* 17: 02-18.

[43] Manandhar, A., Lee, S.C., Ho, K.F., Hung, W.T. and Joshi, G.R. (2010). Identification of Indoor and Outdoor Particulate Concentration and its Chemical Composition in Kathmandu Valley.

[44] National Health and Medical Research Council (NHMRC) (1996). Definition of indoor air. Retrieved November 2010 from www.nhmrc.gov au

[45] Ojima, M. *et al.* (2002). Bacterial contamination of Japanese households and related concern about sanitation. *Environmental Health Research;* 12(1): 41-52.

[46] Okafor, E.E., Hassan, A.R. and Hassan, A.D. (2008): Environmental issues and corporate social responsibility: The Nigeria experience. *J. Hum. Ecol.;* 23(2): 101-107.

[47] Oluwole, O., Ana, G.R., Arinola, G.O., Wiskel, T., & Falusi, A.G., Huo, D., Olopade, O.I. and Olopade, C.O. (2013). Effect of stove intervention on household air pollution

and the respiratory health of women and children in rural Nigeria. *Air Qual Atmos Health;* 6: 553-561.

[48] Peppin, J., Demers, A.M., Mberyo-yaah, F., Jaffer, S., Blais, C. and Somse, P. (2001). Acute lower respiratory tract infection among children hospitalized in Bangui, Central African Republic: towards a new case management algorithum. *Trans R Soc Trop Med Hyg;* 95: 410-417.

[49] Perez, M., Ivan, N., Lucia, G., Pruneda, A., Fernando, D., Lilia, E., Batres, E., Francisco, J.,

[50] Perez, V. and Rebeca Isabel, M. S. (2011).Indoor air pollution in

[51] Mexico The Impact of Air Pollution on Health Economy Environment and Agricultural

[52] Sources.

[53] Pillai, S.D. and Ricke, S.C. (2002).Bioaerosols from municipal and animal wastes: background and contemporary issues. *Can J Microbiol;* 48: 681-696.

[54] Pokhrel, A.K., Smith, K.R. *et al.* (2005). Case – control study of indoor cooking smoke exposure and cataract in Nepal and India. *Int J Epidemiol;* 34(3):702-708.

[55] Pudpong, N., Krassi, R. and Nipapun, K. (2011). Indoor concentrations of PM10 and factors influencing its concentrations in day care centres in Bangkok, Thailand. *Asia Journal of Public Health Journal;* 2(1).

[56] Regalado, J., Perez-Padilla, R., Sansores, R., Paramo, R.J.I., Brauer, M., Pare, P. (2006). The effect of biomass burning on respiratory symptom s and lung function in rural Mexican women. *Am RespirCrit Care Med J;*174(8): 901-905.

[57] Romieu, I., Castro-Giner, F., Kunzli, N. and Sunyer, J. (2008). Air pollution, oxidative stress and dietary supplementation: a review. *EurRespir J.;* 31(1):179-197.

[58] Rosenstreich, D.L. et al. (1997). The role of cockroach allergy and exposure to cockroach allergen in causing morbidity among inner-city children with asthma. *New England Journal of Medicine;* 336(19): 1356-1363.

[59] Saha, A., Rao, N.M., *et al.* (2005). Pulmonary function and fuel use: a population survey. *Respir Res.;* 6:127.

[60] Shrestha, I.L. and Shrestha, S.L. (2005). Indoor air pollution from biomass fuels and respiratory health of the exposed population in Nepalese households. *Int J Occup Environ Health;* 11(2):150-160.

[61] Siddiqui, A.R., Lee, K., Gold, E.B., Bhutta, Z.A. (2005). Eyes and respiratory symptoms among women exposed to wood smoke emitted from indoor cooking: a study from southern Pakistan. Energy for Sustainable Development; 9(3):58-66.

[62] Smith, S. J., Alexander, A., Easterlow, D. (1997). Rehousing as a health intervention: miracle or mirage? *Health and Place*; 11: 271-286.

[63] Smith, K.R., Mehta, S., *et al.* (2004). Indoor smoke from household use of solid fuels. In: Ezzati, M., Lopez, AD., Rodgers, A., Murray, C.J.L. (eds) Comparative quantification of health risks: the global burden of disease due to selected risk factors. *World Health Organization*; 2:1435-1493.

[64] Smith-Severtsen, T., Diaz, E., Pope, D. (2009). Effect of reducing indoor air pollution on women's respiratory symptoms and lung function: the RESPIRE Randomized Trial, Guatemala. *Am J Epidemiol*;170(2): 211-220.

[65] Speirs, J.P., Anderson, A. and Anderson, J.G. (1995). A study of microbial content of domestic kitchen. *International Journal of Environmental Health*; 5(13): 109-122.

[66] Srikanth, P., Sudharsanam, S., Steinberg, R. (2008). Bioaerosols in indoor environment: Composition, health effects and analysis. *Indian J Med Microbiol*; 26: 302-312.

[67] Stanley, A.M., Mbamali, I., Zubairu, I.K., Bustani, S.A., Andrew, S.S. & Joshua, I.A. (2010). Electric Power Generator Noise Level Characterization and Impact on Selected Commercial Areas of Zaria and Kaduna Nigeria, International Postgraduate Conference on Infrastructure and Environment, The Hong Kong Polytechnic University, Hong Kong.

[68] Stetzenbach, L.D., Buttner, M.P. and Cruz, P. (2004). Detection and enumeration of airborne contaminants. *Curr Biotechnol*; 15: 170-174.

[69] Strachan, D.P., Cook, D.G. and Derek, G. (1998). Parental smoking and allergic sensitization in children: longitudinal and case-control studies. *Thorax*; 53(3): 204-212.

[70] Toivola, M., Alm, S., Reponen, T., Kolari, S. and Nevalainen, A. (2002). Personal exposures and micro-environmental concentrations of Particles and Bioaerosols. *Journal of Environment Monitoring*; 4:166.

[71] United States Environmental Protection Agency (USEPA) (1997). Revisions to the national ambient air quality standards for particles matter. *Federal Register*; 62: 38651-38701.

[72] Verreault, D., Moineau, S. and Duchaine, C. (2008). Methods for sampling of airborne viruses. Microbiology and Molecular Biology Reviews; 72(3): 413-444.

[73] WHO (2000). The World Health Report 2000 – Health systems: Improving performance. World Health Organization, Geneva.

[74] World Health Report (2002). Reducing risks, promoting healthy life. Geneva, World

[75] Health Organization (WHO). Retrieved March 4, 2010, from http://www.who.int/whr/2003/chapter1/en/index4.html

[76] World Health Organization (2007). Public Health and the Environment. Geneva: WHO.

[77] Yli-Pirilä, T., Kusnetsov, J., Haatainen, S., Hänninen, M., Jalava, P., Rei-man, M. et al. (2004). Amoebae and other protozoa in material samples from moisture-damaged buildings. Environ *Res*; 96(3): 250-256.

[78] Yli-Pirilä, T., Huttunen, K., Nevalainen , A., Seuri, M., Hirvonen, M. (2007). Effects of co-culture of amoebae with indoor microbes on their cytotoxic and pro-inflammatory potential. *Environ Toxicol*; 22(4): 357-367.

[79] Zock, J.P., Heinrich, J., Jarvis, D., Verlato, G., Norback, D., Plana, E. et al. (2006). Distribution and determinants of house dust mite allergens in Europe: the European Community Respiratory Health Survey II. J Allergy ClinImmunol; 118(3): 682-690.

Section 5

Air Quality and Health

Chapter 25

Health Effects of Metals in Particulate Matter

T.I. Fortoul, V. Rodriguez-Lara, A. Gonzalez-Villalva,
M. Rojas-Lemus, L. Colin-Barenque,
P. Bizarro-Nevares, I. García-Peláez,
M. Ustarroz-Cano, S. López-Zepeda,
S. Cervantes-Yépez, N. López-Valdez,
N. Meléndez-García, M. Espinosa-Zurutuza,
G. Cano-Gutierrez and M.C. Cano-Rodriguez

Additional information is available at the end of the chapter

http://dx.doi.org/10.5772/59749

1. Introduction

There is increased evidence of the association of air pollution and deleterious health effects. Particulate matter (PM) has obtained more attention, especially the small size components (PM_{10}, $PM_{2.5}$, UFP -ultrafine particles-) that carry on their surface different organic and inorganic elements whose composition differ with local and regional variations [1].

PMs might be identified by source, e.g., primary or secondary; combustion products, traffic, or by particle's size (aerodynamic diameter) (PM_{10}, $PM_{2.5}$, UFP -ultrafine particles-). This former parameter is important because the larger the particle the shorter the time it remains suspended in the air and the lower the risk of it being inhaled. Also the smaller the particles the higher the chances of deleterious health effects [1].

1.1. Metals in particulate matter, air pollution and sources

Particles toxic components are complex mixtures of solids or liquids with different characteristics (e.g., mass, number, size, shape, surface area, chemical composition, acidity, solubility). Chemical components may be located on the surface or inside the particle. Considering the source there are natural and anthropogenic emissions. Natural sources include sea salt, volcanic ash, pollens, fungal spores, soil particles, forest fires and wind-blow dust [2]. Anthropogenic sources consist of fossil fuel combustion products, industrial process, mining activities, wood stove burning and cigarette smoking. In urban areas the main source of PM are motor vehicles especially those derived from diesel fuel combustion.

One of the components adhered to the small particles are metals that come from impurities derived from fuel additives and brakes and tires attrition. Transition metals are generated by non-exhaust emissions, impurities in fuel additives, or metallurgic process. Iron (Fe), nickel (Ni), vanadium (V), chromium (Cr) copper (Cu) are considered because its potential to produce Reactive Oxygen Species (ROS) in biological systems [1]. Others such as zinc (Zn) may also exert toxic effects by other mechanisms besides ROS production.

Heavy metals such as cadmium, lead and mercury are some of more common air pollutants emitted by industrial activities, combustion, extraction and processing activities. The diagnosis of health effects for heavy metals might be difficult if there is no previous evidence of the exposure source, while acute exposures usually occur in the workplace and are more easily identified; signs and symptoms differ within metals because each one interacts with different targets such as: specific metal-binding proteins (metallothioneins, transferrin, ferritin, ceruloplasmin) in cases such as Cd, Cu, Hg, Ag, Mn Zn, Al, and Be; membrane carrier-proteins (phosphate and sulphate-transporters), divalent cation-transporters, some examples are V, Cr, Mo, Se. Heavy metals share with transition metals the possibility to exert its toxic effects by the production of ROS [2-4].

2. Metals and its effects on health

A variety of health effects associated with PM exposure began with the increased mortality risk in those cities with high particulate concentrations (PM), later myocardial infarction incidence and high particulate matter emitted by internal combustion engines were reported. Also ultrafine particles emitted from the vehicles, especially diesel, induced oxidative stress in the endothelium, and through the nose these particles penetrate the olfactory bulb an reached other structures from the nervous system, inducing increased inflammatory responses in the brain. Reduced lung function and respiratory diseases are frequently reported. Thrombosis, hearth rate changes, blood pressure modifications have been associated with exposure to particulate matter also [5] and more recently metabolic abnormalities such as altered glucose metabolism have been reported as well. Low birth-weight, infertility, genotoxicity and cancer [6] are also part of the spectrum of alterations associated with PM. Metals are some of the components adhered to particles surface associated with its toxicity [7, 8]. Here we included the health effects of some metals carried by inhaled particles, and its possible mechanisms of damage.

2.1. Respiratory system and metals

Human beings can survive three weeks without food, three days without water but no more than three minutes without air [9]. As Aaron Cohen mentioned, "You can't avoid breathing the air no matter what you are" [10] and this comment is because the recent publications about

http://dx.doi.org/10.5772/59749

the air pollution problems reported in China [11] and the case of a Chinese eight-year-old girl diagnosed with lung cancer, until now the youngest victim of lung cancer [10]. Particulate Matter (PM) is the carrier of metals into the lung structure and its content determines its potential health hazard [7].

The exact mechanisms involved in PM exposures and lung damage have been discussed and mentioned previously, but it seems that oxidative stress, inflammation, and modulation of the immune response are some of the mechanisms proposed. Differences according to the age of the population exposed also have some differences because in children's asthma, reduction in lung growth, allergic rhinitis and respiratory infections are more frequently reported compared with adults. Metals such as Cd, Hg and Au have been associated with autoimmune reactions. Other metals reported on the particles are Fe, Zn, and Ni, and recently with the use of the catalytic converters an increase in the presence of Pt, Pd and Rh in the particles inhaled has been observed. These three elements have been associated with asthma, rhinoconjuntivitis, and dermatitis among occupational exposure workers [7].

Lead has been present in the atmosphere in different concentrations and since tetraethyl lead was reduced as anti-knocking agent in gasolines Pb-concentrations in the air and in children 's blood has decreased [12]. Changes in the ultrastructure of the non-ciliated bronchiolar cell (NCBC) after the inhalation of Pb has been reported along with cell hypertrophy, whorl-like structures and mitochondria cristae disarray [13]. The NCBC was also the target of the combined inhalation of Cd and Pb. A decrease in the cell volume with surface irregularities and the presence of sloughed cells on the epithelial surface were noticed in the exposed group; however at 4-week exposure time, clusters of dividing cells were observed [14]. Differences in the lung concentration of each element, compared with the mixture were reported and could explain the morphological differences reported [15].

Cigarette smoke is a source of direct deposit of metals in lung tissue; because metals are not biodegradable remain in the tissue for long periods of time. In a review from Stavrides [16] chromium, cadmium and nickel are mentioned as carcinogenic and genotoxic metals, indicating that the main effect of these metals as carcinogenic agents is because of altered DNA's repair capacity as a consequence of the oxidative stress. In addition the destruction of the cilia by the gaseous phase of the cigarette smoke facilitates the stagnation of the mucus, whose production increases as a consequence of the irritating effect of the smoke components, resulting in longer periods of cells-metals contact with the increasing opportunity to interact with the genetic material and also to interact with DNA, producing cumulative genetic alterations that could result in lung cancer [16]. Cerium and lanthanum are also reported in the environmental tobacco smoke and may produce inflammation and granulomatosis in lung tissue [17]. Cakmak [18] found association with respiratory effects and the content of Cd, Zn and V in particulate matter ($PM_{2.5}$] and the suggested mechanism was oxidative stress resulting in inflammation and tissue damage.

2.2. Cardiovascular toxic effects of metals

Particulate Suspended matter (PM) exposure raises the risk of developing cardiovascular diseases, both in the short and long term [19, 20]. The proposed mechanisms include inflammation and the induction of hypercoagulability, oxidative stress and endothelial dysfunction [20, 21]. The best evidence that PM have an important role in cardiovascular morbidity and mortality are the interventional studies, in which measures to reduce the levels of PM are taken and a significant decrease in cardiovascular risk is observed [22, 23].

Fine and ultrafine particles are considered the more toxic compared with bigger particles; particularly Diesel exhaust ultrafine particles are dangerous because of its high metals content, causing oxidative stress that leads to endothelial dysfunction and the development of atherosclerosis increasing the risk of heart infarction or stroke [24]. There are associations of elevated concentrations of some metals such as nickel, copper, arsenic, and selenium in fine or ultrafine particles and markers of cardiovascular disease: higher levels of markers of inflammation (C-reactive protein, interleukin-6, and vascular endothelial growth factor) and reduced levels of circulating endothelial progenitor cells (CEPC) suggesting reduced capacity of endothelial repair [25]. Iron overload is directly associated with hypertension, atherosclerosis progression and increased cardiovascular risk [26] Reports that iron chelation with deferoxamine decreases endothelial dysfunction and has been successfully used to reduce cardiovascular risk in diabetic and non-diabetic patients [27]. Lead causes cardiovascular effects even at low doses and it has been associated with hypertension in animals and humans [28]. Peripheral arterial, coronary heart and cerebral vascular diseases had been also associated with Pb exposure, but further studies are needed to establish its causality [29]. Cadmium affects the cardiovascular system too, demonstrated *in vitro* and in animal studies, in addition of epidemiological evidence that Cd is associated with hypertension, promoting atherosclerosis and myocardial infarction. Mercury promotes atherosclerosis and cardiovascular disease in much lower concentrations than those reported for its neurotoxic effects [30].

2.3. Metals and its relationship with cancer

Many epidemiological studies have demonstrated an association between long-term exposure to ambient air pollution and cancer mortality mainly from lung cancer. The risk of cancer is associated with exposure to different pollutants such as nitrogen dioxide (NO_2), Sulphur dioxide (SO_2), particulate matter (PM) and several metals attached to its surface. Metals can promote carcinogenesis, through several pathways such as: producing DNA damage, activating different signaling pathways that lead to tumor progression and promoting inflammation. In the following paragraphs the mechanisms by which some metals promote carcinogenesis are detailed.

Experimental animal studies have demonstrated clearly that cadmium and cadmium compounds by multiple routes of exposure generate cancer at various sites in many animal species and in humans. Cadmium exposures of laboratory animals causes leukemia and lymphoma, local sarcoma and cancer of the adrenal gland, liver, lung, kidney, pancreas, pituitary, prostate and testis. Moreover, cadmium and their compounds are classified by the International Agency for Research on Cancer (IARC) as carcinogens in humans. Cadmium exposure is associated with lung adenocarcinoma, also with prostate, kidney, urinary bladder, pancreas and breast cancer in humans. Several studies have shown that Cd carcinogenicity seems to be crucially mediated by the production of ROS such as hydroxyl radicals, superoxide anions, nitric oxide and hydrogen peroxide and are due to the inactivation of detoxifying enzymes (e.g., catalase, glutathione peroxidase, glutathione reductase and superoxide dismutase) as a consequence of the interaction with thiol groups. Cadmium is also capable of replacing copper and iron in various cytoplasmic and membrane proteins (e.g., ferritin, apoferritin), leading to an increase in the amount of unbound or poorly chelated copper and iron ions inducing oxidative stress via Fenton reactions. Cadmium also produces genotoxicity by the production of DNA single-strand breaks an damage and competes for binding at sites (specifically with a zinc finger motif) that are important in gene regulation, enzyme activity, or maintenance of genomic stability. In addition, this metal modifies the expression of several genes related to carcinogenesis, including intermediate early-response genes such as c-fos, c-jun, and c-myc; stress-response genes such as metallothionein, and heat-shock genes; genes controlling glutathione and related proteins, as well as transcription and translation factors. Also disrupts cell adhesion mediated by E-cadherin and affects the regulation of cell growth and apoptosis causing tumor progression [31-33]. In addition, cadmium has estrogenic effect and may bind to and activate mammary cell estrogen receptors; it also interacts and regulates the transcription of estrogen-dependent genes affecting the synthesis of proteins and/or the activity of cell-signaling pathways in a manner similar to estradiol [34]. In addition to its endocrine effects on mammary tumor cells, cadmium transforms healthy breast epithelial cells into cells with a cancer-like profile through non-hormone-related pathways. Thus, in the presence of cadmium, the cells alter gene expression and DNA changes in DNA methylation (an epigenetic change) that are typical of cells undergoing transformation from healthy to cancerous type [35].

Furthermore, epidemiological data provide increasing evidence that environmental as well as occupational lead exposures may be associated with increased cancer risk. The IARC has classified lead as possible human carcinogen (group 2B) and its inorganic compounds as probable human carcinogens (group 2A). Lead exposure has been associated to increased lung cancer risk. Some studies looking at blood lead levels in the general population have also found a small increased risk of lung cancer in people with higher lead levels. In addition, the majority of the studies found an increased risk of stomach cancer with higher lead exposure, even though the studies did not take into account other factors that could also have been affected stomach cancer risk. There is a stronger association of kidney tumors with lead exposure; brain, lung and bladder cancer have also been linked to lead in different studies, however results are controversial [36]. Several studies have identified the carcinogenic potential of lead, because the genotoxicity of the metal that induces alterations in DNA synthesis, mutations, chromosome aberrations, as well as inhibiting DNA repair or displacing zinc in DNA binding proteins [37]. Lead, also stimulates cell proliferation, induces alterations in gene transcription and causes oxidative damage that promote carcinogenesis [36].

The International Agency for Research on Cancer (IARC) has determined that some nickel compounds are carcinogenic to humans and that metallic nickel may be carcinogenic to humans. The EPA has determined that nickel refinery dust and nickel subsulfide, are human carcinogens. Occupationally exposed people have a higher risk of respiratory tract cancer (nasal sinus and lung cancer mainly) due to inhalation of nickel at their workplace in nickel refineries, nickel-producing or processing plants or using industries. High cancer risk is related to less soluble oxidic and especially sulfidic nickel species in refinery dust. Earlier studies gave already indications that rats during two years to inhalation exposure of nickel subsulfide developed a significant higher number of lung tumors [38]. Mechanisms of nickel carcinogenicity have not been fully elucidated yet. Ionic nickel (Ni^{2+}) is supposed to be the carcinogenic species because it can bind to cellular components such as nuclear proteins (histones and protamines) and DNA. Nickel also induces chromatin condensation modifications, DNA hypermethylation, histone acetylation and gene slicing, which disturb gene expression. Moreover, there is evidence that nickel ions inhibit enzymes required for DNA repair. Furthermore, nickel modulates gene expression by the induction of DNA methylation and/or suppression of histone acetylation [38-41].

Some studies have shown that serum copper levels are elevated in cancer patients and correlate with the severity of the disease and the response to therapies [42, 43]. Copper-chelating drugs have been reported to have antiangiogenic activity in animal models. Other study has shown that cancer cells express higher levels of the copper transporter Ctr1 and that the tumors were sensitive to the reduction in systemic copper levels compared with normal tissues [44]. Pharmacological suppression of systemic copper impairs oxidative phosphorylation and tumor growth, since copper can modulate the proliferation of cancer cells and associated tumor growth. It has been proposed that copper can be a rate-limiting nutrient for tumors, similar to oxygen and glucose. Copper levels in tumors affect cytochrome c oxidase activity, additional bioavailable copper facilitates increased production of ATP, which is consumed to fuel rapid proliferation of cancer cells. Thus, copper may not initiate transformation, but may stimulate proliferation of transformed cells by providing energy needed for cell-cycle progression as proposed Ishida and coworkers [42]. Additionally, copper ions are well suited to facilitate formation of ROS that can damage biomolecules, including DNA and chromatin. This event occur *in vitro* with isolated DNA or chromatin, or by exposure of cultured mammalian cells to copper complexed with various agents. Whether if this is likely to occurs *in vivo* is not well defined. However, copper, can directly bind with high affinity to DNA molecule; this binding can modify the conformational structure of DNA promoting carcinogenesis[45].

Mercury and its compounds mainly methylmercury have been classified as "possibly carcinogenic compounds to humans". Mercury has been associated with lung cancer, genitourinary tract cancer and probably brain cancer risk in occupational exposed personnel, however these results are in controversy because workers might be also exposed to other metals [46]. Mercury promotes carcinogenesis inducing oxidative stress. In addition, mercury compounds are genotoxic, mainly by inhibiting the mitotic spindle and altering DNA repair processes decreasing the incision step. Lead and aluminum can increases the toxicity of mercury. It has been shown that mercury rapidly depletes the immune system and could decrease immune tumor response. Chronic exposure to relatively low levels of mercury may inhibit antioxidant enzymatic activity due to persistent oxidative stress. This phenomenon might represent an important peripheral target for mercury toxicity in exposed populations [33].

Inhalation exposures to aluminum in several cohort studies reveal a relationship with increased cancer incidence and mortality in the aluminum smelting industry [47]. The IARC has classified occupational exposures during aluminum production as a causal factor, with sufficient evidence in humans, for cancers of the lung and the bladder. Nowadays, authors conclude that exposure levels to know health hazards associated with the emissions from primary aluminum production should be studied to establish a clear relationship between inhaled aluminum exposure and cancer [48]. The carcinogenic mechanisms of inhaled aluminum exposure are associated to different compounds an not with the metal *per se*, as it is described in different studies [49].

Chromium is widely used in the industry for the production of stainless steel, chromium plating, and spray-painting. The health effects and toxicity/carcinogenicity of Cr inhalation are primarily related to its oxidation state at the time of exposure [50]. According to epidemiological studies, the hexavalent [Cr(VI) form of this metal, appears to be drastically toxic and carcinogenic, thus it has been classified as carcinogenic to humans by the IARC [51]. The carcinogenicity of the metal is site specific, mainly to the lungs and nasal cavity [50]. The molecular mechanisms of [Cr(VI)-induced carcinogenesis are well studied and characterized, and the main mechanism of chromium carcinogenesis, is the production of free radicals, resulting in the generation of oxidative stress, This stress causes a series of modifications that are directly linked to the establishment of the cancer phenotype. The genetic changes involve Cr-DNA adducts, formation of DNA-protein cross-links, single and double strand DNA breaks [52, 53].

Studies exploring excessive environmental exposure to iron are often limited by poor characterization of the environmental factors and causal relations to effects other than the chemical properties of the iron [54]. Iron is present primarily in two oxidation states, ferrous ions [Fe(II) and ferric ions [Fe(III). Mechanisms by which iron may contribute to tumor induction or progression, includes oxidative damage-induced changes in genetic material as the initial step involved in Fe-induced mutagenesis and carcinogenesis. Other mechanisms are alterations in gene expression consistent with increased iron requirements in proliferating cells and decreased immune surveillance against cancer [55].

In occupationally exposed individuals, inhalation of Mn is a potential important route of exposure [56]. In general, Mn and its inorganic compounds are considered to possess low mutagenic or carcinogenic potential compared with other heavy metals. The experimental evidence on its carcinogenicity does not provide any clear evidence, while the available occupational and environmental epidemiological evidence is equivocal as to whether exposure to inorganic Mn is associated with a significant cancer risk or not [50]. Apparently from *in vitro* data, Mn and some inorganic Mn compounds are cytocytotoxic at differing concentrations depending on the test system, generating ROS *in vitro* and *in vivo*, interfering with DNA polymerases, mitochondrial function and activating some cytokines and MAPK cell signaling cascades. These mechanisms should probably contributes to Mn carcinogenesis, but further studies must be made [57].

Finally, V is a major transition element that is released primarily by the burning of fossil fuels, including petroleum, oil, coal, tar, bitumen, and asphaltite. Among V compounds, V pentoxide is highly toxic [58]. The IARC classified vanadium pentoxide as a possible carcinogen to humans (Group 2B) in 2003 [59]. The pentavalent forms, such as V and vanadate have carcinogenic potential via ROS generation, DNA damage, and activation of hypoxia signaling

[51, 60]. There is *in vivo* preclinical data on cancer chemoprevention and therapy, which provide a rationale for its use in human populations [61].

2.4. Genotoxicity and metals

The term genotoxicity refers to any detrimental change in genetic material, regardless of the mechanism by which the change was caused [62]. The DNA lesions can be classified according to the extent and severity of damage. Repairable damage includes single-stranded breaks, oxidized bases, AP-sites and alkali-labile sites. The damages that result from the incorrect repair are chromosomal rearrangements and sister chromatid exchange. Finally, there are irreparable injuries including gains or losses of whole chromosomes or chromosome fragments (chromosomal aberrations) that are the product of clastogenic or aneugenic events. The consequences of DNA damage include alterations in the three-dimensional conformation, blocking the processes of replication and transcription, deletions, chromosomal instability, cell death and mutagenic events. Thus, genotoxic mutagenic events precede and therefore are the source of cellular malignancy [63]. Studies indicate that nickel induces chromosome aberrations in rats [64]; cadmium [65] and lead [66] causes single strand breaks and alkali-labile sites in mouse cells; in rats uranium causes double strand breaks [67]; iron induces in mice chromosome aberrations and micronuclei; in human beings vanadium produces bases oxidation and micronucleus [68], but in mice produces single strand breaks and micronucleus too [69, 70]. Conversely to others metals, titanium showed no adverse effects on DNA in mice cells [71].

2.5. Neurotoxicity of metals

The brain is vulnerable to oxidative stress damage produced by metals, due to its great metabolic activity, high cellular content of lipids, and low levels antioxidants such as catalase and superoxide dismutase. Air pollution is a mixture of gases and metals associated with particulate matter (PM) [72] that induced olfactory dysfunction, neuroinflammation and elevated markers of neurodegeneration [73] and heavy metals in PM are accumulated in brain tissues [74].

Inhaled nanosized particle (NSP) includes ambient spherical particles < 100 nm deposited directly on olfactory dendritic cilia in the olfactory mucosa. Subsequently uptake and translocated along axons of the olfactory nerve by nasal route via neuronal transsynaptic transport and uptake through the blood brain barrier from systemic circulation, and induced oxidative stress and gene expression in central nervous system in human and experimental animals [75, 76].

Occupational and environmental exposure to neurotoxicants such as iron (Fe), copper (Cu), manganese (Mn), aluminum (Al), zinc (Zn) mercury (Hg), lead (Pb) and vanadium induced oxidative stress [77, 78] that generated accumulation of ROS inducing protein, lipid and deoxyribonucleic acid (DNA) oxidation, and produced neurotoxic changes that are risk factors for development neurodegenerative diseases such a Parkinson (PD), Alzheimer (AD), Huntington (HD), amyotrophic lateral sclerosis (ALS) and transmissible spongiform encephalopathy (TSE), [79, 80] [81, 82].

Protein misfolding is implicated in neurodegenerative diseases, such as amyloid-β precursor protein, in senile plaques and tau in neurofibrillary tangles of AD,α-synuclein in Lewy bodies in sustantia nigra in PK, prion protein in TSE and huntingtin in HD striatum have been connected to neuronal iron homeostatic control [83].

At a molecular level metal dishomeostasis and mitochondria dysfunction in AD and PD and the cytoplasmic predominance of neuronal 8-hydroxy-Guanine supports mitochondria as the major source of ROS responsible for RNA oxidation and might induce DNA oxidation neuronal damage [84, 85].

Atmospheric Mn is present in gasoline additive methylcyclopentadienyl manganese tricarbonyl (MMT) is a putative modulator of dopamine biology (the primary target of Mn neurotoxicity) and workers exposed to airborne Mn are in risk of developing PD known as manganism, an extrapyramidal neurological disease characterized by rigidity action tremor, bradykinesia, memory and cognitive dysfunction. Mn in blood crosses the blood brain barrier and accumulates inside the neuron disrupting the synaptic transmission and inducing glial activation[86].

In previous studies it has been reported that aluminum inhalation induced altered of expressions of glycogen synthase kinase-3 (GSK3) and protein phosphatase (PP1), which help in the regulation of carbohydrate metabolism in the rat's brain [87]. Mercury exposures in male marmoset monkey showed deposits of Hg in ventral horn motor neuron and atrophy of large myelinated motor axon [88]. In our mice model the inhalation of vanadium pentoxide caused morphological and functional alterations in the central nervous system. In the olfactory bulb we showed dendritic spiny loss of granule cells and ultrastructural changes characterized by swelled organelles, disrupted mitochondria and necrotic and apoptotic neuronal death that might correlate with the olfactory dysfunction. Also, the hippocampal formation showed a decrease in dendritic spines and necrosis of the pyramidal layer of CA1 and granule cell of dentate gyrus that could be related with spatial memory impairment at a 4-week exposure. We also found the loss of immunoreactive- tyrosine hydroxylase + in substantia nigra and a decrease in dendritic spine density in the medium striatal spiny neurons at 8-week exposure time. The blood brain barrier (BBB), after the inhalation of vanadium showed cilia loss, cell sloughing and ependymal epithelium detachment from the basal membrane, and the presence of oxidative damage in the choroid plexus, which was confirmed by the presence of 4-hydroxynonenal. The reported alterations were associated with an increase in MMP-9 and MMP-2 activity in the cortex, the olfactory bulb, hippocampus and striatum [89, 90].

2.6. Metals and mental health

Environmental pollution by heavy metals that are produced by the combustion of hydrocarbons is a public health problem, which affects different organs such as central nervous system [91] have resulted in behavior, learning, mental disorders, attention deficit and low mental performance [92-94].

Inhaled metals involved in centrals nervous damage, mental and behavioral disorders are arsenic, lead, cadmium, mercury, manganese and vanadium [95] [96]. The inhalation of these metals induces brain damage and resulted in behavioral disorders associated with the severity of the poisoning, the metal involved, the chemical state of the element as well as the exposure route and the age of the exposure [97]. Inhaled arsenic might cause Guillain-Barre similar syndrome with confusion, irritability, cognitive loss, decrease in verbal responses and paralysis [98].

Lead causes an irreversible reduction in cognitive ability in children resulting in an IQ decrease [99]. Oxidative stress is also the mechanism proposed to induce the damage [100]. Other behavioral effects induced by inhaled-Pb are depression, irritability, bipolar states, mental retardation and cognitive deficit [101].

Cadmium has the ability to replace iron and copper, which induces an increase in the production of ROS via the Fenton reaction, which translates into GABA and serotonin systems alteration, causing irritability, depression, amnesia and cognitive disorders [102].

Mercury mechanism of action are not fully known but it seems that interact molecularly with the antioxidant systems such as GSH, cysteine and melatonin [103]. Apoptosis, necrosis, lysis and phagocytosis have been reported in exposed humans [104]. Other neurotoxic effects reported are weakness, inability to concentrate, lethargy, depression, irritability, blindness, coma and death [105].

Excessive manganese exposure during childhood causes hyperactivity and learning disorders [106]. Studies in animals show that inhaled manganese reaches the central nervous system through the olfactory nerve and by the blood, crossing the blood-brain barrier inducing [107] "manganism" that includes tremor, “manganic madness", schizophrenic symptoms, violent behavior, compulsions, emotional instability, hallucinations, and other psychiatric disorders [108].

Vanadium brain accumulation has been related with behavioral and cognitive disorders, [109] [110]. Memory loss, a decrease in the sense of taste and Parkinson's disease has been reported in vanadium exposure [111, 112].

2.7. Metals and its toxic effects in the eye

Despite the fact that the eye is an important air pollutant target because it is directly exposed to the atmosphere, as well as to the elements that enter into the lungs and further are distributed through the systemic circulation [113], until now, only two studies have been found to approach the effect of atmospheric pollutants inhalation on the retina, where phototransduction process takes place. Such studies were performed in a mice model in which the animals were exposed to vanadium (V) [0.02M] inhalation 1 h twice a week, for 4, 8 and 12-week time periods.

In all exposure times, morphological alterations in the photoreceptor layer (PL) and in the inner and outer nuclear layers were observed, as well as a gradual rhodopsin pigment reduction in the PL and an increase of the oxidative stress biomarker -4-hydroxinonenal- in the PL and in the inner and outer plexiform layers [114].

Additionally, the effect of V exposure was evaluated on the Müller glial cell (MGC), which is the predominant radial glia in the retina, for 4 and 8-week time period. Glial fibrillary acidic protein (GFAP) expression increased at four weeks after the exposure, probably as evidence of reactive gliosis, whereas it was observed a gradual reduction in Glutamine synthetase (GS) expression as exposure time passed. Given that GS is an enzyme whose levels are regulated by its substrates -glutamate (Glu) and ammonium- its reduction might suggest that photoreceptors, that produce most of the Glu in the retina, are degenerating in response to the V toxic insult [113]. This is consistent with the rhodopsin pigment reduction that previously was mentioned, because it evidenced damage to the photoreceptors as a consequence of the increase in the oxidative stress induced by the exposure [114].

2.8. Metals and glucose metabolism

There is limited information about the role of metals in carbohydrate metabolism and glycemic regulation; however there are some studies showing that metals have hypoglycemic or hyperglycemic effects. The relevance of studying these effects is that the exposure to some metals has been related to increased risk of diabetes, but some other metals have shown a hypoglycemic effect and have been studied as a potential treatment of diabetes (as vanadium).

Iron, mercury, nickel and lead are hyperglycemic metals that are also pollutants and will be discussed in this chapter. Iron is an atmospheric pollutant in both urban and industrial sites [115], near iron or steelmaking industries, near petrochemical areas [116], cement mills and in metro systems of many cities [117]. Mercury is a toxic heavy metal widespread and persistent in the environment and it is considered one of the most relevant atmospheric pollutants (Wang et al, 2006). Nickel is a metal released from many industries and it has proved to be toxic at high concentrations and lead is a heavy metal pollutant because it is released to the atmosphere by the burning fossil fuels, industries and mining activities.

Iron is an essential metal for life, but iron overload is a health risk because it is associated to insulin resistance, hyperglycemia and an elevated risk of type 2 Diabetes Mellitus [26, 118, 119]. Several authors have explored the mechanism of this risk; oxidative stress has been implicated because it is associated with insulin resistance and with direct damage on beta pancreatic cells [120].

There is evidence that mercury can cause hyperglycemia because it can directly damage pancreatic beta cells inducing necrosis or apoptosis [121]. In a follow-up study of young people who were exposed to high levels of mercury a higher risk of developing diabetes after 18 years was found [122]. The majority of the associations are related to water pollution, but it is important to evaluate people who have been exposed by atmospheric pollution. On the other hand, there is no evidence of higher levels of mercury in blood of diabetic type 1 or 2. A causal relationship between mercury and pancreatic dysfunction that leads to hyperglycemia and diabetes has been reported, but not all the diabetics have higher levels of this metal [123, 124].

There are reports about the effect of lead on glycaemia regulation but the effect seems to be related to the dose and the compound. Ibrahim and cols in 2012 [125] had reported hyperglycemic effect on rats exposed to different doses of lead acetate, just as Adham et al. did in birds in 2011 [126], however other authors have found in rats hypoglycemia, after low doses of lead dissolved in water [127]. It is necessary to understand the mechanism and the reason for this dual effect.

There are multiple reports of hyperglycemia and insulin resistance after exposure to nickel in different animal models [128, 129], so it is important to consider this element as a possible risk factor to glycemic deregulation.

Some metals decrease the levels of blood glucose such as vanadium, chromium, magnesium and zinc. Proposed mechanisms for this effect include: activation of insulin receptors, increasing insulin sensitivity, and function as cofactors or components of the enzymatic systems involved in glucose metabolism or acting as antioxidants to prevent tissue peroxidation. [130].

Vanadium potentiates the action of insulin and lowers blood glucose levels. Some vanadium compounds have been studied as antidiabetic agents [131]. At first it was thought that vanadium exerted an effect on the glycemia because it inhibits appetite at certain concentrations, but hypoglycemia was observed only minutes after its administration, which is not a period of time sufficient to exert its anorectic effects [132]. The hypoglycemic effect is explained because vanadium inhibits some tyrosine protein phosphatases increasing the phosphorylation levels of various insulin pathway intermediaries. Activation of these signaling pathways results in GLUT transporter translocation to the plasma membrane [133]. Another factor that contributes to the hypoglycemic effect of vanadium is its inhibitory effect on gluconeogenesis because it inhibits the expression of the gluconeogenic enzymes PEPCK and GTPase [134]. There are reports of severe hypoglycemia that may threaten life in vanadium acute intoxication [135]. Further studies are needed to evaluate the hypoglycemic effect of vanadium as an air pollutant or in workers occupationally exposed.

Chromium induces hypoglycemia because it is a promoter of glucose catabolism in muscle cells and adipocytes. Also, it functions as a regulator of glycaemia in different animal and human models, and as an inhibitor of glucogenolysis in muscle cells. The trivalent compounds as chromium picolinate increase insulin activity [136]. There are reports that after the consumption of Chromium based compounds; patients suffering from diabetes improved their sensibility to insulin [137]. The mechanisms studied are: 1) increase the concentrations of the messenger RNA for insulin receptor; 2) making a complex with insulin that has a greater activity in the metabolism of glucose than insulin alone; 3) through the decrease in TNFα, resistin and interleukin 6 concentrations; and 4) increasing the sensitivity of pancreatic β-cells [138, 139]. There are no reports of severe hypoglycemia, but further studies are needed to elucidate the effects of chromium in concentrations inhaled as atmospheric pollutants.

Depletion of magnesium is associated to insulin resistance, hyperglycemia and type 2 Diabetes [140]. Low levels of serum magnesium in diabetics is associated with poor glycemic control [141] and foot ulcers [142]. Zinc has insulinomimetic activity *in vitro* and blood glucose lowering effect *in vivo* [143]. In some studies zinc deficiency has been associated with hyperglycemia and diabetes [141]. Also, zinc supplementation in diabetic patients improves glycemic control [144]. However, a higher zinc concentration has promoted metabolic syndrome (overweight, hypertension and dyslipidemia) in Wistar rats [145] Ugwuja et al, in

http://dx.doi.org/10.5772/59749

2014 [141] reported higher zinc levels in complicated diabetic patients compared with diabetic uncomplicated cases. The meaning of these associations is unclear and needs further studies.

2.9. Metals and its toxic effects on liver

Inhaled air pollutants that travel through the blood, also produce changes in the integrity of liver parenchyma, which leads to a slowly and irreversible liver damage [146] [147]. Acute or chronic liver damage is the usual consequence in the majority of the toxic agents that enter into the organism, because the liver is the main organ that metabolizes xenobiotic agents such as metals. The progressive deleterious events in the liver starts with steatosis ending in hepatocellular carcinoma, passing through chronic hepatitis, fibrosis and cirrhosis, finalizing in liver failure and death [148].

It is important to emphasize that regardless the etiological agent in all types of liver damage there is overwhelming evidence of an increase in free radicals or a decrease in antioxidant defenses [149]. As well, the reactive oxygen and nitrogen species play a crucial role in the induction and progression of the liver diseases.

Arsenic is a metalloid that has been characterized by causing a variety of alterations in the organism [150]. Arsenic crosses lung alveolar membrane and reaches the blood stream; hence it is transported to all the organs, mainly to the liver, in which is metabolized. It has been reported that arsenic induces liver cancer [151]. This is done through the modulation of transcription factors like NF-kB, AP-1 and p53 that promotes liver tumors. Likewise, Arsenic causes liver lipoperoxidation producing large amounts of ROS [152]. Also a decrease in the levels of the superoxide dismutase enzymes (SOD), catalase (CAT) and glutathione peroxidase has been reported [153].

Lead has been characterized by inducing damage by the production of ROS, that increases lipid peroxidation and decreases antioxidant defenses [154]. It has been reported that lead damages the cell membrane of hepatocytes and its DNA [155].

Cadmium causes liver damage mainly by induction ROS inducing lipoperoxidation via Fenton reaction [156]. The increment of ROS induces DNA damage, proteins oxidation and lipoperoxidation. Cadmium replaces iron and copper in the Fenton reaction. It is also capable of moving to zinc from proteins and changing their structure [157]. Chronic liver exposure to cadmium induces liver failure [158].

Mercury toxicity is the consequense of its high affinity to sulphydric groups in proteins and enzymes involved in cell cycle progression [159] [160]. It induces hepatocyte apoptosis causing acute liver failure [161].

Liver is the main organ for the metabolism of iron and it is also the target of its pathological accumulation, as a consequence of a metabolic disease, such as hemochromatosis or because of an increased exposure [162]. Iron accumulates in the hepatocytes in which induces the formation of hydroxyl radicals (.OH) from reduced forms of O_2, ending in oxidative stress [163].

Cooper is associated with Wilson's disease resulting in its liver accumulation, because of the reduced metal elimination by bile duct [164]; air pollution exposure also causes liver accumulation and injury [165]. Copper and iron share the same mechanism of damage causing an increase in reactive species in liver parenchyma ending in fibrosis and cirrhosis [166].

Vanadium is other metal that alters liver function, inducing ROS via Fenton reaction, damaging proteins and altering the genetic material [167]. Inhalation of V_2O_5 induces alterations in liver function tests with an increase in ALT y AST, as well as hepatic megalocytosis [168].

2.10. Pancreas and metals

Pancreatic parenchyma damage by metals has been poorly analyzed, but some reports mention acute and chronic pancreatitis, and cancer [169]. The pathophysiology and etiology of pancreatic damage are still unknown, and usually are lethal [170] [171] [172]. Vanadium induces an increase in pancreatic enzymes, hypertrophic acinar cells, which results in an acute pancreatitis [173].

2.11. Immunotoxicity of metals

Urban populations are often exposed to metals as constituent of particulate matter (PM), one aspect of the myriad toxicities that might arise from these exposures is altered lymphoid system and thus immune responses. Among the metals that when inhaled damage the lymphoid system we can find vanadium, cadmium, mercury, iron, lead, manganese, chromium, copper and arsenic.

Among the main effects of vanadium that our group has reported we can find splenomegaly. Spleens of mice exposed to vanadium showed morphological changes that included an increase in the size of the white pulp, germinal center hyperplasia, and an increase in the size and number of megakaryocytes and CD19+ lymphocytes. In the same study we found a decrease in the mice capacity to star a humoral immune response, when presented to Hepatitis B surface antigen (HBsAg), vanadium exposed mice presented higher antibodies concentrations with lower affinity compared to controls [174]. On the other hand, our group, in the thymus, has reported morphological changes. We found that vanadium exposed mice presented a shift of the normal cortex-medulla relationship, showing a much thinner medulla and the presence of medulla-like areas within the cortex regions. These changes suggest an alteration of the immune response [69, 89]. In addition to these findings we have reported a decrease in the presence of CD11c, a dendritic cell marker, and MHCII an antigen presenting cell marker, in the thymus of vanadium-exposed mice. This study was conducted using two methods, immunohistochemistry and FACS, with similar findings. This hyperplasia downturn could be detrimental for the negative selection of thymocytes, as dendritic cells are closely related to this process, leading to the persistence of self-reactive cells and increased risk of autoimmune diseases [175].

On the other hand, it has been shown that mice exposed to low concentrations of cadmium have an enhanced humoral immune response [176], however, exposing rats to high concentrations of Cd results in the decrease of B and T cell function and impairment of the phagocytic capacity of NK cells [177]. Chronic exposure to Cd increases serum concentrations of diverse pro-inflammatory cytokines such as IL-1β, TNF-α y la IL-6. Chronic exposure has also been associated to splenomegaly, alterations in the histology of the spleen and the appearance of giant cells and fibrosis. In the thymus, atrophy evidenced as a decreased weight of the thymus has been reported with this metal. All these toxic effect can lead to disturbances in the immune selection and response [178].

One of the worst threats mercury inhalation imposes, is the development of autoimmunity in genetically predisposed individuals, chronic exposure is capable of inducing an immunosuppressive state, alongside apoptotic defects that can lead to a syndrome similar to that of Lupus. The mechanism through which this occurs is not entirely understood. It has been proposed that Hg can associate with proteins creating large complexes capable of activating the immune system, this is specially true for molecules present in antigen presenting cells leading to a massive activation of T cells [179]. Hg is not only capable of inducing immunosuppression, it can also induce immunostimulation in both, mice and humans, its exact mechanism is not known yet [180].

Iron is an essential element for metabolic processes occurring in both, human and microbial cells. Therefore it's relationship with immune function is evident There is an hypothesis that the persistence of certain extracellular pathogens in circulation induces an iron restriction in the mononuclear phagocyte system, blocking its phagocytic capacity. As a result changing concentrations of Fe in the system cause by exposure to this metal could alter the immune function [181].

Lead exposure through inhalation can affect the immune system, an increase in circulating concentrations of IL-4 and IFN-γ and leukocytosis have been reported in a murine model [182]. The effects of lead on the immune system have been studied using macrophages. Being present in a diverse range of tissues, any adverse effect on them could be associated with several presentations. It has been shown that lead diminishes the phagocytic capacity of macrophages, which plays a central role in innate immunity, and therefore could redirect the response towards a Th2 or antibody producing response which could in turn favor the development antibody mediated autoimmunity. In the presence of lead NO decreases and with it the macrophages capacity to kill pathogens. Due to Pb the membrane of erythrocytes gets damaged leading to anemia by increasing the rate at with the spleen phagocytes damaged red cells. It has been reported that macrophages can increase their production of TNF-α due to Pb exposure, and with these can damage peripheral tissue. This has been proven in two models; in the first peritoneal macrophages exposed to Pb damaged liver tissue and in the nervous system Pb exposed microglia damaged peripheral neurons [183].

An immunosuppressive state has been documented in workers exposed to Mn inhalation. In an experimental model using rats, Mn decreased circulated populations of CD4+ and CD8+ lymphocytes that could explain the immunosuppressive state reported in the work exposure studies [184].

Other metal with immunotoxic properties is Chromium. Occupational exposure to this metal can lead to an imbalance of the humoral and cellular components of the immune response. In a study conducted on workers exposed to inhaled Cr, it was found that circulating levels of immunoglobulins and complement components were different when compared to their controls. Seric titers of IgG and IgA were lower than those found in the control group, whilst C3 and C4 concentrations were higher. These results suggest that exposure to Cr can produce ill effects on lymphocytes that include inhibition of immunoglobulin secretion and complement activation [185].

Copper inhalation by humans is more frequently found as an occupational exposure. This metal can cause morphological changes in lymphoid tissue. In a murine study Cr inhalation led to an increase in spleen weight and decrease in thymus weight. Morphologically splenomegaly and thymus atrophy were also reported. All of the above could lead to an immunological imbalance [186].

Studies with Arsenic have been conducted using pregnant women that live in rural areas contaminated with this metal. In this study their IgG titers were elevated compared to non-exposed pregnant women from a different region. The study suggest that a higher titer of IgG can increase the pregnant women mortality, however, the overall effect on humoral response that these elevation in IgG can have is still unknown [187].

The immune system is an important and complex system composed of different kind of cells and factors whose function is of the outmost importance in preserving health. This system is on of the most sensible targets of atmospheric pollution. Metals associated with particulate matter can cause an ample specter of immunological disorders. The susceptibility to these disorders, however, can be deeply related to a genetic predisposition in the individual.

2.12. Reprotoxicity of metals

Air pollution has been related to adverse effects on female reproductive health such as infertility, miscarriages, delay menarche [188] and an increased risk of hypertensive disorders during pregnancy [189]. Metals contained in particulate matter could be related with these reproductive alterations [190], because it has been proved that some metals like cadmium, lead, mercury, manganese, chromium and nickel have female reprotoxic effects [191].

Cadmium increased the duration of estrous cycle in rats exposed to 1 mg Cd/m^3 (NTP, 1995); as well as after 6 weeks of exposure to 1 mg Cd/m^3 and after 18 weeks after exposure to 0.16 mg Cd/m^3 (5h/5 days/20 weeks) [192]. Female mice exposed to 230 μg Cd/m^3 daily showed a lower incidence of pregnancies, and a lower level of serum 17-β estradiol [193]. Inhalation of 1 mg Cd/m^3 (5 h daily/5 days weekly/5 months) caused a decreased female rats fertility [194].

Mercury vapor has reproductive effects in occupationally exposed women (dental assistants and dentists), reports included abortion, stillbirth and menstrual disorders (irregularity, painful or hemorrhagic menstrual bleeding) [195]. Polymenorrhoea or oligomenorrhoea [196] and reduced fertility [197] it was also observed in women working in a lamp factory and dental assistants, respectively. In animal models, Davis and coworkers [198] observed longer estrous cycles in rats exposed to 2 mg Hg/m^3 (2h/day/11days).

Female workers at a lead smelter showed an increased frequency of abortions and their child showed low birth weight [199]. In pregnant women living in Mexico City, high levels of lead in maternal blood were related with an increased incidence of abortions [200]. Vigeh and coworkers [201] observed high levels of lead and manganese in pregnant women diagnosed with preeclampsia.

MnO_2 dust inhalation augments the number of pups and decreased their body weight gain, when female rats were exposed preconception [202]. In other study rats transfer manganese to their offspring trough milk, after preconception exposure to manganese inhalation [203].

Inhaled chromium caused menstrual alterations, postnatal hemorrhage and delivery complication in female workers and near-resident women [204]. In female workers at a nickel refinery plant, it was observed an increased incidence of abortions [205]. In other study, high levels of nickel were found in women with endometriosis [206].

Vanadium inhalation caused an increased length of estrous cycle in females exposed to 4.5 mg V/m^3, and exposure to 9 mg V/m^3 reduced the number of females with a normal cycle (NTP, 2002). After inhalation of V_2O_5 0.02M (1h/week/4 weeks) female mice get into anestrous and showed lower serum levels of estradiol and progesterone and an increase in the width of uterine stroma and myometrium [207]; as well as an increase in the lipidic peroxidation in the ovary and a reduced size of secondary and preovulatory follicles [208].

Bucher and coworkers [209] studied the effect of inhalation of copper sulfate (3 mg/m^3, 6 h daily, 5 days per week for 13 weeks) in rats; they find alterations in sperm (decreased motility and abnormal sperm increase) and in testicular weight.

In male rats exposed to inhalation of $MnSO_4$ (3 mg/m^3) for 6 h/day for 7 days a week, Dorman and coworkers (2001) quantified an increase in the concentration of $MnSO_4$ in the testes of treated animals (0.79 ± 0.18 μg / mg of dry tissue) compared with the testes of control animals (0.32 ± 0.04 μg / mg of dry tissue).

In mice exposed by inhalation of lead and cadmium has been reported mitochondrial damage in Sertoli cells of mice exposed to inhalation of lead acetate (0.01 M, 1 h / week/4 weeks) and chloride cadmium (0.006 M, 1h / week / 4 weeks); in addition, mitochondrial alterations were more severe and an detected earlier in animals exposed together to both compounds than in controls [210].

In mice exposed to inhaled vanadium pentoxide (0.02 M) for 1 h twice a week, for 12 weeks, alterations were observed in the cells of the seminiferous tubules: necrosis, pseudo-nuclear inclusions and disruption of cellular junctions [211]; alterations were also found in proteins of the, such as decrease of gamma-tubulin [212] and actin [213].

In humans, it has been proposed that exposure to toxic metals is a risk factor in reproductive health. A study by Akinloye and coworkers [214] indicated that the cadmium concentration in serum and seminal fluid from azoospermic men was higher than in oligozoospermic and control men.

In the case of lead, there is evidence of its reproductive toxicity effects in humans. Occupational exposure to lead causes decreased sperm motility and dysfunction of the sex glands [215].

2.13. Teratogenesis

During the last decade, epidemiologic studies have researched the connection between air pollution and its adverse effects during pregnancy, being found an increase in preterm birth risk, low birth weight and foetus underdevelopment; however, results are contradictory because of different methodological approaches [216]. A study held in California showed the link between the expose to PM2.5 and low birth weight in children born at the end of pregnancy. The particles had sulphide, sulphate, vanadium, iron, manganese, bromine, ammonium and zinc. These particles can affect the fetus weight because of their impact in cardiovascular and respiratory health in the mother, produce oxidative stress and damage the fetus DNA, affecting in its development [217]. An experimental study of mouse in utero exposure to diesel emissions, a pollutant that cause the major number of PM2.5 suspended particles, demonstrated that there is embryonic reabsorption and placental changes such as hemorrhage, necrosis, swelling and oxidative stress. In adulthood it was found propensity to arterial hypertension and cardiac failure in mice that were exposed to diesel during prenatal development [218].

Waste incinerators produce environmental pollutants such as heavy metals, specifically cadmium, lead, mercury, chromium and arsenic. A study held in Cumbria, UK, found and excessive number of perinatal and child mortality because of spina bifida and cardiac malformations near to incineration places [219].

Heavy metals can produce health problems because of oxidative stress (Cd, Cr, Pb, As), neurological damage (Pb, Hg), DNA damage (As, Cr, Cd), changes in the metabolism of glucose (As) and calcium (Cd, Pb) and interfere with essential elements (Cd, Hg) [220]. It is because of this that these metals have a teratogenic potential, which depends also on its placental transportation. The placenta is an active transporter of essential elements, such as calcium, copper, zinc and iron, as well as toxic elements such as cadmium, lead, mercury and nickel. Heavy metals can go through the placental barrier and accumulate in the fetus tissues and amniotic fluid.

Cadmium accumulates in the placenta, and it has been found a correlation between cadmium levels and the expression of the metallothionein, which retains the cadmium and prevents that it reaches the fetus; but its increase blocks the transportation of Zn to the fetus, decreasing the placental permeability to this essential element. Cadmium also affects the synthesis of the placental hormones such as progesterone and leptin, affects the trophoblastic cells migration and induces an early development on the decidua of the endometrial stroma [220]. The maternal exposure to cadmium during pregnancy is linked to preterm birth, intrauterine growth restriction and low birth weight [220, 221]. In experimental models it has demonstrated that it can affect the embryo development and the implantation. Cadmium increases in rats the oxidative stress and decreases the antioxidant enzymes activity [221]. The exposure of rats to cadmium during the organogenesis period produces external and internal malformations, as well as alterations in the ossification. After birth it was observed an alteration in males and females sexual behavior [222, 223]. The exposure to cadmium in human fetal gonads in culture produces a decrease of germinal cells due to apoptosis, but it does not affect the cells proliferation [223].

Lead can easily go through the placental barrier by simple diffusion. Lead can affect calcium-mediated processes in the syncytiotrophoblast; it accumulates there and reduces the cytochrome oxidase activity, an enzyme of the respiratory chain, reducing therefore the ATP production in this cells. Lead produces oxidative damage and induces preterm birth, abortion, intrauterine growth restriction and congenital abnormalities [220]. Chromium that can be found in the polluted air has been linked to neuroblastoma in children [224].

Arsenic has been linked to miscarriage, low birth weight and malformations in populations working or living near foundries in which emissions arsenic can be found. In animal models arsenic produces toxicity in development, since it produces malformations, intrauterine death and intrauterine growth restriction. These malformations include neural tube defects, gonadal and renal agenesis, eye defects and malformations of the ribs [225].

Manganese can be found in the air due to, essentially, diesel combustion. Manganese goes through the placental barrier by simple diffusion. Manganese is an essential nutrient, and that is why it can be found in the tissues and fluids of both, mother and foetus. The lack or excess of manganese affect the prenatal development. Because of the air pollution an excess of manganese has been linked to intrauterine growth restriction during the foetal period and, postnatally, to hyperactive behavior, decrease of intellectual ability and alteration of the psychomotor development [226]. Experimental studies have showed that the excess of manganese has a teratogenic effect, since that produces growth restriction, embryonic death and bone alterations [227].

High levels of iron during the human embryonic period can be teratogenic. Iron is toxic mainly because it produces oxidative stress. Iron catalyzes the production of hydroxyl-free radicals, which destroy cells by lipid peroxidation, enzymes denaturation, carbohydrates depolymerization and ruptures in DNA. Experimental studies in the mouse show that an increase of iron produces histological alterations in the encephalon, as well as spine and ribs malformations [228].

Equally, high levels of nickel affect the mouse embryo development, producing embryonic death, fetal death, and malformations such as hydrocephalus, eyelids alterations, microphthalmia, exophthalmia, clubfoot, umbilical hernia and bone anomalies. The excess in nickel ions can replace other metals required for the structure and functions of enzymes, which get inactivated, and this could be the cause of the embryo and fetus toxicity observed in mice, rats and women [229].

Vanadium causes reproductive damage. It goes through the placenta, constituting complexes with transferrin or albumin, and accumulates in the placenta and the fetal tissues. It has been found that vanadium affects the prenatal development, since it produces embryo mortality, fetus toxicity and teratogenicity in mice, rats, hamsters and chickens [69, 89]. In humans, it has been proved that the exposure to vanadium during pregnancy is linked to low birth weight. In the chicken embryo, the vanadium pentoxide produces embryo mortality and teratogenicity, since it provokes alterations in tubulation and the central nervous system, microphthalmia, abnormalities in the pharyngeal arches and facial development processes or their derivatives, congenital heart defect, limbs malformations and visceral ectopia [89]. Vanadium produces oxidative damage in proteins, lipids and DNA, interferes with DNA repair and affects cellular signaling pathways and cells proliferation [230].

3. Conclusion

Metals enter into the respiratory system adhered to particulate matter, and by this route they reach the systemic circulation. In the blood metals are attached to proteins or ionized entering into the different organs and cells producing a variety of outcomes. The chemical characteristics of the inhaled metals, the length of the exposure, the route, and the physiology of each organ will determinate the metabolism, the affected functions and the possible manifestations, which are resumed in Figure 1. Some metals interact with enzymes inhibiting its actions by the interaction of the metal with the SH group of the enzyme or by displacement of an essential metal cofactor; another interference mechanism is the inhibition of the synthesis of the enzyme, indirectly altering the systemic function; the binding of the metal by certain cytosolic proteins may modify its toxicity. Also metals may interact directly with the components of the cell, and may be accumulated in the lysosomes, or damaging the mitochondria and inhibiting respiratory enzymes leading to cell death. A direct interaction with DNA may produce gene mutations, chromosome aberrations or aneuploidy; these changes could pave the way to proliferation and cancer development [231].

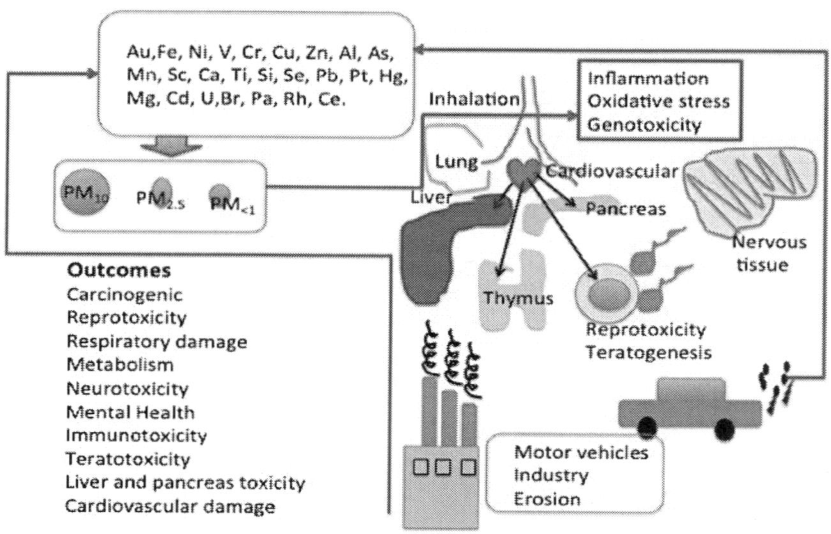

Figure 1. Interaction of metals in particular matter that enters into the respiratory system inducing inflammation, oxidative stress and genotoxicity. The sources and possible outcomes are resumed.

The problem of metals associated with atmospheric particulate matter is not new and there are increasing reports about its health effects, as we have mentioned in the previous sections. Even though the toxic effects of these elements are severe, there are scant specific knowledge about the association of a disease and the inhalation exposure. This problem needs more research in order to count with more information for understanding the mechanisms of damage and to propose measures to control the emissions, decrease the exposure and its adverse effects.

Acknowledgements

Authors thank Alejandra Núñez-Fortoul for reviewing the final English version.

Author details

T.I. Fortoul[1*], V. Rodriguez-Lara[1], A. Gonzalez-Villalva[1], M. Rojas-Lemus[1], L. Colin-Barenque[2], P. Bizarro-Nevares[1], I. García-Peláez[1], M. Ustarroz-Cano[1], S. López-Zepeda[1], S. Cervantes-Yépez[1], N. López-Valdez[1], N. Meléndez-García[1], M. Espinosa-Zurutuza[3], G. Cano-Gutierrez[4] and M.C. Cano-Rodríguez[4]

*Address all correspondence to: fortoul@unam.mx

1 Cellular and Tissular Biology Department, School of Medicine, National Autonomous University of Mexico (UNAM). Mexico City, Mexico

2 Neuromorphology Laboratory, FES Iztacala, National Autonomous University of Mexico (UNAM), Mexico

3 Faculty of Sciences, Biology, National Autonomous University of Mexico (UNAM) Mexico City, Mexico

4 Valle de Mexico University (Coyoacan) School of Health Sciences, Mexico City, Mexico

References

[1] Englert N. Fine particles and human health--a review of epidemiological studies. Toxicology letters. 2004;149(1-3):235-42.

[2] Kelly FJ, Fussell JC. Size, source and chemical composition as determinants of toxicity attributable to ambient particulate matter. Atmos Environ. 2012; 60:504-26.

[3] Bu Changming WL HL, Tang Jing.. Evaluation of Health Effects of Fine Particulate PM2.5: a Review. Advanced Materials Research 2013; 790:441-4.

[4] Goyer R, Clarkson T. Toxic effects of metals. In: Klaassen C, editor. Casarett and Doull's Toxicology The basic science of poisons. 6th Ed ed. New York: McGraw-Hill; 2001.

[5] Anderson JO, Thundiyil JG, Stolbach A. Clearing the air: a review of the effects of particulate matter air pollution on human health. Journal of medical toxicology : official journal of the American College of Medical Toxicology. 2012;8(2):166-75.

[6] Loomis D, Grosse Y, Lauby-Secretan B, El Ghissassi F, Bouvard V, Benbrahim-Tallaa L, et al. The carcinogenicity of outdoor air pollution. Lancet Oncol. 2013;14(13): 1262-3.

[7] Wiseman CL, Zereini F. Airborne particulate matter, platinum group elements and human health: a review of recent evidence. The Science of the total environment. 2009;407(8):2493-500.

[8] Cao S, Duan X, Zhao X, Ma J, Dong T, Huang N, et al. Health risks from the exposure of children to As, Se, Pb and other heavy metals near the largest coking plant in China. The Science of the total environment. 2014;472:1001-9.

[9] Mumford L. The conduct of life. Co Ba, editor. New York: Harcord; 1951.

[10] Kessler R. Air of danger. Sci Am. 2014;311(1):S16-7.

[11] Matus K, Nam KM, Selin NE, Lamsal LN, Reilly JM, Paltsev S. Health damages from air pollution in China. Global Environ Chang. 2012;22(1):55-66.

[12] Soto-Jimenez MF, Flegal AR. Childhood lead poisoning from the smelter in Torreon, Mexico. Environ Res. 2011;111(4):590-6.

[13] Fortoul TI SR, Monca SG, Sanchez IG, Lopez IE, Espejel G, Calderon NL, Saldivar L.. Ultrastructural findings in the murine nonciliated bronchiolar cells (NCBC) after subacute inhalation of lead acetate. Acta Vet Brno 1999;68:51-5.

[14] Fortoul TI, Avila-Costa MR, Espejel-Maya G, Mussali-Galante P, Avila-Casado Mdel C, Hernandez-Serrato MI, et al. Metal mixture inhalation (Cd-Pb) and its effects on the bronchiolar epithelium. An ultrastructural approach. Toxicology and industrial health. 2004;20(1-5):69-75.

[15] Fortoul TI, Saldivar OL, Espejel-Maya G, Bazarro NP, Mussali-Galante P, Avila-Casado Mdel C, et al. Inhalation of cadmium, lead or its mixture Effects on the bronchiolar structure and its relation with metal tissue concentrations. Environmental toxicology and pharmacology. 2005;19(2):329-34.

[16] Stavrides JC. Lung carcinogenesis: pivotal role of metals in tobacco smoke. Free Radic Biol Med. 2006;41(7):1017-30.

[17] Bohlandt A, Schierl R, Diemer J, Koch C, Bolte G, Kiranoglu M, et al. High concentrations of cadmium, cerium and lanthanum in indoor air due to environmental tobacco smoke. Sci Total Environ. 2012;414:738-41.

[18] Cakmak S, Dales R, Kauri LM, Mahmud M, Van Ryswyk K, Vanos J, et al. Metal composition of fine particulate air pollution and acute changes in cardiorespiratory physiology. Environ Pollut. 2014;189:208-14.

[19] Bell ML, Zanobetti A, Dominici F. Evidence on vulnerability and susceptibility to health risks associated with short-term exposure to particulate matter: a systematic review and meta-analysis. Am J Epidemiol. 2013;178(6):865-76.

[20] Martinelli N, Olivieri O, Girelli D. Air particulate matter and cardiovascular disease: a narrative review. Eur J Intern Med. 2013;24(4):295-302.

[21] Franchini M, Guida A, Tufano A, Coppola A. Air pollution, vascular disease and thrombosis: linking clinical data and pathogenic mechanisms. J Thromb Haemost. 2012;10(12):2438-51.

[22] Johnston FH, Hanigan IC, Henderson SB, Morgan GG. Evaluation of interventions to reduce air pollution from biomass smoke on mortality in Launceston, Australia: retrospective analysis of daily mortality, 1994-2007. BMJ. 2013;346:e8446.

[23] Pope CA, 3rd, Ezzati M, Dockery DW. Fine-particulate air pollution and life expectancy in the United States. N Engl J Med. 2009;360(4):376-86.

[24] Miller MR, Shaw CA, Langrish JP. From particles to patients: oxidative stress and the cardiovascular effects of air pollution. Future Cardiol. 2012;8(4):577-602.

[25] Niu J, Liberda EN, Qu S, Guo X, Li X, Zhang J, et al. The role of metal components in the cardiovascular effects of PM2.5. PLoS One. 2013;8(12):e83782.

[26] Jomova K, Valko M. Advances in metal-induced oxidative stress and human disease. Toxicology. 2011;283(2-3):65-87.

[27] Kruszewski M. The role of labile iron pool in cardiovascular diseases. Acta Biochim Pol. 2004;51(2):471-80.

[28] Mordukhovich I, Wright RO, Hu H, Amarasiriwardena C, Baccarelli A, Litonjua A, et al. Associations of toenail arsenic, cadmium, mercury, manganese, and lead with blood pressure in the normative aging study. Environ Health Perspect. 2012;120(1): 98-104.

[29] Flora G, Gupta D, Tiwari A. Toxicity of lead: A review with recent updates. Interdiscip Toxicol. 2012;5(2):47-58.

[30] Alissa EM, Ferns GA. Heavy metal poisoning and cardiovascular disease. J Toxicol. 2011;2011:870125.

[31] Hartwig A. Cadmium and cancer. Met Ions Life Sci. 2013;11:491-507.

[32] Huff J, Lunn RM, Waalkes MP, Tomatis L, Infante PF. Cadmium-induced cancers in animals and in humans. Int J Occup Environ Health. 2007;13(2):202-12.

[33] Koedrith P, Seo YR. Advances in carcinogenic metal toxicity and potential molecular markers. Int J Mol Sci. 2011;12(12):9576-95.

[34] Silva N, Peiris-John R, Wickremasinghe R, Senanayake H, Sathiakumar N. Cadmium a metalloestrogen: are we convinced? J Appl Toxicol. 2012;32(5):318-32.

[35] Benbrahim-Tallaa L, Tokar EJ, Diwan BA, Dill AL, Coppin JF, Waalkes MP. Cadmium malignantly transforms normal human breast epithelial cells into a basal-like phenotype. Environ Health Perspect. 2009;117(12):1847-52.

[36] Silbergeld EK. Facilitative mechanisms of lead as a carcinogen. Mutat Res. 2003;533(1-2):121-33.

[37] Garcia-Leston J, Mendez J, Pasaro E, Laffon B. Genotoxic effects of lead: an updated review. Environ Int. 2010;36(6):623-36.

[38] Schaumloffel D. Nickel species: analysis and toxic effects. J Trace Elem Med Biol. 2012;26(1):1-6.

[39] Costa M. Molecular mechanisms of nickel carcinogenesis. Biol Chem. 2002;383(6): 961-7.

[40] Oller AR, Costa M, Oberdorster G. Carcinogenicity assessment of selected nickel compounds. Toxicol Appl Pharmacol. 1997;143(1):152-66.

[41] Kasprzak KS, Sunderman FW, Jr., Salnikow K. Nickel carcinogenesis. Mutat Res. 2003;533(1-2):67-97.

[42] Ishida S, Andreux P, Poitry-Yamate C, Auwerx J, Hanahan D. Bioavailable copper modulates oxidative phosphorylation and growth of tumors. Proc Natl Acad Sci U S A. 2013;110(48):19507-12.

[43] Linder MC, Moor JR, Wright K. Ceruloplasmin assays in diagnosis and treatment of human lung, breast, and gastrointestinal cancers. J Natl Cancer Inst. 1981;67(2): 263-75.

[44] Ishida S, McCormick F, Smith-McCune K, Hanahan D. Enhancing tumor-specific uptake of the anticancer drug cisplatin with a copper chelator. Cancer Cell. 2010;17(6): 574-83.

[45] Theophanides T, Anastassopoulou J. Copper and carcinogenesis. Crit Rev Oncol Hematol. 2002;42(1):57-64.

[46] Barregard L, Sallsten G, Jarvholm B. Mortality and cancer incidence in chloralkali workers exposed to inorganic mercury. Br J Ind Med. 1990;47(2):99-104.

[47] Gibbs GW, Labreche F. Cancer risks in aluminum reduction plant workers: a review. J Occup Environ Med. 2014;56(5 Suppl):S40-59.

[48] Martin SC, Lariviere C. Community health risk assessment of primary aluminum smelter emissions. J Occup Environ Med. 2014;56(5 Suppl):S33-9.

[49] Gibbs GW, Labreche F, Busque MA, Duguay P. Mortality and cancer incidence in aluminum smelter workers: a 5-year update. J Occup Environ Med. 2014;56(7):739-64.

[50] Lee JC, Son YO, Pratheeshkumar P, Shi X. Oxidative stress and metal carcinogenesis. Free Radic Biol Med. 2012;53(4):742-57.

[51] Galanis A, Karapetsas A, Sandaltzopoulos R. Metal-induced carcinogenesis, oxidative stress and hypoxia signalling. Mutat Res. 2009;674(1-2):31-5.

[52] Salnikow K, Zhitkovich A. Genetic and epigenetic mechanisms in metal carcinogenesis and cocarcinogenesis: nickel, arsenic, and chromium. Chem Res Toxicol. 2008;21(1):28-44.

[53] Hartwig A. Metal interaction with redox regulation: an integrating concept in metal carcinogenesis? Free Radic Biol Med. 2013;55:63-72.

[54] Beguin Y, Aapro M, Ludwig H, Mizzen L, Osterborg A. Epidemiological and non-clinical studies investigating effects of iron in carcinogenesis--a critical review. Crit Rev Oncol Hematol. 2014;89(1):1-15.

[55] Durackova Z. Some current insights into oxidative stress. Physiol Res. 2010;59(4): 459-69.

[56] IEH. Occupational exposure lim- its: Criteria document for manganese and inorganic manganese compounds. 2004.. Available from: Available at: http://www.cranfield.ac.uk/ health/researchareas/environmenthealth/ ieh/ieh%20publications/ w17.pdf.

[57] Valko M, Morris H, Cronin MT. Metals, toxicity and oxidative stress. Curr Med Chem. 2005;12(10):1161-208.

[58] Beyersmann D, Hartwig A. Carcinogenic metal compounds: recent insight into molecular and cellular mechanisms. Arch Toxicol. 2008;82(8):493-512.

[59] (IARC) IAfRoC. Vanadium pentoxide, IARC Monogr. Eval. Carcinog. Risks Hum.. 2003;83:227-43.

[60] Assem FL, Levy LS. A review of current toxicological concerns on vanadium pentoxide and other vanadium compounds: gaps in knowledge and directions for future research. J Toxicol Environ Health B Crit Rev. 2009;12(4):289-306.

[61] Bishayee A, Waghray A, Patel MA, Chatterjee M. Vanadium in the detection, prevention and treatment of cancer: the in vivo evidence. Cancer Lett. 2010;294(1):1-12.

[62] ICH. GfiSR. Genotoxicity testing and data interpretation for pharmaceuticals intended for human use. 2012. Available from: http://www.fda.gov/Drugs/GuidanceComplianceRegulatoryInformation/Guidances/default.htm.

[63] Sorsa M. Biological Monitoring. In: La Ferla F, Lauwerys, Robert R.,Jeanne Mager Stellman editor. Encyclopedia of Occupational Health and safety. Genova: International Labor Organization,; 2011.

[64] Chorvatovicova D, Kovacikova Z. Inhalation exposure of rats to metal aerosol. II. Study of mutagenic effect on alveolar macrophages. J Appl Toxicol. 1992;12(1):67-8.

[65] Valverde M, Fortoul TI, Diaz-Barriga F, Mejia J, del Castillo ER. Induction of genotoxicity by cadmium chloride inhalation in several organs of CD-1 mice. Mutagenesis. 2000;15(2):109-14.

[66] Valverde M, Fortoul TI, Diaz-Barriga F, Mejia J, del Castillo ER. Genotoxicity induced in CD-1 mice by inhaled lead: differential organ response. Mutagenesis. 2002;17(1):55-61.

[67] Monleau M, De Meo M, Paquet F, Chazel V, Dumenil G, Donnadieu-Claraz M. Genotoxic and inflammatory effects of depleted uranium particles inhaled by rats. Toxicol Sci. 2006;89(1):287-95.

[68] Ehrlich VA, Nersesyan AK, Atefie K, Hoelzl C, Ferk F, Bichler J, et al. Inhalative exposure to vanadium pentoxide causes DNA damage in workers: results of a multiple end point study. Environ Health Perspect. 2008;116(12):1689-93.

[69] Fortoul TI, Rodriguez-Lara V, Gonzalez-Villalva A, Rojas-Lemus M, Cano-Gutierrez G, Ustarroz-Cano M, et al. Vanadium Inhalation in a Mouse Model for the Understanding of Air-Suspended Particle Systemic Repercussion. J Biomed Biotechnol. 2011:1-11.

[70] Rojas-Lemus M, Altamirano-Lozano M, Fortoul TI. Sex differences in blood genotoxic and cytotoxic effects as a consequence of vanadium inhalation: micronucleus assay evaluation. Journal of Applied Toxicology. 2014;34(3):258-64.

[71] Lindberg HK, Falck GC, Catalan J, Koivisto AJ, Suhonen S, Jarventaus H, et al. Genotoxicity of inhaled nanosized TiO(2) in mice. Mutat Res. 2012;745(1-2):58-64.

[72] MohanKumar SM, Campbell A, Block M, Veronesi B. Particulate matter, oxidative stress and neurotoxicity. Neurotoxicology. 2008;29(3):479-88.

[73] Calderon-Garcidueñas L, Torres-Jardon R, Kulesza RJ, Park SB, D'Angiulli A. Air pollution and detrimental effects on children's brain. The need for a multidisciplinary approach to the issue complexity and challenges. Front Hum Neurosci. 2014;8:613.

[74] Ljubimova JY GP, Portilla-Arias J, Patil R, Konda B, Paff M, Markman JL, Inoue S, Espinoza A, Chesnokova A, Funari V, Kleinman MT, Holler E, Black KL.. Molecular changes in rat brain due to air nano pollution.. Nanotech. 2012;3:261-4..

[75] Oberdorster G, Oberdorster E, Oberdorster J. Nanotoxicology: an emerging discipline evolving from studies of ultrafine particles. Environ Health Perspect. 2005;113(7):823-39.

[76] Win-Shwe TT, Fujimaki H. Nanoparticles and neurotoxicity. Int J Mol Sci. 2011;12(9): 6267-80.

[77] Colin-Barenque L, Fortoul T. Oxidative stress and metals. In: Fortoul T, editor. Metals and toxicological implications in Health. Kerala: Research Signpost; 2007. p. 15-25.

[78] Jellinger KA. The relevance of metals in the pathophysiology of neurodegeneration, pathological considerations. International review of neurobiology. 2013;110:1-47.

[79] Modgil S, Lahiri DK, Sharma VL, Anand A. Role of early life exposure and environment on neurodegeneration: implications on brain disorders. Transl Neurodegener. 2014;3:9.

[80] Cannon JR, Greenamyre JT. The role of environmental exposures in neurodegeneration and neurodegenerative diseases. Toxicol Sci. 2011;124(2):225-50.

[81] Charlet L, Chapron Y, Faller P, Kirsch R, Stone AT, Baveye PC. Neurodegenerative diseases and exposure to the environmental metals Mn, Pb, and Hg. Coordin Chem Rev. 2012;256(19-20):2147-63.

[82] Kozlowski H, Luczkowski M, Remelli M, Valensin D. Copper, zinc and iron in neurodegenerative diseases (Alzheimer's, Parkinson's and prion diseases). Coordin Chem Rev. 2012;256(19-20):2129-41.

[83] Wong BX, Duce JA. The iron regulatory capability of the major protein participants in prevalent neurodegenerative disorders. Front Pharmacol. 2014;5:31.

[84] Braidy N, Poljak A, Marjo C, Rutlidge H, Rich A, Jayasena T, et al. Metal and complementary molecular bioimaging in Alzheimer's disease. Front Aging Neurosci. 2014;6:138.

[85] Jomova K, Vondrakova D, Lawson M, Valko M. Metals, oxidative stress and neurodegenerative disorders. Mol Cell Biochem. 2010;345(1-2):91-104.

[86] Aschner M, Erikson KM, Herrero Hernandez E, Tjalkens R. Manganese and its role in Parkinson's disease: from transport to neuropathology. Neuromolecular Med. 2009;11(4):252-66.

[87] Singla N, Dhawan DK. Regulatory role of zinc during aluminium-induced altered carbohydrate metabolism in rat brain. J Neurosci Res. 2012;90(3):698-705.

[88] Roos PM, Dencker L. Mercury in the spinal cord after inhalation of mercury. Basic Clin Pharmacol Toxicol. 2012;111(2):126-32.

[89] Fortoul TI, Rodriguez-Lara V, Gonzalez-Villalva A, Rojas-Lemus M, Cano-Gutierrez G, Ustarroz-Cano M, et al. Inhalation of vanadium pentoxide and its toxic effects in a mouse model. Inorg Chim Acta. 2014;420:8-15.

[90] Colín-Barenque L M-HM, Baiza-Gutman LA, Avila-Costa MR, Ordóñez-Librado, JL, Bizarro-Nevares P, Rodriguez-Lara V, Piñón-Zarate G, Rojas-Lemus M,Mussali-Gal-

ante P, Fortoul TI. 2008.. Matrix metalloproteinases 2 and 9 in central nervous system and its modification after vanadium inhalation. J Appl Toxicol 2008;28(6):718-23.

[91] Ritchie GD, Still KR, Alexander WK, Nordholm AF, Wilson CL, Rossi J, 3rd, et al. A review of the neurotoxicity risk of selected hydrocarbon fuels. Journal of toxicology and environmental health Part B, Critical reviews. 2001;4(3):223-312.

[92] Terzano C, Di Stefano F, Conti V, Graziani E, Petroianni A. Air pollution ultrafine particles: toxicity beyond the lung. European review for medical and pharmacological sciences. 2010;14(10):809-21.

[93] Julvez J, Grandjean P. Neurodevelopmental toxicity risks due to occupational exposure to industrial chemicals during pregnancy. Industrial health. 2009;47(5):459-68.

[94] Bellinger DC. Children's cognitive health: the influence of environmental chemical exposures. Alternative therapies in health and medicine. 2007;13(2):S140-4.

[95] Landrigan P, Nordberg M, Lucchini R, Nordberg G, Grandjean P, Iregren A, et al. The Declaration of Brescia on prevention of the neurotoxicity of metals June 18, 2006. American journal of industrial medicine. 2007;50(10):709-11.

[96] Jarup L. Hazards of heavy metal contamination. British medical bulletin. 2003;68:167-82.

[97] Trejo-Acevedo A, Diaz-Barriga F, Carrizales L, Dominguez G, Costilla R, Ize-Lema I, et al. Exposure assessment of persistent organic pollutants and metals in Mexican children. Chemosphere. 2009;74(7):974-80.

[98] Sinczuk-Walczak H, Szymczak M, Halatek T. Effects of occupational exposure to arsenic on the nervous system: clinical and neurophysiological studies. International journal of occupational medicine and environmental health. 2010;23(4):347-55.

[99] Needleman HL, Geiger SK, Frank R. Lead and IQ scores: a reanalysis. Science. 1985;227(4688):701-2, 4.

[100] Baranowska-Bosiacka I, Gutowska I, Rybicka M, Nowacki P, Chlubek D. Neurotoxicity of lead. Hypothetical molecular mechanisms of synaptic function disorders. Neurologia i neurochirurgia polska. 2012;46(6):569-78.

[101] Mason LH, Harp JP, Han DY. Pb neurotoxicity: neuropsychological effects of lead toxicity. BioMed research international. 2014;2014:840547.

[102] Schoeters G, Den Hond E, Zuurbier M, Naginiene R, van den Hazel P, Stilianakis N, et al. Cadmium and children: exposure and health effects. Acta Paediatr Suppl. 2006;95(453):50-4.

[103] Murata K, Grandjean P, Dakeishi M. Neurophysiological evidence of methylmercury neurotoxicity. American journal of industrial medicine. 2007;50(10):765-71.

[104] Ceccatelli S, Dare E, Moors M. Methylmercury-induced neurotoxicity and apoptosis. Chemico-biological interactions. 2010;188(2):301-8.

[105] Magos L, Clarkson TW. Overview of the clinical toxicity of mercury. Annals of clinical biochemistry. 2006;43(Pt 4):257-68.

[106] Roels HA, Bowler RM, Kim Y, Claus Henn B, Mergler D. Hoet P. et al. Manganese exposure and cognitive deficits: a growing concern for manganese neurotoxicity. Neurotoxicology. 2012;33(4):872-80.

[107] Racette BA, Aschner M, Guilarte TR, Dydak U, Criswell SR, Zheng W. Pathophysiology of manganese-associated neurotoxicity. Neurotoxicology. 2012;33(4):881-6.

[108] Avila DS, Puntel RL, Aschner M. Manganese in health and disease. Metal ions in life sciences. 2013;13:199-227.

[109] Barceloux DG. Vanadium. J Toxicol Clin Toxicol. 1999;37(2):265-78.

[110] Zwolak I. Vanadium carcinogenic, immunotoxic and neurotoxic effects: a review of in vitro studies. Toxicology mechanisms and methods. 2014;24(1):1-12.

[111] Ngwa HA, Kanthasamy A, Jin H, Anantharam V, Kanthasamy AG. Vanadium exposure induces olfactory dysfunction in an animal model of metal neurotoxicity. Neurotoxicology. 2014;43:73-81.

[112] Afeseh Ngwa H, Kanthasamy A, Anantharam V, Song C, Witte T, Houk R, et al. Vanadium induces dopaminergic neurotoxicity via protein kinase Cdelta dependent oxidative signaling mechanisms: relevance to etiopathogenesis of Parkinson's disease. Toxicology and applied pharmacology. 2009;240(2):273-85

[113] Cervantes Yépez SE. Efecto de la exposición por inhalación de vanadio en las células gliales de Müller (MGC) de la retina en un modelo murino. México: Universidad Nacional Autónoma de México.; 2014.

[114] Quezada Maldonado EM. Cambios en la histología de la retina de ratones expuestos a la inhalación de pentóxido de vanadio y la participación del estrés oxidante. México: Universidad Nacional Autónoma de México. ; 2013.

[115] Zhou S, Yuan Q, Li W, Lu Y, Zhang Y, Wang W. Trace metals in atmospheric fine particles in one industrial urban city: spatial variations, sources, and health implications. J Environ Sci (China). 2014;26(1):205-13.

[116] dos Anjos Paulino S, Oliveira RL, Loyola J, Minho AS, Arbilla G, Quiterio SL, et al. Trace metals in PM10 and PM 2.5 samples collected in a highly industrialized chemical/petrochemical area and its urbanized surroundings. Bull Environ Contam Toxicol. 2014;92(5):590-5.

[117] Chillrud SN, Grass D, Ross JM, Coulibaly D, Slavkovich V, Epstein D, et al. Steel dust in the New York City subway system as a source of manganese, chromium, and iron exposures for transit workers. J Urban Health. 2005;82(1):33-42.

[118] Fernandez-Real JM, Lopez-Bermejo A, Ricart W. Cross-talk between iron metabolism and diabetes. Diabetes. 2002;51(8):2348-54.

[119] Ferrannini E. Insulin resistance, iron, and the liver. Lancet. 2000;355(9222):2181-2.

[120] Hatunic M, Finucane FM, Brennan AM, Norris S, Pacini G, Nolan JJ. Effect of iron overload on glucose metabolism in patients with hereditary hemochromatosis. Metabolism. 2010;59(3):380-4.

[121] Chang JW, Chen HL, Su HJ, Liao PC, Guo HR, Lee CC. Simultaneous exposure of non-diabetics to high levels of dioxins and mercury increases their risk of insulin resistance. J Hazard Mater. 2011;185(2-3):749-55.

[122] He K, Xun P, Liu K, Morris S, Reis J, Guallar E. Mercury exposure in young adulthood and incidence of diabetes later in life: the CARDIA Trace Element Study. Diabetes Care. 2013;36(6):1584-9.

[123] Serdar MA, Bakir F, Hasimi A, Celik T, Akin O, Kenar L, et al. Trace and toxic element patterns in nonsmoker patients with noninsulin-dependent diabetes mellitus, impaired glucose tolerance, and fasting glucose. Int J Diabetes Dev Ctries. 2009;29(1): 35-40.

[124] Forte G, Bocca B, Peruzzu A, Tolu F, Asara Y, Farace C, et al. Blood metals concentration in type 1 and type 2 diabetics. Biol Trace Elem Res. 2013;156(1-3):79-90.

[125] Ibrahim NM, Eweis EA, El-Beltagi HS, Abdel-Mobdy YE. Effect of lead acetate toxicity on experimental male albino rat. Asian Pac J Trop Biomed. 2012;2(1):41-6.

[126] Adham KG, Al-Eisa NA, Farhood MH. Impact of heavy metal pollution on the hemogram and serum biochemistry of the Libyan jird, Meriones libycus. Chemosphere. 2011;84(10):1408-15.

[127] Novakova J, Lukacinova A, Lovasova E, Cimbolakova I, Racz O, Nistiar F. Lifetime exposure to low doses of lead in rats: Effect on selected parameters of carbohydrate metabolism. Toxicol Ind Health. 2013; 10.1177/0748233713475510

[128] Xu X, Rao X, Wang TY, Jiang SY, Ying Z, Liu C, et al. Effect of co-exposure to nickel and particulate matter on insulin resistance and mitochondrial dysfunction in a mouse model. Part Fibre Toxicol. 2012;9:40.

[129] Das Gupta A, Dhara PC, Dhundasi SA, Das KK. Effect of garlic (Allium sativum) on nickel II or chromium VI induced alterations of glucose homeostasis and hepatic antioxidant status under sub-chronic exposure conditions. J Basic Clin Physiol Pharmacol. 2009;20(1):1-14.

[130] Praveena SP, S. Sameera, K. Trace elements in diabetes mellitus. J Clin Diagn Res. 2013;7(9):1863-5.

[131] Soveid M, Dehghani GA, Omrani GR. Long- term efficacy and safety of vanadium in the treatment of type 1 diabetes. Arch Iran Med. 2013;16(7):408-11.

[132] Yuen VG, Orvig C, McNeill JH. Glucose-lowering effects of a new organic vanadium complex, bis(maltolato)oxovanadium(IV). Can J Physiol Pharmacol. 1993;71(3-4): 263-9.

[133] Mehdi MZ, Pandey SK, Theberge JF, Srivastava AK. Insulin signal mimicry as a mechanism for the insulin-like effects of vanadium. Cell Biochem Biophys. 2006;44(1):73-81.

[134] Marzban L, Rahimian R, Brownsey RW, McNeill JH. Mechanisms by which bis(maltolato)oxovanadium(IV) normalizes phosphoenolpyruvate carboxykinase and glucose-6-phosphatase expression in streptozotocin-diabetic rats in vivo. Endocrinology. 2002;143(12):4636-45.

[135] Boulassel B, Sadeg N, Roussel O, Perrin M, Belhadj-Tahar H. Fatal poisoning by vanadium. Forensic Sci Int. 2011;206(1-3):e79-81.

[136] Mertz W. Chromium in human nutrition: a review. J Nutr. 1993;123(4):626-33.

[137] Lai MH. Antioxidant effects and insulin resistance improvement of chromium combined with vitamin C and E supplementation for type 2 diabetes mellitus. J Clin Biochem Nutr. 2008;43(3):191-8.

[138] Jain SK, Kannan K. Chromium chloride inhibits oxidative stress and TNF-alpha secretion caused by exposure to high glucose in cultured U937 monocytes. Biochem Biophys Res Commun. 2001;289(3):687-91.

[139] Qiao W, Peng Z, Wang Z, Wei J, Zhou A. Chromium improves glucose uptake and metabolism through upregulating the mRNA levels of IR, GLUT4, GS, and UCP3 in skeletal muscle cells. Biol Trace Elem Res. 2009;131(2):133-42.

[140] Garg N, Weinberg J, Ghai S, Bradauskaite G, Nuhn M, Gautam A, et al. Lower magnesium level associated with new-onset diabetes and pre-diabetes after kidney transplantation. J Nephrol 2014;27:339-344.

[141] Ugwuja E NA, Ezenkwa U, Oshim A, Nnabu R, Ogiji E, Ogbanshi M. J. Effects of diabetes complications and glycaemic control on some mineral elements in Nigerians patients with diabetes. Journal of Diabetology, 2014;27(3):339-344.;In press.

[142] Ozgur Keskek S, Kirim S, Karaca A, Saler T. Low serum magnesium levels and diabetic foot ulcers. Pak J Med Sci. 2013;29(6):1329-33.

[143] Adachi Y, Yoshikawa Y, Sakurai H. Antidiabetic zinc(II)-N-acetyl-L-cysteine complex: evaluations of in vitro insulinomimetic and in vivo blood glucose-lowering activities. Biofactors. 2007;29(4):213-23.

[144] Capdor J, Foster M, Petocz P, Samman S. Zinc and glycemic control: a meta-analysis of randomised placebo controlled supplementation trials in humans. J Trace Elem Med Biol. 2013;27(2):137-42.

[145] Taneja SK, Mandal R, Girhotra S. Long term excessive Zn-supplementation promotes metabolic syndrome-X in Wistar rats fed sucrose and fat rich semisynthetic diet. Indian J Exp Biol. 2006;44(9):705-18.

[146] Laing S, Wang G, Briazova T, Zhang C, Wang A, Zheng Z, et al. Airborne particulate matter selectively activates endoplasmic reticulum stress response in the lung and liver tissues. Am J Physiol Cell Physiol. 2010;299(4):C736-49.

[147] Tan HH, Fiel MI, Sun Q, Guo J, Gordon RE, Chen LC, et al. Kupffer cell activation by ambient air particulate matter exposure may exacerbate non-alcoholic fatty liver disease. J Immunotoxicol. 2009;6(4):266-75.

[148] Tarantino G, Capone D, Finelli C. Exposure to ambient air particulate matter and non-alcoholic fatty liver disease. World J Gastroenterol. 2013;19(25):3951-6.

[149] Kovacic P, Somanathan R. Unifying mechanism for metals in toxicity, carcinogenicity and therapeutic action: integrated approach involving electron transfer, oxidative stress, antioxidants, cell signaling and receptors. J Recept Signal Transduct Res. 2010;30(2):51-60.

[150] Casale T, Rosati MV, Ciarrocca M, Samperi I, Andreozzi G, Schifano MP, et al. Assessment of liver function in two groups of outdoor workers exposed to arsenic. Int Arch Occup Environ Health. 2014;87(7):745-52.

[151] Dutta M, Ghosh D, Ghosh AK, Bose G, Chattopadhyay A, Rudra S, et al. High fat diet aggravates arsenic induced oxidative stress in rat heart and liver. Food Chem Toxicol. 2014;66:262-77.

[152] Shi X, Wei X, Koo I, Schmidt RH, Yin X, Kim SH, et al. Metabolomic analysis of the effects of chronic arsenic exposure in a mouse model of diet-induced Fatty liver disease. J Proteome Res. 2014;13(2):547-54.

[153] Xu Z, Wang Z, Li JJ, Chen C, Zhang PC, Dong L, et al. Protective effects of selenium on oxidative damage and oxidative stress related gene expression in rat liver under chronic poisoning of arsenic. Food Chem Toxicol. 2013;58:1-7.

[154] Labudda M., Lead hepatotoxicity: selected aspects of pathobiochemistry. Med Pr. 2013;64(4):565-8.

[155] Mudipalli A. Lead hepatotoxicity & potential health effects. Indian J Med Res. 2007;126(6):518-27.

[156] Fowler BA. Monitoring of human populations for early markers of cadmium toxicity: a review. Toxicol Appl Pharmacol. 2009;238(3):294-300.

[157] Lupo S, Hewitt WR, Rush GF. Cadmium toxicity in the isolated perfused rat liver. Toxicol Lett. 1986;34(1):5-11.

[158] Koyuturk M, Yanardag R, Bolkent S, Tunali S. The potential role of combined antioxidants against cadmium toxicity on liver of rats. Toxicol Ind Health. 2007;23(7): 393-401.

[159] Landrigan PJ, Wright RO, Birnbaum LS. Mercury toxicity in children. Science. 2013;342(6165):1447.

[160] Bernhoft RA. Mercury toxicity and treatment: a review of the literature. Journal of environmental and public health. 2012;2012:460508.

[161] Gibb H, O'Leary KG. Mercury exposure and health impacts among individuals in the artisanal and small-scale gold mining community: a comprehensive review. Environmental health perspectives. 2014;122(7):667-72.

[162] Ackerman Z. Pappo O, Link G, Glazer M, Grozovski M. Liver Toxicity of Thioacetamide is Increased by Hepatocellular Iron Overload. Biol Trace Elem Res. 2014.; 10.1007/s12011-014-0110-9.

[163] Pietrangelo A. Iron-induced oxidant stress in alcoholic liver fibrogenesis. Alcohol. 2003;30(2):121-9.

[164] Uriu-Adams JY, Keen CL. Copper, oxidative stress, and human health. Mol Aspects Med. 2005;26(4-5):268-98.

[165] Pietrangelo A. Metals, oxidative stress, and hepatic fibrogenesis. Seminars in liver disease. 1996;16(1):13-30.

[166] Dong W, Simeonova PP, Gallucci R, Matheson J, Flood L, Wang S. et al. Toxic metals stimulate inflammatory cytokines in hepatocytes through oxidative stress mechanisms. Toxicol Appl Pharmacol. 1998;151(2):359-66.

[167] Dafnis E, Sabatini S. Biochemistry and pathophysiology of vanadium. Nephron. 1994;67(2):133-43.

[168] Cano-Gutierrez G, Acevedo-Nava S, Santamaria A, Altamirano-Lozano M, Cano-Rodriguez MC, Fortoul TI. Hepatic megalocytosis due to vanadium inhalation: participation of oxidative stress. Toxicol Ind Health. 2012;28(4):353-60.

[169] Hall TC, Garcea G, Webb MA, Al-Leswas D, Metcalfe MS, Dennison AR. The socioeconomic impact of chronic pancreatitis: a systematic review. J Eval Clin Pract. 2014;20(3):203-7.

[170] Tan CR, Yaffee PM, Jamil LH, Lo SK, Nissen N, Pandol SJ, et al. Pancreatic cancer cachexia: a review of mechanisms and therapeutics. Front Physiol. 2014;5:88.

[171] Das SL, Singh PP, Phillips AR, Murphy R, Windsor JA, Petrov MS. Newly diagnosed diabetes mellitus after acute pancreatitis: a systematic review and meta-analysis. Gut. 2014;63(5):818-31.

[172] Tong GX, Geng QQ, Chai J, Cheng J, Chen PL, Liang H, et al. Association between pancreatitis and subsequent risk of pancreatic cancer: a systematic review of epide-

miological studies. Asian Pacific journal of cancer prevention : APJCP. 2014;15(12): 5029-34.

[173] Cano-Gutierrez GF-R, C. Montaño Luis, F. Rodriguez-Lara, V. and Fortoul, TI. Pancreatic changes and vanadium inhalation. Current Topics in Toxicology. 2009;6:39-44.

[174] Pinon-Zarate G, Rodriguez-Lara V, Rojas-Lemus M, Martinez-Pedraza M, Gonzalez-Villalva A, Mussali-Galante P, et al. Vanadium pentoxide inhalation provokes germinal center hyperplasia and suppressed humoral immune responses. Journal of immunotoxicology. 2008;5(2):115-22.

[175] Ustarroz-Cano M, Garcia-Pelaez I, Pinon-Zarate G, Herrera-Enriquez M, Soldevila G, Fortoul TI. CD11c decrease in mouse thymic dendritic cells after vanadium inhalation. Journal of immunotoxicology. 2012;9(4):374-80.

[176] Fujimaki H, Shimizu F, Kubota K. Suppression of antibody response in mice by acute exposure to nitrogen dioxide: in vitro study. Environ Res. 1981;26(2):490-6.

[177] Cifone MG, Alesse E, Di Eugenio R, Napolitano T, Morrone S, Paolini R, et al. In vivo cadmium treatment alters natural killer activity and large granular lymphocyte number in the rat. Immunopharmacology. 1989;18(3):149-56.

[178] Liu J, Liu Y, Habeebu SS, Klaassen CD. Metallothionein-null mice are highly susceptible to the hematotoxic and immunotoxic effects of chronic CdCl2 exposure. Toxicol Appl Pharmacol. 1999;159(2):98-108.

[179] Jiang Y, Moller G. In vitro effects of $HgCl_2$ on murine lymphocytes. I. Preferable activation of CD4+ T cells in a responder strain. J Immunol. 1995;154(7):3138-46.

[180] Tchounwou PB, Ayensu WK, Ninashvili N, Sutton D. Environmental exposure to mercury and its toxicopathologic implications for public health. Environ Toxicol. 2003;18(3):149-75.

[181] Nairz M, Haschka D, Demetz E, Weiss G. Iron at the interface of immunity and infection. Front Pharmacol. 2014;5:152.

[182] Boskabaddy MH, Farkhondeh T. Inhaled lead exposure affects tracheal responsiveness and lung inflammation in guinea pigs during sensitization. Biol Trace Elem Res. 2013;154(3):363-71.

[183] Kasten-Jolly J, Lawrence DA. Lead modulation of macrophages causes multiorgan detrimental health effects. J Biochem Mol Toxicol. 2014;28(8):355-72.

[184] Antonini JM, Zeidler-Erdely PC, Young SH, Roberts JR, Erdely A. Systemic immune cell response in rats after pulmonary exposure to manganese-containing particles collected from welding aerosols. J Immunotoxicol. 2012;9(2):184-92.

[185] Qian Q, Li P, Wang T, Zhang J, Yu S, Chen T, et al. Alteration of Th1/Th2/Th17 cytokine profile and humoral immune responses associated with chromate exposure. Occup Environ Med. 2013;70(10):697-702.

[186] Mitra S, Keswani T, Dey M, Bhattacharya S, Sarkar S, Goswami S, et al. Copper-induced immunotoxicity involves cell cycle arrest and cell death in the spleen and thymus. Toxicology. 2012;293(1-3):78-88.

[187] Ser PH, Banu B, Jebunnesa F, Fatema K, Rosy N, Yasmin R, et al. Arsenic exposure increases maternal but not cord serum immunoglobulin G level in Bangladesh. Pediatr Int. 2014.

[188] Veras MM, Caldini EG, Dolhnikoff M, Saldiva PH. Air pollution and effects on reproductive-system functions globally with particular emphasis on the Brazilian population. J Toxicol Environ Health B Crit Rev. 2010;13(1):1-15.

[189] Xu X, Hu H, Ha S, Roth J. Ambient air pollution and hypertensive disorder of pregnancy. J Epidemiol Community Health. 2014;68(1):13-20.

[190] Mendola P, Messer LC, Rappazzo K. Science linking environmental contaminant exposures with fertility and reproductive health impacts in the adult female. Fertil Steril. 2008;89(2 Suppl):e81-94.

[191] Gerhard I, Monga B, Waldbrenner A, Runnebaum B. Heavy metals and fertility. J Toxicol Environ Health A. 1998;54(8):593-611.

[192] Baranski B, Sitarek K. Effect of oral and inhalation exposure to cadmium on the oestrous cycle in rats. Toxicol Lett. 1987;36(3):267-73.

[193] Blum JL, Xiong JQ, Hoffman C, Zelikoff JT. Cadmium associated with inhaled cadmium oxide nanoparticles impacts fetal and neonatal development and growth. Toxicol Sci. 2012;126(2):478-86.

[194] Baranski B. Effect of exposure of pregnant rats to cadmium on prenatal and postnatal development of the young. J Hyg Epidemiol Microbiol Immunol. 1984;29(3):253-62.

[195] Sikorski R, Juszkiewicz T, Paszkowski T, Szprengier-Juszkiewicz T. Women in dental surgeries: reproductive hazards in occupational exposure to metallic mercury. Int Arch Occup Environ Health. 1987;59(6):551-7.

[196] De Rosis F, Anastasio SP, Selvaggi L, Beltrame A, Moriani G. Female reproductive health in two lamp factories: effects of exposure to inorganic mercury vapour and stress factors. Br J Ind Med. 1985;42(7):488-94.

[197] Rowland AS, Baird DD, Weinberg CR, Shore DL, Shy CM, Wilcox AJ. The effect of occupational exposure to mercury vapour on the fertility of female dental assistants. Occup Environ Med. 1994;51(1):28-34.

[198] Davis BJ, Price HC, O'Connor RW, Fernando R, Rowland AS, Morgan DL. Mercury vapor and female reproductive toxicity. Toxicol Sci. 2001;59(2):291-6.

[199] Nordstrom S, Beckman L, Nordenson I. Occupational and environmental risks in and around a smelter in northern Sweden. V. Spontaneous abortion among female employees and decreased birth weight in their offspring. Hereditas. 1979;90(2):291-6.

[200] Borja-Aburto VH, Hertz-Picciotto I, Rojas Lopez M, Farias P, Rios C, Blanco J. Blood lead levels measured prospectively and risk of spontaneous abortion. Am J Epidemiol. 1999;150(6):590-7.

[201] Vigeh M, Yokoyama K, Ramezanzadeh F, Dahaghin M, Sakai T, Morita Y, et al. Lead and other trace metals in preeclampsia: a case-control study in Tehran, Iran. Environ Res. 2006;100(2):268-75.

[202] Lown BA, Morganti JB, D'Agostino R, Stineman CH, Massaro EJ. Effects on the postnatal development of the mouse of preconception, postconception and/or suckling exposure to manganese via maternal inhalation exposure to MnO2 dust. Neurotoxicology. 1984;5(1):119-29.

[203] Yoon M, Nong A, Clewell HJ, 3rd, Taylor MD, Dorman DC, Andersen ME. Lactational transfer of manganese in rats: predicting manganese tissue concentration in the dam and pups from inhalation exposure with a pharmacokinetic model. Toxicol Sci. 2009;112(1):23-43.

[204] Banu SK, Stanley JA, Lee J, Stephen SD, Arosh JA, Hoyer PB, et al. Hexavalent chromium-induced apoptosis of granulosa cells involves selective sub-cellular translocation of Bcl-2 members, ERK1/2 and p53. Toxicol Appl Pharmacol. 2011;251(3):253-66.

[205] Chashschin VP, Artunina GP, Norseth T. Congenital defects, abortion and other health effects in nickel refinery workers. Sci Total Environ. 1994;148(2-3):287-91.

[206] Silva N, Senanayake H, Waduge V. Elevated levels of whole blood nickel in a group of Sri Lankan women with endometriosis: a case control study. BMC Res Notes. 2013;6:13.

[207] Meléndez-García N. Efecto del vanadio en la morfofisiología del útero de ratones CD.. México: Universidad Nacional Autónoma de México. ; 2014.

[208] García-Ibarra F. Efecto del ácido ascórbico sobre las alteraciones morfofisiológicas del ovario inducidas por la inhalación de vanadio en un modelo murino. México: Universidad Nacional Autónoma de México.; 2014.

[209] Bucher JR, Elwell MR, Thompson MB, Chou BJ, Renne R, Ragan HA. Inhalation toxicity studies of cobalt sulfate in F344/N rats and B6C3F1 mice. Fundam Appl Toxicol. 1990;15(2):357-72.

[210] Bizarro P, Acevedo S, Nino-Cabrera G, Mussali-Galante P, Pasos F, Avila-Costa MR, et al. Ultrastructural modifications in the mitochondrion of mouse Sertoli cells after inhalation of lead, cadmium or lead-cadmium mixture. Reproductive toxicology. 2003;17(5):561-6.

[211] Fortoul TI, Bizarro-Nevares P, Acevedo-Nava S, Pinon-Zarate G, Rodriguez-Lara V, Colin-Barenque L, et al. Ultrastructural findings in murine seminiferous tubules as a consequence of subchronic vanadium pentoxide inhalation. Reproductive toxicology. 2007;23(4):588-92.

[212] Mussali-Galante P, Rodriguez-Lara V, Hernandez-Tellez B. Avila-Costa MR, Colin-Barenque L, Bizarro-Nevarez P, et al. Inhaled vanadium pentoxide decrease gamma-tubulin of mouse testes at different exposure times. Toxicol Ind Health. 2005;21(9): 215-22.

[213] Rodriguez-Lara V, Morales-Rivero A, Rivera-Cambas AM, Fortoul TI. Vanadium inhalation induces actin changes in mice testicular cells. Toxicology and industrial health. 2013; DOI: 10.1177/0748233713501364.

[214] Akinloye O, Arowojolu AO, Shittu OB, Anetor JI. Cadmium toxicity a possible cause of male infertility in Nigeria. Reprod Biol. 2006;6(1):17-30.

[215] Naha N, Manna B. Mechanism of lead induced effects on human spermatozoa after occupational exposure. Kathmandu Univ Med J (KUMJ). 2007;5(1):85-94.

[216] Bosetti C, Nieuwenhuijsen MJ, Gallus S, Cipriani S, La Vecchia C, Parazzini F. Ambient particulate matter and preterm birth or birth weight: a review of the literature. Arch Toxicol. 2010;84(6):447-60.

[217] Basu R, Harris M, Sie L, Malig B, Broadwin R, Green R. Effects of fine particulate matter and its constituents on low birth weight among full-term infants in California. Environ Res. 2014;128:42-51.

[218] Weldy CS, Liu Y, Liggitt HD, Chin MT. In utero exposure to diesel exhaust air pollution promotes adverse intrauterine conditions, resulting in weight gain, altered blood pressure, and increased susceptibility to heart failure in adult mice. PLoS One. 2014;9(2):e88582.

[219] Dolk H, Vrijheid M. The impact of environmental pollution on congenital anomalies. Br Med Bull. 2003;68:25-45.

[220] Caserta D, Graziano A, Lo Monte G, Bordi G, Moscarini M. Heavy metals and placental fetal-maternal barrier: a mini-review on the major concerns. Eur Rev Med Pharmacol Sci. 2013;17(16):2198-206.

[221] Al-Gubory KH. Environmental pollutants and lifestyle factors induce oxidative stress and poor prenatal development. Reprod Biomed Online. 2014;29(1):17-31.

[222] Salvatori F, Talassi CB, Salzgeber SA, Spinosa HS, Bernardi MM. Embryotoxic and long-term effects of cadmium exposure during embryogenesis in rats. Neurotoxicol Teratol. 2004;26(5):673-80.

[223] Angenard G, Muczynski V, Coffigny H, Pairault C, Duquenne C, Frydman R, et al. Cadmium increases human fetal germ cell apoptosis. Environ Health Perspect. 2010;118(3):331-7.

[224] Heck JE, Park AS, Qiu J, Cockburn M, Ritz B. An exploratory study of ambient air toxics exposure in pregnancy and the risk of neuroblastoma in offspring. Environ Res. 2013;127 1-6.

[225] Golub MS, Macintosh MS, Baumrind N. Developmental and reproductive toxicity of inorganic arsenic: animal studies and human concerns. J Toxicol Environ Health B Crit Rev. 1998;1(3):199-241.

[226] Lin YY, Hwang YH, Chen PC, Chen BY, Wen HJ, Liu JH, et al. Contribution of gestational exposure to ambient traffic air pollutants to fetal cord blood manganese. Environ Res. 2012;112:1-7.

[227] Gerber GB, Leonard A, Hantson P. Carcinogenicity, mutagenicity and teratogenicity of manganese compounds. Crit Rev Oncol Hematol. 2002;42(1):25-34.

[228] Weinberg ED. Can iron be teratogenic? Biometals. 2010;23(2):181-4.

[229] Saini S, Nair N, Saini MR. Embryotoxic and teratogenic effects of nickel in Swiss albino mice during organogenetic period. Biomed Res Int. 2013;2013:701439.

[230] Fortoul TI., Rodriguez-Lara V., Cano-Gutiérrez G., González Villalba A., Colín-Barenque L., Santamaría A., Ustarroz-Cano M., García-Peláez I., López-Valdez N., Falcón-Rodríguez CI., Pedraza-Chaverri J., Bizarro-Nevares P., Carrillo-Mora P.. Free radicals and health effects. In: Kozyrev D. SV, editor. Handbook of free radicals: formation, types and effects. Hauppauge, NY, USA,: Nova Science Publishers, Inc. ; 2010. p. pp. 263-89..

[231] Cope G, Hodgson E. Classes of toxicants: Use classes. In: Hodgson E, editor. Modern Toxicology. Fourth ed. New Jersey: John Wiley & Sons, Inc; 2010. p. 49-50.

Chapter 26

Molecular Epidemiology and Air Pollution

Pavel Rossner, Jr., Blanka Binkova,
Andrea Rossnerova and Radim J. Sram

Additional information is available at the end of the chapter

http://dx.doi.org/10.5772/59955

1. Introduction

A new and evolving area of research termed molecular epidemiology aims to merge sophisticated and highly sensitive laboratory methods (many of them developed during the recent revolution in molecular biology) with analytical epidemiological methods. Molecular epidemiology bridges from basic research in molecular biology to studies of human cancer causation by combining laboratory measurement of internal dose, biologically effective dose, biological effects and the influence of individual susceptibility with epidemiologic methodologies [1]. The most common view is that the approach represents a natural convergence of molecular biology and epidemiology [2].

The number of biomarkers available for evaluating genetic and cancer risk in humans is quite large. Their utility for human biomonitoring is suggested by the well-known paradigm of environmentally induced cancer, which represents endpoints for assessing the entire spectrum of human-genotoxicant interactions [3]. These biomarkers begin with exposure and include absorption, metabolism, distribution, critical target interaction (i.e. DNA damage and repair), genetic changes and finally disease. Disease is the province of traditional epidemiology. The development of biomarkers has given rise to the field of molecular epidemiology, which uses these biomarkers rather than disease to assess the risk of environmental exposure [4, 5].

The paradigm of environmental cancer starts with exposure. A large number of biomarkers are available now, but to evaluate their sensitivity and to interpret the results obtained we need to know data on exposure. Previously, data on exposure were usually not published in studies using various biomarkers.

It is well established that ambient air pollution is related to human health. Increased exposure to respirable particulate matter (PM) correlates with increased mortality caused by lung cancer and cardiovascular diseases [6-8]. Pope et al. [9] suggested that a long term increase in PM2.5

of 10 μg/m^3 is connected with an 8% increase in lung cancer mortality in adult men. Despite the fact that other factors related to cancer incidence, such as smoking habit or inappropriate diet, are probably stronger influences, the absolute number of cancer cases related to air pollution is high due to the high prevalence of exposure [10].

Although the quantitative health risk related to air pollution is assessed by epidemiological studies [11-14], alternative types of studies are necessary for the purposes of primary prevention. On the level of human populations, such studies are first of all molecular epidemiological studies evaluating quantitative relations between external exposure and measurable biological events (biomarkers). These biomarkers form a chain from exposure to disease [15-17].

One of the most frequently used biomarkers are DNA adducts as biomarker of exposure, quantifying the biologically effective dose of genotoxic compounds that were covalently bound to DNA as a target molecule of carcinogenesis [18-22]. If DNA adducts are not effectively repaired, they might be fixed as mutations during replication. According to the well known scheme of the multi-step process of chemical carcinogenesis, an accumulation of mutations may lead to carcinogenesis. Thus, DNA adduct levels have a direct relation to mutagenesis and carcinogenesis. Data are accumulating about the relation of DNA adducts to environmental exposure to complex mixture components such as carcinogenic polycyclic aromatic hydrocarbons (c-PAHs) [23] and to malignant tumors and other degenerative diseases [24, 25].

The Comet assay (single cell gel electrophoresis, SCGE) in its standard alkaline version allows the detection of DSB (double strand breaks), SSB (single strand breaks), alkali-labile sites and transient repair sites. DSB represent the principal lesion leading to the formation of chromosomal aberrations. The majority of chemical mutagens induce DSB indirectly via the generation of other DNA lesions such as SSB or oxidative damage that may be converted to DSB during DNA replication or repair [26]. When combined with specific bacterial repair enzymes, it identifies a broad spectrum of additional lesions including oxidized purines and pyrimidines [27]. Due to its relative simplicity, its low requirements on the number of analyzed cells as well as its capacity to detect DNA damage independently of the cell cycle, comet assay is believed to serve as an exposure biomarker assay providing information about the biologically effective dose [28].

DNA fragmentation in the mature sperm may be understood as another biomarker of exposure, measured by the sperm chromatin structure assay (SCSA) [29]. The sperm sample is stained with acridine orange, which is a metachromatic DNA dye that fluorescence green when intercalated into native DNA and shifts to a red fluorescence when associated with collapsed single-stranded DNA. These stained samples are measured by flow cytometry [30,31]. Human sperm cells can undergo DNA fragmentation or covalent modification of nucleotides. Both these types of disturbances contribute to infertility. Altered sperm chromatin structure can be significant contributing factor to subfertility that is not revealed by standard spermiogram parameters, because they can range within physiological values. Consequently, fertilization failure, impaired embryo morphology, abnormal development of blastocysts, embryo implantation failure or repeated spontaneous abortions can occur.

Chromosomal aberrations in human peripheral lymphocytes are recognized as a valuable biomarker of effect, probably the only one that has been internationally standardized and validated [32]. While classic cytogenetic analysis (conventional method, CCA) is the method of choice for determining unstable types of aberrations, the fluorescent in situ hybridization technique (FISH) seems to be a rapid, sensitive, and reliable method for the detection of stable structural rearrangements that remain undiminished over time, such as translocations [33-37]. The FISH painting technique appears to be more sensitive than the conventional technique for detecting the genomic frequency of translocations induced by various chemical agents or irradiation [38].

The measurement of micronuclei (MN) in human peripheral blood lymphocytes is one of the methods frequently used in molecular epidemiology. MN represent a measure of both chromosome breakage and chromosome loss. Therefore, an increased frequency of micronucleated cells, used as a biomarker of genotoxic effects, can reflect exposure to agents with clastogenic or aneugenic modes of action [39]. Currently, the MN assay is one of the preferred methods for assessing chromosomal damage as a result of environmental mutagen exposure as well as a tool for genotoxicity testing.

Another important aspect demonstrating the advantages of molecular epidemiology studies over classical epidemiology is the possibility of identifying the genetic susceptibility of individuals to the action of various compounds [17]. Biomarkers of susceptibility mostly take into account the role of genetic makeup of the organism in the response to the exposure to xenobiotics. The role of genetic polymorphisms on the metabolic activation of xenobiotics (oxygenases of cytochromes P450 such as *CYP 1A1*) and also their detoxification (glutathione-S-transferases) is well known. Further biomarkers of individual susceptibility are polymorphisms in genes encoding DNA repair enzymes (*XRCC1, XPD, hOGG1*) [40-42]. One more factor affecting susceptibility to the genotoxic and carcinogenic effects of xenobiotics is the saturation of the organism by vitamins A, C, E, folic acid etc., which are known to play a significant role as free radical scavengers and antioxidant agents and which also affect the synthesis of DNA repair enzymes [43-46].

Inhalation of PM, particularly PM of aerodynamic diameter < 2.5 μm (PM2.5) and smaller, leads to inflammation and subsequent production of reactive oxygen species (ROS) [47]. The production of ROS, that include e.g. the hydroxyl radical, superoxide anion, or hydrogen peroxide, is caused by both the physical effects of PM (PM is phagocyted by macrophages that consequently produce ROS), and the presence of various chemicals on the surface of PM (e.g. metals, PAHs) with pro-oxidant properties. ROS may arise from exogenous or endogenous sources. The latter are mostly physiologic and include various metabolic processes and inflammation, whereas exogenous sources are environmental factors such as smoking, diet [46, 49], ultraviolet radiation, ionizing radiation or exposure to environmental pollution [50].

ROS can attack lipids, proteins and nucleic acids simultaneously [51]. The highly reactive hydroxyl radical reacts with DNA by addition to the double bonds of DNA bases, and by abstraction of a hydrogen atom from the methyl group of thymine and C–H bonds of 2'-deoxyribose [52]. This reaction yields several modified bases but, as a biomarker of oxidative DNA damage, 8-oxodeoxyguanosine (8-oxodG) is most often studied. The oxidized base is highly mutagenic and, if unrepaired, its presence in DNA causes GC>TA transversions. If repaired, 8-oxodG is excreted in urine where it can be assessed by various techniques as a general biomarker of oxidative stress. Urinary levels of 8-oxodG are also believed to reflect the total DNA excision repair capacity of an organism [53]. Other sources of 8-oxodG, including oxidation of the nucleotide pool, may affect urinary levels of 8-oxodG. Numerous reports have indicated that urinary 8-oxodG is not only a biomarker of generalized, cellular oxidative stress, but may also imply the risk of cancer, atherosclerosis and diabetes [51, 54].

Several metabolic pathways of PAH activation have been described [55] and one of them, activation through PAH-*o*-quinones, leads to ROS generation and oxidative stress [56]. The modification of DNA molecules represents the most serious form of impact of ROS on the organism because it may lead to base changes, mutations, and/or DNA breaks. If ROS attack both DNA strands, double-strand DNA breaks may appear. These breaks may lead either to unstable chromosomal aberrations, or, if homologous or non-homologous end-joining repair seals the breaks, to stable chromosomal translocations. Translocations are more serious because they are usually fixed in the genome and may lead to rearrangements of regulatory elements and genes, including oncogenes thus increasing cancer risk [57]. Another, indirect mechanism of DNA double strand breaks induction is associated with DNA adducts formation. Adducts may cause persistent blockage of one DNA strand during its synthesis and uncoupling of the other strand which may result in the formation of double strand breaks [58].

The attack of ROS on lipids leads to lipid peroxidation. This reaction may have potentially serious consequences, as it may damage cellular membrane and inactivate membrane-bound receptors or enzymes. In addition, secondary products of lipid peroxidation, such as aldehydes, are highly reactive and may propagate oxidative stress by reacting with other cellular molecules including proteins [59]. Currently, isoprostanes are considered the most reliable markers of lipid peroxidation. These prostaglandin-like compounds, first described in the 1990s, are formed by free-radical induced peroxidation of arachidonic acid, independent of cyclooxygenase enzymes. Lipid peroxidation products, including isoprostanes, play a role in the pathogenesis of many diseases [60].

New omics biomarkers: mRNA expression. Although the effect of air pollutants on humans may be monitored by the analysis of mRNA expression of individual selected genes [61], the current trend is to use transcriptomics as a tool for studying genome-wide responses of the organism to environmental exposures [62]. It has been concluded that transcriptome is a dynamic entity that is highly responsive to environmental exposures [62]. But studies of the effects of environmental pollutants on gene expression profiles are scarce [63, 64, 65].

2. Czech studies – Fig. 1

The Northern Bohemia was in late eighties one of the most polluted regions in Europe. It was therefore believed that such specific situation is just the location to study the sensitivity of biomarkers to detect genetic damage [15]. As the exposed region was selected the Teplice district, as the control region the Prachatice district.

http://dx.doi.org/10.5772/59955

Later EXPAH project (Effects of polycyclic aromatic hydrocarbons in environmental pollution on exogenous and oxidative DNA damage) tried to evaluate the hypothesis that PAHs are the major source of genotoxic activities of organic mixtures associated with air pollution, one of the studied cities was Prague [66-72].

Ostrava Region is area highly polluted by benzo[a]pyrene, nowadays one of the highest exposure in EU [73].

Molecular epidemiology studies in those regions are examples of the use of biomarkers to identify genetic damage induced by air pollution as well as which biomarkers may be used to evaluate the health risk for exposed populations.

Figure 1. Map of the Czech Republic with the locations of molecular epidemiology studies.

3. Teplice program

The first studies in which biomarkers detected seasonal differences in exposure and genetic damage from environmental pollution were conducted in a population from a highly polluted area in Silesia (Poland) [73-76]. Perera et al. [74] found that the exposure to environmental pollution was associated with a significant increase of DNA adducts, SCE (sister chromatid exchanges) and chromosomal aberrations, as well as with frequency of *ras* oncogene overexpression. In these studies, the investigators were not able to determine the relationship between individual exposure to air pollution and biomarkers of exposure and genetic damage because personal exposure data were not collected.

The methods of molecular epidemiology were widely used in studies on the impact of air pollution to genetic damage in Czechoslovakia. Mining districts in the northwestern region of the Czech Republic were polluted by combustion from power plants and heavy industry, which resulted in one of the worst environmental pollution in Europe [77]. The Teplice Program was initiated by the Czech Ministry of the Environment in 1990 to provide scientifically valid information needed to assess environmental health problems in the Northern Bohemian mining districts. The program was successful thanks to collaboration with U.S. Environmental Protection Agency; it included the air pollution monitoring, human exposure, biomarker, and health effects studies [15]. An air quality monitoring and receptor modeling study conducted in the Teplice district during 1992 to 1994 identified emissions from residential heating and vehicles as the major sources of organic carbon, including c-PAHs. c-PAHs and their nitroderivatives associated with the respirable particle fraction PM10 have been identified as a major source of carcinogenic risk in urban areas [78, 79]. Therefore this situation was understood as a convenient model to check the relationship between c-PAHs in the polluted air and biomarkers.

Polycyclic aromatic hydrocarbons (PAHs) adsorbed onto respirable air particles (< 2.5 μm) are mainly derived from incomplete combustion, including mobile sources such as motor vehicles and stationary sources such as power plants, residential heating etc. Some of these compounds exhibit carcinogenic and/or mutagenic properties [80]. Molecular epidemiology studies using biomarkers of exposure and early biological effects could provide invaluable information about the genotoxic effects of environmental exposure to such PAHs mixtures. The measurement of DNA adducts was selected as a promising biomarker of exposure, since such measurements take into account individual differences in exposure, absorption, distribution, metabolic activation and detoxification of PAHs in the body as well as cell turnover and the repair of DNA damage [22].

The first study compared a group of women working as postal workers or gardeners from a highly polluted Teplice district with the group of women working as postal workers and nursery school teachers in the town of Prachatice, a control district in southern Bohemia in winter 1993/1994. At that time personal exposure to B[a]P in Teplice reached up to 7.5±3.6 ng/m^3. DNA adducts by ^{32}P-postlabeling in WBC (white blood cells) were significantly affected by personal exposure to c-PAHs ($r=0.710$; $p<0.001$). Percentage of DNA in tail measured by Comet assay correlated significantly ($r=0.304$; $p<0.05$) with personal exposures to respirable particles (PM2.5). No effect was observed on the frequency of chromosomal aberrations or SCE [18]. A group of 10 women non-smokers also participated in a follow up study during 5 samplings in November 1992, October 1993-February 1994. Personal exposure to B[a]P during 24 h sampling was during this period between 2.0±1.1 ng/m^3 and 7.5±3.6 ng/m^3. Analyzing data from the follow up study, a significant effect of personal exposure on DNA adduct levels and their relationship with short-term exposure to c-PAHs was found ($r=0.621$, $p<0.001$). No other variables as age, passive smoking and consumption of fried and smoked food during the 24 h of personal exposure monitoring had significant effect on DNA adduct levels [19].

Svecova et al. [81] analyzed the effect of air pollution, particularly PM10, PM2.5, c-PAHs and B[a]P, on urinary levels of 8-oxodG in children from the districts of Teplice and Prachatice.

The urine was collected in the year 2004 from 894 children born in the period 1994-1998 [82]. 8-oxodG was determined by ELISA [83]. Stationary monitoring of PM2.5, PM10 and c-PAHs was done continually during the entire sampling period as well as before this period. Collected data on air pollution allowed to correlate individual urinary 8-oxodG levels with levels of pollutants measured at different times and for different periods before collection of urine samples. In multivariate models, they found that exposure to PM10 and PM2.5 measured in a 3-day interval 4–6 days before sampling, PM10, PM2.5, and B[a]P in a 7-day period before sampling, c-PAHs and B[a]P in a 3-day interval 1–3 and 7–9 days before urine collection were significant factors positively affecting 8-oxodG levels in urine. It may be generalized that PM10, PM2.5, and c-PAHs increase oxidative damage within one week of exposure. Increased level of 8-oxodG was observed also in children exposed to environmental tobacco smoke (ETS). It may be hypothesized that ETS exposure and short-term exposure to fine particles and c-PAHs induce oxidative stress, and therefore may be starting point for respiratory and allergic morbidity in children.

Rubes et al. [84] studied the impact of air pollution in Teplice in the period 1995-1997 to sperm DNA damage repeatedly in the same donors, measuring the sperm chromatin structure assay (SCSA), when the percentage of mature sperm with abnormal chromatin/fragmented DNA was determined and expressed as % DNA fragmentation index (DFI). In the study 36 semen donors participated, 21 men gave seven samples, 10 gave six samples. Air pollution levels over the last 90 days before sampling ranged from 28.7- 67.8 μg/m^3 (for PM10), and from 0.3-7.9 ng/m^3 (for B[a]P). None of other semen outcomes (sperm concentration, semen volume, sperm morphology and sperm motility) showed significant associations with air pollution. Only mean % DFI was significantly associated with exposure ($p<0.05$). It was the first study reporting association between exposure to ambient air pollution and DNA fragmentation in human sperm. Rubes et al. [85] further studied the hypothesis, if men homozygous null for *GSTM1* (*GSTM1-*) are less able to detoxify reactive metabolites of c-PAHs found in air pollution. Using a longitudinal study design, this study revealed a significant association between *GSTM1* null genotype and increased DNA damage in sperm, defined as % DFI. This study shows for the first time that endogenous DNA fragmentation in human sperm can be modulated by polymorphism in *GSTM1*, a gene involved in c-PAH metabolism. Men who are homozygous null for *GSTM1* exhibit increased susceptibility to sperm DNA damage associated with exposure.

4. Prague

The capital city of Prague has become one of the most polluted localities in the Czech Republic, especially due to traffic. Therefore, the effect of exposure to carcinogenic polycyclic aromatic hydrocarbons (c-PAHs) adsorbed onto respirable air particles (PM2.5, <2.5 μm) on DNA adducts and chromosomal aberrations was repeatedly studied in groups of city policemen working in the downtown area as well as in bus drivers [86].

Studied groups of a total of 950 subjects from three studies were used for analyses of associations between c-PAHs exposure, DNA adducts and chromosomal aberrations in 1) city policemen in 2001 (exposed N = 53, controls N = 52) [36], 2) city policemen in 2004 (exposed N = 480, 120 x 4) [87], 3) bus drivers in 2005-2006 (N = 120 x 3, exposed I N = 50, exposed II N = 20, controls N = 50) [83]. All volunteers ad 1 and 2) were male city policemen working in the Prague downtown and spending >8 h outdoors daily. Controls from study 1) were age- and sex-matched healthy males spending >90% of their daily time outdoors and working in a suburban area. Study 3) involved 50 bus drivers working in the center of Prague (I), 20 garagemen (II) and, as controls, 50 healthy administrative workers spending >90% of their daily time indoors.

Ambient air particles (PM10, PM2.5) and c-PAHs (benz[a]anthracene, benzo[b]fluoranthene, benzo[k]fluoranthene, benzo[ghi]perylene, benzo[a]pyrene (B[a]P), chrysene, dibenz[ah]-anthracene and indeno[cd]pyrene) were monitored using VAPS samplers, while personal exposure to c-PAHs was evaluated using personal samplers during working shifts. Quantitative chemical analysis of c-PAHs was performed by HPLC with fluorimetric detection according to the EPA method [19].

DNA adducts were analyzed in lymphocytes by the ^{32}P-postlabeling assay, which was performed according to a standardized procedure – Fig. 2 [22].

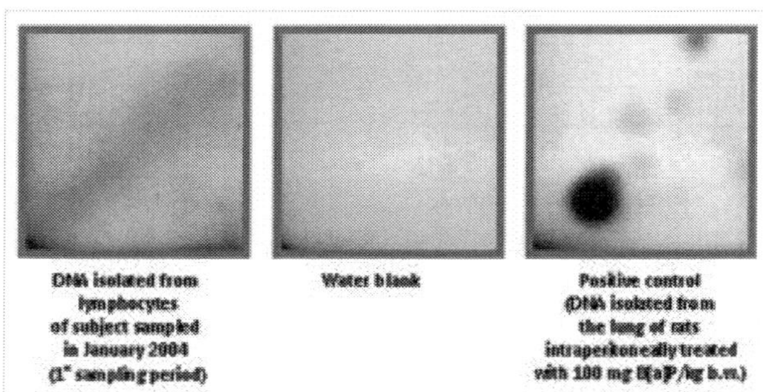

Figure 2. DNA adducts by ^{32}P-postlabeling.

Based on the personal monitoring data, during their working shifts the city policemen were exposed to significantly higher concentrations of c-PAHs as well as B[a]P than the controls (median): 9.7 vs. 5.8 ng/m^3 ($p<0.01$) and 1.6 vs 0.8 ng/m^3 ($p<0.01$), respectively The level of B[a]P-like DNA adducts was higher in the exposed group (0.122±0.036 vs. 0.099±0.035 adducts/ 10^8 nucleotides, p=0.003). The results of multivariate regression analysis showed smoking, vitamin C levels and polymorphisms of the *XPD* repair gene in exon 23 and the *GSTM1* gene to be significant predictors for total DNA adduct levels. Exposure to ambient air pollution, smoking, and polymorphisms of the *XPD* repair gene in exon 6 were significant predictors for B[a]P-like DNA adducts [22].

Using the FISH technique and probes for chromosomes 1 and 4 (Fig. 3), the genomic frequency of translocations calculated as $F_G/100$ was 1.72 and 1.24 for exposed and controls ($p<0.05$), respectively. *CYP1A1*2C (Ile/Ile)*, *XPD 23 (Lys/Lys)*, and *XPD 6 (CC)* genotypes were associated with an increase in the number of aberrant cells as determined by the conventional method. Factors associated with an increased level of translocations determined by FISH included age, smoking, B[a]P-like DNA adducts, folate, and polymorphisms of *CYP1A1*2C, GSTP1, EPHX1, p53* and *MTHFR* genes. Ambient air exposure to c-PAHs significantly increased FISH cytogenetic parameters in nonsmoking city policemen [36].

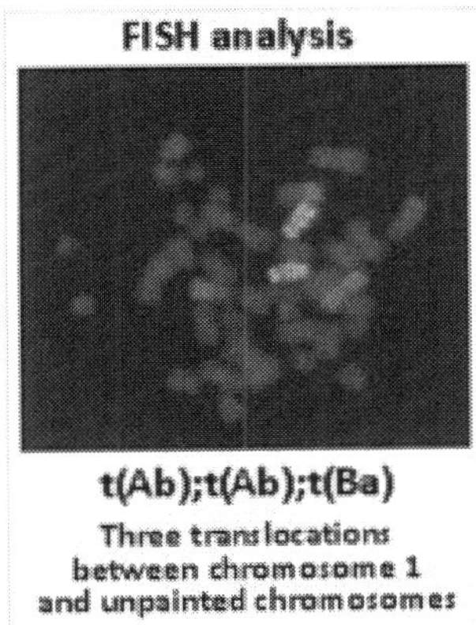

Figure 3. Chromosomal aberrations by fluorescence in situ hybridization (FISH) – whole chromosome painting #1 and #4. (t(Ab); t(Ab); t(Ba); three translocations between chromosome 1 and unpainted chromosomes.

Total DNA adducts, B[a]P-like DNA adducts and the genomic frequency of translocations in the study conducted in the year 2001 were significantly affected by smoking – an effect of air pollution was observed only in nonsmokers [22, 36]. Therefore, later studies used only nonsmokers as volunteers.

The obtained results were confirmed in a subsequent study in which city policemen were sampled in January, March, June, and September 2004. Using personal monitoring, the concentrations of c-PAHs were 1.58 ng/m^3 for B[a]P and 9.07 ng/m^3 for c-PAHs during January, 0.39 ng/m^3 for B[a]P and 3.46 ng/m^3 for c-PAHs during March, 0.18 ng/m^3 for B[a]P and 1.92 ng/m^3 for c-PAHs during June, and 0.45 ng/m^3 for B[a]P and 3.08 ng/m^3 for c-PAHs during September. Total DNA adducts were only slightly elevated in January (2.08 ± 1.60) compared to March (1.66 ± 0.65), June (1.96 ± 1.73) and September (1.77 ± 1.77). B[a]P-like DNA adducts, however, were significantly higher in January than in the March and June sampling periods (0.26 ± 0.14 vs. 0.19 ± 0.12 and 0.22 ± 0.13, respectively; $p<0.0001$ and $p=0.017$), indicating that c-PAH exposure probably plays a crucial role in DNA adduct formation in lymphocytes [80]. In those periods, the mean frequency of translocations measured by FISH (F_G/100) was 1.32±1.07, 0.85±0.95, 0.87±0.81, and 1.08±0.94, respectively. The frequency of chromosomal aberrations determined by CCA was 2.07±1.48, 1.84±1.28, 1.84±1.42.and 1.64±1.46 % AB.C., respectively [86].

In another study bus drivers were sampled in winter 2005, summer 2006 and winter 2006. Using personal monitoring, the concentrations of B[a]P for the exposed group were 1.25 ng/m^3 during winter 2005, 0.20 ng/m^3 during summer 2006 and 1.04 ng/m^3 during winter 2006; for controls the concentrations were 1.75 ng/m^3 during winter 2005, 0.24 ng/m^3 during summer 2006, and 0.75 ng/m^3 during winter 2006. The total DNA adducts in the exposed group were 1.72 ± 0.56 during winter 2005, 1.22 ± 0.45 during summer 2006, and 1.62 ± 0.59 adducts/10^8 nucleotides during winter 2006; in garagemen the totals were 1.24 ± 0.41 during winter 2005, 1.27 ± 0.48 during summer 2006, and 1.70 ± 0.08 adducts/10^8 nucleotides during winter 2006; in controls the totals were 2.15 ± 0.61 during winter 2005, 1.18 ± 0.36 during summer 2006, and 1.90 ± 0.79 adducts/10^8 nucleotides during winter 2006. In those periods the mean frequencies of translocations measured by FISH (F_G/100) were 1.62±1.17, 2.18±1.75, and 1.77±1.31 in the group of bus drivers, 1.20±1.24, 0.88±1.11, and 1.01±0.78 in garagemen, and 1.65±149, 1.34±1.01, and 1.87±1.29 in controls, respectively. The frequencies of chromosomal aberrations determined by CCA were s 1.30±1.15, 1.43±1.01, and 1.30±1.04 % AB.C. in the group of bus drivers, 0.95±0.76, 1.15±1.09, and 1.55±0.97 % AB.C. in garagemen, and 1.17±0.93, 1.50±0.99, and 1.52±1.12 % AB.C. in controls, respectively [86].

The levels of DNA adducts were significantly affected by stationary exposure to B[a]P within the last 30 days before samples collection. Data obtained for biomarkers of exposure and effect were used for pooled analysis. Using multivariate logistic regression, the relationship between personal exposure to B[a]P and DNA adducts measured by ^{32}P-postlabeling was calculated (DNA adducts = 1.042 + B[a]P x 0.077, $p<0.001$, Fig. 4).

These results indicate that c-PAH exposure plays a crucial role in DNA adduct formation in lymphocytes. A similar relationship was observed between personal exposure to B[a]P and the genomic frequency of translocations measured by FISH (F_G/100 = 1.255 + B[a]P x 0.082, $p<0.05$, Fig. 5) [86].

In the subset of 54 city policemen (exposed) and 11 controls (working indoors) genetic damage was analyzed by Comet assay, the sampling was performed during January and September 2004. The exposed group displayed significantly higher levels of unspecified DNA damage than controls during both seasons, oxidative DNA damage was significantly higher in the exposed group only in January. Correlation analysis revealed a strong association in the exposed group between the level of oxidative DNA damage and personal exposure to c-PAHs during January. Regression analysis of the influence of genetic polymorphism on the level of DNA damage suggested higher oxidative DNA damage with variant alleles of genes *CYP1A1*2C (Ile/Val), MTHFR, MS*; variant allele of *p53MspI* polymorphism increased both unspecific and oxidative DNA damage [88].

In the same groups of bus drivers Rossner et al. studied the effect of air pollution to oxidative damage to DNA [83] and oxidative damage to lipids [89]. 8-oxodG in urine was used as the biomarker of DNA oxidative damage. Increased level of urinary 8-oxodG in bus drivers was observed in all three sampling periods, as well as a protective effect of vitamin C on oxidative DNA damage. Multivariate logistic regression analysis identified PM2.5 and PM10 levels, measured by stationary monitoring during a 3-day period before urine collection, as the only

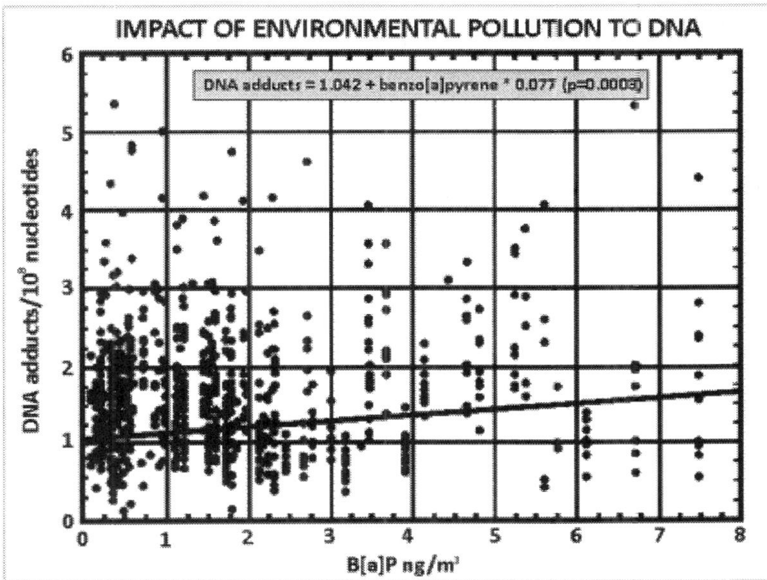

Figure 4. Impact of B[a]P exposure to DNA adducts.

factors significantly affecting 8-oxodG levels, while the levels of c-PAHs had no significant influence [83].

Lipid peroxidation was determined by 15-F_{2t}-isoprostane (15-F_{2t}-IsoP) in urine. It was significantly higher in bus drivers than in controls in both winter seasons. Personal c-PAHs and B[a]P exposure 48 h before sample collection significantly increased 15-F_{2t}-IsoP levels in urine. When data from stationary monitors were used, c-PAHs and B[a]P had a significant effect on 15-F_{2t}-IsoP levels for the 3-day period immediately preceding sample collection, but exposure to PM2.5 and PM10 affected lipid peroxidation at least 25-27 days before sampling [89].

Another study on city policemen in February and May 2007 analyzed the impact of air pollution on the level of micronuclei measured by automated image analysis (MetaSystem Metafer 4) [90] (Fig. 6).

Using multivariate logistic regression, the relationship between personal exposure to B[a]P and micronuclei expressed as MN/1000 cells was calculated (MN = 5.18 + B[a]P x 1.11, p=0.002, Fig. 7).

These results indicate that MN frequencies, when measured by the automated scoring system, are significantly affected by higher levels of air pollutants. Others have published similar observations in general populations living in polluted regions [91] as well as in heavily exposed workers [92, 93]. A recent meta-analysis of data from a 25-year period also indicates increased MN frequencies in environmentally exposed subjects, specifically children. Results of Ross-

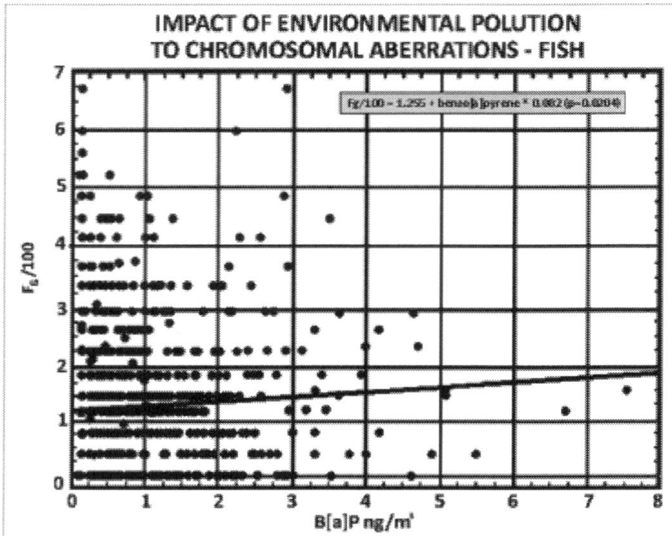

Figure 5. Impact of B[a]P exposure to genomic frequency of translocations (Fg/100).

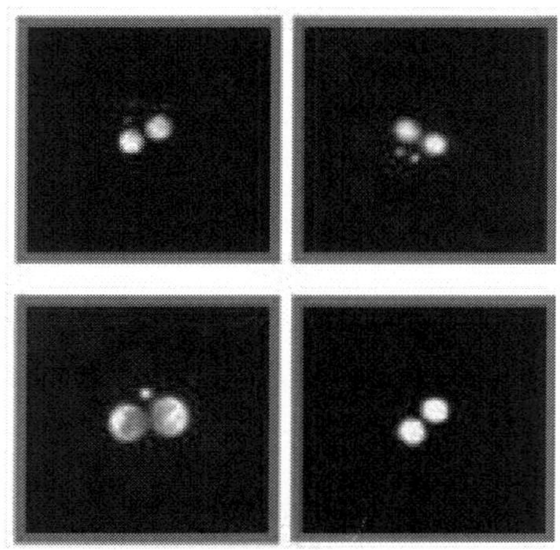

Figure 6. Micronuclei by automated image analysis in binucleated cells (BNC): a) BNC with three MN; b) BNC with two MN; c) BNC with one MN; d) BNC without MN.

nerova et al. [95] also showed that MN frequency was affected by exposure to c-PAHs up to 60 days before sample collection. Concentrations of c-PAHs measured more than 60 days before the collection of samples had no effect on MN formation. Similar results were obtained

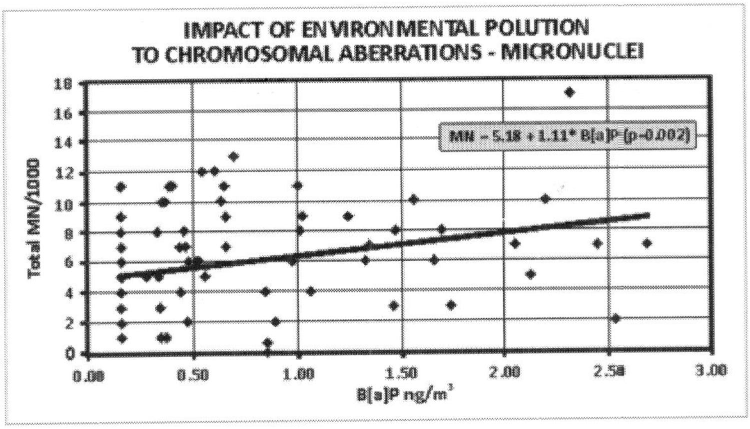

Figure 7. Impact of B[a]P exposure to micronuclei.

using conventional cytogenetic analysis, where the frequency of aberrant cells corresponded to the exposure to chemical carcinogens during the periods of 3 months [95].

The work of Rossnerova et al. [90] was the first human biomonitoring study focused on the measurement of MN by automated image analysis for assessing chromosomal damage as a result of environmental mutagen exposure. The results demonstrate the ability of c-PAHs to increase MN frequency, even if the environmental exposure to c-PAHs occurred up to 60 days before collection of biological material. Further, those findings indicate the ability of the automated image analysis system to analyze easily slides with a low density of cells that would be very laborious to score using visual analysis. Considering the possibility to analyze more BNC more quickly, it may be suggested to analyze 2000 to 3000 binucleated cells in future studies to obtain more statistically powerful data.

Comparing stationary exposure in the center of Prague as well as the personal exposure of city policemen to c-PAHs in February 2007 [90] vs. February 2001 [22] or January 2004 [96], exposure to c-PAHs in February 2007 was lower due to meteorological conditions, e.g. personal exposure to B[a]P was only 1.04±0.76 ng/m^3. It is important to note that even this B[a]P concentration increased the frequency of MN.

In the same study population of 47 city policemen DNA fragmentation in mature sperm by SCSA, relationship between air pollution and genetic polymorphism in metabolic genes (*CYP1A1, EPHX1, GSTM1, GSTP1, GSTT1*), folic acid metabolism genes (*MTR, MTHFR*) and DNA repair genes (*XRCC1, XPD6, XPD23, hOGG1*) was analyzed. DNA fragmentation index (DFI) was significantly higher in February vs. May 2007. Rubes et al. [97] observed that concentration of 1 ng B[a]P/ m^3 induces DNA fragmentation in mature spermatozoa, which may be modulated by a polymorphism in metabolic (*CYP1A1MspI, GSTM1*) and DNA repair genes (*XRCC1, XPD6, XPD23*). Sperm DNA fragmentation seems to be a sensitive biomarker of air pollution.

The effect of exposure to air pollution to biomarkers in newborns was analyzed in two locations with different level of pollution: Prague vs. Ceske Budejovice in winter season 2008/2009. The levels of B[a]P, benzene and PM2.5 for both locations for the years 2008 and 2009 were obtained from the Czech Hydrometeorological Institute. The mean concentration of these pollutants 3 months before birth were calculated to estimate the individual exposure of each mother-newborn pair: B[a]P concentration in Prague was 1.9±0.5 ng/m^3 vs. 3.2±0.2 ng/m^3 in Ceske Budejovice ($p<0.001$), PM2.5 27.0±2.5 μg/m^3 vs. 24.5±0.7 μg /m^3 ($p<0.001$), benzene 2.5±0.5 μg / m^3 vs. 2.1±0.1 μg /m^3, respectively [98].

As biomarkers, umbilical cord blood (UCB) was used to determine DNA adducts by ^{32}P-postlabeling [22] and micronuclei using automated image analysis [90]. DNA adducts were analyzed as B[a]P-like adducts and total adducts, both categories were significantly higher in Ceske Budejovice vs. Prague ($p<0.001$). Higher frequencies of micronuclei in newborns in Ceske Budejovice vs. Prague ($p<0.001$) were observed. Multivariate logistic regression showed a significant impact of 3 months mean B[a]P exposure before birth.

The changes in the transcriptome of newborns from UCB were studied [92]. Non-smoking mothers were selected: 52 from Ceske Budejovice and 35 from Prague. Total RNA was isolated from leukocytes, gene expression profiles were determined by HumanRef-8 Expression BeadChips (Illumina, San Diego, CA, USA) containing 24 526 transcript probes. Genes with / logFC/ > 0.58 (binary logarithm of fold change) and $P<0.01$ were considered as differentially expressed between Ceske Budejovice and Prague. Leukocytes from newborns showed different expression of 104 genes (37 up-regulated and 67 down-regulated genes). Down-regulated biological processes were immune and defense response (*KIR2DL3, KIR3DL3, KIR3DL4, KIR2DS5, KLRC3, CLTA4*), negative regulation of proliferation (*CNDKN1A, CTLA4, TGFBR3*), apoptosis (*PRF1, NR4A2, GZMB, TNFAIP3, PP2R2B, DDIT4*), response to oxygen levels, cell migration, organ regeneration, signal transduction (*RGS1, SOCS1, THBS*) and cell differentiation (*FLT3, ZBTB16*), up-regulated gene encoding *SERPINA1* (which is considered as biomarker of exposure to genotoxic agents). Down regulated signaling pathways were natural killer cell mediated cytotoxicity, antigen processing and presentation, autoimmune thyroid disease, graft vs. host disease, up-regulated MAPK signaling pathway [98].

The results were surprising because air pollution in Prague was understood to be higher than in Ceske Budejovice. Results of analysis of DNA adducts, micronuclei and transcriptome indicate in the same direction the significance of exposure to B[a]P in Ceske Budejovice to induce genetic damage in newborns, when PM2.5 and benzene was higher in Prague. It seems to be a coincidence that changes observed in Ceske Budejovice were observed when exposure to B[a]P was 3.2 ng/m^3, corresponding to the effect of exposure inducing IUGR 2.8 ng/m^3 [100].

5. Ostrava program

The Ostrava Region (OSTR) is an industrial and heavily populated area situated in the easternmost part of the Czech Republic. Concentrations of PM10 (particulate matter < 10 μm) were continuously above 40 μg/m^3 daily average in the years 2002-2011 and considerably higher than urban background in the largest city of CR - Prague. Similarly, population in this region is exposed to high concentrations of PM2.5 (particulate matter < 2.5 μm). Concentrations of B[a]P in the OSTR are the highest in the Czech Republic as well as in the European Union. Standard of 1 ng/m^3/year B[a]P has been exceeded on all OSTR monitoring stations in all years 2004-2011 [72].

Sram et al. [72] investigated the impact of high level of environmental air pollution on selected biomarkers. Exposure was measured as follows: PM2.5 by stationary monitoring, c-PAHs (B[a]P) and VOC (benzene) by personal and stationary monitoring. Personal exposure to c-PAHs was defined using outdoor concentration, ETS exposure (environmental tobacco smoke), indicator of home heating by coal, wood or gas, frequency of exhaust fan use, cooking habits, and commuting by a car [101]. Cotinine in urine, triglycerids, total, HDL and LDL cholesterols, and vitamins A, C, E in plasma were used as life-style indicators.

The following parameters were analyzed: DNA adducts by ^{32}P-postlabeling as biomarkers of effect, chromosomal aberrations by FISH (fluorescent in situ hybridization) and MN as biomarkers of effect, 8-oxo-7,8-dihydro-2'-deoxyguanosine (8-oxodG) as a marker of oxidative DNA damage, 15-F2t-isoprostane (15-F2t-IsoP) as a marker of lipid peroxidation, protein carbonyls as a marker of protein oxidation, and genetic polymorphisms as biomarkers of susceptibility. Sampling was done in winter 2009, summer 2009 and winter 2010. Volunteers were recruited from office workers in Ostrava city, city policemen from Havirov and Karvina (N=98), and in 2010 also from general population of Ostrava-Radvanice (N=28). City policemen from Prague (N=65) served as a control group.

During all sampling periods, the study subjects from OSTR were exposed to significantly higher concentrations of B[a]P and benzene than subjects in Prague as measured by personal monitoring. Taken separately, B[a]P levels were lowest in Prague in 2009, Prague winter 2010 concentrations were about equal to the lower Ostrava 2009 levels, and levels in Ostrava in winter 2010 were 5-fold higher. Despite higher B[a]P air pollution in OSTR during all sampling periods, the levels of B[a]P-like DNA adducts per 10^8 nucleotides were significantly higher in the Ostrava subjects only in winter 2009 (mean ± SD: 0.21 ± 0.06 versus 0.28 ± 0.08 adducts/10^8 nucleotides, $p < 0.001$ for Prague and Ostrava subjects, respectively) (Table 1; controls - unexposed subjects from [22]). During the other two sampling periods, the levels of B[a]P-like DNA adducts were significantly higher in the Prague subjects ($p < 0.001$). Multi-variate analyses done separately for subjects from Ostrava and from Prague, combining all sampling periods in each location, revealed that exposure to B[a]P and PM2.5 significantly increased levels of B[a]P-like DNA adducts only in the Ostrava subjects [102].

Despite several-fold higher concentrations of air pollutants in the Ostrava Region, the levels of stable aberrations (genomic frequency of translocations per 100 cells (F_G/100), percentage of aberrant cells (% AB.C.) were comparable (Table 2; controls - unexposed subjects from [36]).

The frequency of unstable aberrations measured as number of micronuclei was unexpectedly significantly lower in the Ostrava Region subjects in both seasons of 2009. Urinary excretion of 8-oxodG did not differ between locations in either season.

		Exposure		DNA adducts (adducts/10^8 nucleotides)			
	N	B[a]P (ng/m^3)		B[a]P - "like"		Total	
		2009	2010	2009	2010	2009	2010
Prague	60	0.80 ± 0.55	2.86± 1.87	0.21± 0.06	0.25± 0.12	1.30± 0.41	1.37± 0.41
Ostrava	98	2.73± 2.60	14.8± 13.3*	0.28± 0.08	0.16± 0.06	0.37± 0.37	1.03± 0.33*
Controls	42	0.80±0.62		0.10±0.03		0.76±0.20	

* P < 0.05

		Exposure		FISH			
	N	B[a]P (ng/m^3)		% AB.C.		F_G/100	
		2009	2010	2009	2010	2009	2010
Prague	60	0.80 ± 0.55	2.86± 1.87	0.27± 0.18	0.25± 0.15	1.43± 1.15	1.39±1.03
Ostrava	98	2.73± 2.60	14.8± 13.3*	0.26± 0.19	0.22± 0.18	1.44± 1.23	1.25±1.18
Controls	42	0.80±0.62		0.21±0.16		1.13±1.01	

* P < 0.05

Lipid peroxidation measured as levels of 15-F2t-IsoP in blood plasma was elevated in the Ostrava subjects sampled in 2009, similarly increased in Prague samples in 2010 (Table 3). Multivariate analyses conducted separately for subjects from Prague and Ostrava showed a negative association between the frequency of micronuclei and concentrations of B[a]P and PM2.5 in both regions. A positive relationship was observed between lipid peroxidation and air pollution [103]. Plasma levels of 15-F2t-IsoP increased linearly with increasing concentration of B[a]P in the ambient air up to concentration of 9 ng B[a]P/m^3 – higher doses of B[a]P did not increase plasma 15-F2t-IsoP levels any further [103].

		Exposure		Oxidative stress 15-F2t-isoprostane (pg/mL)	
	N	B[a]P (ng/m^3)		15-2Ft-IsoP	
		2009	2010	2009	2010
Prague	60	0.80 ± 0.55	2.86± 1.87	165.9± 41.7	265.5± 104.7*
Ostrava	98	2.73± 2.60	14.8± 13.3*	279.3± 303.6*	279.5± 124.5

* P < 0.05

In contrast to the above results, changes were observed in a group of 4 subjects from Prague who spent 3 weeks in Ostrava just in the period of inversion in winter 2010, when the average daily concentration of B[a]P reported by stationary monitoring was 14.7±13.3 ng/m^3. The frequency of micronuclei in peripheral lymphocytes in those individuals increased approx. 50% (Table 4) [104], and similar increase was observed for genomic frequency of translocations.

Groups	Sampling	N (6000c./person)	Frequency of micronuclei % AB.C. with MN	Frequency of micronuclei MN/1000
Controls	(i)	4	0.81 ± 0.15	8.32 ± 1.63
	(ii)	4	0.80 ± 0.14	8.47 ± 1.55
Exposed	(i)	4	0.74 ± 0.43	7.96 ± 4.92
	(ii)	4	1.14 ± 0.55	12.91 ± 6.49*

* $P < 0.05$

The relationship between exposure to B[a]P and the level of DNA adducts and chromosomal aberrations in winter 2010 in Ostrava inhabitants was surprising, as the results did not correspond with the expected dose effect relationship. Therefore Rossner et al. [61] put forward a hypothesis about a possible adaptive response, indicating that this outcome may be affected by DNA repair. In 64 subjects from Prague and 75 subjects from Ostrava they investigated the levels of oxidative stress markers (8-oxodG, 15-F2t-IsoP, protein carbonyls) and cytogenetic parameters [F_G/100, % AB.C. and acentric fragments (ace)], and their relationship with the expression of genes participating in base excision repair (BER) and non-homologous end-joining (NHEJ) by quantitative PCR. Multivariate analyses revealed that subjects living in Ostrava had increased odds of having above-median levels of *XRCC5* expression (OR; 95% CI: 3.33; 1.03–10.8; q = 0.046). Above-median levels of 8-oxodG were associated with decreased levels of vitamins C (OR; 95% CI: 0.37; 0.16–0.83; p = 0.016) and E (OR; 95% CI: 0.25; 0.08–0.75; p = 0.013), which were elevated in subjects from Ostrava. They suggest that air pollution by c-PAHs affects *XRCC5* expression, which probably protect subjects from Ostrava against the induction of a higher frequency of translocations; elevated vitamin C and E levels in the Ostrava subjects decrease the levels of 8-oxodG. Such changes in gene expression were not observed in the 4 subjects from Prague after 3 weeks stay in Ostrava, their reaction differed from subjects with long residence time in OSTR.

Global gene expression analysis in a group of total 312 exposed subjects from OSTR and 154 controls from Prague [98] was conducted with the aim to characterize molecular response of the organism exposed to heavy air pollution [105]. A combination of geographical and meteorological conditions (a valley affected by frequent atmospheric inversions), heavy industry and the fact that industrial production exists in the OSTR region continually for almost three centuries creates a specific situation suitable for research on environmental air pollution and human health. Given these characteristics a higher number of differentially expressed genes was expected to be found in subjects living in the polluted region. The rationale behind this hypothesis was that the protection of the organism against deleterious effects of air pollution would require greater changes in the transcriptome than in the control subjects. Unexpectedly, despite lower concentrations of air pollutants a higher number of deregulated genes and affected KEGG pathways was found in subjects from Prague. In both locations differences between seasons were observed. The quantitative real-time PCR (qRT-

PCR) analysis showed a significant decrease in expression of *APEX, ATM, FAS, GSTM1, IL1B* and *RAD21* in subjects from Ostrava, in a comparison of winter and summer seasons. In the control subjects, an increase in gene expression was observed for *GADD45A* and *PTGS2*. The Rossner et al. [105] concluded that high concentrations of pollutants in Ostrava do not increase the number of deregulated genes. This may be explained by adaptation of humans to chronic exposure to air pollution. To further explain this phenomenon analyses focused on regulation of mRNA expression are necessary.

For the first time, this study measures the levels of biomarkers in subjects exposed to air pollutants in this region. Simultaneous assessment of oxidative stress markers, DNA adducts, chromosomal aberrations and transcriptomics is a new approach that can bring more clarity to the mechanisms of pollution effects.

6. In vitro studies

A wide variety of in vitro systems was developed in order to study the genotoxicity of chemicals and their mixtures, including complex mixtures of environmental pollutants adsorbed onto respirable air particles (PM2.5). Complex mixtures of organic compounds to which humans are exposed through air pollution are only partially characterized with respect to their chemical composition due to difficulties with chemical analysis of the individual components. Therefore, assays based on biological effects of complex mixture components may be a suitable alternative to a circumstantial chemical analysis. Using rat liver microsomal fraction (S9), it has been repeatedly shown that PAHs formed DNA adducts after metabolic activation by P450 enzymes to diol epoxides. This activation system may be used in acellular assay coupled with ^{32}P-postlabeling to assess genotoxic potential of complex environmental mixtures via the analysis of DNA forming activity of the mixtures in native DNA [106-109].

The first study comparing the biological activities of complex mixtures from urban air particles PM10 was published by Binkova et al. [110]. HiVol samples were collected during the winter (October-March) and summer (April-September) seasons in the years 1993-1994, extracts (EOM-extractable organic matter) were analyzed in several fractions using in vitro acellular assay (calf thymus DNA with/without rat liver microsomal S9 fraction) with DNA adduct analysis by ^{32}P-postlabeling (Teplice district: winter: PM10 69.3 μg/m^3, B[a]P 7.4 ng/m^3, summer: PM10 36.5 μg/m^3, B[a]P 0.8 ng/m^3; Prachatice district: winter: PM10 29.6 μg/m^3, B[a]P 5.4 ng/m^3; summer: PM10 23.6 μg/m^3, B[a]P 0.7 ng/m^3). The highest total DNA adduct levels were observed in the neutral fraction, especially in the aromatic subfraction with metabolic activation, which contained mainly PAHs and their methylderivatives. The major PAH-DNA adducts contributed about 50% of the total DNA adducts resulting from all of the crude extracts using S9-metabolic activation. These results indicate that PAHs are a major source of genotoxic activities of organic mixtures associated with urban air particles.

This approach was later repeatedly used in different localities with different level of PM10 and PM2.5 pollution. When EOM extracted from these particles was analyzed, total PAH-DNA adducts highly correlated with concentrations of B[a]P and c-PAHs [111-113]. All studies showed that a cell-free system in conjunction with the sensitive ^{32}P-postlabeling is a suitable model to detect genotoxic potential of EOMs, particularly those containing c-PAHs, as well as to distinguish between direct and indirect genotoxicants in the complex environmental pollutants. Those results indicate that c-PAHs contribute predominantly to the total genotoxicity of various EOMs. The strong correlation between B[a]P and other c-PAH content in all EOMs tested in these studies ($r = 0.98$; $p<0.001$) suggests that B[a]P may be used as an indicator of other c-PAHs in mixtures.

Topinka et al. [114] used acellular assay for the DNA adduct analysis of EOM according to the size fraction of particulate matter: 1-10 μm, 0.5-1 μm, 0.17-0.5 μm and <0.17 μm and the concentration of c-PAHs. The fraction of 0.5-1μm, that formed 37-46% of total PM mass, was the major carrier of c-PAHs, and induce highest genotoxicity detected as DNA adducts by ^{32}P-postlabeling.

Numerous studies analyzing the effect of c-PAHs, particularly B[*a*]P, on oxidative damage in cell lines *in vitro* or in animals *in vivo* have been published [115-121], but the results are conflicting, probably because of differences in the experimental protocols. Gabelova et al. [115] did not find any significant increase of oxidative DNA damage measured by a single cell gel electrophoresis assay (the Comet assay) in HepG2 cells treated with 7.5 μM B[*a*]P for 2, 24 or 48 h. In the study by Park et al. [116], B[*a*]P at concentrations up to 10 μM induced nonspecific DNA damage measured by the Comet assay after 24 h treatment of HepG2 cells. Lipid peroxidation, analyzed as malondialdehyde (MDA) levels, was increased but showed no dose-response. The authors concluded that oxidative DNA damage is probably related to B[*a*]P toxicity. In the A549 cell line, Garcon et al. [117] tested the effect of B[*a*]P treatment on lipid peroxidation measured as MDA levels. They treated the cells for 72 h with 0.05 μM B[*a*]P but found no increase. They hypothesized that the antioxidant defenses of the cells prevented the induction of lipid peroxidation by B[*a*]P [118, 119].

Rossner et al. [122] investigated the ability of organic extracts of size segregated aerosol particles (EOM; three fractions of PM, aerodynamic diameter 1–10 μm, 0.5–1μm and 0.17–0.5μm) to induce oxidative damage to DNA in an in vitro acellular system of calf thymus (CT) DNA with and without S9 metabolic activation. PM was collected in the Czech Republic at four places with different levels of air pollution. Levels of 8-oxodG tended to increase with decreasing sizes of PM. S9 metabolic activation increased the oxidative capacity of PM.

These results indicate that smaller size fractions are more potent inducers of oxidative damage to DNA. This observation is in agreement with other studies [123-125]. In these studies, however, water-soluble PM extracts were used. Moreover, end-point parameters for measuring the potency of PM to induce oxidative damage differed from Rossner's approach. There are only two other reports that used organic PM extracts in the acellular CTDNA system [124, 126] and only one that tested S9 metabolic activation of extracts [126], but none of these analyzed the oxidative capacity of individual size fractions. Thus, Rossner's results [122]) are probably the first showing that the ability of organic PM extracts to induce oxidative damage to DNA also increases with decreasing sizes of particles. This trend seems to be less pronounced after S9 metabolic activation of EOM. Thus, the presence of PAHs in EOM is probably not the only factor responsible for oxidative damage induction by PM organic extracts.

On the other hand, results of Rossner et al. [122] showed that metabolic activation of PAHs plays at least a partial role in the induction of oxidative damage to DNA because 8-oxodG levels in CT-DNA incubated with S9 fraction were significantly higher than in samples without S9 metabolic activation. Also, they observed a positive correlation between c-PAHs concentrations and 8-oxodG levels induced by PM. This correlation was stronger and statistically significant when PM extracts were incubated with S9 metabolic fraction.

The oxidative capacity of PM extracts increases with increasing levels of air pollution. Smaller size fractions of PM induce higher oxidative damage, which is caused partly by higher content of c-PAHs and partly by other unidentified factors.

7. Discussion

The average personal environmental exposure to approx. 10 ng/m^3 of c-PAHs (or 1.6 ng/m^3 of B[a]P) [86], or approx. twice as high c-PAHs concentrations by stationary monitoring – according to outdoor/indoor ratio) during winter increases DNA adduct formation and decreases repair efficiency, which may be further affected by genetic polymorphisms. This concentration of c-PAHs and its impact on increased DNA adduct formation seems to be important for re-evaluating the risk assessment of c-PAHs. PAH-DNA adduct formation represent one of the key first steps in carcinogenesis [127]. According to Phillips [128] the use of DNA adducts as a measure of exposure, several years prior to the onset (or clinical manifestation) of disease, can identify individuals at higher probability of subsequently developing cancer. Another study demonstrated significant correlations between DNA adduct levels and atherosclerosis [25]. Both examples demonstrate the potential of DNA adducts as biomarkers of risk.

DNA adducts determined by the ^{32}P-postlabeling method are sensitive biomarkers of environmental exposure to c-PAHs, if the study simultaneously includes personal and stationary monitoring, information on life style, determination of cotinine, vitamin and lipid levels, as well as genetic polymorphisms of metabolic and DNA repair genes.

Nowadays it is generally accepted that a high frequency of chromosomal aberrations in peripheral lymphocytes is predictive of an increased risk of cancer [32]. Therefore, it may be hypothesized that environmental exposure to c-PAHs that increases the genomic frequency of translocations represents a significant health risk. Surprising results were observed in the group of city policemen who were examined in January and March: the genomic frequency of translocations decreased similarly as did their exposure to c-PAHs. This reveals that these chromosomal aberrations are not so stable as originally expected [79]. When Binkova et al. [22] studied the relationship between chromosomal aberrations and DNA adducts in the same subjects using FISH, multiple regression analysis indicated that B[a]P-like DNA adducts are a significant predictor of the genomic frequency of translocations. Whole chromosome painting using the FISH technique is more sensitive than the originally used conventional cytogenetic method, which was not affected by the studied concentrations of c-PAHs.

Studies in the Czech Republic [22, 36, 90, 97] suggest that environmental exposure to concentrations higher than 1 ng B[a]P/m^3 represent a risk of DNA damage as indicated by an increase in DNA adducts and an increase in translocations detected by FISH, increase of micronuclei as well as the increase of DNA fragmentation in the mature sperm.

A relationship between DNA adducts (biomarker of exposure) and chromosomal aberrations as determined by FISH (biomarker of effect) was observed for the first time [22].

Studies by Binkova et al. [18, 19, 22] and Topinka et al. [87] suggest that DNA adducts in the lymphocytes of subjects exposed to increased c-PAH levels are an appropriate biomarker of a biologically effective dose, directly indicating whether or not the extent of exposure to these compounds is related to an increased mutagenic and carcinogenic risk. All results indicate that molecular epidemiology studies should be done in a very complex manner, simultaneously using biomarkers of exposure, effect and susceptibility. Air pollution, and specifically c-PAHs, induce genetic damage and may significantly affect human health.

Although PAHs may induce oxidative stress during their metabolism by the formation of reactive quinones, the results of epidemiological studies on 8-oxodG levels after occupational, as well as environmental, PAH exposure are contradictory. Several authors reported a positive correlation between PAH exposure, measured by urinary levels of 1-hydroxypyrene, and 8-oxodG levels, either in lymphocytes or in urine [129-132]. Others did not see any correlation between 8-oxodG levels and urinary 1-hydroxypyrene and/or PAH levels in the air [133-135]. Similarly, Rossner et al. [83] did not find any effect of either personal or stationary c-PAH exposure on 8-oxodG levels in urine. These observations suggest that while c-PAHs may have the potential to induce oxidative DNA damage, other factors probably play a role in the response of the organism to environmental pollution.

Unlike c-PAHs and B[a]P, Rossner et al. [89] did not observe any effect of stationary PM2.5 and PM10 on levels of 15-F_{2t}-IsoP when PM concentrations over a 3-day period immediately preceding urine collection were used in the statistical analysis. Only when the multivariate model was applied did the effect of PM2.5 exposure reach borderline significance. Interestingly, when concentrations of PM2.5 and PM10 over various 3-day periods before sampling were included in the statistical analysis, they found significant effects of PM exposure on 15-F_{2t}-IsoP levels. This observation suggests that 15-F_{2t}-IsoP levels, unlike 8-oxodG, are not markers of immediate effect; rather, they reflect the exposure to pollutants that occurred at least 3-4 weeks before sampling. The possible explanation may be the existence of DNA repair mechanisms that remove damage to DNA shortly after it occurs. Damage to lipids is not repaired; rather the lesions accumulate and thus allow the detection of exposure to pollutants longer period before sampling.

Among other factors that seem to affect the levels of oxidative stress markers, a positive association between cotinine (considered as a marker of passive smoking) and 15-F_{2t}-IsoP levels was observed [136-138], although negative reports have also been published [139].

In vitro studies reveal that the biological activity of PM2.5 may differ according to the sources and the content of c-PAHs. The genotoxic risk may be substantially different when comparing industrial emissions vs. traffic related emissions or emissions from the local heating using coal.

Review of in vitro studies allows to propose the cell free system with CT-DNA to be used as a system to evaluate the biologic activity of organic compounds extracted from particulate matter, detecting genetic damage as DNA adducts or oxidative damage as 8-oxodG, as a first information about a possible genetic risk of air pollution, and the different quality of respirable particles according to the content of c-PAHs adsorbed on their surface as well as oxidative damage induced according to the size of PM.

8. Conclusion

All discussed studies indicate, that DNA adducts, Comet assay and DNA fragmentation in sperm are sensitive biomarkers of exposure to c-PAHs in polluted air, chromosomal aberrations by FISH and micronuclei as biomarkers of effect, and 8-oxodG and 15-F2t-IsoP as biomarkers of oxidative damage.

It seems that when using these biomarkers the dose-effect is seen only in a certain range, probably up to 10 ng B[a]P/m^3.

It is important to identify simultaneously the gene susceptibility, especially the genetic polymorphisms of metabolic genes and genes encoding DNA repair enzymes. DNA damage may be further affected by life style as smoking, ETS, diet – intake of vitamins A, C, E, folic acid, oxidative metabolism by lipid metabolism (triglycerides, cholesterol, HDL, LDL) – it is therefore pertinent to analyze all these endpoints in the biological material in the course of molecular epidemiology studies.

New perspectives may be seen in using the microarray methods, e.g. studying the gene expression of genes coding DNA repair enzymes.

Summing up, molecular epidemiology studies on the environmental exposure to c-PAHs in ambient air should be very complex: determining personal exposure, DNA and oxidative damage, gene susceptibility and life style factors. It will bring new results, which may specify new information important to evaluate properly c-PAHs human health risk.

Acknowledgements

We would like to acknowledge the great help and support of our friends from National Health and Environmental Effects Research Laboratory, US Environmental Protection Agency, Research Triangle Park, NC 27711, USA, especially Drs. Joellen Lewtas, Lawrence W. Reiter, and Sally Perault Darney. Thanks to their support we were able to establish molecular epidemiology methods in the Czech Republic.

Writing of this review was supported by Grant Agency of the Czech Republic (P30113-13458S) and by CITI-SENSE, a Collaborative Project funded by the EU FP7-ENV-2012 (no. 308524).

Author details

Pavel Rossner, Jr., Blanka Binkova, Andrea Rossnerova and Radim J. Sram*

*Address all correspondence to: sram@biomed.cas.cz

Institute of Experimental Medicine AS CR, Videnska, Prague, Czech Republic

References

[1] Perera FP, Whyatt RM. Biomarkers and molecular epidemiology in mutation/cancer research. Mutat Res 1994; 313:117-129

[2] Perera FP, Whyatt RM, Jedrzychowski W, Raugh V, Manchester D, Santella RM, Ortman R. Recent developments in molecular epidemiology. A study of the effects of environmental polycyclic aromatic hydrocarbons on birth outcomes in Poland. Am J Epidemiol 1998; 147: 309-314

[3] Committee on Biological Markers of the National Research Council. Biological markers in environmental health research. Environ Health Perspect 1987; 74:3-9

[4] Albertini RJ, Nicklas JA. Future research directions for evaluating human genetic and cancer risk from environmental exposures. Environ Health Perspect 1996; (suppl 3) 503-510

[5] Albertini RJ. The use and interpretation of biomarkers of environmental genotoxicity in humans. Biotherapy 1998; 11:155-167

[6] Dockery DW, Pope CA 3rd, Xu X, Spengler JD, Ware JH, Fay ME, Ferris BG, Speizer FE. An association between air pollution and mortality in six U.S. cities. N Engl J Med 1993; 329:1753-1759

[7] Pope CA 3rd, Thun MJ, Namboodiri MM, Dockery DW, Evans JS, Speizer FE, Health CW Jr. Particulate air pollution as a predictor of mortality in a prospective study of U.S. adults. Am J Respir Crit Care Med 1995; 151:669-674

[8] Sarnat JA, Schwartz J, Suh HH. Fine particulate air pollution and mortality in 20 U.S. cities. N Engl J Med 2001; 344:1253-1254

[9] Pope CA 3rd, Burnett RT, Thun MJ, Calle EE, Krewski K. Lung cancer, cardiopulmonary mortality, and long-term exposure to fine particulate air pollution. JAMA 2002; 287:1132-1141

[10] Beaglehole R, Bonita R, Kjellström T. Basic epidemiology1993; WHO, Geneve

[11] Seaton A, MacNee W, Donaldson K, Godden D. Particulate air pollution and acute health effects. Lancet 1995; 345:176-178

[12] Luginaah I, Fung KY, Gorey KM, Webster G. Association of ambient air pollution with respiratory hospitalization in a government designated „area concern": the case of Windsor, Ontario. Environ Health Perspect 2005; 113:290-296

[13] Fung KY, Luginaah I, Gorey KM, Webster G. Air pollution and daily hospital admissions for cardiovascular diseases in Windsor, Ontario. Can J Public Health 2005; 96:29-33

[14] Kunzli N, Kaiser R, Medina S, Studnicka M, Chanel O, Filliger P, Herry M, Horak F Jr, Puybonnieux-Texier V, Quenel P, Schneider J, Seethaler R, Vergnaud JC, Sommer H. Public-health impact of outdoor and traffic- related air pollution: European assessment. Lancet 2000; 356:795–801

[15] Sram RJ, Benes I, Binkova B, Dejmek J, Horstman D, Kotesovec F, Otto D, Perreault SD, Rubes J, Selevan SG, Skalik I, Stevens RK, Lewtas. Teplice program—the impact of air pollution on human health. Environ Health Perspect 1996; 104 (suppl 4) 699-714

[16] Bonassi S. Combining environmental exposure and genetic effect measurements in health outcome ssessment. Mutat Res. 1999; 428:177–185

[17] Sram RJ, Binkova B. Molecular epidemiology studies on occupational and environmental exposure to mutagens and carcinogens, 1997–1999. Environ Health Perspect 2000; 108 (suppl 1) 57-70

[18] Binkova B, Lewtas J, Miskova I, Rossner P, Cerna M, Mrackova G, Peterkova K, Mumford J, Meyer S, Sram R. Biomarker studies in Northern Bohemia. Environ Health Perspect 1996; 104 (suppl 3) 591–597

[19] Binkova B, Lewtas J, Miskova I, Lenicek J, Sram R. DNA adducts and personal air monitoring of carcinogenic polycyclic aromatic hydrocarbons in an environmentally exposed population. Carcinogenesis 1995; 16:1037–1046

[20] Binkova B, Topinka J, Mrackova G, Gajdosova D, Vidova P, Stavkova Z, Peterka V, Pilcik T, Rimar V, Dobias L, Farmer PB, Sram RJ. Coke oven workers study, the effect of exposure and GSTM1 and NAT2 genotypes on DNA adduct levels in white blood cells and lymphocytes as determined by 32P-postlabelling. Mutat Res 1998;416:67–84

[21] Phillips DH, Castegnaro M.Standardization and validation of DNA adduct postlabelling methods. R report of interlaboratory trials and production of recommended protocols. Mutagenesis 1999; 14:301-315

[22] Binkova B, Chvatalova I, Lnenickova Z, Milcova A, Farmer PB, Sram RJ. PAH-DNA adducts in environmentally exposed population in relation to metabolic and DNA repair genes polymorphisms. Mutat Res 2007; 620:49-61

[23] Georgiadis P, Topinka J, Stoikidou M, Kaila S, Gioka M, Katsouyanni K, Sram R, Autrup H, Kyrtopoulos SA. AULIS Network. Biomarkers of genotoxicity of air pollution (the AULIS project): bulky DNA adducts in subjects with moderate to low exposures to airborne polycyclic aromatic hydrocarbons and their relationship to environmental tobacco smoke and other parameters. Carcinogenesis 2001; 22:1447–1457

[24] Migliore L, Coppede F. Genetic and environmental factors in cancer and neurodegenerative diseases. Mutat Res 2002; 512:135–153

[25] Binkova B, Smerhovsky Z, Strejc P, Boubelik O, Stavkova Z, Chvatalova I, Sram RJ. DNA-adducts and atherosclerosis, a study of accidental and sudden death males in the Czech Republic. Mutat Res 2002; 501:115–128

[26] Obr G, Pfeiffer P, Savage JR, Johannes C, Goedecke W, Jeppesen P, Natarajan AT, Martinez-Lopez W, Folle GA, Drets ME. Chromosomal aberrations: formation, identification and distribution. Mutat Res 2002; 504:17-36

[27] Collins AR. The comet assay for DNA damage and repair: principles, applications, and limitations. Mol Biotechnol 2004; 26:249-261

[28] Møller P. The alkaline comet assay: towards validation in biomonitoring of DNA damaging exposures. Basic Clin Pahrmacol Toxicol 2006; 98:336-345

[29] Evenson DP, Jost LK. Sperm chromatin structure assay: DNA denaturability. In: Darzynkiewicz Z, Robinson JP, Crissman HA (eds). Methods in Cell Biology, Vol. 42, Flow Cytometry. Acad. Press, Inc, Orlando, FL, 1994; pp. 159-176

[30] Evenson DP, Kasperson K, Wixon RL. Analysis of sperm DNA fragmentation using flow cytometry and other techniques. Society of Reproduction and Fertility supplement. 2007; 65:93-113.

[31] Evenson DP. Sperm chromatin structure assay (SCSA(R)). Methods Mol Biol. 2013; 927: 147-164. doi:10.1007/978-1-62703-038-0_14.

[32] Bonassi S, Norppa H, Ceppi M, Stromberg U, Vermeulen R, Znaor A, Cebulska- Wasilewska A, Fabianova E, Fucic A, Gundy S, Hansteen IL, Knudsen LE, Lazutka J, Rossner P, Sram RJ, Boffetta P. Chromosomal aberration frequency in lymphocytes predicts the risk of cancer: results from a pooled cohort study of 22 358 subjects in 11 countries. Carcinogenesis 2008; 29:1178-1183

[33] Rubes J, Kucharova S, Vozdova M, Musilova P, Zudova Z. Cytogenetic analysis of peripheral lymphocytes in medical personnel by means of FISH. Mutat Res 1998; 412:293-298

[34] Verdorfer I, Neubauer S, Letzer S, Angerer J, Arutyunyan R, Martus P, Wucherer M, Gebhart E. Chromosome painting for cytogenetic monitoring of occupationally exposed and non-exposed groups of human individuals. Mutat Res 2001; 491:97-109

[35] Tucker JD, Moore DH 2nd, Ramsey MJ, Kato P, Langlois RG, Burroughs B, Long L, Garry VF. Multi-endpoint biological monitoring of phosphine workers. Mutat Res 2003; 536:7-14

[36] Sram RJ, Beskid O, Binkova B, Chvatalova I, Lnenickova Z, Milcova A, Solansky I, Tulupova E, Bavorova H, Ocadlikova D, Farmer FB. Chromosomal aberrations in environmentally exposed population in relation to metabolic and DNA repair genes polymoprhisms. Mutat Res 2007; 620:22-33

[37] Sigurdson AJ, Ha M, Hauptmann M, Bhatti P, Sram RJ, Beskid O, Tawn EJ, Whitehouse CA, Lindholm C, Nakano M, Kodama Y, Nakamura N, Vorobtsova I, Oestreicher U, Staphan G, Yong LC, Bauchinger M, Schmid E, Chung HM, Darroudi F, Roy L, Voisin P, Barquinero JF, Livingston G, Blakey D, Hayata I, Zhang W, Wang C, Benett LM, Littlefield LG, Edwards AA, Kleinerman RA, Tucker JD. International study of factors affecting human chromosome translocations. Mutat Res 2008; 652:112–121

[38] Sram RJ, Beskid O, Rossnerova A, Rossner P, Lnenickova Z, Milcova A, Solansky I, Binkova B. Environmental exposure to carcinogenic polycyclic hydrocarbons – the interpretation of cytogenetic analysis by FISH. Toxicol Lett 2007; 172:12-20

[39] Fenech M, Morley AA. Measurement of micronuclei in lymphocytes. Mutat Res 1985; 147:29-36

[40] Tuimala J, Szekely G, Gundy S, Hirvonen A, Norppa H. Genetic polymorphisms of DNA repair and xenobiotic metabolizing enzymes: role in mutagen sensitivity. Carcinogenesis 2002; 23:1003–1008

[41] Thacker J, Zdzienicka Z. The mammalian XRCC genes: their roles in DNA repair and genetic stability. DNA Repair (Amst) 2003; 2:655–672

[42] Kelada SN, Eaton DL, Wang SS, Rothman NR, Khoury MJ. The role of genetic polymorphism in environmental health. Environ Health Perspect 2003; 111:1055–1064

[43] Zijno A, Andreoli C, Leopardi P, Marcon F, Rossi S, Caiola S, Verdina A, Galati R, Cafolla A, Crebelli R. Folate status, metabolic genotype, and biomarkers of genotoxicity in healthy subjects. Carcinogenesis 2003; 24:1097– 1103

[44] Ames BN. DNA damage from micronutrient deficiencies is likely to be a major cause of cancer. Mutat Res 2001; 475:7-20

[45] Fenech M.The role of folic acid and vitamin B12 in genomic stability in humans. Mutat Res 2001; 475:57-67

[46] Fenech M, Ferguson LR. Vitamins/minerals and genomic stability in humans. Mutat Res 2001; 475:1–6

[47] Mazzoli-Rocha F, Fernandes S, Einicker-Lamas M, Zin WA. Roles of oxidative stress in signaling and inflammation induced by particulate matter. Cell Biol Toxicol 2010; 26:481-498

[48] Loft S, Vistisen K, Ewertz M, Tjonneland A, Overvad K, Poulsen HE. Oxidative DNA damage estimated by 8- hydroxydeoxyguanosine excretion in humans: influence of smoking, gender and body mass index. Carcinogenesis 1992; 13:2241-2247

[49] Klaunig JE, Kamendulis LM. The role of oxidative stress in carcinogenesis. Ann Rev Pharmacol Toxicol 2004; 44:239-267

[50] Wu LL, Chiou CC, Chang PY, Wu JT. Urinary 8-OHdG: a marker of oxidative stress to DNA and a risk factor for cancer, atherosclerosis and diabetics. Clin Chim Acta 2004; 339:1-9

[51] Cooke MS, Evans MD, Dizdaroglu M, Lunec J. Oxidative DNA damage: mechanisms, mutation, and disease. Faseb J 2003; 17:1195-1214

[52] Haghdoost S, Czene S, Naslund I, Skog S, Harms-Ringdahl M. Extracellular 8-oxo-dG as a sensitive parameter for oxidative stress in vivo and in vitro. Free Radic Res. 2005; 39:153-162

[53] Sorensen M, Autrup H, Moller P, Hertel O, Jensen SS, Vinzents P Knudsen LE, Loft S. Linking exposure to environmental pollutants with biological efects. Mutat Res 2003; 544:255-271

[54] Nair J, De Flora S, Izzotti A, Bartsch H. Lipid peroxidation-derived etheno-DNA adducts in human atherosclerotic lesions. Mutat Res 2007; 621:95-105

[55] Xue W, Warshawsky D. Metabolic activation of polycyclic and heterocyclic aromatic hydrocarbons and DNA damage: a review. Toxicol Appl Pharmacol 2005; 206:73-93

[56] Penning TM, Ohnishi ST, Ohnishi T, Harvey RG. Generation of reactive oxygen species during the enzymatic oxidation of polycyclic aromatic hydrocarbon trans-dihydrodiols catalyzed by dihydrodiol dehydrogenase. Chem Res Toxicol 1996; 9:84-92

[57] van Gent DC, Hoeijmakers JH, Kanaar R. Chromosomal stability and the DNA double-stranded break connection. Nat Rev Genet 2001; 2:196-206

[58] Bi X, Slater DM, Ohmori H, Vaziri C. DNA polymerase kappa is specifically required for recovery from the benzo[a]pyrene-dihydrodiol epoxide (BPDE)-induced S-phase checkpoint. J Biol Chem 2005; 280: 22343-22355

[59] Slade PG, Williams MV, Brahmbhatt V, Dash A, Wishnok JS, Tannenbaum SR. Proteins modified by the lipid peroxidation aldehyde 9,12-dioxo-10(E)-dodecenoic acid in MCF7 breast cancer cells. Chem Res Toxicol 200; 23:557-567

[60] Montuschi P, Barnes P, Roberts LJ 2nd. Insights into oxidative stress: the isoprostanes. Curr Med Chem 2007;14:703-717

[61] Rossner P Jr, Uhlirova K, Beskid O, Rossnerova A, Svecova V, Sram RJ. Expression of XRCC5 in peripheral blood lymphocytes is upregulated in subjects from a heavily polluted region in the Czech Republic. Mutat Res 2011; 713:76-82

[62] Wild CP, Scalbert A, Herceg Z. Measuring the exposome: a powerful basis for evaluating environmental exposures and cancer risk. Environ Mol Mutagen 2013; 54(7): 480-99. doi:10.1002/em.21777.

[63] van Leeuwen DM, van Herwijnen MH, Pedersen M, Knudsen LE, Kirsch-Volders M, Sram RJ et al. Genome-wide differential gene expression in children exposed to air pollution in the Czech Republic. Mutat Res 2006; 600(1- 2):12-22. doi:10.1016/ j.mrfmmm.2006.05.032.

[64] van Leeuwen DM, Pedersen M, Hendriksen PJ, Boorsma A, van Herwijnen MH, RW Gottschalk, Kirsch-Volders L, Knudsen LE, Sram RJ, Bajak E, van Delft JH, Kleinjans JC. Genomic analysis suggests higher susceptibility of children to air pollution. Carcinogenesis 2008; 29:977–983

[65] De Coster S, van Leeuwen DM, Jennen DG, Koppen G, Den Hond E, Nelen V et al. Gender- specific transcriptomic response to environmental exposure in Flemish adults. Environ Mol Mutagen 2013; 54(7):574- 88. doi:10.1002/em.21774.

[66] Taioli E, Sram RJ, Garte S, Kalina I, Popov TA, Farmer PB. Effects of polycyclic aromatic hydrocarbons (PAHs) in environmental pollution on exogenous and oxidative DNA damage (EXPAH project): Description of the population under study. Mutat Res 2007; 620: 1-6

[67] Garte S, Taioli E, Raimondi S, Paracchini V, Binkova B, Sram RJ, Kalina I, Popov TA, Singh R, Farmer PB. Effects of matabolic genotypes on intermediary biomarkers in subjects exposed to PAHS: Results from the EXPAH study. Mutat Res 2007; 620: 7-15

[68] Taioli E, Sram RJ, Binkova B, Kalina I, Popov TA, Garte S, Farmer PB. Biomarkers of exposure to carcinogenic PAHs and their relationship with environmental factors. Mutat Res 2007; 620: 16-21

[69] Singh R, Kaur B, Kalina I, Popov TA, Georgieva T, Garte S, Binkova B, Sram RJ, Taioli E, Farmer PB. Effects of environmental air pollution on endogenous oxidative DNA damage in humans. Mutat Res 2007; 620: 71- 82

[70] Singh R, Sram RJ, Binkova B, Kalina I, Popov TA, Georgieva T, Garte S, Taioli E, Farmer PB. The relationship between biomarkers of oxidative DNA damage, polycyclic aromatic hydrocarbon DNA adducts, antioxidant status and genetic susceptibility following exposure to environmental air pollution in humans. Mutat Res 2007;620: 83-92

[71] Cebulska-Wasilewska A, Binkova B, Sram RJ, Kalina I, Popov T, Farmer PB. Repair competence assay in studies of the influence of environmental exposure to c-PAHs

on individual susceptibility to induction of DNA damage. Mutat Res 2007; 620: 155-164

[72] Sram RJ, Dostal M, Libalova H, Rossner P Jr, Rossnerova A, Svecova V, Topinka J, Bartonova A. The European hot spot of B[a]P and PM2.5 exposure – The Ostrava Region, Czech Republic: Health research results. ISRN Public Health ID 2013; 416701:1-12

[73] Hemminki K, Grzybowska E, Chorazy M, Twardowska-Saucha K, Sroczynski JW, Putman KL, Randerath K, Phillips DH, Hewer A, Santella RM, Young TL, Perera FP. DNA adducts in humans environmentally exposed to aromatic compounds in an industrial area of Poland. Carcinogenesis 1990; 11:1229-1231

[74] Perera FP, Hemminki K, Grzybowska E, Motykiewicz G, Michalska J, Santella RM, Young TL, Dickey C, Brandt- Rau P, DeVivo I, Blaner W, Tsai WY, Chorazy M. Molecular and genetic damage in humans from environmental pollution in Poland. Nature 1992; 360:256- 258

[75] Grzybowska E, Hemminki K, Chorazy M. Seasonal variations in levels of DNA adducts and X-spots in human populations living in different parts o Poland. Environ Health Perspect 1993; 99:77-81

[76] Grzybowska E, Hemminki K, Szeliga J, Chorazy M. Seasonal variation of aromatic adducts in human lymphocytes and granulocytes. Carcinogenesis 1993; 14:2523-2526

[77] Moldan B, Schnoor L. Czechoslovakia: restoring a critically ill environment. Environ Sci Technol 1992; 26:14-21

[78] Lewtas J, Lewis C, Zweidinger R, Stevens R, Cupitt L. Sources of genotoxicity and cancer risk in ambient air. Pharmacokinetics 1992; 2:288-296

[79] Lioy PJ, Greenberg A. Factors associated with human exposures to polycyclic aromatic hydrocarbons. Toxicol Ind Health 1990; 6:209-223

[80] Lewtas J. Air pollution combustion emissions: characterization of causative agents and mechanisms associated with cancer, reproductive, and cardiovascular effects. Mutat Res 2007; 636:95-133

[81] Svecova V, Rossner P Jr, Dostal M, Topinka J, Solansky I, Sram RJ. Urinary 8-oxodeoxyguanosine levels in children exposed to air pollution. Mutat Res 2009; 662:37-43

[82] Dejmek J Selevan SG, Benes I, Solansky I, Sram RJ. Fetal growth and maternal exposure to particulate matter during pregnancy. Environ Health Perspect 1999; 107:475-480

[83] Rossner P Jr, Svecova V, Milcova A, Lnenickova Z, Solansky I, Sram RJ. Seasonal variability of oxidative stress markers in city bus drivers. Part I. Oxidative damage to DNA. Mutat Res 2008; 642:14-20

[84] Rubes J, Selevan SG, Evenson DP, Zudova D, Vozdova M, Zudova Z, Robbins WA, Perreault SD. Episodic air pollution is associated with increased DNA fragmentation in human sperm without other changes in semen quality. Hum Reprod 2005; 20:2776-2783

[85] Rubes J, Selevan SG, Sram RJ, Evenson DP, Perreault SD. GSMT1 genotype influences the susceptibility of men to sperm DNA damage associated with exposure to air pollution. Mutat Res 2007; 625:20-28

[86] Sram RJ, Binkova B, Beskid O, Milcova A, Rossner P, Rossner P Jr, Rossnerova A, Solansky I, Topinka J. Biomarkers of exposure and effect – interpretation in human risk assessment. Air Qual Atmos Health 2011; 4:161-167

[87] Topinka J, Sevastyanova O, Binkova B, Chvatalova I, Milcova A, Lnenickova Z, Novakova Z, Solansky I, Sram RJ. Biomarkers of air pollution exposure – a study of policemen in Prague. Mutat Res 2007; 624:9-17

[88] Novotna B, Topinka J, Solansky I, Chvatalova I, Lnenickova Z, Sram RJ. Impact of air pollution and genotype variability on DNA damage in Prague policemen. Toxicol Lett 2007; 172:37-47

[89] Rossner P Jr, Svecova V, Milcova A, Lnenickova Z, Solansky I, Sram RJ. Seasonal variability of oxidative stress markers in city bus drivers. Part II. Oxidative damage to lipids and proteins. Mutat Res 2008; 642:21-27

[90] Rossnerova A, Spatova M, Rossner P, Solansky I, Sram RJ. The impact of air pollution on the levels of micronuclei measured by automated image analysis. Mutat Res 2009; 669:42-47

[91] Ishikawa H, Tian Y, Piao F, Duan Z, Zhang M, Ma M, Li H, Yamamoto H, Matsumoto Y, Sakai S, Cui J, Yamauchi T, Yokoyama K. Genotoxic damage in female residents exposed to environmental air pollution in Shenyang city, China. Cancer Lett 2006; 240:29-35

[92] Pavanello S, Kapka L, Siwinska E, Mielzynska D, Bolognesi C, Clonfero E. Micronuclei related to anti-B[a]PDE- DNA adduct in peripheral blood lymphocytes of heavily polycyclic aromatic hydrocarbon-exposed nonsmoking coke-oven workers and controls. Cancer Epidemiol Biomarkers Prev 2008; 17:2795–2799

[93] Liu AL, Lu WQ, Wang ZZ, Chen WH, Lu WH, Yuan J, Nan PH, Sun JV, Zou YL, Zhou LH, Zhang C, Wu TC. Elevated levels of urinary 8-hydroxy-2 -deoxyguanosine, lymphocytic micronuclei, and serum glutathione S- transferase in workers exposed to coke oven emissions. Environ Health Perspect 2006; 114:673–677

[94] Neri M, Ugolini D, Bonassi S, Fucic A, Holland N, Knudsen LE, Sram RJ, Ceppi M, Bocchini V, Merlo DF. Children's exposure to environmental pollutants and bio-

markers of genetic damage. II. Results of a comprehensive literature search and meta-analysis. Mutat Res 2006; 612:14–39

[95] Sram RJ, Rossner P, Smerhovsky Z. Cytogenetic analysis and occupational health in the Czech Republic. Mutat Res 2004; 566:21–48

[96] Sram RJ, Beskid O, Rossnerova A, Rossner P, Lnenickova Z, Milcova A, Solansky I, Binkova B. Environmental exposure to carcinogenic polycyclic aromatic hydrocarbons – the interpretation of cytogenetic analysis by FISH. Toxicol Lett 2007; 172:12-20

[97] Rubes J, Rybar R, Prinosilova P, Veznik Z, Chvatalova I, Solansky I, Sram RJ. Genetic polymorphisms influence the susceptibility of men to sperm DNA damage associated with exposure to air pollution. Mutat Res 2010; 683:9-15

[98] Sram RJ, Binkova B, Dostal M, Merkerova-Dostalova M, Libalova H, Milcova A, Rossner P, Rossnerova A, Schmuczerova J, Svecova V, Topinka J, Votavova H. Health impacts of air pollution to children. Int J Hyg Environm Health 2013; 216 (5) 533-540.

[99] Votavova, H., Dostalova-Merkerova, M., Krejcik, Z., Fejglova, K., Vasikova, A., Pastorkova, A., Tabashidze, N., Topinka, J., Balascak, I., Sram, R.J., Brdicka, R., 2012. Deregulation of gene expression induced by environmental tobacco smoke exposure in pregnancy. Nicotine Tob Res 2012; 14 (9) 1073-1082.

[100] Dejmek, J., Solansky, I., Benes, I., Lenicek, J., Sram, R.J. The impact of polycyclic aromatic hydrocarbons and fine particles on pregnancy outcome. Environ Health Perspect 2000; 108:1159-1164.

[101] Svecova V, Topinka J, Solansky I, Rossner P Jr, Sram RJ. Personal exposure to carcinogenic polycyclic aromatic hydrocarbons in the Czech Republic. J Expo Sci Environ Epidemiol 2013; 23:350-355

[102] Rossner P Jr, Svecova V, Schmuczerova J, Milcova A, Tabashidze N, Topinka J, Pastorkova A, Sram RJ. Analysis of biomarkers in a Czech population exposed to heavy air pollution. Part I. Bulky DNA adducts. Mutagenesis 2013; 28:89-95

[103] Rossner P Jr, Rossnerova A, Spatova M, Beskid O, Uhlirova K, Libalova H, Solanksy I, Topinka J, Sram RJ. Analysis of biomarkers in a Czech population exposed to heavy air pollution. Part II. Chromosomal aberrations and oxidative stress. Mutagenesis 2013; 28:97-106

[104] Rossnerova A, Spatova M, Shunk Ch, Sram RJ. Automated scoring of lymphocyte micronuclei by the MetaSystems Metafer image cytometry system and its application in studies of human mutagen sensitivity and biodosimetry of genotoxic exposure. Mutagenesis 2011; 26:169-175

[105] Rossner P, Jr., Tulupova E, Rossnerova A, Libalova H, Gmuender H, Svecova V et al. Gene expression profiling in populations exposed to different levels of respirable air particles. Mutagenesis. 2014b; submitted.

[106] Adams SP, Laws GM, Storer RD, DeLuca JG, Nichols WW. Detection of DNA damage induced by human carcinogens in acellular assays: potential application for determining genotoxic mechanisms. Mutat Res 1996; 368:235–248

[107] Reddy MV, Blackburn GR, Schreiner CA, Mackerer CR. Correlation of mutagenic potencies of various petroleum oils and oil coal tar mixtures with DNA adduct levels in vitro. Mutat Res 1997; 378:89–95

[108] Smith WA, Arif JM, Gupta RC. Effect of cancer chemopreventive agents on microsome-mediated DNA adduction of the breast carcinogen dibenzo[a,l]pyrene. Mutat Res 1998; 412:307–314

[109] Pohjola SK, Lappi M, Honkanen M, Savela K. Comparison of mutagenicity and calf thymus DNA adducts formed by the particulate and semivolatile fractions of vehicle exhausts. Environ Mol Mutagen 2003; 42:26- 36

[110] Binkova B, Vesely D, Vesela D, Jelinek R, Sram RJ. Genotoxicity and embryotoxicity of urban air particulate matter collected during winter and summer period in two different districts of the Czech Republic. Mutat Res 1999; 440:45-58

[111] Binkova B, Cerna M, Pastorkova A, Jelinek R, Benes I, Novak J, Sram RJ. Biological activities of organic compounds adsorbed onto ambient air particles: comparison between the cities of Teplice and Prague during the summer and winter seasons 2000-2001. Mutat Res 2003; 525:43-59

[112] Binkova B, Topinka J, Sram RJ, Sevastyanova O, Novakova Z, Schmuczerova J, Kalina I, Popov T, Farmer PB. In vitro genotoxicity of PAH mixtures and organic extract from air particles. Part I: Acellular assay. Mutat Res 2007; 620:114-122

[113] Topinka J, Rossner P Jr, Milcova A, Schmuczerova J, Svecova V, Sram RJ. DNA adducts and oxidative DNA damage induced by organic extracts from PM2.5 in an acellular assay. Toxicol Lett 2011; 202:186-192

[114] Topinka J, Milcova A, Schmuczerova J, Krouzek J, Hovorka J. Ultrafine particles are not major carriers of carcinogenic PAHs and their genotoxicity in size-segregated aerosols. Mutat Res 2013; 754:1-6

[115] Gabelova A, Valovicova Z, Labaj J, Bacova G, Binkova B, Farmer PB. Assessment of oxidative DNA damage formation by organic complex mixtures from airborne particles PM(10). Mutat Res 2007; 620:135-144

[116] Park SY, Lee SM, Ye SK, Yoon SH, Chung MH, Choi J. Benzo[a]pyrene-induced DNA damage and p53 modulation in human hepatoma HepG2 cells for the identification

of potential biomarkers for PAH monitoring and risk assessment Toxicol Lett 2006; 167:27- 33

[117] Garcon G, Garry S, Gosset P, Zerimech F, Martin A, Hannothiaux M, Shirali P. Benzo[a]pyrene-coated onto Fe_2O_3 particles-induced lung tissue injury: role of free radiacals. Cancer Lett 2001; 167:7-15

[118] Garcon G, Zerimech F, Hannothiaux M, Gosset P, Martin A, Marez T, Shirali P. Antioxidant defense disruption by polycyclic aromatic hydrocarbons-coated onto Fe_2O_3 particles in human lung cells (A549). Toxicology 2001;166(3):129-137

[119] Kim KB, Lee BM. Oxidative stress to DNA, protein, and antioxidant enzymes (superoxide dismutase and catalase) in rats treated with benzo[a]pyrene. Cancer Lett 1997; 113:205-212

[120] Mauthe RJ, Cook VM, Coffin SL, Baird WM. Exposure of mammalian cell cultures to benzo[a]pyrene and light results in oxidative DNA damage as measured by 8-hydroxydeoxyguanosine formation. Carcinogenesis 1995; 16:133-137

[121] Danielsen PH, Loft S, Moller P. DNA damage and cytotoxicity in type II lung epithelial (A549) cell cultures after exposure to diesel exhaust and urban street particles. Part Fibre Toxicol 2008; 5:6

[122] Rossner P Jr, Topinka J, Hovorka J, Milcova A, Schmuczerova J, Krouzek J, Sram RJ. An acellular assay to assess the genotoxicity of complex mixtures of organic pollutants bound on size segregated aerosol. Part II. Oxidative damage to DNA. Toxicol Lett 2010; 198:312-316

[123] Cho AK, Sioutas C, Miguel AH, Kumagai Y, Schmitz DA, Singh M, Eiguren-Fernandez A, Froines JR. Redox activity of airborne particulate matter at different sites in the Los Angeles. Basin Environ Res 2005; 99:40–47

[124] Healey K, Lingard JJ, Tomlin AS, Hughes A, White KL, Wild CP, Routledge MN. Genotoxicity of size- fractionated samples of urban particulate matter. Environ Mol Mutagen 2005; 45:380–387

[125] Shi T, Duffin R, Borm PJ, Li H, Weishaupt C, Schins RP. Hydroxyl-radical- dependent DNA damage by ambient particulate matter from contrasting sampling locations. Environ Res 2006; 101:18-24

[126] de Kok TM, Hogervorst JG, Briede JJ, van Herwijnen MH, Maas LM, Moonen EJ, Driece, HA, Kleinjans JC. Genotoxicity and physiochemical characteristics of traffic-related ambient particulate matter. Environ Mol Mutagen 2005; 46:71-80

[127] Gammon MD, Sagiv SK, Eng SM, Shantakumar S, Gaudet MM, Teitelbaum SL, Britton JA, Terry MB, Wang LW, Wang Q, Stellman SD, Beyea J, Hatch M, Kabat GC, Wolff MS, Levin B, Neugut AI, Santella RM. Polycyclic aromatic hydrocarbon-DNA adducts and breast cancer:a pooled analysis. Arch Environ Health 2004; 59:640–649

[128] Phillips DH. DNA adducts as markers of exposure and risk. Mutat Res 2005; 577:384–392

[129] Nilsson R, Nordlinder R, Moen BE, Ovrebo S, Bleie K, Skorve AH, Hollund BE,Tagesson C. Increased urinary excretion of 8-hydroxydeoxyguanosine in engine room personnel exposed to polycyclic aromatic hydrocarbons. Occup Environ Med 2004; 61:692-696

[130] Hu CW, Wu MT, Chao MR, Pan CH, Wang CJ, Swenberg JA, Wu KY. Comparison of analyses of urinary 8- hydroxy-2'-deoxyguanosine by isotope-dilution liquid chromatography with electrospray tandem mass spectrometry and by enzyme-linked immunosorbent assay. Rapid Commun Mass Spectrom 2004; 18:505- 510

[131] Chuang CY, Lee CC, Chang YK, Sung FC. Oxidative DNA damage estimated by urinary 8- hydroxydeoxyguanosine: influence of taxi driving, smoking and areca chewing. Chemosphere 2003; 52:1163- 1171

[132] Marczynski B, Rihs H, Rossbach B, Holzer J, Angerer J, Scherenberg M, Hoffmann G, Bruning T, Wilhelm M. Analysis of 8-oxo-7,8-dihydro-2'-deoxyguanosine and DNA strand breaks in white blood cells of occupationally exposed workers: comparison with ambient monitoring, urinary metabolites and enzyme polymorphisms. Carcinogenesis 2002; 23:273-281

[133] Marczynski B, Preuss R, Mensing T, Angerer J, Seidel A, El MA,Wilhelm M, Bruning T. Genotoxic risk assessment in white blood cells of occupationally exposed workers before and after alteration of the polycyclic aromatic hydrocarbon (PAH) profile in the production material: comparison with PAH air and urinary metabolite levels. Int Arch Occup Environ Health 2005; 78:97-108

[134] Zhang J, Ichiba M, Hanaoka T, Pan G, Yamano Y, Hara K, Takahashi K, Tomokuni K. Leukocyte 8- hydroxydeoxyguanosine and aromatic DNA adduct in coke-oven workers with polycyclic aromatic hydrocarbon exposure. Int Arch Occup Environ Health 2003;76:499-504

[135] Casado A, De LN, Lopez-Fernandez E, Sanchez A, Jimenez JA. Lipid peroxidation, occupational stress and aging in workers of a prehospital emergency service. Eur J Emerg Med 2006; 13:165-171

[136] Block G, Dietrich M, Norkus EP, Morrow JD, Hudes M, Caan B, Packer L. Factors associated with oxidative stress in human populations. Am J Epidemiol 2002; 156:274-285

[137] Kato T, Inoue T, Morooka T, Yoshimoto N, Node K. Short-term passive smoking causes endothelial dysfunction via oxidative stress in nonsmokers. Can J Physiol Pharmacol 2006; 84:523-529

[138] Ahmadzadehfar H, Oguogho A, Efthimiou Y, Kritz H, Sinzinger H. Passive cigarette smoking increases isoprostane formation. Life Sci 2006; 78:894-897

[139] Kitano S, Hisatomi H, Hibi N, Kawano K, Harada S. Improved method of plasma 8-Isoprostane measurement and association analyses with habitual drinking and smoking. World J Gastroenterol 2006; 12:5846-5852